PARTIAL DIFFERENTIAL EQUATIONS OF MATHEMATICAL PHYSICS

BY

H. BATEMAN, M.A., PH.D.

Late Fellow of Trinity College, Cambridge;
Professor of Mathematics, Theoretical Physics
and Aeronautics, California Institute of Technology,
Pasadena, California

CAMBRIDGE UNIVERSITY PRESS

1969

CAMBRIDGE UNIVERSITY PRESS
Cambridge, New York, Melbourne, Madrid, Cape Town, Singapore, São Paulo, Delhi

Cambridge University Press
The Edinburgh Building, Cambridge CB2 8RU, UK

Published in the United States of America by Cambridge University Press, New York

www.cambridge.org
Information on this title: www.cambridge.org/9780521041218

First published 1932
First American Edition 1944
Paperback Edition 1959
Reprinted 1960, 1964, 1969
Re-issued in this digitally printed version 2008

A catalogue record for this publication is available from the British Library

ISBN 978-0-521-04121-8 hardback
ISBN 978-0-521-09163-3 paperback

Dedicated

to

MY MOTHER

CONTENTS

CHAPTER I

THE CLASSICAL EQUATIONS

CHAPTER II

APPLICATIONS OF THE INTEGRAL THEOREMS OF GREEN AND STOKES

CHAPTER III

TWO-DIMENSIONAL PROBLEMS

CHAPTER IV

CONFORMAL REPRESENTATION

CHAPTER V

EQUATIONS IN THREE VARIABLES

CHAPTER VI

POLAR CO-ORDINATES

CHAPTER VII

CYLINDRICAL CO-ORDINATES

CHAPTER VIII

ELLIPSOIDAL CO-ORDINATES

CHAPTER IX

PARABOLOIDAL CO-ORDINATES

CHAPTER X

TOROIDAL CO-ORDINATES

CHAPTER XI

DIFFRACTION PROBLEMS

CHAPTER XII

NON-LINEAR EQUATIONS

PREFACE

In this book the analysis has been developed chiefly with the aim of obtaining exact analytical expressions for the solution of the boundary problems of mathematical physics.

In many cases, however, this is impracticable, and in recent years much attention has been devoted to methods of approximation. Since these are not described in the text with the fullness which they now deserve, a brief introduction has been written in which some of these methods are sketched and indications are given of portions of the text which will be particularly useful to a student who is preparing to use these methods.

No discussion has been given of the partial differential equations which occur in the new quantum theory of radiation because these have been well treated in several recent books, and an adequate discussion in a book of this type would have greatly increased its size. It is thought, however, that some of the analysis may prove useful to students of the new quantum theory.

Some abbreviations and slight departures from the notation used in recent books have been adopted. Since the L-notation for the generalised Laguerre polynomial has been used recently by different writers with slightly different meanings, the original T-notation of Sonine has been retained as in the author's *Electrical and Optical Wave Motion*. It is thought, however, that a standardised L-notation will eventually be adopted by most writers in honour of the work of Lagrange and Laguerre.

The abbreviations "eit" and "eif" used in the text might be used with advantage in the new quantum theory, together with some other abbreviations, such as "eil" for eigenlevel and "eiv" for eigenvector.

The Heaviside Calculus and the theory of integral equations are only briefly mentioned in the text; they belong rather to a separate subject which might be called the Integral Equations of Mathematical Physics. Accounts of the existence theorems of potential theory, Sturm-Liouville expansions and ellipsoidal harmonics have also been omitted. Many excellent books have however appeared recently in which these subjects are adequately treated.

I feel indeed grateful to the Cambridge University Press for their very accurate work and intelligent assistance during the printing of this book.

H. BATEMAN

November 1931

INTRODUCTION

THE differential equations of mathematical physics are now so numerous and varied in character that it is advisable to make a choice of equations when attempting a discussion.

The equations considered in this book are, I believe, all included in some set of the form

$$\frac{\delta F}{\delta y_1} = 0, \quad \frac{\delta F}{\delta y_2} = 0, \quad \dots \quad \frac{\delta F}{\delta y_m} = 0,$$

where the quantities on the left-hand sides of these equations are the variational derivatives* of a quantity F, which is a function of l independent variables $x_1, \dots x_l$, of m dependent variables $y_1, \dots y_m$ and of the derivatives up to order n of the y's with respect to the x's. The meaning of a variational derivative will be gradually explained.

The first property to be noted is that the variational derivatives of a function F all vanish identically when the function can be expressed in the form

$$F = \frac{dG_1}{dx_1} + \frac{dG_2}{dx_2} + \dots + \frac{dG_l}{dx_l},$$

where each of the functions G_s is a function of the x's and y's and of the derivatives up to order $n-1$ of the y's with respect to the x's. The notation d/dx_s is used here for a complete differentiation with respect to x_s when consideration is taken of the fact that F is not only an explicit function of x_s but also an explicit function of quantities which are themselves functions of x_s.

Another statement of the property just mentioned is that the variational derivatives of F vanish identically when the expression

$$F\, dx_1\, dx_2 \dots dx_l$$

is an exact differential.

In the case when there is only one independent variable x and only one dependent variable y, whose derivatives up to order n are respectively $y', y'', \dots y^{(n)}$, the condition that Fdx may be an exact differential is readily found to be

$$\frac{\partial F}{\partial y} - \frac{d}{dx}\left(\frac{\partial F}{\partial y'}\right) + \frac{d^2}{dx^2}\left(\frac{\partial F}{\partial y''}\right) - \dots + (-)^n \frac{d^n}{dx^n}\left(\frac{\partial F}{\partial y^{(n)}}\right) = 0.$$

Now the quantity on the left-hand side of this equation is indeed the variational derivative of F with respect to y and will be denoted by the symbol $\delta F/\delta y$.

* For a systematic discussion of variational derivatives reference may be made to the papers of Th. de Donder in *Bulletin de l'Académie Royale de Belgique, Classe des Sciences* (5), t. xv (1929–30). In some cases a set of equations must be supplemented by another to give all the equations in a set of the variational form.

In the case when F is of the form

$$yN_x(z) - zM_x(y),$$

where z is a function of x and $M_x(y)$, $N_x(z)$ are linear differential expressions involving derivatives up to order n and coefficients of these derivatives which are functions of x with a suitable number of continuous derivatives, we can say that the differential expressions $M_x(y)$, $N_x(z)$ are adjoint when $\delta F/\delta y \equiv 0$ for all forms of the function z. When z is chosen to be a solution of the differential equation $N_x(z) = 0$ the expression $zM_x(y)\,dx$ is an exact differential and so z is an integrating factor of the differential equation adjoint to $N_x(z) = 0$. The relation between two adjoint differential equations is, moreover, a reciprocal one.

The idea of adjoint differential expressions was introduced by Lagrange and extended by Riemann to the case when there is more than one independent variable. Further extensions have been made by various writers for the case when there are several dependent variables*. Adjoint differential expressions and adjoint differential equations are now of great importance in mathematical analysis.

A second important property of the variational derivatives may be introduced by first considering a simple integral

$$I = \int_\alpha^\beta F(x, y, y', y'', \dots y^{(n)})\,dx,$$

and its first variation

$$\delta I = \int_\alpha^\beta \frac{\partial F}{\partial y}\,\delta y\,dx + \int_\alpha^\beta \frac{\partial F}{\partial y'}\,\delta y'\,dx + \dots + \int_\alpha^\beta \frac{\partial F}{\partial y^{(n)}}\,\delta y^{(n)}\,dx.$$

Integrating the $(s+1)$th term s times by parts, making use of the equations

$$\delta y' = \frac{d}{dx}(\delta y), \quad \delta y'' = \frac{d}{dx}(\delta y'), \text{ etc.},$$

it is readily seen that the portion of δI which still remains under the sign of integration is

$$\int_\alpha^\beta \frac{\delta F}{\delta y}\,\delta y\,dx.$$

It is readily understood now why the name "variational derivative" is used. The variational derivative is of fundamental importance in the Calculus of Variations because the Eulerian differential equation for a variational problem involving an integral of the above form is obtained by equating the variational derivative to zero.

This rule is capable of extension, and rules for writing down the variational derivatives of a function F in the general case when there are

* See for instance J. Kürschák, *Math. Ann.* Bd. LXII, S. 148 (1906); D. R. Davis, *Trans. Amer. Math. Soc.* vol. XXX, p. 710 (1928).

l independent variables and m dependent variables can be derived at once from the rules of § 2·42 for the derivation of the Eulerian equations.

Since our differential equations are always associated with variational problems, direct methods of solving these problems are of great interest.

The important method of approximation invented by Lord Rayleigh* and developed by W. Ritz† is only briefly mentioned in the text, though it has been used by Ritz‡, Timoshenko§ and many other writers‖ to obtain approximate solutions of many important problems. An adequate discussion by means of convergence theorems is rather long and difficult, and has been omitted from the text largely for this reason and partly because important modifications of the method have recently been suggested which lead more rapidly to the goal and furnish means of estimating the error of an approximation.

In Ritz's method a boundary problem for a differential equation $D(u) = 0$ is replaced by a variation problem in which a certain integral I is to be made a minimum, the unknown function u being subject to certain supplementary conditions which are usually linear boundary conditions and conditions of continuity. The function u_a, used by Ritz as an approximation for u, is not generally a solution of the differential equation, but it does satisfy the boundary conditions for all values of the arbitrary constants which it contains. The result is that when an integral I_a is calculated from u_a in the way that I is to be calculated from u, the integral I_a is greater than the minimum value I_m of I, even when the arbitrary constants in u_a are chosen so as to make I_a as small as possible. This means that I_m is approached from above by integrals of the type I_a.

Now it was pointed out by R. Courant¶ that I_m can often be approached from below by integrals I_b calculated from approximation functions u_b which satisfy the differential equation but are subject to less restrictive supplementary conditions. If, for instance, u is required to be zero on the boundary, the boundary condition may be loosened by merely requiring u_b to give a zero integral over the boundary in each of the cases in which it is first multiplied by a function v_s belonging to a certain finite set. This idea has been developed by Trefftz** who uses the arithmetical mean of I_a and I_b

* *Phil. Trans.* A, vol. CLXI, p. 77 (1870); *Scientific Papers*, vol. I, p. 57.

† W. Ritz, *Crelle*, Bd. CXXXV, S. 1 (1908); *Œuvres*, pp. 192–316 (Gauthier-Villars, Paris, 1911).

‡ W. Ritz, *Ann. der Phys.* (4), Bd. XXVIII, S. 737 (1909).

§ S. Timoshenko, *Phil. Mag.* (6), vol. XLVII, p. 1093 (1924); *Proc. London Math. Soc.* (2), vol. XX, p. 398 (1921); *Trans. Amer. Soc. Civil Engineers*, vol. LXXXVII, p. 1247 (1924); *Vibration Problems in Engineering* (D. Van Nostrand, New York, 1928).

‖ See especially M. Plancherel, *Bull. des Sciences Math.* t. XLVII, pp. 376, 397 (1923), t. XLVIII, pp. 12, 58, 93 (1924); *Comptes Rendus*, t. CLXIX, p. 1152 (1919); R. Courant, *Acta Math.* t. XLIX, p. 1 (1926); K. Friedrichs, *Math. Ann.* Bd. XCVIII, S. 205 (1927–8).

¶ R. Courant, *Math. Ann.* Bd. XCVII, S. 711 (1927).

** E. Trefftz, *Int. Congress of Applied Mechanics, Zürich* (1926), p. 135; *Math. Ann.* Bd. C, S. 503 (1928).

as a close approximation for I_m, and uses the difference of I_a and I_b as an upper bound for the error in this method of approximation. This method is simplified by a choice of functions v_s which will make it possible to find simple solutions of the differential equation for the loosened boundary conditions. Sometimes it is not the boundary conditions but the conditions of continuity which should be loosened, and this makes it advisable not to lose interest in a simple solution of a differential equation because it does not satisfy the requirements of continuity suggested by physical conditions.

In order that Ritz's method may be used we must have a sequence of functions which satisfy the boundary conditions and conditions of continuity peculiar to the problem in hand. It is advantageous also if these functions can be chosen so that they form an orthogonal set. To explain what is meant by this we consider for simplicity the case of a single independent variable x. The functions $\psi_1(x)$, $\psi_2(x)$, ..., defined in an interval $a \leqslant x \leqslant b$, are then said to form a normalised orthogonal set when the orthogonal relations

$$\begin{aligned}\int_a^b \psi_m(x)\,\psi_n(x)\,dx &= 0, \quad m \neq n,\\ &= 1, \quad m = n\end{aligned}$$

are satisfied for each pair of functions of the set. This definition is readily extended to the case of several independent variables and functions defined in a domain R of these variables; the only difference is that the simple integrals are replaced by integrals over the domain of definition. The definition may be extended also to complex functions $\psi_n(x)$ of the form $\alpha_n(x) + i\beta_n(x)$, where $\alpha_n(x)$ and $\beta_n(x)$ are real. The orthogonal relations are then of type

$$\begin{aligned}\int_a^b \psi_m(x)\,\overline{\psi_n(x)}\,dx \equiv \int_a^b [\alpha_m(x) + i\beta_m(x)]\,[\alpha_n(x) - i\beta_n(x)]\,dx &= 0, \quad m \neq n,\\ &= 1, \quad m = n.\end{aligned}$$

Many types of orthogonal functions are studied in this book. The trigonometrical functions sin (nx), cos (nx) with suitable factors form an orthogonal set for the interval $(0, 2\pi)$, the Legendre functions $P_n(x)$, with suitable normalising factors, form an orthogonal set for the interval $(-1, 1)$, while in Chapter IX sets of functions are obtained which are orthogonal in an infinite interval. The functions of Laplace, which form the complete system of spherical harmonics considered in Chapter VI, give an orthogonal set of functions for the surface of a sphere of unit radius, and it is easy to construct functions which are orthogonal in the whole of space. In Chapter IV methods are explained by which sets of normalised orthogonal functions may be associated with a given curve or with a given area. In many cases functions suitable for use in Ritz's method of approximation are furnished by the Lamé products defined in Chapters III–XI. These products are important, then, for both the exact and the approximate solution of problems. It was shown by Ritz, moreover, that sometimes the functions occurring in the exact solution of one problem may be used in the approximate solution of another; the functions giving the deflection of a clamped bar were in fact used in the form of products to represent the approximate deflection of a clamped rectangular plate.

Early writers* using Ritz's method were content to indicate the degree of approximation obtainable by applying the method to problems which could be solved exactly and comparing the approximate solution with the exact solution. This plan is somewhat unsatisfactory because the examples chosen may happen to be particularly favourable ones. Attempts have, however, been made by Kryloff† and others to estimate the error when an approximating function of order n, say

$$U_a = \psi_0(x) + c_1\psi_1(x) + c_2\psi_2(x) + \dots + c_n\psi_n(x),$$

is substituted in the integral to be minimised and the coefficients c_s are chosen so as to make the resulting algebraic expression a minimum. Attempts have been made also to determine the order n needed to make the error less than a prescribed quantity ϵ.

In Ritz's method a boundary problem for a given differential equation must first of all be replaced by a variation problem. There are, however, modifications of Ritz's method in which this step is avoided. If, for instance, the differential equation is a variational equation $\delta F/\delta u = 0$, the same set of equations for the determination of the constants c_s is obtained by substituting the expression U_a for u directly in the equation

$$\int_a^b \delta u \,.\, (\delta F/\delta u)\, dx = 0,$$

and equating to zero the coefficients of the variations δc_s.

This method has been recommended by Hencky‡ and Goldsbrough§; it has the advantage of indicating a reason why in the limit the function u_a should satisfy the differential equation.

Another method, proposed by Boussinesq|| many years ago, has been called the method of least squares. If the differential equation is

$$L_x(u) = f(x),$$

and $a \leqslant x \leqslant b$ is the range in which it is to be satisfied with boundary

* See, for instance, M. Paschoud, *Sur l'application de la méthode de W. Ritz*: Thèse (Gauthier-Villars, Paris, 1914).

† N. Kryloff, *Comptes Rendus*, t. CLXXX, p. 1316 (1925), t. CLXXXVI, p. 298 (1928); *Annales de Toulouse* (3), t. XIX, p. 167 (1927).

‡ H. Hencky, *Zeits. für angew. Math. u. Mech.* Bd. VII, S. 80 (1927).

§ G. R. Goldsbrough, *Phil. Mag.* (7), vol. VII, p. 333 (1929).

|| J. Boussinesq, *Théorie de la chaleur*, t. I, p. 316.

conditions at the ends, the constants c_s in an approximating function u_a, which satisfies these boundary conditions, are chosen so as to make the integral

$$\int_a^b [L_x(u_a) - f(x)]^2\,dx$$

as small as possible. The accuracy of this method has been studied by Kryloff* who believes that Ritz's method and the method of least squares are quite comparable in usefulness. The method of least squares is, of course, closely allied to the well-known method of approximating to a function $f(x)$ by a finite series of orthogonal functions

$$c_1\psi_1(x) + c_2\psi_2(x) + \ldots + c_n\psi_n(x),$$

the coefficients c_s being chosen so that the integral†

$$\int_a^b [f(x) - c_1\psi_1(x) - c_2\psi_2(x) - \ldots - c_n\psi_n(x)]^2\,dx$$

may be as small as possible. The conditions for a minimum lead to the equations

$$c_s = \int_a^b f(s)\,\psi_s(x)\,ds \quad (s = 1, 2, \ldots n).$$

For an account of such methods of approximation reference may be made to recent books by Dunham Jackson‡, S. Bernstein§ and de la Vallée Poussin‖.

In the discussion of the convergence of methods of approximation there is an inequality due to Bouniakovsky and Schwarz which is of fundamental importance. If the functions $f(x)$, $g(x)$ and the parameter c are all real, the integral

$$\int_a^b [f(x) + cg(x)]^2 = A + 2cH + c^2B$$

is never negative and so $AB - H^2 \geqslant 0$. This gives the inequality

$$\int_a^b [f(x)]^2\,dx \int_a^b [g(x)]^2\,dx \geqslant \int_a^b [f(x)\,g(x)]^2\,dx.$$

There is a similar inequality for two complex functions $f(x)$, $g(x)$ and

* N. Kryloff, *Comptes Rendus*, t. CLXI, p. 558 (1915); t. CLXXXI, p. 86 (1925). See also Krawtchouk, *ibid.* t. CLXXXIII, pp. 474, 992 (1926).

† G. Plarr, *Comptes Rendus*, t. XLIV, p. 984 (1857); A. Toepler, *Anzeiger der Kais. Akad. zu Wien* (1876), p. 205.

‡ D. Jackson, "The Theory of Approximation," *Amer. Math. Soc. Colloquium publications*, vol. XI (1930).

§ S. Bernstein, *Leçons sur les propriétés extrémales et la meilleure approximation des fonctions analytiques d'une variable réelle* (Gauthier-Villars, Paris, 1926).

‖ C. J. de la Vallée Poussin, *Leçons sur l'approximation des fonctions d'une variable réelle* (*ibid.* 1919).

their conjugates $\overline{f(x)}$, $\overline{g(x)}$. Indeed, if c and $\bar{c}$ are conjugate complex quantities, the integral

$$\int_a^b [f(x) + cg(x)]\,[\overline{f(x)} + \overline{cg(x)}]\,dx$$

is never negative. Writing

$$f = l + im, \quad g = p + iq, \quad c = \xi + i\eta,$$

where l, m, p, q, ξ, η are all real, the integral may be written in the form

$$A(\xi^2 + \eta^2) + 2B\xi + 2C\eta + D = \frac{1}{A}[(A\xi + B)^2 + (A\eta + C)^2 + AD - B^2 - C^2],$$

where
$$A = \int_a^b (p^2 + q^2)\,dx = \int_a^b g\bar{g}\,dx,$$
$$D = \int_a^b (l^2 + m^2)\,dx = \int_a^b f\bar{f}\,dx,$$
$$B + iC = \int_a^b f\bar{g}\,dx, \quad B - iC = \int_a^b \bar{f}g\,dx.$$

Since the integral is never negative, we have the inequality

$$AD \geqslant B^2 + C^2,$$

which may be written in the form*

$$\int_a^b |f|^2\,dx \int_a^b |\bar{g}|^2\,dx \geqslant \left|\int_a^b f\bar{g}\,dx\right|^2.$$

In this inequality the functions f and $\bar{g}$ may be regarded as arbitrary integrable functions. This inequality and the analogous inequality for finite sums are used in § 4·81.

In the approximate treatment of problems in vibration the natural frequencies are often computed with the aid of isoperimetric variation problems. Ritz's method is now particularly useful. If, for instance, the differential equation is

$$\frac{d^2u}{dx^2} + \lambda u = 0,$$

and the end conditions $u(a) = 0, \quad u(b) = 0,$

the aim is to make the integral

$$\int_a^b \left(\frac{du}{dx}\right)^2 dx$$

a minimum when the integral

$$\int_a^b [u(x)]^2\,dx$$

* This inequality is called Schwarz's inequality by E. Schmidt, *Rend. Palermo*, t. xxv, p. 58 (1908).

has an assigned value. This is accomplished by replacing u by a finite series in both integrals and reducing the problem to an algebraic problem. It was noted by Rayleigh that very often a single term in the series will give a good approximation to the frequency of the fundamental frequency of vibration. To obtain approximate values of the frequencies of overtones it is necessary, however, to use a series of several terms and then the work becomes laborious as it is necessary to solve an algebraic equation of high order. Many other methods of approximating to the frequencies of overtones are now available.

Trefftz has recently introduced a new method of approximating to the solution of a differential equation, in which the original variation problem $\delta I = 0$ is replaced by a modified variation problem $\delta I\ (\epsilon) = 0$ in such a way that the desired solution u can be expressed in the form

$$2\epsilon u = \{I\ (\epsilon)\}_{\min} - \{I\ (-\ \epsilon)\}_{\min}.$$

This method, combined with Trefftz's method of estimating the error of approximation to an integral such as $I\ (\epsilon)$, can lead to an estimate of the error involved in a computation of u. In the problem of the deflection of a clamped plate under a given distribution of load, the function $I\ (\epsilon)$ represents the potential energy when a concentrated load ϵ is placed at the point where the deflection u is required.

Courant has shown that the rapidity of convergence of a method of approximation can often be improved by modifying the variational problem, introducing higher derivatives in such a way that the Eulerian equation of the problem is satisfied whenever the original differential equation is satisfied. This device is useful also in applications of Trefftz's method.

An entirely different method of approximation is based on the use of difference equations which in the limit reduce to the differential equation of a problem. The early writers were content to adopt the principle, usually called Rayleigh's principle, that it is immaterial whether the limiting process is applied to the difference equations or their solutions. Some attempts have been made recently to justify this principle* and also to justify the use of a similar principle in the treatment of problems of the Calculus of Variations by a direct method, due to Euler, in which an integral is replaced by a finite sum. An example indicating the use of partial difference equations and finite sums is discussed in § 2·33.

* See the paper by N. Bogoliouboff and N. Kryloff, *Annals of Math.* vol. XXIX, p. 255 (1928). Many references to the literature are contained in this paper. In particular the method is discussed by R. B. Robbins, *Amer. Journ.* vol. XXXVII, p. 367 (1915).

CHAPTER I

THE CLASSICAL EQUATIONS

§ **1·11.** *Uniform motion.* It seems natural to commence a study of the differential equations of mathematical physics with a discussion of the equation

$$\frac{d^2x}{dt^2} = 0,$$

which is the equation governing the motion of a particle which moves along a straight line with uniform velocity. It may be thought at first that this equation needs no discussion because the general solution is simply

$$x = At + B,$$

where A and B are arbitrary constants, but in mathematical physics a differential equation is almost always associated with certain supplementary conditions, and it is this association which presents the most interesting problems.

A similar differential equation

$$\frac{d^2y}{dx^2} = 0$$

describes an essential property of a straight line, when x and y are interpreted as rectangular co-ordinates, and its solution

$$y = mx + c$$

is the familiar equation of a straight line: the property in question is that the line has a constant direction, the direction or slope of the line being specified by the constant m. For some purposes it is convenient to regard the line as a ray of light, especially as the conditions for the reflection and refraction of rays of light introduce interesting supplementary or boundary conditions, and there is the associated problem of geometrical foci of a system of lenses or reflecting surfaces.

If a ray starts from a point Q on the axis of the system and is reflected or refracted at the different surfaces of the optical system it will, after completely traversing the system, be transformed into a second ray which meets the axis of the system in a point Q, which is called the geometrical focus of Q. The problem is to find the condition that a given point Q may be the geometrical focus of another given point Q.

This problem is generally treated by an approximate method which illustrates very clearly the mathematical advantages gained by means of simplifying assumptions. It is assumed that the angle between the ray and the axis is at all times small, so that it can be represented at any time by dy/dx.

Let $y^2 = 4ax$ give an approximate representation of a refracting surface in the immediate neighbourhood of the point (0, 0) on the axis. If y is a small quantity of the first order the value of x given by this equation can be regarded as a small quantity of the second order if a is of order unity. Neglecting quantities of the second order we may regard x as zero and may denote the slope of the normal at (x, y) by

$$-\frac{dx}{dy} = \frac{y}{2a}.$$

Now let suffixes 1 and 2 refer to quantities relating to the two sides of the refracting surface. Since the angle between a ray and the normal to the refracting surface is approximately $dy/dx + y/2a$, the law of refraction is represented by the equations

$$\mu_2\left(\frac{dy_2}{dx} + \frac{y_2}{2a}\right) = \mu_1\left(\frac{dy_1}{dx} + \frac{y_1}{2a}\right),$$

$$y_2 = y_1.$$

Denoting by $[u]$ the discontinuity $u_2 - u_1$ in a quantity u, we have the boundary conditions

$$\mu_2\left[\frac{dy}{dx}\right] = (\mu_1 - \mu_2)\left(\frac{dy_1}{dx} + \frac{y_1}{2a}\right),$$

$$[y] = 0.$$

Dropping the suffixes we see that these boundary conditions are of type

$$\left[\frac{dy}{dx}\right] = A\frac{dy}{dx} + By,$$

$$[y] = 0,$$

where A and B are constants which may be either positive or negative.

In the case of a moving particle, which for the moment we shall regard as a billiard ball, a supplementary condition is needed when the ball strikes another ball, which for simplicity is supposed to be moving along the same line. If u_1, u_2 are the velocities of the f[illegible] ball before and after collision, U_1, U_2 those of the second ball before and after collision, the laws of impact give

$$u_2 - U_2 = -e(u_1 - U_1),$$

$$mu_2 + MU_2 = mu_1 + MU_1,$$

where e is the coefficient of restitution and m, M are the masses of the two balls. Regarding U_1 as known and eliminating U_2 we have

$$(M + m)\,u_2 = (m - eM)\,u_1 + M(1 + e)\,U_1.$$

Replacing $u_2 - u_1$ by $[dx/dt]$ we have the boundary conditions for the collision

$$[x] = 0, \qquad (M + m)\left[\frac{dx}{dt}\right] = M(1 + e)\left(U_1 - \frac{dx_1}{dt}\right).$$

These hold for the place $x = x_1$ where the collision occurs, x being the coordinate of the centre of the colliding ball.

The boundary conditions considered so far may be included in the general conditions

$$\left[\frac{dy}{dx}\right] = A\frac{dy}{dx} + By + C,$$

$$[y] = 0,$$

where A, B and C are constants associated with the particular boundary under consideration.

§ **1·12**. Other types of boundary condition occur in the theory of uniform fields of force.

A field of force is said to be uniform when the vector E which specifies the field strength is the same in magnitude and direction for each point of a certain domain D. Taking the direction to be that of the axis of x the field strength E may be derived from a potential $\dot{V}$ of type $V - Ex$ by means of the equation

$$E = -\frac{dV}{dx},$$

V being an arbitrary constant. This potential V satisfies the differential equation

$$\frac{d^2V}{dx^2} = 0$$

throughout the domain D.

Boundary conditions of various types are suggested by physical considerations. At the surface of a conductor V may have an assigned value. At a charged surface $\left[\frac{dV}{dx}\right]$ may have an assigned value, while there may be a surface at which $[V]$ has an assigned value (contact difference of potential).

With boundary conditions of the types that have already been considered many interesting problems may be formulated. We shall consider only two.

§ **1·13**. *Problem* 1. To find a solution of $d^2y/dx^2 = 0$ which satisfies the conditions

$$y = 0 \text{ when } x = 0 \text{ and when } x = 1; \quad [dy/dx] = -1 \text{ when } x = \xi;$$
$$[y] = 0.$$

The first condition is satisfied by writing

$$\begin{aligned} y &= Ax & x &< \xi \\ &= B(1-x) & x &> \xi. \end{aligned}$$

The condition $[y] = 0$ gives

$$A\xi = B(1-\xi),$$

and the condition $[dy/dx] = -1$ gives

$$A + B = 1.$$

$$\therefore\ A = 1 - \xi, \qquad B = \xi.$$

Hence
$$y = g(x, \xi) = x(1-\xi) \qquad (x \leqslant \xi)$$
$$= \xi(1-x) \qquad (x \geqslant \xi).$$

This function is called the Green's function for the differential expression d^2y/dx^2, on account of its analogy to a function used by George Green in the theory of electrostatics.

It may be remarked in the first place that a solution of type $P + Qx$ which satisfies the conditions $y = a$ when $x = 0$, $y = b$ when $x = 1$, is given by the formula

$$y = a\left(\frac{\partial g}{\partial \xi}\right)_{\xi=0} - b\left(\frac{\partial g}{\partial \xi}\right)_{\xi=1}.$$

Secondly, it will be noticed that $g(x, \xi)$ is a symmetrical function of x and ξ; in other words $g(x, \xi) = g(\xi, x)$.

A third property is obtained by considering a solution of $d^2y/dx^2 = 0$ which is a linear combination of a number of such Green's functions, for example,

$$y = \sum_{s=1}^{n} f_s g(x, \xi_s),$$

where $f_1, f_2, f_3, \ldots$ are arbitrary constants. The derivative dy/dx drops by an amount f_1 at ξ_1, by an amount f_2 at ξ_2, and so on.

Let us now see what happens when we increase the number of points $\xi_1, \xi_2, \xi_3, \ldots$ and proceed to a limit so that the sum is replaced by an integral

$$y = \int_0^1 g(x, \xi) f(\xi)\, d\xi \qquad \ldots\ldots\text{(A)}$$

$$= \int_0^x (1-x)\,\xi f(\xi)\, d\xi + \int_x^1 x(1-\xi) f(\xi)\, d\xi.$$

We find on differentiating that

$$\frac{dy}{dx} = -\int_0^x \xi f(\xi)\, d\xi + \int_x^1 (1-\xi) f(\xi)\, d\xi,$$

$$\frac{d^2y}{dx^2} = -xf(x) - (1-x)f(x) = -f(x),$$

the function $f(x)$ being supposed to be continuous in the interval $(0, 1)$. It thus appears that the integral is no longer a solution of the differential equation $d^2y/dx^2 = 0$, but is a solution of the non-homogeneous equation

$$\frac{d^2y}{dx^2} + f(x) = 0.$$

Conversely, if the function $f(x)$ is continuous in the interval $(0, 1)$ a solution of this differential equation and the boundary conditions, $y = 0$ when $x = 0$ and when $x = 1$, is given by the formula (A); this formula, moreover, represents a function which is continuous in the interval and has continuous first and second derivatives in the interval. Such a function will be said to be continuous (D, 2), or of class C'' (Bolza's notation).

§ **1·14**. *Problem* 2. To find a solution of $d^2y/dx^2 = 0$ and the supplementary conditions

$$y = 0 \text{ when } x = 0 \text{ and when } x = 1; \quad \left.\begin{aligned}[y] &= 0\\ n\,[dy/dx] + k^2 y &= 0\end{aligned}\right\} \text{ at } x = s/n,$$

where
$$s = 1, 2, 3, \ldots n - 1.$$

Let
$$y = A_s x + B_s, \qquad s - 1 < nx < s,$$

then the supplementary conditions give $B_1 = 0$, $A_n + B_n = 0$,

$$(A_2 - A_1)\frac{1}{n} + B_2 - B_1 = 0, \qquad n\,(A_2 - A_1) + k^2\,(A_1/n + B_1) \;= 0,$$

$$(A_3 - A_2)\frac{2}{n} + B_3 - B_2 = 0, \qquad n\,(A_3 - A_2) + k^2\,(2A_2/n + B_2) = 0,$$

$$(A_4 - A_3)\frac{3}{n} + B_4 - B_3 = 0, \qquad n\,(A_4 - A_3) + k^2\,(3A_3/n + B_3) = 0,$$

$$\ldots\ldots\ldots\ldots\ldots\ldots \qquad \ldots\ldots\ldots\ldots\ldots\ldots$$

$$\therefore\; nB_2 = A_1 - A_2, \quad nB_3 = A_1 + A_2 - 2A_3, \ldots$$

$$nB_s = (A_1 + A_2 + \ldots A_s) - sA_s,$$

$$n^2\,(A_{s+1} - A_s) + k^2\,(A_1 + A_2 + \ldots A_s) = 0,$$

$$n^2\,(A_{s+1} - 2A_s + A_{s-1}) + k^2 A_s = 0, \qquad s > 1.$$

This difference equation may be solved by writing $k^2 = 2n^2\,(1 - \cos\theta)$.

$$\therefore\; A_s = A_1 \cos(s-1)\,\theta + K \sin(s-1)\,\theta,$$

where K is a constant to be determined. Now

$$A_2 = A_1 + 2A_1\,(\cos\theta - 1) = A_1\,(2\cos\theta - 1),$$

therefore
$$K \sin\theta = A_1\,(\cos\theta - 1),$$

and so
$$A_s = A_1\,\frac{\sin s\theta - \sin\overline{s-1}\theta}{\sin\theta},$$

$$A_1 + A_2 + \ldots A_s = A_1\,\frac{\sin s\theta}{\sin\theta},$$

$$n\,(A_n + B_n) = A_1 + \ldots A_n = A_1\,\frac{\sin n\theta}{\sin\theta}.$$

The condition $0 = A_n + B_n$ is satisfied if $n\theta = r\pi$, where r is an integer. In the limit when $n = \infty$ this condition becomes

$$k = \lim_{n\to\infty} 2n \sin\frac{r\pi}{2n} = r\pi,$$

and this is exactly the condition which must be satisfied in order that the differential equation

$$\frac{d^2y}{dx^2} + k^2y = 0 \qquad \text{......(B)}$$

may possess a non-trivial solution which satisfies the boundary conditions $y = 0$ when $x = 0$ and when $x = 1$. The general solution of this equation is, in fact,

$$y = P \cos kx + Q \sin kx,$$

where P and Q are arbitrary constants. To make $y = 0$ when $x = 0$ we choose $P = 0$. The condition $y = 0$ when $x = 1$ is then satisfied with $Q \neq 0$ only if $\sin k = 0$, i.e. if $k = r\pi$.

The exceptional values of k^2 of type $(r\pi)^2$ are called by the Germans "Eigenwerte" of the differential equation (B) and the prescribed boundary conditions. A non-trivial solution $Q \sin (kx)$ which satisfies the boundary conditions is called an "Eigenfunktion." These words are now being used in the English language and will be needed frequently in this book. To save printing we shall make use of the abbreviation eit for Eigenwert and eif for Eigenfunktion. The conventional English equivalent for Eigenwert is characteristic or proper value and for Eigenfunktion proper function.

The theorem which has just been discussed tells us that the differential equation (B) and the prescribed boundary conditions have an infinite number of real eits which are all simple inasmuch as there is only one type of eif for each eit. The eits are, moreover, all positive.

The quantities
$$k_r^2 = \left(2n \sin \frac{r\pi}{2n}\right)^2$$

may be regarded as eits of the differential equation $d^2y/dx^2 = 0$ and the preceding set of boundary conditions. These eits are also positive, and in the limit $n \to \infty$ they tend towards eits of the differential equation (B) and the associated boundary conditions. The solution y corresponding to k is, for $s - 1 < nx < s$,

$$y = A_1 \operatorname{cosec}\left(\frac{r\pi}{n}\right)\left[\left(x - \frac{s}{n}\right)\left(\sin\frac{sr\pi}{n} - \sin\frac{\overline{s-1}\,.\,r\pi}{n}\right) + \frac{1}{n}\sin\frac{sr\pi}{n}\right], \qquad \text{......(C)}$$

and it is interesting to study the behaviour of this function as $n \to \infty$ to see if the function tends to the limit

$$\frac{A_1}{r\pi}\sin(r\pi x).$$

Let us write
$$A_0 = A_1\left(\frac{r\pi}{n}\right)\operatorname{cosec}\left(\frac{r\pi}{n}\right),$$

$$F_0(x) = \frac{A_0}{r\pi}\sin(r\pi x), \qquad F_1(x) = \frac{A_1}{r\pi}\sin(r\pi x),$$

and let us use $F(x)$ to denote the function (C) which represents a polygon with straight sides inscribed in the curve $y = F_0(x)$.

The closeness of the approximation of $F(x)$ to $F_1(x)$ can be inferred from the uniform continuity of $F_0(x)$.

Given any small quantity ϵ we can find a number $n(\epsilon)$ such that for any number n greater than $n(\epsilon)$ we have the inequality

$$\left| F_0(x) - F_0\left(\frac{s}{n}\right) \right| < \epsilon$$

for any point x in the interval $s - 1 \leqslant nx \leqslant s$ and for any value of s in the set 1, 2, 3, ... n. In particular

$$\left| F_0\left(\frac{s-1}{n}\right) - F_0\left(\frac{s}{n}\right) \right| < \epsilon.$$

Now $F(x) = F_0\left(\frac{s}{n}\right) + (s - nx)\left[F_0\left(\frac{s-1}{n}\right) - F_0\left(\frac{s}{n}\right)\right] \quad (s - 1 \leqslant nx < s).$

Therefore $$| F(x) - F_0(x) | < \epsilon + \epsilon.$$

On the other hand

$$| F_0(x) - F_1(x) | = | F_0(x) | \left[1 - \frac{n}{r\pi}\sin\left(\frac{r\pi}{n}\right)\right].$$

Therefore $$\leqslant A_1\left[\frac{r\pi}{n}\operatorname{cosec}\frac{r\pi}{n} - 1\right],$$

$$| F(x) - F_1(x) | \leqslant 2\epsilon + A_1\left[\frac{r\pi}{n}\operatorname{cosec}\frac{r\pi}{n} - 1\right].$$

But when ϵ is given we can also choose a number $m(\epsilon)$ such that for $n > m(\epsilon)$ we have the inequality

$$\left| A_1\left[\frac{r\pi}{n}\operatorname{cosec}\frac{r\pi}{n} - 1\right] \right| < \epsilon.$$

Consequently, by choosing n greater than the greater of the two quantities $n(\epsilon)$ and $m(\epsilon)$, if they are not equal, we shall have

$$| F(x) - F_1(x) | \leqslant 3\epsilon.$$

This inequality shows that as $n \to \infty$, $F(x)$ tends uniformly to the limit $F_1(x)$.

This method of obtaining a solution of the equation

$$\frac{d^2y}{dx^2} + k^2y = 0$$

from a solution of the simpler equation $d^2y/dx^2 = 0$ by a limiting process, can be extended so as to give solutions of other differential equations and specified boundary conditions, but the question of convergence must always be carefully considered.

§ **1·15**. *Fourier's theorem.* It seems very natural to try to find a solution of the equation

$$\frac{d^2y}{dx^2} + f(x) = 0,$$

and a prescribed set of supplementary conditions by expanding $f(x)$ in a series of solutions of $\frac{d^2y}{dx^2} + k^2y = 0$ and the prescribed supplementary conditions, because if

$$f(x) = \Sigma b_n \sin nx \qquad \text{......(A)}$$

the differential equation is formally satisfied by the series

$$y = \Sigma b_n \frac{\sin nx}{n^2},$$

and if the original series is uniformly convergent the two differentiations term by term of the last series can be justified. When the function $f(x)$ is continuous it is not necessary to postulate uniform convergence because Lusin has proved that if the series (A) converge at all points of an interval I to the values of a continuous function $f(x)$ then the series (A) is integrable term by term in the interval I. Unfortunately it has not been proved that an arbitrary continuous function can be expanded in a trigonometrical series. Indeed, we are faced with the question of the possibility of expanding a given function $f(x)$ in a trigonometrical series of type (A). This question is usually made more definite by stipulating the range of values of x for which the representation of $f(x)$ is required and the type of function $f(x)$ to which the discussion will be limited. A mathematician who starts out to find an expansion theorem for a perfectly arbitrary function will find after mature consideration that the programme is too ambitious*, as there are functions with very peculiar properties which make trouble for the mathematician who seeks complete generality. It is astonishing, however, that a function represented by a trigonometrical series is not of an exceedingly restricted type but has a wide degree of generality, and after the discussions of the subject by the great mathematicians of the eighteenth century it came as a great surprise when Fourier pointed out that a trigonometrical series could represent a function with a discontinuous derivative, and even a discontinuous function if a certain convention were adopted with regard to the value at a point of discontinuity. In Fourier's work the coefficients were derived by a certain rule now called Fourier's rule, though indications of it are to be found in the writings of Clairaut, Euler and d'Alembert. In the case of the sine-series the rule is that

$$b_n = \frac{2}{\pi}\int_0^\pi f(x) \sin nx\, dx,$$

and the range in which the representation is required is that of the interval $(0 < x < \pi)$. When the range is $(0 < x < 2\pi)$ and the complete trigonometrical series

$$f(x) = \tfrac{1}{2}a_0 + \sum_{n=1}^{\infty} a_n \cos nx + \sum_{n=1}^{\infty} b_n \sin nx$$

* For the history of the subject see Hobson's *Theory of Functions of a Real Variable* and Burkhardt's Report, *Jahresbericht der Deutschen Math. Verein*, vol. x (1908).

is to be used for the representation, Fourier's rule takes the form

$$a_n = \frac{1}{\pi}\int_0^{2\pi} f(x)\cos nx\,dx,$$

$$b_n = \frac{1}{\pi}\int_0^{2\pi} f(x)\sin nx\,dx, \qquad \text{......(B)}$$

and the coefficients a_n, b_n are called the Fourier constants of the function $f(x)$.

Unless otherwise stated the symbol $f(x)$ will be used to denote a function which is single-valued and bounded in the interval $(0, 2\pi)$ and defined outside this interval by the equation $f(x + 2\pi) = f(x)$.

For some purposes it is more convenient to use the range $(-\pi < u < \pi)$ and the variable $u = 2\pi - x$. If $f(x) = F(u)$ the coefficients in the expansion of $F(u)$ in a trigonometrical series of Fourier's type are given by formulae exactly analogous to (B) except that the limits are $-\pi$ and π instead of 0 and 2π.

The advantage of using the interval $(-\pi, \pi)$ instead of the interval $(0, 2\pi)$ is that if $F(u)$ is an odd function of u, i.e. if $F(-u) = -F(u)$, the coefficients a_n are all zero, and if $F(u)$ is an even function of u, i.e. if $F(-u) = F(u)$, the coefficients b_n are all zero. In one case the series becomes a sine-series and in the other case a cosine-series.

The possibility of the expansion of $f(x)$ in a Fourier series is usually established for a function of limited variation*, that is a function such that the sum

$$\sum_{s=0}^{n-1} |f(x_{s+1}) - f(x_s)|$$

is bounded and $\leqslant N$, say, for all sets of points of subdivision $x_1, x_2, \ldots x_{n-1}$ dividing the interval $(0, 2\pi)$ up into n parts and for all finite integral values of n. Such a function is also called a function of limited total fluctuation and a function of bounded variation.

In addition to this restriction on $f(x)$ it is also supposed that the integrals in the expressions for the coefficients exist in the ordinary sense†. In the case when the integral representing a_n is an improper integral it is assumed that the integral

$$\int_0^{2\pi} |f(x)|\,dx \qquad \text{......(C)}$$

is convergent. If x is any interior point of the interval $(0, 2\pi)$ it can be shown that when the foregoing conditions are satisfied the series is convergent and its sum is

$$\lim_{\epsilon \to 0} \tfrac{1}{2}[f(x+\epsilon) + f(x-\epsilon)]$$

* Whittaker and Watson's *Modern Analysis*, 3rd ed. p. 175.

† That is, in the Riemann sense. There are corresponding theorems for the cases in which other definitions of integral (such as those of Stieltjes and Lebesgue) are used.

when the limits of $f(x \pm \epsilon)$ exist, i.e. with a convenient notation

$$\tfrac{1}{2}[f(x+0)+f(x-0)] = \bar{f}(x), \text{ say.}$$

When the function $f(x)$ is continuous in an interval $(\alpha \leqslant x \leqslant \beta)$ contained in the interval $(0, 2\pi)$, is of limited variation in the last interval and the other conditions relating to the coefficients are satisfied, it can be shown* that the series is uniformly convergent for all values of x for which $\alpha + \delta \leqslant x \leqslant \beta - \delta$, where δ is any positive number independent of x.

When the conditions of continuity and limited variation are dropped and the function $f(x)$ is subject only to the conditions relating to the existence of the integrals in the formulae for the coefficients and the convergence of the integral (C), there is a theorem due to Fejér, which states that†

$$\bar{f}(x) = \lim_{m \to \infty} \frac{1}{m} \{A_0 + S_1(x) + S_2(x) + \dots S_{m-1}(x)\},$$

where $A_0 = \tfrac{1}{2}a_0,\quad A_n(x) = a_n \cos nx + b_n \sin nx,\quad S_m(x) = \sum_{n=0}^{m} A_n(x).$

This means that the series is summable in the Cesàro sense by the simple method of averaging which is usually denoted by the symbol $(C, 1)$.

This is a theorem of great generality which can be used in applied mathematics in place of Fourier's theorem. It is assumed, of course, that the limits

$$\lim_{\epsilon \to 0} f(x+\epsilon) = f(x+0), \quad \lim_{\epsilon \to 0} f(x-\epsilon) = f(x-0)$$

exist‡.

§ **1·16**. *Cesàro's method of summation*§. Let

$$s_n = u_1 + u_2 + \dots + u_n,$$

$$nS_n = s_1 + s_2 + \dots + s_n,$$

then, if $S_n \to S$ as $n \to \infty$, the infinite series

$$\sum_{n=1}^{\infty} u_n \qquad \text{......(1)}$$

is said to be summable $(C, 1)$ with a Cesàro sum S.

For consistency of the definition of a sum it must be shown that when the series (1) converges to a sum s, we have $s = S$. To do this we choose a positive integer n, such that

$$|s_{n+p} - s_n| < \epsilon \qquad \text{......(2)}$$

for all positive integral values of p. This is certainly possible when s exists and we have in the limit

$$|s - s_n| < \epsilon. \qquad \text{......(3)}$$

* Whittaker and Watson, *Modern Analysis*, 3rd ed. p. 179.

† *Ibid.* p. 169.

‡ When $f(x+0) = f(x-0)$ this implies that $f(x)$ is continuous at the point x.

§ *Bull. des Sciences Math.* (2), t. XIV, p. 114 (1890). See also Bromwich's *Infinite Series*.

Now let ν be an integer greater than n and let C_m be defined by the equation

$$\nu C_{m+1} = \nu - m, \qquad \text{......(4)}$$

then

$$S_\nu = c_1 u_1 + c_2 u_2 + \ldots + c_\nu u_\nu. \qquad \text{......(5)}$$

But $c_1 > c_2 > \ldots > c_\nu > 0$, hence it follows from (2) that

$$| c_{n+1} u_{n+1} + c_{n+2} u_{n+2} + \ldots + c_\nu u_\nu | < \epsilon c_{n+1},$$

i.e.

$$| S_\nu - (c_1 u_1 + c_2 u_2 + \ldots + c_n u_n) | < \epsilon c_{n+1}.$$

Making $\nu \to \infty$ we see that if S be any limit of S_ν

$$| S - s_n | < \epsilon. \qquad \text{......(6)}$$

Combining (3) and (6) we find that

$$| S - s | < 2\epsilon.$$

Since ϵ is an arbitrary small positive quantity it follows that $S = s$ and so the sequence S_ν has only one limit s.

§ **1·17**. *Fejér's theorem.* Let us now write $u_1 = A_0$, $u_{n+1} = A_n(x)$, then, by using the expressions for the cosines as sums of exponentials, it is readily found that*

$$S_m(x) = \frac{1}{2\pi}\int_{-\pi}^{\pi} \frac{\sin^2 m\theta}{m \sin^2\theta} f(t)\,dt, \qquad \text{......(A)}$$

where $2\theta = | x - t |$. Now the integrand is a periodic function of t of period 2π, consequently we may also write

$$S_m(x) = \frac{1}{2\pi}\int_{x-\pi}^{x+\pi} \frac{\sin^2 m\theta}{m \sin^2\theta} f(t)\,dt$$

$$= \frac{1}{\pi}\int_0^{\frac{1}{2}\pi} \frac{\sin^2 m\theta}{m \sin^2\theta} [f(x + 2\theta) + f(x - 2\theta)]\,d\theta.$$

Furthermore, since

$$\frac{\sin^2 m\theta}{\sin^2\theta} = m + 2(m-1)\cos 2\theta + 2(m-2)\cos 4\theta + \ldots + 2\cos 2(m-1)\theta$$

it is readily seen that

$$\int_0^{\frac{1}{2}\pi} \frac{\sin^2 m\theta}{m \sin^2\theta} = \tfrac{1}{2}\pi.$$

Writing $\phi_x(\theta) = f(x + 2\theta) + f(x - 2\theta) - 2\bar{f}(x)$,

and making use of the last equation, we find that

$$S_m(x) - \bar{f}(x) = \frac{1}{\pi}\int_0^{\frac{1}{2}\pi} \frac{\sin^2 m\theta}{m \sin^2\theta}\phi_x(\theta)\,d\theta,$$

where $\phi_x(\theta) \to 0$ as $\theta \to 0$.

Now if ϵ is any small positive quantity we can choose a number δ so that

$$| \phi_x(\theta) | < \epsilon$$

* The details of the analysis are given in Whittaker and Watson's *Modern Analysis*.

whenever $0 < \theta < \delta$, and if ϵ is independent of m the number δ may be regarded as independent of m.

Writing for brevity

$$\sin^2 m\theta = m \sin^2 \theta P(\theta), \quad \pi = 2\alpha$$

and noting that $P(\theta)$ is never negative, we have

$$\left| \int_0^\alpha P(\theta)\,\phi_x(\theta)\,d\theta \right| \leqslant \int_0^\delta P(\theta)\,|\,\phi_x(\theta)\,|\,d\theta + \int_\delta^\alpha P(\theta)\,|\,\phi_x(\theta)\,|\,d\theta$$

$$< \epsilon \int_0^\delta P(\theta)\,d\theta + \frac{1}{m\sin^2\delta}\int_\delta^\alpha |\,\phi_x(\theta)\,|\,d\theta$$

$$< \epsilon \int_0^\alpha P(\theta)\,d\theta + \frac{1}{m\sin^2\delta}\int_0^\alpha |\,\phi_x(\theta)\,|\,d\theta. \quad \text{...(B)}$$

Let us now suppose that $\int_{-\pi}^{\pi} |\,f(t)\,|\,dt$ exists, then

$$\int_0^\alpha |\,\phi_x(\theta)\,|\,d\theta$$

also exists, and by choosing a sufficiently large value of m we can make

$$\alpha\epsilon m \sin^2\delta > \int_0^\alpha |\,\phi(\theta)\,|\,d\theta.$$

This makes the second integral on the right of (B) less than $\alpha\epsilon$, which is also the value of the first integral. Therefore

$$|\,S_m(x) - \bar{f}(x)\,| < 2\alpha\epsilon/\pi = \epsilon;$$

consequently $\qquad S_m(x) \to \bar{f}(x)$ as $m \to \infty$.

When $f(x)$ is continuous throughout the interval $(-\pi \leqslant x \leqslant \pi)$ all the foregoing requirements are satisfied and in addition $\bar{f}(x) = f(x)$; consequently, in this case, $S_n(x) \to f(x)$, and this is true for each point x of the interval.

This celebrated theorem was discovered by Fejér*. The conditions of the theorem are certainly satisfied when the range $(-\pi \leqslant x \leqslant \pi)$ can be divided up into a finite number of parts in each of which $f(x)$ is bounded and continuous. Such a function is said to be continuous bit by bit (Stückweise stetig); the Cesàro sum for the Fourier series is then $\bar{f}(x)$ at any point of the range, $\bar{f}(x)$ and $f(x)$ being the same except at the points of subdivision.

§ **1·18**. *Parseval's theorem.* Let the function $f(x)$ be continuous bit by bit in the interval $(-\pi, \pi)$ and let its Fourier constants be a_n, b_n; it will then be shown that

$$\int_{-\pi}^{\pi} [f(x)]^2\,dx = \pi\left[\tfrac{1}{2}a_0^2 + \sum_{n=1}^{\infty}(a_n^2 + b_n^2)\right]. \quad \text{......(A)}$$

* *Math. Ann.* Bd. LVIII, S. 51 (1904).

We shall find it convenient to sum the series (A) by the Cesàro method*. This will give the correct value for the sum because the inequality

$$\int_{-\pi}^{\pi} [f(x)]^2\,dx - \pi\left[\tfrac{1}{2}a_0{}^2 + \sum_{n=1}^{m}(a_n{}^2 + b_n{}^2)\right]$$
$$= \int_{-\pi}^{\pi}\left[f(x) - \tfrac{1}{2}a_0 - \sum_{n=1}^{m} A_n(x)\right]^2 dx > 0$$

indicates that the series is convergent.

To find the sum $(C, 1)$ we have to find the limit of S_m where, by a simple extension of 1·17 (A),

$$S_m = \frac{1}{2\pi}\int_{-\pi}^{\pi}\int_{-\pi}^{\pi}\frac{\sin^2 m\theta}{m\sin^2\theta} f(x) f(t)\,dx\,dt,$$

2θ being equal to $|x - t|$.

Since the region of integration can be divided up into a finite number of parts in each of which the integrand is a continuous function of x and t, the double integral exists and can be transformed into a repeated integral in which x and θ are the new independent variables. The region for which θ lies between θ_0 and $\theta_0 + d\theta$, while x lies between x_0 and $x_0 + dx$ consists of two equal parts†; sometimes two, sometimes one and sometimes none of these parts lie within the region of integration. When this is taken into consideration the correct formula for the transformation of the integral is found to be

$$S_m = \frac{1}{2\pi}\int_0^{\pi} P(\theta)\,d\theta\left[\int_{-\pi}^{\pi-2\theta} f(x) f(x+2\theta)\,dx + \int_{2\theta-\pi}^{\pi} f(x) f(x-2\theta)\,dx\right]. \qquad \text{......(B)}$$

In the derivation of this result Fig. 1 will be found to be helpful. The lines M_1M_2, M_3M_4 are those on which θ has an assigned value, while N_1N_2, N_3N_4 are lines on which θ has a different assigned value. It will be noticed that a line parallel to the axis of t meets M_1M_2, M_3M_4 either once or twice, while it meets N_1N_2, N_3N_4 either once or not at all.

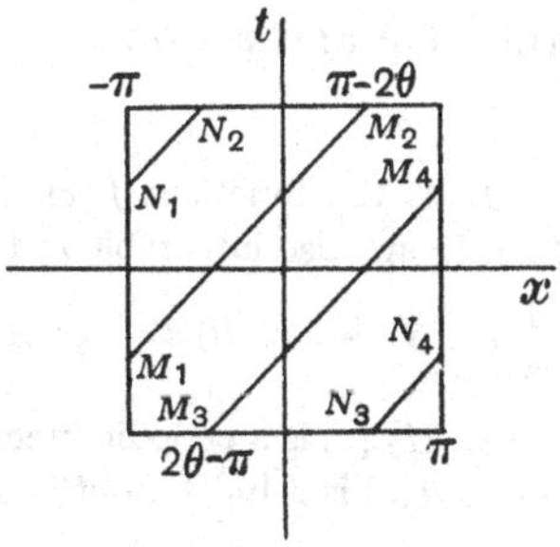

Fig. 1.

Applying the theorem of § 1·17 to (B) we get

$$\lim_{m\to\infty} S_m = \int_{-\pi}^{\pi} f(x)\bar{f}(x)\,dx,$$

and when $f(x)$ is defined to be $\bar{f}(x)$ this result gives (A).

* This is the plan adopted in Whittaker and Watson's *Modern Analysis*, p. 181. The present proof, however, differs from that given in *Modern Analysis*, which is for the case in which $f(x)$ is bounded and integrable.

† It will be noted that the Jacobian of the transformation has a modulus equal to two.

The theorem (A) was first proved by Liapounoff*; the present investigation is a modification of that given by Hurwitz†.

Now let $F(x)$ be a second function which is continuous bit by bit in the interval $-\pi \leqslant x \leqslant \pi$ and let A_n, B_n be its Fourier constants. Applying the foregoing theorem to $F(x)+f(x)$ and $F(x)-f(x)$, we obtain

$$\int_{-\pi}^{\pi} [F(x)+f(x)]^2\,dx = \pi\left[\tfrac{1}{2}(A_0+a_0)^2 + \sum_{n=1}^{\infty}\{(A_n+a_n)^2+(B_n+b_n)^2\}\right],$$

$$\int_{-\pi}^{\pi} [F(x)-f(x)]^2\,dx = \pi\left[\tfrac{1}{2}(A_0-a_0)^2 + \sum_{n=1}^{\infty}\{(A_n-a_n)^2+(B_n-b_n)^2\}\right].$$

Subtracting, we obtain the important formula

$$\int_{-\pi}^{\pi} f(x)\,F(x)\,dx = \pi\left[\tfrac{1}{2}A_0 a_0 + \sum_{n=1}^{\infty}(A_n a_n + B_n b_n)\right],$$

which is usually called Parseval's theorem, though Parseval's derivation of the formula was to some extent unsatisfactory.

In the modern theory, when Lebesgue integrals are used, the theorem is usually established for the case in which the functions $f(x)$, $F(x)$, $[f(x)]^2$ and $[F(x)]^2$ are integrable in the sense of Lebesgue. There is also a converse theorem which states that when the series (A) converges there is a function $f(x)$ with a_n and b_n as Fourier constants which is such that $[f(x)]^2$ is integrable and equal to the sum of the series. This theorem was first proved by Riesz and Fischer. Several proofs of the theorem are given in a paper by W. H. Young and Grace Chisholm Young‡. The theorem has also been extended by W. H. Young§, the complete theorem being also an extension of Parseval's theorem. A general form of Parseval's theorem has been used to justify the integration term by term of the product of a function and a Fourier series.

ADDITIONAL RESULTS

1. If the functions $f(x)$, $F(x)$ are integrable in the sense of Lebesgue, and $[f(x)]^2$, $[F(x)]^2$ are also integrable in the same sense, then‖

$$\frac{1}{\pi}\int_{-\pi}^{\pi} f(t+x)\,F(t)\,dt = \tfrac{1}{2}a_0A_0 + \sum_{n=1}^{\infty}(a_nA_n+b_nB_n)\cos nx - \sum_{n=1}^{\infty}(a_nB_n-b_nA_n)\sin nx.$$

2. If $f(x)$ is a periodic function of period 2π which is integrable in the sense of Lebesgue, and if $g(x)$ is a function of bounded variation which is such that the integral

$$\int_0^{\infty} |\,g(x)\,|\,dx$$

* *Comptes Rendus*, t. CXXVI, p. 1024 (1898).

† *Math. Ann.* Bd. LVII, S. 429 (1903).

‡ *Quarterly Journal*, vol. XLIV, p. 49 (1913).

§ *Comptes Rendus*, t. CLV, pp. 30, 472 (1912); *Proc. Roy. Soc. London*, A, vol. LXXXVII, p. 331 (1912); *Proc. London Math. Soc.* (2), vol. XII, p. 71 (1912). See also F. Hausdorff, *Math. Zeits.* Bd. XVI, S. 163 (1923).

‖ W. H. Young, *Comptes Rendus*, t. CLV, p. 30 (1912); *Proc. Roy. Soc. London*, A, vol. LXXXVII, p. 331 (1912).

is convergent, then the value of the integral

$$\int_0^\infty f(x)\, g(x)\, dx$$

may be calculated by replacing $f(x)$ by its Fourier series and integrating formally term by term. In particular, the theorem is true for a positive function $g(x)$ which decreases steadily as x increases and is such that the first integral is convergent.

[W. H. Young, *Proc. London Math. Soc.* (2), vol. IX, pp. 449, 463 (1910); vol. XIII, p. 109 (1913); *Proc. Roy. Soc.* A, vol. XXXV, p. 14 (1911). G. H. Hardy, *Mess. of Math.* vol. LI, p. 186 (1922).]

§ **1·19**. *The expansion of the integral of a bounded function which is continuous bit by bit.* If in Parseval's theorem we put

$$F(x) = 1, \quad -\pi < x < z, \quad F(x) = 0, \quad z < x < \pi,$$

we have

$$A_0 = \frac{1}{\pi}\int_{-\pi}^{\pi} F(x)\, dx = \frac{1}{\pi}\int_{-\pi}^{z} dx = \frac{z+\pi}{\pi},$$

$$A_n = \frac{1}{\pi}\int_{-\pi}^{z} \cos nx \,.\, dx = \frac{1}{n\pi}[\sin nz],$$

$$B_n = \frac{1}{\pi}\int_{-\pi}^{z} \sin nx \,.\, dx = \frac{1}{n\pi}[\cos n\pi - \cos nz],$$

and we have the result that

$$\int_{-\pi}^{z} f(x)\, dx = \tfrac{1}{2}a_0 (z+\pi) + \sum_{n=1}^{\infty} \frac{1}{n}[a_n \sin nz + b_n (\cos n\pi - \cos nz)]. \qquad \text{......(A)}$$

Now the function $\frac{1}{2}a_0 z$ can be expanded in the Fourier series

$$-a_0 \sum_{n=1}^{\infty} \frac{1}{n} \cos n\pi \sin nz,$$

hence the integral on the left of (A) can be expanded in a convergent trigonometrical series. To show that this is the Fourier series of the function we must calculate the Fourier constants.

Now

$$\frac{1}{\pi}\int_{-\pi}^{\pi} \sin nz\, dz \int_{-\pi}^{z} f(x)\, dx = -\frac{1}{n\pi}\cos n\pi \int_{-\pi}^{\pi} f(x)\, dx + \frac{1}{\pi}\int_{-\pi}^{\pi} \frac{dz}{n} f(z) \cos nz = \frac{a_n}{n} - \frac{a_0}{n}\cos n\pi,$$

$$\frac{1}{\pi}\int_{-\pi}^{\pi} \cos nz\, dz \int_{-\pi}^{z} f(x)\, dx = -\frac{1}{n\pi}\int_{-\pi}^{\pi} dz \sin nz\, f(z) = -\frac{1}{n} b_n,$$

$$\frac{1}{\pi}\int_{-\pi}^{\pi} dz \int_{-\pi}^{z} f(x)\, dx = \int_{-\pi}^{\pi} f(x)\, dx - \frac{1}{\pi}\int_{-\pi}^{\pi} z f(z)\, dz$$

$$\pi a_0 + \sum_{n=1}^{\infty} \frac{1}{n} b_n \cos n\pi,$$

by Parseval's theorem. Hence the coefficients are precisely the Fourier constants and so the integral of a function which is continuous bit by bit can be expanded in a Fourier series. This means that a continuous periodic function with a derivative continuous bit by bit can be expanded in a Fourier series.

Proofs of this theorem differing from that in the text are given by Hilbert-Courant, *Methoden der Mathematischen Physik*, Bd. I (1924), and by M. G. Carman, *Bull. Amer. Math. Soc.* vol. XXX, p. 410 (1924).

It should be noticed that equation (A) shows that when $f(x)$ can be expanded in a Fourier series this series can be integrated term by term. A more general theorem of this type is proved by E. W. Hobson, *Journ. London Math. Soc.* vol. II, p. 164 (1927).

Fourier's theorem may be extended to functions which become infinite in certain ways in the interval $(0, 2\pi)$. When the number of singularities is limited the singularities may be removed one by one by subtracting from $f(x)$ a simple function $h_s(x)$ with a singularity of the same type. This process is continued until we arrive at a function

$$g(x) = f(x) - \sum_{s=1}^{r} h_s(x)$$

which does not become infinite in the interval $(0, 2\pi)$. The problem then reduces to the discussion of the Fourier series associated with each of the functions $h_s(x)$.

§ **1·21**. *The bending of a beam.* We shall now consider some boundary problems for the differential equation $d^4y/dx^4 = 0$, which is the natural one to consider after $d^2y/dx^2 = 0$ from the historical standpoint and on account of the variety of boundary conditions suggested by mechanical problems.

The quantity y will be regarded here as the deflection from the equilibrium position of the central axis of a long beam at a point Q whose distance from one end is x. The beam will be assumed to have the same cross-section at all points of its length and to be of uniform material, also the deflection at each point will be regarded as small. The physical properties of the beam needed for the simple theory of flexure are then represented simply by the value of a certain quantity B which is called the flexural rigidity and which may be calculated when the form of the cross-section and the elasticity of the material of the beam are known. We are not interested at this stage in the calculation of B and shall consequently assume that the value of B for a given beam is known. The fundamental hypothesis on which the theory is based is that when the beam is bent by external forces there is at each point x of the central axis a resisting couple proportional to the curvature of the beam which just balances the bending moment introduced by the external forces. When the flexure takes place in the plane of xy this resisting couple has a moment which can be set equal to Bd^2y/dx^2 and the fundamental equation for the bending moment is

$$M = Bd^2y/dx^2.$$

The origin of the bending moment will be better understood when it is remarked that the bending moment M is associated with a transverse shearing force S by the equation

$$-S = dM/dx.$$

Fig. 2.

When the beam is so light that its weight may be disregarded, this shearing force S is constant along any portion of the beam that does not contain a point of support or point of attachment of a weight. If we have a simple cantilever OA built into a wall at O and carrying a weight W at the point B the shearing force S is zero from A to B and is W from B to O, while M is zero from A to B and equal to Wx between B and O. At the point O the fact that the beam is built in or clamped implies that $y=0$ and $dy/dx = 0$, consequently the equation

$$Bd^2y/dx^2 = -Wx + Wb$$

gives $y = -Wx^3/6B + Wbx^2/2B.$

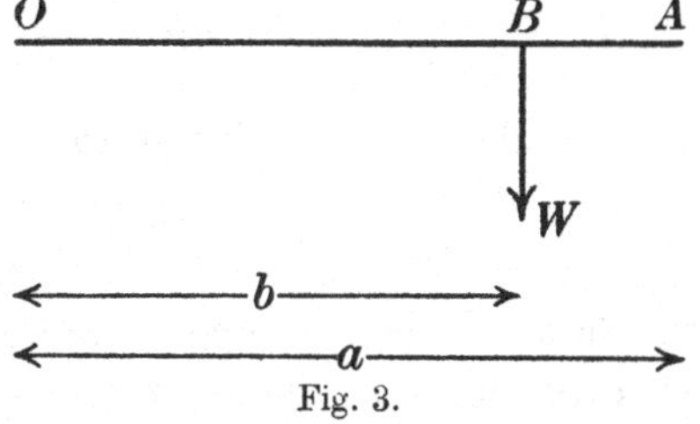

Fig. 3.

This holds for $x < b$. For $x > b$ the differential equation for y is

$$Bd^2y/dx^2 = 0,$$

and so

$$y = mx + c.$$

The quantities y and dy/dx are supposed to be continuous at B and so we have the equations

$$mb + c = Wb^3/3B, \qquad m = Wb^2/2B$$

which give $c = -Wb^3/6B$. The deflection of B is $Wb^3/3B$ and is seen to be proportional to the force W. The deflection of A is also proportional to W.

§ **1·22**. Let us next consider the deflection of a beam of length l which is clamped at both ends $x = 0$, $x = l$ and which carries a concentrated load W at the point $x = \xi$.

We have the equations

$x < \xi$	$x > \xi$
$S = T + W,$	$S = T$, say,
$-M = (T+W)x + N = -Bd^2y/dx^2,$	$-M = Tx + W\xi + N = -Bd^2y/dx^2,$
$\frac{1}{2}(T+W)x^2 + Nx = -Bdy/dx,$	$\frac{1}{2}Tx^2 - \frac{1}{2}Tl^2 + (W\xi + N)(x-l) = -Bdy/dx,$
$\frac{1}{6}(T+W)x^3 + \frac{1}{2}Nx^2 = -By,$	$\frac{1}{6}T(x^3 - l^3) - \frac{1}{2}Tl^2(x-l) + \frac{1}{2}(W\xi + N)(x-l)^2 = -By,$

where T and N are constants to be determined. M has been made continuous at $x = \xi$, but we have still to make y and dy/dx continuous. This gives the equations

$$\tfrac{1}{2}(W\xi^2 - Tl^2) = Wl\xi + Nl,$$

$$\tfrac{1}{3}(W\xi^3 + Tl^3) = \tfrac{1}{2}(-W + T)\,l^2\xi + \tfrac{1}{2}Nl\,(2\xi - l) + W\xi^2 l.$$

Therefore $\quad Tl^3 = W\xi^2(2\xi - 3l), \qquad l^2 N = W\xi\,(2\xi l - l^2 - \xi^2),$

$$By = -\frac{1}{6l^3}Wx^2(l-\xi)^2[x(l+2\xi) - 3\xi l] \qquad x < \xi$$

$$= -\frac{1}{6l^3}W\xi^2(l-x)^2[\xi(l+2x) - 3xl] \qquad x > \xi.$$

This solution will be written in the form $By = -Wg(x, \xi)$ and the function $g(x, \xi)$ will be called a Green's function for the differential expression d^4y/dx^4 and the prescribed boundary conditions.

If
$$By = -\int_0^l g(x, \xi)\, w(\xi)\, d\xi$$

it is found on differentiation that y is a solution of the differential equation

$$B\frac{d^4y}{dx^4} = w(x),$$

the function $w(x)$ being supposed to be continuous in the range $(0, l)$. This solution corresponds to the case of a distributed load of amount $w\,dx$ for a length dx. When w is independent of x the expression found for y is

$$By = \frac{w}{24}x^2(l-x)^2.$$

It should be noticed that the Green's function $g(x, \xi)$ is a symmetrical function of x and y, its first two derivatives are continuous at $x = \xi$, but the third derivative is discontinuous, in fact

$$\left[\frac{\partial^3}{\partial x^3}g(x, \xi)\right] = -1.$$

The reactions at the ends of the clamped beam with concentrated load are found by calculating the shear S. When $x < \xi$ we have

$$S = W(l-\xi)^2(l+2\xi)/l^3,$$

and this is equal in magnitude to the reaction at $x = 0$. The reaction at $x = l$ is similarly

$$R = W(3l - 2\xi)\,\xi^2/l^3.$$

The deflection of the point $x = \xi$ is

$$y_\xi = W\xi^3(l-\xi)^3/3l^3B,$$

when $\xi = l/2$ this amounts to $Wl^3/192B$.

In the case of the uniformly loaded beam the reactions at the ends are respectively $\tfrac{1}{2}W$ and $\tfrac{1}{2}W$ as we should expect. The deflection of the middle point is $Wl^4/384B$.

§ **1·23**. When the beam is pin-jointed at both ends, M is zero there and the boundary conditions are

$$y = 0, \quad d^2y/dx^2 = 0 \text{ for } x = 0 \text{ and } x = a.$$

When there is a concentrated load W at $x = \xi$ and the beam is of negligible weight, the solution is $By = Wk(x, \xi)$, where

$$\begin{aligned} k(x, \xi) &= x(a - \xi)(x^2 + \xi^2 - 2a\xi)/6a \qquad x \leqslant \xi \\ &= \xi(a - x)(x^2 + \xi^2 - 2ax)/6a \qquad x \geqslant \xi. \end{aligned}$$

The reactions at the supports are

$$\begin{aligned} R_0 &= W(1 - \xi/a) \quad \text{at } x = 0, \\ R_a &= W\xi/a \qquad\quad \text{at } x = a. \end{aligned}$$

As before, the deflection corresponding to a distributed load of density $w(x)$ is

$$By = \int_0^a k(x, \xi)\, w(\xi)\, d\xi,$$

and when w is constant

$$By = w(x^4 - 2ax^3 + a^3x)/24.$$

The reactions at the supports are in this case

$$R_0 = \frac{wa}{2} \qquad \text{at } x = 0,$$

$$R_a = \frac{wa}{2} \qquad \text{at } x = a.$$

In the case of a beam of length l clamped at the end $x = 0$ and pin-jointed at the end $x = l$, the solution for the case of a concentrated load W at $x = \xi$ is

$$\begin{aligned} 12Bl^3y &= Wx^2(l - \xi)[(\xi^2 - 2\xi l - 2l^2)(x - 3l) - 6l^3] \qquad x < \xi, \\ &= W\xi^2(l - x)[(x^2 - 2xl - 2l^2)(\xi - 3l) - 6l^3] \qquad x > \xi. \end{aligned}$$

The deflection at $x = \xi$ is now

$$y = W\xi^3(l - \xi)^2(4l - \xi)/12Bl^3 = 7Wl^3/768B$$

when $\xi = l/2$, and the reaction at $x = l$ is

$$R = W\xi^2(3l - \xi)/2l^3.$$

If, on the other hand, we consider a beam which is clamped at the end $x = 0$ but is free at the end $x = l$ except for a concentrated load P which acts there, we have, at the point $x = \xi$,

$$By_\xi = P\xi^2(3l - \xi)/6, \qquad \text{......(I)}$$

while at the point $x = l$

$$By_l = Pl^3/3.$$

Hence

$$R = Wy_\xi/y_l.$$

If the original beam is acted on by a number of loads of type W we have, for the reaction at the end $x = l$,

$$y_l R = \Sigma W_\xi y_\xi.$$

On account of this relation the curve (I) is called the "influence line" of the original beam. Much use is made now of influence lines in the theory of structures*.

There are three reciprocal theorems analogous to $g(x, \xi) = g(\xi, x)$ which are fundamental in the theory of influence lines. These theorems, which are due to Maxwell and Lord Rayleigh, may be stated as follows:

Consider any elastic structure with ends fixed or hinged, or with one end fixed and the other hinged, to an immovable support, then

(1) The displacement at any point A due to a load P applied at any point B is equal to the displacement at B due to the same load P placed at A instead of B.

(2) If the displacement at any point A is prevented by a load P at A with displacement y_B at B under a load Q, and alternatively if a load Q_1 at B prevents displacement at B with displacement y_A at A under a load P, then if $y_A = y_B$, P must equal Q_1.

(3) If a force Q acts at any point B producing displacements y_B at B and y_A at any other point A, and if a second force P is caused to act at A but in the opposite direction to Q reducing the displacement at B to zero, then $Q/P = y_A/y_B$.

In these three relationships it is supposed that the displacements are in the directions of the acting forces.

Proofs of these relations and some applications will be found in a paper by C. E. Larard, *Engineering*, p. 287 (1923).

§ **1·24**. Let us next consider a continuous beam with supports at A, B and C. The bending moment M at any point in AB or BC is the sum of bending moment M_1 of a beam which is pin-jointed at ABC and of the moment M_2 caused by the fixing moments at the supports. Let us take B as origin and let l_2 denote the length BC.

For a beam which is pin-jointed at B and C we have

$$M_1 = \tfrac{1}{2}w\,(l_2 x - x^2),$$

while for a weightless beam with fixing moments M_B, M_C at B and C respectively, we have

$$M_2 = -M_B - (M_C - M_B)\,x/l_2.$$

Hence $$M = B_2 d^2y/dx^2 = \tfrac{1}{2}w_2 l_2 x - \tfrac{1}{2}w_2 x^2 - M_B - (M_C - M_B)\,x/l_2.$$

Integrating, we have

$$B_2 dy/dx = \tfrac{1}{4}w_2 l_2 x^2 - w_2 x^3/6 - M_B x - x^2\,(M_C - M_B)/2l_2 - B_2 i_B,$$

where i_B is the value of dy/dx at $x = 0$. Integrating again,

$$B_2 y = w_2 l_2 x^3/12 - w_2 x^4/24 - x^2 M_B/2 - x^3\,(M_C - M_B)/6l_2 - B_2 i_B x.$$

When $x = l_2$, $y = -y_2$, say

$$6B_2 i_B = \tfrac{1}{4}w_2 l_2^3 - 2M_B l_2 - M_C l_2 + 6l_2^{-1} y_2.$$

* See especially Spofford, *Theory of Structures*; D. B. Steinman, *Engineering Record* (1916); G. E. Beggs, *International Engineering*, May (1922).

Similarly, by considering the span BA, taking B again as origin, but in this case taking x as positive when measured to the left,

$$-6B_1 i_B = \tfrac{1}{4}w_1 l_1^3 - 2M_B l_1 - M_A l_1 + 6l_1^{-1}y_1.$$

Eliminating i_B we obtain the equation

$$B_2 l_1 (M_A + 2M_B) + B_1 l_2 (M_C + 2M_B) = \tfrac{1}{4}(w_1 l_1^3 B_2 + w_2 l_2^3 B_1) + 6(B_2 l_1^{-1} y_1 + B_1 l_2^{-1} y_2).$$

This is the celebrated equation of three moments which was given in a simpler form by Clapeyron* and subsequently extended for the general case by Heppel†, Weyrauch‡, Webb§ and others||.

The reaction at B is the sum of the shears on the two sides of B and is therefore

$$R_B = \frac{w_2 l_2}{2} + \frac{M_B - M_C}{l_2} + \frac{w_1 l_1}{2} + \frac{M_B - M_A}{l_1}.$$

Similarly for the other supports.

§ **1·25**. When a light beam or thin rod originally in a vertical position is acted upon by compressive forces P at its ends (Fig. 4) the equation for the bending moment is

$$M = Bd^2y/dx^2 = -Py,$$

or

$$d^2y/dx^2 + k^2y = 0,$$

where

$$k^2 = P/B,$$

and if $y = 0$ when $x = 0$ and when $x = a$, the solution is $y = A \sin kx$, where $\sin ka = 0$ or $A = 0$.

If $ak < \pi$, the analysis indicates that $A = 0$. A solution with $A \neq 0$ becomes possible when $ak = \pi$. The corresponding load $P = B\pi^2/a^2$ is called Euler's critical load for a rod pinned at its ends. When P is given there is a corresponding critical length $a = P^{\frac{1}{2}}/B^{\frac{1}{2}}\pi$.

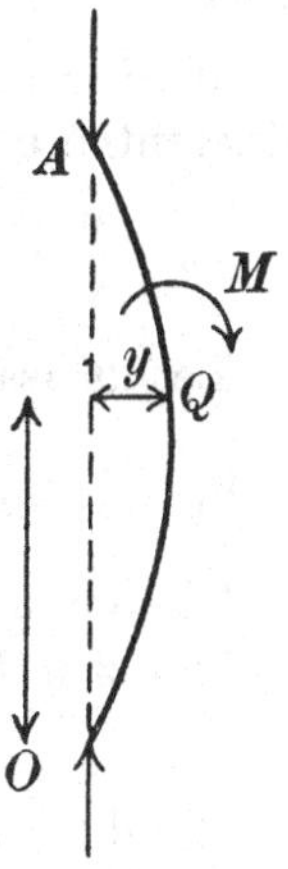

Fig. 4.

To obtain these critical values experimentally great care must be taken to eliminate initial curvature of the rod and bad centering of the loads. The formula of Euler has been confirmed by the experiments of Robertson. In general practice, however, the crippling load P_c is found to be less than the critical load P_0 given by Euler's formula, and many formulae for struts have been proposed. For these reference must be made to books on Elasticity and the Strength of Materials.

In the case of a strut clamped at both ends there is an unknown couple M_0 acting at each end. The equation is now

$$Bd^2y/dx^2 + Py = M_0,$$

* E. Clapeyron, *Comptes Rendus*, t. XLV, p. 1076 (1857).

† J. M. Heppel, *Proc. Inst. Civil Engineers*, vol. XIX, p. 625 (1859–60).

‡ Weyrauch, *Theorie der continuierlichen Träger*, pp. 8–9.

§ R. R. Webb, *Proc. Camb. Phil. Soc.* vol. VI (1886). (Case $B_1 \neq B_2$.)

|| M. Lévy, *Statique graphique*, t. II (Paris, 1886). (Case $y_1 \neq 0$, $y_2 \neq 0$.)

and the solution is of type

$$Py = M_0 + \alpha \cos kx + \beta \sin kx.$$

The boundary conditions $y = 0$, $dy/dx = 0$ at $x = a$ and $x = 0$ give

$$\beta = 0, \quad \alpha = -M_0, \quad M_0 \sin ka = 0, \quad M_0(1 - \cos ka) = 0.$$

Hence either $M_0 = 0$ or $\sin(ka/2) = 0$. The critical load is now given by the equation $ka = 2\pi$ and is $P_0 = 4B\pi^2/a^2$.

When the load P reaches the critical value the rod begins to buckle, and for a discussion of the equilibrium for a load greater than P_0 a theory of curved rods is needed.

In the case of a heavy horizontal beam of weight w per unit length and under the influence of longitudinal forces p at its ends, the equation satisfied by the bending moment M is

$$\frac{d^2M}{dx^2} + k^2M = w,$$

where

$$k^2 = P/B.$$

If $M = M_0$ when $x = 0$ and $M = M_1$ when $x = a$, the solution of the differential equation is

$$\left(\frac{w}{k^2} - M\right)\sin ka = \left(\frac{w}{k^2} - M_0\right)\sin k(a - x) + \left(\frac{w}{k^2} - M_1\right)\sin kx.$$

Let us assume that M_0 and M_1 are both positive and write

$$M_0 = \frac{2w}{k^2}\sin^2\theta, \quad M_1 = \frac{2w}{k^2}\sin^2\phi, \quad kx = \alpha, \quad k(a - x) = \beta, \quad ka = \alpha + \beta.$$

The equation which determines the points of zero bending moment (points of inflexion) is

$$\sin(\alpha + \beta) = \cos 2\theta \,.\, \sin\beta + \cos 2\phi \,.\, \sin\alpha.$$

We shall show that if α and β are both positive this equation implies that $\alpha + \beta \geqslant 2(\theta + \phi)$ and so determines a certain minimum length which must not be exceeded if there are to be two real points of inflexion.

Let us regard θ and ϕ as variable quantities connected by the last equation and ask when $\theta + \phi$ is a maximum. Writing $z = \theta + \phi$ we have

$$\frac{dz}{d\theta} = 1 + \frac{d\phi}{d\theta}, \qquad 0 = \sin 2\theta \sin\beta + \sin 2\phi \sin\alpha \frac{d\phi}{d\theta},$$

$$\frac{d^2z}{d\theta^2} = \frac{d^2\phi}{d\theta^2} = -\frac{2\sin\beta}{\sin\alpha \sin^2 2\phi}\left[\cos 2\theta \sin 2\phi - \sin 2\theta \cos 2\phi \frac{d\phi}{d\theta}\right].$$

When $2\theta = \alpha$, we have $2\phi = \beta$, and these values of θ and ϕ give a zero value of $\frac{dz}{d\theta}$ and a negative value of $\frac{d^2z}{d\theta^2}$, they therefore give us a maximum value of z, and so for ordinary values of θ and ϕ we have the inequality

$$2(\theta + \phi) \leqslant \alpha + \beta.$$

The position of the points of inflexion is of some practical interest because, in the first place, A. R. Low* has pointed out that instability is determined by the usual Eulerian formula for a pin-jointed strut of a length equal to the distance between the points of inflexion, if these lie on the beam, and secondly, if any splicing is to be done, the flanges should be spliced at one of the points where the bending moment is zero†.

When P is negative and so represents a pull we may put $p^2 = -P/B$, and the solution is

$$\left(\frac{w}{p^2} + M\right) \sinh pa = \left(\frac{w}{p^2} + M_0\right) \sinh p (l - x) + \left(\frac{w}{p^2} + M_1\right) \sinh px.$$

If we write

$$M_0 = \frac{2w}{p^2} \sinh^2 \theta, \quad M_1 = \frac{2w}{p^2} \sinh^2 \phi, \quad px = \alpha, \quad p (a - x) = \beta, \quad pa = \alpha + \beta,$$

a value of x for which $M = 0$ is determined by the equation

$$\sinh (\alpha + \beta) = \cosh 2\theta \sinh \beta + \cosh 2\phi \sinh \alpha.$$

This equation implies that

$$\alpha + \beta \geqslant 2 (\theta + \phi).$$

For a continuous beam acted on by longitudinal forces at the points of support there is an equation analogous to the equation of three moments which is obtained by a method similar to that used in obtaining the ordinary equation of three moments. We give only an outline of the analysis.

Case 1. $\quad k^2 = P_2/B_2, \quad BC = b, \quad 2\beta = bk, \quad y_A = y_B = y_C = 0,$

$$\frac{d^2M}{dx^2} + k^2 M = w_2, \quad M = M_B, \quad x = 0, \quad M = M_C, \quad x = b.$$

Therefore

$$M = \frac{M_C - M_B}{2} \frac{\sin kz}{\sin \beta} + \frac{M_C + M_B}{2} \frac{\cos kz}{\cos \beta} + \frac{w_2}{k^2}\left(1 - \frac{\cos kz}{\cos \beta}\right), \quad z = x - l/2,$$

$$B_2 y = \frac{M_B - M_C}{2k^2} \frac{\sin kz}{\sin \beta} - \frac{M_B + M_C}{2k^2} \frac{\cos kz}{\cos \beta} + \frac{w}{k^4}\left(\frac{\cos kz}{\cos \beta} - 1\right) + \frac{w_2}{2k^2} x (x - b) + \frac{M_c}{bk^2} x + \frac{M_b}{bk^2} (b - x),$$

$$B_2 \left(\frac{dy}{dx}\right)_0 = \frac{M_B - M_C}{2k\beta} (\beta \cot \beta - 1) - \frac{M_B + M_C}{2k} \tan \beta + \frac{w_2}{k^3} (\tan \beta - \beta)$$

$$= -2bM_B\, \phi (\beta) - bM_C f (\beta) + \frac{w_2 b^3}{4} \psi (\beta),$$

where $\quad f (\beta) = \dfrac{3}{2} \dfrac{2\beta \operatorname{cosec} 2\beta - 1}{\beta^2}, \quad \phi (\beta) = \dfrac{3}{4} \dfrac{1 - 2\beta \cot 2\beta}{\beta^2},$

$$\psi (\beta) = 3 \frac{\tan \beta - \beta}{\beta^3}.$$

* *Aeronautical Journal*, vol. XVIII, p. 144 (April, 1914). See also J. Perry, *Phil. Mag.* (March, 1892); A. Morley, *ibid.* (June, 1908); L. N. G. Filon, *Aeronautics*, p. 282 (Sept. 1919).

† H. Booth, *Aeronautical Journal*, vol. XXIV, p. 563 (1920).

There is a corresponding equation for the bay BA which is of length a, if $h^2 = P_1/B_1$, $2\alpha = ah$, the equation of three moments has the form

$$\frac{aM_A}{B_1} f(\alpha) + \frac{bM_C}{B_2} f(\beta) + 2M_B \left\{ \frac{a}{B_1} \phi(\alpha) + \frac{b}{B_2} \phi(\beta) \right\} = \frac{w_1 a^3}{4B_1} \psi(\alpha) + \frac{w_2 b^3}{4B_2} \psi(\beta).$$

Case 2. $P < 0$. The corresponding equations are

$$k^2 = -P_2/B_2, \quad h^2 = -P_1/B_1, \quad 2\alpha = ah, \quad 2\beta = bk,$$

$$\frac{aM_A}{B_1} F(\alpha) + \frac{bM_C}{B_2} F(\beta) + 2M_B \left\{ \frac{a}{B_1} \Phi(\alpha) + \frac{b}{B_2} \Phi(\beta) \right\}$$
$$= \frac{w_1 a^3}{4B_1} \Psi(\alpha) + \frac{w_2 b^3}{4B_2} \Psi(\beta),$$

$$F(\alpha) = \frac{3}{2} \frac{1 - 2\alpha \operatorname{cosech} 2\alpha}{\alpha^2}, \quad \Phi(\alpha) = \frac{3}{4} \frac{2\alpha \coth 2\alpha - 1}{\alpha^2},$$

$$\Psi(\alpha) = 3 \frac{\alpha - \tanh \alpha}{\alpha^3}.$$

The functions $f(\alpha)$, $F(\alpha)$, etc., have been tabulated by Berry* who has also given a complete exposition of the analysis. These equations are much used in the design of airplanes built of wood.

EXAMPLES

1. Find the crippling load for a rod which is clamped at one end and pinned at the other.

2. Prove that in the case of a uniform light beam of length a with a concentrated load W at $x = \xi$ the solution can be written in the form

$$M = \frac{2Wa}{\pi^2} \left(\sin \frac{\pi x}{a} \sin \frac{\pi \xi}{a} + \frac{1}{2^2} \sin \frac{2\pi x}{a} \sin \frac{2\pi \xi}{a} + \dots \right),$$

$$y = \frac{2Wa^3}{B\pi^4} \left(\sin \frac{\pi x}{a} \sin \frac{\pi \xi}{a} + \frac{1}{2^4} \sin \frac{2\pi x}{a} \sin \frac{2\pi \xi}{a} + \dots \right),$$

when the beam is pinned at both ends. The corresponding formulae for a uniformly distributed load are

$$w(x) = \frac{4w}{\pi} \left(\sin \frac{\pi x}{a} + \frac{1}{3} \sin \frac{3\pi x}{a} + \frac{1}{5} \sin \frac{5\pi x}{a} + \dots \right),$$

$$M = \frac{4wa^2}{\pi^3} \left(\sin \frac{\pi x}{a} + \frac{1}{3^3} \sin \frac{3\pi x}{a} + \frac{1}{5^3} \sin \frac{5\pi x}{a} + \dots \right),$$

$$y = \frac{4wa^4}{B\pi^5} \left(\sin \frac{\pi x}{a} + \frac{1}{3^5} \sin \frac{3\pi x}{a} + \frac{1}{5^5} \sin \frac{5\pi x}{a} + \dots \right).$$

[Timoshenko and Lessells *Applied Elasticity*, p. 230.]

3. Find the form of a strut pinned at its ends and eccentrically loaded at its ends with compressional loads P.

4. The Green's function for the differential expression†

$$\frac{d^2}{dx^2} \left[B(x) \frac{d^2 u}{dx^2} \right],$$

* *Trans. Roy. Aeronautical Soc.* (1919). The tables are given also in Pippard and Pritchard's *Aeroplane Structures*, App. I (1919).

† Examples 4–6 are taken from a paper by A. Myller, *Diss. Göttingen* (1906).

and the end conditions $u(0) = u(1) = u'(0) = u'(1) = 0$ is

$$G(x, \xi) = -\frac{1}{4}\int_0^1 \frac{|x-z|\,|\xi-z|}{B(z)}\,dz + \frac{\gamma}{4(\alpha\gamma-\beta^2)}\phi(\xi)\phi(x)$$
$$+\frac{\alpha}{4(\alpha\gamma-\beta^2)}\psi(\xi)\psi(x) - \frac{\beta}{4(\alpha\gamma-\beta^2)}[\psi(\xi)\phi(x)+\phi(\xi)\psi(x)],$$

where $$\phi(x) = \int_0^1 |x-z|\frac{dz}{B(z)}, \quad \psi(x) = \int_0^1 |x-z|\frac{z\,dz}{B(z)},$$
$$\alpha = \int_0^1 \frac{dz}{B(z)}, \quad \beta = \int_0^1 \frac{z\,dz}{B(z)}, \quad \gamma = \int_0^1 \frac{z^2dz}{B(z)}.$$

5. If in the last example the boundary conditions are $u(0) = u'(0) = u''(1) = u'''(1) = 0$ the Green's function is

$$G(x, \xi) = -\frac{1}{4}\int_0^1 \frac{|x-z|\,|\xi-z|}{B(z)}\,dz - \frac{1}{4}[\xi\phi(x) + x\phi(\xi)] + \frac{1}{4}[\psi(x)+\psi(\xi)]$$
$$-\frac{\alpha x\xi}{4} - \frac{\beta}{4}(x+\xi) - \gamma.$$

6. When the end conditions are $u(0) = u''(0) = u(1) = u''(1) = 0$, the Green's function is $G(x, \xi)$, where

$$4G(x, \xi) = -\int_0^1 \frac{|x-z|\,|\xi-z|\,dz}{B(z)} + \xi\phi(x) + x\phi(\xi) + (1-2\xi)\psi(x) + (1-2x)\psi(\xi)$$
$$-(\alpha - 4\beta + 4\gamma)x\xi - (\beta - 2\gamma)(x+\xi) - \gamma.$$

§ **1·31**. *Free undamped vibrations.* Whenever a particle performs free oscillations in a straight line under the influence of a restoring force proportional to the distance from a fixed point on the line the equation of motion is

$$m\ddot{x} = -x\mu,$$

where m is the mass of the particle and $x\mu$ is the restoring force. Writing $\mu = k^2m$, we have

$$\ddot{x} + k^2x = 0, \qquad \text{......(I)}$$

an equation which has already been briefly considered. The general solution is

$$x = A\cos(kt) + B\sin(kt),$$

where A and B are arbitrary constants. Writing $k = 2\pi n$, $A = a\sin\theta$, $B = a\cos\theta$ we have

$$x = a\sin(2\pi nt + \theta).$$

The quantity a specifies the amplitude, n the frequency and $2\pi nt + \theta$ the phase of the oscillation. The angle θ gives the phase at time $t = 0$. The period of vibration T may be found from the equations

$$kT = 2\pi, \quad nT = 1.$$

This type of vibration is called simple-harmonic vibration because it is of fundamental importance in the theory of sound. The vibrations of solid bodies which are almost perfectly rigid are often of this type, thus the end of a prong of a tuning fork which has been properly excited moves in a manner which may be described approximately by an equation of this type. The harmonic vibrations of the tuning fork produce corresponding vibrations in the surrounding air which are of audible frequency if

$$24 < n < 24000.$$

The range of frequencies used in music is generally

$$40 < n < 4000.$$

The differential equation (I) may be replaced by two simultaneous equations of the first order

$$\dot{x} + ky = 0, \quad \dot{y} - kx = 0 \qquad \text{......(II)}$$

which imply that the point Q with rectangular co-ordinates (x, y) moves in a circle with uniform speed ka. We have, in fact, the equation

$$x\dot{x} + y\dot{y} = 0,$$

which signifies that $x^2 + y^2$ is a constant which may be denoted by a^2. There is also an equation

$$\dot{x}^2 + \dot{y}^2 = k^2 (x^2 + y^2) = k^2a^2,$$

which indicates that the velocity has the constant magnitude ka. The solution of the simultaneous equations may be expressed in the form

$$x = a \cos \alpha, \quad y = a \sin \alpha,$$

where $$\alpha = 2\pi nt + \theta - \frac{\pi}{2}.$$

Simultaneous equations of type (II) describe the motion of a particle which is under the influence of a deflecting force perpendicular to the direction of motion and proportional to the velocity of the particle. The equations of motion are really

$$\ddot{x} + k\dot{y} = 0, \quad \ddot{y} - k\dot{x} = 0,$$

but an integration with respect to t and a suitable choice of the origin of co-ordinates reduces them to the form (II). The equations may also be written in the form

$$\dot{u} + kv = 0, \quad \dot{v} - ku = 0,$$

where (u, v) are the component velocities.

If the deflecting force mentioned above is the deflecting force of the earth's rotation the deflection is to the right of a horizontal path in the northern hemisphere and to the left in the southern hemisphere. If the angle ϕ represents the latitude of the place and ω the angular velocity of the earth's rotation, the quantity k is given by the formula

$$k = 2\omega \sin \phi.$$

When the resistance of the air can be neglected, the suspended mass M of a pendulum performs simple harmonic oscillations after it has been slightly displaced from its position of equilibrium. The vertical motion is now so small that it may be neglected and the acceleration may, to a first approximation, be regarded as horizontal and proportional to the horizontal component of the pull P of the string. We thus have the equation of motion

$$Ml\ddot{x} = -Px = -Mgx,$$

where l is the length of the string and g the acceleration of gravity. The mass of the string is here neglected. With this simplifying assumption the pendulum is called a simple pendulum. In dealing with connected systems of simple pendulums it is convenient to use the notation (l, M) for a simple pendulum whose string is of length l and whose bob is of mass M (Fig. 5).

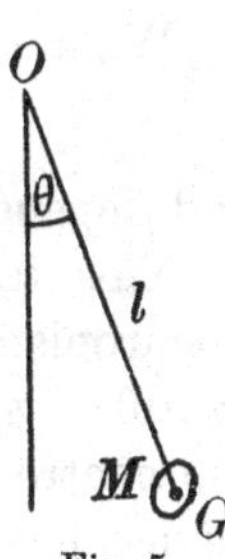

Fig. 5.

If the string and suspended mass are replaced by a rigid body free to swing about a horizontal axis through the point O, the equation of motion is approximately

$$I\ddot{\theta} = -Mgh\theta,$$

where I is the moment of inertia of the body about the horizontal axis through O and h is the depth of the centre of mass below the axis in the equilibrium position in which the centre of mass is in the vertical plane through O. Writing $Mhg = Ik^2$ the equation of motion becomes

$$\ddot{\theta} + k^2\theta = 0,$$

and the period of vibration is $2\pi/k$, a quantity which is independent of the angle through which the pendulum oscillates.

This law was confirmed experimentally by Galileo, who showed that the times of vibration of different pendulums were proportional to the square roots of their lengths. The isochronism of the pendulum for small oscillations was also discovered by him but had been observed previously by others. When the pendulum swings through an angle which is not exceedingly small it is better to use the more accurate equation

$$\ddot{\theta} + k^2 \sin\theta = 0,$$

which may be derived by resolving along the tangent to the path of the centre of gravity G or by differentiating the energy equation

$$\tfrac{1}{2}I\dot{\theta}^2 = Mgh(\cos\theta - \cos\alpha),$$

which is written down on the supposition that the velocity of G is zero when $\theta = \alpha$. With the aid of the substitution

$$\sin(\tfrac{1}{2}\theta) = \sin(\tfrac{1}{2}\alpha)\sin\phi,$$

this equation may be written in the form

$$\dot{\phi}^2 = k^2[1 - \sin^2\tfrac{1}{2}\alpha\sin^2\phi].$$

As θ varies from $-\alpha$ to α, ϕ varies from $-\dfrac{\pi}{2}$ to $\dfrac{\pi}{2}$, and so the time of a swing from one extreme position ($\theta = -\alpha$) to the next extreme position ($\theta = \alpha$) is

$$k^{-1}\int_{-\frac{\pi}{2}}^{\frac{\pi}{2}} (1 - \sin^2 \tfrac{1}{2}\alpha \sin^2 \phi)^{-\frac{1}{2}}\, d\phi.$$

When α is small the period T is given approximately by the formula

$$kT = 2\pi\,(1 + \tfrac{1}{4}\sin^2 \tfrac{1}{2}\alpha)$$

and depends on α, so that there is not perfect isochronism.

This fact was recognised by Huygens who discovered that perfect isochronism could theoretically be secured by guiding the string (or other flexible suspension) with the aid of a pair of cycloidal cheeks so as to make the centre of gravity describe a cycloidal instead of a circular arc. This device has not, however, proved successful in practice as it introduces errors larger than those which it is supposed to remove*. More practicable methods of securing isochronism with a pendulum have been described by Phillips†.

§ **1·32.** Simultaneous equations of type

$$L\ddot{x} + M\ddot{y} + Lm^2x = 0,$$
$$M\ddot{x} + N\ddot{y} + Nn^2y = 0,$$

in which L, M, N, m, n are constants, occur in many mechanical and electrical problems. When the coefficient M is zero the co-ordinates x and y oscillate in value independently with periods $2\pi/m$ and $2\pi/n$ respectively, but when $M \neq 0$ the assumption

$$x = p^2MAe^{ipt}, \quad y = LA\,(m^2 - p^2)\,e^{ipt}$$

gives the equation

$$p^4\,(1 - \gamma^2) - p^2\,(m^2 + n^2) + m^2n^2 = 0,$$

where

$$\gamma^2 = M^2/LN.$$

This quantity γ is called the coefficient of coupling‡.

When $m \neq n$ we have

$$p^2 - m^2 = \frac{\gamma^2p^4}{p^2 - n^2},$$

and when γ is small the value of p which is close to m is given approximately by the equation

$$p^2 - m^2 = \frac{\gamma^2m^4}{m^2 - n^2} = k^2\gamma^2, \text{ say.}$$

A simple harmonic oscillation of the x-co-ordinate, with a period close to the free period $2\pi/m$, is accompanied by a similar oscillation of the

* See R. A. Sampson's article on "Clocks and Time-Keeping" in *Dictionary of Applied Physics*, vol. III.

† *Comptes Rendus*, t. CXII, p. 177 (1891).

‡ See, for instance, E. H. Barton and H. Mary Browning, *Phil. Mag.* (6), vol. XXXIV, p. 246 (1917).

y-co-ordinate with the same period but opposite phase. The amplitude of the y-oscillation is proportional to γ^2. Now let p_1 be the greater of the two values of p. If $m \geqslant n$ we have $p_1 > m$ but if $m < n$ we have $p_2 < m$. The effect of the coupling is thus to lower the frequency of the gravest mode of vibration and to raise the frequency of the other mode of simple harmonic vibration. If $m = n$ the equation for p^2 gives

$$p^2 = m^2 \pm \gamma p^2,$$

and the effect of the coupling is to make the periods of the two modes unequal. In the general case we can say that the effect of the coupling is to increase the difference between the periods. The periods may, in fact, be represented geometrically by the following construction:

Let OA, OB represent the squares of the free periods, the points O, A, B being on a straight line. Now draw a circle Γ on AB as diameter and let a larger concentric circle cut the line OAB in U and V; the distances OU, OV then represent the squares of the periods when there is coupling. If a tangent from O to the circle Γ touches this circle at T and meets the larger circle in the points M and L the coefficient of coupling is represented by the ratio TL/TO (Fig. 6).

So long as O lies outside the larger circle it is evident that the difference between the periods is increased by the coupling, but when $\gamma > 1$ the point O lies within the larger circle and the difference between the periods decreases to zero as the radius CU of this circle increases without limit. There is thus some particular value of the coupling for which the difference between the periods has the original value, both periods being greater than before.

When $\gamma = 1$ the equations of motion may be written in the forms

$$L\ddot{x} + M\ddot{y} + Lm^2x = 0, \quad L\ddot{x} + M\ddot{y} + Mn^2y = 0,$$

and imply that $Lm^2x = Mn^2y$. There is now only one period of vibration. The cases $\gamma \geqslant 1$ are not of much physical interest as the values of the constants are generally such that $M^2 < LN$, this being the condition that the kinetic energy may be always positive.

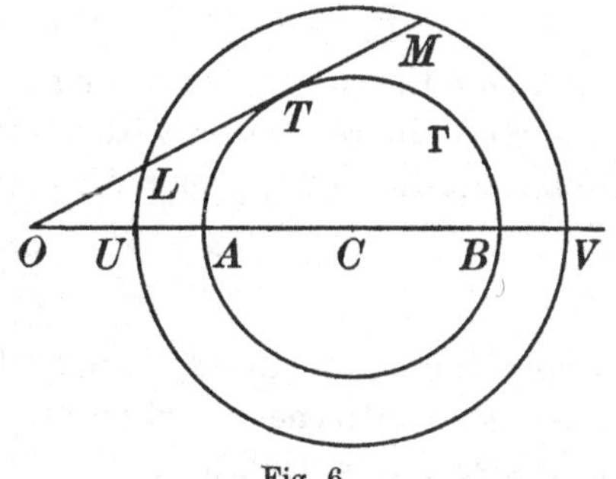

Fig. 6.

Equations of the present type occur in electric circuit theory when resistances are neglected. In the case of a simple circuit of self-induction L and capacity C furnished, say, by a Leyden jar in the circuit, the charge Q on the inside of the jar fluctuates in accordance with the equation

$$L\ddot{Q} + \frac{Q}{C} = 0$$

when the discharge is taking place. The period of the oscillations is thus

$$T = 2\pi / \sqrt{(LC)}.$$

This is the result obtained by Lord Kelvin in 1857 and confirmed by the experiments of Fedderson in 1857. The oscillatory character of the discharge had been suspected by Joseph Henry from observations on the magnetization of needles placed inside a coil in a discharging circuit.

In the case of two coupled circuits (L_1, C_1), (L_2, C_2) the mutual induction M needs to be taken into consideration and the equations for free oscillations are

$$L_1\ddot{Q}_1 + M\ddot{Q}_2 + \frac{Q_1}{C_1} = 0,$$

$$L_2\ddot{Q}_2 + M\ddot{Q}_1 + \frac{Q_2}{C_2} = 0.$$

§ **1·33**. *The Lagrangian equations of motion.* Consider a mechanical system consisting of l material points of which a representative one has mass m and co-ordinates x, y, z at time t. Using square brackets to denote a summation over these material points, we may express d'Alembert's principle in the Lagrangian form

$$[m\,(\ddot{x}\,\delta x + \ddot{y}\,\delta y + \ddot{z}\,\delta z)] = [X\,\delta x + Y\,\delta y + Z\,\delta z],$$

where δx, δy, δz are arbitrary increments of the co-ordinates which are compatible with the geometrical conditions limiting the freedom of motion of the system. On account of these conditions, the number of degrees of freedom is a number N, which is less than $3l$, and it is advantageous to introduce a set of "generalised" co-ordinates $q_1, q_2, \ldots q_N$ which are independent in the sense that any infinitesimal variation δq_s of q_s is compatible with the geometrical conditions. These conditions may, indeed, be expressed in the form

$$x = f\,(q_1, q_2, \ldots q_N, t), \quad y = g\,(q_1, q_2, \ldots q_N, t), \quad z = h\,(q_1, q_2, \ldots q_N, t).$$

Using the sign Σ to denote a summation from 1 to N, a prime to denote a partial differentiation with respect to t and a suffix s to denote a partial differentiation of x, y or z with respect to q_s, we have the equations

$$\dot{x} = x' + \Sigma x_s \dot{q}_s, \quad \delta x = \Sigma x_s \delta q_s,$$

$$[X\,\delta x + Y\,\delta y + Z\,\delta z] = \Sigma Q^{(s)}\,\delta q_s,$$

where the quantities $Q^{(s)}$ may be called generalised force components associated with the co-ordinates q. The first of these equations shows that x_s is also the partial derivative of $\dot{x}$ with respect to $\dot{q}_s$ and so if the kinetic energy of the system is T, where

$$2T = [m\,(\dot{x}^2 + \dot{y}^2 + \dot{z}^2)],$$

we shall have

$$\frac{\partial T}{\partial \dot{q}_s} = [m\,(\dot{x}x_s + \dot{y}y_s + \dot{z}z_s)] = p_s,$$

where p_s is the generalised component of momentum. Since

$$\frac{\partial \dot{x}}{\partial q_r} = \frac{\partial x'}{\partial q_r} + \Sigma\, x_{rs}\dot{q}_s = \frac{dx_r}{dt} = \dot{x}_r,$$

and

$$\frac{d}{dt}(\dot{x}x_s) = \ddot{x}x_s + \dot{x}\dot{x}_s,$$

we have

$$[m\,(\ddot{x}\,\delta x + \ddot{y}\,\delta y + \ddot{z}\,\delta z)] = \Sigma\delta q_s\left[\frac{d}{dt}m\,(\dot{x}x_s + \dot{y}y_s + \dot{z}z_s) - m\,(\dot{x}\dot{x}_s + \dot{y}\dot{y}_s + \dot{z}\dot{z}_s)\right]$$

$$= \Sigma\delta q_s\left\{\frac{d}{dt}\left(\frac{\partial T}{\partial \dot{q}_s}\right) - \frac{\partial T}{\partial q_s}\right\},$$

and so Lagrange's principle may be written in the form

$$\Sigma\delta q_s\left\{\frac{d}{dt}\left(\frac{\partial T}{\partial \dot{q}_s}\right) - \frac{\partial T}{\partial q_s}\right\} = \Sigma Q^{(s)}\delta q_s.$$

On account of the arbitrariness of the increments δq_s this relation gives the Lagrangian equations of motion

$$\frac{d}{dt}\left(\frac{\partial T}{\partial \dot{q}_s}\right) - \frac{\partial T}{\partial q_s} = Q^{(s)}.$$

If there is a potential energy function V, which can be expressed simply in terms of the generalised co-ordinates q, we may write

$$Q^{(s)} = -\,V_s \equiv -\,\partial V/\partial q_s, \quad L = T - V,$$

and the equations of motion take the simple form

$$\frac{d}{dt}\left(\frac{\partial L}{\partial \dot{q}_s}\right) - \frac{\partial L}{\partial q_s} = 0. \qquad \text{......(A)}$$

The quantity L is called the Lagrangian function.

Introducing the reciprocal function

$$\bar{T} = \Sigma\left(\dot{q}_s\frac{\partial T}{\partial \dot{q}_s}\right) - T,$$

we have

$$\frac{d\bar{T}}{dt} = \Sigma\left\{\dot{q}_s\frac{d}{dt}\left(\frac{\partial T}{\partial \dot{q}_s}\right) + \ddot{q}_s\frac{\partial T}{\partial \dot{q}_s}\right\} - \Sigma\left(\ddot{q}_s\frac{\partial T}{\partial \dot{q}_s} + \dot{q}_s\frac{\partial T}{\partial q_s}\right)$$

$$= \Sigma\dot{q}_s\left\{\frac{d}{dt}\left(\frac{\partial T}{\partial \dot{q}_s}\right) - \frac{\partial T}{\partial q_s}\right\} = -\,\Sigma\dot{q}_s V_s = -\frac{dV}{dt}.$$

Hence we have the energy equation

$$\bar{T} + V = \text{constant}.$$

When the functions f, g and h do not contain the time explicitly we have on account of Euler's theorem for homogeneous functions

$$2T = \Sigma\dot{q}_s\frac{\partial T}{\partial \dot{q}_s}, \quad \bar{T} = T.$$

The Lagrangian equations of motion may be replaced by another set of equations for the quantities p and q. For this purpose we introduce the Hamiltonian function H defined by

$$H\,(q_1, q_2, \ldots q_N, p_1, \ldots p_N) = -\,L + \Sigma p_s\dot{q}_s.$$

If we always consider H as a function of the quantities $\dot{q}_s$ and p_s but L as a function of q_s and $\dot{q}_s$, we have

$$\frac{\partial H}{\partial p_i} = - \Sigma \frac{\partial L}{\partial \dot{q}_s}\frac{\partial \dot{q}_s}{\partial p_i} + \Sigma p_s \frac{\partial \dot{q}_s}{\partial p_i} + \dot{q}_i,$$

$$\frac{\partial H}{\partial q_i} = - \frac{\partial L}{\partial q_i} - \Sigma \frac{\partial L}{\partial \dot{q}_s}\frac{\partial \dot{q}_s}{\partial q_i} + \Sigma p_s \frac{\partial \dot{q}_s}{\partial q_i}.$$

Thus

$$\frac{\partial H}{\partial p_i} = \dot{q}_i, \quad \frac{\partial H}{\partial q_i} = - \frac{\partial L}{\partial q_i},$$

consequently the equations of motion can be expressed in the Hamiltonian form

$$\frac{dq_i}{dt} = \frac{\partial H}{\partial p_i}, \quad \frac{dp_i}{dt} = - \frac{\partial H}{\partial q_i}.$$

Systems of equations whose solutions represent superposed simple harmonic vibrations are derived from the Lagrangian equations of motion of a dynamical system

$$\frac{d}{dt}\left(\frac{\partial T}{\partial \dot{q}_s}\right) - \frac{\partial T}{\partial q_s} = - \frac{\partial V}{\partial q_s},$$

$$s = 1, 2, \ldots N,$$

whenever the kinetic energy T, and the potential energy V, can be expressed for small displacements and velocities in the forms

$$2T = \sum_{m=1}^{N} \sum_{n=1}^{N} a_{mn} \dot{q}_m \dot{q}_n,$$

$$2V = \sum_{m=1}^{N} \sum_{n=1}^{N} c_{mn} q_m q_n$$

respectively, where the constant coefficients a_{mn} and c_{mn} are such that T and V are positive whenever the quantities q_m, $\dot{q}_m$ do not all vanish. For such a system the equations (A) give the differential equations

$$\sum_{n=1}^{N} (a_{mn}\ddot{q}_n + c_{mn} q_n) = 0,$$

$$m = 1, 2, \ldots N.$$

Multiplying by u_m, where u_m is a constant to be determined, and summing with respect to m, the resulting equation is of type

$$\ddot{v} + k^2 v = 0, \qquad \ldots\ldots(B)$$

if the quantities u_m are such that for each value of n

$$\sum_{m=1}^{N} u_m c_{mn} = k^2 \sum_{m=1}^{N} u_m a_{mn} = k^2 b_n, \text{ say.}$$

The corresponding value of v is then

$$v = \sum_{n=1}^{N} b_n q_n.$$

Now when the quantities u_m are eliminated from the linear homogeneous equations

$$\sum_{m=1}^{N} (c_{mn} - k^2 a_{mn})\, u_m = 0, \qquad \text{......(C)}$$

we obtain an algebraic equation of the Nth degree for k^2. With the usual method of elimination this equation is expressed by the vanishing of a determinant and may be written in the abbreviated form

$$| \, c_{mn} - k^2 a_{mn} \, | = 0.$$

To show that the values of k^2 given by this equation are all real and positive, we substitute $k^2 = h + ij$, $u_m = v_m + iw_m$ in equation (C). Equating the real and imaginary parts, we have

$$\sum_{m=1}^{N} (c_{mn} - ha_{mn})\, v_m + j \sum_{m=1}^{N} a_{mn} w_m = 0,$$

$$\sum_{m=1}^{N} (c_{mn} - ha_{mn})\, w_m - j \sum_{m=1}^{N} a_{mn} v_m = 0.$$

Multiplying these equations by w_n and $-v_n$ respectively, adding and summing, we find that

$$j \sum_{m,n} a_{mn} (w_m w_n + v_m v_n) = 0.$$

The factor multiplying j is a positive quadratic form which vanishes only when the quantities v_n, w_n are all zero, hence we must have $j = 0$ and this means that k^2 is necessarily real. That k is necessarily positive is seen immediately from the equation

$$\sum_{m,n} c_{mn} u_m u_n = k^2 \sum_{m,n} a_{mn} u_m u_n,$$

which involves two positive quadratic forms.

If $u_m(k_1)$, $u_m(k_2)$ are values of u_m corresponding to two different values of k we have the equations

$$\sum_{m=1}^{N} (c_{mn} - k_1^2 a_{mn})\, u_m(k_1) = 0,$$

$$\sum_{m=1}^{N} (c_{mn} - k_2^2 a_{mn})\, u_m(k_2) = 0.$$

Multiplying these by $u_n(k_2)$, $u_n(k_1)$ respectively and subtracting we find that

$$(k_2^2 - k_1^2) \sum_{m,n} a_{mn} u_m(k_1)\, u_n(k_2) = 0. \qquad \text{......(D)}$$

Denoting the constant b_n associated with the value k by $b_n(k)$, we see from the last equation that if $k_2 \neq k_1$,

$$\sum_{n=1}^{N} b_n(k_1)\, u_n(k_2) = 0.$$

On the other hand,

$$\sum_{n=1}^{N} b_n(k_1)\, u_n(k_1) = \sum_{m,n} a_{mn} u_m(k_1)\, u_n(k_1),$$

and is an essentially positive quantity which may be taken without loss of generality to be unity since the quantities $u_n(k_1)$ contain undetermined constant factors as far as the foregoing analysis is concerned.

Using the symbol $v(k, t)$ to denote the function v corresponding to a definite value of k, we observe that if

$$q_m = \sum_{s=1}^{N} v(k_s, t)\, \lambda_{ms},$$

we must have

$$\sum_{s=1}^{N} b_n(k_s)\, \lambda_{ms} = 0 \qquad n \neq m$$
$$= 1 \qquad n = m.$$

Multiplying by $u_n(k_r)$ and summing with respect to n we find that

$$\lambda_{mr} = u_m(k_r).$$

Hence

$$q_m = \sum_{s=1}^{N} u_m(k_s)\, v(k_s, t).$$

This expresses the solution of our system of differential equations in terms of the simple harmonic vibrations determined by the equations of type (B). The analysis has been given for the simple case in which the roots of the equation for k^2 are all different but extensions of the analysis have been given for the case of multiple roots.

The relation (D) may be regarded as an orthogonal relation in generalised co-ordinates. When

$$a_{mn} = 0, \quad m \neq n, \quad a_{mm} = 1,$$

the relation takes the simpler form

$$\sum_{m=1}^{N} u_m(k_p)\, u_m(k_q) = 0 \qquad p \neq q.$$

§ **1·34**. An interesting mechanical device for combining automatically any number of simple harmonic vibrations has been studied by A. Garbasso*.

A small table of mass m is supported by four light strings of equal length l so that it remains horizontal as it swings like a pendulum. The table is attached at various points to n simple pendulums (l_s, m_s), $s = 1, 2, \ldots n$. Each string is regarded as light and is supposed to oscillate in a vertical plane and remain straight as the apparatus oscillates.

Specifying the configuration of the apparatus by the angular variables $\theta_0, \theta_1, \ldots \theta_n$ we have in a small oscillation

$$T = \frac{1}{2}\left[m_0 l_0^2 \dot{\theta}_0^2 + \sum_{s=1}^{n} m_s (l_0\dot{\theta}_0 + l_s\dot{\theta}_s)^2\right],$$

$$V = \frac{1}{2} g \left[m_0 l_0^2 \theta_0^2 + \sum_{s=1}^{n} m_s (l_0\theta_0^2 + l_s\theta_s^2)\right].$$

* *Vorlesungen über Spektroskopie*, p. 65; *Torino Atti*, vol. XLIV, p. 223 (1908–9). The case in which $n = 2$ has been studied in connection with acoustics by Barton and Browning, *Phil. Mag.* (6), vol. XXXIV, p. 246 (1917); vol. XXXV, p. 62 (1918); vol. XXXVI, p. 36 (1918) and by C. H. Lees, *ibid.* vol. XLVIII (1924).

The equations of motion are

$$m_0 l_0 \ddot{\theta}_0 + \sum_{s=1}^{n} m_s (l_0 \ddot{\theta}_0 + l_s \ddot{\theta}_s) + g\left(m_0 + \sum_{s=1}^{n} m_s\right) \theta_0 = 0,$$

$$l_0 \ddot{\theta}_0 + l_s \ddot{\theta}_s + g\theta_s = 0,$$

$$(s = 1, 2, \dots n).$$

Writing $m_s = c_s m_0$, $\sum_{s=1}^{n} c_s = c$, the equation for k^2 is in this case

$$\begin{vmatrix} g(1+c) - l_0 k^2 & -gc_1 & -gc_2 & \dots & -gc_n \\ -l_0 k^2 & g - l_1 k^2 & 0 & \dots & 0 \\ -l_0 k^2 & 0 & g - l_2 k^2 & \dots & 0 \\ \dots & \dots & \dots & \dots & \dots \\ -l_0 k^2 & 0 & 0 & \dots & g - l_n k^2 \end{vmatrix} = 0,$$

or

$$\begin{vmatrix} g - l_0 k^2 & -gc_1 & -gc_2 & \dots & -gc_n \\ -(l_0 + l_1) k^2 + g & g - l_1 k^2 & 0 & \dots & \dots \\ \dots & \dots & \dots & \dots & \dots \\ -(l_0 + l_n) k^2 + g & 0 & \dots & \dots & g - l_n k^2 \end{vmatrix} = 0.$$

Expanding the determinant we obtain the equation

$$\left\{g - l_0 k^2 + g \sum_{s=1}^{n} \frac{c_s [g - k^2 (l_0 + l_s)]}{g - l_s k^2}\right\} f(k^2) = 0,$$

where

$$f(k^2) = \prod_{s=1}^{n} (g - l_s k^2).$$

Now $\quad (l_0 - l_s)[g - (l_0 + l_s) k^2] = l_0 (g - l_0 k^2) - l_s (g - l_s k^2),$

consequently the equation for k^2 can be written in the form

$$f(k^2) \left\{(g - l_0 k^2) \left[1 + g l_0 \sum_{s=1}^{n} \frac{c_s}{(l_0 - l_s)(g - l_s k^2)}\right] - g \sum_{s=1}^{n} \frac{c_s l_s}{l_0 - l_s}\right\} = 0.$$

If the mass of each pendulum is so small in comparison with that of the table that we may neglect terms of the second order in the quantities c_s, the equation may be written in the form

$$\left(g - l_0 k^2 - g \sum_{s=1}^{n} \frac{c_s l_s}{l_0 - l_s}\right) \prod_{s=1}^{n} \left(g - l_s k^2 + \frac{g c_s l_0}{l_0 - l_s}\right).$$

Hence the periods of the normal vibrations are approximately

$$T_0 = 2\pi\sqrt{(l_0/g)} \left[1 + \frac{1}{2} \sum_{s=1}^{n} \frac{c_s l_s}{l_0 - l_s}\right],$$

$$T_s = 2\pi\sqrt{(l_s/g)} \left[1 - \frac{1}{2} \frac{c_s l_0}{l_0 - l_s}\right],$$

$$(s = 1, 2, \dots n).$$

If $l_0 > l_s$ the period of the sth pendulum is decreased by attaching it to the table. If $l_0 < l_s$ the period is increased.

EXAMPLES

1. A simple pendulum (b, N) is suspended from the bob of another simple pendulum (a, M) whose string is attached to a fixed point. Prove that the equations of motion for small oscillations are

$$(M+N)\,a^2\ddot{\theta} + Nab\ddot{\phi} + (M+N)\,ga\theta = 0,$$
$$Nab\ddot{\theta} + Nb^2\ddot{\phi} + Ngb\phi = 0,$$

where θ and ϕ are the angles which the strings make with the vertical.

2. Prove that the coefficient of coupling of the compound pendulum in the last example is given by

$$\gamma^2 = \frac{N}{M+N}.$$

3. Prove that it is not possible for the centre of gravity of the two bobs to remain fixed in a simple type of oscillation.

4. A simple pendulum (l, M) is suspended from the bob of a lath pendulum which is treated as a rigid body with a moment of inertia different from that of the bob. Find the equations of motion and the coefficient of coupling.

5. A simple pendulum (l, M) is attached to a point P of an elastic lath pendulum which is clamped at its lower end and carries a bob of mass N at its upper end. At time t the horizontal displacements of M, N and P are y, z and αz respectively, α being regarded as constant. By adopting the simplifying assumption that a horizontal component force F at P gives N the same horizontal acceleration as a force αF acting directly on N, obtain the equations of motion

$$lM\ddot{y} + Mg\,(y - \alpha z) = 0, \quad lN\ddot{z} + lNn^2 z = Mg\alpha\,(y - \alpha z),$$

and show that the coefficient of coupling is given by the equation

$$\gamma^2 = \frac{Mm^2\alpha^2}{Nn^2 + Mm^2\alpha},$$

where
$$lm^2 = g.$$

[L. C. Jackson, *Phil. Mag.* (6), vol. XXXIX, p. 294 (1920).]

6. Two masses m and m' are attached to friction wheels which roll on two parallel horizontal steel bars. A third mass M, which is also attached to friction wheels which roll on a bar midway between the other two, is constrained to lie midway between the other two masses by a light rigid bar which passes through holes in swivels fixed on the upper part of the masses. The masses m and m' are attached to springs which introduce restoring forces proportional to the displacements from certain equilibrium positions. Find the equations of motion and the coefficient of coupling.

This mechanical device has been used to illustrate mechanically the properties of coupled electric circuits. [See Sir J. J. Thomson, *Electricity and Magnetism*, 3rd ed. p. 392 (1904); W. S. Franklin, *Electrician*, p. 556 (1916).]

7. Two simple pendulums (l_1, M_1), (l_2, M_2) hang from a carriage of mass M which, with the aid of wheels, can move freely along a horizontal bar. Prove that the equations of motion are

$$(M + M_1 + M_2)\,\ddot{x} + M_1 l_1 \ddot{\theta}_1 + M_2 l_2 \ddot{\theta}_2 = 0,$$
$$\ddot{x} + l_1\ddot{\theta}_1 + g l_1\theta_1 = 0,$$
$$\ddot{x} + l_2\ddot{\theta}_2 + g l_2\theta_2 = 0.$$

Hence show that the quantities θ_1, θ_2 can be regarded as analogous to electric potential differences at condensers of capacities $M_1 g$ and $M_2 g$, the quantities $M_1 l_1 g \dot{\theta}_1$ and $M_2 l_2 g \dot{\theta}_2$ as analogous to electric currents in circuits, the quantities

$$\frac{M + M_2}{M_1 g^2 (M_1 + M_2 + M)} \quad \text{and} \quad \frac{M + M_1}{M_2 g^2 (M_1 + M_2 + M)}$$

as analogous to coefficients of self-induction and $[(M_1 + M_2 + M) g^2]^{-1}$ as analogous to a coefficient of mutual induction.

[T. R. Lyle, *Phil. Mag.* (6), vol. xxv, p. 567 (1913).]

8. A simple pendulum of length l, when hanging vertically, bisects the horizontal line joining the knife edges. When the pendulum oscillates it swings freely until the string comes into contact with one of the knife edges and then the bob swings as if it were suspended by a string of length h. Assuming that the motion is small and that in a typical quarter swing

$$l\ddot{\theta} + g\theta = 0 \quad \text{for} \quad 0 < t < \tau,$$

$$h\ddot{\theta} + g\theta = 0 \quad \text{for} \quad \tau < t < T,$$

prove that the quarter period T is given by the equation

$$m \cot n (T - \tau) = n \tan m\tau,$$

where

$$g = lm^2 = hn^2.$$

§ **1·35**. *Some properties of non-negative quadratic forms**. Let

$$g = \sum_{1,1}^{n,n} g_{rs} x_r x_s$$

be a quadratic form of the real variables $x_1, \dots x_n$, which is negative for no set of values of these variables, then there are n linear forms

$$\sum_1^n p_{rs} x_s$$

with real coefficients p_{rs} such that

$$g = \sum_{r=1}^{n} \left(\sum_{s=1}^{n} p_{rs} x_s \right)^2.$$

This identity gives the relation

$$g_{ik} = \sum_1^n p_{ri} p_{rk}$$

which can be regarded as a parametric representation of the coefficients in a non-negative form.

This result may be obtained by first noting that g_{ss} is not negative, for g_{ss} is the value of g when $x_r = 0$, $r \neq s$ and $x_s = 1$. If the coefficients g_{rs} are not all zero the coefficients g_{ss} are not all zero, because if they were and if, say, $g_{12} \gtrless 0$ a negative value of g could be obtained by choosing $x_1 = 1$, $x_2 = \mp 1$, $x_3 = \dots x_n = 0$.

We may, then, without loss of generality assume that there is at least one coefficient g_{11} of the set g_{ss} which is positive.

* L. Fejér, *Math. Zeits.* Bd. I, S. 70 (1918).

Writing $$p_{11}p_{1s} = g_{1s} \quad s = 1, 2, \dots n,$$
$$z_1 = \sum_{s=1}^{n} p_{1s}x_s,$$
$$g^{(1)} = g - z_1^2,$$
it is easily seen that the quadratic form $g^{(1)}$ does not depend on x_1. $g^{(1)}$ is moreover non-negative because if it were negative for any set of values of $x_2, x_3, \dots x_n$ we could obtain a negative value of g by choosing x_1 so that $z_1 = 0$.

Since $g^{(1)}$ is non-negative it either vanishes identically or the coefficient of at least one of the quantities $x_2^2, x_3^2, \dots x_n^2$ in $g^{(1)}$ must be positive. Let us suppose that $g_{22}^{(1)}$ is positive and write
$$p_{22}\, p_{2r} = g_{2r}^{(1)} \qquad r = 2, 3, \dots n,$$
$$z_2 = \sum_{r=2}^{n} p_{2r}x_r,$$
$$g^{(2)} = g^{(1)} - z_2^2.$$

Continuing this process it is found that $g = z_1^2 + z_2^2 + \dots z_n^2$, where the linear forms $z_1, z_2, \dots z_n$ are not all zero; it is also found that none of the quantities $g_{11}, g_{22}^{(1)}, g_{33}^{(2)}, \dots$ are negative and that all these quantities, except the first, are ratios of leading diagonal minors in the determinant
$$| g_{rs} |,$$
and are not all zero.

Now let $$h = \sum_{1,1}^{n,n} h_{ik}x_i x_k$$
be a second non-negative form, and let
$$h_{ik} = \sum_{1}^{n} q_{si}q_{sk}$$
be its parametric representation, then*
$$\sum_{1,1}^{n,n} g_{ik}h_{ik} \equiv \sum_{1,1}^{n,n} \left(\sum_{\sigma=1}^{n} p_{r\sigma}q_{s\sigma} \right)^2.$$

If $\theta, y_1, y_2, \dots y_n$ are arbitrary real quantities,
$$\sum_{1}^{n} (x_s - \theta y_s)^2$$
is never negative. Regarding this as a quadratic expression in θ it is readily seen that the quadratic form
$$h = \sum_{1}^{n} x_s^2 \sum_{1}^{n} y_s^2 - \left(\sum_{1}^{n} x_s y_s\right)^2$$
is non-negative. This result, which was known to Cauchy and Bessel, is frequently called Schwarz's inequality as Schwarz obtained a similar inequality for integrals.

* L. Fejér, *Math. Zeits.* Bd. I, S. 70 (1918).

Using this particular form of h in Fejér's inequality we obtain the result that

$$\sum_{1,1}^{n,n} g_{rs} y_r y_s \leqslant \sum_{1}^{n} g_{rr} \sum_{1}^{n} y_s^2.$$

For further properties of quadratic forms the reader is referred to Bromwich's tract, *Quadratic Forms*, Cambridge (1906), to Bôcher's *Algebra*, Macmillan and Co. (1907), and to Dickson's *Modern Algebraic Theories*, Sanborn and Co., Chicago (1926).

§ **1·36**. *Hermitian forms.* Let $\bar{z}$ denote the complex quantity conjugate to a complex quantity z and let the complex coefficients c_{rs} be such that

$$\bar{c}_{rs} = c_{sr},$$

the bilinear form

$$\sum_{1,1}^{n,n} c_{rs} z_r \bar{z}_s$$

is then Hermitian. If $c_{lm} = a_{lm} + ib_{lm}$, $z_m = r_m e^{i\theta_m}$, where a_{lm}, b_{lm}, r_m, θ_m are real quantities, we have

$$a_{lm} = a_{ml}, \quad b_{lm} = -b_{ml},$$

and the Hermitian form can be expressed as a quadratic form

$$\sum_{1,1}^{n,n} p_{lm} r_l r_m,$$

where

$$p_{lm} = a_{lm} \cos(\theta_l - \theta_m) + b_{lm} \sin(\theta_l - \theta_m) = p_{ml}.$$

The positive definite Hermitian forms which are positive whenever at least one of the quantities $z_1, z_2, \dots z_n$ is different from zero are of special interest. In this case the associated quadratic form is positive for all non-vanishing sets of values of $r_1, r_2, \dots r_n$ and for all values of $\theta_1, \theta_2, \dots \theta_n$. An important property of a Hermitian form is that the associated secular equation

$$|c_{rs} - \lambda\delta_{rs}| = 0$$
$$\delta_{rs} = 1 \quad r = s$$
$$= 0 \quad r \neq s$$

has only real roots. When the form is positive these roots are all positive. The proof of this theorem may be based upon analysis very similar to that given in § 1·33.

EXAMPLES

1. If $F(t) \geqslant 0$ for $-\pi \leqslant t \leqslant \pi$ and

$$F(t) = \sum_{\nu=-\infty}^{\infty} c_\nu e^{i\nu t}, \quad \bar{c}_\nu = c_{-\nu},$$

$$H_n = \sum_{1,1}^{n,n} c_{l-m} z_l \bar{z}_m \qquad n = 1, 2, 3 \dots,$$

then $H_n \geqslant 0$. This has been shown by Carathéodory and Toeplitz to be a necessary and sufficient condition that $F(t) \geqslant 0$. See *Rend. Palermo*, t. XXXII, pp. 191, 193 (1911).

2. If
$$f(t) = \int_{-\infty}^{\infty} e^{itx}\,\omega(x)\,dx,$$
where
$$\bar{\omega}(x) = \omega(-x)$$
and
$$H_n = \sum_{1,1}^{n,n} \omega(x_l - x_m)\,\zeta_l\bar{\zeta}_m,$$
Mathias has shown that when $f(t) \geqslant 0$ we have $H_n \geqslant 0$ for any choice of real parameters $x_1, x_2, \ldots x_n$ and of the complex numbers $\zeta_1, \zeta_2, \ldots \zeta_n$. See *Math. Zeits.* Bd. XVI, p. 103 (1923). The analysis depends upon Fourier's inversion formula which is studied in § 3·12 and it appears that, with suitable restrictions on the function $\omega(x)$, the inequality $H_n \geqslant 0$ is the necessary and sufficient condition that $f(t) \geqslant 0$. Mathias gives two methods of choosing a function $\omega(x)$ which will make $H_n \geqslant 0$. The correctness of these should be verified by the reader.

(1) If the functions $\chi(x)$, $\bar{\chi}(x)$ are such that when x has any real value $\chi(x)$ and $\bar{\chi}(x)$ are conjugate complex quantities, the function
$$\omega(x) = \int_{-\infty}^{\infty} \chi(\sigma + x)\,\bar{\chi}(\sigma - x)\,d\sigma$$
will make $H_n \geqslant 0$.

(2) If the positive constants λ_ν and the functions $\chi_\nu(x)$ are such that
$$\sum_{\nu=1}^{\infty} \frac{1}{\lambda_\nu}\,\chi_\nu(x)\,\bar{\chi}_\nu(y)$$
is a function of $x - y$, say $\omega(x - y)$, then this function $\omega(x)$ will make $H_n \geqslant 0$.

§ **1·41**. *Forced oscillations.* When a particle, which is normally free to oscillate with simple harmonic motion about a position of equilibrium, is acted upon by a periodic force varying with the time like sin (pt), the equation of motion takes the form
$$\ddot{x} + k^2x = A \sin pt.$$

Writing $x = z + C \sin pt$, where C is a constant to be determined, we find that if we choose C so that
$$(k^2 - p^2)\,C = A, \qquad \ldots\ldots(\text{A})$$
the equation for z takes the form
$$\ddot{z} + k^2z = 0.$$

The motion thus consists of a free oscillation superposed on an oscillation with the same period as the force. In other words the motion is partly original and partly imitation. It should be noticed, however, that if $p^2 > k^2$ the imitation is not perfect because there is a difference in phase. The difference between the case $p^2 > k^2$ and the case $p^2 < k^2$ is beautifully illustrated by giving a simple harmonic motion to the point of suspension of a pendulum.

When $p^2 = k^2$ the quantity C is no longer determined by equation (A) and the solution is best obtained by the method of integrating factors which may be applied to the general equation
$$\ddot{x} + k^2x = F(t).$$

Multiplying successively by the integrating factors cos kt and sin kt and integrating, we find that if $x = c$, $\dot{x} = u$ when $t = \tau$, we have

$$\dot{x}\cos kt + kx \sin kt = u \cos k\tau + kc \sin k\tau + \int_\tau^t F(s) \cos ks \,.\, ds,$$

$$\dot{x}\sin kt - kx \cos kt = u \sin k\tau - kc \cos k\tau + \int_\tau^t F(s) \sin ks \,.\, ds,$$

$$kx = u \sin k(t-\tau) + kc \cos k(t-\tau) + \int_\tau^t F(s) \sin k(t-s)\, ds,$$

$$\dot{x} = u \cos k(t-\tau) - kc \sin k(t-\tau) + \int_\tau^t F(s) \cos k(t-s)\, ds,$$

the function $F(s)$ being supposed to be integrable over the range τ to t.

In particular, if the particle starts from rest at the time t we have at any later time

$$kx = \int_\tau^t F(s) \sin k(t-s)\, ds,$$

$$\dot{x} = \int_\tau^t F(s) \cos k(t-s)\, ds.$$

When $F(s) = A \sin ks$ and $\tau = 0$ we find that

$$2kx = t \cos kt - k^{-1} \sin kt,$$

and the oscillations in the value of x increase in magnitude as t increases. This is a simple case of resonance, a phenomenon which is of considerable importance in acoustics. In engineering one important result of resonance is the whirling of a shaft which occurs when the rate of rotation has a critical value corresponding to one of the natural frequencies of lateral vibration of the shaft. For a useful discussion of vibration problems in engineering the reader is referred to a recent book on the subject by S. Timoshenko, D. Van Nostrand Co., New York (1928).

By choosing the unit of time so that $k = 1$ the mathematical theory may be illustrated geometrically with the aid of the curve whose radius of curvature, ρ, is given by the equation $\rho = a \sin \omega\psi$, where ψ is the angle which the tangent makes with a fixed line. Using p now to denote the length of the perpendicular from the origin to the tangent, we have the equation

$$\rho = \frac{d^2p}{d\psi^2} + p = a \sin \omega\psi. \qquad \text{......(B)}$$

The quantity p thus represents a solution of the differential equation, and by suitably choosing the position of the origin the arbitrary constants in the solution can be given any assigned real values. In this connection it should be noticed that $\frac{dp}{d\psi}$ has a simple geometrical meaning (Fig. 7).

When $\omega = 1$ the equation (B) is that of a cycloid, while epicycloids and hypocycloids are obtained by making ω different from unity. The intrinsic equation of these curves is in fact

$$s = \frac{4(a+b)b}{a} \sin \frac{a\psi}{a+2b},$$

where a is the radius of the fixed circle and b the radius of the rolling circle which contains the generating point.

Epicycloids $b > 0$.

Pericycloids and hypocycloids $b < 0$.

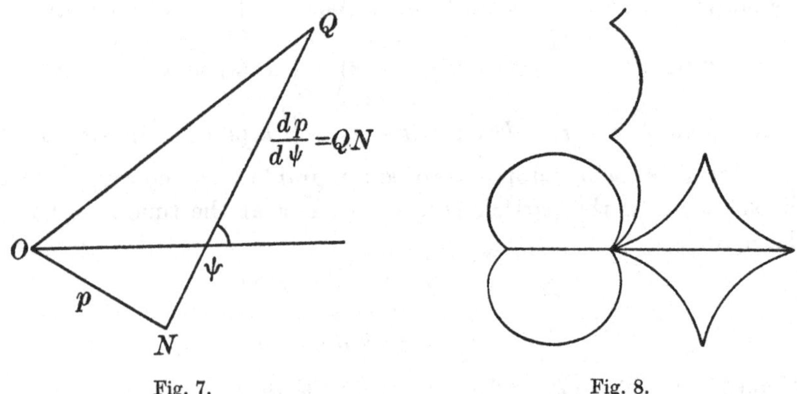

Fig. 7. Fig. 8.

§ **1·42**. The effect of a transient force in producing forced oscillations is best studied by putting $\tau = -\infty$ and assuming that c and u are both zero, then

$$kx = \int_0^\infty F(t-\sigma)\sin k\sigma \, . \, d\sigma,$$

$$\dot{x} = \int_0^\infty F(t-\sigma)\cos k\sigma \, . \, d\sigma.$$

Let us consider first of all the case when

$$F(t) = e^{-h|t|} \qquad h > 0.$$

In this case $F'(t)$ is discontinuous at time $t = 0$ in a way such that

$$F'(-0) - F'(+0) = 2h.$$

The solution which is obtained by supposing that x, $\dot{x}$ and $\ddot{x}$ are con tinuous at time $t = 0$ is

$$kx = \int_0^t e^{-h(t-\sigma)} \sin k\sigma d\sigma + \int_t^\infty e^{h(t-\sigma)} \sin k\sigma d\sigma$$

$$= \frac{2h \sin kt}{k^2 + h^2} + \frac{k}{k^2 + h^2} e^{-ht} \qquad t > 0$$

$$= \frac{k}{k^2 + h^2} e^{ht} \qquad t < 0,$$

$$\dot{x} = \frac{2h \cos kt - he^{-ht}}{k^2 + h^2} \qquad t > 0$$

$$= \frac{h}{k^2 + h^2} e^{ht} \qquad t < 0.$$

It will be noticed that $\ddot{x}$ is discontinuous at $t = 0$.

It is clear from these equations that x increases with t until t reaches the first positive root of the transcendental equation

$$2\cos kt = e^{-ht}.$$

The corresponding value of x is then

$$\frac{2(h\sin kt + k\cos kt)}{k(k^2+h^2)}.$$

As t increases beyond the critical value x begins to oscillate in value, and as $t \to \infty$ there is an undamped residual oscillation given by

$$kx = \frac{2h\sin kt}{k^2+h^2}.$$

A general formula for the displacement x at time t in the residual oscillation produced by a transient force may be obtained by putting $t = \infty$ in the upper limit of the integral (t being retained in the integrand)*. This gives

$$kx = \int_{-\infty}^{\infty} F(s)\sin k(t-s)\,ds.$$

In particular, if

$$F(s) = \int_0^b \cos ms\,.\,dm = \frac{\sin bs}{s},$$

we have

$$kx = \sin kt\int_{-\infty}^{\infty}\cos ks\sin bs\frac{ds}{s} - \cos kt\int_{-\infty}^{\infty}\sin ks\sin bs\frac{ds}{s};$$

the second integral vanishes and we may write

$$\begin{aligned} 2kx &= \sin kt\int_{-\infty}^{\infty}[\sin(b+k)s + \sin(b-k)s]\frac{ds}{s} \\ &= 2\pi\sin kt \qquad \text{if } b > k > 0 \\ &= 0 \qquad \text{if } 0 < b < k \\ &= \pi\sin kt \qquad \text{if } b = k > 0. \end{aligned}$$

There is thus a residual oscillation only when $b \geqslant k$.

EXAMPLES

1\. If $$F(s) = \int_a^b \cos ms\,.\,dm,$$
there is a residual oscillation only when k lies within the range $a \leqslant k \leqslant b$. Extend this result by considering cases when

$$F(s) = \int_a^b \cos ms\,.\,\phi(m)\,dm, \quad F(s) = \int_a^b \sin ms\,.\,\phi(m)\,dm,$$

$\phi(m)$ being a suitable arbitrary function.

2\. Determine the residual oscillation in the case when

$$F(s) = (c^2+s^2)^{-1}.$$

* Cf. H. Lamb, *Dynamical Theory of Sound*, p. 19.

3. If $F(s) = se^{-h|s|}$, where $h > 0$ and x is chosen to be a solution of the differential equation and the supplementary conditions $x = 0$, $\dot{x} = 0$, when $t = -\infty$, x and $\dot{x}$ continuous at $t = 0$, the residual oscillation is given by

$$x = -\frac{4h \cos kt}{(k^2 + h^2)^2}.$$

If $k^2 > h^2$ there is a negative value of t for which $\dot{x} = 0$, but if $k^2 < h^2$ there is no such value.

4. If
$$\ddot{x} + k^2 x = x\omega(t),$$
where $\omega(t)$ is zero when $t = \infty$ and is bounded for other real values of t, the solution for which x and $\dot{x}$ are initially zero can be regarded as the residual oscillation of a simple pendulum disturbed by a transient force.

5. If $0 < a^2 < A(t) < b^2$, the differential equation
$$\ddot{x} + A(t)\, x = 0 \qquad \text{......(C)}$$
is satisfied only by oscillating functions. Prove that the interval between two consecutive roots of the equation $x = 0$ lies between π/a and π/b.

[Let y be a solution of $\ddot{y} + b^2 y = 0$ which is positive in the interval $\tau_1 < t < \tau_2$, then
$$\Big[\dot{x}y - \dot{y}x\Big]_{\tau_1}^{\tau_2} = \int_{\tau_1}^{\tau_2} [b^2 - A(t)]\, xy\, dt.$$

Let us now suppose that it is possible for x to be of one sign (positive, say) in the interval $\tau_1 \leqslant t \leqslant \tau_2$ and zero at the ends of the interval. We are then led to a contradiction because the suppositions make $\dot{x}$ positive near τ_1 and negative near τ_2, they thus make the left-hand side negative (or zero) and the right-hand side positive. Hence the interval between two consecutive roots of x must be greater than any range in which y is positive, that is, greater than y/b. In a similar way it may be shown that if z is a solution of $\ddot{z} + a^2 z = 0$ the interval between two consecutive roots of the equation $z = 0$ is greater than any range in which $x > 0$.]

This is one of the many interesting theorems relating to the oscillating functions which satisfy an equation of type (C). For further developments the reader is referred to Bôcher's book, *Leçons sur les méthodes de Sturm*, Gauthier-Villars, Paris (1917) and his article, "Boundary problems in one dimension," *Fifth International Congress of Mathematicians, Proceedings*, vol. I, p. 163, Cambridge (1912).

6. Prove that $x^2 + \dot{x}^2$ remains bounded as $t \to \infty$ but may not have a definite limiting value, x being any solution of (C). [M. Fatou, *Comptes Rendus*, t. 189, p. 967 (1929).]

The generality of this result has recently been questioned. See Note I, Appendix.

§ **1·43**. *Motion with a resistance proportional to the velocity.* Let us first of all discuss the motion of a raindrop or solid particle which falls so slowly that the resistance to its motion through the air varies as the first power of the velocity. This is called Stokes' law of resistance; it will be given in a precise form in the section dealing with the motion of a sphere through a viscous fluid; for the present we shall use simply an unknown constant coefficient k and shall write the equation of motion in the form
$$m\, dv/dt = m'g - kv, \qquad \text{......(A)}$$
where m is the apparent mass of the body when it moves in air, m' is the reduced mass when the buoyancy of the air is taken into consideration and g is the acceleration of gravity. The solution of this equation is
$$kv = m'g + Be^{-\frac{kt}{m}},$$

where B is a constant depending on the initial conditions. If $v = 0$ when $t = 0$ we have

$$kv = m'g\,(1 - e^{-\frac{kt}{m}}).$$

To find the distance the body must fall to acquire a specified velocity v we write $v = \dot{x}$,

$$mv\,dv/dx = m'g - kv,$$

$$k^2x = mm'g \log\left(\frac{m'g}{m'g - kv}\right) - kmv.$$

If V denote the terminal velocity the equation may be written in the form

$$m'gx = mV^2 \log \frac{V}{V - v} - mVv.$$

Let us now consider the case in which the particle moves in a fluctuating vertical current of air. Let $f'(t)$ be the upward velocity of the air at time t, v the velocity of the particle relative to the ground and $u = v + f'(t)$ the relative velocity. On the supposition that the resistance is proportional to the relative velocity the equation of motion is

$$m\,dv/dt = m'g - ku = m'g - kv - kf'(t).$$

If $v = 0$ when $t = 0$ the solution is

$$v = \frac{m'g}{k}(1 - e^{-\frac{kt}{m}}) - \frac{k}{m}\int_0^t f'(t_0)\,dt_0\, e^{-\frac{k}{m}(t - t_0)}$$

The last integral may be written in the form

$$\int_0^t f'(t - \tau)\, e^{-\frac{k\tau}{m}}\, d\tau,$$

which is very useful for a study of its behaviour when t is very large.

The distance traversed in time t is given by the equation

$$x = \frac{m'gt}{k} - \frac{mm'g}{k^2}(1 - e^{-\frac{kt}{m}}) - \frac{k}{m}\int_0^t f(t - \tau)\, e^{-\frac{k\tau}{m}}\, d\tau,$$

the constant in $f(t)$ being chosen so that $f(0) = 0$. In particular if

$$f'(t) = \frac{m'g}{k} + c\cos pt, \quad f(t) = \frac{m'gt}{k} + \frac{c}{p}\sin pt,$$

we have
$$v = \frac{ck}{m^2p^2 + k^2}[ke^{-\frac{kt}{m}} - mp\sin pt - k\cos pt],$$

$$x = -\frac{ck}{p\,(m^2p^2 + k^2)}[k\sin pt - mp\cos pt + mpe^{-\frac{kt}{m}}].$$

As $t \to \infty$ we are left with a simple harmonic oscillation which is not in the same phase as the air current.

It should be emphasised that this law of resistance is of very limited application as there is only a small range of velocities and radius of particle for which Stokes' law is applicable. It should be mentioned that the product of the radius and the velocity must have a value lying in a certain range if the law is to be valid.

An equation similar to (A) may be used to describe the course of a unimolecular chemical reaction in which only one substance is being transformed. If the initial concentration of the substance is a and at time t altogether x gram-molecules of the substance have been transformed, the concentration is then $a - x$ and the law of mass action gives

$$\frac{dx}{dt} = k(a - x),$$

$$x = a(1 - e^{-kt}),$$

the coefficient k being the rate of transformation of unit mass of the substance.

Simultaneous equations involving only the first derivatives of the variables and linear combinations of these variables occur in the theory of consecutive unimolecular chemical reactions. If at the end of time t the concentrations of the substances A, B and C are x, y and z respectively and the reactions are represented symbolically by the equations

$$A \to B, \quad B \to C,$$

the equations governing the reactions are

$$dx/dt = -k_1 x, \quad dy/dt = k_1 x - k_2 y, \quad dz/dt = k_2 y.$$

A more general system of linear equations of this type occurs in the theory of radio-active transformations. Let $P_0, P_1, \ldots P_n$ represent the amounts of the substances $A_0, A_1, \ldots A_n$ present at time t, then the law of mass action gives

$$\frac{dP_0}{dt} = -\lambda_0 P_0,$$

$$\frac{dP_1}{dt} = \lambda_0 P_0 - \lambda_1 P_1,$$

$$\frac{dP_2}{dt} = \lambda_1 P_1 - \lambda_2 P_2,$$

$$\cdots\cdots\cdots\cdots\cdots\cdots$$

$$\frac{dP_n}{dt} = \lambda_{n-1} P_{n-1} - \lambda_n P_n,$$

where the coefficients λ_s are constants. In my book on differential equations this system of equations is solved by the method of integrating factors. This method is elementary but there is another method* which, though more recondite, is more convenient to use.

Let us write
$$p_s(x) = \int_0^\infty e^{-xt} P_s(t)\, dt, \qquad \text{......(B)}$$

then
$$\int_0^\infty e^{-xt} \frac{dP_s}{dt}\, dt = -P_s(0) + x p_s(x),$$

* H. Bateman, *Proc. Camb. Phil. Soc.* vol. xv, p. 423 (1910).

and so the system of differential equations gives rise to the system of linear algebraic equations

$$xp_0(x) - P_0(0) = -\lambda_0 p_0(x),$$
$$xp_1(x) - P_1(0) = \lambda_0 p_0(x) - \lambda_1 p_1(x),$$
$$xp_2(x) - P_2(0) = \lambda_1 p_1(x) - \lambda_2 p_2(x),$$
$$\dots\dots\dots\dots\dots\dots\dots\dots\dots$$
$$xp_n(x) - P_n(0) = \lambda_{n-1} p_{n-1}(x) - \lambda_n p_n(x),$$

from which the functions $p_0(x), p_1(x), \dots p_n(x)$ may at once be derived.

If $P_1(0) = P_2(0) = \dots = P_n(0) = 0$, i.e. if there is only one substance initially, and if $P_0(0) = Q$, we have

$$p_0(x) = \frac{Q}{x + \lambda_0}, \quad p_1(x) = \frac{\lambda_0 Q}{(x + \lambda_0)(x + \lambda_1)},$$
$$\dots\dots\dots\dots\dots\dots\dots\dots\dots$$
$$p_n(x) = \frac{\lambda_0 \lambda_1 \dots \lambda_{n-1} Q}{(x + \lambda_0)(x + \lambda_1) \dots (x + \lambda_n)}.$$

To derive $P_s(t)$ from $p_s(x)$ we simply express $p_s(x)$ in partial fractions

$$p_s(x) = \frac{c_0}{x + \lambda_0} + \dots \frac{c_s}{x + \lambda_s}.$$

The corresponding function $P_s(t)$ is then given by

$$P_s(t) = c_0 e^{-\lambda_0 t} + \dots c_s e^{-\lambda_s t},$$

for this is evidently of the correct form and the solution of the system of differential equations is unique. The uniqueness of the function $P_s(t)$ corresponding to a given function $p_s(x)$ can also be inferred from Lerch's theorem which will be proved in § 6·29.

If some of the quantities λ_m are equal there may be terms of type

$$\Gamma(x + 1) \frac{C_{m, \kappa}}{(x + \lambda_m)^{\kappa + 1}}$$

in the representation of $p_s(x)$ in partial fractions. In this case the corresponding term in $P_s(t)$ is

$$C_{m, \kappa} t^{\kappa} e^{-\lambda_m t}.$$

Such a case arises in the discussion of a system of linear differential equations occurring in the theory of probability*.

§ **1·44**. *The equation of damped vibrations.* A mechanical system with one degree of freedom may be represented at time t by a single point P which moves along the x-axis and has a position specified at this instant by the co-ordinate x. This point P, which may be called the image of the system, may in some cases be a special point of the system, provided that the path of such a point is to a sufficient approximation rectilinear. The

* H. Bateman, *Differential Equations*, p. 45.

mechanical system may also in special cases be just one part of a larger system; it may, for instance, be one element of a string or vibrating body on which attention is focussed. To obtain a simple picture of our system and to fix ideas we shall suppose that P is the centre of mass of a pendulum which swings in a resisting medium.

The motion of the point P is then similar to that of a particle acted upon by forces which depend in value on t, x, $\dot{x}$ and possibly higher derivatives. For simplicity we shall consider the case in which the force F is a linear function of x and $\dot{x}$,

$$F = f(t) - h(t)\,x - 2k(t)\,\dot{x}.$$

In the case when $h(t)$ and $k(t)$ are constants the equation of motion takes the simple form

$$\ddot{x} + 2k\dot{x} + n^2 x = f(t). \qquad \text{......(A)}$$

The motion of the particle is in this case retarded by a frictional force proportional to the velocity. If it were not for this resistance the free motion of the particle would be a simple harmonic vibration of frequency $n/2\pi$. The effect of the resistance when $n^2 > k^2$ is to reduce the free motion to a damped oscillation of type

$$x = Ae^{-kt} \sin(pt + \epsilon), \qquad \text{......(B)}$$

where A and ϵ are arbitrary constants and

$$p^2 = n^2 - k^2.$$

The period of this damped oscillation may be defined as the interval between successive instants at which x is a maximum and is $2\pi/p$. One effect of the resistance, then, is to lengthen the period of free oscillations. It may be noted that the interval between successive instants at which $x = 0$ is $2\pi/p$. The time range $0 \leqslant t < \infty$ may, then, be divided up into intervals of this length. The sign of x changes as t passes from one interval to the next and so the point P does in fact oscillate. Points P and P' of two intervals in which x has the same sign may be said to correspond if their associated times t, t', are connected by the relation

$$pt' = pt + 2m\pi,$$

where m is an integer. We then have

$$x' = xe^{-k(t'-t)} = xe^{-2km\pi/p}.$$

The positive constant k is seen, then, to determine the rate of decay of the oscillations.

When $n^2 = k^2$ the free motion is given by

$$x = (A + Bt)\,e^{-kt},$$

where A and B are arbitrary constants. In this case x vanishes at a time t given by $B = k(A + Bt)$, thus $|x|$ increases to a maximum value and then decreases rapidly to zero. The motion of a dead-beat galvanometer needle may be represented by an equation of type (A) with $n^2 = k^2$.

When $k^2 > n^2$ the free motion is of type

$$x = Ae^{-ut} + Be^{-vt},$$

where u and v are the roots of the equation

$$z^2 - 2kz + n^2 = 0.$$

In this case the general value of x is obtained by the addition of two terms each of which represents a simple subsidence, the logarithmic decrement of which is u for the first and v for the second. The time τ which is needed for the value of x in one of these subsidences to fall to half value is given by the equation

$$\tfrac{1}{2} = e^{-u\tau}.$$

In the case of the damped oscillation (B) the quantity Ae^{-kt} can be regarded as the amplitude at time t. In an interval of time τ this diminishes in the ratio $r:1$, where $r = e^{k\tau}$.

Putting $\tau = 1$, we have $k = \log r$, whence the name logarithmic decrement usually given to k. Instead of considering the logarithmic decrement per unit time we may, in the case of a damped vibration, consider the logarithmic decrement per period or per half-period*. It is $k\pi/p$ for the half-period.

When $f(t) = C \sin mt$, where C and m are constants, the solution of the differential equation (A) is composed of a particular integral of type

$$x = C\frac{(n^2 - m^2)\sin mt - 2km\cos mt}{(n^2 - m^2)^2 + 4k^2m^2} \qquad \text{......(I)}$$

and a complementary function of type (B). The particular integral is obtained most conveniently by the symbolical method in which we write $D = d/dt$ and make use of the fact that $D^2 f = -m^2 f$. The operator D is treated as an algebraic quantity in some of the steps

$$x = \frac{f(t)}{D^2 + 2kD + n^2} = \frac{f(t)}{n^2 - m^2 + 2kD}$$
$$= \frac{n^2 - m^2 - 2kD}{(n^2 - m^2)^2 - 4k^2D^2} f(t) = \frac{n^2 - m^2 - 2kD}{(n^2 - m^2)^2 + 4k^2m^2} f(t).$$

If $x = 0$, $\dot{x} = 0$ when $t = a$ the unknown constants in the complementary function may be determined and we find that

$$px = \int_a^t e^{-k(t-\tau)} \sin p(t - \tau) f(\tau)\, d\tau. \qquad \text{......(C)}$$

This result may be obtained directly from the differential equation by using the integrating factors $e^{kt} \sin pt$ and $e^{kt} \cos pt$.

We thus obtain the equations

$$px\, e^{kt} \sin pt + (\dot{x} + kx)\, e^{kt} \cos pt = \int_a^t e^{k\tau} \cos p\tau \,.\, f(\tau)\, d\tau,$$
$$-px\, e^{kt} \cos pt + (\dot{x} + kx)\, e^{kt} \sin pt = \int_a^t e^{k\tau} \sin p\tau \,.\, f(\tau)\, d\tau,$$

* E. H. Barton and E. M. Browning, *Phil. Mag.* (6), vol. XLVII, p. 495 (1924).

from which (C) is immediately derived. We also obtain the formula

$$\dot{x} + kx = \int_a^t e^{-k(t-\tau)} \cos p(t-\tau) f(\tau)\, d\tau.$$

Introducing an angle ϵ defined by the equation

$$\tan \epsilon = \frac{2km}{n^2 - m^2}$$

the particular integral (I) may be expressed in the form

$$x = A \sin (mt - \epsilon),$$

where the amplitude A is given by the formula

$$A^2 [(n^2 - m^2)^2 + 4k^2m^2] = C^2.$$

This is the forced oscillation which remains behind when the time t is so large that the free oscillations have died down. The amplitude A is a maximum when m is such that $m^2 = n^2 - 2k^2 = p^2 - k^2$. We then have

$$A_{\max} = C/2kp.$$

Writing $A = aA_{\max}$ it is easily seen that when m is nearly equal to n and k/n is so small that its square can be neglected we have the approximate formula*

$$k = a \mid n - m \mid (1 - a^2)^{-\frac{1}{2}}.$$

This formula has been used to determine the damping of forced oscillations of a steel piano wire.

It should be noticed that a differential equation of type (A) may be obtained from the pair of equations

$$\dot{u} + ku + nv = 0,$$
$$\dot{v} + kv - nu = 0,$$

where k and n are constants. These represent the equations of horizontal motion of a particle under the influence of the deflecting force of the earth's rotation and a frictional force proportional to the velocity. These equations give

$$u\dot{u} + v\dot{v} + k(u^2 + v^2) = 0,$$
$$u^2 + v^2 = q^2 e^{-2kt},$$

hence k is the logarithmic decrement for the velocity.

The equation of damped vibrations has some interesting applications in seismology and in fact in any experimental work in which the motion of the arms of a balance is recorded mechanically.

The motion of a horizontal or vertical seïsmograph subjected to displacements of the ground in a given direction, say $x = f(t)$, can be represented by an equation of form

$$\ddot{\theta} + 2k\dot{\theta} + n^2\theta + \ddot{x}/l = 0,$$

* Florence M. Chambers, *Phil. Mag.* (6), vol. XLVIII, p. 636 (1924).

where θ is the deviation of the instrument, k a constant which depends upon the type of damping and l the reduced pendulum length*.

The motion of a dead-beat galvanometer, coupled with the seismograph, is governed by an equation of type

$$\ddot{\phi} + 2m\dot{\phi} + m^2\phi + h\theta = 0,$$

where ϕ is the angle of deviation of the galvanometer and h and m are constants of the instrument.

To get rid as soon as possible of the natural oscillations of the pendulum, introduced by the initial circumstances, it is advisable to augment the damping of the instrument, driving it if possible to the limit of aperiodicity and making it dead beat. By doing this a more truthful record of the movement of the ground is obtained.

When $n^2 = k^2$ the solution of the equation of forced motion

$$\ddot{x} + 2k\dot{x} + k^2x = f(t)$$

is
$$x = \int_a^t (t - \tau)\, e^{-k(t-\tau)} f(\tau)\, d\tau + (A + Bt)\, e^{-kt}.$$

When t is large the second term is negligible and the lower limit of the integral may, to a close approximation, be replaced by 0, $-\infty$, or any other instant from which the value of $f(t)$ is known.

When k is large the second term is negligible even when t has moderate values and if, for such values of t, $f(t)$ is represented over a certain range with considerable accuracy by $C \sin mt$, the value of x is given approximately by the formula

$$x = \frac{f(t)}{D^2 + 2kD + k^2} = \frac{C \sin mt}{k^2 - m^2 + 2kD}$$

$$= \frac{C}{(k^2 - m^2)^2 + 4k^2m^2}\left[(k^2 - m^2)\sin mt - 2km\cos mt\right]. \quad \ldots(\text{I}')$$

When k is large in comparison with m a good approximation is given by

$$k^2x = C \sin mt = f(t),$$

and the factor of proportionality k^2 is independent of m, consequently, if a number of terms were required to give a good representation of $f(t)$ within a desired range of values of t, the record of the instrument would still give a faithful representation, on a certain definite scale, of the variation of the force.

When k and m are of the same order of magnitude this is no longer true, consequently, if the "high harmonics" occur to a marked degree in the representation of $f(t)$ by a series of sine functions, the record of the instrument may not be a true picture of the force†.

* B. Galitzin, "The principles of instrumental seismology," *Fifth International Congress of Mathematicians, Proceedings*, vol. I, p. 109 (Cambridge, 1912).

† If $m = k/10$ the solution of the differential equation is approximately $k^2x = \cdot 99 \sin(mt - \epsilon)$, where ϵ is the circular measure of an angle of about 16° 59′. When $m = k/5$ the solution is approximately $k^2x = \cdot 96 \sin(mt - \epsilon)$, where ϵ is the circular measure of an angle of about 31° 47′.

When $n \neq k$ the formula (I′) shows that if k is large in comparison with both m and n the solution is given approximately by the formula

$$2kmx = -C \cos mt$$

which may be written in the form

$$2k\dot{x} = f(t).$$

This result may be obtained directly from the differential equation by neglecting the terms $\ddot{x}$ and n^2x in comparison with $2k\dot{x}$. In this case the velocity $\dot{x}$ gives a faithful record of the force on a certain scale.

Finally, if n^2 is large in comparison with k and m, formula (I) gives simply

$$n^2x \sim f(t),$$

and the instrument gives a faithful record of the force when the natural vibrations have died down.

§ **1·45**. *The dissipation function.* The equation of damped vibrations

$$m\ddot{x} + k\dot{x} + \mu x = f(t)$$

may be written in the form of a Lagrangian equation of motion

$$\frac{\partial}{\partial t}\left(\frac{\partial T}{\partial \dot{x}}\right) + \frac{\partial F}{\partial \dot{x}} + \frac{\partial V}{\partial x} = f(t),$$

where $$T = \tfrac{1}{2}m\dot{x}^2, \quad F = \tfrac{1}{2}k\dot{x}^2, \quad V = \tfrac{1}{2}\mu x^2.$$

Regarding m as the mass of a particle whose displacement at time t is x, T may be regarded as the kinetic energy, V as the potential energy and F as the dissipation function introduced by the late Lord Rayleigh*. The function F is defined for a system containing a number of particles by an equation of type

$$F = \tfrac{1}{2}\Sigma\,(k_x\dot{x}^2 + k_y\dot{y}^2 + k_z\dot{z}^2),$$

where k_x, k_y, k_z are the coefficients of friction, parallel to the axes, for the particle x, y, z. Transforming to general co-ordinates $q_1, q_2, q_3, \ldots q_n$ we may write

$$T = \tfrac{1}{2}\,[11]\,\dot{q}_1{}^2 + \ldots [12]\,\dot{q}_1\dot{q}_2,$$
$$F = \tfrac{1}{2}\,(11)\,\dot{q}_1{}^2 + \ldots (12)\,\dot{q}_1\dot{q}_2,$$
$$V = \tfrac{1}{2}\,\{11\}\,q_1{}^2 + \ldots \{12\}\,q_1q_2,$$

where the coefficients $[rs]$, (rs), $\{rs\}$ are of such a nature that the quadratic forms T, F, V are essentially positive, or rather, never negative. These coefficients are generally functions of the co-ordinates $q_1, \ldots q_n$, but if we are interested only in small oscillations we may regard $q_1, \ldots q_n, \dot{q}_1, \ldots \dot{q}_n$ as small quantities and in the expansions of the coefficients in ascending powers of $q_1, \ldots q_n$ it will be necessary only to retain the constant terms if we agree to neglect terms of the third and higher orders in $q_1, \ldots q_n$, $\dot{q}_1, \ldots \dot{q}_n$.

* *Proc. London Math. Soc.* (1), vol. IV, p. 357 (1873).

The generalised Lagrangian equations of motion are now

$$\frac{d}{dt}\left(\frac{\partial T}{\partial \dot{q}_m}\right) - \frac{\partial T}{\partial q_m} + \frac{\partial F}{\partial \dot{q}_m} + \frac{\partial V}{\partial q_m} = Q_m,$$

where Q_m is the generalised force associated with the co-ordinate q_m. Since T is supposed to be approximately independent of the quantities q_1, q_2, ... q_n the second term may be omitted.

Using $\overline{rs}$ as an abbreviation for the quadratic operator

$$[rs]\frac{d^2}{dt^2} + (rs)\frac{d}{dt} + \{rs\}$$

the equations of motion assume the linear form

$$\overline{11}q_1 + \overline{12}q_2 + \ldots = Q_1,$$
$$\overline{21}q_1 + \overline{22}q_2 + \ldots = Q_2. \qquad \text{......(E)}$$

Since $[rs] = [sr]$, $(rs) = (sr)$, $\{rs\} = \{sr\}$, it follows that $\overline{rs} = \overline{sr}$.

§ **1·46**. *Rayleigh's reciprocal theorem.* Let a periodic force Q_s equal to $A_s \cos pt$ act on our mechanical system and produce a forced vibration of type

$$q_r = KA_s \cos(pt - \epsilon),$$

where K is the coefficient of amplitude and ϵ the retardation of phase. The reciprocal theorem asserts that if the system be acted on by the force $Q_r = A \cos pt$, the corresponding forced vibration for the co-ordinate q_s will be

$$q_s = KA_r \cos(pt - \epsilon).$$

Let D denote the determinant

$$\begin{vmatrix} \overline{11} & \overline{12} & \overline{13} & \ldots \\ \overline{21} & \overline{22} & \overline{23} & \ldots \\ \overline{31} & \overline{32} & \overline{33} & \ldots \\ \ldots & \ldots & \ldots & \ldots \end{vmatrix}$$

and let $\widehat{rs}$ denote its partial derivative with respect to the constituent $\overline{rs}$ when no recognition is made of the relation $\overline{rs} = \overline{sr}$ and when all the constituents are treated as algebraic quantities. This means that $\widehat{rs}$ is the cofactor of $\overline{rs}$ in the determinant operator D.

Solving the equations (E) like a set of linear algebraic equations on the assumption that $D \neq 0$, we obtain the relations

$$Dq_1 = \widehat{11}\,Q_1 + \widehat{21}\,Q_2 + \ldots \widehat{n1}\,Q_n,$$
$$Dq_2 = \widehat{12}\,Q_1 + \widehat{22}\,Q_2 + \ldots \widehat{n2}\,Q_n,$$
$$\ldots\ldots\ldots\ldots\ldots\ldots\ldots\ldots\ldots\ldots\ldots\ldots$$

From a property of determinants we may conclude that since $\overline{rs} = \overline{sr}$ we have also $\widehat{rs} = \widehat{sr}$. Thus the component displacement q_r due to a force Q_s is given by

$$Dq_r = \widehat{rs}\,Q_s.$$

Similarly, the component displacement q_s due to a force Q_r is given by

$$Dq_s = \widehat{sr}\,Q_r.$$

Distinguishing the second case by a dash affixed to the various quantities, we write

$$Q_s = A_s e^{ipt}, \quad Q_r' = A_r' e^{ipt},$$

where the coefficients A_s, A_r' may without loss of generality be supposed to be real. If they were complex but had a real ratio they could be made real by changing the initial time from which t is measured.

Expressing the solution in the form

$$q_r = A_s \frac{\widehat{rs}}{D} e^{ipt}, \quad q_s' = A_r' \frac{\widehat{sr}}{D} e^{ipt},$$

and defining the forced vibration as the particular integral obtained by replacing d/dt in each of the operators by ip, we obtain the relation

$$A_r' q_r = A_s q_s'$$

which gives reciprocal relations for both amplitude and phase.

In the statical case the quantities $[rs]$, (rs), are all zero and D, $\overline{rs}$, $\widehat{rs}$ are simply constants. Rayleigh then gives two additional theorems corresponding to those already considered in § 2.

(2) Suppose that only two forces Q_1, Q_2 act, then

$$\left.\begin{aligned} Dq_1 &= \widehat{11}\,Q_1 + \widehat{21}\,Q_2, \\ Dq_2 &= \widehat{12}\,Q_1 + \widehat{22}\,Q_2. \end{aligned}\right\} \qquad \text{......(F)}$$

If $q_1 = 0$ we have

$$\widehat{12}\,Dq_2 = [\widehat{12}^2 - \widehat{11}\,\widehat{22}]\,Q_1.$$

From this we conclude that if q_2 is given an assigned value a it requires the same force to keep $q_1 = 0$ as would be required if the force Q_2 is to keep $q_2 = 0$ when q_1 has the assigned value a.

(3) Suppose, first, that $Q_1 = 0$, then the equations (F) give

$$q_1 : q_2 = \widehat{21} : \widehat{22}.$$

Secondly, suppose $q_2 = 0$, then

$$Q_2 : Q_1 = -\,\widehat{12} : \widehat{22}.$$

Thus, when Q_2 acts alone, the ratio of the displacements q_1, q_2 is $-\,Q_2/Q_1$, where Q_1, Q_2 are the forces necessary to keep $q_2 = 0$.

§ **1·47**. *Fundamental equations of electric circuit theory.* A system of equations analogous to the system (E) occurs in the theory of electric circuits. This theory may be based on Kirchhoff's laws.

(1) The total impressed electromotive force (E.M.F.) taken around any closed circuit in a network is equal to the drop of electric potential expressed as the sum of three parts due respectively to resistance, induction and capacity.

Thus, if we consider an elementary circuit consisting of a resistance element of resistance R, an inductance element of self-induction (or inductance) L and a capacity element of capacity (capacitance) C, all in series, and suppose that an E.M.F. of amount E is applied to the circuit, Kirchhoff's first law states that at any instant of time

$$RI + L\frac{dI}{dt} + \frac{Q}{C} = E, \qquad \text{......(I)}$$

where I is the current in the circuit and $Q = \int I\, dt$.

The fall of potential due to resistance is in fact represented by RI where R is the resistance of the circuit (Ohm's law), the drop due to inductance is $L dI/dt$ and the drop across the condenser is Q/C.

There is an associated energy equation

$$RI^2 + \frac{d}{dt}(\tfrac{1}{2}LI^2) + \frac{d}{dt}(Q^2/2C) = EI,$$

in which RI represents the rate at which electrical energy is being converted into heat, while the second and third terms represent rates of increase of magnetic energy and electrical energy respectively. The right-hand side represents the rate at which the impressed E.M.F. is delivering energy to the circuit, while the left-hand side is the rate at which energy is being absorbed by the circuit. The inductance element and the condenser may be regarded as devices for storing energy, while the resistance is responsible for a dissipation of energy since the energy converted into Joule's heat is eventually lost by conduction and radiation of heat or by conduction and convection if the circuit is in a moving medium.

If we regard Q as a generalised co-ordinate, we may obtain the equation (I) by writing

$$T = \tfrac{1}{2}L\dot{Q}^2, \quad F = \tfrac{1}{2}R\dot{Q}^2, \quad V = Q^2/2C,$$

$$\frac{\partial}{\partial t}\left(\frac{\partial T}{\partial \dot{Q}}\right) + \frac{\partial F}{\partial \dot{Q}} + \frac{\partial V}{\partial Q} = E.$$

(2) In the case of a network the sum of the currents entering any branch point in the network is always zero.

If we consider a general form of network possessing n independent circuits, Kirchhoff's second law leads to the system of equations

$$\sum_{s=1}^{n} \left(L_{rs}\frac{d^2Q_s}{dt^2} + R_{rs}\frac{dQ_s}{dt} + \Gamma_{rs}Q_s\right) = E_r, \qquad \text{......(II)}$$

$$(r = 1, 2, \ldots n),$$

where E_r is the E.M.F. applied to the rth circuit, L_{ss}, R_{ss}, C_{ss} denote the total inductance, resistance and capacitance in series in the circuit s, while L_{rs}, R_{rs}, C_{rs} denote the corresponding mutual elements between circuits r and s. We have written Γ_{rs} for the reciprocal of C_{rs} and Q_s for $\int I_s dt$, where I_s is the current in the sth circuit or mesh.

An elaborate study of these equations in connection with the modern applications to electrotechnics has been made by J. R. Carson*.

In discussing complicated systems of resistances, inductances and capacities it will be convenient to use the symbol (LRC) for an inductance L, a resistance R and a capacity C in series. If a transmission line running between two terminals T and T' divides into two branches, one of which contains (LRC) and the other $(L'R'C')$, and if the two branches subsequently reunite before the terminal T' is reached, the arrangement will be represented symbolically by the scheme

$$(T) - \frac{(LRC)}{(L'R'C')} - (T').$$

In electrotechnics a mechanical system with a period constant n and a damping constant k is frequently used as the medium between the quantity to be studied (which actuates the oscillograph) and the record. The oscillograph is usually critically damped ($k = n$) so as to give a faithful record over a limited range of frequencies but even then the usefulness of the instrument is very limited as the range for accurate results is given roughly by the inequality $10m \leqslant n$.

A method of increasing the working range of such an oscillograph has been devised recently by Wynn-Williams†. If an E.M.F. of amount E acts between two terminals T and T' and the aim is to determine the variation of E, the usual plan is to place the oscillograph O in line with T and T' so that T, O and T' are in series. In our notation the arrangement is

$$T - O - T'.$$

Instead of this Wynn-Williams proposes the following scheme

$$(T) - (L_1 R_1 C_1) - \frac{(LRC)}{(O)} - (T')$$

in which L, R and C are chosen so that $2k = R/L$, $n^2 = 1/CL$ and L_1, R_1, C_1 are chosen so that L_1, C_1 are small and R_1 is such that there is a relation $(k + k_1)^2 = n^2 + n_1^2$, where $2k_1 = R_1/L$, $n_1^2 = 1/LC_1$.

Putting $L_1 = 0$ and writing $\dot{q}$ for the current flowing between T and T', x for the reading of the oscillograph, F for the back E.M.F. of the system (LRC), we have

$$\left[LD^2 + (R + R_1)D + \frac{1}{C} + \frac{1}{C_1}\right] q = E,$$

$$\left[LD^2 + RD + \frac{1}{C}\right] q = F,$$

$$[D^2 + 2kD + n^2]\, x = F.$$

Therefore
$$x = \frac{E}{D^2 + 2(k + k_1)D + n^2 + n_1^2}.$$

* *Electric circuit theory and operational calculus* (McGraw Hill, 1926).

† *Phil. Mag.* vol. L, p. 1 (1925).

Hence when the reading of the oscillograph is used to determine E the oscillograph behaves as if its period constant were $(n^2 + n_1^2)^{\frac{1}{2}}$ instead of n and its damping constant $k + k_1$ instead of k. By choosing $n_1^2 = 24n^2$, $k_1 = 4k = 4n$ we have

$$N = (n^2 + n_1^2)^{\frac{1}{2}} = 5n, \quad K = k + k_1 = 5k = 5n,$$

thus we obtain an amplitude scale which is the same as that of an oscillograph with a period constant $5n$ and a damping constant $5k$.

§ **1·48**. *Cauchy's method of solving a linear equation**. Let us suppose that we need a particular solution of the linear differential equation

$$\phi(D)\, u = f(t), \qquad \text{......(I)}$$

where $$\phi(D) = a_0 D^n + a_1 D^{n-1} + \dots a_n,$$

and D denotes the operator d/dt. The coefficients a_s are either constants or functions of t. For convenience we shall write $\alpha(t) = 1/a_0$.

If $(v_1, v_2, \dots v_n)$ are distinct solutions of the homogeneous equation

$$\phi(D)\, v = 0$$

the coefficients $C_1, C_2, \dots C_n$ in the general solution

$$v = C_1 v_1 + C_2 v_2 + \dots C_n v_n$$

may be chosen so that v satisfies the initial conditions

$$v(\tau) = v'(\tau) = \dots v^{(n-2)}(\tau) = 0, \quad v^{(n-1)}(\tau) = \alpha(\tau) f(\tau),$$

where $v^{(s)}(t)$ denotes the sth derivative of $v(t)$ and τ is the initial value of t. We denote this solution by the symbol $v(t, \tau)$ and consider the integral

$$u(t) = \int_0^t v(t, \tau)\, d\tau.$$

Assuming that the differentiations under the integral sign can be made by the rule of Leibnitz, we have

$$D^s u = \int_0^t D^s v(t, \tau)\, d\tau, \quad s = 1, 2, \dots n-2, n-1;$$

the terms arising from the upper limit vanishing on account of the properties of the function v. On the other hand

$$D^n u = \int_0^t D^n v(t, \tau)\, d\tau + \alpha(t) f(t),$$

and so $$\phi(D)\, u = f(t) + \int_0^t \phi(D)\, v \,.\, d\tau = f(t).$$

This particular solution is characterised by the properties

$$u(0) = u'(0) = \dots u^{(n-1)}(0) = 0, \quad u^{(n)}(0) = \alpha(0) f(0).$$

If, when $f(t) = 1$, $v(t, \tau) = \psi(t, \tau)$, the general value for an arbitrary

* Except for some slight modifications this presentation follows that of F. D. Murnaghan, *Bull. Amer. Math. Soc.* vol. XXXIII, p. 81 (1927).

function $f(t)$ is seen to be $f(t)\,\psi(t,\tau)$ and so we may write $u(t)$ in the form

$$u(t) = \int_0^t \psi(t,\tau)\, f(t)\, d\tau. \qquad \text{......(II)}$$

Introducing the notation

$$\xi(t,\tau) = \int_\tau^t \psi(t,s)\, ds,$$

we have

$$\frac{\partial}{\partial\tau}\xi(t,\tau) = -\psi(t,\tau),$$

and the integral in (II) may be integrated by parts giving

$$u(t) = f(0)\,\xi(t,0) + \int_0^t \xi(t,\tau)\, f'(\tau)\, d\tau.$$

This is the mathematical statement of the Boltzmann-Hopkinson principle of superposition, according to which we are able to build up a particular solution of equation (I) from a corresponding particular solution $\xi(t,\tau)$ for the case in which

$$f(t) = 1 \quad t > \tau,$$
$$= 0 \quad t < \tau.$$

When the coefficients in the polynomial $\phi(D)$ are all constants we may write

$$\phi(D) = a_0 (D - r_1)(D - r_2) \ldots (D - r_n),$$

where $r_1, r_2, \ldots r_n$ are the roots of the algebraic equation $\phi(x) = 0$. Taking first the case in which these roots are all distinct, we write

$$v_s = e^{r_s(t-\tau)} \quad s = 1, 2, \ldots n.$$

The equations to determine the constants C are then

$$C_1 + \ldots C_n = 0,$$
$$r_1 C_1 + \ldots r_n C_n = 0,$$
$$r_1^{n-1} C_1 + \ldots r_n^{n-1} C_n = af(t).$$

We may solve for C by multiplying these equations respectively by the coefficients of the successive powers of x in the expansion

$$(x - r_2)(x - r_3) \ldots (x - r_n) = b_0 + b_1 x + \ldots b_{n-1} x^{n-1}.$$

Since $b_{n-1} = 1$ we find that

$$af(\tau) = C_1 (r_1 - r_2)(r_1 - r_3) \ldots (r_1 - r_n) = aC_1\phi'(r_1),$$

and the other coefficients may be determined in a similar way.

Writing k_r for the reciprocal of $\phi'(r_s)$ we have

$$v(t,\tau) = f(\tau) \sum_{s=1}^{\infty} k_s e^{r_s(t-\tau)} = f(\tau)\,\psi(t,\tau),$$

and if ρ_s denotes the reciprocal of r_s

$$\xi(t,\tau) = \sum_{s=1}^{n} \rho_s k_s \left[e^{r_s(t-\tau)} - 1\right].$$

Now
$$\frac{1}{\phi(D)} = \sum_{s=1}^{n} \frac{k_s}{D - r_s},$$

and so
$$[\phi(0)]^{-1} = -\sum_{s=1}^{n} \rho_s k_s,$$

$$\xi(t, \tau) = \frac{1}{\phi(0)} + \sum_{s=1}^{n} \frac{e^{r_s(t-\tau)}}{r_s \phi'(r_s)}.$$

When there is a double root $r_1 = r_2$, we write

$$v_1 = e^{r_1(t-\tau)}, \quad v_2 = (t - \tau)\, e^{r_1(t-\tau)},$$

$v_3, \ldots v_n$ being the same as before. The equations to determine the constants C are now

$$C_1 + 0 \,.\, C_2 + C_3 + \ldots C_n = 0,$$
$$r_1 C_1 + 1 \,.\, C_2 + r_3 \,.\, C_3 + \ldots r_n \,.\, C_n = 0,$$
$$r_1^{n-1} C_1 + (n-1)\, r_2^{n-1} C_2 + r_3^{n-1} C_3 + \ldots r_n^{n-1} C_n = af(t).$$

Writing
$$(r - \lambda) F(r) \equiv (r - \lambda)(r - r_3) \ldots (r - r_n)$$
$$= c_0 + c_1 r + \ldots c_{n-1} r^{n-1}$$

and multiplying the equations by $c_0, c_1, \ldots c_{n-1}$ respectively we find that, since $C_{n-1} = 1$,

$$C_1 (r_1 - \lambda) F(r_1) + C_2 [F(r_1) + (r_1 - \lambda) F'(r_1)] = af(\tau).$$

The quantity λ is at our disposal. Let us first write $\lambda = r_1$, we then have

$$C_2 F(r_1) = af(\tau).$$

Simplifying the preceding equation with the aid of this relation we obtain

$$C_1 F(r_1) + C_2 F'(r_1) = 0.$$

Writing
$$G(r) = a/F(r),$$

we have
$$C_1 = f(\tau)\, G'(r_1), \quad C_2 = f(\tau)\, G(r_1).$$

These are just the constants obtained by writing

$$\frac{f(\tau)}{\phi(r)} = \frac{C_1}{r - r_1} + \frac{C_2}{(r - r_1)^2} + \frac{C_3}{r - r_3} + \ldots \frac{C_n}{r - r_n},$$

and a similar rule holds in the case of a multiple root of any order or any number of multiple roots. Thus in the case of a triple root,

$$\frac{f(\tau)}{\phi(r)} = \frac{C_1}{r - r_1} + \frac{C_2}{(r - r_1)^2} + \frac{2!\, C_3}{(r - r_1)^3} + \ldots.$$

§ **1·49**. *Heaviside's expansion.* The system of differential equations (II) of § 1·47 may be written in the form

$$\sum_{a=1}^{n} a_{ra} Q_a = E_r(t),$$

where the a's are analogous to the operators $\overline{rs}$ of § 1·45.

Denoting the determinant $| a_{rs} |$ by $\phi(D)$ and using A_{rs} to denote the co-factor of the constituent a_{rs} in this determinant, we have

$$\phi(D)\, Q_s = \sum_{r=1}^{n} A_{rs} E_r(t).$$

To obtain an expansion for Q_s we first solve the equation

$$\phi(D)\, y_s(t) = 1$$

with the supplementary conditions

$$y_s(0) = y_s'(0) = \dots y_s^{(2n-1)}(0) = 0,$$

then
$$x_{rs} = A_{sr} y_s(t)$$
is a particular solution of the equation $\phi(D)\, x_r = A_{sr} \,.\, 1$ for which

$$\dot{x}_r(0) = x_r'(0) = 0;$$

and by the expansion theorem of § 1·48,

$$\cdot\, x_{rs}(t, 0) = A_{sr} y_s(t, 0)$$
$$= \sum_{a=1}^{2n} \frac{A_{sr}(r_a)}{(r_a)\, \phi'(r_a)} e^{r_a t} + \frac{\angle_{s0}(0)}{\phi(0)},$$

which is Heaviside's expansion formula. The corresponding formula for $Q_s(t)$ is

$$Q_s(t) = \sum_{r=1}^{n} \left[E_r(0)\, x_{sr}(t, 0) + \int_0^t E_r'(\tau)\, x_{sr}(t, \tau)\, d\tau \right],$$

and this particular solution satisfies the conditions

$$Q_s(0) = Q_s'(0) = 0.$$

§ **1·51.** *The simple wave-equation.* There are a few partial differential equations which occur so frequently in physical problems that they may be called classical. The first of these is the simple wave-equation

$$\frac{\partial^2 V}{\partial t^2} = c^2 \frac{\partial^2 V}{\partial x^2} \qquad \text{......(I)}$$

which occurs in the theory of a vibrating string and also in the theory of the propagation of plane waves which travel without change of form. These waves may be waves of sound, elastic waves of various kinds, waves of light, electromagnetic waves and waves on the surface of water. In each case the constant c represents the velocity of propagation of a phase of a disturbance. The meaning of phase may be made clear by considering the particular solution

$$V = \sin(x - ct)$$

which shows that V has a constant value whenever the angle $x - ct$ has a constant value. This angle may be called the phase angle, it is constant for a moving point whose x-co-ordinate is given by an equation of type $x = ct + a$, where a is a constant. This point moves in one direction with uniform velocity c. There is also a second particular solution

$$V = \sin(x + ct)$$

for which the phase angle $x + ct$ is constant for a point which moves with velocity c in the direction for which x decreases. These solutions may be generalised by multiplying the argument $x \pm ct$ by a frequency factor $2\pi\nu/c$, where ν is a constant called the frequency, by adding a constant γ to the new phase angle and by multiplying the sine by a factor A to represent the amplitude of a travelling disturbance. In this way the particular solution is made more useful from a physical standpoint because it involves more quantities which may be physically measurable. In some cases these quantities may be more or less determined by the supplementary conditions which go with the equation when it is derived from physical principles or hypotheses.

Usually this particular equation is derived by the elimination of the quantity U from two equations

$$\frac{\partial V}{\partial t} = \alpha \frac{\partial U}{\partial x}, \quad \frac{\partial U}{\partial t} = \beta \frac{\partial V}{\partial x} \qquad \text{......(A)}$$

involving the quantities U and V, the coefficients α and β being constants. The constant c is now given by the equation

$$c^2 = \alpha\beta.$$

It should be noticed that if V is eliminated instead of U, the equation obtained for U is

$$\frac{\partial^2 U}{\partial t^2} = c^2 \frac{\partial^2 U}{\partial x^2},$$

and is of the same type as that obtained for V. This seems to be a general rule when the original equations are linear homogeneous equations of the first order with constant coefficients, however many equations there may be. The rule breaks down, however, when the coefficients are functions of the independent variables. If, for instance, α and β are functions of x the resulting equations are respectively

$$\frac{\partial^2 V}{\partial t^2} = \alpha \frac{\partial}{\partial x}\left(\beta \frac{\partial V}{\partial x}\right),$$

$$\frac{\partial^2 U}{\partial t^2} = \beta \frac{\partial}{\partial x}\left(\alpha \frac{\partial U}{\partial x}\right).$$

These equations may be called *associated equations*. Partial differential equations of this type occur in many physical problems. If, for instance, y denotes the horizontal deflection of a hanging chain which is performing small oscillations in a transverse direction, the equation of vibration is

$$\frac{\partial^2 y}{\partial t^2} = g \frac{\partial}{\partial x}\left(x \frac{\partial y}{\partial x}\right),$$

where g is the acceleration of gravity and x is the vertical distance above the free end. Equations of the above type occur also in the theory of the propagation of shearing waves in a medium stratified in horizontal plane layers, the physical properties of the medium varying with the depth.

§ 1·52. The differential equation (I) was solved by d'Alembert who showed that the solution can be expressed in the form

$$V = f(x - ct) + g(x + ct),$$

where $f(z)$ and $g(z)$ are arbitrary functions of z with second derivatives $f''(z)$, $g''(z)$ that are continuous for some range of the real variable z. A solution of type $f(x - ct)$ will be called a "primary solution," a term which will be extended in § 1·92 to certain other partial differential equations.

To illustrate the way in which primary solutions can be used to solve a physical problem we consider the transverse vibrations of a fine string or the shearing vibration of a building*.

The co-ordinate x is supposed to be in the direction of the undisturbed string and in the vertical direction for the building, the co-ordinate y is taken to represent the transverse displacement. If A denotes the area of the cross-section, which is a horizontal section in the case of the building, and ρ the density of the material, the momentum of the slice $A\,dx$ is $M\,dx$, where $M = \rho A \dfrac{\partial y}{\partial t}$. The slice is acted upon by two shearing forces acting in a transverse direction and by other forces acting in a "vertical" direction, i.e. in the direction of the undisturbed string. Denoting the shearing force on the section x by S, that on the section $x + dx$ is $S + \dfrac{\partial S}{\partial x}\,dx$. The difference is $\dfrac{\partial S}{\partial x}\,dx$, and so the equation of motion is

$$\frac{\partial M}{\partial t} = \frac{\partial S}{\partial x}.$$

We now adopt the hypothesis that when the displacement y is very small

$$S = A\mu \frac{\partial y}{\partial x},$$

where μ is a constant which represents the rigidity of the material in the case of the building and the tension in the case of the string. According to this hypothesis if ρ and A are also constants

$$\frac{\partial^2 y}{\partial t^2} = c^2 \frac{\partial^2 y}{\partial x^2},$$

where $c^2 = \mu/\rho$. The expressions for M and S also give the equation

$$\rho \frac{\partial S}{\partial t} = \mu \frac{\partial M}{\partial x},$$

and so we have two equations of the first order connecting the quantities M and S; these equations imply that M and S satisfy the same partial differential equation as y.

* The shearing vibrations of a building have been discussed by K. Suyehiro, *Journal of the Institute of Japanese Architects*, July (1926).

In the case of the building one of the boundary conditions is that there is no shearing force at the top of the building, therefore $S = 0$ when $x = h$. Assuming that

$$y = f(x - ct) + g(x + ct)$$

the condition is $\dfrac{\partial y}{\partial x} = 0$ when $x = h$, and so

$$0 = f'(h - ct) + g'(h + ct).$$

This condition may be satisfied by writing

$$y = \phi(ct + x - h) + \phi(ct - x + h),$$

where $\phi(z)$ is an arbitrary function.

A motion of the ground ($x = 0$) which will give rise to a motion of this kind is obtained by putting $x = 0$ in the above equation.

Denoting the motion of the ground by $y = F(t)$ we have the equation

$$F(t) = \phi(ct - h) + \phi(ct + h) \qquad \text{......(A)}$$

for the determination of the function $\phi(z)$.

In the case of the string the end $x = l$ may be stationary. We therefore put $y = 0$ for $x = l$ and obtain the equation

$$0 = f(l - ct) + g(l + ct)$$

which is satisfied by

$$y = \psi(ct + x - l) - \psi(ct + l - x),$$

where ψ is another arbitrary function. If the motion of the end $x = 0$ is prescribed and is $y = G(t)$ we have the equation

$$G(t) = \psi(ct - l) - \psi(ct + l) \qquad \text{......(B)}$$

for the determination of the function $\psi(z)$.

If, on the other hand, the initial displacement and velocity are prescribed, say

$$y = \theta(x), \quad \frac{\partial y}{\partial t} = \chi(x)$$

when $t = 0$, we have the equations

$$\theta(x) = f(x) + g(x), \quad \chi(x) = c\,[g'(x) - f'(x)]$$

which give

$$2cf'(x) = c\theta'(x) - \chi(x),$$
$$2cg'(x) = c\theta'(x) + \chi(x),$$

and the solution takes the form

$$y = \tfrac{1}{2}[\theta(x - ct) + \theta(x + ct)] + \frac{1}{2c}\int_{x-ct}^{x+ct} \chi(\tau)\,d\tau.$$

If in the preceding case both ends of the string are fixed, the equation (B) implies that $\psi(x)$ is a periodic function of period $2l$, the corresponding time interval being $2l/c$. Submultiples of these periods are, of course, admissible, and the inference is that a string with its ends fixed can perform oscillations in which any state of the system is repeated after every time interval of length $2lm/nc$, where m and n are integers, n being a constant for this type of oscillation.

In the case of the building the ground can remain fixed in cases when $\phi(z)$ is a periodic function of period $4h$ such that

$$\phi(z + 2h) = -\phi(z).$$

It should be noticed that the conditions of periodicity may be satisfied by writing

$$\psi(z) = \sin \frac{m\pi z}{l}, \quad \phi(z) = \sin [(n + \tfrac{1}{2}) \pi z/h],$$

where m and n are integers. Thus in the case of the string with fixed ends there are possible vibrations of type

$$y = a_m \sin \frac{m\pi x}{l} \cos \frac{m\pi ct}{l},$$

and in the case of the building with a free top and fixed base there are possible vibrations of type

$$y = b_m \sin \left[(n + \tfrac{1}{2}) \frac{\pi x}{h} \right] \cos \left[(n + \tfrac{1}{2}) \frac{\pi ct}{h} \right].$$

These motions may be generalised by writing for the case of the string

$$y = \sum_{m=1}^{s} a_m \sin \frac{m\pi x}{l} \cos \frac{m\pi ct}{l},$$

where the coefficients a_m are arbitrary constants. For complete generality we must make s infinite, but for the present we shall treat it as a finite constant. The total kinetic energy of the string is

$$\int_0^l \tfrac{1}{2}\rho A \left(\frac{\partial y}{\partial t}\right)^2 dx = \sum_{m=1}^{s} \frac{m^2\pi^2c^2}{4l^2} a_m{}^2 \sin^2 \frac{m\pi ct}{l},$$

since we have
$$\int_0^l \sin \frac{m\pi x}{l} \sin \frac{n\pi x}{l} \, dx = 0 \qquad n \neq m$$
$$= l/2 \qquad n = m.$$

Since the kinetic energy is the sum of the kinetic energies of the motions corresponding to the individual terms of the series, these terms are supposed to represent independent natural vibrations of the string. These are generally called the normal vibrations.

The solution for the vibrating building can also be generalised so as to give

$$y = \sum_{n=1}^{s} b_n \sin \left[(n + \tfrac{1}{2}) \frac{\pi x}{h} \right] \cos \left[(n + \tfrac{1}{2}) \frac{\pi ct}{h} \right],$$

and the kinetic energy is in this case

$$\sum_{n=1}^{s} \frac{(2n + 1)^2 \pi^2 c^2}{16l} b_n{}^2 \sin^2 \left[(n + \tfrac{1}{2}) \frac{\pi ct}{h} \right],$$

for now we have corresponding relations

$$\int_0^h \sin\left[(n+\tfrac{1}{2})\frac{\pi x}{h}\right]\sin\left[(m+\tfrac{1}{2})\frac{\pi x}{h}\right]dx = 0 \qquad m \neq n$$
$$= h/2 \qquad m = n.$$

Such relations are called orthogonal relations.

§ 1·53. In the case of the equation of the transverse vibrations of a string there is a type of solution which can be regarded as fundamental.

Let us suppose that the point $x = a$ is compelled to move with a simple harmonic motion*

$$y = \beta \cos(pt + \alpha),$$

where α, β and p are arbitrary real constants. If the ends $x = 0$, $x = l$ remain fixed, it is easily seen that the differential equation

$$\frac{\partial^2 y}{\partial t^2} = c^2 \frac{\partial^2 y}{\partial x^2},$$

and the conditions $y = 0$ at the ends may be satisfied by writing $y = y_1$ for $0 \leqslant x \leqslant a$ and $y = y_2$ for $a \leqslant x \leqslant l$, where

$$y_1 = \beta \operatorname{cosec}(\lambda a) \sin(\lambda x) \cos(pt + \alpha),$$
$$y_2 = \beta \operatorname{cosec} \lambda(l - a) \sin \lambda(l - x) \cos(pt + \alpha),$$

and $\lambda c = p$. The case of a periodic force $F = F_0 \cos(pt + \alpha)$ concentrated on an infinitely short length of the string may be deduced by writing down the condition that the forces on the element must balance, the inertia being negligible. This condition is

$$F = Py_1' - Py_2' \qquad \text{for } x = a,$$

where P is the pull of the string. Substituting the values of y_1' and y_2' we get

$$cF_0 = p\beta P\,[\cot \lambda a + \cot \lambda(l - a)].$$

Therefore
$$\beta = \frac{F_0}{P\lambda} \operatorname{cosec} \lambda l \sin \lambda a \sin \lambda(l - a).$$

The solution can now be written in the form

$$y = \frac{F}{P} g(x, a),$$

where
$$g(x, a) = \frac{\sin \lambda x \sin \lambda(l - a)}{\lambda \sin \lambda l} \qquad 0 \leqslant x \leqslant a$$
$$= \frac{\sin \lambda(l - x) \sin \lambda a}{\lambda \sin \lambda l} \qquad a \leqslant x \leqslant l.$$

This function $g(x, a)$ is a solution of the differential equation

$$\frac{d^2 g}{dx^2} + \lambda^2 g = 0 \qquad \text{......(A)}$$

* Rayleigh, *Theory of Sound*, vol. I, p. 195.

and satisfies the boundary conditions $g = 0$ when $x = 0$ and when $x = l$. It is continuous throughout the interval $0 \leqslant x \leqslant l$, but its first derivative is discontinuous at the point $x = a$ and indeed in such a manner that

$$\lim_{s \to 0} \left[\frac{\partial}{\partial x} g(x - s, a) - \frac{\partial}{\partial x} g(x + s, a) \right]_{x=a} = 1.$$

The function $g(x, a)$ is called a Green's function for the differential expression $\frac{d^2u}{dx^2} + \lambda^2 u$, it possesses the remarkable property of symmetry expressed by the relation

$$g(x, a) = g(a, x).$$

This is a particular case of the general reciprocal theorem proved by Maxwell and the late Lord Rayleigh.

It should be noticed that the Green's function does not exist when λ has a value for which $\sin \lambda l = 0$, that is, a value for which the equation (A) possesses a solution $g = \sin \lambda x$ which satisfies the boundary conditions and is continuous $(D, 1)$ throughout the range $(0 \leqslant x \leqslant l)$.

A fundamental property of the Green's function $g(x, a)$ is obtained by solving the differential equation

$$\frac{d^2u}{dx^2} + \lambda^2 u = f(x)$$

by the method of integrating factors. Assuming that y is continuous $(D, 2)$ in the interval $(0, l)$ and that the function $f(x)$ is continuous in this interval, the result is that

$$u = u(0) \left(\frac{\partial g}{\partial a} \right)_{a=0} - u(l) \left(\frac{\partial g}{\partial a} \right)_{a=l} - \int_0^l g(x, a) f(a) \, da,$$

where $u(0)$ and $u(l)$ are assigned values of u at the ends. If these values are both zero y is expressed simply as a definite integral involving the Green's function and $f(a)$.

§ **1·54**. The torsional oscillations of a circular rod are very similar in character to the shearing oscillations of a building. Let us consider a straight rod of uniform cross-section, the centroids of the sections by planes $x =$ constant, perpendicular to the length of the rod, being on a straight line which we take as axis of x. Let us assume that the section at distance x from the origin is twisted through an angle θ relative to the section at the origin. It is on account of the variation of θ with x that an element of the rod must be regarded as strained. The twist per unit length at the place x is defined to be

$$\tau = \frac{\partial \theta}{\partial x};$$

it vanishes when θ is constant throughout the element bounded by the planes x and $x + dx$, i.e. when this element is simply in a displaced position just as if it had been rotated like a rigid body.

The torque which is transmitted from element to element across the plane x is assumed to be $K\mu\tau$, where μ is an elastic constant for the material (the modulus of rigidity) and K is a quantity which depends upon the size and shape of the cross-section and has the same dimensions as I, the moment of inertia of the area about the axis of x.

Let ρ denote the density of the material, then the moment of inertia about the axis of x of the element previously considered is $\rho I\,dx$ and the angular momentum is $\rho I\,\dot{\theta}\,dx$.

Equating the rate of change of angular momentum to the difference between the torques transmitted across the plane faces of the element, we obtain the equation of motion

$$\rho I \frac{\partial^2\theta}{\partial t^2} = \mu K \frac{\partial^2\theta}{\partial x^2}$$

which holds in the case when the rod is entirely free or is acted upon by forces and couples at its ends. In this case the differential equation must be combined with suitable end conditions.

A simple case of some interest is that in which the end $x = 0$ is tightly clamped, whilst the motion of the other end $x = a$ is prescribed.

§ **1·55**. The same differential equation occurs also in the theory of the longitudinal vibrations of a bar or of a mass of gas.

Consider first the case of a bar or prism whose generators are parallel to the axis of x. Let $x + \xi$ denote the position at time t of that cross-section whose undisturbed position is x, then ξ denotes the displacement of this cross-section. An element of length, δx, is then altered to $\delta\,(x + \xi)$, or $(1 + \xi')\,\delta x$, where the prime denotes differentiation with respect to x. Equating this to $(1 + e)\,\delta x$ we shall call e the strain. The strain is thus the ratio of the change in length to the original length of the element and is given by the formula

$$e = \frac{\partial\xi}{\partial x}.$$

According to Hooke's law stress is proportional to strain for small displacements and strains. The total force acting across the sectional area A in a longitudinal direction is therefore $F = EeA$, where E is Young's modulus of elasticity for the material of which the rod is composed. The stress across the area is simply Ee.

The momentum of the portion included between the two sections with co-ordinates x and $x + \delta x$ is $M\,\delta x$, where $M = \rho A \dfrac{\partial\xi}{\partial t}$ and ρ is the density of the material. The equation of motion is then

$$\frac{\partial M}{\partial t} = \frac{\partial F}{\partial x},$$

or
$$\rho A \frac{\partial^2 \xi}{\partial t^2} = \frac{\partial}{\partial x}\left[EA \frac{\partial \xi}{\partial x}\right].$$

When the material is homogeneous and the rod is of uniform section the equation is
$$\frac{\partial^2 \xi}{\partial t^2} = c^2 \frac{\partial^2 \xi}{\partial x^2},$$
where $c^2 = E/\rho$. Since the modulus E for most materials is about two or three times the modulus of rigidity μ, longitudinal waves travel much more rapidly than shearing waves and the frequency of the fundamental mode of vibration is higher for longitudinal oscillations than it is for shearing oscillations. In the case of a thin rod shearing oscillations would not occur alone but would be combined with bending, and the motion is different.

The fundamental frequency for the lateral oscillations is, however, much lower than that for the longitudinal oscillations. Let us next consider the propagation of plane waves of sound in a direction parallel to the axis of x.

Let $v_0 = A\,\delta x$ be the initial volume of a disc-shaped mass of the gas through which the sound travels, $v = A\delta\,(x + \xi)$ the volume of the same mass at time t. We then have
$$v = v_0\,(1 + e),$$
where e is now the dilatation. If ρ_0 is the original density and ρ the density of the mass at time t, we may write
$$\rho = \rho_0\,(1 + s),$$
where s is the condensation, it is the ratio of the increment of density to the original density. Since $\rho v = \rho_0 v_0$ we have
$$(1 + s)\,(1 + e) = 1,$$
and if s and e are both small we may write
$$s = -\,e = -\,\frac{\partial \xi}{\partial x}.$$

To obtain the equation of motion we assume that the pressure varies with the density according to some definite law such as the adiabatic law
$$\frac{p}{p_0} = \left(\frac{\rho}{\rho_0}\right)^{\gamma},$$
where p_0 is the pressure corresponding to the density γ and is a constant which is different for different gases.

This law holds when there is no sensible transfer of heat between adjacent portions of the gas. Such a state of affairs corresponds closely to the facts, since in the case of vibration of audible frequency the condensations and rarefactions of our disc-shaped mass of gas follow one another with a frequency of 500 or more per second.

For small values of s we may write
$$p = p_0\,(1 + \gamma s).$$

The equation of motion is now

$$\frac{\partial M}{\partial t} = \frac{\partial F}{\partial x},$$

where $$M = \rho_0 A \frac{\partial \xi}{\partial t}, \quad F = -Ap.$$

Substituting the values of p and s we obtain the equation

$$\frac{\partial^2 \xi}{\partial t^2} = c^2 \frac{\partial^2 \xi}{\partial x^2}$$

in which $$c^2 = \gamma \frac{p_0}{\rho_0} = \left(\frac{dp}{d\rho}\right)_0.$$

For sound waves in a tube closed at both ends the boundary conditions are $\xi = 0$ when $x = 0$ and when $x = l$. The solution is just the same as the solution of the problem of transverse vibration of a string with fixed ends.

For sound waves in a pipe open at both ends and for the longitudinal vibrations of a bar free at both ends we have the boundary conditions

$$\frac{\partial \xi}{\partial x} = 0$$

when $x = 0$ and when $x = l$, which express that there is no stress at the ends. The normal modes of vibration are now of type

$$\xi = C_m \cos \frac{m\pi x}{l} \cos\left(\frac{m\pi ct}{l}\right),$$

where C_m is an arbitrary constant and m is an integer. This solution is of type

$$\xi = \phi\,(x + ct) + \phi\,(ct - x),$$

and may be interpreted to mean that the progressive waves represented by $\xi_1 = \phi\,(ct - x)$ are reflected at the end $x = 0$ with the result that there is a superposed wave represented by $\xi_2 = \phi\,(ct + x)$.

There is a different type of reflection at a closed end of a tube (or fixed end of a rod), as may be seen from the solution

$$\xi = \phi\,(ct - x) - \phi\,(ct + x),$$

which makes $\xi = 0$ when $x = 0$.

Reflection at a boundary between two different fluid media or between two parts of a bar composed of different materials may be treated by introducing the boundary condition that the stress and the velocity must be continuous at the boundary.

If progressive waves represented by $\xi_0 = a_0\phi\,(t - x/c)$ approach the boundary $x = 0$ from the negative side and give rise to a reflected wave $\xi_1 = a_1\phi\,(t + x/c)$ and a transmitted wave $\xi_2 = a_2\phi\,(t - x/c')$, the boundary conditions are

$$\frac{\partial \xi_0}{\partial t} + \frac{\partial \xi_1}{\partial t} = \frac{\partial \xi_2}{\partial t}, \quad \kappa\,(s_0 + s_1) = \kappa' s_2,$$

where $\kappa = \gamma p_0$ and $\kappa' = \gamma' p_0$, the constants γ and γ' referring respectively to the media on the negative and positive sides of the origin. The equilibrium pressure p_0 is the same for both media.

Now $$\frac{\partial \xi_0}{\partial t} = cs_0, \quad \frac{\partial \xi_1}{\partial t} = -cs_1, \quad \frac{\partial \xi_2}{\partial t} = c's_2,$$

hence, when $x = 0$,

$$cs_0 = a_0\phi'(t), \quad -cs_1 = a_1\phi'(t), \quad c's_2 = a_2\phi'(t),$$

and $$c(s_0 - s_1) = c's_2, \quad \kappa(s_0 + s_1) = \kappa' s_2.$$

Therefore $$s_1 = \frac{\kappa' c - \kappa c'}{\kappa' c + \kappa c'} s_0, \quad s_2 = \frac{2\kappa c}{\kappa' c + \kappa c'} s_0,$$

$$a_1 = \frac{\kappa c' - \kappa' c}{\kappa' c + \kappa c'} a_0, \quad a_2 = \frac{2\kappa c'}{\kappa' c + \kappa c'} a_0.$$

§ **1·56**. The simple wave-equation occurs also in an approximate theory of long waves travelling along a straight canal, with horizontal bed and parallel vertical sides, the axis of x being parallel to the vertical sides and in the bed (see Lamb's *Hydrodynamics*, Ch. VIII).

Let b be the breadth of the canal and h the depth of the fluid in an initial state at time $t = 0$ when the fluid is at rest and its surface horizontal. We shall denote the density of the fluid by ρ and the pressure at a point (x, y, z) by p. The motion is investigated on the assumption that p is approximately the same as the hydrostatic pressure due to the depth below the free surface. This means that we write

$$p = p_0 + g\rho(h + \eta - y), \qquad \text{......(I)}$$

where p_0 is the external pressure, which is supposed to be uniform, η is the elevation of the free surface above its undisturbed position and g is the acceleration of gravity. One consequence of this assumption is that there is no vertical acceleration, in other words, the vertical acceleration is neglected in making this approximation.

If, in fact, we consider a small element of fluid bounded by horizontal and vertical planes parallel to the planes of reference, the axis of y being vertically upwards, the equations of motion are

$$\rho\alpha \,.\, \delta x \delta y \delta z = -\frac{\partial p}{\partial x} \delta x \,.\, \delta y \delta z,$$

$$\rho\beta \,.\, \delta x \delta y \delta z = -\frac{\partial p}{\partial y} \delta y \,.\, \delta z \delta x - \rho g \delta x \delta y \delta z,$$

$$\rho\gamma \,.\, \delta x \delta y \delta z = -\frac{\partial p}{\partial z} \delta z \,.\, \delta x \delta y,$$

where α, β, γ are the component accelerations. With the above assumption we have $\beta = \gamma = 0$, and so

$$\rho\alpha = -\frac{\partial p}{\partial x}.$$

The assumption of no vertical acceleration is not equivalent to the assumption (I), because an arbitrary function of x, z and t could be added to the right-hand side of (I) and the equations of motion would still give no vertical acceleration.

Equation (I) gives $$\rho\alpha = -g\rho\frac{\partial\eta}{\partial x}.$$

This expression for α is independent of y, consequently, since g is assumed to be constant, the acceleration α is the same for all particles in a vertical plane perpendicular to the axis of x. The horizontal velocity u depends on x and t only.

Now let ξ be the total displacement from their initial position of the particles which at time t occupy the vertical plane x. Each particle is supposed to have moved horizontally through a distance ξ, but actually some of the particles will have moved slightly upwards or downwards as well.

If $$AX = \xi,\ A'X' = \xi + \frac{\partial\xi}{\partial x}\delta x,$$
the fluid which occupies the region $QQ'X'X$ is supposed to have initially occupied the region $PP'A'A$.

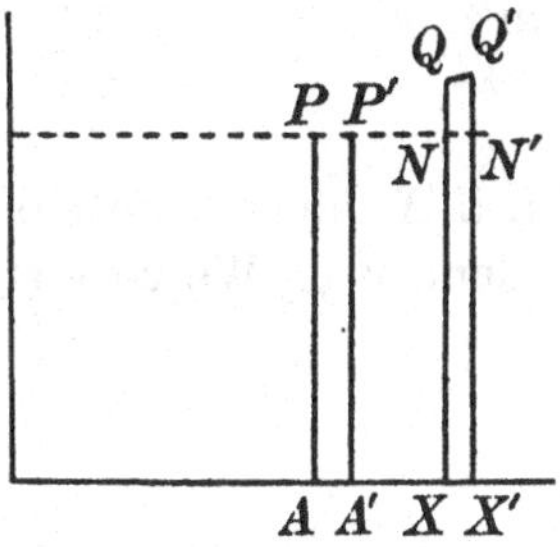

Fig. 9.

Equating the amount of fluid in the region $QQ'N'N$ to the difference of the amounts in the regions $PNXA$, $P'N'X'A'$ we obtain the equation of continuity

$$-\frac{\partial}{\partial x}(\xi hb)\,\delta x = \eta b\,\delta x,$$

or

$$\eta = -h\frac{\partial\xi}{\partial x}. \qquad \text{......(II)}$$

A second equation is obtained by writing $\alpha = \frac{\partial u}{\partial t}$. This is approximately true in the case of infinitely small motions, the exact equation being

$$\alpha = \frac{\partial u}{\partial t} + u\frac{\partial u}{\partial x}.$$

Writing $$\xi = \int_0^t u\,dt,$$

we have $$\frac{\partial^2\xi}{\partial t^2} = \frac{\partial u}{\partial t} = \alpha = -g\frac{\partial\eta}{\partial x}. \qquad \text{......(III)}$$

The equations (II) and (III) now give the wave-equations

$$\frac{\partial^2\xi}{\partial t^2} = c^2\frac{\partial^2\xi}{\partial x^2},\quad \frac{\partial^2\eta}{\partial t^2} = c^2\frac{\partial^2\eta}{\partial x^2},$$

where $$c^2 = gh.$$

When, in addition to gravity, the fluid is acted upon by small disturbing forces with components (X, Y) per unit mass of the fluid, the assumption that the pressure is approximately equal to the hydrostatic pressure leads to the equation

$$p = p_0 + \rho \int_y^{h+\eta} (g - Y)\, dy.$$

This gives
$$\frac{\partial p}{\partial x} = \rho (g - Y) \frac{\partial \eta}{\partial x} - \rho \int_y^{h+\eta} \frac{\partial Y}{\partial x}\, dy,$$

and the equation of horizontal motion

$$\rho \frac{\partial u}{\partial t} = \rho X - \frac{\partial p}{\partial x}$$

indicates that in general u depends on y as well as on x and t.

With, however, the simplifying assumptions that Y is small compared with g and that $h \frac{\partial Y}{\partial x}$ is small in comparison with X the equation takes the form

$$\frac{\partial u}{\partial t} = X - g \frac{\partial \eta}{\partial x},$$

and, if X depends only on x and t, this equation indicates that u is independent of y. We may then proceed as before and obtain the equations

$$\frac{\partial^2 \xi}{\partial t^2} = c^2 \frac{\partial^2 \xi}{\partial x^2} + X,$$

$$\frac{\partial^2 \eta}{\partial t^2} = c^2 \frac{\partial^2 \eta}{\partial x^2} - h \frac{\partial X}{\partial x}.$$

EXAMPLES

1. An elastic bar of length l has masses m_0, m_1 at the ends $x = 0$, $x = l$ respectively. Prove that the terminal conditions are

$$EA \frac{\partial \xi}{\partial x} = m_0 \frac{\partial^2 \xi}{\partial t^2} \quad \text{when } x = 0,$$

$$EA \frac{\partial \xi}{\partial x} = -m_1 \frac{\partial^2 \xi}{\partial t^2} \quad \text{when } x = l.$$

Prove that the possible frequencies of vibration are given by the equation

$$(1 - \mu_0 \mu_1 \theta^2) \tan \theta + (\mu_0 + \mu_1)\, \theta = 0,$$

where
$$c^2 m_0 = lAE\mu_0, \quad c^2 m_1 = lAE\mu_1, \quad \theta = nl,$$

and $nc/2\pi$ is the number of vibrations per second.

2. If a prescribed vibration $\xi = C \cos nt$ is maintained at the end $x = 0$ of a straight pipe which is closed at the end $x = l$ the vibration at the place x is given by

$$\xi = C \operatorname{cosec} \frac{nl}{c} \sin \frac{n(l - x)}{c} \cos nt.$$

Obtain the corresponding solution for the case in which the end $x = l$ is open.

3. Discuss the longitudinal oscillations of a weighted bar whose upper end is fixed.

4. If $La = \log\left\{1 + v\frac{ct+x}{ca}\right\}, \quad L\beta = \log\left\{1 + v\frac{ct-x}{ca}\right\}, \quad L = \log\frac{c+v}{c-v}$

and A is an arbitrary constant, the function

$$y = A\,[\sin 2s\pi a - \sin 2s\pi\beta]$$

satisfies the differential equation

$$\frac{\partial^2 y}{\partial t^2} = c^2\frac{\partial^2 y}{\partial x^2},$$

and the end conditions $y = 0$ when $x = 0$ and when $x = a + vt$. Prove also that when $v \to 0$,

$$y \to 2A\sin\frac{s\pi x}{a}\cos\frac{s\pi ct}{a}.$$

[T. H. Havelock, *Phil. Mag.* vol. XLVII, p. 754 (1924).]

5. Prove that if $y = 0$ when $x = 0$ and $x = vt$,

$$y = f(x), \quad \dot{y} = g(x), \text{ when } t = t_0,$$

a solution of $\frac{\partial^2 y}{\partial t^2} = c^2\frac{\partial^2 y}{\partial x^2}$ is given by

$$y = \sum_{n=1}^{\infty} \sin\left[\frac{na}{2}\log\frac{ct+x}{ct-x}\right]\left\{A_n\cos\frac{na\tau}{2} + B_n\sin\frac{na\tau}{2}\right\},$$

where

$$a\log\frac{c+v}{c-v} = 2\pi,$$

$$exp\,(\tau) = \frac{t^2}{t_0^2} - \frac{x^2}{c^2t_0^2}, \quad F(ct_0 + x) = \tfrac{1}{2}f(x) + \frac{1}{2c}\int_0^x g(x)\,dx, \quad -vt_0 < x < vt_0,$$

$$A_n = \frac{2a}{\pi}\int_{-vt_0}^{vt_0} F(ct_0 + x)\sin(na\omega)\frac{dx}{ct_0 + x},$$

$$B_n = -\frac{2a}{\pi}\int_{-vt_0}^{vt_0} F(ct_0 + x)\cos(na\omega)\frac{dx}{ct_0 + x},$$

and it is supposed that

$$f(-x) = -f(x), \quad g(-x) = -g(x).$$

[E. L. Nicolai, *Phil. Mag.* vol. XLIX, p. 171 (1925).]

§ 1·61. *Conjugate functions and systems of partial differential equations.* If in equations ((A) § 1·51) we write $\alpha = 1$, $\beta = -1$ and use the variable y in place of t we obtain the equations

$$\frac{\partial U}{\partial x} = \frac{\partial V}{\partial y}, \quad \frac{\partial U}{\partial y} = -\frac{\partial V}{\partial x}$$

satisfied by two conjugate functions U and V. In this case both functions satisfy the two-dimensional form of Laplace's equation

$$\frac{\partial^2 V}{\partial x^2} + \frac{\partial^2 V}{\partial y^2} = 0.$$

This equation is important in hydrodynamics and in electricity and magnetism.

The equations (A) may be generalised in another way by writing

$$\left.\begin{aligned} \theta\frac{\partial V}{\partial t} &= \alpha\frac{\partial U}{\partial x} + \gamma\frac{\partial V}{\partial x} + \lambda U + \mu V, \\ \phi\frac{\partial U}{\partial t} &= \beta\frac{\partial V}{\partial x} + \delta\frac{\partial U}{\partial x} + \sigma U + \tau V, \end{aligned}\right\} \quad \text{......(A)}$$

where α, β, γ, δ, θ, ϕ, λ, μ, σ, τ are arbitrary constants. In particular, the equations

$$\frac{\partial U}{\partial t} = \kappa \frac{\partial V}{\partial x}, \quad \frac{\partial U}{\partial x} = V$$

lead to the equation

$$\frac{\partial U}{\partial t} = \kappa \frac{\partial^2 U}{\partial x^2},$$

which is the equation for the conduction of heat in one direction when U is interpreted as the temperature and κ as the diffusivity. The same equation occurs in the theory of diffusion. It should be noticed that the quantity V satisfies the same equation.

Again, if we write

$$\frac{\partial V}{\partial x} = L \frac{\partial U}{\partial t} + RU,$$

$$\frac{\partial U}{\partial x} = C \frac{\partial V}{\partial t} + SV,$$

and interpret V as electric potential, U as electric current, we obtain the differential equation

$$\frac{\partial^2 V}{\partial x^2} = CL \frac{\partial^2 V}{\partial t^2} + (RC + SL) \frac{\partial V}{\partial t} + RSV = 0$$

which governs the propagation of an electric current in a cable*. The coefficients have the following meanings:

R	L	C	S
resistance	inductance	capacity	leakance

all per unit of length of the cable. The quantity U satisfies the same differential equation as V. This differential equation may be reduced to a canonical form by introducing the new dependent variables u, v, defined by the equations

$$u = Ue^{Rt/L}, \quad v = Ve^{Rt/L}.$$

These variables satisfy the equations

$$\frac{\partial v}{\partial x} = L \frac{\partial u}{\partial t},$$

$$\frac{\partial u}{\partial x} = C \frac{\partial v}{\partial t} + (S - CR/L)\, v,$$

and the canonical equations of propagation are Heaviside's equations

$$\frac{\partial^2 u}{\partial x^2} = CL \frac{\partial^2 u}{\partial t^2} + (SL - CR) \frac{\partial u}{\partial t},$$

$$\frac{\partial^2 v}{\partial x^2} = CL \frac{\partial^2 v}{\partial t^2} + (SL - CR) \frac{\partial v}{\partial t}.$$

These equations are of the simple type (I) if

$$SL = CR.$$

In this case a wave can be propagated along the cable without distortion.

* Cf. J. A. Fleming, *The Propagation of Electric Currents in Telephone and Telegraph Circuits*, ch. v.

When dealing with the general equations (A) it is advantageous to use algebraic symbols for the differential operators and to write

$$\frac{\partial}{\partial t} = D_t, \quad \frac{\partial}{\partial x} = D_x;$$

the differential equations may then be written symbolically in the form

$$(\theta D_t - \gamma D_x - \mu)\, V = (\alpha D_x + \lambda)\, U, \quad (\phi D_t - \delta D_x - \sigma)\, U = (\beta D_x + \tau)\, V.$$

The first equation may be satisfied by writing

$$U = (\theta D_t - \gamma D_x - \mu)\, W, \quad V = (\alpha D_x + \lambda)\, W, \qquad \text{......(B)}$$

where W is a new dependent variable. Substituting in the second equation we obtain the following equation for W,

$$[(\theta D_t - \gamma D_x - \mu)(\phi D_t - \delta D_x - \sigma) - (\alpha D_x + \lambda)(\beta D_x + \tau)]\, W = 0,$$

which, when written in full, has the form

$$\theta\phi \frac{\partial^2 W}{\partial t^2} - (\theta\delta + \phi\gamma)\frac{\partial^2 W}{\partial t \partial x} + (\gamma\delta - \alpha\beta)\frac{\partial^2 W}{\partial x^2} - (\theta\sigma + \phi\mu)\frac{\partial W}{\partial t}$$
$$- (\alpha\tau + \beta\lambda + \gamma\sigma + \delta\mu)\frac{\partial W}{\partial x} + (\mu\sigma - \lambda\tau)\, W = 0.$$

When this equation has been solved the variables U and V may be determined with the aid of equations (B). It is easily seen that U and V satisfy the same equation as W.

The equation for W is said to be hyperbolic, parabolic or elliptic according as the roots of the quadratic equation

$$\theta\phi X^2 - (\theta\delta + \phi\gamma)\, X + \gamma\delta - \alpha\beta = 0$$

are real and distinct, equal or imaginary. In this classification the coefficients α, β, γ, δ, θ, ϕ, λ, μ, σ, τ are supposed to be all real, the simple wave-equation is then of hyperbolic type, the equation of the conduction of heat of parabolic type and Laplace's equation of elliptic type. The telegraphic equation is generally of hyperbolic type, but if either $C = 0$ or $L = 0$ it is of parabolic type and the canonical equation is of the same form as the equation of the conduction of heat.

The foregoing analysis requires modification if the coefficients α, β, γ, δ, θ, ϕ, λ, μ, σ, τ are functions of x and t, because then the operators $\alpha D_x + \lambda$ and $\theta D_t - \gamma D_x - \mu$ are not commutative in general, and so the first equation cannot usually be satisfied by means of the substitution (B). If, however, the conditions

$$\alpha \frac{\partial \theta}{\partial x} = 0,$$

$$\alpha \frac{\partial \gamma}{\partial x} = \gamma \frac{\partial \alpha}{\partial x} - \theta \frac{\partial \alpha}{\partial t},$$

$$\alpha \frac{\partial \mu}{\partial x} = \gamma \frac{\partial \lambda}{\partial x} - \theta \frac{\partial \lambda}{\partial t}$$

are satisfied the operators are commutative (permutable) and a differential equation may be obtained for W. In this case the variables U and V do not necessarily satisfy the same partial differential equation. This is easily seen by considering the simple case when the first equation is $U = \theta \partial V/\partial t$ and β and τ are independent of t.

Differential operators which are not permutable play an interesting part in the new mechanics.

§ **1·62**. For some purposes it is useful to consider the partial difference equations which are analogous to partial differential equations in which we are interested. The notation which is now being used in Germany is the following*:

$$u\,(x+h,\,y) - u\,(x,\,y) = hu_x, \quad u\,(x,\,y+h) - u\,(x,\,y) = hu_y,$$

$$u\,(x,\,y) - u\,(x-h,\,y) = hu_{\bar{x}}, \quad u\,(x,\,y) - u\,(x,\,y-h) = hu_{\bar{y}},$$

$$u\,(x+h,\,y) - 2u\,(x,\,y) + u\,(x-h,\,y) = h^2 u_{x\bar{x}} = h^2 u_{\bar{x}x}.$$

The equations
$$u_x = v_{\bar{y}}, \quad u_y = -\,v_{\bar{x}}$$
are analogous to those satisfied by conjugate functions since they imply that
$$u_{x\bar{x}} + v_{y\bar{y}} = 0, \quad v_{x\bar{x}} + v_{y\bar{y}} = 0.$$

The equations
$$u_{\bar{x}} = v_y, \quad u_{\bar{y}} = v_x$$
give the equations
$$u_{x\bar{x}} = u_{\bar{y}}, \quad v_{x\bar{x}} = v_{\bar{y}}$$
analogous to the equation of the conduction of heat.

§ **1·63**. The simultaneous equations from which the final partial differential equation is derived need not be always of the first order. In the theory of the transverse vibrations of a thin rod the primary equations are†
$$M = EA\kappa^2 \frac{\partial^2 \eta}{\partial x^2}, \quad \rho A \left(\frac{\partial^2 \eta}{\partial t^2} - \kappa^2 \frac{\partial^4 \eta}{\partial x^2 \partial t^2} \right) = - \frac{\partial^2 M}{\partial x^2},$$
where η is the lateral displacement, M the bending moment, A the sectional area, x the radius of gyration of the area of the cross-section about an axis through its centre of gravity, ρ the density and E the Young's modulus of the material. The resulting equation
$$\frac{\partial^2 \eta}{\partial t^2} - \kappa^2 \frac{\partial^4 \eta}{\partial x^2 \partial t^2} + \frac{E\kappa^2}{\rho} \frac{\partial^4 \eta}{\partial x^4} = 0 \qquad \text{......(I)}$$
is of the fourth order. The equation is usually simplified by the omission of the second term. This process of approximation needs to be carefully justified because it will be noticed that the term omitted involves a derivative of the fourth order, that is a derivative of the highest order. Now there is a danger in omitting terms involving derivatives of the highest

* See an article by R. Courant, K. Friedrichs and H. Lewy, *Math. Ann.* Bd. c, S. 32 (1928).

† Cf. H. Lamb, *Dynamical Theory of Sound*, p. 121.

order because their coefficients are small. This may be illustrated in a very simple way by considering the equation

$$\frac{d\eta}{dx} = \nu \frac{d^2\eta}{dx^2}, \qquad \text{......(II)}$$

where ν is small. The solution is of type

$$\eta = A + Be^{x/\nu},$$

where A and B are constants. When the term on the right of (II) is omitted the solution is simply $\eta = A$. When x and ν are both small and positive the term $Be^{x/\nu}$, which is omitted in the foregoing method of approximation, may be really the dominant term. In this example all the terms involving derivatives of the highest order have been omitted, and as a general rule this is more dangerous than the omission of only some of the terms as in the case of the vibrating rod. The omission of the second term from the rod equation seems to be quite justifiable when the rod is very thin. When the rod is thick Timoshenko's theory* shows that there is a term giving the correction for shear which is at least as important as the second term of the usual equation (I).

This point relating to the danger of omitting terms involving derivatives of the highest order comes up again in hydrodynamics when the question of the omission of some or all of the viscous terms comes under consideration. The omission of all the viscous terms lowers the order of the equations and requires a modification of the boundary conditions. This does not lead to very good results. On the other hand, in Prandtl's theory of the boundary layer some of the viscous terms are retained, the boundary condition of no slipping at the surface of a solid body is also retained and the results are found to be fairly satisfactory.

EXAMPLE

Prove that the equations

$$\frac{\partial v}{\partial x} = a\frac{\partial u}{\partial x} + b\frac{\partial u}{\partial y},$$

$$\frac{\partial v}{\partial y} = c\frac{\partial u}{\partial x} + d\frac{\partial u}{\partial y}$$

give an equation of the second order which is elliptic, parabolic or hyperbolic according as $(a-d)^2 + 4bc$ is less than, equal to or greater than zero.

[E. Picard, *Compt. Rend.* t. CXII, p. 685 (1891).]

§ **1·71.** *Potentials and stream-functions.* The classical equations are of great mathematical interest and have played an important part in the

* *Phil. Mag.* (6), vol. XLI, p. 744 (1921). The equation used by Timoshenko is of type

$$E\kappa^2\frac{\partial^4\eta}{\partial x^4} + \rho\frac{\partial^2\eta}{\partial t^2} - \rho\kappa^2\left(1 + \frac{E}{\alpha\mu}\right)\frac{\partial^4\eta}{\partial x^2\partial t^2} + \frac{\rho^2\kappa^2}{\alpha\mu}\frac{\partial^4\eta}{\partial t^4} = 0,$$

where μ is the modulus of rigidity and α is a constant which depends upon the shape of the cross-section. For the equation of resisted vibrations see Note II, Appendix.

development of mathematical analysis by suggesting fruitful lines of investigation. It can be truly said that the modern theory of functions owes its origin largely to a study of these equations. The theory of functions of a complex variable is associated, for instance, with the theory of conjugate functions and the solutions of Laplace's equation.

If, for instance, we write

$$\phi + i\psi = f(x + iy) = f(z),$$

where $f(z)$ is an analytic function* and ϕ and ψ are real when x and y are real, we have, for points in the domain for which $f(z)$ is analytic,

$$\frac{\partial \phi}{\partial x} + i\frac{\partial \psi}{\partial x} = f'(z),$$

$$\frac{\partial \phi}{\partial y} + i\frac{\partial \psi}{\partial y} = if'(z),$$

where $f'(z)$ denotes the derivative of $f(z)$.

These equations give

$$\frac{\partial \phi}{\partial y} + i\frac{\partial \psi}{\partial y} = i\left[\frac{\partial \phi}{\partial x} + i\frac{\partial \psi}{\partial x}\right].$$

Equating the real and imaginary parts of the two sides of this equation, we see that

$$\frac{\partial \phi}{\partial x} = \frac{\partial \psi}{\partial y} = u, \text{ say,} \qquad \text{......(A)}$$

$$\frac{\partial \phi}{\partial y} = -\frac{\partial \psi}{\partial x} = v, \text{ say.}$$

These relations between the derivatives of two conjugate functions ϕ and ψ are called Cauchy's relations because they play a fundamental part in Cauchy's theory of functions of a complex variable. The relations can also be given many very interesting physical interpretations.

The simplest from a physical standpoint is, perhaps, that in which u and v are regarded as the component velocities in the plane of x, y of a particle of a fluid in two-dimensional motion, the particle in question being the particular one which happens* to be at the point (x, y) at time t. If u and v are independent of t the motion is said to be "steady" and a curve along which it is constant may be regarded as a "stream-line" or "line of flow" of the particles of fluid. The condition that a particle of the fluid should move along such a line is, in fact, expressed by the differential equations

$$\frac{dx}{u} = \frac{dy}{v} = dt \qquad \text{......(B)}$$

which give

$$v\,dx - u\,dy = 0,$$

that is $d\psi = 0$ or $\psi =$ constant.

* The reader is supposed to possess some knowledge of the properties of analytic functions.

Another way of looking at the matter is to calculate the "flux" across any line AP from right to left. This is expressed by the integral

$$\int \left(v \frac{dx}{ds} - u \frac{dy}{ds} \right) ds = - \int d\psi = \psi_A - \psi_P,$$

where ds denotes an element of length of AP and the suffix is used to indicate the point at which ψ is calculated. It is clear from this equation that there is no flow across a line AP along which ψ is constant.

The conjugate function ϕ is called the "velocity potential" and was first introduced by Euler. The curves on which ϕ is constant are called "equipotential curves." The function ψ is called the stream-function or current function, it was used in a general manner by Earnshaw.

It must be understood that the fluid motion which is represented by such simple formulae is of an ideal character and is only a very rough approximation to a real motion of a fluid. A study of this type of fluid motion serves, however, as a good introduction to the difficult mathematical analysis connected with the studies of actual fluid motions. It will be worth while, then, to make a few remarks on the peculiarities of this ideal type of fluid motion.

In the first place, it should be noticed that the expression $udx + vdy$ is an exact differential $d\phi$, and so the integral

$$\int (udx + vdy)$$

represents the difference between the values of ϕ at the ends of the path of integration. If the function ϕ is one-valued the integral round a closed curve is zero, but if ϕ is many-valued the integral may not vanish. The value of the integral in such a case is called the circulation round the closed curve. It is different from zero in the case when

$$\phi + i\psi = i \log z = i (\log r + i\theta)$$

and the curve is a circle whose centre is at the origin. In this case

$$\phi = - \theta, \quad \psi = \log r,$$

and it is easily seen that the circulation Γ defined by the integral

$$\Gamma = \int udx + vdy = \int d\phi = - \int_0^{2\pi} d\theta$$

is equal to $- 2\pi$. The fluid motion for which

$$\phi + i\psi = - A \log z,$$

where A is a constant, is said to be that due to a vortex of strength Γ when A is an imaginary quantity $\frac{i\Gamma}{2\pi}$. If, on the other hand, A is real, the motion is said to be due to a source if $- A$ is positive and due to a sink if $- A$ is negative. The flow in the last two cases is radial.

Since the stream-function in the last two cases is $-A\theta$ and is not one-valued, the flux across a circle whose centre is at O is $-2\pi A$.

The flow due to a vortex, source or sink at a point other than the origin may be represented in the same way by simply interpreting r and θ as polar co-ordinates relative to the point in question.

Since the equations expressing u and v in terms of ϕ and ψ are linear, the component velocities for the flow due to any number of vortices, sources and sinks may be derived from the complex potential

$$\phi + i\psi = \frac{1}{2\pi}\Sigma(\alpha_s - i\beta_s)\log[x - x_s + i(y - y_s)],$$

where the constants α_s, β_s specify the strengths of the source and vortex associated with the point (x_s, y_s). The word source is used here in a general sense to include both source and sink.

One further remark may be made regarding the motion if we are interested in the career of a particular particle of fluid. If x_0, y_0 are the initial co-ordinates of this particle at time t these quantities at time t will be functions of x, y and t

$$x_0 = f(x, y, t), \quad y_0 = g(x, y, t),$$

but functions of such a nature that the equations (B) are satisfied when x_0 and y_0 are regarded as constant. We have then

$$u\frac{\partial f}{\partial x} + v\frac{\partial f}{\partial y} + \frac{\partial f}{\partial t} = 0, \quad u\frac{\partial g}{\partial x} + v\frac{\partial g}{\partial y} + \frac{\partial g}{\partial t} = 0 \quad \text{......(C, D)}$$

and any quantity h which can be expressed in the form $h = F(x_0, y_0)$ will be a solution of the equation

$$\frac{\partial h}{\partial t} + u\frac{\partial h}{\partial x} + v\frac{\partial h}{\partial y} = 0,$$

and will be constant throughout the motion. We shall write this equation in the form $dh/dt = 0$ and shall call dh/dt the complete time derivative of h. When the motion is steady we evidently have $d\psi/dt = 0$.

The equations (C) and (D) may be solved for u and v if $\dfrac{\partial(f, g)}{\partial(x, y)} = 1$ and give expressions

$$u = \frac{\partial(f, g)}{\partial(y, t)}, \quad v = -\frac{\partial(f, g)}{\partial(x, t)}$$

which satisfy the equation

$$\frac{\partial u}{\partial x} + \frac{\partial v}{\partial y} = 0$$

on account of

$$\frac{\partial(x_0, y_0)}{\partial(x, y)} = \frac{\partial(f, g)}{\partial(x, y)} = 1.$$

This last equation expresses that the area occupied by a group of particles remains constant during the motion. To obtain a solution of this equation we take x and x_0 as new independent variables, then

$$\frac{\partial y}{\partial x_0} = \frac{\partial(x, y)}{\partial(x, x_0)} = \frac{\partial(x, y)}{\partial(x, x_0)}\frac{\partial(x_0, y_0)}{\partial(x, y)} = \frac{\partial(x_0, y_0)}{\partial(x, x_0)},$$

and so
$$\frac{\partial y}{\partial x_0} = -\frac{\partial y_0}{\partial x}.$$

This means that $ydx - y_0 dx_0$ is an exact differential and so we may write

$$y = \frac{\partial F}{\partial x}, \quad y_0 = -\frac{\partial F}{\partial x_0},$$

where $F = F(x, x_0, t)$ and t is regarded as constant. If, however, we allow t to vary and use brackets to denote derivatives when x, y and t are regarded as independent variables, we have

$$v = \frac{dy}{dt} = u\frac{\partial^2 F}{\partial x^2} + \frac{\partial^2 F}{\partial x \partial t}, \quad u = \frac{dx}{dt},$$

$$0 = -\frac{dy_0}{dt} = u\frac{\partial^2 F}{\partial x \partial x_0} + \frac{\partial^2 F}{\partial x_0 \partial t},$$

$$0 = \left(\frac{\partial y}{\partial x}\right) = \frac{\partial^2 F}{\partial x^2} + \frac{\partial^2 F}{\partial x \partial x_0}\left(\frac{\partial x_0}{\partial x}\right),$$

$$1 = \left(\frac{\partial y}{\partial y}\right) = \frac{\partial^2 F}{\partial x \partial x_0}\left(\frac{\partial x_0}{\partial y}\right),$$

$$\left(\frac{\partial^2 F}{\partial x \partial t}\right) = \frac{\partial^2 F}{\partial x \partial t} + \frac{\partial^2 F}{\partial x_0 \partial t}\left(\frac{\partial x_0}{\partial x}\right) = \frac{\partial^2 F}{\partial x \partial t} + u\frac{\partial^2 F}{\partial x^2} = v,$$

$$\left(\frac{\partial^2 F}{\partial y \partial t}\right) = \frac{\partial^2 F}{\partial x_0 \partial t}\left(\frac{\partial x_0}{\partial y}\right) = -u.$$

Hence we may write $\psi = -\dfrac{\partial F}{\partial t}$ and obtain a convenient expression for the stream-function.

Another physical interpretation of the functions ϕ and ψ is obtained by regarding ϕ as the electric potential and u, v as the components of the electric field strength due to a set of fictitious point charges, or, if we prefer a three-dimensional interpretation, to a system of uniform line charges on lines perpendicular to the plane of x, y. The curves ϕ = constant are then sections by this plane of the equipotential surfaces ϕ = constant, while the curves ψ = constant are the "lines of force" in the plane of x, y. For brevity we shall sometimes think in terms of the fictitious point charges and call a curve ϕ = constant an "equipotential."

Again, ϕ may be interpreted as a magnetic potential of a system of magnetic line charges (fictitious magnetic point charges) or of electric currents of uniform intensity flowing along wires of infinite length at right angles to the plane of x, y. The curves ψ = constant are again lines of force, a line of force being defined by the equations

$$\frac{dx}{u} = \frac{dy}{v}.$$

In all cases the lines of force are the orthogonal trajectories of the equipotentials, as may be seen immediately from the relation

$$\frac{\partial \phi}{\partial x}\frac{\partial \psi}{\partial x}+\frac{\partial \phi}{\partial y}\frac{\partial \psi}{\partial y}=0,$$

which is a consequence of Cauchy's relations.

For any number of electric or magnetic line charges perpendicular to the plane of x, y we have by definition

$$\phi + i\psi = \Sigma 2\mu_s \log [x - x_s + i\,(y - y_s)],$$

where μ_s is the density per unit length of the electricity, or magnetism as the case may be, on the line which passes through the point (x, y). It must be understood, of course, that when ϕ is the electric potential we consider only electric charges and when ϕ is the magnetic potential we consider only magnetic charges. When the number of terms in the series is finite we can certainly write

$$\phi + i\psi = f\,(x + iy) = f\,(z),$$

where $f\,(z)$ is a function which is analytic except at the points $z = z_s$.

When in the foregoing equation μ_s is regarded as a purely imaginary quantity, ϕ may be interpreted as the magnetic potential of a system of electric currents flowing along wires perpendicular to the plane of x, y. If $\mu_s = iC_s$ the current along the wire x_s, y_s is of strength C_s and flows in the positive direction, i.e. the direction associated with the axes Ox, Oy by the right-handed screw rule.

When a potential function ϕ is known it is sometimes of interest to determine the curves along which the associated force (or velocity) has either a constant magnitude or direction. This may be done as follows. We have

$$u - iv = \frac{\partial \phi}{\partial x} - i\frac{\partial \phi}{\partial y} = f'\,(x + iy),$$

$$\log\,(u - iv) = \log f'\,(x + iy) = \Phi + i\Psi, \text{ say,}$$

where $$\Phi = \tfrac{1}{2}\log\,(u^2 + v^2), \quad \Psi = \pi - \tan^{-1}(v/u).$$

The curves Φ = constant are clearly curves along which the magnitude $(u^2 + v^2)^{\frac{1}{2}}$ of the force or velocity is constant, while Ψ = constant is the equation of a curve along which the direction of the force is constant. The functions Φ and Ψ are clearly solutions of Laplace's equations, i.e.

$$\frac{\partial^2 \Phi}{\partial x^2} + \frac{\partial^2 \Phi}{\partial y^2} = 0.$$

A function Φ which satisfies this equation is called a logarithmic potential to distinguish it from the ordinary Newtonian potential which occurs in the theory of attractions. The electric and magnetic potentials of line charges are thus logarithmic potentials.

A logarithmic potential Φ is said to be regular in a domain D if

$$\Phi, \quad \frac{\partial \Phi}{\partial x}, \quad \frac{\partial \Phi}{\partial y}, \quad \frac{\partial^2 \Phi}{\partial x^2}, \quad \frac{\partial^2 \Phi}{\partial x \partial y} \quad \text{and} \quad \frac{\partial^2 \Phi}{\partial y^2}$$

are continuous functions of x and y for all points of D. If D is a region which extends to infinity it is further stipulated that

$$\lim_{r \to \infty} \phi(x, y) = C, \quad \lim_{r \to \infty} r \frac{\partial \phi}{\partial x} = \lim_{r \to \infty} r \frac{\partial \phi}{\partial y} = 0,$$

$$(r^2 = x^2 + y^2),$$

where C is a finite quantity which may be zero. In this sense the potential of a single line charge is not regular at infinity.

Still another physical interpretation of conjugate functions is obtained by writing

$$X_x = -Y_y = \phi, \quad X_y = Y_x = \psi.$$

Cauchy's relations then give

$$\frac{\partial X_x}{\partial x} + \frac{\partial X_y}{\partial y} = 0,$$

$$\frac{\partial Y_x}{\partial x} + \frac{\partial Y_y}{\partial y} = 0.$$

These are the equations for the equilibrium of an elastic solid when there are no body forces and the stress is two-dimensional. The quantities (X_x, X_y) are interpreted as the component stresses across a plane through (x, y) perpendicular to the axis of x, while (Y_x, Y_y) are the component stresses across a plane perpendicular to the axis of y. The relation $X_y = Y_x$ is quite usual but the relation $X_x + Y_y = 0$ indicates that the distribution of stress is of a special character. A stress system satisfying this condition can, however, be obtained by writing

$$X_x = -Y_y = \tfrac{1}{2}(u^2 - v^2), \quad Y_x = X_y = uv,$$

for these equations give

$$\frac{\partial X_x}{\partial x} + \frac{\partial X_y}{\partial y} = u\left(\frac{\partial u}{\partial x} + \frac{\partial v}{\partial y}\right) - v\left(\frac{\partial v}{\partial x} - \frac{\partial u}{\partial y}\right) = 0,$$

$$\frac{\partial Y_x}{\partial x} + \frac{\partial Y_y}{\partial y} = u\left(\frac{\partial v}{\partial x} - \frac{\partial u}{\partial y}\right) + v\left(\frac{\partial u}{\partial x} + \frac{\partial v}{\partial y}\right) = 0.$$

The fact that the various potentials ϕ and ψ which have been considered so far are solutions of Laplace's equation

$$\frac{\partial^2 \phi}{\partial x^2} + \frac{\partial^2 \phi}{\partial y^2} = 0$$

is a consequence of the circumstance that they have been defined as sums of quantities that are individually solutions of this equation. No physical principle has been used except a principle of superposition which states that when the individual terms give quantities with a physical meaning, the sum will give a quantity with a similar physical meaning. In the analysis of many physical problems such a superposition of individual effects is not strictly applicable, for the sources of a disturbance cannot be supposed to act independently, each source may, in fact, be modified by the presence of the others or may modify the mode of propagation of the disturbance produced by another. Such interactions will be left out of consideration at present, for our aim is not to formulate at the outset a complete theory of physical phenomena but to gradually make the student familiar with the mathematical processes which have been used successfully in the gradual discovery of the laws of physical phenomena.

In applied mathematics the student has always found the formulation of the fundamental equations of a problem to be a matter of some difficulty. Some men have been very successful in formulating simple equations because, by a kind of physical instinct, they have known what to neglect. The history of mathematical physics shows that in many cases this so-called physical instinct is not a safe guide, for terms which have been neglected may sometimes determine the mathematical behaviour of the true solution. In recent years the tendency has been to try to work with partial differential equations and their solutions without the feeling of orthodoxy which is created by a derivation of the equations that is regarded for the time being as fully satisfactory. The mathematician now feels that it is only by a comparison of the inferences from his equations with the results of experiment and the inferences from slightly modified equations that he can ascertain whether his equations are satisfactory or not. In the present state of physics the formulation of equations has not the air of finality that it had a few years ago.

This does not mean, however, that the art of formulating equations should be neglected, it means rather that mathematicians should also include amongst their special topics of study the processes which lead to the most interesting partial differential equations of physics. These processes are of various kinds. Besides the process of elimination from equations of the first order there are the methods of the Calculus of Variations and methods which depend upon the use of line, surface and volume integrals. Mathematically, the direct process of elimination is the simplest and will be given further consideration in § 1·82.

§ **1·72**. *Geometrical properties of equipotentials and lines of force.* When the potential ϕ is a single-valued function of x and y there cannot be more than one equipotential curve through a given point P in the (x, y) plane. An equipotential curve $\phi = \phi_0$ may, however, cross itself at a point and have a multiple point of any order at a point P_0. In such a case the tangents at the multiple point are arranged like the radii from the centre to the corners of a regular polygon. To see this, let us take the origin at P_0, then the terms of lowest degree in the Taylor expansion of $\phi - \phi_0$ are of type

$$c^n e^{nia} (x + iy)^n + c^n e^{-nia} (x - iy)^n,$$

where n is an integer and c and α are constants. In polar co-ordinates $x = r \cos \theta$, $y = r \sin \theta$, these terms become

$$2c^n r^n \cos n (\theta + \alpha),$$

and the directions of the n tangents are given by $\cos n (\theta + \alpha)$. The possible values of $n (\theta + \alpha)$ are thus $\pi/2, 3\pi/2, \ldots (n - \frac{1}{2}) \pi$, the angle between consecutive tangents being π/n.

Since $\cos n (\theta + \alpha)$ is positive for some values of θ and negative for others, the function ϕ cannot have a maximum or minimum value at a point, for this point may be chosen for origin and the expansion shows that there are points in the immediate neighbourhood of the origin for which $\phi > \phi_0$, and also points for which $\phi < \phi_0$.

By means of the transformation

$$x' - iy' = k^2 (x + iy)^{-1},$$
$$x' + iy' = k^2 (x - iy)^{-1},$$

which represents an inversion with respect to a circle of radius k and centre at the origin, an equipotential curve of a system of line charges is transformed into an equipotential curve of another system of line charges.

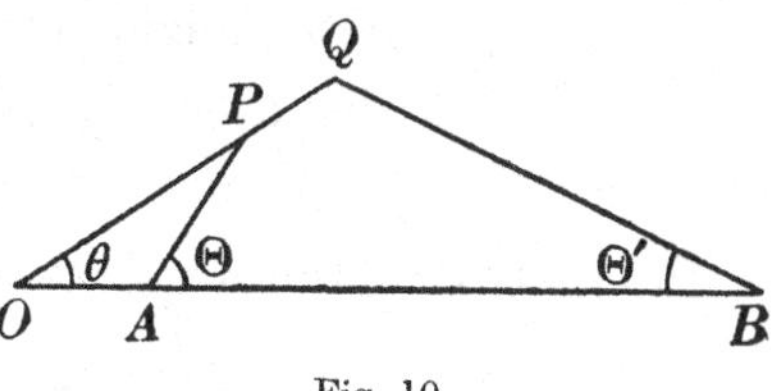

Fig. 10.

In polar co-ordinates we have

$$r' = k^2/r, \quad \theta' = \theta.$$

If in Fig. 10 Q corresponds to P and B to A, we have

$$R/r = R'/b,$$

where $AP = R$, $OP = r$, $BQ = R'$, $OB = b$.

For a number of points A and the corresponding points B

$$\Sigma \mu_s \log (R_s/r) = \Sigma \mu_s (\log R' - \log b).$$

An equipotential system of curves represented by the equation

$$\Sigma \mu_s \log R' = C$$

is thus transformed into an equipotential system represented by the equation

$$\Sigma \mu_s \log R_s - \log r \Sigma \mu_s = C - \Sigma \mu_s \log b.$$

A line charge at B is seen to correspond to a line charge of equal strength at A and another one of opposite sign at O which may be supposed to correspond to a line charge at infinity sufficient to compensate the charge at B.

An equipotential curve with a multiple point at O inverts into an equipotential which goes to infinity in the directions of the tangents at the multiple point. This indicates that the directions in which an equipotential goes to infinity are parallel to the radii from the centre to the corners of a regular polygon.

In the simple case of two equal line charges at the points $(c, 0)$, $(-c, 0)$ the equipotentials are

$$\log R_1 + \log R_2 = \text{constant},$$

or

$$R_1 R_2 = a^2,$$

where a is constant for each equipotential. These curves are Cassinian ovals with the polar equation

$$r^4 + c^4 - 2r^2c^2 \cos 2\theta = a^4.$$

When $a = c$ we obtain the lemniscate $r^2 = c^2 \cos 2\theta$ with a double point at the origin. The tangents at the double point are perpendicular.

Inverting we get the equipotentials for two equal line charges of strength $+1$ at the points $(b, 0)$, $(-b, 0)$, where $bc = k^2$ and a line charge of strength -2 at the origin. The equipotentials are now

$$\log R_1' + \log R_2' - 2 \log r' = \text{constant},$$

or

$$c^2 R_1' R_2' = a^2 r'^2.$$

Dropping the primes we have the polar equation

$$c^2 (r^4 + c^4 - 2r^2c^2 \cos 2\theta) = a^2 r^4$$

of a system of bicircular quartic curves. When $a = c$ we obtain the rectangular hyperbola $r^2 \cos 2\theta = c^2$ which is the inverse of the lemniscate. The rectangular hyperbola goes to infinity in two perpendicular directions.

It is easily seen that lines of force invert into lines of force. In Fig. 10, if we denote the angles POA, PAB, QBO by θ, Θ and Θ' respectively, we have the relation

$$\Theta - \theta = \Theta'.$$

Hence the lines of force represented by the equation

$$\Sigma \mu_s \Theta_s' = \text{constant}$$

transform into the lines of force represented by the equation

$$\Sigma \mu_s \Theta_s - \theta \Sigma \mu_s = \text{constant}.$$

In particular, the lines of force of two equal line charges

$$\Theta_1 + \Theta_2 = \text{constant},$$

being rectangular hyperbolas, invert into the family of lemniscates represented by

$$\Theta_1' + \Theta_2' - 2\Theta' = \text{constant},$$

and these are the lines of force of two equal line charges of strength $+1$ and a single line charge of strength -2 at O.

At a point of equilibrium in a gravitational, electrostatic or magnetic field, the first derivatives of the potential vanish and so the equipotential curve through the point has a double point or multiple point. A similar remark applies to a curve $\psi =$ constant, but this curve cannot strictly be regarded as a single line of force for, if we consider any branch which passes through the point of equilibrium without change of direction, the force is in different directions on the two sides of the point of equilibrium and the neighbouring lines of force avoid the point of equilibrium by turning through large angles in a short distance. This is exemplified in the case of two equal masses or charges when the equipotentials are Cassinian ovals which include a lemniscate with a double point at the point of equilibrium. The lines of force are then rectangular hyperbolas, the system including one pair of perpendicular lines which cross at the point of equilibrium.

In plotting equipotential curves and lines of force for a given system of line charges it is very useful to know the position of the points of equilibrium, since the properties just mentioned can be employed to indicate the behaviour of the lines of force. At a point of stagnation in an irrotational two-dimensional flow of an inviscid fluid the component velocities vanish and so the first derivatives of the velocity potential and stream-function are zero. The properties of the equipotentials and stream-lines at a point of stagnation are, then, similar to those of equipotential and lines of force at a point of equilibrium. There is, however, one important difference between the two cases. In the electric problem the field is often bounded by a conductor, i.e. an equipotential surface, while in the hydrodynamical problem the field of flow is generally bounded by some solid body whose profile in the plane $z = 0$ is a stream-line. A point of stagnation frequently lies on the boundary of the body and two coincident stream-lines may be supposed to meet and divide there, running round the body in opposite directions and reuniting at the back of the body when the profile is a simple closed curve.

A point on a conductor may be a point of equilibrium if the conductor's profile is a curve with a double point with perpendicular tangents or if it consists of two curves cutting one another orthogonally at all their common points. It should be remarked, however, that the force at a double point may be either zero or infinite; it is zero when the double point represents a pit or dent in the curve, but is infinite when the double point represents a peak. This may be exemplified by the equations $\phi = x^2 - y^2$, $\psi = 2xy$. If the field lies in the region $x > 0$, $y^2 < x^2$, the force is zero at O and there is a single line of force through O, namely, $y = 0$ (Fig. 11). If, on the other hand, the field is outside the region $x < 0$, $x^2 > y^2$, and

$$\phi = \frac{x^2 - y^2}{(x^2 + y^2)^2}, \quad \psi = \frac{2xy}{(x^2 + y^2)^2},$$

the force at the origin is infinite for most methods of approach and there are three lines of force through the origin (Fig. 12).

Similarly, when two conductors meet at any angle less than π, but a submultiple of π, the angle being measured outside the conductor. The point of intersection is a point of equilibrium and we have the approximate expression

$$\phi = 2c^n r^n \cos n\,(\theta + \alpha)$$

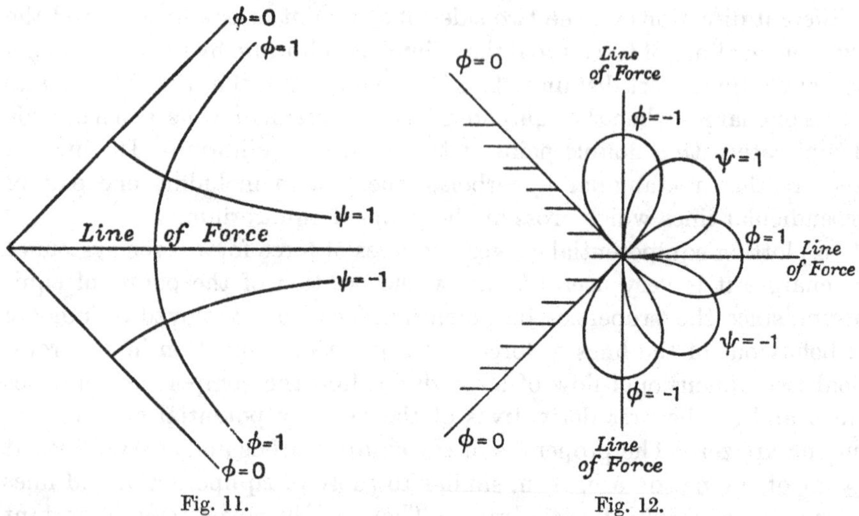

Fig. 11. Fig. 12.

for the value of ϕ in the neighbourhood of the point, the equation of the conductor in the neighbourhood of the point being $n\,(\theta + \alpha) = \pm\,\pi/2$ and the field being in the region $-\dfrac{\pi}{2} < n\,(\theta + \alpha) < \dfrac{\pi}{2}$. The angle is in this case π/n and the radial force $\dfrac{\partial\phi}{\partial r}$ varies initially according to the $(n - 1)$th power of the distance as a point recedes from the position of equilibrium.

The corresponding approximate expression for ψ is

$$\psi = 2c^n r^n \sin n\,(\theta + \alpha)$$

and there is a single line of force $\theta = -\,\alpha$ which lies within the field, this being its equation in the immediate neighbourhood of the point O.

There is another simple transformation which is sometimes useful for deriving the equipotentials and lines of force of one set of line charges from those of another. This is the transformation

$$z' = z + a^2/z$$

which gives two values of z for each value of z'. Let these be z and $\bar{z}$, then $z\bar{z} = a^2$. Similarly, let z_1 and $\bar{z}_1$ correspond to z_1', then

$$\begin{aligned} z' - z_1' &= z - z_1 + a^2/z - a^2/z_1 = (z - z_1)\,(1 - a^2/zz_1) \\ &= (z - z_1)\,(1 - \bar{z}_1/z) = (z - z_1)\,(z - \bar{z}_1)/z. \end{aligned}$$

Taking the moduli we obtain a relation

$$r_1' = r_1\bar{r}_1/r,$$

where $\quad r_1' = |\,z' - z_1'\,|,\quad r_1 = |\,z - z_1\,|,\quad \bar{r}_1 = |\,z - \bar{z}_1\,|,\quad r = |\,z\,|.$

Similarly, if z_2 and $\bar{z}_2$ correspond to z_2' we have, with a similar notation,

$$r_2' = r_2\bar{r}_2/r,$$

and so

$$\frac{r_2'}{r_1'} = \frac{r_2\bar{r}_2}{r_1\bar{r}_1}.$$

The transformation thus enables us to derive the equipotentials for four charges $(1, 1, -1, -1)$ from the equipotentials for two charges $(1, -1)$ and a similar remark holds for the lines of force, as may be seen by equating the arguments on the two sides of the equation

$$\frac{z' - z_1'}{z' - z_2'} = \frac{(z - z_1)(z - \bar{z}_1)}{(z - z_2)(z - \bar{z}_2)}.$$

This is just one illustration of the advantages of a transformation.

A general theory of such transformations will be developed in Chapter III.

Some geometrical properties of equipotential curves and lines of force may be obtained by using the idea of imaginary points. The pair of points with co-ordinates $(a \pm i\beta, b \mp i\alpha)$ are said to be the anti-points of the pair with co-ordinates $(a \pm \alpha, b \pm \beta)$, the upper or lower sign being taken throughout. Denoting the two pairs by F_1, F_2; S_1, S_2 respectively, we can say that if S_1 and S_2 are the real foci of an ellipse, then F_1 and F_2 are the imaginary foci. F_1 and F_2 can also be regarded as the imaginary points of intersection of the coaxial system of circles having S_1 and S_2 as limiting points.

If the co-ordinates of F_1 and F_2 are (x_1, y_1), (x_2, y_2) respectively and those of S_1, S_2 are (ξ_1, η_1), (ξ_2, η_2) respectively, we have

$$\begin{aligned} x_1 + iy_1 &= a + ib + \alpha + i\beta = \xi_1 + i\eta_1, \\ x_1 - iy_1 &= a - ib - \alpha + i\beta = \xi_2 - i\eta_2, \\ x_2 + iy_2 &= a + ib - \alpha - i\beta = \xi_2 + i\eta_2, \\ x_2 - iy_2 &= a - ib + \alpha - i\beta = \xi_1 - i\eta_1. \end{aligned}$$

If now $\quad u + iv = f(x + iy), \quad u - iv = f(x - iy),$

and S_1, S_2 lie on a curve $u =$ constant, we have

$$f(\xi_1 + i\eta_1) + f(\xi_1 - i\eta_1) = f(\xi_2 + i\eta_2) + f(\xi_2 - i\eta_2).$$

The foregoing relations now show that

$$f(x_1 + iy_1) + f(x_2 - iy_2) = f(x_2 + iy_2) + f(x_1 - iy_1),$$

and this means that F_1, F_2 lie on a curve $v =$ constant.

When the imaginary points on a curve $v =$ constant admit of a simple geometrical representation or description, the foregoing result may be sometimes used to find the curves $u =$ constant. If the curve $v =$ constant is a hyperbola, the imaginary points in which a family of parallel lines meet the curve have geometrical properties which are sufficiently well known to enable us to find the anti-points of each pair of points of intersection.

These anti-points lie on a confocal ellipse which is a curve of the family $u =$ constant. By taking lines in different directions the different ellipses of the family $u =$ constant are obtained. Similarly, by taking a set of parallel chords of an ellipse and the anti-points of the two points of intersection of each chord, it turns out that these anti-points all lie on a confocal hyperbola, and by taking families of lines with different directions the different hyperbolas of the confocal family may be obtained.

In this case the relations are particularly simple. In the general case when one curve of the family $u =$ constant is given there will be, presumably, a family of lines whose imaginary intersections with this curve are pairs of points with anti-points lying on one curve of the family $v =$ constant, but these lines cannot be expected, in general, to be parallel, and a simple description of the family is wanting.

EXAMPLES

1. If a family of circles gives a set of equipotential curves, the circles are either concentric or coaxial.

2. Equipotentials which form a family of parallel curves must be either straight lines or circles.

[Proofs of these propositions will be found in a paper by P. Franklin, *Journ. of Math. and Phys. Mass. Inst. of Tech.* vol. VI, p. 191 (1927).]

§ **1·81**. *The classical partial differential equations for Euclidean space.* Passing now to the consideration of some partial differential equations in which the number of independent variables is greater than two we note here that the most important equations are Laplace's equation

$$\frac{\partial^2 V}{\partial x^2} + \frac{\partial^2 V}{\partial y^2} + \frac{\partial^2 V}{\partial z^2} = 0, \qquad \text{......(A)}$$

the wave-equation

$$\frac{\partial^2 V}{\partial x^2} + \frac{\partial^2 V}{\partial y^2} + \frac{\partial^2 V}{\partial z^2} = \frac{1}{c^2}\frac{\partial^2 V}{\partial t^2}, \qquad \text{......(B)}$$

the equation of the conduction of heat

$$\frac{\partial V}{\partial t} = \kappa\left(\frac{\partial^2 V}{\partial x^2} + \frac{\partial^2 V}{\partial y^2} + \frac{\partial^2 V}{\partial z^2}\right), \qquad \text{......(C)}$$

the equation for the conduction of electricity

$$c^2\left(\frac{\partial^2 E}{\partial x^2} + \frac{\partial^2 E}{\partial y^2} + \frac{\partial^2 E}{\partial z^2}\right) = K\mu\frac{\partial^2 E}{\partial t^2} + \sigma\mu\frac{\partial E}{\partial t}, \qquad \text{......(D)}$$

and the wave-equation of Schrödinger's theory of wave-mechanics. This last equation takes many different forms and we shall mention here only the simple form of the equation in which the dependence of ψ on the time has already been taken into consideration. The reduced equation is then

$$\frac{\partial^2 \psi}{\partial x^2} + \frac{\partial^2 \psi}{\partial y^2} + \frac{\partial^2 \psi}{\partial z^2} + \frac{8\pi^2}{h^2}(E - V)\psi = 0, \qquad \text{......(E)}$$

where V is a function of x, y and z and E is a constant to be determined.

In these equations κ represents the diffusivity or thermometric conductivity of the medium, K the specific inductive capacity, μ the permeability, and σ the electric conductivity of the medium. The quantities c and h are universal constants, c being the velocity of light in vacuum and h being Planck's constant which occurs in his theory of radiation.

Laplace's equation, which for brevity may be written in the form

$$\nabla^2 V = 0,$$

may be obtained in various ways from a set of linear equations of the first order. One set,

$$X = \frac{\partial V}{\partial x}, \quad Y = \frac{\partial V}{\partial y}, \quad Z = \frac{\partial V}{\partial z}, \quad \frac{\partial X}{\partial x} + \frac{\partial Y}{\partial y} + \frac{\partial Z}{\partial z} = 0, \quad \text{......(F)}$$

occurs naturally in the theory of attractions, V being the gravitational potential and X, Y, Z the components of force per unit mass. The last equation is then a consequence of Gauss's theorem that the surface integral of the normal force is zero for any closed surface not containing any attracting matter.

The same equations occur also in hydrodynamics, the potential V being replaced by the velocity potential ϕ and the quantities X, Y, Z by the component velocities u, v, w. The equation is then the equation of continuity of an incompressible fluid.

The electric and magnetic interpretations of X, Y, Z and V are similar to the gravitational except that the electric (or magnetic) potential is usually taken to be $-V$ when X, Y, Z are the force intensities.

As in the two-dimensional theory, Laplace's equation is satisfied by the potential V because by the principle of superposition V is expressed as the sum of a number of elementary potentials each of which happens to be a solution of Laplace's equation, the elementary potential being of type

$$V = [(x - x')^2 + (y - y')^2 + (z - z')^2]^{-\frac{1}{2}} = 1/R.$$

When V is interpreted as the electrostatic potential this elementary potential is regarded as that of a unit point charge at the point (x', y', z'); when V is interpreted as a magnetic potential the elementary potential is that of a unit magnetic pole. In the theory of gravitation the elementary potential is that of unit mass concentrated at the point (x, y, z). A more general expression for a potential is

$$V = \Sigma m_s \, [(x - x_s)^2 + (y - y_s)^2 + (z - z_s)^2]^{-\frac{1}{2}},$$

where the coefficient m_s is a measure of the strength of the charge, pole or mass concentrated at the point (x_s, y_s, z_s). If we write ϕ in place of V, where ϕ is a velocity potential for a fluid motion in three dimensions, the elementary potential is that of a source and the coefficient m_s can be interpreted as the strength of the source at (x_s, y_s, z_s). Sources and sinks are useful in hydrodynamics as they give a convenient representation of the disturbance produced by a body when it is placed in a steady stream.

§ 1·82. *Systems of partial differential equations of the first order which lead to the classical equations.* When we introduce algebraic symbols

$$D_x \equiv \frac{\partial}{\partial x}, \quad D_y \equiv \frac{\partial}{\partial y}, \quad D_z \equiv \frac{\partial}{\partial z}$$

for the differential operators the equations (F) of § 1·81 become

$$\begin{gathered} X - D_x V = 0, \\ Y - D_y V = 0, \\ Z - D_z V = 0, \\ D_x X + D_y Y + D_z Z = 0, \end{gathered}$$

and the algebraic eliminant

$$\begin{vmatrix} 1 & 0 & 0 & -D_x \\ 0 & 1 & 0 & -D_y \\ 0 & 0 & 1 & -D_z \\ D_x & D_y & D_z & 0 \end{vmatrix} = 0$$

is simply

$$D_x^2 + D_y^2 + D_z^2 = 0. \qquad \text{......(G)}$$

If, on the other hand, we consider the set of equations

$$\left.\begin{aligned} \frac{\partial w}{\partial y} - \frac{\partial v}{\partial z} - \frac{\partial s}{\partial x} &= 0, \\ \frac{\partial u}{\partial z} - \frac{\partial w}{\partial x} - \frac{\partial s}{\partial y} &= 0, \\ \frac{\partial v}{\partial x} - \frac{\partial u}{\partial y} - \frac{\partial s}{\partial z} &= 0, \\ \frac{\partial u}{\partial x} + \frac{\partial v}{\partial y} + \frac{\partial w}{\partial z} &= 0, \end{aligned}\right\} \qquad \text{......(H)}$$

which give

$$\nabla^2 u = \nabla^2 v = \nabla^2 w = \nabla^2 s = 0,$$

the corresponding algebraic equations

$$\begin{aligned} -D_z v + D_y w - D_x s &= 0, \\ D_z u \qquad - D_x w - D_y s &= 0, \\ -D_y u + D_x v \qquad - D_z s &= 0, \\ D_x u + D_y v + D_z w \qquad &= 0 \end{aligned}$$

give the eliminant

$$\begin{vmatrix} 0 & -D_z & D_y & -D_x \\ D_z & 0 & -D_x & -D_y \\ -D_y & D_x & 0 & -D_z \\ D_x & D_y & D_z & 0 \end{vmatrix} = 0,$$

which is equivalent to

$$(D_x^2 + D_y^2 + D_z^2)^2 = 0. \qquad \text{......(I)}$$

These examples show that the problem of finding a set of linear equations of the first order which will lead to a given partial differential equation of higher order admits a variety of solutions which may be classified by noting the power of the complete differential operator (in this case $(\nabla^2)^2$) which is represented by the algebraic eliminant written in the form of a determinant.

It is known that Laplace's equation also occurs in the theory of elasticity. If u, v, w denote the components of the displacements and X_x, Y_y, Z_z, Y_z, Z_x, X_y the component stresses the equations for the case of no body forces are

$$\left.\begin{aligned}\frac{\partial X_x}{\partial x}+\frac{\partial X_y}{\partial y}+\frac{\partial X_z}{\partial z}=0,\\ \frac{\partial Y_x}{\partial x}+\frac{\partial Y_y}{\partial y}+\frac{\partial Y_z}{\partial z}=0,\\ \frac{\partial Z_x}{\partial x}+\frac{\partial Z_y}{\partial y}+\frac{\partial Z_z}{\partial z}=0,\end{aligned}\right\} \qquad \text{......(J)}$$

and if the substance is isotropic the relations between stress and strain take the form

$$\left.\begin{aligned}X_x&=\lambda\Delta+2\mu\frac{\partial u}{\partial x},\\ Y_y&=\lambda\Delta+2\mu\frac{\partial v}{\partial y},\\ Z_z&=\lambda\Delta+2\mu\frac{\partial w}{\partial z},\\ Y_z=Z_y&=\mu\left(\frac{\partial w}{\partial y}+\frac{\partial v}{\partial z}\right),\\ Z_x=X_z&=\mu\left(\frac{\partial u}{\partial z}+\frac{\partial w}{\partial x}\right),\\ X_y=Y_x&=\mu\left(\frac{\partial v}{\partial x}+\frac{\partial u}{\partial y}\right),\end{aligned}\right\} \qquad \text{......(K)}$$

where
$$\Delta=\frac{\partial u}{\partial x}+\frac{\partial v}{\partial y}+\frac{\partial w}{\partial z}.$$

The equations obtained by eliminating X_x, Y_y, Z_z, Y_z, Z_x, X_y are

$$\begin{aligned}\mu\nabla^2 u&=(\lambda+\mu)\frac{\partial\Delta}{\partial x},\\ \mu\nabla^2 v&=(\lambda+\mu)\frac{\partial\Delta}{\partial y},\\ \mu\nabla^2 w&=(\lambda+\mu)\frac{\partial\Delta}{\partial z},\end{aligned}$$

and, except in the case when $\lambda+2\mu=0$, a case which is excluded because λ and μ are positive constants when the substance is homogeneous, these equations imply that Δ is a solution of Laplace's equation. The algebraic eliminant is in this case

$$(D_x^2 + D_y^2 + D_z^2)^3 = 0. \qquad \text{......(L)}$$

It is easily seen that the quantities X_x, Y_y, Z_z, Y_z, Z_x, X_y, u, v, w are all solutions of the equation of the fourth order

$$\nabla^2\nabla^2 u = 0,$$

i.e.

$$\nabla^4 u = 0,$$

which may be called the elastic equation. The algebraic equation obtained by eliminating the twelve quantities X_x, X_y, X_z, Y_x, Y_y, Y_z, Z_x, Z_y, Z_z, u, v, w from the twelve equations (J) and (K) by means of a determinant is also equivalent to (L).

The question naturally arises whether as many as four equations are necessary for the derivation of Laplace's equation from a set of equations of the first order. The answer seems to be yes or no according as we do or do not require all the quantities occurring in the linear equations of the first order to be real. Thus, if we write $U = u - iv$, $V = w + is$, where u, v, w, s are the quantities satisfying the equations (H), it is easily seen that

$$\frac{\partial U}{\partial x} + i\frac{\partial U}{\partial y} + \frac{\partial V}{\partial z} = 0, \quad \frac{\partial V}{\partial x} - i\frac{\partial V}{\partial y} - \frac{\partial U}{\partial z} = 0,$$

and these equations imply that

$$\nabla^2 U = 0, \quad \nabla^2 V = 0.$$

The algebraic eliminant is in this case simply (G).

It should be noticed that if we write ict in place of y the two-dimensional wave-equation

$$\frac{\partial^2 V}{\partial x^2} + \frac{\partial^2 V}{\partial z^2} = \frac{1}{c^2}\frac{\partial^2 V}{\partial t^2}$$

may be derived from the two equations

$$\frac{\partial U}{\partial x} + \frac{\partial V}{\partial z} + \frac{1}{c}\frac{\partial U}{\partial t} = 0, \quad \frac{\partial V}{\partial x} - \frac{\partial U}{\partial z} - \frac{1}{c}\frac{\partial V}{\partial t} = 0$$

which have real coefficients. The wave-equation (B) may also be derived from two linear equations of the first order

$$\frac{\partial U}{\partial x} + i\frac{\partial U}{\partial y} = \frac{\partial V}{\partial z} + \frac{1}{c}\frac{\partial V}{\partial t},$$

$$\frac{\partial V}{\partial x} - i\frac{\partial V}{\partial y} = \frac{1}{c}\frac{\partial U}{\partial t} - \frac{\partial U}{\partial z},$$

but in this case the coefficients are not all real. The algebraic eliminant is in this case simply

$$c^2(D_x^2 + D_y^2 + D_z^2) - D_t^2 = 0.$$

To obtain the wave-equation from a set of linear equations of the first order with only real coefficients we may use the set of eight linear equations,

$$\frac{\partial\gamma}{\partial y}-\frac{\partial\beta}{\partial z}=\frac{\partial X}{\partial s}-\frac{\partial T}{\partial x},\quad \frac{\partial Z}{\partial y}-\frac{\partial Y}{\partial z}=\frac{\partial\tau}{\partial x}-\frac{\partial\alpha}{\partial s},$$

$$\frac{\partial\alpha}{\partial z}-\frac{\partial\gamma}{\partial x}=\frac{\partial Y}{\partial s}-\frac{\partial T}{\partial y},\quad \frac{\partial X}{\partial z}-\frac{\partial Z}{\partial x}=\frac{\partial\tau}{\partial y}-\frac{\partial\beta}{\partial s},$$

$$\frac{\partial\beta}{\partial x}-\frac{\partial\alpha}{\partial y}=\frac{\partial Z}{\partial s}-\frac{\partial T}{\partial z},\quad \frac{\partial Y}{\partial x}-\frac{\partial X}{\partial y}=\frac{\partial\tau}{\partial z}-\frac{\partial\gamma}{\partial s},$$

$$\frac{\partial X}{\partial x}+\frac{\partial Y}{\partial y}=\frac{\partial T}{\partial s}-\frac{\partial Z}{\partial z},\quad \frac{\partial\alpha}{\partial x}+\frac{\partial\beta}{\partial y}=\frac{\partial\tau}{\partial s}-\frac{\partial\gamma}{\partial z},$$

in which for convenience s has been written in place of ct.

These equations imply that X, Y, Z, T, α, β, γ and τ are all solutions of the wave-equation. The algebraic eliminant is now

$$\Theta^4 \equiv (D_x^2 + D_y^2 + D_z^2 - D_s^2)^4 = 0.$$

If in the foregoing equations we put $T = \tau = 0$ we obtain a set of equations very similar to that which occurs in Maxwell's electromagnetic theory. The eight equations may be divided into two sets of four and an algebraic eliminant may be obtained by taking three equations from each set and eliminating the six quantities X, Y, Z, α, β, γ. There are altogether sixteen possible eliminants but they are all of type $\Theta^2 L = 0$, where the last factor L is obtained by multiplying a term from the first of the two rows

$$\begin{matrix} D_x & D_y & D_z & D_s \\ D_x & D_y & D_z & D_s \end{matrix}$$

by a term from the second.

§ 1·91. *Primary solutions.* Let $f(\xi_1, \xi_2, \ldots \xi_m)$ be a homogeneous polynomial of the degree in its m arguments $\xi_1, \xi_2, \ldots \xi_m$ and let each of the quantities D_s that is used to denote an operator $\partial/\partial x_s$ be treated as an algebraic quantity when successive operations are performed. The equation

$$f(D_1, D_2, \ldots D_m)\, u = 0 \qquad \ldots\ldots(A)$$

is then a linear homogeneous partial differential equation of a type which frequently occurs in physics. An equation such as

$$D_1^2 w = D_2 w$$

may be included among equations of the foregoing type by writing

$$u = e^{x_3} . w,$$

and noting that u satisfies the equation

$$(D_1^2 - D_2 D_3)\, u = 0.$$

A solution of the form

$$u = F(\theta_1, \theta_2, \ldots \theta_s)$$

in which $\theta_1, \theta_2, \ldots \theta_s$ are particular functions of $x_1, x_2, x_3, \ldots x_m$, and F is an arbitrary function of the parameters $\theta_1, \theta_2, \theta_3, \ldots \theta_s$, will be called a *primary solution*. An arbitrary function will be understood here to be a function which possesses an appropriate number of derivatives which are all continuous in some region R. Such a function will be said to be continuous (D, n) when derivatives up to order n are specified as continuous.

It can be shown that the general equation (A) always possesses primary solutions of type

$$u = F(\theta), \qquad \ldots\ldots\text{(B)}$$

where

$$\theta = c_1 x_1 + c_2 x_2 + \ldots c_m x_m, \qquad \ldots\ldots\text{(C)}$$

and $c_1, c_2, \ldots c_m$ are constants satisfying the relation

$$f(c_1, c_2, \ldots c_m) = 0. \qquad \ldots\ldots\text{(D)}$$

This relation may be satisfied in a variety of ways and when a parametric representation

$$\left.\begin{aligned} c_1 &= C_1\ (\alpha_1, \alpha_2, \ldots \alpha_{m-2}), \\ c_2 &= C_2\ (\alpha_1, \alpha_2, \ldots \alpha_{m-2}), \\ &\ldots\ldots\ldots\ldots\ldots\ldots\ldots\ldots \\ c_m &= C_m\ (\alpha_1, \alpha_2, \ldots \alpha_{m-2}), \end{aligned}\right\} \qquad \ldots\ldots\text{(E)}$$

is known for the co-ordinates of points on the variety whose equation is represented by (D), the formulae (B) and (C) will give a family of primary solutions.

When $m = 2$ there is generally no family of primary solutions but simply a number of types, thus in the case of the equation

$$(D_1^2 - D_2^2)\, u = 0$$

there are the two types

$$u = F(x_1 + x_2), \quad u = F(x_1 - x_2).$$

Primary solutions may be generalised by summing or integrating with respect to a parameter after multiplication by an arbitrary function of the parameter. Thus in the case of Laplace's equation we have a family of primary solutions $V = F(\theta) \,.\, G(\alpha)$, where

$$\theta = z + ix \cos \alpha + iy \sin \alpha,$$

and α is an arbitrary parameter. Generalisation by the above method leads to a solution which may be further generalised by summation over a number of arbitrary functional forms for $F(\theta)$ and $G(\alpha)$ and we obtain Whittaker's solution*

$$V = \int_0^{2\pi} W(z + ix \cos \alpha + iy \sin \alpha, \alpha)\, d\alpha,$$

which may also be obtained directly by making the arbitrary function F a function of α as well as of θ.

The primary solutions (B) are not the only primary solutions of

* *Math. Ann.* vol. LVII, p. 333 (1903); Whittaker and Watson, *Modern Analysis*, ch. XVIII.

Laplace's equation, for it was shown by Jacobi* that if θ is defined by the equation

$$\theta = x\xi(\theta) + y\eta(\theta) + z\zeta(\theta), \qquad \text{......(F)}$$

where $\xi(\theta)$, $\eta(\theta)$ and $\zeta(\theta)$ are functions connected by the relation

$$[\xi(\theta)]^2 + [\eta(\theta)]^2 + [\zeta(\theta)]^2 = 0,$$

then $V = F(\theta)$ is a solution of Laplace's equation.

This is easily verified because if†

$$M = 1 - x\xi'(\theta) - y\eta'(\theta) - z\zeta'(\theta),$$

we have

$$M\frac{\partial\theta}{\partial x} = \xi(\theta),$$

$$M\frac{\partial^2\theta}{\partial x^2} - \left(\frac{\partial\theta}{\partial x}\right)^2[x\xi''(\theta) + y\eta''(\theta) + z\zeta''(\theta)] = \xi'(\theta)\frac{\partial\theta}{\partial x}.$$

These equations give

$$\left(\frac{\partial\theta}{\partial x}\right)^2 + \left(\frac{\partial\theta}{\partial y}\right)^2 + \left(\frac{\partial\theta}{\partial z}\right)^2 = 0,$$

$$\xi'(\theta)\frac{\partial\theta}{\partial x} + \eta'(\theta)\frac{\partial\theta}{\partial y} + \zeta'(\theta)\frac{\partial\theta}{\partial z} = 0,$$

and so

$$\nabla^2\theta = 0, \quad \nabla^2\{F(\theta)\} = 0.$$

This theorem is easily generalised. If $c_1(\alpha)$, $c_2(\alpha)$, ... $c_m(\alpha)$ are functions connected by the identical relation (D) the quantity θ defined by the equation

$$\theta = x_1c_1(\theta) + x_2c_2(\theta) + \dots x_mc_m(\theta) \qquad \text{......(G)}$$

is such that $u = F(\theta)$ is a primary solution of equation (A).

Since $v = \partial u/\partial x_1$ is also a solution of the same differential equation it follows that if $G(\theta)$ is an arbitrary function and

$$M = 1 - x_1c_1'(\theta) - x_2c_2'(\theta) - \dots x_mc_m'(\theta)$$

the expression

$$v = M^{-1}G(\theta)$$

is a second solution of the differential equation. The reader who is familiar with the principles of contour integration will observe that this solution may be expressed as a contour integral

$$v = \frac{1}{2\pi i}\int_C \frac{G(\alpha)\,d\alpha}{\alpha - x_1c_1(\alpha) - x_2c_2(\alpha) - \dots x_mc_m(\alpha)},$$

where C is a closed contour enclosing that particular root of equation (G) which is used as the argument of the function $G(\theta)$.

It is easy to verify that the contour integral is a solution of the differential equation because the integrand is a primary solution for all values of the parameter α and has been generalised by the method already suggested.

* *Journal für Math.* vol. XXXVI, p. 113 (1848); *Werke*, vol. II, p. 208.

† We use primes to denote differentiations with respect to θ.

In this method of generalisation by integration with respect to a parameter the limits of integration are generally taken to be constants or the path of integration is taken to be a closed contour in the complex plane. It is possible, however, to still obtain a solution of the differential equation when the limits of integration are functions of the independent variables of type θ. Thus the integral

$$V = \int_0^\theta W(z + ix\cos\alpha + iy\sin\alpha, \alpha)\, d\alpha$$

satisfies Laplace's equation $\nabla^2 V = 0$ when θ is defined by an equation of type (F) where

$$\frac{\xi(\alpha)}{i\cos\alpha} = \frac{\eta(\alpha)}{i\sin\alpha} = \frac{\zeta(\alpha)}{1}.$$

When the equation (A) possesses primary solutions of type $u = F(\theta)$ and no primary solutions of type $u = F(\theta, \phi)$ it will be said to be of the first grade. When it possesses primary solutions of type $u = F(\theta, \phi)$ and no primary solutions of type $u = F(\theta, \phi, \psi)$ it will be said to be of the second grade and so on.

The equation $\partial^2 u/\partial x\partial y = 0$ is evidently of the first grade because the general solution is $u = F(x) + G(y)$, where F and G are arbitrary functions. The primary solutions are in this case $F(x)$ and $G(y)$.

Laplace's equation $\nabla^2(u) = 0$ is also of the first grade but the equation

$$\frac{\partial u}{\partial x} + \frac{\partial u}{\partial y} + \frac{\partial u}{\partial z} = 0$$

is of the second grade because the general solution is of type

$$u = F(y - z, z - x).$$

There is, of course, a primary solution of type

$$u = F(y - z, z - x, x - y),$$

where $F(\theta, \phi, \psi)$ is an arbitrary function of the three arguments θ, ϕ, ψ, but these arguments are not linearly independent; indeed, since

$$\theta + \phi + \psi = 0,$$

a function of θ, ϕ and ψ, is also a function of θ and ϕ. In the foregoing definition of the grade of the equation it must be understood, then, that the parameters θ, ϕ, ψ, etc., are supposed to be functionally independent.

The differential equation

$$f(D_1, D_2, D_3, D_4)\, u = 0$$

has not usually a grade higher than one. If, in particular, an attempt is made to find a solution of type

$$u = F(\theta, \phi),$$

where

$$\theta = x_1\xi_1 + x_2\xi_2 + x_3\xi_3 + x_4\xi_4,$$

$$\phi = x_1\eta_1 + x_2\eta_2 + x_3\eta_3 + x_4\eta_4,$$

it is found that a number of equations must be satisfied. These equations imply that

$$f\,(a\xi_1 + b\eta_1,\, a\xi_2 + b\eta_2,\, a\xi_3 + b\eta_3,\, a\xi_4 + b\eta_4) = 0,$$

where a and b are arbitrary parameters and this means that all points of the line

$$\frac{x_1}{a\xi_1 + b\eta_1} = \frac{x_2}{a\xi_2 + b\eta_2} = \frac{x_3}{a\xi_3 + b\eta_3} = \frac{x_4}{a\xi_4 + b\eta_4}$$

lie on the surface whose equation is

$$f\,(x_1, x_2, x_3, x_4) = 0.$$

When $f\,(x_1, x_2, x_3, x_4)$ has a linear factor of the first degree or is itself of the first degree the equation (A) is of grade 3. In particular the equation

$$(D_1 + D_2 + D_3 + D_4)\, u = 0$$

possesses the general solution

$$u = F\,(x_1 - x_2,\, x_1 - x_3,\, x_1 - x_4),$$

and so is of grade 3. An equation with m independent variables which, by a simple change of variables, can be written in the form

$$\frac{\partial}{\partial z_m}\, g\left(\frac{\partial}{\partial z_1}, \frac{\partial}{\partial z_2}, \ldots \frac{\partial}{\partial z_m}\right) u = 0,$$

is said to be reducible. Such an equation is evidently of grade $m - 1$. It is likely that whenever the number of independent variables is m and the grade $m - 1$ the equation is reducible. The wave-equation

$$\Box^2 u \equiv \frac{\partial^2 u}{\partial x^2} + \frac{\partial^2 u}{\partial y^2} + \frac{\partial^2 u}{\partial z^2} - \frac{1}{c^2}\frac{\partial^2 u}{\partial t^2} = 0$$

is of grade 2 because there is a primary solution of type

$$u = F\,(\theta, \phi),$$

where

$$\theta = x \cos \alpha + y \sin \alpha + iz, \quad \phi = x \sin \alpha - y \cos \alpha + ct.$$

This solution may be generalised so as to give a solution

$$u = \int_0^{2\pi} F\,(\theta, \phi, \alpha)\, d\alpha,$$

analogous to Whittaker's solution of Laplace's equation.

The equations

$$\frac{\partial Q_z}{\partial y} - \frac{\partial Q_y}{\partial z} + \frac{i}{c}\frac{\partial Q_x}{\partial t} = 0,$$

$$\frac{\partial Q_x}{\partial z} - \frac{\partial Q_z}{\partial x} + \frac{i}{c}\frac{\partial Q_y}{\partial t} = 0,$$

$$\frac{\partial Q_y}{\partial x} - \frac{\partial Q_x}{\partial y} + \frac{i}{c}\frac{\partial Q_z}{\partial t} = 0,$$

$$\frac{\partial Q_x}{\partial x} + \frac{\partial Q_y}{\partial y} + \frac{\partial Q_z}{\partial z} = 0,$$

which may be written in the abbreviated form*

$$\operatorname{curl} Q + \frac{i}{c}\frac{\partial Q}{\partial t} = 0, \quad \operatorname{div} Q = 0,$$

and which give the simple equations of Maxwell

$$\operatorname{curl} H = \frac{1}{c}\frac{\partial E}{\partial t}, \quad \operatorname{div} E = 0,$$

$$\operatorname{curl} E = -\frac{1}{c}\frac{\partial H}{\partial t}, \quad \operatorname{div} H = 0,$$

when the vector Q is replaced by $H + iE$, where E and H are real, may be satisfied by writing

$$Q = \int_0^{2\pi} F(\theta, \phi, \alpha)\, q(\alpha)\, d\alpha,$$

where $q(\alpha)$ is a vector with components $\cos\alpha$, $\sin\alpha$, i, respectively and $F(\theta, \phi, \alpha)$ is an arbitrary function of its arguments.

EXAMPLES

1. Let ξ, η, ζ, τ be functions of α, β, γ connected by the relation

$$\xi^2 + \eta^2 + \zeta^2 + \tau^2 = 1,$$

and let $\quad X = \tau x - \zeta y + \eta z - \xi t + u, \quad Z = -\eta x + \xi y + \tau z - \zeta t + w,$

$$Y = \zeta x + \tau y - \xi z - \eta t + v, \quad T = \xi x + \eta y + \zeta z + \tau t + s.$$

Prove that if the integration extends over a suitable fixed region the definite integral

$$V = \iiint f[X, Y, Z, T; \alpha, \beta, \gamma]\, d\alpha\, d\beta\, d\gamma$$

satisfies the differential equation

$$\frac{\partial^2 V}{\partial x^2} + \frac{\partial^2 V}{\partial y^2} + \frac{\partial^2 V}{\partial z^2} + \frac{\partial^2 V}{\partial t^2} = \frac{\partial^2 V}{\partial u^2} + \frac{\partial^2 V}{\partial v^2} + \frac{\partial^2 V}{\partial w^2} + \frac{\partial^2 V}{\partial s^2}.$$

2. If $\quad V = F(A, B, C, D, E)$

is a solution of the equation

$$\frac{\partial^2 V}{\partial A^2} + \frac{\partial^2 V}{\partial B^2} + \frac{\partial^2 V}{\partial C^2} + \frac{\partial^2 V}{\partial D^2} = \frac{\partial^2 V}{\partial E^2}$$

when considered as a function of A, B, C, D and E; then, when

$$A = 2(xs + yw - zv - tu), \qquad C = 2(zs + xv - yu - tw),$$

$$B = 2(ys + zu - xw - tv), \qquad D = 2(ts + xu + yv + zw),$$

$$E = x^2 + y^2 + z^2 + t^2 + u^2 + v^2 + w^2 + s^2,$$

the function V is a solution of

$$\frac{\partial^2 V}{\partial x^2} + \frac{\partial^2 V}{\partial y^2} + \frac{\partial^2 V}{\partial z^2} + \frac{\partial^2 V}{\partial t^2} = \frac{\partial^2 V}{\partial u^2} + \frac{\partial^2 V}{\partial v^2} + \frac{\partial^2 V}{\partial w^2} + \frac{\partial^2 V}{\partial s^2}$$

when considered as a function of x, y, z, t, u, v, w, s.

* We use the symbol Q to denote the vector with components Q_x, Q_y, Q_z respectively. This abbreviated form is due to H. Weber and L. Silberstein.

§ 1·92. *The partial differential equation of the characteristics.* It is easily seen that when $\theta = c_1x_1 + c_2x_2 + \dots c_mx_m$ and $c_1, c_2, \dots c_m$ are constants satisfying the equation $f(c_1, c_2, \dots c_m) = 0$, the function $F(\theta) = u$ is not only a primary solution of the equation $f(D_1, D_2, \dots D_m)\,u = 0$ but it is also a solution of the equation

$$f(D_1u, D_2u, \dots D_mu) = 0. \qquad \dots\dots(A)$$

This partial differential equation of the first order is usually called the partial differential equation of the characteristics of the equation

$$f(D_1, D_2, \dots D_m)\,u = 0. \qquad \dots\dots(B)$$

In particular, the quantity $u = \theta$ is a solution of this differential equation and the locus $\theta(x_1, x_2, \dots x_m) =$ constant is a characteristic or characteristic locus of the partial differential equation.

A characteristic locus can generally be distinguished from other loci of type $\phi(x_1, x_2, \dots x_m) =$ constant by the property that it is a locus of "singularities" or "discontinuities" of some solution of the differential equation. If we adopt this definition of a characteristic locus $\theta =$ constant it is clear that $\theta =$ constant is a characteristic locus whenever there is a solution of the equation which involves in some explicit manner an arbitrary function $F(\theta)$, for the function $F(\theta)$ can be given a form which will make the solution discontinuous on the characteristic locus.

Thus the quantity $u = e^{-1/\theta}$ is a solution of the differential equation (B) when $\theta \neq 0$ and is discontinuous at each point of the characteristic locus $\theta = 0$. It should be observed that this function and all its derivatives on the side $\theta > 0$ of the locus $\theta = 0$ are zero for $\theta = 0$. The function $u = e^{-1/\theta^2}$ possesses a similar property and the additional one that the derivatives on the side $\theta < 0$ of the locus $\theta = 0$ are also zero. From these remarks it is evident that if there is a solution of the partial differential equation (B) which satisfies the condition that u and its derivatives up to order $n - 1$ have assigned values on the locus $\phi(x_1, x_2, \dots x_m) =$ constant and so gives the solution of the problem of Cauchy for the equation, this solution is not unique when $\phi \equiv \theta$ because a second solution may be obtained by adding to the former one a solution such as e^{-1/θ^2} which vanishes and has zero derivatives at all points of the locus. This property of a lack of uniqueness of the solution of the Cauchy problem for the locus $\theta(x_1, x_2, \dots x_m) = 0$ is the one which is usually used to define the characteristic loci of a partial differential equation and can be used in the case when the equation does not possess primary solutions. Since, however, we are dealing at present with equations having primary solutions the simpler definition of θ as the argument of a primary solution or other arbitrary function occurring in a solution will serve the purpose quite well.

An equation with a solution involving an arbitrary function explicitly (not under the sign of integration) will be called a *basic equation.*

Let us now write $p_1 = D_1 u$, $p_2 = D_2 u$, ... so that the partial differential equation for the characteristics may be written in the form

$$f(p_1, p_2, \dots p_m) = 0.$$

The curves defined by the differential equations

$$\frac{dx_1}{\frac{\partial f}{\partial p_1}} = \frac{dx_2}{\frac{\partial f}{\partial p_2}} = \dots \frac{dx_m}{\frac{\partial f}{\partial p_m}} \qquad \text{......(C)}$$

are called the bicharacteristics* of the equation; they are the characteristics of the equation (A) according to the theory of partial differential equations of the first order.

When $p_1, p_2, \dots$ are eliminated from these equations it is found that

$$F(dx_1, dx_2, \dots dx_m) = 0,$$

where $F(x_1, x_2, \dots x_m) = 0$ is the equation reciprocal to $f(p_1, p_2, \dots p_m) = 0$ in the sense of the theory of reciprocal polars.

In mathematical physics the loci of type $u =$ constant, where u satisfies the equation (A), frequently admit of an interesting interpretation as wave-surfaces. The curves given by the equations (C) associated with the function u are interpreted as the rays associated with the system of wave-surfaces.

In the particular case when the partial differential equation of the characteristics is

$$\left(\frac{d\theta}{dt}\right)^2 = V^2\left\{\left(\frac{\partial\theta}{\partial x}\right)^2 + \left(\frac{\partial\theta}{\partial y}\right)^2 + \left(\frac{\partial\theta}{\partial z}\right)^2\right\}, \qquad \text{......(D)}$$

where
$$\frac{d\theta}{dt} = \frac{\partial\theta}{\partial t} + u\frac{\partial\theta}{\partial x} + v\frac{\partial\theta}{\partial y} + w\frac{\partial\theta}{\partial z},$$

and u, v and w are constants representing the velocity of a medium and V is another constant representing the velocity of propagation of waves in the medium, the differential equations of the bicharacteristics are

$$\frac{dx}{u\frac{d\theta}{dt} - V^2\frac{\partial\theta}{\partial x}} = \frac{dy}{v\frac{d\theta}{dt} - V^2\frac{\partial\theta}{\partial y}} = \frac{dz}{w\frac{d\theta}{dt} - V^2\frac{\partial\theta}{\partial z}} = \frac{dt}{\frac{d\theta}{dt}}$$

and the equation obtained by eliminating $\frac{\partial\theta}{\partial x}, \frac{\partial\theta}{\partial y}, \frac{\partial\theta}{\partial z}, \frac{\partial\theta}{\partial t}$ is

$$(dx - u\,dt)^2 + (dy - v\,dt)^2 + (dz - w\,dt)^2 = V^2dt^2.$$

This result is of considerable interest in the theory of sound and may be extended so as to be applicable to the case in which u, v, w and V are functions of x, y, z and t.

It may be remarked that if we have a solution of (D) in the form of a complete integral

$$\theta = t - \tau - g(x, y, z, \alpha, \beta),$$

* See J. Hadamard's *Propagation des Ondes*. The theory is illustrated by the analysis of § 19.

in which τ, α and β are arbitrary constants, the rays may be obtained by combining the foregoing equation with the equations

$$\frac{\partial g}{\partial \alpha} = 0, \quad \frac{\partial g}{\partial \beta} = 0.$$

The characteristics of a set of linear equations of the first order may be defined to be the characteristics of the partial differential equation obtained by eliminating all the dependent variables except one. The relation of the primary solutions of this equation to the dependent variables in the set of equations of the first order is a question of some interest which will now be examined.

Let us first consider the equations

$$\frac{\partial u}{\partial x} = \frac{\partial v}{\partial y}, \quad \frac{\partial u}{\partial y} = \frac{\partial v}{\partial x}, \qquad \text{......(E)}$$

which lead to the equation

$$\frac{\partial^2 u}{\partial x^2} = \frac{\partial^2 u}{\partial y^2}. \qquad \text{......(F)}$$

In this case the quantity $w = u + v$ satisfies a linear equation of the first order

$$\frac{\partial w}{\partial x} = \frac{\partial w}{\partial y},$$

and this equation possesses the primary solution $w = F(x + y)$ which is also a primary solution of the equation (F).

Similarly the quantity $z = u - v$ satisfies the equation

$$\frac{\partial z}{\partial x} + \frac{\partial z}{\partial y} = 0,$$

which possesses a primary solution $z = G(x - y)$ which is also a primary solution of the equation (F).

To generalise this result we consider a set of m linear partial differential equations of the first order,

$$\left.\begin{aligned} L_{11} u_1 + L_{12} u_2 + \ldots L_{1m} u_m &= 0, \\ L_{21} u_1 + L_{22} u_2 + \ldots L_{2m} u_m &= 0, \\ \ldots\ldots\ldots\ldots\ldots\ldots\ldots\ldots\ldots\ldots \\ L_{m1} u_1 + L_{m2} u_2 + \ldots L_{mm} u_m &= 0, \end{aligned}\right\} \qquad \text{......(G)}$$

where L_{pq} denotes a linear operator of type

$$(p, q, 1) D_1 + (p, q, 2) D_2 + \ldots (p, q, m) D_m,$$

where the coefficients (p, q, r) are constants.

Multiplying these equations by coefficients $b_1, b_2, \ldots b_m$ respectively, the resulting equation is of the form

$$L(a_1 u_1 + a_2 u_2 + \ldots a_m u_m) = 0 \qquad \text{......(H)}$$

if the constants $b_1, b_2, \ldots b_m; a_1, a_2, \ldots a_m$ are of such a nature that

$$\left.\begin{array}{l} b_1 L_{11} + b_2 L_{21} + \ldots b_m L_{m1} \equiv a_1 L, \\ b_1 L_{12} + b_2 L_{22} + \ldots b_m L_{m2} \equiv a_2 L, \\ \cdots\cdots\cdots\cdots\cdots\cdots\cdots\cdots \\ b_1 L_{1m} + b_2 L_{2m} + \ldots b_m L_{mm} \equiv a_n L, \end{array}\right\} \quad \ldots\ldots(\mathrm{I})$$

and the operator L is of the form

$$L = l_1 D_1 + l_2 D_2 + \ldots l_m D_m,$$

where the operator coefficients $l_1, l_2, \ldots l_m$ are constants to be determined.

Equating the coefficients of the operator D in the identities (I) we obtain

$$\Sigma b_p (p, q; r) = a_q l_r.$$

This equation indicates that if $z_1, z_2, \ldots z_m;\ y_1, y_2, \ldots y_m$ are arbitrary quantities, the bilinear form

$$\Sigma\Sigma\Sigma b_p (p, q; r)\, y_q z_r$$

can be resolved into linear factors

$$\Sigma a_q y_q \times \Sigma l_r z_r.$$

When the coefficients $b_1, \ldots b_m$ can be chosen so that the bilinear form breaks up in this way the two factors will give the required coefficients $a_1, \ldots a_m;\ l_1, \ldots l_m$ and the partial differential equations will give an expression for $a_1 u_1 + \ldots a_m u_m$ which may be called a primary solution of the set of linear partial differential equations of grade $m - 1$. When such a solution exists the system is said to be reducible. The problem of finding when a set of equations is reducible is thus reduced to an algebraic problem.

Now let Ω denote the determinant

$$\begin{vmatrix} L_{11} & L_{12} & \ldots L_{1m} \\ L_{21} & L_{22} & \ldots L_{2m} \\ \cdots & \cdots & \cdots \\ L_{m1} & L_{m2} & \ldots L_{mm} \end{vmatrix}$$

and let $\Lambda_{11}, \ldots \Lambda_{1m}$ denote the co-factors of the constituents $L_{11}, L_{12}, \ldots L_{1m}$ respectively. If we write

$$u_1 = \Lambda_{11} v, \quad u_2 = \Lambda_{12} v, \quad u_m = \Lambda_{1m} v$$

it is easily seen that the last $m - 1$ equations of the set are all formally satisfied, and since

$$\Omega = L_{11}\Lambda_{11} + L_{12}\Lambda_{12} + \ldots L_{1m}\Lambda_{1m},$$

the first equation is formally satisfied if v is a solution of the partial differential equation

$$\Omega v = 0,$$

which is of order m. Since

$$\Omega u_1 = \Omega \Lambda_{11} v = \Lambda_{11} \Omega v = 0,$$

the quantities $u_1, u_2, \ldots u_m$ are all solutions of the same partial differential equation.

It should be noticed that

$$a_1 u_1 + \dots a_m u_m = (a_1 \Lambda_{11} + a_2 \Lambda_{12} + \dots a_m \Lambda_{1m})\, v,$$

consequently

$$L\,(a_1 u_1 + \dots a_m u_m) = L\,(a_1 \Lambda_{11} + a_2 \Lambda_{12} + \dots a_m \Lambda_{1m})\, v.$$

The equation $L\,(a_1 u_1 + \dots a_m u_m) = 0$ will be a consequence of the equation $\Omega v = 0$ if the operator Ω breaks up into two factors L and

$$(a_1 \Lambda_{11} + a_2 \Lambda_{12} + \dots a_m \Lambda_{1m})$$

of which one, L, is linear. The set of linear equations is thus reducible when the equation $\Omega v = 0$ is reducible.

It is clear from this result that we cannot generally expect a set of linear homogeneous equations of type (G) to possess primary solutions of grade $m - 1$.

The equations do, however, generally possess primary solutions of grade 1. To see this we try

$$u_1 = f_1(\theta), \quad u_2 = f_2(\theta), \dots u_m = f_m(\theta).$$

Substituting in the set of equations we obtain the set of linear equations

$$f_1'(\theta)\, L_{11}\theta + f_2'(\theta)\, L_{12}\theta + \dots f_m'(\theta)\, L_{1m}\theta = 0,$$
$$f_1'(\theta)\, L_{21}\theta + f_2'(\theta)\, L_{22}\theta + \dots f_m'(\theta)\, L_{2m}\theta = 0,$$
$$\dots\dots\dots\dots\dots\dots\dots\dots\dots\dots\dots\dots$$

from which the quantities $f_1'(\theta), f_2'(\theta), \dots f_m'(\theta)$ may be eliminated. The resulting equation,

$$\begin{vmatrix} L_{11}\theta & L_{12}\theta & \dots & L_{1m}\theta \\ L_{21}\theta & L_{22}\theta & \dots & L_{2m}\theta \\ \dots & \dots & \dots & \dots \\ L_{m1}\theta & L_{m2}\theta & \dots & L_{mm}\theta \end{vmatrix} = 0,$$

is no other than the partial differential equation of the characteristics of the equation $\Omega u = 0$.

§ 1·93. *Primary solutions of the second grade.* We have already seen that the wave-equation possesses primary solutions of type $F(\theta, \phi)$ which may be called primary solutions of the second grade. The result already obtained may be generalised by saying that if $l_0, m_0, n_0, p_0, l_1, m_1, n_1, p_1$ are quantities independent of x, y, z and t and connected by the relations

$$\left.\begin{aligned} l_0^2 + m_0^2 + n_0^2 &= p_0^2, \\ l_1^2 + m_1^2 + n_1^2 &= p_1^2, \\ l_1 l_0 + m_1 m_0 + n_1 n_0 &= p_1 p_0, \end{aligned}\right\} \qquad \dots\dots(\mathrm{J})$$

the quantities

$$\theta = l_0 x + m_0 y + n_0 z - p_0 ct, \quad \phi = l_1 x + m_1 y + n_1 z - p_1 ct$$

are such that the function $u = F(\theta, \phi)$ is a solution of $\Box^2 u = 0$.

This result may be generalised still further by making the coefficients l_0, m_0, etc., functions of two parameters σ, τ and forming the double contour integral

$$u = -\frac{1}{4\pi^2}\iint \frac{f(\sigma, \tau)\, d\sigma d\tau}{(l_0 x + m_0 y + n_0 z - p_0 ct - g_0)(l_1 x + m_1 y + n_1 z - p_1 ct - g_1)},$$

where $f(\sigma, \tau)$, $g_0(\sigma, \tau)$, $g_1(\sigma, \tau)$ are arbitrary functions of their arguments.

This integral will generally be a solution of the wave-equation and the value of the integral which is suggested by the theory of the residues of double integrals is

$$u = J^{-1} f(\alpha, \beta),$$

in which α, β satisfy

$$h_1(\alpha, \beta) = 0, \qquad h_0(\alpha, \beta) = 0, \qquad \text{......(K)}$$

where

$$h_1(\sigma, \tau) = x l_1(\sigma, \tau) + y m_1(\sigma, \tau) + z n_1(\sigma, \tau) - ct p_1(\sigma, \tau) - g_1(\sigma, \tau),$$
$$h_0(\sigma, \tau) = x l_0(\sigma, \tau) + y m_0(\sigma, \tau) + z n_0(\sigma, \tau) - ct p_0(\sigma, \tau) - g_0(\sigma, \tau),$$

and J is the value when $\sigma = \alpha$, $\tau = \beta$ of the Jacobian

$$J = \frac{\partial(h_0, h_1)}{\partial(\sigma, \tau)}.$$

This result, which may be extended to any linear equation with a two-parameter family of primary solutions of the second grade, will now be verified for the case of the wave-equation. It should be remarked that the method gives us a solution of the wave-equation of type

$$u = \gamma f(\alpha, \beta),$$

where γ is a particular solution of the wave-equation. Such a solution will be called a *primitive solution*; it is easily verified that the parameters α and β occurring in a primitive solution are such that the function $v = F(\alpha, \beta)$ is a solution of the partial differential equation of the characteristics

$$\left(\frac{\partial v}{\partial x}\right)^2 + \left(\frac{\partial v}{\partial y}\right)^2 + \left(\frac{\partial v}{\partial z}\right)^2 = \frac{1}{c^2}\left(\frac{\partial v}{\partial t}\right)^2. \qquad \text{......(L)}$$

Instead of considering the wave-equation it is more advantageous to consider the set of partial differential equations comprised in the vector equations

$$\operatorname{curl} Q + \frac{i}{c}\frac{\partial Q}{\partial t} = 0, \quad \operatorname{div} Q = 0 \qquad \text{......(M)}$$

and to look for a primitive solution of these equations of type

$$Q = q f(\alpha, \beta)$$

in which f is an arbitrary function of the two parameters α and β, which are certain functions of x, y, z and t, and the vector q is a particular solution of the set of equations.

Substituting in the equations (L) we find that since f is arbitrary α and β must satisfy the equations in ω,

$$c\nabla\omega \times q = -iq\partial\alpha/\partial t, \qquad q \cdot \nabla\omega = 0,$$

which indicate that

$$\left.\begin{aligned} q_x &= \kappa\frac{\partial(\alpha,\beta)}{\partial(y,z)} = \frac{i\kappa}{c}\frac{\partial(\alpha,\beta)}{\partial(x,t)},\\ q_y &= \kappa\frac{\partial(\alpha,\beta)}{\partial(z,x)} = \frac{i\kappa}{c}\frac{\partial(\alpha,\beta)}{\partial(y,t)},\\ q_z &= \kappa\frac{\partial(\alpha,\beta)}{\partial(x,y)} = \frac{i\kappa}{c}\frac{\partial(\alpha,\beta)}{\partial(z,t)},\end{aligned}\right\} \qquad \text{......(N)}$$

where κ is some multiplier. To solve these equations we take α, β, x, y as new independent variables and write the equation connecting α and β in the form

$$\frac{\partial(\alpha,\beta,x,t)}{\partial(y,z,x,t)} = \frac{i}{c}\frac{\partial(\alpha,\beta,y,z)}{\partial(x,t,y,z)},$$

$$\frac{\partial(\alpha,\beta,y,t)}{\partial(z,x,y,t)} = \frac{i}{c}\frac{\partial(\alpha,\beta,z,x)}{\partial(y,t,z,x)},$$

$$\frac{\partial(\alpha,\beta,z,t)}{\partial(x,y,z,t)} = \frac{i}{c}\frac{\partial(\alpha,\beta,x,y)}{\partial(z,t,x,y)}.$$

Now multiply each of the Jacobians by $\dfrac{\partial(x,y,z,t)}{\partial(\alpha,\beta,x,y)}$ and make use of the multiplication theorem for Jacobians. We then obtain a set of equations similar to the above but with $\partial(\alpha,\beta,x,y)$ in each denominator. The new equations reduce to the form

$$\frac{\partial t}{\partial y} = -\frac{i}{c}\frac{\partial z}{\partial x}, \quad \frac{\partial t}{\partial x} = \frac{i}{c}\frac{\partial z}{\partial y}, \quad \frac{\partial(z,t)}{\partial(x,y)} = \frac{i}{c}.$$

The first two of these equations are analogous to the equations connecting conjugate functions t and iz/c, consequently we may write

$$z - ct = \mathscr{F}[x + iy, \alpha, \beta],$$
$$z + ct = \mathscr{G}[x - iy, \alpha, \beta].$$

Substituting in the third equation, we find that

$$\mathscr{F}'\mathscr{G}' = -1,$$

where in each case the prime denotes a derivative with respect to the first argument. Evidently $\mathscr{F}'$ must be independent of $x + iy$ and $\mathscr{G}'$ independent of $x - iy$. The general solution is thus determined by equations of the form

$$z - ct = \phi(\alpha,\beta) + (x + iy)\,\theta(\alpha,\beta),$$
$$z + ct = \psi(\alpha,\beta) - (x - iy)\,[\theta(\alpha,\beta)]^{-1},$$

where θ, ϕ, ψ are arbitrary functions of α and β which are continuous $(D, 1)$ in some domain of the complex variables α and β.

For some purposes it is more convenient to write the equations in the equivalent form

$$z - ct = \phi(\alpha,\beta) + (x + iy)\,\theta(\alpha,\beta),$$
$$\theta(\alpha,\beta)\,(z + ct) = \chi(\alpha,\beta) - (x - iy).$$

These equations are easily seen to be of the type (K) and may indeed be regarded as a canonical form of (K). When the expressions for q are substituted in the equations (M) it is easily seen that κ is a function of α and β. Since Q already contains an arbitrary function of α and β we may without loss of generality take $\kappa = 1$.

A case of particular interest arises when

$$\phi = \zeta(\alpha) - c\tau(\alpha) - [\xi(\alpha) + i\eta(\alpha)]\,\theta,$$
$$\psi = \zeta(\alpha) + c\tau(\alpha) + [\xi(\alpha) - i\eta(\alpha)]\,\theta^{-1},$$
$$\theta = \beta,$$

where $\xi(\alpha)$, $\eta(\alpha)$, $\zeta(\alpha)$ and $\tau(\alpha)$ are real arbitrary functions of α which are continuous (D, 2). We then have

$$\beta = \frac{z - \zeta(\alpha) - c[t - \tau(\alpha)]}{x - \xi(\alpha) + i[y - \eta(\alpha)]} = -\frac{x - \xi(\alpha) - i[y - \eta(\alpha)]}{z - \zeta(\alpha) + c[t - \tau(\alpha)]},$$

and so α is defined by the equation

$$[x - \xi(\alpha)]^2 + [y - \eta(\alpha)]^2 + [z - \zeta(\alpha)]^2 = c^2[t - \tau(\alpha)]^2.$$

We may without loss of much generality take $\tau(\alpha) = \alpha$ and use τ as variable in place of α. Let us now regard $\xi(\tau)$, $\eta(\tau)$, $\zeta(\tau)$ as the coordinates of a point S moving with velocity v which is a function of τ. For the sake of simplicity we shall suppose that for each value of τ we have the inequality $v^2 < c^2$, which means that the velocity of S is always less than the velocity of light. We shall further introduce the inequality $\tau < t$. This is done to make the value of τ associated with a given space-time point (x, y, z, t) unique*.

To prove that it is unique we describe a sphere of radius $c(t - \tau)$ with its centre at the point occupied by S at the instant τ. As τ varies we obtain a family of spheres ranging from the point sphere corresponding to $\tau = t$ to a sphere of infinite radius corresponding to $\tau = -\infty$.

Now, since $v^2 < c^2$ it is easily seen that no two spheres intersect. Each sphere is, in fact, completely surrounded by all the spheres that correspond to earlier times τ. There is consequently only one sphere through each point of space and so the value of τ corresponding to (x, y, z, t) is unique. The corresponding position of S may be called the effective position of S relative to (x, y, z, t).

In calculating the Jacobians τ may be treated as constant in the differentiations of β. Now

$$x - \xi(\tau) = M\frac{\partial\alpha}{\partial x}, \quad z - \zeta(\tau) = M\frac{\partial\alpha}{\partial z},$$

$$y - \eta(\tau) = M\frac{\partial\alpha}{\partial y}, \quad c^2(t - \tau) = -M\frac{\partial\alpha}{\partial t},$$

* Proofs of this theorem have been given by A. Liénard, *L'éclairage électrique*, t. XVI, pp. 5, 53, 106 (1898); A. W. Conway, *Proc. London Math. Soc.* (2), vol. I (1903); G. A. Schott, *Electromagnetic Radiation* (Cambridge, 1912).

where

$$M = [x - \xi(\tau)]\,\xi'(\tau) + [y - \eta(\tau)]\,\eta'(\tau) + [z - \zeta(\tau)]\,\zeta'(\tau) - c^2(t - \tau),$$

and primes denote derivatives with respect to τ. We thus find that

$$\frac{\partial(\alpha, \beta)}{\partial(y, z)} = -\frac{i}{2M}(1 - \beta^2),$$

$$\frac{\partial(\alpha, \beta)}{\partial(z, x)} = -\frac{1}{2M}(1 + \beta^2),$$

$$\frac{\partial(\alpha, \beta)}{\partial(x, y)} = -\frac{i\beta}{M}.$$

The ratios of the Jacobians thus depend only on β and we have the general result that the function

$$u = M^{-1} f(\alpha, \beta)$$

is a solution of the wave-equation.

When the point (ξ, η, ζ) is stationary and at the origin of co-ordinates this result tells us that if f is an arbitrary function which is continuous $(D, 2)$ in some domain of the variables α, β and if $r^2 = x^2 + y^2 + z^2$ the function

$$u = \frac{1}{r} f\left[t - \frac{r}{c}, \; \frac{z - r}{x + iy}\right]$$

is a solution of the wave-equation. There is a corresponding primitive solution of type

$$u = \frac{1}{r} g\left[t + \frac{r}{c}, \; \frac{z - r}{x + iy}\right]$$

obtained by changing the sign of t and using another arbitrary function.

In the case of the wave-function $M^{-1} f(\alpha, \beta)$ the parameter α may be called a phase-parameter because it determines the phase of a disturbance which reaches the point (x, y, z) at time t when the function f is periodic in α. The parameter β is on the other hand a ray-parameter because a given complex value of β determines the direction of a ray when α is given.

It is easily deduced from the equations (N) that α and β satisfy the differential equation of the characteristics

$$\left(\frac{\partial \alpha}{\partial x}\right)^2 + \left(\frac{\partial \alpha}{\partial y}\right)^2 + \left(\frac{\partial \alpha}{\partial z}\right)^2 = \frac{1}{c^2}\frac{\partial \alpha}{\partial t^2}, \qquad \text{......(O)}$$

and that

$$\frac{\partial \alpha}{\partial x}\frac{\partial \beta}{\partial x} + \frac{\partial \alpha}{\partial y}\frac{\partial \beta}{\partial y} + \frac{\partial \alpha}{\partial z}\frac{\partial \beta}{\partial z} = \frac{1}{c^2}\frac{\partial \alpha}{\partial t}\frac{\partial \beta}{\partial t}. \qquad \text{......(P)}$$

It follows that the quantity $v = F(\alpha, \beta)$ is also a solution of (L).

An interesting property of this equation (O) is that if α is any solution and we depart from the space-time point (x, y, z, t) in a direction and velocity defined by the equations

$$\frac{dx}{\dfrac{\partial \alpha}{\partial x}} = \frac{dy}{\dfrac{\partial \alpha}{\partial y}} = \frac{dz}{\dfrac{\partial \alpha}{\partial z}} = -\frac{c^2\,dt}{\dfrac{\partial \alpha}{\partial t}} = ds, \text{ say,}$$

then α and its first derivatives are unaltered in value as we follow the moving point. We have in fact

$$d\alpha = \frac{\partial\alpha}{\partial x}dx + \frac{\partial\alpha}{\partial y}dy + \frac{\partial\alpha}{\partial z}dz + \frac{\partial\alpha}{\partial t}dt$$
$$= \left[\left(\frac{\partial\alpha}{\partial x}\right)^2 + \left(\frac{\partial\alpha}{\partial y}\right)^2 + \left(\frac{\partial\alpha}{\partial z}\right)^2 - \frac{1}{c^2}\left(\frac{\partial\alpha}{\partial t}\right)^2\right] ds$$
$$= 0,$$

$$d\left(\frac{\partial\alpha}{\partial x}\right) = \frac{\partial^2\alpha}{\partial x^2}dx + \frac{\partial^2\alpha}{\partial x\partial y}dy + \frac{\partial^2\alpha}{\partial x\partial z}dz + \frac{\partial^2\alpha}{\partial x\partial t}dt$$
$$= \left[\frac{\partial\alpha}{\partial x}\frac{\partial^2\alpha}{\partial x^2} + \frac{\partial\alpha}{\partial y}\frac{\partial^2\alpha}{\partial x\partial y} + \frac{\partial\alpha}{\partial z}\frac{\partial^2\alpha}{\partial x\partial z} - \frac{1}{c^2}\frac{\partial\alpha}{\partial t}\frac{\partial^2\alpha}{\partial x\partial t}\right] ds$$
$$= 0.$$

Also, if α and β are connected by an equation of type (P),

$$d\beta = \frac{\partial\beta}{\partial x}dx + \frac{\partial\beta}{\partial y}dy + \frac{\partial\beta}{\partial z}dz + \frac{\partial\beta}{\partial t}dt$$
$$= \left(\frac{\partial\alpha}{\partial x}\frac{\partial\beta}{\partial x} + \frac{\partial\alpha}{\partial y}\frac{\partial\beta}{\partial y} + \frac{\partial\alpha}{\partial z}\frac{\partial\beta}{\partial z} - \frac{1}{c^2}\frac{\partial\alpha}{\partial t}\frac{\partial\beta}{\partial t}\right) ds$$
$$= 0.$$

The equations (O and P) thus indicate that the path of the particle which moves in accordance with these equations is a straight line described with uniform velocity c and is, moreover, a ray for which β is constant.

§ **1·94**. *Primitive solutions of Laplace's equation.* As a particular case of the above theorem we have the result that the function

$$V = \frac{1}{r} f\left(\frac{z - r}{x + iy}\right)$$

is a primitive solution of Laplace's equation. This is not the only type of primitive solution, for the following theorem has been proved*.

In order that Laplace's equation may be satisfied by an expression of the form $V = \gamma f(\theta)$, in which the function f is arbitrary, the quantity θ must either be defined by an equation of the form

$$[x - \xi(\theta)]^2 + [y - \eta(\theta)]^2 + [z - \zeta(\theta)]^2 = 0,$$

or by an equation of the form

$$xl(\theta) + ym(\theta) + zn(\theta) = p(\theta),$$

where l, m, n are either constants or functions of θ connected by the relation

$$l^2 + m^2 + n^2 = 0.$$

The most general value of γ is in each case of the form

$$\gamma = \gamma_1 a(\theta) + \gamma_2 b(\theta),$$

* See my *Differential Equations*, p. 202.

where γ_1 and γ_2 are particular values of γ, whose ratio is not simply a function of θ. In the first case we may take

$$\gamma_1 = w^{-\frac{1}{2}}, \quad \gamma_2 = w_1^{-\frac{1}{2}},$$

where $\quad w = [x - \xi(\theta)]\,\lambda(\theta) + [y - \eta(\theta)]\,\mu(\theta) + [z - \zeta(\theta)]\,\nu(\theta),$

$$w_1 = [x - \xi(\theta)]\,\lambda_1(\theta) + [y - \eta(\theta)]\,\mu_1(\theta) + [z - \zeta(\theta)]\,\nu_1(\theta),$$

and $\lambda, \mu, \nu, \lambda_1, \mu_1, \nu_1$ are two independent sets of three functions of θ which satisfy relations of type

$$\lambda^2 + \mu^2 + \nu^2 = 0,$$

$$\lambda(\theta)\,\xi'(\theta) + \mu(\theta)\,\eta'(\theta) + \nu(\theta)\,\zeta'(\theta) = 0.$$

In the second case we may take $\gamma_1 = 1$ and define γ_2 by the equation

$$\gamma_2^{-1} = xl'(\theta) + ym'(\theta) + zn'(\theta) - p'(\theta).$$

If in the first theorem we choose $\xi = 0$, $\eta = i\theta$, $\zeta = 0$, we have

$$\theta = \frac{r^2}{x + iy}, \quad \lambda + i\mu = 0, \quad \nu = 0,$$

$$w = x + iy,$$

and the theorem tells us that the function

$$V = (x + iy)^{-\frac{1}{2}} f\left(\frac{r^2}{x + iy}\right)$$

is a primitive solution of Laplace's equation. If we write $x + iy = t$, $x - iy = 4s$ this theorem tells us that the function

$$V = t^{-\frac{1}{2}} f(4s + z^2/t)$$

is a primitive solution of the equation

$$\frac{\partial^2 V}{\partial s\,\partial t} + \frac{\partial^2 V}{\partial z^2} = 0.$$

§ **1·95**. *Fundamental solutions**. The equations with primary and primitive solutions have been called basic because it is believed that solutions of a differential equation with the same characteristics as a basic equation can be derived from solutions of the basic equation by some process of integration or summation in which singularities of these solutions of the basic equation fill the whole of the domain under consideration.

This point will be illustrated by a consideration of Laplace's equation as our basic equation.

We have seen that there is a primitive solution of type

$$V = \frac{1}{r} f\left(\frac{z - r}{x + iy}\right).$$

By a suitable choice of the function f we obtain a primitive solution

* These are also called elementary solutions. See Hadamard, *Propagation des Ondes.*

with singularities at isolated points and along isolated straight lines issuing from isolated singular points. The particular solution

$$V = 1/r$$

has the single isolated point singularity $x = 0$, $y = 0$, $z = 0$. Let us take this particular solution as the starting-point and generalise it by forming a volume integral

$$V = \iiint [(x-\xi)^2 + (y-\eta)^2 + (z-\zeta)^2]^{-\frac{1}{2}} F(\xi, \eta, \zeta)\, d\xi\, d\eta\, d\zeta, \quad \text{......(A)}$$

over a portion of space which we shall call the domain $\mathscr{D}$.

When the point (x, y, z) is in the domain $\mathscr{D}$ this integral is not a solution of Laplace's equation but is generally a solution of the equation

$$\nabla^2 V + 4\pi F(x, y, z) = 0, \quad \text{......(B)}$$

provided suitable limitations are imposed upon the function F.

Now the function F is at our disposal and in most cases it can be chosen so as to represent the terms which make the given differential equation differ from the basic equation of Laplace. It is true that this choice of F does not give us a formula for the solution of the given equation but gives us instead an integro-differential equation for the determination of the solution. Yet the point is that when this equation has been solved the desired solution is expressed by means of the formula (A) in terms of primitive solutions of the basic equation.

A solution of the basic equation which gives by means of an integral a solution of the corresponding equation, such as (B), in which the additional term is an arbitrary function of the independent variables, is called a *fundamental solution*. Rules for finding fundamental solutions have been given by Fredholm and Zeilon. In some cases the solution which is called fundamental seems to be unique and the theory is simple. In other cases difficulties arise. In any case much depends upon the domain $\mathscr{D}$ and the supplementary conditions that are imposed upon the solution.

When the basic equation is the wave-equation the question of a fundamental solution is particularly interesting. There are, indeed, two solutions,

$$V = \frac{1}{r} f(t - r/c)$$

and

$$V = \frac{1}{2r}\left[\frac{1}{r - ct} + \frac{1}{r + ct}\right] = \frac{1}{r^2 - c^2t^2},$$

which may be regarded as natural generalisations of the fundamental solution $1/r$ of Laplace's equation. The former seems to be the most useful as is shown by a famous theorem due to Kirchhoff.

In the case of the equation of the conduction of heat the solution which is regarded as fundamental is

$$V = t^{-\frac{3}{2}} e^{-\frac{x^2}{4\kappa t}}$$

when the equation is taken in the form

$$\frac{\partial V}{\partial t} = \kappa \nabla^2 V,$$

and is

$$V = t^{-\frac{1}{2}} e^{-\frac{x^2}{4\kappa t}}$$

when the equation is taken in the simpler form

$$\frac{\partial V}{\partial t} = \kappa \frac{\partial^2 V}{\partial x^2}.$$

The equation of heat conduction is not a basic equation but may be transformed into a basic equation by the introduction of an auxiliary variable in a manner already mentioned. Thus the basic equation derived from

$$\frac{\partial V}{\partial t} = \kappa \frac{\partial^2 V}{\partial x^2}$$

is

$$\frac{\partial^2 W}{\partial s \partial t} = \frac{\partial^2 W}{\partial x^2}, \quad W = V e^{s/\kappa},$$

and this equation possesses the primitive solution

$$W = t^{-\frac{1}{2}} F\left[\frac{x^2}{4t} - s\right],$$

of which

$$W = t^{-\frac{1}{2}} \exp\left[\frac{s}{\kappa} - \frac{x^2}{4\kappa t}\right]$$

is a special case.

The theory of fundamental solutions is evidently closely connected with the theory of primitive solutions but some principles are needed to guide us in the choice of the particular primitive solution which is to be regarded as fundamental. The necessary principles are given by some general theorems relating to the transformation of integrals which are forms or developments of the well-known theorems of Green and Gauss. These theorems will be discussed in Chapter II. An entirely different discussion of the fundamental solutions of partial differential equations with constant coefficients has been given recently by G. Herglotz, *Leipziger Berichte*, vol. LXXVIII, pp. 93, 287 (1926) with references to the literature.

EXAMPLES

1. Prove that the equation

$$\frac{\partial V}{\partial t} = \frac{\partial^4 V}{\partial x^4}$$

is satisfied by the two definite integrals

$$V = 4\int_0^\infty e^{-xs}(\cos xs - \sin xs)\, e^{-4ts^4}\, ds,$$

$$V = \int_0^\infty v(s, t)\, v(x, s)\, ds,$$

where $v(x, t) = t^{-\frac{1}{2}} e^{-x^2/4t}$.

Show also that the two integrals represent the same solution.

2. Prove that this solution can be expanded in the form

$$V = V_0 - V_1 + V_2,$$

where
$$V_0 = \Gamma(\tfrac{1}{4})(4t)^{-\frac{1}{4}}\left[1 - \frac{1}{4!}\frac{x^4}{4t} + \frac{1.5}{8!}\left(\frac{x^4}{4t}\right)^2 + \ldots\right],$$

$$V_1 = \Gamma(\tfrac{1}{2})\, x t^{-\frac{1}{2}}\left[1 - \frac{1}{5!}\frac{x^4}{2t} + \frac{1.3}{9!}\left(\frac{x^4}{2t}\right)^2 - \ldots\right],$$

$$V_2 = 2\Gamma(\tfrac{3}{4})\, x^2 (4t)^{-\frac{3}{4}}\left[\frac{1}{2!} - \frac{1.3}{6!}\left(\frac{x^4}{4t}\right) + \frac{1.3.7}{(10)!}\left(\frac{x^4}{4t}\right)^2 \ldots\right].$$

3. Show also that

$$V_0 = 4\int_0^\infty e^{-4ts^4} \cos sx \cosh sx\, ds.$$

$$V_1 = 4\int_0^\infty e^{-4ts^4} [\sin sx \cosh sx + \cos sx \sinh sx]\, ds,$$

$$V_2 = 4\int_0^\infty e^{-4ts^4} \sin sx \sinh sx\,.\, ds.$$

4. Prove that there is a fourth solution

$$V_3 = x^3 t^{-1}\left[\frac{1}{3!} - \frac{1}{7!}\frac{x^4}{t} + \frac{1.2}{(11)!}\left(\frac{x^4}{t}\right)^2 - \ldots\right] = 4\int_0^\infty e^{-4ts^4} [\sin sx \cosh sx - \cos sx \sinh sx]\, ds.$$

5. If $V(x, t)$ is a solution of the equation

$$\frac{\partial V}{\partial t} = \frac{\partial^s V}{\partial x^s} \qquad s \geq 1,$$

the quantity

$$y_n(t) = \frac{1}{\Gamma(1+n)}\int_0^\infty x^n V(x, t)\, dx$$

is generally a solution of the set of equations

$$y_n'(t) = y_{n-s}(t) \qquad n > s - 1.$$

In particular, if $s = 2$ and $V(x, t)$ is the function $v(x, t)$ of Ex. 1, the corresponding function $y_n(t)$ is

$$y_n(t) = \frac{t^{\frac{n}{2}}}{\Gamma\left(1 + \frac{n}{2}\right)} = \frac{t^{\frac{n}{s}}}{\Gamma\left(1 + \frac{n}{s}\right)}.$$

This may be called the fundamental solution, and when the second form is adopted s may have any positive integral value. In particular, when $s = 4$, this function is derivable from the function $v(x, t)$ of Ex. 1, p. 113.

CHAPTER II

APPLICATIONS OF THE INTEGRAL THEOREMS OF GAUSS AND STOKES

§ **2·11.** In the following investigations much use will be made of the well-known formulae

$$\int_C (u\,dx + v\,dy + w\,dz) = \iint \left[l\left(\frac{\partial w}{\partial y} - \frac{\partial v}{\partial z}\right) + m\left(\frac{\partial u}{\partial z} - \frac{\partial w}{\partial x}\right) + n\left(\frac{\partial v}{\partial x} - \frac{\partial u}{\partial y}\right)\right] dS,$$

$$\iint_S (lX + mY + nZ)\, dS = \iiint \left(\frac{\partial X}{\partial x} + \frac{\partial Y}{\partial y} + \frac{\partial Z}{\partial z}\right) dx\,dy\,dz \qquad \text{......(A)}$$

for the transformation of line and surface integrals into surface and volume integrals respectively. In these equations l, m, n are the direction cosines of the normal to the surface element dS, the normal being drawn in a direction away from the region over which the volume integral is taken or in a direction which is associated with the direction of integration round the closed curve C by the right-handed screw rule.

The functions u, v, w, X, Y, Z occurring in these equations will be supposed to be continuous over the domains under consideration and to possess continuous first derivatives of the types required*. The equations may be given various vector forms, the simplest being those in which u, v, w are regarded as the components of a vector q and X, Y, Z the components of a vector F. The equations are then

$$\left.\begin{aligned} \int q \,.\, \vec{ds} &= \int (\text{curl } q) \,.\, \vec{dS}, \\ \int F \,.\, \vec{dS} &= \int (\text{div } F)\, d\tau \quad (d\tau \equiv dx \,.\, dy \,.\, dz), \end{aligned}\right\} \qquad \text{......(B)}$$

where $\vec{ds}$ now stands for a vector of magnitude ds and the direction of the tangent to the curve C, while $\vec{dS}$ represents a vector of magnitude dS and the direction of the outward-drawn normal. The dot is used to indicate a scalar product of two vectors. Another convenient notation is

$$\left.\begin{aligned} \int q_t\, ds &= \int (\text{curl } q)_n\, dS, \\ \int F_n\, dS & \quad \int (\text{div } F)\, d\tau, \end{aligned}\right\} \qquad \text{......(C)}$$

* See for instance Goursat-Hedrick, *Mathematical Analysis*, vol. I, pp. 262, 309. Some interesting remarks relating to the proofs of the theorems will be found in a paper by J. Carr, *Phil. Mag.* (7), vol. IV, p. 449 (1927). The first theorem is well discussed by W. H. Young, *Proc. London Math. Soc.* (2), vol. XXIV, p. 21 (1926); and by O. D. Kellogg, *Foundations of Potential Theory*, Springer, Berlin (1929), ch. IV.

where the suffixes t and n are used to denote components in the direction of the tangent and normal respectively.

If we write $Z = v$, $Y = -w$, $X = 0$; $X = w$, $Z = -u$, $Y = 0$; $Y = u$, $X = -v$, $Z = 0$ in succession we obtain three equations which may be written in the vector form

$$\int (q \times \vec{dS}) = -\int (\text{curl}\, q)\, d\tau, \qquad \text{......(D)}$$

where the symbol $\times$ is used to denote a vector product.

Again, if we write successively $X = p$, $Y = Z = 0$; $Y = p$, $Z = X = 0$; $Z = p$, $X = Y = 0$, we obtain three equations which may be written in the vector form

$$\int (p)\, \vec{dS} = \int (\nabla p)\, d\tau, \qquad \text{......(E)}$$

where ∇p denotes the vector with components $\frac{\partial p}{\partial x}, \frac{\partial p}{\partial y}, \frac{\partial p}{\partial z}$ respectively.

§ 2·12. To obtain physical interpretations of these equations we shall first of all regard u, v, w as the component velocities of a particle of fluid which happens to be at the point (x, y, z) at time t. The quantities ξ, η, ζ defined by the equations

$$\xi = \frac{\partial w}{\partial y} - \frac{\partial v}{\partial z}, \quad \eta = \frac{\partial u}{\partial z} - \frac{\partial w}{\partial x}, \quad \zeta = \frac{\partial v}{\partial x} - \frac{\partial u}{\partial y}$$

may then be regarded as the components of the vorticity.

The line integral in (A) is called the circulation round the closed curve C and the theorem tells us that this is equal to the surface integral of the normal component of the vorticity. When there is a velocity potential ϕ we have

$$u = \frac{\partial \phi}{\partial x}, \quad v = \frac{\partial \phi}{\partial y}, \quad w = \frac{\partial \phi}{\partial z}$$

(in vector notation $q = \nabla\phi$) and $\xi = \eta = \zeta = 0$, the circulation round a closed curve is then zero so long as the conditions for the transformation of the line integral into a surface integral are fulfilled. The circulation is not zero when

$$\phi = \tan^{-1}(y/x),$$

and the curve C is a simple closed curve through which the axis of z passes once without any intersection. The axis of z is then a line of singularities for the functions u and v. The value of the integral is 2π, for ϕ increases by 2π in one circuit round the axis of z. The velocity potential

$$\phi = (\Gamma/2\pi) \tan^{-1}(y/x)$$

may be regarded as that of a simple line vortex along the axis of z, the strength of the vortex being represented by the quantity Γ which is supposed to be constant. Γ represents the circulation round a closed curve which goes once round the line vortex.

If we write $X = \rho u$, $Y = \rho v$, $Z = \rho w$, where ρ is the density of the fluid, the surface integral in (A) may be interpreted as the rate at which the mass of the fluid within the closed surface S is decreasing on account of the flow across the surface S. If fluid is neither created nor destroyed within the surface this decrease of mass is also represented by

$$-\frac{\partial}{\partial t}\int \rho\, d\tau = -\int \frac{\partial \rho}{\partial t}\, d\tau.$$

The two expressions are equal when the following equation is satisfied at each place (x, y, z) and at each time t,

$$\frac{\partial \rho}{\partial t} + \frac{\partial}{\partial x}(\rho u) + \frac{\partial}{\partial y}(\rho v) + \frac{\partial}{\partial z}(\rho w) = 0.$$

This is the equation of continuity of hydrodynamics. There is a similar equation in the theory of electricity when ρ is interpreted as the density of electricity and u, v, w as the component velocities of the electricity which happens to be at the point (x, y, z) at time t. When ρ is constant the equation of continuity takes the simple form

$$\frac{\partial u}{\partial x} + \frac{\partial v}{\partial y} + \frac{\partial w}{\partial z} = 0$$

(in vector notation $\operatorname{div} q = 0$). This simple form may be used also when $d\rho/dt = 0$, where d/dt stands for the hydrodynamical operator

$$\frac{d}{dt} = \frac{\partial}{\partial t} + u\frac{\partial}{\partial x} + v\frac{\partial}{\partial y} + w\frac{\partial}{\partial z},$$

a fluid for which $d\rho/dt = 0$ is said to be incompressible.

When p is interpreted as fluid pressure the equation (E) indicates that as far as the components of the total force are concerned the effect of fluid pressure on a surface is the same as that of a body force which acts at the point (x, y, z) and is represented in magnitude and direction by the vector $-\nabla p$, the sign being negative because the force acts inwards and not outwards relative to each surface element. Putting $q = pr$ in equation (D), where r is the vector with components x, y, z, we have an equation

$$\int (r \times p\, ds) = -\int (\operatorname{curl} pr)\, d\tau = \int (r \times \nabla p)\, d\tau,$$

which indicates that the foregoing distribution of body force gives the same moments about the three axes of co-ordinates as the set of forces arising from the pressures on the surface S. The body forces are thus completely equivalent to the forces arising from the pressures on the surface elements. This result is useful for the formulation of the equations of hydrodynamics which are usually understood to mean that the mass multiplied by the acceleration of each fluid element is equal to the total body force. If in addition to the body force arising from the pressure there is a body force F whose components per unit mass are X, Y, Z for a particle which is at (x, y, z) at time t, the equations of hydrodynamics may be written in the vector form

$$\rho \frac{dq}{dt} = - \nabla p + \rho F.$$

When viscosity and turbulence are neglected the body force often can be derived from a potential Ω so that $F = \nabla \Omega$. The hydrodynamical equations then take the simple form

$$\frac{dq}{dt} = - \frac{1}{\rho} \nabla p + \nabla \Omega,$$

which implies that in this case there is an acceleration potential if ρ is a constant or a function of p. When in addition there is a velocity potential ϕ the equations may be written in the form

$$\nabla \left[\frac{\partial \phi}{\partial t} + \tfrac{1}{2} q^2 + \int \frac{dp}{\rho} - \Omega \right],$$

and imply that

$$\int \frac{dp}{\rho} + \frac{\partial \phi}{\partial t} + \tfrac{1}{2} q^2 = \Omega + f(t),$$

where $f(t)$ is some function of t. This may be regarded as an equation for the pressure, when $\Omega = 0$ it indicates that the pressure is low where the velocity is high.

§ **2·13.** *The equation of the conduction of heat.* When different parts of a body are at different temperatures, energy in the form of heat flows from the hotter parts to the colder and a state of equilibrium is gradually established in which the temperature is uniformly constant throughout the body, if the different parts of the body are relatively at rest and do not participate in an unequal manner in heat exchanges with other bodies. When, however, a steady supply of heat is maintained at some place in the body, the steady state which is gradually approached may be one in which the temperature varies from point to point but remains constant at each point.

A hot body is not like a pendulum swinging in air and performing a series of damped oscillations as the position of equilibrium is approached, it is more like a pendulum moving in a very viscous fluid and approaching its position of equilibrium from one side only. The steady state appears, in fact, to be approached without oscillation.

These remarks apply, of course, to the phenomenon of conduction of heat when there is no relative motion (on a large scale) of different parts of the body. When a liquid is heated, a state of uniform temperature is produced largely by convection currents in which part of the fluid moves from one place to another and carries heat with it. There are convection currents also in the atmosphere and these are responsible not only for the diffusion of heat and water vapour but also for a transportation of momentum which is responsible for the diurnal variation of wind velocity and other phenomena.

A third process by which heat may be lost or gained by a body is by the emission or absorption of radiation. This process will be treated here as a surface phenomenon so that the laws of emission and absorption are expressed as boundary conditions; the propagation of the radiation in the intervening space between two bodies or between different parts of the same body is considered in electromagnetic theory. The mechanism of the emission or absorption is not fully understood and is best described by means of the quantum theory and the theory of the electron. The use of a simple boundary condition avoids all the difficulty and is sufficiently accurate for most mathematical investigations. In many problems, however, radiation need not be taken into consideration at all.

The fundamental hypothesis on which the mathematical theory of the conduction of heat is based is that the rate of transfer of heat across a small element dS of a surface of constant temperature (i.e. an isothermal surface) is represented by

$$- K \frac{\partial \theta}{\partial n} dS, \qquad \text{......(A)}$$

where K is the thermal conductivity of the substance, θ is the temperature in the neighbourhood of dS, and $\frac{\partial}{\partial n}$ denotes a differentiation along the outward-drawn normal to dS. The negative sign in this expression simply expresses the fact that the flow of heat is from places of higher to places of lower temperature. The rate of transfer of heat across any surface element $d\sigma$ in time dt may be denoted by $f_\nu d\sigma dt$, where the quantity f_ν is called the flux of heat across the element and the suffix ν is used to indicate the direction of the normal to the element.

Let us now consider a small tetrahedron $DABC$ whose faces DBC, DCA, DAB, ABC are normal respectively to the directions Ox, Oy, Oz, Ow, where the first three lines are parallel to the axes of co-ordinates. Denoting the area ABC by Δ, the areas DBC, DCA, DAB are respectively $w_x\Delta$, $w_y\Delta$, $w_z\Delta$, where w_x, w_y, w_z are the direction cosines of Ow.

Wher Δ is very small the rate at which heat is being gained by the tetrahedron at time t is approximately

$$(w_x f_x + w_y f_y + w_z f_z - f_w)\,\Delta.$$

This must be equal to $Vc\rho \frac{d\theta}{dt}$, where V is the volume of the tetrahedron, c the specific heat of the material and ρ its density. Now $V = \frac{1}{3}p\Delta$, where p is the perpendicular distance of D from the plane ABC, hence

$$w_x f_x + w_y f_y + w_z f_z - f_w = \tfrac{1}{3} pc\rho \frac{d\theta}{dt}$$

and so tends to zero as p tends to zero.

When DAB is an element of an isothermal surface we may use the additional hypothesis that f_x and f_y are both zero and the equation gives

$$f_w = w_z f_z = -Kw_z \frac{\partial \theta}{\partial z} = -K \frac{\partial \theta}{\partial w}.$$

The law (A) thus holds not simply for an isothermal surface but for any surface separating two portions of the same material. The vector $\Delta\theta$ whose components are $\frac{\partial \theta}{\partial x}, \frac{\partial \theta}{\partial y}, \frac{\partial \theta}{\partial z}$ is called the temperature gradient at the point (x, y, z) at time t.

Let us now consider a portion of the body bounded by a closed surface S. Assuming that f_x, f_y, f_z and their partial derivatives with respect to x, y and z are continuous functions of x, y and z for all points of the region bounded by S, the rate at which this region is gaining heat on account of the fluxes across its surface elements is

$$+\iint K \frac{\partial \theta}{\partial n}\, dS.$$

Transforming this into a volume integral and equating the result to

$$\iiint c\rho \frac{d\theta}{dt}\, dx\, dy\, dz,$$

we have the equation

$$\iiint \left[c\rho \frac{d\theta}{dt} - \frac{\partial}{\partial x}\left(K \frac{\partial \theta}{\partial x}\right) - \frac{\partial}{\partial y}\left(K \frac{\partial \theta}{\partial y}\right) - \frac{\partial}{\partial z}\left(K \frac{\partial \theta}{\partial z}\right)\right] dx\, dy\, dz.$$

This must hold for any portion of the material that is bounded by a simple closed surface and this condition is satisfied if at each point

$$c\rho \frac{d\theta}{dt} - \operatorname{div}(K\nabla\theta) = 0.$$

If the body is at rest we can write $\frac{\partial \theta}{\partial t}$ in place of $\frac{d\theta}{dt}$, but if it is a moving fluid the appropriate expression for $\frac{d\theta}{dt}$ is

$$\frac{d\theta}{dt} = \frac{\partial \theta}{\partial t} + u \frac{\partial \theta}{\partial x} + v \frac{\partial \theta}{\partial y} + w \frac{\partial \theta}{\partial z},$$

where u, v and w are the component velocities of the medium.

In most mathematical investigations the medium is stationary and the quantities c, K and ρ are constant in both space and time and the equation takes the simple form

$$\frac{\partial \theta}{\partial t} = \kappa \nabla^2 \theta,$$

in which κ is a constant called the diffusivity*. If at the point (x, y, z) there is a source of heat supplying in time dt a quantity $F(x, y, z, t)\, dx\, dy\, dz\, dt$

* This name was suggested by Lord Kelvin. A useful table of the quantities K, c, ρ and x is given in Ingersoll and Zobel's *Mathematical Theory of Heat Conduction* (Ginn & Co., 1913).

of heat to the volume element $dxdydz$, a term $F(x, y, z, t)$ must be added to the right-hand side of the equation.

A similar equation occurs in the theory of diffusion; it is only necessary to replace temperature by concentration of the diffusing substance in order to obtain the derivation of the equation of diffusion. The quantity of diffusing substance conducted from place to place now corresponds to the amount of heat that is being conducted. The theory of diffusion of heat was developed by Fourier, that of a substance by Fick. In recent times a theory of non-Fickian diffusion has been developed in which the coefficient K is not a constant. Reference may be made to the work of L. F. Richardson*.

§ **2·14**. An equation similar to the equation of the conduction of heat has been used recently by Tuttle† in a theory of the drying of wood. It is known that when different parts of a piece of wood are at different moisture contents, moisture transfuses from the wetter to the drier regions; Tuttle therefore adopts the fundamental hypothesis that the rate at which transfusion takes place transversely with respect to the wood fibres or elements is proportional to the slope of the moisture gradient.

This assumption leads to the equation

$$\frac{\partial \theta}{\partial t} = h^2 \frac{\partial^2 \theta}{\partial x^2},$$

where θ is moisture content expressed as a percentage of the oven-dry weight of the wood and h^2 is a constant for the particular wood and may be called the transfusivity (across the grain) of the species of wood under consideration.

From actual data on the distribution of moisture in the heartwood of a piece of Sitka spruce after five hours' drying at a temperature of 160° F. and in air with a relative humidity of 30 %, Tuttle finds by a computation that h^2 is about 0·0053, where lengths are measured in inches, time in hours and moisture content in percentage of dry weight of wood.

The actual boundary conditions considered in the computation were

$$\theta = 0 \text{ at } x = 0, \quad \theta = 0 \text{ at } x = 1, \quad \theta = \theta_0 \text{ when } t = 0.$$

A more complete theory of drying has been given recently by E. E. Libman‡ in his theory of porous flow. He denotes the mass of fluid per unit mass of dry material by v and calls it the moisture density. The symbols ρ, σ, τ are used to denote the densities of moist material, dry material and fluid respectively and β is used to denote the coefficient of compressibility of the moist material.

* *Proc. Roy. Soc. London*, vol. CX, p. 709 (1926).

† F. Tuttle, *Journ. of the Franklin Inst.* vol. CC, p. 609 (1925).

‡ E. E. Libman, *Phil. Mag.* (7), vol. IV, p. 1285 (1927).

The rate of gain of fluid per unit mass of dry material in the volume V is the rate of increase of $\bar{v}$, where $\bar{v}$ is the average value of v in V. If w is the mass of dry material in volume V of moist material and f_n = mass of fluid flowing in unit time across unit area normal to the direction n we have

$$\frac{\partial \bar{v}}{\partial t} = -\frac{1}{w}\iint f_n dS = -\frac{1}{w}\iiint \operatorname{div} f \,.\, dx\,dy\,dz,$$

therefore

$$\frac{\partial v}{\partial t} = -\frac{V}{w}\operatorname{div} f. \qquad \text{......(B)}$$

Now, the mass of fluid in the volume V is $w\bar{v}$ and the total mass of material in V is $w\bar{v} + w$ and is also ρV, hence

$$w = \frac{\rho}{1+\bar{v}} V,$$

and (B) gives the equation

$$\frac{\partial v}{\partial t} = -\frac{1+v}{\rho}\operatorname{div} f$$

for the interior of the porous body.

If $E\,dS$ denotes the mass of fluid evaporating in unit time from a small area dS of the boundary of the porous body the boundary condition is

$$f_n = E.$$

The flow of fluid in a porous material may be regarded as the sum of three separate flows due respectively to capillarity, gravity and a pressure gradient caused by shrinkage. We therefore write, for the case in which the z axis is vertical and p is the pressure,

$$f_n = -K\frac{\partial v}{\partial n} - kg\tau\frac{\partial z}{\partial n} - k\frac{\partial p}{\partial n},$$

where K and k are constants characteristic of the material.

Consider now a small element of volume $\dfrac{1+v}{\rho}\delta w$ at the point $P(x,y,z)$, the associated mass of dry material being δw and the volume per unit mass of dry material

$$V = \frac{1+v}{\rho}.$$

Then

$$\frac{dV}{dv} = \frac{d}{dv}\left(\frac{1+v}{\rho}\right),$$

$$\frac{dp}{dv} = \frac{dp}{dV}\frac{dV}{dv} = \frac{dp}{dV}\frac{d}{dv}\left(\frac{1+v}{\rho}\right).$$

But by definition

$$\beta = -\frac{1}{V}\frac{dV}{dp},$$

therefore

$$\frac{dp}{dv} = -\frac{1}{V\beta}\frac{d}{dv}\left(\frac{1+v}{\rho}\right) = -\frac{1}{\beta}\frac{d}{dv}\left(\log\frac{1+v}{\rho}\right),$$

$$\frac{\partial p}{\partial n} = -\frac{1}{\beta}\frac{d}{dv}\left(\log\frac{1+v}{\rho}\right)\cdot\frac{\partial v}{\partial n}.$$

Putting $$\frac{d\phi}{dv} = K - \frac{k}{\beta}\frac{d}{dv}\left(\log\frac{1+v}{\rho}\right),$$

we have $$f_n = -\frac{\partial\phi}{\partial n} - kg\tau\frac{\partial z}{\partial n},$$

or $$f_x = -\frac{\partial\phi}{\partial x}, \quad f_y = -\frac{\partial\phi}{\partial y}, \quad f_z = -\frac{\partial\phi}{\partial z} - kg\tau,$$

$$\operatorname{div} f = -\nabla^2\phi,$$

and so $$\frac{\rho}{1+v}\frac{\partial v}{\partial t} = \nabla^2\phi,$$

while the boundary condition takes the form

$$\frac{\partial\phi}{\partial n} + E + kg\tau\frac{\partial z}{\partial n} = 0.$$

It should be mentioned that in the derivation of this equation the material has been assumed to be isotropic.

In the special case of no shrinkage we have

$$\rho = \sigma(1+v), \quad \frac{d\phi}{dv} = K, \quad \phi = Kv + \text{const.},$$

and the equation for v becomes

$$\frac{\partial v}{\partial t} = \frac{K}{\sigma}\nabla^2 v,$$

which is similar in form to the equation of the conduction of heat. The boundary condition is

$$K\frac{\partial v}{\partial n} + E + kg\tau\frac{\partial z}{\partial n} = 0.$$

§ 2·15. *The heating of a porous body by a warm fluid**. A warm fluid carrying heat is supposed to flow with constant velocity into a tube which contains a porous substance such as a solid body in a finely divided state. For convenience we shall call the fluid steam and the porous substance iron. The steam is initially at a constant temperature which is higher than that of the iron. The problem is to determine the temperatures of the iron and steam at a given time and position on the assumption that the specific heats of the iron and steam are both constant and that there are no heat exchanges between the wall of the tube and either the iron or steam, no heat exchanges between different particles of steam and no heat exchanges between different particles of iron. The problem is, of course, idealised by these simplifying assumptions. We make the further assumption that the velocity of the steam is the same all over the cross-section of the pipe. This, too, would not be quite true in actual practice.

Let U be the temperature of the iron at a place specified by a co-ordinate x measured parallel to the axis of the pipe, V the corresponding temperature

* A. Anzelius, *Zeits. f. ang. Math. u. Mech.* Bd. VI, S. 291 (1926).

of the steam. These quantities will be regarded as functions of x and t only. This is approximately true if the pipe is of uniform section so that the cross-sectional area is a constant quantity A.

Let us now consider a slice of the pipe bounded by the wall and two transverse planes x and $x + dx$. At times t and $t + dt$ the heat contents of the iron contained in this slice are respectively

$$uUA\,dx \quad \text{and} \quad u\left(U + \frac{\partial U}{\partial t}\,dt\right)A\,dx,$$

where u is the quantity of heat necessary to raise the temperature of unit volume of the iron through unit temperature. Thus the quantity of heat imparted to the iron in the slice in time dt is

$$dQ_1 = uA\,\frac{\partial U}{\partial t}\,dt\,dx.$$

Similarly, at time t the heat content of the vapour in the slice is $vVA\,dx$ and at time $t + dt$ it is $v\left(V + \frac{\partial V}{\partial t}\,dt\right)A\,dx$, where v is a quantity analogous to u.

With the steam flowing across the plane x in time dt a quantity of heat $vVAc\,dt$ is brought into the slice where c is the constant velocity of flow. In the same time a quantity of heat $v\left(V + \frac{\partial V}{\partial x}\,dx\right)Ac\,dt$ leaves the slice across the plane $x + ax$. The steam has thus conveyed to the iron a quantity of heat

$$dQ_2 = -\,v\left(\frac{\partial V}{\partial t} + c\,\frac{\partial V}{\partial x}\right)A\,dt\,dx.$$

In accordance with the law of heat transfer that is usually adopted the quantity of heat transferred from the steam to the iron in the slice in time dt is

$$dQ_3 = k\,(V - U)\,A\,dt\,dx,$$

where k is the heat transfer factor for iron and steam. We thus have the equations

$$v\left(\frac{\partial V}{\partial t} + c\,\frac{\partial V}{\partial x}\right) = k\,(U - V),$$

$$u\,\frac{\partial U}{\partial t} = k\,(V - U).$$

With the notation $a = k/cv$, $b = k/cu$ and the new variables

$$\xi = ax, \quad \tau = b\,(ct - x),$$

the equations become

$$\frac{\partial V}{\partial \xi} = U - V, \quad \frac{\partial U}{\partial \tau} = V - U.$$

These equations imply that the quantity $\Delta\,(\xi, \tau)$ defined by

$$\Delta\,(\xi, \tau) = e^{\xi+\tau}\,(V - U)$$

is a solution of the partial differential equation

$$\frac{\partial^2 \Delta}{\partial \xi\,\partial \tau} = \Delta. \qquad \text{......(A)}$$

The supplementary conditions which will be adopted are

$$U(\xi, 0) = U_1, \quad V(0, \tau) = V_1,$$

where U_1 and V_1 are constants. The equation (A) then gives

$$U(0, \tau) = V_1 - (V_1 - U_1) e^{-\tau},$$
$$V(\xi, 0) = U_1 + (V_1 - U_1) e^{-\xi},$$

and so the supplementary conditions for the quantity Δ are

$$\Delta(\xi, 0) = \Delta(0, \tau) = V_1 - U_1 = W, \text{ say.}$$

§ **2·16**. *Solution by the method of Laplace.* The equation (A) may be solved by a method of successive approximations by writing

$$\Delta = \Delta_0 + \Delta_1 + \Delta_2 + \dots,$$

where $\Delta_0 = W$ and
$$\frac{\partial^2 \Delta_n}{\partial \xi \partial \tau} = \Delta_{n-1}.$$

This gives
$$\Delta = W I_0 [2\sqrt{(\xi\tau)}],$$
$$V - U = W e^{-(\xi+\tau)} I_0 [2\sqrt{(\xi\tau)}],$$
$$V = V_1 - W e^{-b(ct-x)} \int_0^{ax} e^{-s} I_0 [2\sqrt{\{bs(ct-x)\}}]\, ds,$$
$$U = U_1 + W e^{-ax} \int_0^{b(ct-x)} e^{-s} I_0 [2\sqrt{(axs)}]\, ds.$$

For $x > ct$ the solution has no physical meaning but for such values of x the iron has not yet been reached by the steam and so $U = U_1$.

As $t \to \infty$ we should have $U \to V_1$, $V \to V_1$; this condition is easily seen to be satisfied, for our formula for $V - U$ indicates that $V - U \to 0$ and $U \to V_1$ because

$$\int_0^\infty e^{-s} I_0 [2\sqrt{(axs)}]\, ds = e^{ax}.$$

The properties of the solution might be used, however, to infer the value of this integral.

EXAMPLE

Prove that if
$$E(\tau) = \sum_0^\infty \frac{\tau^n}{(n!)^3},$$
the differential equation
$$\frac{\partial^3 V}{\partial x \partial y \partial z} = V$$
is satisfied by
$$V = \int_0^y \int_0^z \phi(v, w) E\{x(y-v)(z-w)\}\, dv\, dw$$
$$+ \int_0^z \int_0^x \psi(w, u) E\{y(z-w)(x-u)\}\, dw\, du$$
$$+ \int_0^x \int_0^y \chi(u, v) E\{z(x-u)(y-v)\}\, du\, dv$$
$$+ \int_0^x P(u) E\{(x-u) yz\}\, du + \int_0^y Q(v) E\{(y-v) zx\}\, dv$$
$$+ \int_0^z R(w) E\{(z-w) xy\}\, dw + SE(xyz).$$

[T. W. Chaundy, *Proc. London Math. Soc.* (2), vol. XXI, p. 214 (1923).]

§ 2·21. *Riemann's method.* Let $L(u)$ be used to denote the differential expression

$$\frac{\partial^2 u}{\partial x \partial y} + a\frac{\partial u}{\partial x} + b\frac{\partial u}{\partial y} + cu,$$

where a, b, c are continuous $(D, 1)$ in a region R in the (x, y) plane. The adjoint expression $\bar{L}(v)$ is defined by the relation

$$vL(u) - u\bar{L}(v) = \frac{\partial M}{\partial x} + \frac{\partial N}{\partial y},$$

where M and N are certain quantities which can be expressed in terms of u, v and their first derivatives. Appropriate forms for $\bar{L}$, M and N are

$$\bar{L}(v) = \frac{\partial^2 v}{\partial x \partial y} - \frac{\partial}{\partial x}(av) - \frac{\partial}{\partial y}(bv) + cv,$$

$$M = auv + \frac{1}{2}\left(v\frac{\partial u}{\partial y} - u\frac{\partial v}{\partial y}\right) = \frac{1}{2}\frac{\partial}{\partial y}(uv) - uP(v), \text{ say,}$$

$$N = buv + \frac{1}{2}\left(v\frac{\partial u}{\partial x} - u\frac{\partial v}{\partial x}\right) = \frac{1}{2}\frac{\partial}{\partial x}(uv) - uQ(v), \text{ say,}$$

where $\quad P(v) = \dfrac{\partial v}{\partial y} - av, \quad Q(v) = \dfrac{\partial v}{\partial x} - bx.$

Now if C is a closed curve which lies entirely within the region R and if both u and v are continuous $(D, 1)$ in R, we have by the two-dimensional form of Green's theorem

$$\int (lM + mN)\, ds = \iint \left(\frac{\partial M}{\partial x} + \frac{\partial N}{\partial y}\right) dx\, dy = \iint [vL(u) - u\bar{L}(v)]\, dx\, dy,$$

where l, m are the direction cosines of the normal to the curve C and the double integral is taken over the area bounded by C, and so will be expressed in terms of the values of u and its normal derivative at points of the curve Γ, for when u is known its tangential derivative is known and $\dfrac{\partial u}{\partial x}$ and $\dfrac{\partial u}{\partial y}$ can be expressed in terms of the normal and tangential derivatives. If (x_0, y_0) are the co-ordinates of the point A the function v which enables us to solve the foregoing problem may be written in the form

$$v = g(x, y; x_0, y_0),$$

and may be called a Green's function of the differential expression $L(u)$.

This theorem will now be applied in the case where the curve C consists of lines XA, AY parallel to the axes of x and y respectively and a curve Γ joining the points Y and X.

Using letters instead of particular values of the variable of integration to denote the end points of each integral, we have when $L(u) = 0$, $\bar{L}(v) = 0$,

$$\int_A^X N\, dx - \int_Y^A M\, dy = -\int_Y^X (lM + mN)\, ds.$$

Now $$-\int_Y^A M\,dy = \tfrac{1}{2}[(uv)_Y - (uv)_A] + \int_Y^A uP(v)\,dy,$$

and $$\int_A^X N\,dx = \tfrac{1}{2}[(uv)_X - (uv)_A] - \int_A^X uQ(v)\,dx,$$

therefore $$(uv)_A = \tfrac{1}{2}[(uv)_X + (uv)_Y] + \int_Y^X (lM + mN)\,ds$$
$$+ \int_Y^A uP(v)\,dy - \int_A^X uQ(v)\,dx.$$

If now the function v can be chosen so that $P(v) = 0$ on AY and $Q(v) = 0$ on AX, the value of u at the point A will be given by the formula

$$(uv)_A = \tfrac{1}{2}[(uv)_X + (uv)_Y] + \int_Y^X (lM + mN)\,ds;$$

It should be noticed that if u is not a solution of $L(u) = 0$ but a solution of

$$L(u) + f(x, y) = 0$$

the corresponding expression for u is

$$(uv)_A = \tfrac{1}{2}[(uv)_X + (uv)_Y] + \int_Y^X (lM + mN)\,ds + \iint vf(x, y)\,dx\,dy.$$

An interesting property of the function g may be obtained by considering the case when the curve Γ consists of a line YB parallel to AX and a line BX parallel to YA. We then have

$$\int_Y^X (lM + mN)\,ds = \int_B^X M\,dy - \int_Y^B N\,dx,$$

also $$M = -\frac{1}{2}\frac{\partial}{\partial y}(uv) + v\bar{P}(u), \quad \bar{P}(u) = \frac{\partial v}{\partial y} + au,$$

$$N = -\frac{1}{2}\frac{\partial}{\partial x}(uv) + v\bar{Q}(u), \quad \bar{Q}(u) = \frac{\partial v}{\partial x} + bu.$$

we have $$\int_Y^B N\,dx = \tfrac{1}{2}[(uv)_Y - (uv)_B] + \int_Y^B v\bar{Q}(u)\,dx,$$

$$\int_B^X M\,dy = \tfrac{1}{2}[(uv)_B - (uv)_X] + \int_B^X v\bar{P}(u)\,dy.$$

Hence $$(uv)_A = (uv)_B - \int_Y^B v\bar{Q}(u)\,dx + \int_B^X v\bar{P}(u)\,dy.$$

Now let a function $u = h(x, y; x_1, y_1)$ be supposed to exist such that $L(u) = 0$, $\bar{P}(u) = 0$ on BX, $\bar{Q}(u) = 0$ on BY; the co-ordinates x_1, y_1 being those of B. The formula then gives

$$(uv)_A = (uv)_B.$$

Choosing the arbitrary constant multipliers which occur in the general expressions for g and h, in such a way that

$$g(x_0, y_0; x_0, y_0) = 1, \quad h(x_1, y_1; x_1, y_1) = 1,$$

the preceding relation can be written in the form

$$h(x_0, y_0; x_1, y_1) = g(x_1, y_1; x_0, y_0).$$

When considered as a function of (x, y) the Green's function g satisfies the adjoint equation $\bar{L}(v) = 0$, but when considered as a function of (x_0, y_0) it satisfies the original equation expressed in the variables x_0, y_0.

EXAMPLE

Prove that the Green's function for the differential equation

$$\frac{\partial^2 u}{\partial x \partial y} = \tfrac{1}{4}k^2 u$$

is

$$I_0[k\surd(x - x_0)(y - y_0)],$$

and obtain Laplace's formula

$$u = \int_0^y I_0\{k\surd[x(y-s)]\}\phi(s)\,ds + \int_0^x I_0\{k\surd[y(x-s)]\}\psi(s)\,ds$$

for a solution which satisfies the conditions

$$\frac{\partial u}{\partial x} = \psi(x) \quad \text{when } y = 0,$$

$$\frac{\partial u}{\partial y} = \phi(x) \quad \text{when } x = 0.$$

§ 2·22. Solution of the equation

$$L(z) \equiv \frac{\partial^2 z}{\partial x^2} - \frac{\partial z}{\partial y} = f(x, y). \qquad \text{.....(A)}$$

Let the curve B consist of a line A_1A_2 parallel to the axis of x and two curves C_1, C_2 starting from A_1, A_2 respectively and running in an upward direction from the line A_1A_2. Let S denote the realm bounded by the portion of B below a line γ parallel to the axis of x and the portion of B which lies between C_1 and C_2. When γ is replaced by the parallel lines γ_0, γ' the corresponding realms will be denoted by S_0 and S' respectively. The portions of B below the lines γ, γ_0, γ' will be denoted by β, β_0 and β' respectively. The equation of A_1A_2 will be taken to be $y = y_1$.

We shall now suppose that γ and γ' both lie below γ_0 and that z is a solution of (A) which is regular in S_0, regularity meaning that z, $\frac{\partial z}{\partial x}$, $\frac{\partial z}{\partial y}$ and $\frac{\partial^2 z}{\partial x^2}$ are continuous functions of x and y in the realm S_0.

The differential expression adjoint to $L(z)$ is $\bar{L}(t)$, where

$$\bar{L}(t) = \frac{\partial^2 t}{\partial x^2} + \frac{\partial t}{\partial y},$$

and we have the identity

$$t[L(z) - f(x, y)] - z\bar{L}(t) \equiv \frac{\partial}{\partial x}\left[t\frac{\partial z}{\partial x} - z\frac{\partial t}{\partial x}\right] - \frac{\partial}{\partial y}[tz] - tf.$$

Hence if. $$L(z) = f(x, y),$$
and $$\bar{L}(t) = 0,$$
we have $$\frac{\partial}{\partial x}\left[t\frac{\partial z}{\partial x} - z\frac{\partial t}{\partial x}\right] - \frac{\partial}{\partial y}[tz] - tf = 0.$$

Let us now write $\zeta = z(\xi, \eta)$, $\tau = t(\xi, \eta)$ and integrate the last equation over the region S', then

$$\int_{\gamma'} \tau\zeta d\xi = \int_{\beta'}\left[\tau\zeta d\xi + \left(\tau\frac{\partial\zeta}{\partial\xi} - \zeta\frac{\partial\tau}{\partial\xi}\right)d\eta\right] - \iint_{S'} \tau f d\xi d\eta.$$

In this equation we write

$$\tau(\xi, \eta) = T[x, y; \xi, \eta] \equiv (y-\eta)^{-\frac{1}{2}} \exp[-(x-\xi)^2/(y-\eta)],$$

and we take γ' to be a line which lies just below the line γ which passes through the point (x, y). Our aim now is to find the limiting value of the integral on the left as $\gamma' \to \gamma$.

By means of the substitution $\xi = x + 2u\sqrt{(y-\eta)}$ this integral is transformed into

$$2\int_{u_1}^{u_2} z[x + 2u(y-\eta)^{\frac{1}{2}}, \eta]\, e^{-u^2} du.$$

If the equations of the curves C_1, C_2 are respectively

$$x = c_1(y), \quad x = c_2(y),$$

the limits of the integral are respectively

$$u_1 = [c_1(\eta) - x]/2\sqrt{(y-\eta)}, \quad u_2 = [c_2(\eta) - x]/2\sqrt{(y-\eta)}.$$

If the point (x, y) lies within S we have

$$u_1 \to -\infty, \quad u_2 \to +\infty \text{ as } \gamma' \to \gamma \text{ and } \eta \to y;$$

if it lies outside S we have

$$u_1 \to u_2 \to \pm\infty \text{ as } \gamma' \to \gamma \text{ and } \eta \to y.$$

Finally, if the point (x, y) lies on either C_1 or C_2 one limit is zero, thus we may have either $u_1 \to 0$, $u_2 \to \infty$, or $u_1 \to -\infty$, $u_2 \to 0$.

The limiting value obtained by putting $\eta = y$ in the integral is $2z(x, y)\sqrt{\pi}$ in the first case, zero in the second case and $z(x, y)\sqrt{\pi}$ in the third. Hence when the limiting value is actually attained we have the formula

$$z(x, y)[2\sqrt{\pi}, 0 \text{ or } \sqrt{\pi}] = \int_{\beta}\left[T\zeta d\xi + \left(T\frac{\partial\zeta}{\partial\xi} - \zeta\frac{\partial T}{\partial\xi}\right)d\eta\right] - \iint_S Tfd\xi d\eta.$$

The transition to the limit has been carefully examined by Levi*, Goursat† and Gevrey‡. The last named has imposed further conditions

* E. E. Levi, *Annali di Matematica* (3), vol. XIV, p. 187 (1908).

† E. Goursat, *Traité d'Analyse*, t. III, p. 310.

‡ M. Gevrey, *Journ. de Mathématiques* (6), t. IX, p. 305 (1913). See also Wera Lebedeff, *Diss. Göttingen* (1906); E. Holmgren, *Arkiv för Matematik, Astronomi och Fysik*, Bd. III (1907), Bd. IV (1908); G. C. Evans, *Amer. Journ. Math.* vol. XXXVII, p. 431 (1915).

in order to establish the formula in the case when (x, y) lies on either C_1 or C_2. His conditions are that, if $\phi(s)$ is continuous,

$$\lim_{\eta \to y} [c_p(y) - c_p(\eta)](y - \eta)^{-\frac{1}{2}} = 0 \quad (p = 1, 2),$$

that

$$\int_\eta^y [c_p(y) - c_p(s)](y - s)^{-\frac{3}{2}} \phi(s)\, ds \,.\, \exp[-\{c_p(y) - c_p(s)\}^2/4(y - s)]$$

should exist and that the functions $c_1(y)$, $c_2(y)$ be of bounded variation.

It may be remarked that the line integrals in this formula are particular solutions of the equation $\frac{\partial z}{\partial y} = \frac{\partial^2 z}{\partial x^2}$, while the integral

$$z(x, y) = -\frac{1}{2\sqrt{\pi}} \iint_S T(x, y; \xi, \eta) f(\xi, \eta)\, d\xi\, d\eta$$

is a particular solution of the equation (A). Sufficient conditions that this may be true have been given by Levi and less restrictive conditions have been formulated by Gevrey. The properties of the integrals

$$I(x, y) = \int_C T(x, y; \xi, \eta)\, \phi(\eta)\, d\eta, \quad J(x, y) = \int_C \frac{\partial T}{\partial x} \phi(\eta)\, d\eta$$

have also been studied, where C is a curve running from a point on the line $y = y_1$ to a point on the line $y = y$. It appears that when the point P crosses the curve C at a point P_0 the integral J suffers a discontinuity indicated by the formula

$$\lim_{P \to P_0} (J_P - J_{P_0}) = \pm \phi_{P_0} \sqrt{\pi},$$

the sign being + or − according as P approaches P_0 from the right or the left of the curve C.

In this formula ϕ denotes any continuous function and a suffix P is used to denote the value of a function of position at the point P.

A Green's function for the region S may be defined by the formula

$$G(x, y; \xi, \eta) = T(x, y; \xi, \eta) - H(x, y; \xi, \eta),$$

where $H(x, y; \xi, \eta)$, which satisfies the equation $\frac{\partial^2 H}{\partial \xi^2} + \frac{\partial H}{\partial \eta} = 0$, is zero on γ when considered as a function of ξ and η, which is regular and which takes the same values as $T(x, y; \xi, \eta)$ on the curves C_1 and C_2. The function G satisfies the two equations $\frac{\partial^2 G}{\partial x^2} = \frac{\partial G}{\partial y}$, $\frac{\partial^2 G}{\partial \xi^2} + \frac{\partial G}{\partial \eta} = 0$, is zero when $x = c_1(y)$, when $\xi = c_1(\eta)$, when $x = c_2(y)$ and when $\xi = c_2(\eta)$, and is positive in S. With the aid of this function a formula

$$2z(x, y)\sqrt{\pi} = -\int_{C_1 + C_2} z(\xi, \eta) \frac{\partial G}{\partial \xi} d\eta + \int_{A_1 A_2} G(x, y; \xi, y_1)\, z(\xi, y_1)\, d\xi$$
$$- \iint_S G(x, y; \xi, \eta) f(\xi, \eta)\, d\xi\, d\eta$$

may be given for a solution of (A) which takes assigned values on β. The problem of determining G is reduced by Gevrey to the solution of some integral equations.

Fundamental solutions of the equations

$$\frac{\partial z}{\partial y}=\frac{\partial^3 z}{\partial x^3}, \quad \frac{\partial^2 z}{\partial y^2}=\frac{\partial^3 z}{\partial x^3}$$

have been obtained by H. Block* and have been used by E. Del Vecchio† to obtain solutions of the equations

$$\frac{\partial z}{\partial y}-\frac{\partial^3 z}{\partial x^3}=f(x, y), \quad \frac{\partial^2 z}{\partial y^2}-\frac{\partial^3 z}{\partial x^3}=f(x, y).$$

EXAMPLES

1. Show by means of the substitutions

$$\xi - x = 2\sqrt{(st)}, \quad y - \eta = t$$

that the integral

$$\iint_S (x-\xi)^p (y-\eta)^{-q} \exp[-(x-\xi)^2/4(y-\eta)] f(\xi, \eta)\, d\xi\, d\eta$$

has a meaning when $p+1>0$ and $p-2q+3>0$, f being an integrable function.

[E. E. Levi.]

2. Show that by means of a transformation of variables

$$x' = x'(x, y), \quad y' = \pm y$$

the parabolic equation $\quad \dfrac{\partial^2 z}{\partial x^2}+a\dfrac{\partial z}{\partial x}+b\dfrac{\partial z}{\partial y}+cz+f=0$

may be reduced to the canonical form

$$\frac{\partial^2 z}{\partial x'^2}-\frac{\partial z}{\partial y'}=a\frac{\partial z}{\partial x'}+cz+f.$$

Show also that the term $\dfrac{\partial u}{\partial x'}$ may be removed by making a substitution of type $z=uv$ and that the term involving u will disappear at the same time if

$$\frac{\partial^2 a}{\partial x'^2}=a\frac{\partial a}{\partial x'}\frac{\partial a}{\partial y'}+2\frac{\partial c}{\partial x'}.$$

3. If a, b and c are continuous functions in a region R a solution of

$$\frac{\partial^2 z}{\partial x^2}+a\frac{\partial z}{\partial x}+b\frac{\partial z}{\partial y}+cz=0 \qquad (b<0)$$

which is regular in R can have neither a positive maximum nor a negative minimum. Hence show that there is only one solution of the equation which is regular in R and has assigned values at points of a closed curve C lying entirely within R.

[M. Gevrey.]

§ 2·23. *Green's theorem for a general linear differential equation of the second order.* Let the independent variables $x_1, x_2, \dots x_m$ be regarded as rectangular co-ordinates in a space of m dimensions. The derivatives of

* *Arkiv f. Mat., Ast. och Fysik*, vol. VII (1912), vol. VIII (1913), vol. IX (1913).

† *Mem. d. R. Accad. d. Sc. di Torino* (2), vol. LXVI (1916).

a function u with respect to the co-ordinates may be indicated by suffixes written outside a bracket, thus $(u)_2$ stands for $\frac{\partial u}{\partial x_2}$ and $(u)_{23}$ for $\frac{\partial^2 u}{\partial x_2 \partial x_3}$.

We now consider the differential equation

$$L(u) \equiv \sum_{r=1}^{m} \sum_{s=1}^{m} A_{rs}(u)_{rs} + \sum_{r=1}^{m} B_r(u)_r + Fu = 0,$$

$$A_{rs} = A_{sr},$$

where the coefficients A_{rs}, B_r, F are functions of $x_1, x_2, \ldots x_m$.

The expression $\bar{L}(v)$ adjoint to $L(u)$ is

$$\bar{L}(v) = \sum_{r=1}^{m} \sum_{s=1}^{m} (A_{rs}v)_{rs} - \sum_{r=1}^{m} (B_r v)_r + Fv,$$

and we have the identity

$$vL(u) - u\bar{L}(v) = \sum_{r=1}^{m} (Q_r)_r,$$

where $$Q_r = -u \sum_{s=1}^{m} A_{rs}(v)_s + v \sum_{s=1}^{m} A_{rs}(u)_s + uv\left[B_r - \sum_{s=1}^{m} (A_{rs})_s\right].$$

The n-dimensional form of the theorem for transforming a surface integral into a volume integral may be written in the form

$$\iint \sum_{r=1}^{m} (Q_r)_r \, dx_1 \ldots dx_m = -\iint \sum_{r=1}^{m} n_r Q_r dS,$$

where $n_1, n_2, \ldots n_m$ are the direction cosines of the normal to the hyper-surface S, the normal being drawn into the region of integration. Hence we have the equation

$$\iint [vL(u) - u\bar{L}(v)] \, dx_1 \ldots dx_n = -\iint \{v D_n u - uD_n v - uvP_n\} \, dS,$$

where $$D_n(u) = \sum_{r=1}^{m} \sum_{s=1}^{m} n_s A_{rs}(u)_r,$$

$$P_n = \sum_{r=1}^{m} \left[\sum_{s=1}^{m} (A_{rs})_s - B_r\right].$$

Let us write $$\sum_{s=1}^{m} n_s A_{rs} = \Lambda \nu_r,$$

where $\nu_1, \nu_2, \ldots \nu_m$ are the direction cosines of a line which may be called the conormal*, then

$$D_n(u) = \Lambda \sum_{r=1}^{m} \nu_r (u)_r = \Lambda (u)_\nu.$$

§ **2·24**. *The characteristics of a partial differential equation of the second order*. Let the values of the first derivatives $(u)_1, (u)_2, \ldots (u)_m$ be given at points of the hypersurface $\theta(x_1, x_2, \ldots x_m) = 0$. If $dx_1, dx_2, \ldots dx_m$ are increments connected by the equation

$$(\theta)_1 dx_1 + (\theta)_2 dx_2 + \ldots (\theta)_m dx_m = 0,$$

* This is a term introduced by R. d'Adhémar.

and if $(\theta)_1 \neq 0$, we may regard the increments $dx_2, dx_3, \dots dx_m$ as arbitrary, and since

$$d\,[(u)_p] = (u)_{p1}\,dx_1 + (u)_{p2}\,dx_2 + \dots (u)_{pm}\,dx_m,$$

$$(\theta)_1\,d\,[(u)_p] = [(\theta)_1\,(u)_{p2} - (\theta)_2\,(u)_{p1}]\,dx_2 + \dots [(\theta)_1\,(u)_{pm} - (\theta)_m\,(u)_{p1}]\,dx_m,$$

the quantities $$(\theta)_1\,(u)_{ps} - (\theta)_s\,(u)_{p1}$$

may be regarded as known. Similarly the quantities

$$(\theta)_1\,(u)_{1p} - (\theta)_p\,(u)_{11}$$

may be regarded as known and so the quantities

$$(\theta)_1\,(\theta)_1\,(u)_{ps} - (\theta)_p\,(\theta)_s\,(u)_{11}$$

may be regarded as known. Substituting the values of $(u)_{ps}$ in the partial differential equation

$$L\,(u) = \sum_{r=1}^{m} \sum_{s=1}^{m} A_{rs}\,(u)_{rs} + \sum_{r=1}^{m} B_r\,(u)_r + Fu = 0, \qquad \text{......(I)}$$

we see that we have a linear equation to determine $(u)_{11}$ in which the coefficient of $(u)_{11}$ is

$$A \equiv \sum_{r=1}^{m} \sum_{s=1}^{m} A_{rs}\,(\theta)_r\,(\theta)_s. \qquad \text{......(II)}$$

If this quantity is different from zero the equation determines $(u)_{11}$ uniquely, but if the quantity A is zero the equation fails to determine $(u)_{11}$ and the derivatives (u) are likewise not determined. In this case the hypersurface $\theta\,(x_1, x_2, \dots x_m) = 0$ is called a characteristic and the differential equation $A = 0$ is called the partial differential equation of the characteristics.

The equations of Cauchy's characteristics for this partial differential equation of the first order are

$$\frac{dx_1}{\Sigma A_{1s}\,(\theta)_s} = \frac{dx_2}{\Sigma A_{2s}\,(\theta)_s} = \dots = \frac{dx_m}{\Sigma A_{ms}\,(\theta)_s},$$

and these are called the bicharacteristics of the original partial differential equation. All the bicharacteristics passing through a point $(x_1^0, x_2^0, \dots x_m^0)$ generate a hypersurface or conoid with a singular point at $(x_1^0, x_2^0, \dots x_m^0)$. When all the quantities A_{pq} are constants this conoid is identical with the characteristic cone which is tangent to all the characteristic hypersurfaces through the point $(x_1^0, x_2^0, \dots x_m^0)$.

For the theory of characteristics of equations of higher order reference may be made to papers by Levi* and Sannia†. These authors have also considered multiple characteristics and Sannia gives a complete classification of linear partial differential equations in two variables of orders up to 5.

* E. E. Levi, *Ann. di Mat.* (3 *a*), vol. XVI, p. 161 (1909).

† G. Sannia, *Mem. d. R. Acc. di Torino* (2), vol. LXIV (1914); vol. LXVI (1916)

§ 2·25. *The classification of partial differential equations of the second order.* A partial differential equation with real coefficients is said to be of elliptic type when the quadratic form

$$\Sigma\Sigma A_{rs} X_r X_s$$

is always positive except when $X_1 = X_2 = \ldots = X_n = 0$.

The use of the words elliptic, hyperbolic and parabolic seems natural in the case $n = 2$, the term to be used depending upon the nature of the conic

$$A_{11} X_1^2 + 2A_{12} X_1 X_2 + A_{22} X_2^2 = 1.$$

For a non-linear equation $F(r, s, t, p, q, z) = 0$

$$\left[r = \frac{\partial^2 z}{\partial x^2},\ s = \frac{\partial^2 z}{\partial x \partial y},\ t = \frac{\partial^2 z}{\partial y^2},\ p = \frac{\partial z}{\partial x},\ q = \frac{\partial z}{\partial y} \right],$$

there is a similar classification depending on the nature of the quadratic form

$$X_1^2 \frac{\partial F}{\partial r} + 2X_1 X_2 \frac{\partial F}{\partial s} + X_2^2 \frac{\partial F}{\partial t}.$$

When $n > 2$ the classification is not so simple; for instance, when $n = 3$ it may be based on the different types of quadric surface and it is known that there are two different types of hyperboloid.

The word ellipsoidal might be used in this case instead of elliptic, but it seems better to use the same term for all values of n because the important question from the standpoint of the theory of partial differential equations is whether the equation is or is not of elliptic type. For an equation of elliptic type the characteristics are all imaginary and this fact has a marked influence on the properties of the solutions of the equation. When $n = 2$ typical equations of the three types are

$$\frac{\partial^2 u}{\partial x^2} + \frac{\partial^2 u}{\partial y^2} + a \frac{\partial u}{\partial x} + b \frac{\partial u}{\partial y} + cu = 0 \text{ (elliptic)},$$

$$\frac{\partial^2 u}{\partial x \partial y} + a \frac{\partial u}{\partial x} + b \frac{\partial u}{\partial y} + cu = 0 \text{ (hyperbolic)},$$

$$\frac{\partial^2 u}{\partial x^2} + a \frac{\partial u}{\partial x} + b \frac{\partial u}{\partial y} + cu = 0 \text{ (parabolic)}.$$

A notable difference between elliptic and hyperbolic equations arises when a solution is required to assume prescribed values at points of a closed curve and be regular within the curve. For illustration let us consider the case when the curve is the circle $x^2 + y^2 = 1$. If the boundary condition is $V = \sin 2n\theta$ when $x = \cos\theta$, $y = \sin\theta$, where n is a positive integer, there is no solution of the equation $\dfrac{\partial^2 V}{\partial x \partial y} = 0$ which is continuous $(D, 1)$ and single-valued within and on the circle*, but there is a regular

* When $n = 1$ there is a solution $V = 2y(1 - y^2)^{\frac{1}{2}}$ which satisfies the boundary condition but its derivative $\dfrac{\partial V}{\partial y}$ is infinite on the circle.

solution of $\frac{\partial^2 V}{\partial x^2}+\frac{\partial^2 V}{\partial y^2}=0$, namely, $V=r^{2n}\sin 2n\theta$. On the other hand, if the boundary condition is $V=\sin(2n+1)\theta$, there is a solution of $\frac{\partial^2 V}{\partial x \partial y}=0$ of type $V=f(y)$ which satisfies the conditions and is single-valued and continuous in the circle, but this solution is not unique because $V=1-x^2-y^2$ is a solution of $\frac{\partial^2 V}{\partial x \partial y}=0$ which is zero on the circle and single-valued and continuous inside the circle.

When the solution of a problem is not unique or when there is some uncertainty regarding the existence of a solution the problem may be regarded as not having been formulated correctly. An important property of the boundary problems of mathematical physics is that the correct formulation of the problem is indicated by the physical requirements in nearly every case.

§ 2·26. *A property of equations of elliptic type.* Picard*, Bernstein† and Lichtenstein‡ have shown that the solutions of certain general differential equations of elliptic type cannot have maximum or minimum values in the interior of a region within which they are regular. This property, which has been known for a long time for the case of Laplace's equation, has been proved recently in the following elementary way§.

Let
$$L(u)\equiv\sum_{1,1}^{n,n} A_{\mu\nu}(u)_{\mu\nu}+\sum_{1}^{n} B_\nu (u)_\nu$$
be a partial differential equation of the second order whose coefficients $A_{\mu\nu}$, B_ν are continuous functions of the co-ordinates $(x_1, x_2, \dots x_n)$ of a point P of an n-dimensional region T. For convenience we shall sometimes use a symbol such as $u(P)$ to denote a quantity which depends on the co-ordinates of the point P. We can then state the following theorem:

If $u(P)$ is continuous $(D, 2)$ and satisfies the inequality $L(u)\geqslant 0$ everywhere in T, an inequality of type $u(P)\leqslant u(P_0)$ can only be satisfied throughout T, where P_0 is a fixed internal point, when the inequality reduces to the equality $u(P)=u(P_0)$. Similarly, if $L(u)\leqslant 0$ throughout T, the inequality $u(P)\geqslant u(P_0)$ in T implies that $u(P)=u(P_0)$.

The proof will be given for the case $n=2$ so that we can use the familiar terminology of plane geometry, but the method is perfectly general.

Let us suppose that $L(u)\geqslant 0$ in T and that $u(P_0)=M$, while $u(P)\leqslant M$ if P is in T.

If $u\not\equiv M$ there will be a circle C within T such that at some point P of its boundary, say at P_1, we have $u(P_1)=M$, whilst in the interior of the circle $u<M$.

* E. Picard, *Traité d'Analyse*, t. II, 2nd ed., p. 29 (Paris, 1905).

† S. Bernstein, *Math. Ann.* Bd. LIX, S. 69 (1904).

‡ L. Lichtenstein, *Palermo Rend.* t. XXXIII, p. 211 (1912); *Math. Zeitschr.* vol. XX, p. 205 (1924).

§ E. Hopf, *Berlin. Sitzungsber.* S. 147 (1927).

Let K be a circular realm of radius R whose circular boundary touches C internally at P, then with the exception of the point P_1, we have everywhere in K the inequality $u < M$.

Next let a circle K_1 of radius $R_1 < R$ be drawn so as to lie entirely within T. The boundary of K_1 then consists of an arc S_i (the end points included) which belongs to K and an arc S_0 which does not belong to K. On S_i we have the inequality $u \leqslant M - \epsilon$, where ϵ is a suitable small quantity, while on S_0 we have $u \leqslant M$. ......(A)

We now choose the centre of K as origin and consider the function

$$h(P) = e^{-\alpha r^2} - e^{-\alpha R^2},$$

where $r^2 = x^2 + y^2$ and $\alpha > 0$. If $x_1 = x$, $x_2 = y$ and

$$L(u) = Au_{xx} + 2Bu_{xy} + Cu_{yy} + Du_x + Eu_y,$$

a simple calculation gives

$$e^{\alpha r^2} L(h) = 4\alpha^2 (Ax^2 + 2Bxy + Cy^2) - 2\alpha (A + C + Dx + Ey).$$

Since the equation is of elliptic type we have in the interior and on the boundary of K

$$Ax^2 + 2Bxy + Cy^2 \geqslant k > 0,$$

where k is a suitable constant. By choosing a sufficiently large value of α we can make

$$L(h) > 0$$

in K_1 and so

$$L(u + \delta h) > 0$$

if $\delta > 0$. We have, moreover, $h(P) < 0$ when P is on S_0, $h(P_1) = 0$. ......(B)

We now put $v(P) = u(P) + \delta . h(P)$, $\delta > 0$, where δ is also chosen so small that, in view of (A), we have $v < M$ on S_i. On account of (A) and (B) we have further $v < M$ on S_0. Hence $v < M$ on the whole of the boundary of K_1 and at the centre we have $v = M$. Thus v should have a maximum value at some point in the interior of K_1. This, however, may be shown to be incompatible with the inequality $L(v) \geqslant 0$, for at a place where v is a maximum we have by the usual rule of the differential calculus

$$v_{xx}\lambda^2 + 2v_{xy}\lambda\mu + v_{yy}\mu^2 \leqslant 0$$

for arbitrary real values of λ and μ. Now, by hypothesis,

$$A\lambda^2 + 2B\lambda\mu + B\mu^2 \geqslant 0,$$

therefore by the theorem of Paraf and Fejér (§ 1·35)

$$Av_{xx} + 2Bv_{xy} + Bv_{yy} \leqslant 0.$$

But the expression on the left-hand side is precisely $L(v)$ since $v_x = v_y = 0$, and so we have $L(v) \leqslant 0$ which is incompatible with $L(v) > 0$.

The case in which $L(u) \leqslant 0$, $u(P) \geqslant u(P_0)$ can be treated in a similar way.

In particular, if $L(u) = 0$ in T, where u is not constant, neither of the inequalities $u(P) \leqslant u(P_0)$, $u(P) \geqslant u(P_0)$ can hold throughout T when P is an internal point. This means that $u(P)$ cannot have a maximum or minimum value in the interior of a region T within which it is regular.

This theorem has been extended by Hopf to the case in which the functions A, B, C, D, E are not continuous throughout T but are bounded functions such that an inequality of type

$$A\lambda^2 + 2B\lambda\mu + C\mu^2 \geqslant N(\lambda^2 + \mu^2) > 0$$

holds, with a suitable value of the constant N, for all real values of λ and μ and for all points P in T.

The work of Picard has also been generalised by Moutard* and Fejér†. The latter gives the theorem the following form:

Let

$$\sum_{1,\dots 1}^{n,\dots n} a_{ik}(x_1, x_2, \dots x_n)(u)_{ik} + \sum_{1}^{n} b_r(x_1, x_2, \dots x_n)(u)_r + c(x_1, \dots x_n)\,u = 0,$$

$$a_{ik}(x_1, x_2, \dots x_n) \equiv a_{ki}(x_1, x_2, \dots x_n)$$

be a homogeneous linear partial differential equation of the second order with real independent variables $x_1, x_2, \dots x_n$ and a real unknown function $u(x_1, x_2, \dots x_n)$. The coefficients

$$a_{ik}(x_1, x_2, \dots x_n), \quad b_r(x_1, x_2, \dots x_n), \quad c(x_1, x_2, \dots x_n)$$

are all real functions which can be expanded in convergent power series of types

$$c(x_1, x_2, \dots x_n) = c + c_1x_1 + \dots c_nx_n + c_{11}x_1^2 + \dots,$$

$$b_r(x_1, x_2, \dots x_n) = b_r + b_{r1}x_1 + \dots b_{rn}x_n + b_{r11}x_1^2 + \dots.$$

$$a_{ik}(x_1, x_2, \dots x_n) = a_{ik} + a_{ik1}x_1 + \dots,$$

for

$$|x_1| \leqslant z_1, \quad |x_2| \leqslant z_2, \dots |x_n| \leqslant z_n,$$

where $z_1, z_2, \dots z_n$ are suitable constants. Then, if

$$\sum_{1,1}^{n,n} a_{ik}y_iy_k \geqslant 0$$

for all real values of $y_1, y_2, \dots y_n$, that is, if the quadratic form is non-negative, and if $c < 0$, the differential equation has no solution which is regular at the origin and has there either a negative minimum or a positive maximum. If, however, the quadratic form is not negative, that is, if

$$\sum_{1,1}^{n,n} a_{ik}\eta_i\eta_k < 0$$

for some set of values $\eta_1, \eta_2, \dots \eta_k$, there is always a solution regular at the origin which, if $c < 0$, has either a negative minimum or a positive maximum. Thus when $c < 0$ the requirement that the quadratic form

* Th. Moutard, *Journ. de l'École Polytechnique*, t. LXIV, p. 55 (1894); see also A. Paraf, *Annales de Toulouse*, t. VI, H, p. 1 (1892).

† *Loc. cit.*

should not be of the non-negative type is a necessary and sufficient condition for the existence of a negative minimum or positive maximum at the origin for some regular solution of the differential equation.

§ 2·31. *Green's theorem for Laplace's equation.* Let us now write in equation (A) of § 2·11

$$X = U\frac{\partial V}{\partial x}, \quad Y = U\frac{\partial V}{\partial y}, \quad Z = U\frac{\partial V}{\partial z},$$

the equation then takes the form

$$\int \left(U\frac{\partial V}{\partial n}\right) dS = \int U\nabla^2 V\, d\tau + \int \left(\frac{\partial U}{\partial x}\frac{\partial V}{\partial x} + \frac{\partial U}{\partial y}\frac{\partial V}{\partial y} + \frac{\partial U}{\partial z}\frac{\partial V}{\partial z}\right) d\tau,$$

where the symbol $\frac{\partial V}{\partial n}$ is used for the normal component of ∇V,

$$\frac{\partial V}{\partial n} = (\nabla V)_n \equiv l\frac{\partial V}{\partial x} + m\frac{\partial V}{\partial y} + n\frac{\partial V}{\partial z}.$$

Interchanging U and V we have likewise

$$\int \left(V\frac{\partial U}{\partial n}\right) dS = \int V\nabla^2 U\, d\tau + \int \left(\frac{\partial U}{\partial x}\frac{\partial V}{\partial x} + \frac{\partial U}{\partial y}\frac{\partial V}{\partial y} + \frac{\partial U}{\partial z}\frac{\partial V}{\partial z}\right) d\tau.$$

Subtracting we obtain Green's theorem,

$$\int \left(U\frac{\partial V}{\partial n} - V\frac{\partial U}{\partial n}\right) dS = \int (U\nabla^2 V - V\nabla^2 U)\, d\tau.$$

In this equation the functions U and V are supposed to be continuous $(D, 2)$ within the region over which the volume integrals are taken. This supposition is really too restrictive but it will be replaced later by one which is not quite so restrictive. If the functions U and V are solutions of the same differential equation and one which has the same characteristic as Laplace's equation, an interesting result is obtained. In particular, if

$$\left.\begin{aligned} \nabla^2 U + k^2 U &= 0, \\ \nabla^2 V + k^2 V &= 0, \end{aligned}\right\} \qquad \text{......(B)}$$

where k is either a constant or a function of x, y and z, the volume integral vanishes and we have the relation

$$\int \left(U\frac{\partial V}{\partial n} - V\frac{\partial U}{\partial n}\right) dS = 0.$$

In the special case when the surface S is a sphere and

$$U = f_m(r)\, Y_m\left(\frac{x}{r}, \frac{y}{r}, \frac{z}{r}\right), \quad V = f_n(r)\, Y_n\left(\frac{x}{r}, \frac{y}{r}, \frac{z}{r}\right),$$

where Y_m and Y_n are functions which are continuous over the sphere and $f_m(r), f_n(r)$ are functions such that $f_m(r) f_n'(r) \neq f_m'(r) f_n(r)$ on the sphere we obtain the important integral relation

$$\int Y_m Y_n dS = 0,$$

which implies that the functions Y_m form an orthogonal system.

An appropriate set of such functions will be constructed in § 6·34.

When k is a constant the equations (B) may be derived from the wave-equations $\Box^2 u = 0$, $\Box^2 v = 0$ by supposing that u and v have the forms

$$u = U \sin(kct + \alpha), \quad v = V \sin(kct + \beta)$$

respectively. When k is real these wave-functions are periodic. When $k = 0$, U and V are solutions of Laplace's equation.

The equation may also be derived from the equation of the conduction of heat,

$$\frac{\partial v}{\partial t} = \kappa \nabla^2 v, \qquad \text{......(C)}$$

by supposing that this possesses a solution of type $v = e^{-\kappa k^2 t} V(x, y, z)$.

Green's theorem is particularly useful for proofs of the uniqueness of the solution of a boundary problem for one of our differential equations. Suppose, for instance, that we wish to find a solution of Laplace's equation which is continuous $(D, 2)$ within the region bounded by the surface S and which takes an assigned value $F(x, y, z)$ on the boundary of S. If there are two such solutions U and V the difference $W = U - V$ will be a solution of Laplace's equation which is zero on the boundary and continuous $(D, 2)$ within the region bounded by S. Green's theorem now gives

$$0 = \int W \frac{\partial W}{\partial n} dS = \int \left[\left(\frac{\partial W}{\partial x}\right)^2 + \left(\frac{\partial W}{\partial y}\right)^2 + \left(\frac{\partial W}{\partial z}\right)^2 \right] d\tau,$$

and this equation implies that

$$\frac{\partial W}{\partial x} = 0, \quad \frac{\partial W}{\partial y} = 0, \quad \frac{\partial W}{\partial z} = 0;$$

W is consequently constant and therefore equal to its boundary value zero. Hence $U = V$ and the solution of the problem is unique. A similar conclusion may be drawn if the boundary condition is $\frac{\partial W}{\partial n} = 0$ or $\frac{\partial W}{\partial n} + hW = 0$, where h is positive. If the equation is $\nabla^2 V + \lambda V = 0$ instead of Laplace's equation the foregoing argument still holds when λ is negative, for we have the additional term $-\lambda \int W^2 d\tau$ on the right-hand side. The argument breaks down, however, when λ is positive.

In the case of the equation of heat conduction (C) there are some similar theorems relating to the uniqueness of solutions. If possible, let there be two independent solutions v_1, v_2 of the equation $\frac{\partial v}{\partial t} = \kappa \nabla^2 v$ and the supplementary conditions

$v = f(x, y, z)$ for $t = 0$ for points within S,

$v = \phi(x, y, z, t)$ on S $(t \geqslant 0)$,

v continuous $(D, 2)$ within region bounded by S.

Let $V = v_1 - v_2$, then $V = 0$ for $t = 0$ within S, $V = 0$ on S.

Putting
$$2I = \int V^2 d\tau,$$
we have
$$\frac{\partial I}{\partial t} = \int V \frac{\partial V}{\partial t} d\tau = \kappa \int V(\nabla^2 V)\, d\tau$$
$$= \kappa \int V \frac{\partial V}{\partial n} dS - \kappa \int \left[\left(\frac{\partial V}{\partial x}\right)^2 + \left(\frac{\partial V}{\partial y}\right)^2 + \left(\frac{\partial V}{\partial z}\right)^2\right] d\tau.$$

Since $V = 0$ on S the first integral vanishes, and so
$$\frac{\partial I}{\partial t} = -\kappa \int \left[\left(\frac{\partial V}{\partial x}\right)^2 + \left(\frac{\partial V}{\partial y}\right)^2 + \left(\frac{\partial V}{\partial z}\right)^2\right] d\tau \quad (t \geqslant 0).$$

But $I = 0$ for $t = 0$, therefore $I \leqslant 0$, but on the other hand the integral for I indicates that $I \geqslant 0$, consequently we must have $I = 0$, $V = 0$.

These theorems prove the uniqueness of solutions of certain boundary problems but they do not show that such solutions exist. Many existence theorems have been established by the methods of advanced analysis and the literature on this subject is now very extensive.

§ **2·32.** *Green's functions.* The solution of a problem in which a solution of Laplace's equation or a periodic wave-function is to be determined from a knowledge of its behaviour at certain boundaries can be made to depend on that of another problem—the determination of the appropriate Green's function*.

Let $G(x, y, z; x_1, y_1, z_1)$ be a solution of $\nabla^2 G + k^2 G = 0$ with the following properties: It is finite and continuous $(D, 2)$ with respect to either x, y, z or x_1, y_1, z_1 in a region bounded by a surface S, except in the neighbourhood of the point (x_1, y_1, z_1) where it is infinite like $R^{-1} \cos kR$ as $R \to 0$, R being the distance between the points (x, y, z) and (x_1, y_1, z_1). At the surface S, some boundary condition such as (1) $G = 0$, (2) $\partial G/\partial n = 0$, or (3) $\partial G/\partial n + hG = 0$ is satisfied, h being a positive constant.

Adopting the notation of Plemelj† and Kneser‡, we shall denote the values of a function $\phi(\xi, \eta, \zeta)$ at the points (x, y, z), (x_1, y_1, z_1) respectively by the symbols $\phi(0)$ and $\phi(1)$.

When a function like the Green's function depends upon the co-ordinates of both points it will be denoted by a symbol such as $G(0, 1)$. The importance of the Green's function depends chiefly upon the following theorem:

Let U be a solution of
$$\nabla^2 U + k^2 U + 4\pi f(x, y, z) = 0, \qquad \ldots\ldots(A)$$

* G. Green, *Math. Papers*, p. 31.

† *Monatshefte für Math. u. Phys.* Bd. xv, S. 337 (1904).

‡ A. Kneser, *Die Integralgleichungen und ihre Anwendungen in der mathematischen Physik* (Vieweg, Brunswick, 1911).

which is finite and continuous $(D, 2)$ throughout the interior of a region $\mathscr{D}$ bounded by a surface S and let $f(x, y, z)$ be a function which is finite and continuous throughout $\mathscr{D}$. We shall also allow $f(x, y, z)$ to be finite and continuous throughout parts of the region and zero elsewhere.

Applying Green's theorem to the region between a small sphere whose centre is at (x_1, y_1, z_1) and the surface S, we have

$$\int (U\nabla^2 G - G\nabla^2 U)\, d\tau_0 = -\int\left(U\frac{\partial G}{\partial n} - G\frac{\partial U}{\partial n}\right) d\Sigma_0 + \int\left(U\frac{\partial G}{\partial n} - G\frac{\partial U}{\partial n}\right) dS.$$

Now $\nabla^2 G = -k^2 G$ and the first integral on the right may be found by a simple extension of the analysis already used in a similar case when $G = 1/R$, consequently we have the equation

$$4\pi U(1) = 4\pi \int G(0, 1) f(0)\, d\tau_0 + \int\left(U\frac{\partial G}{\partial n} - G\frac{\partial U}{\partial n}\right) dS_0. \quad \text{......(B)}$$

If U satisfies the same boundary conditions as G on the surface S the surface integral vanishes and we have*

$$U(1) = +\int G(0, 1) f(0)\, d0. \qquad \text{......(C)}$$

If, on the other hand, $f(x, y, z) = 0$ and $G = 0$ on S we have

$$4\pi U(1) = \int U(0)\frac{\partial G}{\partial n}\, dS_0, \qquad \text{......(D)}$$

the value of U is thus determined completely and uniquely by its boundary values. Similarly, if the boundary condition is $\frac{\partial G}{\partial n} = 0$ on S we have

$$4\pi U(1) = -\int \frac{\partial U}{\partial n} G(0, 1)\, dS_0, \qquad \text{......(E)}$$

and the value of U is determined by the boundary values of $\partial U/\partial n$.

Finally, if the boundary condition satisfied by G is $\frac{\partial G}{\partial n} + hG = 0$, we have

$$4\pi U(1) = -\iint\left(\frac{\partial U}{\partial n} + hU\right) G(0, 1)\, dS_0, \qquad \text{......(F)}$$

and U is expressed in terms of the boundary values of $\frac{\partial U}{\partial n} + hU$.

If $g(x_2, y_2, z_2; x, y, z)$ is the Green's function for the same boundary condition as $G(0, 1)$ but for the value σ of k, we must also surround the

* It has not been proved that whenever the function f is finite and continuous throughout D the formula (C) gives a solution of (A). Petrini has shown in fact that when f is merely continuous the second derivatives of the integral may not exist or may not be finite. *Acta Math.* t. XXXI, p. 127 (1908). It should be remarked that Gauss in 1840 derived Poisson's equation (§ 2·61) on the supposition that the density function f is continuous $(D, 1)$. With this supposition (C) does give a solution of (A). Poisson's equation and the solution of (A) are usually derived now for the case of a function f which satisfies a Holder condition. See Kellogg's *Foundations of Potential Theory*, ch. VI.

point (x_2, y_2, z_2) by a small sphere when we apply Green's theorem with $U(x, y, z) = g(2, 0)$. We then obtain the equation

$$g(2, 1) = G(2, 1) - \frac{(k^2 - \sigma^2)}{4\pi} \int g(2, 0)\, G(0, 1)\, d\tau_0. \qquad \text{......(G)}$$

This may be regarded as an integral equation for the determination of $g(2, 0)$ when $G(0, 1)$ is known or for the determination of $G(0, 1)$ when $g(2, 1)$ is known. In some cases the Green's function for Laplace's equation $(k = 0)$ can be found and then the integral equation can be used to calculate $g(2, 0)$ or to establish its existence. The Green's function for Laplace's equation, when it exists, is unique, for if $G(0, 1)$, $H(0, 1)$ were two different Green's functions the function

$$V(0) = G(0, 1) - H(0, 1)$$

would be continuous $(D, 2)$ throughout the region bounded by the surface S and satisfy the boundary condition that was assigned, but such a function is known to be zero.

For small values of σ^2 the function $g(2, 0)$ can be obtained by expanding it in the form

$$g(2, 0) = a(2, 0) + \sigma^2 b(2, 0) + \dots. \qquad \text{......(H)}$$

The first term is the corresponding Green's function for Laplace's equation and is known, the other terms may be obtained successively by substituting the series in the integral equation (with $k = 0$) and equating coefficients of the different powers of σ^2. So long as the series converges this method gives a unique value of $g(2, 0)$. The value of $g(2, 0)$, if it exists, will certainly not be unique when σ^2 has a singular or characteristic value for which the "homogeneous integral equation" has a solution ϕ which is different from zero. In this case

$$4\pi\phi(1) = (\sigma^2 - k^2) \int \phi(0)\, G(0, 1)\, d\tau_0, \qquad \text{......(I)}$$

and the formula (C) indicates that this function $\phi(0) = U(x, y, z)$ is a solution of $\nabla^2 U + \sigma^2 U = 0$, which satisfies the assigned boundary conditions and the other conditions imposed on U. The solutions of this type are of great importance in many branches of mathematical physics, particularly in the theory of vibrations, and have been discussed by many writers.

The characteristic values of σ^2 are called Eigenwerte by the Germans and the corresponding functions ϕ Eigenfunktionen. These terms are now being used by American writers, but it seems worth while to shorten them and use eit in place of Eigenwert and eif in place of Eigenfunktion. The same terms may be used also in connection with the homogeneous integral equation (I). In discussing this equation it is convenient, however, to put $k = 0$, so that G becomes the Green's function for Laplace's equation and the assigned boundary conditions. Denoting this function by the symbol $4\pi K(0, 1)$ we have the integral equation

$$\phi(1) = \sigma^2 \int \phi(0) K(0, 1) d\tau_0$$

for the determination of the solution of $\nabla^2\phi + \sigma^2\phi = 0$ and the assigned boundary conditions, that is, for the determination of the eifs and eits. The function $K(0, 1)$ is called the kernel of the integral equation; it has the important property of symmetry expressed by the equation

$$K(0, 1) = K(1, 0).$$

This may be seen as follows.

If we put $4\pi f(0) = (\sigma^2 - k^2) g(0, 2)$ in the formula (B) and proceed as before, Green's theorem gives

$$g(1, 2) = G(2, 1) - \frac{(k^2 - \sigma^2)}{4\pi} \int g(0, 2) G(0, 1) d\tau_0.$$

Putting $\sigma = k$ and comparing this equation with the previous one we obtain the desired relation. When $k \neq 0$ the relation gives

$$G(1, 2) = G(2, 1).$$

When the boundary condition is $\frac{\partial\phi}{\partial n} = 0$ this result is equivalent to one given by Helmholtz in the theory of sound. If $\psi(0)$ is an eif corresponding to an eit ν^2 which is different from σ^2 we have

$$\psi(1) = \nu^2 \int \psi(0) K(0, 1) d\tau_0,$$

and, if the order of integration can be changed,

$$\sigma^2 \int \phi(1) \psi(1) d\tau_1 = \sigma^2\nu^2 \iint \psi(0) K(0, 1) \phi(1) d\tau_0 d\tau_1$$

$$= \nu^2 \int \phi(0) \psi(0) d\tau_0 = \nu^2 \int \phi(1) \psi(1) d\tau_1.$$

Hence the eifs ϕ and ψ satisfy the orthogonal relation

$$\int \phi(1) \psi(1) d\tau_1 = 0 \qquad \nu^2 \neq \sigma^2.$$

This result may be used to prove that the eits σ^2 are all real. If, indeed, σ^2 were a complex quantity $\alpha + i\beta$ the corresponding eif $\phi(0)$ would also be a complex quantity $\chi(0) + i\omega(0)$, and since K is real the function $\psi(0) = \chi(0) - i\omega(0)$ would be an eif corresponding to the eit $\nu^2 = \alpha - i\beta$, and the orthogonal relation

$$0 = \int \phi(0) \psi(0) d\tau_0 = \int \{[\chi(0)]^2 + [\omega(0)]^2\} d\tau_0$$

would be satisfied. This, however, is impossible because the integrand is either zero for all values of the variables or positive for some, but it is zero only when

$$\chi(0) = \omega(0) = 0, \quad \phi(0) = 0.$$

To prove that the eits are all positive we make use of the equation

$$\sigma^2 \int U^2 d\tau = - \int U \nabla^2 U \, d\tau = \int \left[\left(\frac{\partial U}{\partial x} \right)^2 + \left(\frac{\partial U}{\partial y} \right)^2 + \left(\frac{\partial U}{\partial z} \right)^2 \right] d\tau.$$

The Green's function is usually found in practice by finding the eifs and eits directly from the differential equation and then writing down a suitable expansion for G in terms of these eifs. The question of convergence is, however, a difficult one which needs careful study. The method has been used with considerable success by Heine in his *Kugelfunktionen*, by Hilbert and his co-workers, by Sommerfeld, Kneser and Macdonald.

§ **2·33**. *Partial difference equations.* The partial difference equations analogous to the partial differential equations satisfied by conjugate functions are

$$u_x = v_{\bar{y}}, \quad u_y = - v_{\bar{x}},$$

and these lead to the equations of § 1·62

$$u_{x\bar{x}} + u_{y\bar{y}} = 0, \quad v_{x\bar{x}} + v_{y\bar{y}} = 0,$$

which are analogous to Laplace's equation. These difference equations have been used in recent years to find approximate solutions of Laplace's equation when certain boundary conditions are prescribed* and also to establish the existence of a solution corresponding to prescribed boundary conditions.

Let us consider, for instance, a square whose sides are $x = \pm 2h$, $y = \pm 2h$ and let us introduce the abbreviations

$$a = h, \quad b = 2h, \quad \alpha = - h, \quad \beta = - 2h, \quad u(x, y) = (xy),$$
$$u(b, y) = (y), \quad u(\beta, y) = (\bar{y}); \quad u(x, b) = [x], \quad u(x, \beta) = [\bar{x}],$$

then we have eight non-homogeneous equations (n-h-equations)

$$-(0a) - (\alpha 0) + 4(\alpha a) = (\bar{a}) + [\alpha], \quad (0a) + (a0) - 4(aa) = (a) + [a],$$
$$-(\alpha 0) - (0\alpha) + 4(\alpha\alpha) = (\bar{\alpha}) + [\bar{\alpha}], \quad (0\alpha) + (a0) - 4(a\alpha) = (\alpha) + [\bar{a}],$$
$$(\alpha a) + (00) + (\alpha\alpha) - 4(\alpha 0) = -(\bar{0}), \quad (aa) + (00) + (a\alpha) - 4(a0) = -(0),$$
$$(\alpha a) + (00) + (aa) - 4(0a) = -[0], \quad (\alpha\alpha) + (00) + (a\alpha) - 4(0\alpha) = -[\bar{0}],$$

and one homogeneous equation (h-equation)

$$(0a) + (a0) + (0\alpha) + (\alpha 0) = 4(00).$$

The first step in the solution is to eliminate the quantities (aa), (αa), $(a\alpha)$, $(\alpha\alpha)$ which do not occur in the h-equation. This gives the equations

$$4(00) + (0a) + (0\alpha) - 14(\alpha 0) + (\bar{a}) + (\bar{\alpha}) + [\alpha] + [\bar{\alpha}] + 4(\bar{0}) = 0,$$
$$4(00) + (0a) + (0\alpha) - 14(a0) + (a) + (\alpha) + [a] + [\bar{a}] + 4(0) = 0,$$
$$4(00) + (\alpha 0) + (a0) - 14(0a) + (\bar{a}) + [\alpha] + (a) + [a] + 4[0] = 0,$$
$$4(00) + (\alpha 0) + (a0) - 14(0\alpha) + (\alpha) + [\bar{\alpha}] + (\bar{\alpha}) + [\bar{a}] + 4[\bar{0}] = 0.$$

* L. F. Richardson, *Phil. Trans.* A, vol. ccx, p. 307 (1911); *Math. Gazette* (July, 1925).

Adding these equations we have

$$-16(00) + 12(a0) + 12(0a) + 12(\alpha 0) + 12(0\alpha)$$
$$= 2(a) + 2(\bar{a}) + 2(\alpha) + 2(\bar{\alpha}) + 2\lfloor a\rfloor + 2\lceil\bar{a}\rceil + 2\lfloor\alpha\rfloor + 2\lceil\bar{\alpha}\rceil$$
$$+ 4(0) + 4(\bar{0}) + 4[0] + 4[\bar{0}].$$

Combining this with the homogeneous equation we see that the quantity on the right-hand side of the last equation is equal to $+32(00)$ and so the quantity (00) is obtained uniquely.

Similarly, if the sides of the square are $x = \pm 3h$, $y = \pm 3h$ there are 16 n-h-equations and 9 h-equations which may be solved by first eliminating the quantities which do not occur in the h-equations. We have then to solve 9 linear equations in order to obtain the remaining quantities, but these 9 equations may be treated in exactly the same way as the previous set of 9 linear equations, quantities being eliminated which do not occur in the central equation. In this way a value is finally found for (00).

A similar method may be used for a more general type of square network or lattice. Let the four points $(x + h, y)$, $(x - h, y)$, $(x, y + h)$, $(x, y - h)$ be called the neighbours of the point (x, y) and let the lattice L consist of interior points P, each of which has four neighbours belonging to the lattice, and boundary points Q, each of which has at least one neighbour belonging to the lattice and at least one neighbour which does not belong to the lattice. A chain of lattice points $A_1, A_2, \ldots A_{n+1}$ is said to be connected when A_{s+1} is one of the neighbours of A_s for each value of s in the series $1, 2, \ldots n$. A lattice L is said to be connected when any two of its points belong to a connected chain of lattice points, whether the two points are interior points or boundary points. The lattice has a simple boundary when any two boundary points belong to a chain for which no two consecutive points are internal points and no internal point P is consecutive to two boundary points having the same x or the same y as P.

The solubility of the set of linear equations represented by the equation

$$u_{x\bar{x}} + u_{y\bar{y}} = 0 \qquad \text{......(A)}$$

for such a lattice may be inferred from the fact that this set of linear equations is associated with a certain quadratic form

$$h^2 \Sigma (u_x{}^2 + u_y{}^2),$$

where the summation extends over all the lattice points, and a difference quotient associated with a boundary point is regarded as zero when a point not belonging to the lattice would be needed for its definition. This summation can, by the so-called Green's formula, be expressed in the form

$$-h^2 \sum_P u(u_{x\bar{x}} + u_{y\bar{y}}) - h \sum_Q uR(u), \qquad \text{......(B)}$$

where the boundary expression $R(u)$ associated with a boundary point u_Q is defined by the equation

$$hR(u_Q) = u_1 + u_2 + \ldots u_s - su_Q,$$

where $u_1, u_2, \ldots u_s$ are the s neighbouring points of u_Q ($s \leqslant 3$). Since $u_{x\bar{x}} + u_{y\bar{y}} = 0$ the quadratic form can be expressed in terms of boundary values. If there were two solutions of the partial difference equation with the same boundary values, the foregoing identity could be applied to their difference $u - v$, and since the boundary values of $u - v$ are all zero the identity would give the relation

$$\Sigma \left[(u_x - v_x)^2 + (u_y - v_y)^2\right] = 0,$$

which implies that $u_x - v_x = 0$, $u_y - v_y = 0$ for all points (x, y) of the lattice; consequently, since $u - v$ is zero on the boundary it must be zero throughout the lattice.

There is another identity

$$0 = h^2 \sum_P (vu_{x\bar{x}} + vu_{y\bar{y}} - uv_{x\bar{x}} - uv_{y\bar{y}}) + h \sum_Q \left[vR(u) - uR(v)\right]$$

which, when applied to the case in which $u_{x\bar{x}} + u_{y\bar{y}} = 0$ and the boundary value of v is zero, gives

$$\begin{aligned}\Sigma \left[(u_x + v_x)^2 + (u_y + v_y)^2\right] &= -\sum_P (u+v)(v_{x\bar{x}} + v_{y\bar{y}}) - h^{-1} \sum_Q (u) \left[R(v) + R(u)\right] \\ &= -h^{-1} \sum_Q \left[vR(u) - uR(v)\right] + \Sigma (v_x^2 + v_y^2 + u_x^2 + u_y^2) \\ &\quad + h^{-1} \sum_Q uR(u) - h^{-1} \sum_Q \left[uR(u) + uR(v)\right] \\ &= \Sigma (v_x^2 + v_y^2 + u_x^2 + u_y^2) \\ &\geqslant \Sigma (u_x^2 + u_y^2),\end{aligned}$$

the transformations being made with the aid of Green's formula (B).

This equation shows that the solution of $u_{x\bar{x}} + u_{y\bar{y}} = 0$ and the prescribed boundary condition gives the least possible value to the quadratic form. The system of linear equations $u_{x\bar{x}} + u_{y\bar{y}} = 0$ can, indeed, be obtained by writing down the conditions that the quadratic form should be a minimum when the boundary values of u are assigned.

With a change of notation the quadratic form may be written in the form

$$\sum_{m=1}^{N} \sum_{n=1}^{N} c_{mn} u_m u_n - 2 \sum_{n=1}^{N} a_n u_n + b,$$

where the quadratic form is never negative. The corresponding set of linear equations

$$\begin{aligned} c_{11} u_1 + c_{12} u_2 + \ldots c_{1N} u_N &= a_1, \\ \ldots\ldots\ldots\ldots\ldots\ldots\ldots\ldots\ldots\ldots \\ c_{N1} u_1 + c_{N2} u_2 + \ldots c_{NN} u_N &= a_N \end{aligned}$$

has a determinant $| c_{mn} |$ which is not zero and so can be solved.

For the sake of illustration we consider a lattice in which the internal lattice points are represented in the diagram by the corresponding values of the variable u and the boundary lattice points by corresponding values denoted by v's.

$$\begin{array}{cccc} & v_0 & v_1 & \\ v_{10} & u_0 & u_1 & v_2 \\ v_9 & u_3 & u_2 & v_3 \\ v_8 & u_4 & u_5 & v_4 \\ & v_7 & v_6 & \end{array}$$

The quadratic form is in this case

$$(v_0-u_0)^2+(v_1-u_1)^2+(u_0-u_3)^2+(u_1-u_2)^2+(u_3-u_4)^2+(u_2-u_5)^2+(u_4-v_7)^2,$$
$$(u_5-v_6)^2+(u_1-u_0)^2+(u_0-v_{10})^2+(u_2-u_3)^2+(v_2-u_1)^2+(u_3-v_9)^2,$$
$$(v_3-u_2)^2+(u_4-v_8)^2+(u_5-u_4)^2+(v_4-u_5)^2,$$

and it is easy to see that the equations obtained by differentiating with respect to u_1, u_2 respectively are

$$4u_1 = u_0 + u_2 + v_1 + v_2, \quad 4u_2 = u_1 + u_3 + u_5 + v_3,$$

and are of the required type. The quadratic form is, moreover, equal to the sum of the quantities

$$u_0(4u_0-u_1-u_3-v_{10}-v_0)+u_1(4u_1-u_0-u_2-v_1-v_2)+u_2(4u_2-u_1-u_3-u_5-v_3),$$
$$u_3(4u_3-u_0-u_2-u_4-v_9)+u_4(4u_4-u_3-u_5-v_7-v_8)+u_5(4u_5-u_2-u_4-v_6-v_4),$$
$$v_0(v_0-u_0)+v_1(v_1-u_1)+v_2(v_2-u_1)+v_3(v_3-u_2)+v_4(v_4-u_5)$$
$$+v_6(v_6-u_5)+v_7(v_7-u_4)+v_8(v_8-u_4)+v_9(v_9-u_3)+v_{10}(v_{10}-u_0).$$

§ **2·34**. *The limiting process*†. We assume that G is a simply connected region in the xy-plane with a boundary Γ formed of a finite number of arcs with continuously turning tangents. If v is an integrable function defined within G we shall use the symbol $G\{v\}$ to denote the integral of v over the area G and a similar notation will be used for integrals of v over portions of G which are denoted by capital letters.

Let G_h be the lattice region associated with the mesh-width h and the region G, and let the symbol $G_h[v]$ be used to denote the sum of the values of v over the lattice points of G. Also let the symbol $\Gamma_h(v)$ be used for the sum of the values of v over the boundary points which form the boundary Γ_h of G_h. This notation will be used also for a portion of G_h denoted by a capital letter and for the lattice region $G_h{}^*$ belonging to a partial region G^* of G.

Now let $f(x, y)$ be a given function which is continuous $(D, 2)$ in a region enclosing G and let $u(x, y)$ be the solution of (A) which takes the same value as $f(x, y)$ at the boundary points of G_h. We shall prove that as $h \to 0$ the function $u_h(x, y)$ converges towards a function $u(x, y)$ which

† R. Courant, K. Friedrichs and H. Lewy, *Math. Ann.* vol. c, p. 32 (1928). See also J. le Roux, *Journ. de Math.* (6), vol. x, p. 189 (1914); R. G. D. Richardson, *Trans. Amer. Math. Soc.* vol. xviii, p. 489 (1917); H. B. Phillips and N. Wiener, *Journ. Math. and Phys. Mass. Inst. Tech.* vol. ii, p. 105 (1923).

satisfies the partial differential equation $\nabla^2 u = 0$ and takes the same value as $f(x, y)$ at each of the points of Γ. We shall further show that for any region lying entirely within G the difference quotients of u_h of arbitrary order tend uniformly towards the corresponding partial derivatives of $u(x, y)$.

In the convergence proof it is convenient to replace the boundary condition $u = f$ on Γ by the weaker requirement that

$$S_r\{(u - f)^2\} \to 0 \quad \text{as } r \to 0,$$

where S_r is that strip of G whose points are at a distance from Γ smaller than r.

The convergence proof depends on the fact that for any partial region G^* lying entirely within G, the function $u_h(x, y)$ and each of its difference quotients remains bounded and uniformly continuous as $h \to 0$, where uniform continuity is given the following meaning:

There is for any of these functions $w_h(x, y)$ a quantity $\delta(\epsilon)$, depending only on the region and not on h, such that if $w_h^{(P)}$ denote the value of the function at the point P we have the inequality

$$| w_h^{(P)} - w_h^{(P_1)} | < \epsilon$$

whenever the two lattice points P and P_1 of the lattice region G_h lie in the same partial region and are separated by a distance less than $\delta(\epsilon)$.

As soon as the foregoing type of uniform continuity has been established we can in a well-known manner† select from our functions u_h a partial sequence of functions which tend uniformly in any partial region G^* towards a limit function $u(x, y)$ while the difference quotients of u_h tend uniformly towards that of $u(x, y)$ differential coefficients. The limit function then possesses derivatives of order n in any partial region G^* of G and satisfies $\nabla^2 u = 0$ in this region. If we can also show that u satisfies the boundary condition we can regard it as the solution of our boundary problem for the region G. Since this solution is uniquely determined, it appears then that not only a partial sequence of the functions u_h but this sequence of functions itself possesses the desired convergence property as $h \to 0$.

The uniform continuity of our quantities may be established by proving the following lemmas:

(1) As $h \to 0$ the sums $h^2 G_h[u^2]$ and $h^2 G_h[u_x^2 + u_y^2]$ remain bounded.

(2) If $w = w_h$ satisfies the difference equation (A) at a lattice point of G_h and if, as $h \to 0$ the sum $h^2 G_h^*[w^2]$, extended over a lattice region G_h^* associated with a partial region G^* of G, remains bounded, then for any fixed partial region G^{**} lying entirely within G^* the sum

$$h^2 G_h^{**}[(w_x^2 + w_y^2)]$$

† See for instance, Kellogg's *Foundations of Potential Theory*, p. 265. The theorem to be used is known as Ascoli's theorem; it is discussed in § 4·45.

over the lattice region G_h^{**} associated with G^{**}, likewise remains bounded as $h \to 0$. When this is combined with (1) it follows that, because all the difference quotients w of the function u_h also satisfy the difference equation (A), each of the sums $h^2 G_h^* [w^2]$ is bounded.

(3) From the boundedness of these sums it follows that the difference quotients themselves are bounded and uniformly continuous as $h \to 0$.

The proof of (1) follows from the fact that the functional values u_h are themselves bounded. For the greatest (or least) value of the function is assumed on the boundary† and so tends towards a prescribed finite value. The boundedness of the sum $h^2 G_h [u_x^2 + u_y^2]$ is an immediate consequence of the minimum property of our lattice function which gives in particular

$$h^2 G_h [u_x^2 + u_y^2] \leqslant h^2 G_h [f_x^2 + f_y^2],$$

but as $h \to 0$ the sum on the right tends to $G\left\{\left(\frac{\partial f}{\partial x}\right)^2 + \left(\frac{\partial f}{\partial y}\right)^2\right\}$, which, by hypothesis, exists.

To prove (2) we consider the sum $h^2 Q_1 [w_x^2 + w_{\bar{x}}^2 + w_y^2 + w_{\bar{y}}^2]$, where the summation extends over all the interior points of a square Q_1. Now Green's formula gives

$$h^2 Q_1 [w_x^2 + w_{\bar{x}}^2 + w_y^2 + w_{\bar{y}}^2] = \sum_1 (w^2) - \sum_0 (w^2),$$

where Σ_1 and Σ_0 are respectively the boundary of Q_1 and the square boundary of the lattice points lying within Σ_1.

We now consider a series of concentric squares $Q_0, Q_1, \ldots Q_N$ with the boundaries $\Sigma_0, \Sigma_1, \ldots \Sigma_N$. Applying our formula to each of these squares and observing that we have always

$$2h^2 Q_0 [w_x^2 + w_y^2] \leqslant h^2 Q_k [w_x^2 + w_{\bar{x}}^2 + w_y^2 + w_{\bar{y}}^2] \qquad (k \geqslant 1),$$

we obtain

$$2h^2 Q_0 [w_x^2 + w_y^2] \leqslant \sum_{k+1} (w^2) - \sum_k (w^2) \qquad (0 \leqslant k < n).$$

Adding n inequalities of this type we obtain

$$2nh^2 Q_0 [w_x^2 + w_y^2] \leqslant \sum_n (w^2) - \sum_0 (w^2) \leqslant \sum_n (w^2).$$

Summing this inequality from $n = 1$ to $n = N$ we get

$$N^2 h^2 Q_0 [w_x^2 + w_y^2] \leqslant Q_N [w^2].$$

Diminishing the mesh-width h we can make the squares Q_0 and Q_N converge towards two fixed squares lying within G and having corresponding sides separated by a distance a. In this process $Nh \to a$ and we have independently of the mesh-width

$$h^2 Q_0 [w_x^2 + w_y^2] \leqslant \frac{h^2}{a^2} Q_N [w^2].$$

† On account of equation (A) the value of u_h at an internal point is the mean of the values at the four neighbouring points and so cannot be greater than all of them, consequently the greatest value of u_h cannot occur at an internal point.

With a sufficiently small value of h this inequality holds with another constant a for any two partial regions of G, one of which lies entirely within the other. Hence the surmise in (2) is proved.

To prove that u_h and all its partial difference quotients w remain bounded and uniformly continuous as $h \to 0$ we consider a rectangle R with corners P_0, Q_0, P, Q and with sides P_0Q_0, PQ which are x-lines† of length a. Denoting these lines by the symbols X_0, X respectively we start from the formula

$$w^{Q_0} - w^{P_0} = hX(w_x) + h^2R[w_{xy}]$$

and the inequality

$$|w^{Q_0} - w^{P_0}| \leqslant hX(|w_x|) + h^2R[|w_{xy}|] \qquad \text{......(C)}$$

which is a consequence of it. We now let X vary continuously between an initial position X_1 at a distance b from X_0 and a final position X_2 at a distance $2b$ from X_0 and sum the $(b/h) + 1$ inequalities (C) associated with X's which pass through lattice points. We thus obtain the inequality

$$|w^{P_0} - w^{Q_0}| \leqslant \frac{h^2}{b+h} R_2[|w_x|] + h^2R_2[|w_{xy}|],$$

where the summations on the right are extended over the whole rectangle $P_0Q_0P_2Q_2$. By Schwarz's inequality it then follows that

$$|w^{P_0} - w^{Q_0}| \leqslant (2a/b)^{\frac{1}{2}}(h^2R_2[w_x^2])^{\frac{1}{2}} + (2ab)^{\frac{1}{2}}(h^2R_2[w_{xy}^2])^{\frac{1}{2}}.$$

Since, by hypothesis, the sums which occur here multiplied by h^2 remain bounded, it follows that as $a \to 0$ the difference $|w^{P_0} - w^{Q_0}| \to 0$ independently of the mesh-width, since for each partial region G^* of G the quantity b can be held fixed. Consequently, the uniform continuity of $w = w_h$ is proved for the x-direction. Similarly, it holds for the y-direction and so also for any partial region G^* of G. The boundedness of the function w_h in G^* finally follows from its uniform continuity and the boundedness of $h^2G^*[w_h^2]$.

By this proof we establish the existence of a partial sequence of functions u_h which converge towards a limit function $u(x, y)$ and are, indeed, continuous, together with all their difference quotients, in the sense already explained, for each inner partial region of G. This limit function $u(x, y)$ is thus continuous (D, n) throughout G, where n is arbitrary, and it satisfies the potential equation

$$\frac{\partial^2 u}{\partial x^2} + \frac{\partial^2 u}{\partial y^2} = 0.$$

In order to prove that the solution fulfils the boundary condition formulated above we shall first of all establish the inequality

$$h^2S_{r,h}[v^2] \leqslant Ar^2h^2S_{r,h}[v_x^2 + v_y^2] + Brh\Gamma_h(v^2), \qquad \text{......(D)}$$

where $S_{r,h}$ is that part of the lattice region G_h which lies within the

† This term is used here to denote lines parallel to the axis of x.

boundary strip S_r, which is bounded by Γ and another curve Γ_r. The constants A, B depend only on the region and not on the function v or the mesh-width h.

To do this we divide the boundary Γ of G into a finite number of pieces for which the angle of the tangent with either the x or y-axis is greater than 30°. Let γ, for instance, be a piece of Γ which is sufficiently steep (in the above sense) relative to the x-axis. The x-lines through the end-points of the piece γ cut out on Γ_r a piece γ_r and together with γ and γ_r enclose a piece s_r of the boundary strip S_r. We use the symbol $s_{r,h}$ to denote the portion of G_h contained in s_r and denote the associated portion of the boundary Γ_h by γ_h.

We now imagine an x-line to be drawn through a lattice point P_h of $S_{r,h}$. Let it meet the boundary γ_h in a point $\bar{P}_h$. The portion of this x-line X_h which lies in $s_{r,h}$ we call $p_{r,h}$. Its length is certainly smaller than cr, where the constant c depends only on the smallest angle of inclination of a tangent of γ to the x-axis.

Now between the values of v at P_h and $\bar{P}_h$ we have the relation

$$v^{P_h} = v^{\bar{P}_h} \pm hX_h\,(v_x).$$

Squaring both sides and applying Schwarz's inequality, we obtain

$$(v^{P_h})^2 \leqslant 2\,(v^{\bar{P}_h})^2 + 2crhp_{r,h}\,(v_x^2).$$

Summing with respect to P_h in the x-direction, we get

$$hp_{r,h}\,(v^2) \leqslant 2cr\,(v^{\bar{P}_h})^2 + 2c^2r^2hp_{r,h}\,(v_x^2).$$

Summing again in the y-direction we obtain the relation

$$hs_{r,h}\,[v^2] \leqslant 2cr\Gamma_h\,(v^{\bar{P}_h})^2 + 2c^2r^2S_{r,h}\,[v_x^2].$$

Writing down the inequalities associated with the other portions of Γ and adding all the inequalities together we obtain the desired inequality (D).

By similar reasoning we can also establish the inequality

$$h^2G_h\,[v^2] \leqslant c_1h\Gamma_h\,(v^2) + c_2h^2G_h\,[v_x^2 + v_y^2]$$

in which the constants c_1, c_2 depend only on the region G and not on the mesh division.

We now put $v_h = u_h - f_h$ so that $v_h = 0$ on Γ_h.

Then, since $h^2G_h\,[v_x^2 + v_y^2]$ remains bounded as $h \to 0$, we obtain from (D)

$$(h^2/r)\,S_{r,h}\,[v^2] \leqslant \kappa r, \qquad \text{......(E)}$$

where κ is a constant which does not depend on the function v or the mesh-width. Extending the sum on the left to the difference $S_{\nu,h} - S_{\rho,h}$ of two boundary strips, the inequality (E) still holds with the same constant κ and we can pass to the limit $h \to 0$.

From the inequality (D) we then get

$$(1/r)\,S\,[v^2] \leqslant \kappa r,$$

where $S = S_r - S_\rho$ and $v = u - f$. Now letting $\rho \to 0$, we obtain the inequality

$$(1/r)\, S_r\, [v^2] \equiv (1/r)\, S_r\, [(u-f)^2] \leqslant \kappa r,$$

which signifies that the limit function satisfies the prescribed boundary condition.

§ 2·41. *The derivation of physical equations from a variational principle.* A concise expression may be given to the principles from which an equation or set of equations is derived by using the ideas of the "Calculus of Variations*." This expression is useful for several purposes. In the first place a few methods are now available for the direct solution of problems in the "Calculus of Variations" and these can sometimes be used with advantage when the differential equations are hard to solve. Secondly, when an integral's first variation furnishes the desired physical equations the expression under the integral sign may be used with advantage to obtain a transformation of the physical equations to a new set of co-ordinates, for the transformation of the integral is generally much easier than the transformation of the differential equations and the transformed equations can generally be derived from the transformed integral by the methods of the "Calculus of Variations," that is, by the Eulerian rule

To illustrate the method we consider the variation of the integral

$$I = \frac{1}{2}\iiint \left[\left(\frac{\partial V}{\partial x}\right)^2 + \left(\frac{\partial V}{\partial y}\right)^2 + \left(\frac{\partial V}{\partial z}\right)^2\right] dx\,dy\,dz$$

when the dependent variable V is alone varied and its variation is chosen so that it vanishes on the boundary of the region of integration. We have

$$\delta I = \iiint \left(\frac{\partial V}{\partial x}\,\delta\frac{\partial V}{\partial x} + \frac{\partial V}{\partial y}\,\delta\frac{\partial V}{\partial y} + \frac{\partial V}{\partial z}\,\delta\frac{\partial V}{\partial z}\right) dx\,dy\,dz.$$

Now by a fundamental property of the signs of variation and differentiation

$$\delta\frac{\partial V}{\partial x} = \frac{\partial}{\partial x}(\delta V), \text{ etc.}$$

Hence
$$\delta I = \iiint \left[\frac{\partial V}{\partial x}\frac{\partial}{\partial x}(\delta V) + \ldots\right] dx\,dy\,dz$$

$$= \iint \delta V \frac{\partial V}{\partial n}\, dS - \iiint \delta V \,.\, \nabla^2 V\, dx\,dy\,dz.$$

The surface integral vanishes because $\delta V = 0$ on the boundary, consequently the first variation δI vanishes altogether if V satisfies everywhere the differential equation

$$\nabla^2 V = 0.$$

The condition $\delta V = 0$ on the boundary means that as far as the possible variations of V are concerned V is specified on the boundary. It is easily

* The reader may obtain a clear grasp of the fundamental ideas from the monograph of G. A. Bliss, "Calculus of Variations," *The Carus Mathematical Monographs* (1925).

seen that a function V with the specified boundary values gives a smaller value of I when it is a solution of $\nabla^2 V = 0$, regular within the region, than if it is any other regular function having the assigned values on the boundary.

In the foregoing analysis it is tacitly assumed that $\nabla^2 V$ exists and is such that the transformation from the volume integral to the surface integral is valid. If V is *assumed* to be continuous $(D, 2)$ there is no difficulty but, as Du Bois-Reymond pointed out*, it is not evident that a function V which makes $\delta I = 0$ does have second derivatives. This difficulty, which has been emphasised by Hadamard† and Lichtenstein‡, has been partly overcome by the work of Haar §. There are in fact some sufficient conditions which indicate when the derivation of the differential equation of a variation problem is permissible.

For the corresponding variation problem in one dimension there is a very simple lemma which leads immediately to the desired result. The variation problem is

$$\delta \int_{x_1}^{x_2} \left(\frac{dV}{dx}\right)^2 dx = 0,$$

where x_1 and x_2 are constants and δV is supposed to be zero for $x = x_1$ and for $x = x_2$.

Writing $\dfrac{dV}{dx} = M$, $\delta V = U$ we have to show that if

$$\int_{x_1}^{x_2} M \frac{dU}{dx} = 0 \qquad \text{......(A)}$$

for all admissible functions U which satisfy the conditions

$$U(x_1) = U(x_2) = 0 \qquad \text{......(B)}$$

then M is a constant (Du Bois-Reymond's Lemma).

To prove the lemma we consider the particular function

$$U(x) = (x_2 - x) \int_{x_1}^{x} M(\xi)\, d\xi - (x - x_1) \int_{x}^{x_2} M(\xi)\, d\xi,$$

which satisfies (B) and gives at any point x where $M(x)$ is continuous

$$\begin{aligned} \frac{dU}{dx} &= (x_2 - x_1)\, M(x) - \int_{x_1}^{x_2} M(\xi)\, d\xi \\ &= (x_2 - x_1)\, [M(x) - c], \text{ say.} \end{aligned}$$

* P. du Bois-Reymond, *Math. Ann.* vol. XV, pp. 283, 564 (1879).

† J. Hadamard, *Comptes Rendus*, vol. CXLIV, p. 1092 (1907).

‡ L. Lichtenstein, *Math Ann.* vol. LXIX, p. 514 (1910).

§ A. Haar, *Journ. für Math.* Bd. CXLIX, S. 1 (1919); *Szeged Acta*, t. III, p. 224 (1927). Haar shows that in the case of a two-dimensional variation problem the equation $\delta I = 0$ leads to a pair of simultaneous equations of the first order in which there is an auxiliary function W whose elimination would lead to the Eulerian differential equation if the necessary differentiations could be performed. Many inferences may, however, be derived directly from the simultaneous equations without an appeal to the Eulerian equation.

We shall now assume that $M(x)$ is continuous bit by bit (piecewise continuous) so that this equation holds in the interval (x_1, x_2) except possibly at a finite number of points. The functions $M(x)$, $\frac{dU}{dx}$ are then undoubtedly integrable over the range (x_1, x_2) and, on account of the end conditions (B) satisfied by $U(x)$, we may write (A) in the form

$$\int_{x_1}^{x_2} [M(x) - c] \frac{dU}{dx} dx = 0.$$

With the value adopted for U this equation becomes

$$\int_{x_1}^{x_2} [M(x) - c]^2 dx = 0,$$

and implies that $M(x) = c$, hence $\frac{dM}{dx} = 0$ and $\frac{d^2V}{dx^2} = 0$.

An extension of this analysis to the three-dimensional case is difficult. To avoid this difficulty it is customary* to limit the variation problem and to consider only functions that are continuous $(D, 2)$ throughout the region of integration. The function V and the comparison function $V + U$ are supposed to belong to the *field of functions* with the foregoing property. The problem is to find, if possible, a function V of the field such that δI is zero whenever U belongs to the field and is zero on the boundary of the region of integration.

Even when the problem is presented in this restricted form a lemma is needed to show that V necessarily satisfies the differential equation. We have, in fact, to show that if

$$\iiint U . \nabla^2 V \, dx dy dz = 0,$$

for all admissible functions U, then $\nabla^2 V = 0$.

The nature of the proof may be made clear by considering the one-dimensional case. We then have the equations

$$\int_{x_1}^{x_2} \phi(x) \, U(x) \, dx = 0,$$

$$U(x_1) = 0, \quad U(x_2) = 0,$$

and the conditions:

$U(x)$ is continuous $(D, 2)$, $\qquad \phi(x)$ is continuous in (x_1, x_2).

Since $U(x)$ is otherwise arbitrary we may choose the particular function

$$U(x) = (x - a)^4 (b - x)^4 \qquad x_1 < a < x < b < x_2$$
$$= 0 \qquad \text{otherwise.}$$

If $\phi(x)$ were not zero throughout the interval (x_1, x_2) it would have a definite sign (positive, say) in some interval (a, b) contained within (x_1, x_2),

* See, for instance, Hilbert-Courant, *Methoden der Mathematischen Physik*, vol. I, p. 165.

but this is impossible because with the above form of $U(x)$ the integral $\int_{x_1}^{x_2} \phi(x) U(x) dx$ is positive.

To extend this lemma to the three-dimensional problem it is sufficient to consider a function $U(x, y, z)$ which has a form such as

$$(x - a_1)^4 (b_1 - x)^4 (y - a_2)^4 (b_2 - y)^4 (z - a_3)^4 (b_3 - z)^4,$$

within a small cube with (a_1, a_2, a_3), (b_1, b_2, b_3) as ends of a diagonal, the value of U outside the cube being zero.

In this way it can be shown that a field function V for which $\delta I = 0$ is *necessarily* a solution of $\nabla^2 V = 0$. The foregoing analysis does not prove, however, that such a function exists.

Similar analysis may be used to derive the equation $\nabla^2\phi + k^2\phi = 0$ from a variational principle in which

$$\delta \iiint L\, dx\, dy\, dz = 0,$$

where $$L = \frac{1}{2}\left[\left(\frac{\partial\phi}{\partial x}\right)^2 + \left(\frac{\partial\phi}{\partial y}\right)^2 + \left(\frac{\partial\phi}{\partial z}\right)^2 - k^2\phi^2\right].$$

When the potential ϕ is of the form $\frac{1}{r}\cos kr$ the volume integral is finite although the integral $\iiint k^2\phi^2$ is not*.

EXAMPLES

1. If $$I = \iint \left[\left(\frac{\partial V}{\partial x}\right)^2 - \left(\frac{\partial V}{\partial y}\right)^2\right] dx\, dy,$$
the equation $\delta I = 0$ may be satisfied by making $V = f(x + y) + g(x - y)$ where f and g have first derivatives but not necessarily second derivatives. [Hadamard.]

2. The variation problem
$$\delta \iint F(V_x, V_y, x, y)\, dx\, dy = 0$$
leads to the simultaneous equations
$$W_x = \frac{\partial F}{\partial V_y}, \quad W_y = -\frac{\partial F}{\partial V_x},$$
the suffixes x, y denoting differentiations with respect to these variables. [A. Haar.]

§ **2·42**. *The general Eulerian rule*. To formulate the general rule for finding the equations which express that the first variation of an integral is zero we consider the variation of an integral

$$I = \iint \dots \int L\, dx_1 dx_2 \dots dx_n,$$

where L is a function of certain quantities and their derivatives. For

* See, for instance, the remarks made by J. Lennard-Jones, *Proc. London Math. Soc.* vol. XX, p. 347 (1922).

brevity we use suffixes 1, 2, etc. to denote derivatives with respect to x_1, x_2, etc. If there are m quantities u, v, w, ... which are varied independently except for certain conditions at the boundary of the region of integration, there are m Eulerian equations which are all of type

$$0=\frac{\partial L}{\partial u}-\sum_{s=1}^{n}\frac{\partial}{\partial x_s}\left(\frac{\partial L}{\partial u_s}\right)+\frac{1}{2!}\sum_{s=1}^{n}\sum_{t=1}^{n}\frac{\partial^2}{\partial x_s\partial x_t}\left(\frac{\partial L}{\partial u_{st}}\right)\epsilon_{st}$$
$$-\frac{1}{3!}\sum_{r=1}^{n}\sum_{s=1}^{n}\sum_{t=1}^{n}\frac{\partial^3}{\partial x_r\partial x_s\partial x_t}\left(\frac{\partial L}{\partial u_{rst}}\right)\epsilon_{rst}+\ldots; \qquad \ldots\text{(A)}$$

these are often called the Euler-Lagrange differential equations, but for brevity we shall call them simply the Eulerian equations.

If $l_1, l_2, \ldots l_n$ are the direction cosines of the normal to an element of the boundary, the boundary conditions are of types

$$0=\sum_{s=1}^{n} l_s\left[\frac{\partial L}{\partial u_s}-\frac{1}{2!}\Sigma\frac{\partial}{\partial x_t}\left(\epsilon_{st}\frac{\partial L}{\partial u_{st}}\right)+\frac{1}{3!}\Sigma\frac{\partial^2}{\partial x_r\partial x_t}\left(\epsilon_{rst}\frac{\partial L}{\partial u_{rst}}\right)-\ldots\right],$$

$$0=\sum_{s=1}^{n} l_s\left[2!\,\epsilon_{st}\frac{\partial L}{\partial u_{st}}-3!\sum_{r=1}^{n}\epsilon_{rst}\frac{\partial}{\partial x_r}\left(\frac{\partial L}{\partial u_{rst}}\right)+\ldots\right],$$

....

$$0=\sum_{s=1}^{n} l_s\left[3!\,\epsilon_{stw}\frac{\partial L}{\partial u_{stw}}-4!\sum_{r=1}^{n}\epsilon_{rstw}\frac{\partial}{\partial x_r}\left(\frac{\partial L}{\partial u_{rstw}}\right)+\ldots\right].$$

There are m boundary conditions of the first type, mn boundary conditions of the second type, $\frac{1}{2}mn(n-1)$ boundary conditions of the third type, and so on. In these equations the coefficients ϵ_{st}, ϵ_{rst} are constants which are defined as follows:

$$\begin{aligned}\epsilon_{st} &= 1 \qquad s\neq t,\\ &= 2 \qquad s=t,\\ \epsilon_{rst} &= 1 \qquad r\neq s\neq t,\\ &= 2 \qquad r=s\neq t,\\ &= 6 \qquad r=s=t.\end{aligned}$$

The equations (A) are obtained by subjecting the integral to repeated integrations by parts until one part of the integral is an integral over the boundary and the other part is of type

$$\iint\ldots\int[U\,\delta u+V\,\delta v+W\,\delta w+\ldots]\,dx_1\ldots dx_n.$$

The equations

$$U=0,\quad V=0,\quad W=0,\ldots$$

are then the Eulerian differential equations*, while the boundary integral is

$$\iint dS\,[\bar{U}\,\delta u+\Sigma\,\bar{U}_t\delta u_t+\Sigma\,\bar{U}_{rt}\delta u_{rt}+\ldots],$$

and the boundary conditions are

$$\bar{U}=0,\quad \bar{U}_t=0\quad (t=1,2,\ldots),\quad \bar{U}_{rt}=0\quad (r=1,2,\ldots;\,t=1,2,\ldots).$$

* For general properties of the Eulerian equations see Ex. 2, p. 183, and the remarks at the end of the chapter.

Typical integrations by parts are

$$\delta u_{12}\frac{\partial L}{\partial u_{12}}=\frac{\partial}{\partial x_1}\left[\delta u_2\frac{\partial L}{\partial u_{12}}\right]-\delta u_2\frac{\partial}{\partial x_1}\left(\frac{\partial L}{\partial u_{12}}\right)$$
$$=\frac{\partial}{\partial x_1}\left[\delta u_2\frac{\partial L}{\partial u_{12}}\right]-\frac{\partial}{\partial x_2}\left[\delta u\frac{\partial}{\partial x_1}\left(\frac{\partial L}{\partial u_{12}}\right)\right]+\delta u\frac{\partial^2}{\partial x_1\partial x_2}\left(\frac{\partial L}{\partial u_{12}}\right)$$
$$=\frac{\partial}{\partial x_2}\left[\delta u_1\frac{\partial L}{\partial u_{12}}\right]-\frac{\partial}{\partial x_1}\left[\delta u\frac{\partial}{\partial x_2}\left(\frac{\partial L}{\partial u_{12}}\right)\right]+\delta u\frac{\partial^2}{\partial x_1\partial x_2}\left(\frac{\partial L}{\partial u_{12}}\right),$$
$$\delta u_{11}\frac{\partial L}{\partial u_{11}}=\frac{\partial}{\partial x_1}\left[\delta u_1\frac{\partial L}{\partial u_{11}}\right]-\frac{\partial}{\partial x_1}\left[\delta u\frac{\partial}{\partial x_1}\left(\frac{\partial L}{\partial u_{11}}\right)\right]+\delta u\frac{\partial^2}{\partial x_1^2}\left(\frac{\partial L}{\partial u_{11}}\right).$$

The reason for the introduction of the factors ϵ_{st} is now apparent.

When L depends only on a single quantity u and its first derivatives the Eulerian equation is of the second order. The variation problem is then said to be *regular* when this partial differential equation is of elliptic type. The distinction between regular and irregular variation problems becomes apparent when terms involving the square of δu are retained and the sign of the sum of these terms is investigated (Legendre's rule).

When a variation problem is irregular it is not certain that the boundary conditions suggested by the variation problem will be equivalent to those which are indicated by physical considerations.

For a physically correct variation problem a direct method of solution is often advantageous. The well-known method of Rayleigh and Ritz is essentially a method of approximation in which the unknown function is approximated by a finite series of functions, each of which satisfies the specified boundary conditions. The coefficients in the series are chosen so as to make $\delta I = 0$ when each coefficient is varied. The problem is thus reduced to an algebraic problem.

§ **2·431**. *The transformation of physical equations.* In searching for simple solutions of the partial differential equations of physics it is often useful to transform the equations to a new set of co-ordinates and to look for solutions which are simple functions of these co-ordinates. The necessary transformations can be made without difficulty by the rules of tensor analysis and the absolute calculus, but sometimes they may be obtained very conveniently by transforming to the new co-ordinates the integral which occurs in a variational problem from which they are derived. The principle which is used here is that the Eulerian equations which are derived from the transformed integral must be equivalent to the Eulerian equations which were derived from the original integral because each set of equations means the same thing, namely, that the first variation of the integral is zero. A formal proof of the general theorem of the covariance of the Eulerian equations can, of course, be given*, but in this book we shall

* L. Koschmieder, *Math. Zeits.* Bd. XXIV, S. 181, Bd. XXV, S. 74 (1926); Hilbert and Courant, *l.c.* p. 193.

regard this property of covariance as a postulate. It is well known, of course, that the postulate leads to the Lagrangian equations of motion in the simple case when the integral is of type

$$\int L dt,$$

where
$$L = f[q_1, q_2, \dots q_n; \dot{q}_1, \dot{q}_2, \dots \dot{q}_n]$$
$$= T - V,$$

the Lagrangian equations being of type

$$0 = \frac{\partial L}{\partial q_s} - \frac{d}{dt}\left(\frac{\partial L}{\partial \dot{q}_s}\right).$$

The quantity T here denotes the kinetic energy and V the potential energy. V is a function of the co-ordinates which specify a configuration of the dynamical system, while T is a positive quantity which depends on both the q's and their rates of change, which are denoted here by $\dot{q}$'s.

In the simple case when

$$T = \tfrac{1}{2}\sum_1^n \sum_1^n a_{rs}\dot{q}_r\dot{q}_s,$$

$$V = \tfrac{1}{2}\sum_1^n \sum_1^n c_{rs}q_r q_s,$$

where the coefficients a_{rs}, c_{rs} are constants, the covariance of the equations

$$\Sigma a_{rs}\ddot{q}_r = \Sigma c_{rs}q_r$$

is easily confirmed by considering a linear transformation of type

$$Q_1 = l_{11}q_1 + \dots l_{1n}q_n,$$
$$\dots\dots\dots\dots\dots\dots\dots\dots\dots$$
$$Q_n = l_{n1}q_1 + \dots l_{nn}q_n,$$

in which the coefficients l_{rs} are constants.

The advantage of making a transformation is well illustrated by this case, because when the transformation is chosen so that the expressions for T and V take the forms

$$T = \tfrac{1}{2}\Sigma A_s\dot{Q}_s^2,$$
$$V = \tfrac{1}{2}\Sigma C_s Q_s^2,$$

respectively, the Eulerian equations are simply

$$A_s\ddot{Q}_s + C_s Q_s = 0,$$

and indicate that there are solutions of type

$$Q_s = \alpha_s \cos(n_s t + \beta_s), \quad (n_s^2 A_s = C_s)$$

where α_s and β_s are arbitrary constants. These co-ordinates Q_s are called the normal co-ordinates for the dynamical problem.

Our object now is to see if there are corresponding sets of co-ordinates associated with a partial differential equation.

§ **2·432.** To transform Laplace's equation to new co-ordinates ξ, η, ζ such that

$$dx^2 + dy^2 + dz^2 = a\,d\xi^2 + b\,d\eta^2 + c\,d\zeta^2 + 2f\,d\eta\,d\zeta + 2g\,d\zeta\,d\xi + 2h\,d\xi\,d\eta,$$

where a, b, c, f, g, h, are functions of ξ, η, ζ, we use suffixes x, y, z to denote differentiations with respect to x, y, z and suffixes 1, 2, 3 to denote differentiations with respect to ξ, η, ζ. We then have

$$V_x = V_1\xi_x + V_2\eta_x + V_3\zeta_x,$$

$$dx\,dy\,dz = \frac{\partial(x, y, z)}{\partial(\xi, \eta, \zeta)}\,d\xi\,d\eta\,d\zeta = \frac{1}{J}\,d\xi\,d\eta\,d\zeta, \text{ say,}$$

$$\xi_x = \frac{\partial(\xi, y, z)}{\partial(x, y, z)} = \frac{\partial(\xi, y, z)}{\partial(\xi, \eta, \zeta)}\frac{\partial(\xi, \eta, \zeta)}{\partial(x, y, z)} = J(y_1z_2 - y_2z_1),$$

$$\begin{aligned}\xi_x^2 + \xi_y^2 + \xi_z^2 &= J^2[(y_2z_3 - y_3z_2)^2 + (z_2x_3 - z_3x_2)^2 + (x_2y_3 - x_3y_2)^2]\\ &= J^2[(x_2^2 + y_2^2 + z_2^2)(x_3^2 + y_3^2 + z_3^2)\\ &\qquad - (x_2x_3 + y_2y_3 + z_2z_3)^2]\\ &= J^2(bc - f^2) = J^2A, \text{ say,}\end{aligned}$$

$$\eta_x\zeta_x + \eta_y\zeta_y + \eta_z\zeta_z = J^2(gh - af) = J^2F, \text{ say.}$$

Therefore

$$\begin{aligned}I &= \tfrac{1}{2}\iiint J\,d\xi\,d\eta\,d\zeta\,[AV_1^2 + BV_2^2 + CV_3^2 + 2FV_2V_3 + 2GV_3V_1 + 2HV_1V_2]\\ &= \iiint L\,d\xi\,d\eta\,d\zeta.\end{aligned}$$

By Euler's rule $\delta I = 0$ when

$$\frac{\partial}{\partial\xi}\left(\frac{\partial L}{\partial V_1}\right) + \frac{\partial}{\partial\eta}\left(\frac{\partial L}{\partial V_2}\right) + \frac{\partial}{\partial\zeta}\left(\frac{\partial L}{\partial V_3}\right) = 0.$$

The new form of Laplace's equation is thus

$$\begin{aligned}DV \equiv \frac{\partial}{\partial\xi}\left[J\left(A\frac{\partial V}{\partial\xi} + H\frac{\partial V}{\partial\eta} + G\frac{\partial V}{\partial\zeta}\right)\right] + \frac{\partial}{\partial\eta}\left[J\left(H\frac{\partial V}{\partial\xi} + B\frac{\partial V}{\partial\eta} + F\frac{\partial V}{\partial\zeta}\right)\right]\\ + \frac{\partial}{\partial\zeta}\left[J\left(G\frac{\partial V}{\partial\xi} + F\frac{\partial V}{\partial\eta} + C\frac{\partial V}{\partial\zeta}\right)\right] = 0.\end{aligned}$$

If the original integral is

$$\tfrac{1}{2}\iiint[V_x^2 + V_y^2 + V_z^2 - \lambda V^2]\,dx\,dy\,dz, \qquad \text{......(A)}$$

the transformed integral is

$$\iiint L'\,d\xi\,d\eta\,d\zeta, \qquad \text{......(B)}$$

where $\qquad L' = L - \tfrac{1}{2}\lambda V^2/J,$

and so the equation $\qquad \nabla^2 V + \lambda V = 0,$

which is derived from (A) by Euler's rule, transforms into the equation

$$DV + \lambda V/J = 0$$

which is derived from (B) by Euler's rule. This shows that $\nabla^2 V$ transforms into

$$J \,.\, DV,$$

where

$$J^2 \begin{vmatrix} a & h & g \\ h & b & f \\ g & f & c \end{vmatrix} = 1.$$

This result was given by Jacobi* with the foregoing derivation. The particular case in which

$$dx^2 + dy^2 + dz^2 = \frac{d\xi^2}{h_1^{\,2}} + \frac{d\eta^2}{h_2^{\,2}} + \frac{d\zeta^2}{h_3^{\,2}}$$

was worked out by Lamé. The result is that

$$\nabla^2 V = h_1 h_2 h_3 \left[\frac{\partial}{\partial \xi}\left(\frac{h_1}{h_2 h_3}\frac{\partial V}{\partial \xi}\right) + \frac{\partial}{\partial \eta}\left(\frac{h_2}{h_3 h_1}\frac{\partial V}{\partial \eta}\right) + \frac{\partial}{\partial \zeta}\left(\frac{h_3}{h_1 h_2}\frac{\partial V}{\partial \zeta}\right)\right].$$

This result is of great importance and will be used in the succeeding chapters to find potential functions and wave-functions which are simple functions of polar co-ordinates, cylindrical co-ordinates and other co-ordinates which form an orthogonal system.

In the special case when

$$dx^2 + dy^2 + dz^2 = \kappa^2 (d\xi^2 + d\eta^2 + d\zeta^2),$$

Laplace's equation becomes

$$\frac{\partial}{\partial \xi}\left(\kappa \frac{\partial V}{\partial \xi}\right) + \frac{\partial}{\partial \eta}\left(\kappa \frac{\partial V}{\partial \eta}\right) + \frac{\partial}{\partial \zeta}\left(\kappa \frac{\partial V}{\partial \zeta}\right) = 0,$$

and implies that $\kappa^{\frac{1}{2}} V$ is a solution of

$$\frac{\partial^2 U}{\partial \xi^2} + \frac{\partial^2 U}{\partial \eta^2} + \frac{\partial^2 U}{\partial \zeta^2} = 0$$

if $\kappa^{\frac{1}{2}}$ is a solution of this equation.

Inversion is one transformation which satisfies the requirements, for in this case

$$x = \xi/R^2, \quad y = \eta/R^2, \quad z = \zeta/R^2,$$
$$R^2 = \xi^2 + \eta^2 + \zeta^2,$$
$$\kappa^2 = R^{-4}.$$

The inference is that if $F(x, y, z)$ is a solution of Laplace's equation, the function†

$$\frac{1}{r} F\left(\frac{x}{r^2}, \frac{y}{r^2}, \frac{z}{r^2}\right)$$

is also a solution. Another transformation which satisfies the requirements is

$$x' = \frac{ax}{y + iz}, \quad y' = \frac{r^2 - a^2}{2(y + iz)}, \quad z' = \frac{r^2 + a^2}{2i(y + iz)}.$$

* *Journ. für Math.* vol. XXXVI, p. 113 (1848). See also J. Larmor, *Camb. Phil. Trans.* vol. XII, p. 455 (1884); vol. XIV, p. 128 (1886). H. Hilton, *Proc. London Math. Soc.* (2), vol. XIX, Records of Proceedings, vii (1921). Some very general transformation formulae are given by V. Volterra, *Rend. Lincei*, ser. 4, vol. v, pp. 599, 630 (1889).

† This result was given by Lord Kelvin in 1845.

In this case

$$dx'^2 + dy'^2 + dz'^2 = \frac{a^2}{(y + iz)^2}[dx^2 + dy^2 + dz^2],$$

and we have the result that if $F(x, y, z)$ is a solution of Laplace's equation,

$$(y + iz)^{-\frac{1}{2}} F\left[\frac{ax}{y + iz}, \frac{r^2 - a^2}{2(y + iz)}, \frac{r^2 + a^2}{2i(y + iz)}\right]$$

is also a solution.

These two results may be extended to Laplace's equation in a space of n dimensions

$$\frac{\partial^2 V}{\partial x_1^2} + \frac{\partial^2 V}{\partial x_2^2} + \dots \frac{\partial^2 V}{\partial x_n^2} = 0.$$

If $F(x_1, x_2, \dots x_n)$ is a solution of this equation, and if

$$r^2 = x_1^2 + x_2^2 + \dots x_n^2,$$

then

$$r^{2-n} F\left(\frac{x_1}{r^2}, \frac{x_2}{r^2}, \dots \frac{x_n}{r^2}\right)$$

is also a solution*, and

$$(x_1 + ix_2)^{1-\frac{n}{2}} F\left[\frac{r^2 - a^2}{2(x_1 + ix_2)}, \frac{r^2 + a^2}{2i(x_1 + ix_2)}, \frac{ax_3}{x_1 + ix_2}, \dots \frac{ax_n}{x_1 + ix_2}\right]$$

is a second solution. We shall now use this to obtain Brill's theorem.

Putting $x_1 + ix_2 = t$, $x_1 - ix_2 = s$, the differential equation becomes

$$4\frac{\partial^2 V}{\partial s \partial t} + \frac{\partial^2 V}{\partial x_3^2} + \dots \frac{\partial^2 V}{\partial x_n^2} = 0,$$

and the result is that if $R^2 = x_3^2 + \dots x_n^2$, and if

$$F(s, t, x_3, \dots x_n)$$

is a solution, then

$$t^{1-\frac{n}{2}} F\left(\frac{st + R^2}{t}, -\frac{a^2}{t}, \frac{ax_3}{t}, \dots \frac{ax_n}{t}\right)$$

is also a solution. Now a particular solution is given by

$$V = e^{-\frac{s}{4\kappa}} U(t, x_3, x_4, \dots x_n),$$

where $U(t, x_3, x_4, \dots x_n)$ is a solution of the equation of heat conduction

$$\frac{\partial U}{\partial t} = \kappa\left[\frac{\partial^2 U}{\partial x_3^2} + \frac{\partial^2 U}{\partial x_4^2} + \dots \frac{\partial^2 U}{\partial x_n^2}\right],$$

which is suitable for a space of $n - 2$ dimensions. The inference is that if U is one solution of this equation, the function

$$t^{1-\frac{n}{2}} e^{-\frac{R^2}{4\kappa t}} U\left(-\frac{a^2}{t}, \frac{ax_3}{t}, \dots \frac{ax_n}{t}\right)$$

* The first result is given by Bôcher, *Bull. Amer. Math. Soc.* vol. IX, p. 459 (1903).

is a second solution. When U is a constant the theorem gives us the particular solution

$$t^{1-\frac{n}{2}} e^{-\frac{R^2}{4\kappa t}}$$

which may be regarded as fundamental.

EXAMPLES

1. Prove that if

$$\begin{gathered} \alpha = z - ct, \quad \beta = x + iy, \quad a = z + ct, \quad b = x - iy, \\ \alpha' = z' + ct', \quad \beta' = x' + iy', \quad a' = z' - ct', \quad b' = x' - iy, \end{gathered}$$

the relations

$$\begin{aligned} \alpha' (la - u\beta - p) &= - n\alpha + w\beta + r + \beta' (- ma + v\beta + q), \\ \alpha' (ua + lb - e) &= - wa - nb + g - \beta' (va + mb - f), \\ a' (- ma + v\beta + q) &= ha + j\beta + k - b' (la - u\beta - p), \\ a' (va + mb - f) &= ja - hb + s + b' (ua + lb - e), \end{aligned}$$

in which $l, m, n, u, v, w, f, g, h, p, q, r, s, e, j, k$ are arbitrary constants, lead to a relation of type

$$dx'^2 + dy'^2 + dz'^2 - c^2 dt'^2 = \lambda^2 (dx^2 + dy^2 + dz^2 - c^2 dt^2).$$

2. Prove that if

$$z - ct = \phi + (x + iy)\, \theta,$$
$$(z + ct)\, \theta = \psi - (x - iy),$$

the relations of Ex. 1 give

$$z' - ct' = \phi' + \theta' (x' + iy'),$$
$$\theta' (z' + ct') = \psi' - (x' - iy'),$$

where

$$\begin{aligned} \chi\theta &= w\theta' - v\phi' + u\psi' + j, \\ \chi\phi &= r\theta' - q\phi' + p\psi' + k, \\ \chi\psi &= g\theta' - f\phi' + e\psi' - s, \\ \chi &= n\theta' - m\phi' + l\psi' - h. \end{aligned}$$

§ 2·51. *The equations for the equilibrium of an isotropic elastic solid.* Let u, v, w be the components of the small displacement of a particle, originally at x, y, z, when a solid body is slightly deformed, and let X, Y, Z be the components of the body force per unit mass. We consider the variation of the integral

$$I = \iiint L\, dx\, dy\, dz,$$

where $L = S - W$, with

$$W = \rho\, (uX + vY + wZ),$$

$$\begin{aligned} 2S = (\lambda + 2\mu)\, (u_x + v_y + w_z)^2 + \mu\, [(w_y + v_z)^2 + (u_z + w_x)^2 \\ + (v_x + u_y)^2 - 4v_y w_z - 4w_z u_x - 4u_x v_y], \end{aligned}$$

λ and μ being positive constants. The quantity S may be regarded as the strain energy per unit volume, while W is the work done by the body forces per unit volume. The density ρ is supposed to be constant.

We now wish L to be a minimum subject to the condition that the values of u, v and w are specified at the boundary of the solid. The Eulerian equations of the "Calculus of Variations" give

$$\frac{\partial X_x}{\partial x} + \frac{\partial X_y}{\partial y} + \frac{\partial X_z}{\partial z} + \rho X = 0,$$

$$\frac{\partial Y_x}{\partial x} + \frac{\partial Y_y}{\partial y} + \frac{\partial Y_z}{\partial z} + \rho Y = 0,$$

$$\frac{\partial Z_x}{\partial x} + \frac{\partial Z_y}{\partial y} + \frac{\partial Z_z}{\partial z} + \rho Z = 0,$$

where $$X_x = 2\mu u_x + \lambda (u_x + v_y + w_z), \quad Y_z = Z_y = \mu (w_y + v_z),$$
$$Y_y = 2\mu v_y + \lambda (u_x + v_y + w_z), \quad Z_x = X_z = \mu (u_z + w_x),$$
$$Z_z = 2\mu w_z + \lambda (u_x + v_y + w_z), \quad X_y = Y_x = \mu (v_x + u_y).$$

The quantities X_x, Y_y, Z_z, Y_z, Z_x, X_y, are called the six components of stress, and the quantities

$$e_{xx} = u_x, \quad e_{yy} = v_y, \quad e_{zz} = w_z,$$
$$e_{yz} = w_y + v_z, \quad e_{zx} = u_z + w_x, \quad e_{xy} = v_x + u_y,$$

are called the six components of strain. In terms of these quantities $2S$ may be expressed in the form

$$2S = X_x e_{xx} + Y_y e_{yy} + Z_z e_{zz} + Y_z e_{yz} + Z_x e_{zx} + X_y e_{xy},$$

while the relations between the components of stress and strain are

$$X_x = 2\mu e_{xx} + \lambda\Delta, \quad Y_z = Z_y = \mu e_{yz},$$
$$Y_y = 2\mu e_{yy} + \lambda\Delta, \quad Z_x = X_z = \mu e_{zx},$$
$$Z_z = 2\mu e_{zz} + \lambda\Delta, \quad X_y = Y_x = \mu e_{xy},$$
$$\Delta = u_x + v_y + w_z = e_{xx} + e_{yy} + e_{zz}.$$

The relations may also be written in the form

$$E e_{xx} = X_x - \sigma (Y_y + Z_z),$$
$$E e_{yy} = Y_y - \sigma (Z_z + X_x),$$
$$E e_{zz} = Z_z - \sigma (X_x + Y_y).$$

The coefficient E is Young's modulus, the number σ is Poisson's ratio, and μ is the modulus of rigidity. The quantity Δ is the dilatation and $-\Delta$ the cubical compression. When

$$X_x = Y_y = Z_z = -p, \quad Y_z = Z_x = X_y = 0,$$

we have $$e_{xx} = e_{yy} = e_{zz} = -p/(3\lambda + 2\mu),$$
$$e_{yz} = e_{zx} = e_{xy} = 0,$$
$$-\Delta = p/(\lambda + \tfrac{2}{3}\mu),$$

hence the quantity k defined by the equation

$$k = \lambda + \tfrac{2}{3}\mu$$

is called the modulus of compression. The different elastic constants are connected by the equations

$$E = \frac{\mu(3\lambda + 2\mu)}{\lambda + \mu}, \quad \sigma = \frac{\lambda}{2(\lambda+\mu)}, \quad k = \frac{E}{3 - 6\sigma}.$$

On account of the equations of equilibrium the expression for δI may be written in the form

$$\iiint \left[\frac{\partial}{\partial x}(X_x \delta u + Y_x \delta v + Z_x \delta w) + \frac{\partial}{\partial y}(X_y \delta u + Y_y \delta v + Z_y \delta w) \right.$$
$$\left. + \frac{\partial}{\partial z}(X_z \delta u + Y_z \delta v + Z_z \delta w)\right] dx\,dy\,dz,$$

and may be transformed into the surface integral

$$\iint [X_\nu \delta u + Y_\nu \delta v + Z_\nu \delta w]\, dS,$$

where
$$X_\nu = lX_x + mX_y + nX_z,$$
$$Y_\nu = lY_x + mY_y + nY_z,$$
$$Z_\nu = lZ_x + mZ_y + nZ_z.$$

The quantities X_ν, Y_ν, Z_ν are called the components of the surface traction across the tangent plane to the surface at a point under consideration. In many problems of the equilibrium of an elastic solid these quantities are specified and the expressions for the displacements are to be found.

The equations of motion of an elastic solid may be obtained by regarding $-\frac{\partial^2 u}{\partial t^2}$, $-\frac{\partial^2 v}{\partial t^2}$, $-\frac{\partial^2 w}{\partial t^2}$ as the components of an additional body force per unit mass. The equations are thus of type

$$\frac{\partial X_x}{\partial x} + \frac{\partial X_y}{\partial y} + \frac{\partial X_z}{\partial z} + \rho X = \rho \frac{\partial^2 u}{\partial t^2}.$$

§ 2·52. *The equations of motion of an inviscid fluid.* Let us consider the variation of the integral

$$I = \iiiint L\, dx\,dy\,dz\,dt,$$

where
$$L = \rho\left[\frac{\partial \phi}{\partial t} + \alpha\frac{\partial \beta}{\partial t} + \tfrac{1}{2}(u^2 + v^2 + w^2)\right] + f(\rho),$$

and
$$u = \frac{\partial \phi}{\partial x} + \alpha\frac{\partial \beta}{\partial x}, \quad v = \frac{\partial \phi}{\partial y} + \alpha\frac{\partial \beta}{\partial y}, \quad w = \frac{\partial \phi}{\partial z} + \alpha\frac{\partial \beta}{\partial z}. \quad \text{......(A)}$$

Varying the quantities ϕ, α, β and ρ in such a manner that the variations of ϕ and β vanish on a boundary of the region of integration wherever particles of fluid cross this boundary, the Eulerian equations give

$$\frac{\partial \rho}{\partial t} + \frac{\partial}{\partial x}(\rho u) + \frac{\partial}{\partial y}(\rho v) + \frac{\partial}{\partial z}(\rho w) = 0, \qquad \text{......(B)}$$

$$\frac{d\alpha}{dt} = 0, \quad \frac{d\beta}{dt} = 0, \qquad \text{......(C)}$$

$$\frac{\partial \phi}{\partial t} + \alpha \frac{\partial \beta}{\partial t} + \tfrac{1}{2}(u^2 + v^2 + w^2) + f'(\rho) = 0,$$

where
$$\frac{d}{dt} = \frac{\partial}{\partial t} + u\frac{\partial}{\partial x} + v\frac{\partial}{\partial y} + w\frac{\partial}{\partial z}.$$

If $p = \rho f'(\rho) - f(\rho)$ it is readily seen that

$$\frac{du}{dt} = -\frac{1}{\rho}\frac{\partial p}{\partial x}, \quad \frac{dv}{dt} = -\frac{1}{\rho}\frac{\partial p}{\partial y}, \quad \frac{dw}{dt} = -\frac{1}{\rho}\frac{\partial p}{\partial z}, \qquad \text{......(D)}$$

where
$$\int \frac{dp}{\rho} + \frac{\partial \phi}{\partial t} + \alpha \frac{\partial \beta}{\partial t} + \tfrac{1}{2}(u^2 + v^2 + w^2) = F(t). \qquad \text{......(E)}$$

If p is interpreted as the pressure, the last equation is the usual pressure equation of hydrodynamics for the case when there are no body forces acting. The quantities u, v, w are the component velocities and ρ is the density of the fluid at the point x, y, z. The equation (B) is the equation of continuity and the equations (D) the dynamical equations of motion. The relation $p = \rho f'(\rho) - f(\rho)$ implies that the fluid is a so-called barotropic fluid in which the density is a function of the pressure. It should be noticed that with this expression for the pressure the formula for L becomes

$$L = F(t) - p$$

when use is made of the relation (E).

The foregoing analysis is an extension of that given by Clebsch*. The fact that L is closely related to the expression for the pressure recalls to memory some remarks made by R. Hargreaves† in his paper "A pressure integral as a kinetic potential." The equations of hydrodynamics may also be obtained by writing

$$L = \rho\left[\frac{d\phi}{dt} + \alpha\frac{d\beta}{dt} - \tfrac{1}{2}(u^2 + v^2 + w^2)\right] + f(\rho),$$

and varying ϕ, α, β, u, v, w and ρ independently.

The equations (A) are then obtained by considering the variations of u, v and w. These equations give the following expressions for the components of vorticity:

$$\xi = \frac{\partial w}{\partial y} - \frac{\partial v}{\partial z} = \frac{\partial(\alpha, \beta)}{\partial(y, z)},$$

$$\eta = \frac{\partial u}{\partial z} - \frac{\partial w}{\partial x} = \frac{\partial(\alpha, \beta)}{\partial(z, x)},$$

$$\zeta = \frac{\partial v}{\partial x} - \frac{\partial u}{\partial y} = \frac{\partial(\alpha, \beta)}{\partial(x, y)}.$$

* *Crelle's Journ.* vol. LVI (1859). † *Phil. Mag.* vol. XVI, p. 436 (1908).

These equations indicate that $\alpha = $ constant, $\beta = $ constant are the equations of a vortex line. Now the equations (C) tell us that α and β remain constant during the motion of a particle of fluid, consequently a vortex line moves with the fluid and always contains the same particles.

It should be noticed that in these variational problems no restrictions need be imposed on the small variations $\delta\phi$, $\delta\beta$ at a boundary which is not crossed by particles of the fluid because the integrated terms, derived by the integration by parts, vanish automatically at such a boundary of the region of integration on account of the equation which expresses that fluid particles once on the boundary remain on the boundary.

§ **2·53**. *The equations of vortex motion and Liouville's equation.* Let us consider the variation of the integral

$$I = \tfrac{1}{2}\iint (u^2 + v^2 + \dot{s}^2)\, dx\, dy, \qquad \ldots\ldots\text{(A)}$$

where
$$u = -\frac{\partial\psi}{\partial y}, \quad v = \frac{\partial\psi}{\partial x}, \quad \dot{s} = \frac{\partial(\psi, s)}{\partial(x, y)},$$

the expressions for u and v being chosen so as to satisfy the equation of continuity,

$$\frac{\partial u}{\partial x} + \frac{\partial v}{\partial y} = 0,$$

for the two-dimensional motion of an incompressible fluid.

Varying the integral by giving ψ and s arbitrary variations which vanish at the boundary of the region of integration, we obtain the two equations

$$\frac{\partial(\dot{s}, \psi)}{\partial(x, y)} = 0,$$

$$\frac{\partial^2\psi}{\partial x^2} + \frac{\partial^2\psi}{\partial y^2} + \frac{\partial}{\partial x}\left(\dot{s}\,\frac{\partial s}{\partial y}\right) - \frac{\partial}{\partial y}\left(\dot{s}\,\frac{\partial s}{\partial x}\right) = 0.$$

The first of these gives
$$\dot{s} = g(\psi),$$

where $g(\psi)$ is an arbitrary function which, when the region of integration extends to infinity, must be such that the integral I has a meaning. This requirement usually means that u, v and $\dot{s}$ must vanish at infinity. With the foregoing expression for $\dot{s}$ the second equation takes the form

$$\frac{\partial^2\psi}{\partial x^2} + \frac{\partial^2\psi}{\partial y^2} + g(\psi)\, g'(\psi) = 0, \qquad \ldots\ldots\text{(B)}$$

which is no other than Lagrange's fundamental equation for two-dimensional steady vortex motion. In the special case when $g(\psi) = \lambda e^{h\psi}$, where λ and h are constants, the equation becomes

$$\frac{\partial^2\psi}{\partial x^2} + \frac{\partial^2\psi}{\partial y^2} + \lambda^2 h e^{2\psi h} = 0.$$

This equation, which also occurs in Richardson's theory of the space charge of electricity round a glowing wire*, has been solved by Liouville†, the complete solution being given by

$$\lambda^2 h^2 e^{2h\psi} (\sigma^2 + \tau^2 + 1)^2 = \left\{\left(\frac{\partial \sigma}{\partial x}\right)^2 + \left(\frac{\partial \sigma}{\partial y}\right)^2\right\},$$

where σ and τ are real functions of x and y defined by the equation $\sigma + i\tau = F(x + iy)$ and $F(z)$ is an arbitrary function.

Special forms of F which lead to useful results have been found by G. W. Walker‡. In particular, if $r^2 = x^2 + y^2$, there is a solution of type

$$e^{-h\psi} = \frac{\lambda hr}{2}\left\{\left(\frac{r}{a}\right)^n + \left(\frac{a}{r}\right)^n\right\} = \frac{\lambda}{\dot{s}},$$

and when $n = 1$ the component velocities are given by the expressions

$$u = \frac{2y}{h(a^2 + r^2)}, \quad v = -\frac{2x}{h(a^2 + r^2)}, \qquad \text{......(C)}$$

$$\lambda = 2/ah, \qquad \dot{s} = \frac{2}{h(a^2 + r^2)},$$

which are very like those for a line vortex but have the advantage that they do not become infinite at the origin. If we write

$$\dot{s} = \frac{ds}{dt} = u\frac{\partial s}{\partial x} + v\frac{\partial s}{\partial y},$$

the quantity s may be defined by the equation

$$s = -a \tan^{-1}(y/x),$$

and has a simple geometrical meaning. The quantity $\dot{s}$ may also be interpreted as the velocity of an associated point on the circle $r = a$ which is the locus of points at which the velocity is a maximum.

It should be noticed that if we use the variational principle

$$\delta \iint (u^2 + v^2 - \dot{s}^2)\, dx\, dy = 0, \qquad \text{......(D)}$$

the corresponding equation is

$$\frac{\partial^2 \psi}{\partial x^2} + \frac{\partial^2 \psi}{\partial y^2} = g(\psi)\, g'(\psi), \qquad \text{......(E)}$$

and the solution corresponding to (C) is of type

$$u = -\frac{2y}{h(a^2 - r^2)}, \quad v = \frac{2x}{h(a^2 - r^2)}.$$

This gives an infinite velocity on the circle $r = a$.

* O. W. Richardson, *The Emission of Electricity from hot bodies*, Longmans (1921), p. 50. The differential equation was formulated explicitly by M. v. Laue, *Jahrbuch d. Radioaktivität u. Elektronik*, vol. xv, pp. 205, 257 (1918).

† *Liouville's Journal*, vol. XVIII, p. 71 (1853).

‡ G. W. Walker, *Proc. Roy. Soc. London*, vol. XCI, p. 410 (1915); *Boltzmann Festschrift*, p. 242 (1904).

Other solutions of (B) which give infinite velocities have been discussed by Brodetsky*. It seems that the variational principle (A) may have the advantage over (D) in giving solutions of greater physical interest. It should be noticed that if a boundary of the region of integration is a stream-line $\psi =$ constant, it is not necessary for δs to be zero on this boundary.

When the motion is in three dimensions an appropriate variation principle is $\delta I = 0$, where

$$I = \tfrac{1}{2}\iiint (u^2 + v^2 + w^2 \pm \dot{s}^2)\,dx\,dy\,dz,$$

and the upper or lower sign is chosen according as the vortex motion is of the first or second type. To satisfy the equation of continuity when the fluid is incompressible and the density uniform, we may put

$$u = \frac{\partial(\sigma, \tau)}{\partial(y, z)}, \quad v = \frac{\partial(\sigma, \tau)}{\partial(z, x)}, \quad w = \frac{\partial(\sigma, \tau)}{\partial(x, y)}, \quad \dot{s} = \frac{\partial(s, \sigma, \tau)}{\partial(x, y, z)}.$$

A set of equations of motion is now obtained by varying σ, τ and s in such a way that their variations vanish on the boundary of the region of integration. These equations are

$$\frac{\partial(\dot{s}, \sigma, \tau)}{\partial(x, y, z)} = 0,$$

$$\xi\frac{\partial\sigma}{\partial x} + \eta\frac{\partial\sigma}{\partial y} + \zeta\frac{\partial\sigma}{\partial z} = \pm\frac{\partial(s, \dot{s}, \sigma)}{\partial(x, y, z)},$$

$$\xi\frac{\partial\tau}{\partial x} + \eta\frac{\partial\tau}{\partial y} + \zeta\frac{\partial\tau}{\partial z} = \pm\frac{\partial(s, \dot{s}, \tau)}{\partial(x, y, z)},$$

and are equivalent to the equations

$$\frac{d}{dt}(\dot{s}) = 0, \quad \xi = \pm\frac{\partial(s, \dot{s})}{\partial(y, z)}, \quad \eta = \pm\frac{\partial(s, \dot{s})}{\partial(z, x)}, \quad \zeta = \pm\frac{\partial(s, \dot{s})}{\partial(x, y)},$$

which imply that

$$u = \frac{\partial\chi}{\partial x} \mp \dot{s}\frac{\partial s}{\partial x}, \quad v = \frac{\partial\chi}{\partial y} \mp \dot{s}\frac{\partial s}{\partial y}, \quad w = \frac{\partial\chi}{\partial z} \mp \dot{s}\frac{\partial s}{\partial z}.$$

These equations give

$$\frac{du}{dt} = -\frac{1}{\rho}\frac{\partial p}{\partial x}, \quad \frac{dv}{dt} = -\frac{1}{\rho}\frac{\partial p}{\partial y}, \quad \frac{dw}{dt} = -\frac{1}{\rho}\frac{\partial p}{\partial z},$$

where the pressure p is given by the equation

$$\frac{p}{\rho} + \tfrac{1}{2}(u^2 + v^2 + w^2 \pm \dot{s}^2) = \text{constant}.$$

The equation of continuity may also be derived from the variation problem by adopting Lagrange's method of the variable multiplier. In

* S. Brodetsky, *Proceedings of the International Congress for Applied Mechanics*, p. 374 (Delft, 1924).

this method I is modified by adding $\lambda\left(\frac{\partial u}{\partial x}+\frac{\partial v}{\partial y}+\frac{\partial w}{\partial z}\right)$ to the quantity within brackets in the integrand. The quantities λ, u, v, w are then varied independently. It is better, however, to further modify I by an integration by parts of the added terms. The variation problem then reduces to the type already considered in § 2·52.

§ **2·54**. *The equilibrium of a soap film*. The equilibrium of a soap film will be discussed here on the hypothesis that there is a certain type of surface energy of mechanical type associated with each element of the surface. This energy will be called the tension-energy and will be represented by the integral

$$\iint T\,dS$$

taken over the portion of surface under consideration, T being a constant, called the surface tension. This constant is not dependent in any way upon the shape and size of the film but it does depend upon the temperature. It should be emphasised that a soap film must be considered as having two surfaces which are endowed with tension-energy. The tension-energy is not, moreover, the only type of surface energy; perhaps it would be better to say film energy; for there is also a type of thermal energy associated with the film, and from the thermodynamical point of view it is generally necessary to consider the changes of both mechanical and thermal energy when the film is stretched.

For mechanical purposes, however, useful results can be obtained by using the hypothesis that when a film stretched across a hole or attached to a wire is in equilibrium under the forces of tension alone, the total tension-energy is a minimum.

Assuming, then, as our expression for the total tension-energy E

$$E = 2T\iint (1 + z_x^2 + z_y^2)^{\frac{1}{2}}\,dx\,dy,$$

the z-co-ordinate of a point on the surface or rim being regarded as a function of x and y, the Eulerian equation of the Calculus of Variations gives

$$\frac{\partial}{\partial x}[z_x/H] + \frac{\partial}{\partial y}[z_y/H] = 0, \quad H = (1 + z_x^2 + z_y^2)^{\frac{1}{2}}.$$

This is the differential equation of a minimal surface.

When the film is subject to a difference of pressure on the two sides and the fluid on one side of the film is in a closed vessel whose pressure is p_1 while the pressure on the other side of the film is p_2, there is pressure-energy $(p_1 - p_2)\,V$ associated with the vessel closed by the film, where V is the volume of this vessel. Writing V in the form

$$V = V_0 + \tfrac{1}{3}\int \varpi\,dS,$$

where V_0 is a constant and ϖ is the perpendicular from the origin to the surface element dS, we consider the variation of the integral

$$\iint H\,dx\,dy\,[2T + \tfrac{1}{3}\varpi(p_1 - p_2)].$$

Now $$\varpi H = z - xz_x - yz_y,$$

and so the differential equation of the problem is

$$0 = p_1 - p_2 + 2T\left[\frac{\partial}{\partial x}(z_x/H) + \frac{\partial}{\partial y}(z_y/H)\right].$$

This differential equation may be interpreted by noting that the co-ordinates of a point on the normal at (x, y) are

$$\xi = x - Rz_x/H, \quad \eta = y - Rz_y/H,$$

where R is the distance of the point from (x, y). If now two consecutive normals intersect at this point, we have

$$0 = d\xi = dx - Rd\,(z_x/H), \quad 0 = d\eta = dy - Rd\,(z_y/H),$$

for $dR = 0$. Expanding in the form

$$0 = dx\left[1 - R\frac{\partial}{\partial x}(z_x/H)\right] - dy\,R\frac{\partial}{\partial y}(z_y/H),$$

$$0 = -\,dx\,R\frac{\partial}{\partial x}(z_x/H) + dy\left[1 - R\frac{\partial}{\partial y}(z_y/H)\right],$$

and eliminating dx, dy, we obtain as our equation for R

$$0 = 1 - R\left[\frac{\partial}{\partial x}(z_x/H) + \frac{\partial}{\partial y}(z_y/H)\right] + R^2\frac{\partial\,(z_x/H,\, z_y/H)}{\partial\,(x, y)}.$$

If R_1 and R_2 are the roots, we have

$$\frac{1}{R_1} + \frac{1}{R_2} = \frac{\partial}{\partial x}(z_x/H) + \frac{\partial}{\partial y}(z_y/H).$$

The quantities R_1 and R_2 are called the principal radii of curvature. A minimal surface is thus characterised by the equation $\frac{1}{R_1} + \frac{1}{R_2} = 0$ and a surface of a soap film subject to a constant pressure-difference on its two sides is shaped in accordance with the equation

$$\frac{1}{R_1} + \frac{1}{R_2} = \text{constant}.$$

When the film is subjected to only a small difference of pressure and is stretched across a hole in a thin flat plate we can, to a sufficient approximation, put $H = 1$ in this equation. The resulting equation is

$$\frac{\partial^2 z}{\partial x^2} + \frac{\partial^2 z}{\partial y^2} = K,$$

where K is a constant and the boundary condition is $z = 0$ on the rim.

Now the same differential equation and boundary conditions occur in a number of physical problems and a soap-film method of solving such problems in engineering practice was suggested by Prandtl and has been much developed by A. A. Griffith and G. I. Taylor*. The most important problems of this type are:

(1) The torsion of a prism (Saint-Venant's theory).

(2) The flow of a viscous liquid under pressure in a straight pipe.

These problems will now be considered.

EXAMPLES

1. The forces acting on the rim of a soap film of tension T are equivalent to a force F at the origin and a couple G. Prove that

$$F = \int 2T\,(n \times ds),$$

$$G = \int 2T\,[r \times (n \times ds)],$$

where the vector ds denotes a directed element of the rim and the vector n is a unit vector along the normal to the surface of the film. Show by transforming these integrals into surface integrals that the force and couple are equivalent to a system of normal forces, the force normal to the element dS being of magnitude

$$2T\left(\frac{1}{R_1} + \frac{1}{R_2}\right) = 2T\,(C_1 + C_2), \text{ say.}$$

2. The surface of a film closing up a vessel of volume V can be regarded as one of a family of surfaces for which $C_1 + C_2$ is C, a constant. If within a limited region of space there is just one surface of this family that can be associated with each point by some uniform rule and if S' is another surface through the rim of the hole, ϵ the angle which this surface makes at a point (x, y, z) with the surface of the family through this point, the area of the outer surface of the film is $\iint \cos\epsilon\,.\,dS'$. Hence show that the area of the new surface is greater than that of the film if it encloses the same volume.

3. If
$$u = z_x, \quad v = z_y, \quad q^2 = u^2 + v^2,$$
show that the variation problem
$$\delta \iint G(q)\,dx\,dy = 0$$
leads to the partial differential equation
$$(c^2 - u^2)\frac{\partial^2 z}{\partial x^2} - 2uv\frac{\partial^2 z}{\partial x \partial y} + (c^2 - v^2)\frac{\partial^2 z}{\partial y^2} = 0,$$
where
$$c^2\,[qG''(q) - G'(q)] = q^2\,G'(q).$$

Show also that the two-dimensional adiabatic irrotational flow of a compressible fluid leads to an equation of this type for the velocity potential z, the function $G(q)$ being given by the equation

$$G(q) = [2a^2 + (\gamma - 1)(U^2 - q^2)]^{\frac{\gamma}{\gamma-1}},$$

where U, a and γ are constants.

* See ch. VII of the *Mechanical Properties of Fluids* (Blackie & Son, Ltd., 1923).

§ **2·55**. *The torsion of a prism.* Assuming that the material of the prism is isotropic, we take the axis of z in the direction of the generators of the surface and consider a distortion in which a point (x, y, z) is displaced to a new position $(x + u, y + v, z + w)$, where

$$u = -\tau yz, \quad v = \tau zx, \quad w = \tau\phi,$$

and ϕ is a function of x and y to be determined. The constant τ is called the twist. This distortion is supposed to be produced by terminal couples applied in a suitable manner to the end faces. The portion of the surface generated by lines parallel to the axis of z (the mantle) is supposed to be free from stress. These are the simplifying assumptions of Saint-Venant.

It is easily seen that

$$\frac{\partial u}{\partial x} = \frac{\partial v}{\partial y} = \frac{\partial w}{\partial z} = \frac{\partial v}{\partial x} + \frac{\partial u}{\partial y} = 0,$$

$$e_{zx} = \frac{\partial u}{\partial z} + \frac{\partial w}{\partial x} = \tau\left(\frac{\partial \phi}{\partial x} - y\right),$$

$$e_{yz} = \frac{\partial w}{\partial y} + \frac{\partial v}{\partial z} = \tau\left(\frac{\partial \phi}{\partial y} + x\right).$$

Hence, if $\quad Z_x = \mu e_{zx}, \quad Z_y = \mu e_{yz}, \quad X_x = Y_y = Z_z = X_y = 0,$

the equations of equilibrium

$$\frac{\partial Z_x}{\partial z} = 0, \quad \frac{\partial Z_y}{\partial z} = 0, \quad \frac{\partial Z_x}{\partial x} + \frac{\partial Z_y}{\partial y} = 0$$

show that Z_x and Z_y are independent of z and that

$$\frac{\partial}{\partial x}\left(\frac{\partial \phi}{\partial x} - y\right) + \frac{\partial}{\partial y}\left(\frac{\partial \phi}{\partial y} + x\right) = 0,$$

or

$$\frac{\partial^2 \phi}{\partial x^2} + \frac{\partial^2 \phi}{\partial y^2} = 0.$$

The boundary condition of no stress on the mantle gives

$$lZ_x + mZ_y = 0,$$

where $(l, m, 0)$ are the direction cosines of the normal to the mantle at the point (x, y, z).

Let us now introduce the function ψ conjugate to ϕ, then

$$\frac{\partial \phi}{\partial x} = \frac{\partial \psi}{\partial y}, \quad \frac{\partial \phi}{\partial y} = -\frac{\partial \psi}{\partial x},$$

$$y = \frac{\partial}{\partial y}(\tfrac{1}{2}r^2), \quad x = \frac{\partial}{\partial x}(\tfrac{1}{2}r^2),$$

where $r^2 = x^2 + y^2$. The boundary condition may consequently be written in the form

$$l\frac{\partial \chi}{\partial y} - m\frac{\partial \chi}{\partial x} = 0,$$

or

$$\frac{\partial \chi}{\partial s} = 0,$$

where $\chi = \psi - \frac{1}{2}r^2$ and ds is a linear element of the cross-section. This equation signifies that χ is constant over the boundary and so the problem may be solved by determining a potential function ψ which is regular within the prism and which takes a value differing by a constant from $\frac{1}{2}r^2$ on the mantle of the prism. Without loss of generality this constant may be taken to be zero if there is only one mantle.

It should be noticed that the function χ satisfies the equation

$$\frac{\partial^2 \chi}{\partial x^2} + \frac{\partial^2 \chi}{\partial y^2} = -2,$$

and, with the above choice of the constant, is zero on the mantle when this is unique. It is often more convenient to work with the function χ, especially as

$$Z_x = \mu\tau\left(\frac{\partial \phi}{\partial x} - y\right) = \mu\tau\frac{\partial \chi}{\partial y},$$

$$Z_y = \mu\tau\left(\frac{\partial \phi}{\partial y} + x\right) = -\mu\tau\frac{\partial \chi}{\partial x}.$$

Since χ vanishes on the mantle it is evident that

$$\iint Z_x dx dy = 0, \quad \iint Z_y dx dy = 0.$$

The tractions on a cross-section are thus statically equivalent to a couple about the axis of z of moment

$$M = \iint (xZ_y - yZ_x)\, dx dy = -\mu\tau\iint\left(x\frac{\partial \chi}{\partial x} + y\frac{\partial \chi}{\partial y}\right) dx dy.$$

Integrating by parts we find that

$$M = 2\mu\tau\iint \chi dx dy.$$

The direction of the tangential traction (Z_x, Z_y) across the normal section of the prism by a plane z = constant is that of the tangent to the curve χ = constant which passes through the point. The curves χ = constant may thus be called "lines of shearing stress." The magnitude of the traction is $\mu\tau\frac{\partial \chi}{\partial n}$, where $\frac{\partial \chi}{\partial n}$ is the derivative of χ in a direction normal to the line of shearing stress.

In the case of a circular prism

$$\chi = \tfrac{1}{2}(a^2 - r^2),$$

and in the case of an elliptic prism

$$\chi = C\left(1 - \frac{x^2}{a^2} - \frac{y^2}{b^2}\right),$$

where a and b are the semi-axes of the ellipse and

$$C\left(\frac{1}{a^2} + \frac{1}{b^2}\right) = 1.$$

§ **2·56.** *Flow of a viscous liquid along a straight tube.* Consider the motion of the portion contained between the cross-sections $z = z_1$ and $z = z_1 + h$. If A is the area of the cross-section and ρ the density of the fluid, the equation of motion is

$$\rho A h \frac{\partial u}{\partial t} = A (p_1 - p_2) - D,$$

where p_1 and p_2 are the pressures at the two sections and D is the total frictional drag at the curved surface of the tube. If u is the velocity of flow in the direction of the axis of z, u will be independent of z if the fluid is incompressible and so we may write

$$u = u (x, y, t).$$

We now introduce the hypothesis that there is a constant coefficient of viscosity μ such that

$$D = - \int_{z_1}^{z_1+h} dz \int \mu \frac{\partial u}{\partial n} ds,$$

where $\frac{\partial}{\partial n}$ denotes a differentiation in the direction of the normal to the surface of the tube. Transforming the surface integral into a volume integral, we have the equation of motion

$$\rho A h \frac{\partial u}{\partial t} = A (p_1 - p_2) + \iint h\mu \left(\frac{\partial^2 u}{\partial x^2} + \frac{\partial^2 u}{\partial y^2}\right) dx\, dy.$$

Since h is arbitrary this may be written in the form

$$\rho \frac{\partial u}{\partial t} = - \frac{\partial p}{\partial z} + \mu \left(\frac{\partial^2 u}{\partial x^2} + \frac{\partial^2 u}{\partial y^2}\right).$$

When the motion is steady this equation takes the form

$$\frac{\partial^2 u}{\partial x^2} + \frac{\partial^2 u}{\partial y^2} = - 2K, \qquad \text{......(I)}$$

where $- 2K = \frac{1}{\mu}\frac{\partial p}{\partial z}$ and can be regarded as a constant, because u is independent of z. This is the equation used by Stokes and Boussinesq.

In the case of an elliptic tube

$$u = CK \left(1 - \frac{x^2}{a^2} - \frac{y^2}{b^2}\right),$$

where

$$C\left(\frac{1}{a^2} + \frac{1}{b^2}\right) = 1.$$

For an annular tube bounded by the cylinders $r = a$, $r = b$ we may write

$$u = \tfrac{1}{2}K (a^2 - r^2) + \tfrac{1}{2}K (b^2 - a^2) \log (b/r).$$

The total flux Q is in this case

$$Q = 2\pi \int_a^b u r\, dr = \frac{\pi K a^4}{4} \left\{s^4 - \frac{(s^2 - 1)^2}{\log s}\right\}, \qquad (s = b/a)$$

and so the average velocity is

$$U = \frac{Ka^2}{4}\left\{\frac{s^4}{s^2-1} - \frac{s^2-1}{\log s}\right\}.$$

If there is no pressure gradient the equation of variable flow is

$$\frac{\partial u}{\partial t} = \nu\left(\frac{\partial^2 u}{\partial x^2} + \frac{\partial^2 u}{\partial y^2}\right),$$

where $\nu = \mu/\rho$. This equation is the same as the equation of the conduction of heat in two dimensions. The fluid may be supposed in particular to lie above a plane $z = 0$ which has a prescribed motion, or to lie between two parallel planes with prescribed motions parallel to their surfaces.

The simple type of steady motion of a viscous fluid which is given by the equation (I) does not always occur in practice. The experiments of Osborne Reynolds, Stanton and others have shown that when a viscous fluid flows through a straight pipe of circular section there is a certain critical velocity (which is not very definite) above which the flow becomes irregular or turbulent and is in no sense steady. From dimensional reasoning it has been found advantageous to replace the idea of a critical velocity by that of a critical dimensionless quantity or Reynolds number formed from a velocity, a length and the kinematic viscosity ν of the fluid. In the case of flow through a pipe the velocity V may be taken to be the mean velocity over the cross-section, the length, the diameter of the pipe (d). For steady "laminar flow" the ratio Vd/ν must not exceed about 2300.

In the case of the motion of air past a sphere a similar Reynolds number may be defined in which d is the diameter of the sphere. In order that the drag may be proportional to the velocity V the ratio Vd/ν must be very small.

EXAMPLES

1. In viscous flow between parallel planes $x = \pm a$ the velocity is given by an equation of type

$$u = c\,(1 - x^2/a^2),$$

where c is the maximum velocity. Prove that the mean velocity is two-thirds of the maximum.

2. In a screw velocity pump the motion of the fluid is roughly comparable with that of a viscous liquid between two parallel planes one of which moves parallel to the other and drags the fluid along, although there is a pressure gradient resisting the flow. Calculate the efficiency of the pump and find when it is greatest.

Work out the distribution of velocity and the efficiency when the machine acts as a motor, that is, when the fluid is driven by the pressure and causes the motion of the upper plate.

[Rowell and Finlayson, *Engineering*, vol. CXXVI. p. 249 (1928).]

3. The Eulerian equation associated with the variation problem

$$\delta I \equiv \delta \iint \left[\left(\frac{\partial u}{\partial x}\right)^2 + \left(\frac{\partial u}{\partial y}\right)^2 + 2uf(x, y)\right] dx\,dy = 0,$$

$$\frac{\partial^2 u}{\partial x^2} + \frac{\partial^2 u}{\partial y^2} = f(x, y).$$

L. Lichtenstein [*Math. Ann.* vol. LXIX, p. 514, 1910] has shown that when $f(x, y)$ is merely continuous there may be a function u which makes $\delta I = 0$ and does not satisfy the Eulerian equation.

§ 2·57. *The vibration of a membrane.* Let T be the tension of the membrane in the state of equilibrium and w the small lateral displacement of a point of the membrane from the plane in which the membrane is situated when in a state of equilibrium, the vibrations which will be considered are supposed to be so small that any change in area produced by the deflections w does not produce any appreciable percentage variation of T. The quantity T is thus treated as constant and the potential energy

$$T \iint \left[1 + \left(\frac{\partial w}{\partial x}\right)^2 + \left(\frac{\partial w}{\partial y}\right)^2\right]^{\frac{1}{2}} dx\,dy$$

is replaced by the approximate expression

$$V = T \iint \left[1 + \frac{1}{2}\left(\frac{\partial w}{\partial x}\right)^2 + \frac{1}{2}\left(\frac{\partial w}{\partial y}\right)^2\right] dx\,dy.$$

Let $\rho\,dx\,dy$ be the mass of the element $dx\,dy$. The equation of motion of the membrane will be obtained by considering the variation of

$$\int_{t_1}^{t_2} (E - V)\,dt,$$

where

$$E = \frac{1}{2}\iint \rho \left(\frac{\partial w}{\partial t}\right)^2 dx\,dy, \quad V = T\,dx\,dy + \tfrac{1}{2}T \iint \left[\left(\frac{\partial w}{\partial x}\right)^2 + \left(\frac{\partial w}{\partial y}\right)^2\right] dx\,dy.$$

The integral to be varied is thus

$$I = \frac{1}{2}\iiint \left[\rho\left(\frac{\partial w}{\partial t}\right)^2 - T\left(\frac{\partial w}{\partial x}\right)^2 - T\left(\frac{\partial w}{\partial y}\right)^2\right] dx\,dy,$$

where $w = 0$ on the boundary curve for all values of t. The Eulerian equation of the Calculus of Variations gives

$$\frac{\partial^2 w}{\partial t^2} = c^2\left[\frac{\partial^2 w}{\partial x^2} + \frac{\partial^2 w}{\partial y^2}\right], \qquad \text{......(A)}$$

where
$$c^2 = T/\rho.$$

This is the equation of a vibrating membrane. The equation occurs also in electromagnetic theory and in the theory of sound. In the case when w is of the form

$$w = \sin \kappa y \,.\, v\,(x, t),$$

the function v satisfies the equation

$$\frac{\partial^2 v}{\partial t^2} = c^2\left(\frac{\partial^2 v}{\partial x^2} - \kappa^2 v\right),$$

which is of the same form as the equation of telegraphy.

It should be noticed that a corresponding variation principle

$$\delta \iiiint \left[\rho\left(\frac{\partial w}{\partial t}\right)^2 - T\left(\frac{\partial w}{\partial x}\right)^2 - T\left(\frac{\partial w}{\partial y}\right)^2 - T\left(\frac{\partial w}{\partial z}\right)^2\right] dx\,dy\,dz\,dt = 0$$

gives rise to the familiar wave-equation

$$\frac{\partial^2 w}{\partial t^2} = c^2 \left(\frac{\partial^2 w}{\partial x^2} + \frac{\partial^2 w}{\partial y^2} + \frac{\partial^2 w}{\partial z^2} \right),$$

which governs the propagation of sound in a uniform medium and the propagation of electromagnetic waves. A function w which satisfies this equation is called a wave-function.

Love has shown* that the equation (A) occurs in the theory of the propagation of a simple type of elastic wave.

Taking the positive direction of the axis of z upwards and the axis of x in the direction of propagation, we assume that the transverse displacement v is given by the equation

$$v = Y(z) \cos (pt - fx).$$

The components of stress across an area perpendicular to the axis of y are

$$Y_x = \mu \frac{\partial v}{\partial x}, \quad Y_y = 0, \quad Y_z = \mu \frac{\partial v}{\partial z}$$

respectively and so the equation of motion

$$\rho \frac{\partial^2 v}{\partial t^2} = \frac{\partial Y_x}{\partial x} + \frac{\partial Y_y}{\partial y} + \frac{\partial Y_z}{\partial z}$$

takes the form

$$\rho \frac{\partial^2 v}{\partial t^2} = \frac{\partial}{\partial x}\left(\mu \frac{\partial v}{\partial x}\right) + \frac{\partial}{\partial z}\left(\mu \frac{\partial v}{\partial z}\right). \quad \text{......(B)}$$

When ρ and μ are constants this is the same as the equation of a vibrating membrane, but when ρ and μ are functions of z the equation is of a type which has been considered by Meissner†.

Transverse waves of this type have been called by Jeffreys‡ "Love waves," they are of some interest in connection with the interpretation of the surface waves which are observed after an earthquake.

It may be mentioned that the general equation (B) may be obtained by considering the variation of the integral

$$I = \frac{1}{2} \iiint \left[\rho \left(\frac{\partial v}{\partial t}\right)^2 - \mu \left(\frac{\partial v}{\partial x}\right)^2 - \mu \left(\frac{\partial v}{\partial z}\right)^2 \right] dx\,dz\,dt,$$

and an extension can be made to the case in which ρ and μ are functions of x, z and t.

§ 2·58. *The electromagnetic equations.* Consider the variation of the integral

$$\iiiint L\,dx\,dy\,dz\,dt,$$

* *Some Problems of Geodynamics*, p. 160 (Cambridge University Press, 1911).

† *Proceedings of the Second International Congress for Applied Mathematics* (Zürich, 1926).

‡ *The Earth*, p. 165 (Cambridge University Press, 1924).

where $$2L = H_x^2 + H_y^2 + H_z^2 - E_x^2 - E_y^2 - E_z^2,$$

and
$$\left.\begin{aligned} H_x &= \frac{\partial A_z}{\partial y} - \frac{\partial A_y}{\partial z}, & E_x &= -\frac{\partial A_x}{\partial t} - \frac{\partial \Phi}{\partial x}, \\ H_y &= \frac{\partial A_x}{\partial z} - \frac{\partial A_z}{\partial x}, & E_y &= -\frac{\partial A_y}{\partial t} - \frac{\partial \Phi}{\partial y}, \\ H_z &= \frac{\partial A_y}{\partial x} - \frac{\partial A_x}{\partial y}, & E_z &= -\frac{\partial A_z}{\partial t} - \frac{\partial \Phi}{\partial z}. \end{aligned}\right\} \quad \text{......(A)}$$

If the variations of A_x, A_y, A_z and Φ vanish at the boundary of the region of integration, the Eulerian equations give

$$\frac{\partial H_z}{\partial y} - \frac{\partial H_y}{\partial z} = \frac{\partial E_x}{\partial t},$$
$$\frac{\partial H_x}{\partial z} - \frac{\partial H_z}{\partial x} = \frac{\partial E_y}{\partial t},$$
$$\frac{\partial H_y}{\partial x} - \frac{\partial H_x}{\partial y} = \frac{\partial E_z}{\partial t},$$
$$\frac{\partial E_x}{\partial x} + \frac{\partial E_y}{\partial y} + \frac{\partial E_z}{\partial z} = 0.$$

In vector notation these equations may be written

$$\text{curl } \mathbf{H} = \frac{\partial \mathbf{E}}{\partial t}, \quad \text{div } \mathbf{E} = 0, \qquad \text{......(B)}$$

and equations (A) take the form

$$\mathbf{H} = \text{curl } \mathbf{A}. \quad \mathbf{E} = -\frac{\partial \mathbf{A}}{\partial t} - \nabla\Phi. \qquad \text{......(C)}$$

These equations imply that

$$\text{curl } \mathbf{E} = -\frac{\partial \mathbf{H}}{\partial t}, \quad \text{div } \mathbf{H} = 0. \qquad \text{......(D)}$$

The two sets of equations (B) and (D) are the well-known equations of Maxwell for the propagation of electromagnetic waves in the ether; the unit of time has, however, been chosen so that the velocity of light is represented by unity. The foregoing analysis is due essentially to Larmor.

Writing $\mathbf{Q} = \mathbf{H} + i\mathbf{E}$, the two sets of equations may be combined into the single set of equations

$$\text{curl } \mathbf{Q} = -i\frac{\partial \mathbf{Q}}{\partial t}, \quad \text{div } \mathbf{Q} = 0. \qquad \text{......(E)}$$

By analogy with (C) we may seek a solution for which

$$\mathbf{Q} = -i \text{ curl } \mathbf{L} = -\frac{\partial \mathbf{L}}{\partial t} - \nabla\Lambda. \qquad \text{......(F)}$$

The relations between $\mathbf{L}$ and Λ may be satisfied by writing

$$\mathbf{L} = \frac{\partial \mathbf{G}}{\partial t} + i \text{ curl } \mathbf{G} + \nabla K, \quad \Lambda = -\text{div } \mathbf{G} - \frac{\partial K}{\partial t}, \qquad \text{......(G)}$$

where $\mathbf{G}$ is a complex vector of type $\mathbf{\Gamma} + i\mathbf{\Pi}$, while $\mathbf{\Gamma}$ and $\mathbf{\Pi}$ are real 'Hertzian vectors' whose components all satisfy the wave-equation

$$\Box u \equiv \nabla^2 u - \frac{\partial^2 u}{\partial t^2} = 0. \qquad \text{......(H)}$$

When we differentiate to find an expression for $\mathbf{Q}$ in terms of $\mathbf{G}$ and K the terms involving K cancel and we find the Righi-Whittaker formulae

$$\left.\begin{aligned} \mathbf{Q} &= \operatorname{curl}\left(\operatorname{curl}\mathbf{G} - i\frac{\partial \mathbf{G}}{\partial t}\right), \\ \mathbf{H} &= \operatorname{curl}\left(\operatorname{curl}\mathbf{\Gamma} + \frac{\partial \mathbf{\Pi}}{\partial t}\right), \\ \mathbf{E} &= \operatorname{curl}\left(\operatorname{curl}\mathbf{\Pi} - \frac{\partial \mathbf{\Gamma}}{\partial t}\right). \end{aligned}\right\} \qquad \text{......(I)}$$

If $\mathbf{L} = \mathbf{B} + i\mathbf{A}$, $\Lambda = \Psi + i\Phi$, where $\mathbf{A}$, $\mathbf{B}$, Φ and Ψ are real, we have

$$\left.\begin{aligned} \mathbf{H} &= \quad \operatorname{curl}\mathbf{A} = -\frac{\partial \mathbf{B}}{\partial t} - \nabla\Psi, \\ \mathbf{E} &= -\operatorname{curl}\mathbf{B} = -\frac{\partial \mathbf{A}}{\partial t} - \nabla\Phi, \end{aligned}\right\} \qquad \text{......(J)}$$

$$\left.\begin{aligned} \mathbf{A} = \frac{\partial \mathbf{\Pi}}{\partial t} + \operatorname{curl}\mathbf{\Gamma}, \quad \mathbf{B} = \frac{\partial \mathbf{\Gamma}}{\partial t} - \operatorname{curl}\mathbf{\Pi}, \\ \Phi = -\operatorname{div}\Pi, \quad \Psi = -\operatorname{div}\Gamma, \end{aligned}\right\} \qquad \text{......(K)}$$

where $\mathbf{A}$, $\mathbf{B}$, Φ and Ψ are wave-functions which are connected by the identical relations

$$\operatorname{div}\mathbf{A} + \frac{\partial \Phi}{\partial t} = 0, \quad \operatorname{div}\mathbf{B} + \frac{\partial \Psi}{\partial t} = 0. \qquad \text{......(L)}$$

The corresponding formulae for the case in which the unit of time is not chosen so that the velocity of light is unity are obtained from the foregoing by writing ct in place of t wherever t occurs.

If we write $\mathbf{Q}' = e^{i\theta}\mathbf{Q}$, where θ is a constant, it is evident that the vector $\mathbf{Q}'$ satisfies the same differential equations as $\mathbf{Q}$ and can therefore be used to specify an electromagnetic field ($\mathbf{E}'$, $\mathbf{H}'$) associated with the original field ($\mathbf{E}$, $\mathbf{H}$). It will be noticed that the function L' for this associated field is not the same as L, for

$$L' = H'^2 - E'^2 = (H^2 - E^2)\cos 2\theta - 2\,(E \,.\, H)\sin 2\theta.$$

Also $\quad (E' \,.\, H') = (H^2 - E^2)\sin 2\theta + 2\,(E \,.\, H)\cos 2\theta.$

There are, however, certain quantities which are the same for the two fields. These quantities may be defined as follows:

$$\left.\begin{aligned} W &= \tfrac{1}{2}(E^2 + H^2), \\ S_x &= E_y H_z - E_z H_y = G_x, \\ X_x &= E_x^2 + H_x^2 - W, \\ Y_z &= E_y E_z + H_y H_z = Z_y, \\ &\ldots\ldots\ldots\ldots\ldots\ldots \end{aligned}\right\} \qquad \text{......(M)}$$

It is interesting to note that these quantities may be arranged so as to form an orthogonal matrix*

$$\begin{matrix} X_x & X_y & X_z & iG_x \\ Y_x & Y_y & Y_z & iG_y \\ Z_x & Z_y & Z_z & iG_z \\ iS_x & iS_y & iS_z & W \end{matrix}$$

We have, in fact, the relations

$$\left.\begin{aligned} X_x^2 + X_y^2 + X_z^2 - G_x^2 &= I, \\ W^2 - S_x^2 - S_y^2 - S_z^2 &= I, \\ X_x Y_x + X_y Y_y + X_z Y_z - G_x G_y &= 0, \\ X_x S_x + X_y S_y + X_z S_z + G_x W &= 0, \\ \text{where} \qquad I = \tfrac{1}{4}(H^2 - E^2)^2 &+ (E\,.\,H)^2. \end{aligned}\right\} \qquad \text{......(N)}$$

§ 2·59. *The conservation of energy and momentum in an electromagnetic field.* It follows from the field equations (B) and (D) that the sixteen components of the orthogonal matrix satisfy the equations

$$\left.\begin{aligned} \frac{\partial X_x}{\partial x} + \frac{\partial X_y}{\partial y} + \frac{\partial X_z}{\partial z} - \frac{\partial G_x}{\partial t} &= 0, \\ \frac{\partial Y_x}{\partial x} + \frac{\partial Y_y}{\partial y} + \frac{\partial Y_z}{\partial z} - \frac{\partial G_y}{\partial t} &= 0, \\ \frac{\partial Z_x}{\partial x} + \frac{\partial Z_y}{\partial y} + \frac{\partial Z_z}{\partial z} - \frac{\partial G_z}{\partial t} &= 0, \\ \frac{\partial S_x}{\partial x} + \frac{\partial S_y}{\partial y} + \frac{\partial S_z}{\partial z} + \frac{\partial W}{\partial t} &= 0. \end{aligned}\right\} \qquad \text{......(A)}$$

Regarding S_x, S_y, S_z for the moment as the components of a vector S and using the suffix n to denote the component along the outward-drawn normal to a surface element $d\sigma$ of a surface σ, we have

$$\begin{aligned} \iint S_n d\sigma &= \iiint \operatorname{div} S\,.\,d\tau \qquad (d\tau = dx\,dy\,dz) \\ &= -\iiint \frac{\partial W}{\partial t}\,d\tau \\ &= -\frac{\partial}{\partial t}\iiint W\,d\tau. \end{aligned}$$

In this equation the region of integration is supposed to be such that the derivatives of S and W in which we are interested are continuous functions of x, y, z and t. This will certainly be the case if the field vectors and their first derivatives are continuous functions of x, y, z and t.

* H. Minkowski, *Gött. Nachr.* (1908).

Let us now regard W as the density of electromagnetic energy and S as a vector specifying the flow of energy, then the foregoing equation can be interpreted to mean that the energy gained or lost by the region enclosed by σ is entirely accounted for by the flow of energy across the boundary. This is simply a statement of the Principle of the Conservation of Energy for the electromagnetic energy in the ether.

The equations involving G_x, G_y, G_z may be regarded as expressing the Principle of the Conservation of Momentum. We shall, in fact, regard G_x as the density of the x-component of electromagnetic momentum and $(-X_x, -X_y, -X_z)$ as the components of a vector specifying the flow of the x-component of electromagnetic momentum.

The vector S is generally called Poynting's vector as it was used to describe the energy changes by J. H. Poynting in 1884. The vector G was introduced into electromagnetic theory by Abraham and Poincaré.

In the case of an electrostatic field

$$\iiint W\,d\tau = \frac{1}{2}\iiint \left[\left(\frac{\partial\phi}{\partial x}\right)^2 + \left(\frac{\partial\phi}{\partial y}\right)^2 + \left(\frac{\partial\phi}{\partial z}\right)^2\right] d\tau$$

$$= -\frac{1}{2}\iiint \phi\nabla^2\phi\,d\tau + \frac{1}{2}\iint \phi\frac{\partial\phi}{\partial n}\,d\sigma$$

$$= \frac{1}{2}\iiint \phi\rho\,dx\,dy\,dz,$$

if there are only volume charges and the first integral is taken over all space, for then the surface integral may be taken over a sphere of infinite radius and may be supposed to vanish when the total amount of electricity is finite and there is no electricity at infinity. It should be noticed that in the present system of units Poisson's equation takes the form

$$\nabla^2\phi + \rho = 0,$$

where ρ is the density of electricity. When there are charged surfaces an integral of type

$$-\frac{1}{2}\iint \phi\frac{\partial\phi}{\partial n}\,d\sigma$$

must be added to the right-hand side for each charged surface.

The new expression for the total energy may be written in the form

$$U = \tfrac{1}{2}\Sigma e\phi.$$

This may be derived from first principles if it is assumed that $\phi\,\delta e$ is the work done in bringing up a small charge δe from an infinite distance without disturbing other charges. Now suppose that each charge in an electrostatic field is built up gradually in this way and that when an inventory is taken at any time each carrier of charge has a charge equal to λ times the final amount and a potential equal to λ times the final potential. λ being the same for all carriers. As λ increases from λ to $\lambda + d\lambda$ the work done on the system is

$$dU = \Sigma\,(\lambda\phi)\,e d\lambda.$$

Integrating with respect to λ between 0 and 1 we get

$$U = \tfrac{1}{2}\Sigma e\phi.$$

The carriers mentioned in the proof may be conducting surfaces capable (within limits) of holding any amount of electricity. If the carriers are taken to be atoms or molecules there is the difficulty that, according to experimental evidence, the charge associated with a carrier can only change by integral multiples of a certain elementary charge ϵ. For this reason it seems preferable to start with the assumption that W represents the density of electromagnetic energy.

On account of the symmetrical relations

$$Y_z = Z_y,\ \text{etc.},\quad S_x = G_x,\ \text{etc.},$$

we can supplement the relations (A) by six additional equations of types

$$\frac{\partial}{\partial x}(yZ_x - zY_x) + \frac{\partial}{\partial y}(yZ_y - zY_y) + \frac{\partial}{\partial z}(yZ_z - zY_z) - \frac{\partial}{\partial t}(yG_z - zG_y) = 0,$$

$$\frac{\partial}{\partial x}(xS_x + tX_x) + \frac{\partial}{\partial y}(xS_y + tX_y) + \frac{\partial}{\partial z}(xS_z + tX_z) + \frac{\partial}{\partial t}(xW - tG_x) = 0.$$

The equations of the first type may be supposed to express the Principle of the Conservation of Angular Momentum. We shall, in fact, regard $yG_z - zG_y$ as the density of the x-component of angular momentum and $(zY_x - yZ_x,\ zY_y - yZ_y,\ zY_z - yZ_z)$ as the components of a vector which specifies the flow of the x-component of angular momentum.

The equations of the second type are not so easily interpreted. We shall, however, regard $xW - tG_x$ as the density of the moment of electromagnetic energy with respect to the plane $x = 0$. This quantity is, in fact, analogous to Σmx, a quantity which occurs in the definition of the centre of mass of a system of particles. Here and in the relation $S = G$ we have an indication of Einstein's relation

$$(\text{Energy}) = (\text{Mass})\ (\text{square of the velocity of light})$$

which is of such importance in the theory of relativity.

The quantities $(xS_x + tX_x,\ xS_y + tX_y,\ xS_z + tX_z)$ will be regarded as the components of a vector which specifies the flow of the moment of electromagnetic energy with respect to the plane $x = 0$. The equation may, then, be interpreted to mean that there is conservation of the moment with respect to the plane $x = 0$. There is, in fact, a striking analogy with the well-known principle that the centre of mass of an isolated mechanical system remains fixed or moves uniformly along a straight line*.

* A. Einstein, *Ann. d. Physik* (4), Bd. xx, S. 627 (1906); G. Herglotz, *ibid.* Bd. xxxvi, S. 493 (1911); E. Bessel Hagen, *Math. Ann.* Bd. lxxxiv, S. 258 (1921).

EXAMPLES

1. Prove that when there are no external forces the equations of motion of an incompressible inviscid fluid of uniform density give the following equations which express the principles of the conservation of momentum and angular momentum, the motion being two-dimensional:

$$\frac{\partial}{\partial t}(\rho u) + \frac{\partial}{\partial x}(\rho u^2) + \frac{\partial}{\partial y}(\rho uv) + \frac{\partial p}{\partial x} = 0,$$

$$\frac{\partial}{\partial t}(\rho v) + \frac{\partial}{\partial x}(\rho uv) + \frac{\partial}{\partial y}(\rho v^2) + \frac{\partial p}{\partial y} = 0,$$

$$\frac{\partial}{\partial t}[\rho(vx - uy)] + \frac{\partial}{\partial x}[\rho u(vx - uy) - yp] + \frac{\partial}{\partial y}[\rho v(vx - uy) + xp] = 0.$$

Hence show that the following integrals vanish when the contour of integration does not contain any singularities of the flow or any body which limits the flow, the motion being steady:

$$\int \rho(v + iu)(ul + vm)\,ds + \int p(m + il)\,ds,$$

$$\int \rho(xv - yu)(ul + vm)\,ds + \int p(xm - yl)\,ds.$$

When the contour does contain a body limiting the flow the integrals round the contour are equal to corresponding integrals round the contour of the body.

2. Let $u(x_1, x_2, \ldots x_n)$ be a function which is to be determined by a variational principle $\delta I = 0$, where

$$I = \iint \ldots \int f(x_1, x_2, \ldots x_n, u, u_1, u_2, \ldots u_n)\,dx_1 dx_2 \ldots dx_n$$

and $u_r = \dfrac{\partial u}{\partial x_r}$. Suppose further that I is unaltered in value by the continuous group of transformations whose infinitesimal transformation is

$$X_r = x_r + \Delta x_r, \quad U = u + \Delta u, \quad (r = 1, 2, \ldots n),$$

and let

$$\overline{\delta u} = \Delta u - \sum_{r=1}^{n} u_r \Delta x_r,$$

$$\psi = \frac{\partial f}{\partial u} - \sum_{r=1}^{n} \frac{\partial}{\partial x_r}\left(\frac{\partial f}{\partial u_r}\right),$$

then

$$\psi\,\overline{\delta u} = \sum_{r=1}^{n} \frac{\partial B_r}{\partial x_r},$$

where

$$B_r = -f\Delta x_r - \frac{\partial f}{\partial u_r}\overline{\delta u} \qquad (r = 1, 2, \ldots n).$$

When the function u satisfies the Eulerian equation $\psi = 0$ the foregoing result gives a set of equations of conservation. [E. Noether, *Gött. Nachr.* p. 238 (1918).]

3. If $n = 2$, $f = (u_1^2 - au_2^2 + \beta u^2)\,e^{\gamma x_2}$ where a, β and γ are arbitrary constants, we may write $\Delta x_1 = \epsilon_1$, $\Delta x_2 = \epsilon_2$, $\Delta u = -\frac{1}{2}\gamma\epsilon_2 u$ where ϵ_1 and ϵ_2 are two independent small quantities whose squares and products may be neglected. Hence show that the differential equation $u_{11} = au_{22} + \gamma a u_2 + \beta u$ leads to two equations of conservation

$$D_1\{2u_1u_2 + \gamma u u_1\} = (D_2 + \gamma)\{u_1^2 + au_2^2 + \beta u^2 + \gamma a u u_2\}$$

$$D_1\{u_1^2 + au_2^2 - \beta u^2\} = (D_2 + \gamma)\{2au_1u_2\}$$

where $D_1 \equiv \dfrac{\partial}{\partial x_1}$, $D_2 \equiv \dfrac{\partial}{\partial x_2}$. [E. T. Copson, *Proc. Edin. Math. Soc.* vol. XLII, p. 61 (1924).]

§ **2·61.** *Kirchhoff's formula.* This theorem relates to the equation

$$\square^2 u + \sigma(x, y, z, t) = 0, \qquad \text{......(A)}$$

and to integrals of type

$$u = \int r^{-1} f(t - r/c)\, F(x_0, y_0, z_0)\, d\tau_0.$$

Let us suppose that throughout a specified region of space and a specified interval of time, u and its differential coefficients of the first order are continuous functions of x, y, z and t; let us suppose also that the differential coefficients of the second order such as $\frac{\partial^2 u}{\partial t^2}, \frac{\partial^2 u}{\partial x^2}$ and the quantity σ are finite and integrable.

Let Q be any point (x_0, y_0, z_0) which need not be in the specified region of space and consider the function v derived from u by substituting $t - r/c$ in place of t, r denoting the distance from Q of any point (x, y, z) in the specified region. It is easy to verify that v satisfies the partial differential equation

$$\nabla^2 v + \frac{2r}{c}\left\{\frac{\partial}{\partial x}\left(\frac{\bar{x}}{r^2}\frac{\partial v}{\partial t}\right) + \frac{\partial}{\partial y}\left(\frac{\bar{y}}{r^2}\frac{\partial v}{\partial t}\right) + \frac{\partial}{\partial z}\left(\frac{\bar{z}}{r^2}\frac{\partial v}{\partial t}\right)\right\} + [\sigma] = 0,$$

$$(\bar{x} = x - x_0, \quad \bar{y} = y - y_0, \quad \bar{z} = z - z_0),$$

where $[\sigma]$ denotes the function derived from σ by substituting $t - r/c$ in place of t.

We now multiply the above equation by $\frac{1}{r}$ and integrate it throughout a volume lying entirely within the specified region of space. The volume integral can then be split into two parts, one of which can be transformed immediately into an integral taken over the boundary of this region. Let the point Q be outside the region of integration, then we have

$$-\iint\left[v\frac{\partial}{\partial n}\left(\frac{1}{r}\right) - \frac{1}{r}\frac{\partial v}{\partial n} - \frac{2}{cr}\frac{\partial r}{\partial n}\frac{\partial v}{\partial t}\right] dS + \int \frac{[\sigma]}{r}\, d\tau.$$

When Q lies within the region of integration the volume may be supposed to be bounded externally by a closed surface S_1 and internally by a small closed surface S_2 surrounding the point Q. Passing to the limit by contracting S_2 indefinitely the value of the integral taken over S_2 is eventually

$$v_Q \iint \frac{\partial}{\partial n}\left(\frac{1}{r}\right) dS = 4\pi v_Q,$$

where v_Q denotes the value of v at Q and this is the same as the value of u at Q. Hence in this case

$$4\pi u_Q = \int \frac{[\sigma]}{r}\, d\tau - \iint\left[v\frac{\partial}{\partial n}\left(\frac{1}{r}\right) - \frac{1}{r}\frac{\partial v}{\partial n} - \frac{2}{cr}\frac{\partial v}{\partial t}\frac{\partial r}{\partial n}\right] dS.$$

Now
$$\frac{\partial v}{\partial n} = \left[\frac{\partial u}{\partial n}\right] - \frac{\partial r}{\partial n}\left[\frac{\partial u}{\partial t}\right],$$

hence finally we have Kirchhoff's formula*

$$4\pi u_Q = \int \frac{[\sigma]}{r}\,d\tau - \iint \left\{ [u]\frac{\partial}{\partial n}\left(\frac{1}{r}\right) - \frac{1}{r}\left[\frac{\partial u}{\partial n}\right] - \frac{1}{cr}\frac{\partial r}{\partial n}\left[\frac{\partial u}{\partial t}\right] \right\} dS,$$

where a square bracket $[f]$ indicates that the quantity f is to be calculated at time $t - r/c$. When the point Q lies outside the region of integration the value of the integral is zero instead of u_Q.

When u and σ are independent of t the formula becomes

$$4\pi u_Q = \int \frac{\sigma}{r}\,d\tau - \iint \left\{ u\frac{\partial}{\partial n}\left(\frac{1}{r}\right) - \frac{1}{r}\frac{\partial u}{\partial n} \right\},$$

and the equation for u is

$$\nabla^2 u + \sigma(x, y, z) = 0.$$

If we make the surface S_1 recede to infinity on all sides the surface integrals can in many cases be made to vanish. We may suppose, for instance, that in distant regions of space the function u has been zero until some definite instant t_0. The time $t - r/c$ then always falls below t_0 when r is sufficiently large and so all the quantities in square brackets vanish. The surface integral also vanishes when u and $\frac{\partial u}{\partial t}$ become zero at infinity and tend to zero as $r \to \infty$ in such a way that u is of order r^{-1} and $\frac{\partial u}{\partial t}, \frac{\partial u}{\partial r}$ of order r^{-2}. In such cases we have the formula

$$4\pi u_Q = \int \frac{[\sigma]}{r}\,d\tau, \qquad \text{......(B)}$$

where the integral is extended over all the regions in which the integrand is different from zero.

If $[\sigma]$ exists only within a number of finite regions which do not extend to infinity the function u_Q defined by this integral possesses the property that $u_Q \to 0$ like r_0^{-1} as $r_0 \to \infty$, r_0 being the distance from the origin of co-ordinates, but it is not always true that $\frac{\partial u_Q}{\partial t}$ is of order r_0^{-2}. To satisfy this condition we may, however, suppose that $\frac{\partial u}{\partial t}$ is zero for values of t less than some value t_0. Then if r is sufficiently large $\left[\frac{\partial u}{\partial t}\right]$ is zero because $- r/c$ falls below t_0.

Wave-potentials of type (B) are called retarded potentials; the analysis shows that they satisfy the equation (A) and that the surface integral

$$\iint \left\{ [u]\frac{\partial}{\partial n}\left(\frac{1}{r}\right) - \frac{1}{r}\left[\frac{\partial u}{\partial n}\right] - \frac{1}{cr}\frac{\partial r}{\partial n}\left[\frac{\partial u}{\partial t}\right] \right\} dS$$

* G. Kirchhoff, *Berlin. Sitzungsber.* S. 641 (1882); *Wied. Ann.* Bd. XVIII (1883); *Ges. Abh.* Bd. II, S. 22. The proof given in the text is due substantially to Beltrami, *Rend. Lincei* (5), t. IV (1895), and is given in a paper by A. E. H. Love, *Proc. London Math. Soc.* (2), vol. I, p. 37 (1903). An extension of Kirchhoff's formula which is applicable to a moving surface has been given recently by W. R. Morgans, *Phil. Mag.* (7), vol. IX, p. 141 (1930).

represents a solution of the wave-equation except for points on the surface S, for this integral survives when we put $\sigma = 0$. It should be noticed, however, that when we put $\sigma = 0$ the quantities $[u]$, $\left[\frac{\partial u}{\partial n}\right]$ $\left[\frac{\partial u}{\partial t}\right]$ become those relating to a wave-function u which is supposed in our analysis to exist and to satisfy the postulated conditions. When the quantities $[u]$, $\left[\frac{\partial u}{\partial n}\right]$, $\left[\frac{\partial u}{\partial t}\right]$ are chosen arbitrarily but in such a way that the surface integral exists it is not clear from the foregoing analysis that the surface integral represents a solution of the wave-equation. If, however, the quantities $[u]$, $\left[\frac{\partial u}{\partial n}\right]$, $\left[\frac{\partial u}{\partial t}\right]$ possess continuous second derivatives with respect to the time the integrand is a solution of the wave-equation for each point on the surface. It can, in fact, be written in the form

$$\frac{\partial}{\partial n}\left[\frac{1}{r}f\left(t-\frac{r}{c}\right)\right]-\frac{1}{r}g\left(t-\frac{r}{c}\right),$$

where $[u] = f\left(t-\frac{r}{c}\right)$, $\left[\frac{\partial u}{\partial n}\right] = g\left(t-\frac{r}{c}\right)$. Now $\frac{\partial}{\partial n}$ stands for

$$l\frac{\partial}{\partial x}+m\frac{\partial}{\partial y}+n\frac{\partial}{\partial z},$$

where l, m and n are constants as far as x, y and z are concerned and each term such as $\frac{\partial}{\partial x}\left[\frac{1}{r}f\left(t-\frac{r}{c}\right)\right]$ is a solution of the wave-equation; consequently the whole integrand is a solution of the wave-equation and it follows that the surface integral itself is a solution of the wave-equation.

In the special case when σ and u are independent of t we have the result that when σ satisfies conditions sufficient to ensure the existence and finiteness of the second derivatives of V (see § 2·32) the integral

$$V = \int \frac{\sigma}{r}\,d\tau$$

is a solution of Poisson's equation

$$\nabla^2 V + 4\pi\sigma(x, y, z) = 0,$$

and the integral
$$U = \iint \left\{u\frac{\partial}{\partial n}\left(\frac{1}{r}\right)-\frac{1}{r}\frac{\partial u}{\partial n}\right\} dS$$

is a solution of Laplace's equation.

§ **2·62**. *Poisson's formula.* When the surface S_1 is a sphere of radius ct with its centre at the point Q, $[u]$ denotes the value of u at time $t = 0$ and Kirchhoff's formula reduces to Poisson's formula*

$$u = \frac{\partial}{\partial t}(t\bar{f}) + t\bar{g},$$

* The details of the transformation are given by A. E. H. Love, *Proc. London Math. Soc.* (2), vol. I, p. 37 (1903).

where $\bar{f}$, $\bar{g}$ denote the mean values of f, g respectively over the surface of a sphere of radius ct having the point (x, y, z) as centre and u is a wave-function which satisfies the initial conditions

$$u = f(x, y, z), \quad \frac{\partial u}{\partial t} = g(x, y, z),$$

when $t = 0$.

If we make use of the fact that each of the double integrals in Poisson's formula is an even function of t we may obtain the relation*

$$\frac{1}{2\tau}\int_{-\tau}^{\tau} u(x, y, z, t)\, dt = \bar{f}.$$

This relation may be written in the more general form

$$\frac{1}{2\tau}\int_{t-\tau}^{t+\tau} u(x, y, z, s)\, ds$$

$$= \frac{1}{4\pi}\int_0^{\pi}\int_0^{2\pi} u(x + c\tau\sin\theta\cos\phi, y + c\tau\sin\theta\sin\phi, z + c\tau\cos\theta, t)\sin\theta\, d\theta\, d\phi.$$

When u is independent of the time this equation reduces to Gauss's well-known theorem relating to the mean value of a potential function over a spherical surface.

If $u(x, y, z, s)$ is a periodic function of s of period 2τ, where τ is independent of x, y and z, the function on the left-hand side is a solution of Laplace's equation, for if

$$V = c^2\int_{t-\tau}^{t+\tau} u(x, y, z, s)\, ds,$$

we have

$$\nabla^2 V = c^2\int_{t-\tau}^{t+\tau} \nabla^2 u\, ds = \int_{t-\tau}^{t+\tau}\frac{\partial^2 u}{\partial s^2}\, ds = 0.$$

It then follows that the double integral on the right-hand side is also a solution of Laplace's equation.

If in Poisson's formula the functions f and g are independent of z the formula reduces to Parseval's formula for a cylindrical wave-function. Since we may write

$$c^2t^2\sin\theta\, d\theta\, d\phi = d\sigma\, .\, \sec\theta = ct\,(c^2t^2 - \rho^2)^{-\frac{1}{2}}\, d\sigma,$$

where $d\sigma$ is an element of area in the xy-plane and ρ the distance of the centre of this element from the projection of the centre of the sphere, we find that

$$2\pi\, .\, u(x, y, t) = \frac{\partial}{\partial t}\iint d\sigma\, .\, (c^2t^2 - \rho^2)^{-\frac{1}{2}} f(x + \xi, y + \eta)$$

$$+ \iint d\sigma\, .\, (c^2t^2 - \rho^2)^{-\frac{1}{2}} g(x + \xi, y + \eta),$$

where $d\sigma = d\xi\, d\eta$ and the integration extends over the interior of the circle

$$\xi^2 + \eta^2 = c^2t^2.$$

* Cf. Rayleigh's *Sound*, Appendix.

This formula indicates that the propagation of cylindrical waves as specified by the equation $\Box^2 u = 0$ is essentially different in character from that of the corresponding spherical waves. In the three-dimensional case the value of a wave-function $u(x, y, z, t)$ at a point (x, y, z) at time t is completely determined by the values of u and $\frac{\partial u}{\partial t}$ over a concentric sphere of radius $c\tau$ at time $t - \tau$. If a disturbance is initially localised within a sphere of radius a then at time t the only points at which there is any disturbance are those situated between two concentric spheres of radii $ct + a$ and $ct - a$ respectively, for it is only in the case of such points that the sphere of radius ct with the point as centre will have a portion of its surface within the sphere of radius a. This means that the disturbance spreads out as if it were propagated by means of spherical waves travelling with velocity c and leaving no residual disturbance as they travel along.

In the two-dimensional case, on the other hand, the value of $u(x, y, t)$ at a point (x, y) at time t is not determined by the values of u and $\frac{\partial u}{\partial t}$ over a concentric circle of radius $c\tau$ at time $t - \tau$. To find $u(x, y, t)$ we must know the values of u and $\frac{\partial u}{\partial t}$ over a series of such circles in which τ varies from zero to some other value τ_1. If the initial disturbance at time $t = 0$ is located within a circle of radius a, all that we can say is that the disturbance at time t is located within a circle of radius $ct + a$ and not simply within the region between two concentric circles of radii $ct + a$, $ct - a$ respectively. Hence as waves travel from the initial region of disturbance with velocity c they leave a residual disturbance behind.

The essential difference between the two cases may be attributed to the fact that in the three-dimensional case the wave-function for a source is of type $r^{-1} f(t - r/c)$, while in the two-dimensional case it is of type*

$$\int_\rho^\infty f\left[t - \frac{s}{c}\right](s^2 - \rho^2)^{-\frac{1}{2}}\,ds = \int_0^\infty f\left[t - \frac{\rho}{c}\cosh\alpha\right]d\alpha.$$

This statement may be given a physical meaning by regarding the wave-function as the velocity potential for sound waves in a homogeneous atmosphere, a source being a small spherical surface which is pulsating uniformly in a radial direction.

If

$$\begin{aligned} f(t) &= 0, \quad (t < T_0) \\ &= 1, \quad (T_1 > t > T_0) \\ &= 0, \quad (t > T_1), \end{aligned}$$

we have

$$\begin{aligned} \int_0^\infty f\left[t - \frac{\rho}{c}\cosh\alpha\right]d\alpha &= 0, \quad (ct < cT_0 + \rho) \\ &= \cosh^{-1}[c(t - T_0)/\rho], \quad (cT_0 + \rho < ct < cT_1 + \rho) \\ &= \cosh^{-1}[c(t - T_0)/\rho] - \cosh^{-1}[c(t - T_1)/\rho], \\ &\qquad (cT_0 + \rho < ct, \quad cT_1 + \rho < ct). \end{aligned}$$

* Cf. H. Lamb, *Hydrodynamics*, 2nd ed. p. 474.

EXAMPLES

1. A wave-function u is required to satisfy the following initial conditions for $t = 0$

$$u = f(x, y), \quad \frac{\partial u}{\partial t} = 0 \quad \text{when } z = 0,$$

$$u = 0, \quad \frac{\partial u}{\partial t} = 0 \quad \text{when } z \neq 0.$$

Prove that u is zero when $z^2 > c^2t^2$ and when $z^2 < c^2t^2$ $u = \bar{f}$ where $\bar{f}$ denotes the mean value of the function f round that circle in the plane $z = 0$ whose points are at a distance ct from the point (x, y, z).

2. If in Ex. 1 the plane $z = 0$ is replaced by the sphere $r = a$, where $r^2 = x^2 + y^2 + z^2$, the wave-function u is equal to $\frac{a}{r}\bar{f}$ when there is a circle (on the sphere) whose points are all at distance ct from (x, y, z) and is otherwise zero.

§ 2·63. *Helmholtz's formula.* When a wave-function is a periodic function of t, Kirchhoff's formula may be replaced by the simpler formula of Helmholtz.

Putting
$$u = U(x, y, z)\, e^{ikct}$$
the wave-equation gives
$$\nabla^2 U + k^2 U = 0.$$

Applying Green's theorem to the space bounded by a surface S and a small sphere surrounding the point (x_1, y_1, z_1) we obtain formula (A)

$$4\pi U(x_1, y_1, z_1) = -\iint U(x, y, z) \frac{\partial}{\partial n}(R^{-1}e^{-ikR})\, dS + \iint \frac{\partial U}{\partial n} R^{-1}e^{-ikR}\, dS,$$

where
$$R^2 = (x - x_1)^2 + (y - y_1)^2 + (z - z_1)^2,$$
and the normal is supposed to be drawn out of the space under consideration. This space can extend to infinity and the theorem still holds provided $U \to 0$ like $Ar^{-1}e^{-ikr}$ as $r \to \infty$, r being the distance of the point (x, y, z) from the origin. It is permissible, of course, for U to become zero more rapidly than this.

A solution of the more general equation
$$\nabla^2 U + k^2 U + \omega(x, y, z) = 0$$
is obtained by adding the term
$$\iiint R^{-1}e^{-ikR}\, \omega(x, y, z)\, dx\, dy\, dz$$
to the right-hand side of (A) and it is chiefly in this case that we want to integrate over all space and obtain a formula in which $U(x_1, y_1, z_1)$ is represented by this last integral.

In the two-dimensional case when u is independent of z, the function to be used in place of $R^{-1}e^{-ikR}$ is derived from the function

$$u = \int_0^\infty f\left(t - \frac{\rho}{c}\cosh\alpha\right) d\alpha,$$

already mentioned. Writing $u = Ue^{ikct}$ as before, the elementary potential function satisfying $\nabla^2 U + k^2 U = 0$ is $K_0(ik\rho)$, where $K_0(ik\rho)$ is defined by the equation

$$K_0(ik\rho) = \int_0^\infty e^{-ik\rho \cosh \alpha} \, d\alpha.$$

This is a function associated with the Bessel functions. For large values of R we have

$$K_0(iR) \doteq -i(\pi/2R)^{\frac{1}{2}} e^{-i\left(R-\frac{\pi}{4}\right)},$$

while for small values of R

$$K_0(iR) + \log(R/2)$$

is finite. The two-dimensional form of Green's theorem gives

$$2\pi U(x_1, y_1) = -\int \left[K_0(ik\rho) \frac{\partial U}{\partial n} - U \frac{\partial}{\partial n} K_0(ik\rho) \right] ds, \quad \text{......(B)}$$

where $$\rho^2 = (x - x_1)^2 + (y - y_1)^2,$$

ds is an element of the boundary curve and n denotes a normal drawn into the region in which the point (x_1, y_1) is situated.

A solution of the more general equation

$$\nabla^2 U + k^2 U + \omega(x, y) = 0$$

is likewise obtained by adding the term

$$\frac{1}{2\pi} \iint K_0(ik\rho) \, \omega(x, y) \, dx \, dy$$

to the right-hand side of (B). When $k = 0$ the corresponding theorem is that a solution of the equation

$$\nabla^2 U + \omega(x, y) = 0$$

is given by $$2\pi U = -\iint \log \rho \, \omega(x, y) \, dx \, dy.$$

§ **2·64**. *Volterra's method**. Let us consider the two-dimensional wave-equation

$$L(u) \equiv \frac{\partial^2 u}{\partial z^2} - \frac{\partial^2 u}{\partial x^2} - \frac{\partial^2 u}{\partial y^2} = f(x, y, z)$$

in which $z = ct$. If the problem is to determine the value of u at an arbitrary point (ξ, η, ζ) from a knowledge of the values of u and its derivatives at points of a surface S, we write

$$X = x - \xi, \quad Y = y - \eta, \quad Z = z - \zeta,$$

and construct the characteristic cone $X^2 + Y^2 = Z^2$ with its vertex at the point (ξ, η, ζ). We shall denote this point by P and the cone by the symbol Γ.

* *Acta Math.* t. XVIII, p. 161 (1894); *Proc. London Math. Soc.* (2), vol. II, p. 327 (1904); *Lectures at Clark University*, p. 38 (1912).

Volterra's method is based on the fact that there is a solution of the wave-equation which depends only on the quantity Z/R, where

$$R^2 = X^2 + Y^2.$$

This solution, v, may, moreover, be chosen so that it is zero on the characteristic cone Γ. The solution may be found by integrating the fundamental solution $(Z^2 - X^2 - Y^2)^{-\frac{1}{2}}$ with respect to Z and is $\cosh^{-1} w$ where $w = Z/R$. Since $w = 1$ on Γ it is easily seen that $v = 0$ on Γ.

For this wave-equation the directions of the normal n and the conormal ν are connected by the equations

$$\cos(\nu x) = \cos(nx), \quad \cos(\nu y) = \cos(ny), \quad \cos(\nu z) = -\cos(nz).$$

At points of Γ the conormal is tangential to the surface and since v is zero on Γ, $\frac{\partial v}{\partial \nu}$ is also zero. The function v is infinite, however, when $R = 0$ and a portion of this line lies in the region bounded by the cone Γ and the surface S. We shall exclude this line from our region of integration by means of a cylinder C, of radius ϵ, whose axis is the line $R = 0$. We now apply the appropriate form of Green's theorem, which is

$$\iiint \{vL(u) - uL(v)\}\, d\tau = \iint_S \left(u\frac{\partial v}{\partial \nu} - v\frac{\partial u}{\partial \nu}\right) dS + \iint_\Gamma \left(u\frac{\partial v}{\partial \nu} - v\frac{\partial u}{\partial \nu}\right) dS,$$

to the region outside C and within the realm bounded by S and Γ. On account of the equations satisfied by u and v the foregoing equation reduces simply to

$$\iint_{C+S} \left(u\frac{\partial v}{\partial \nu} - v\frac{\partial u}{\partial \nu}\right) dS = -\iiint vf\, d\tau.$$

On C we have

$$dS = \epsilon\, d\phi\, dz, \quad \frac{\partial v}{\partial \nu} = \frac{\partial v}{\partial R} = -Z/R\,(Z^2 - R^2)^{\frac{1}{2}},$$

and since

$$\lim_{\epsilon \to 0} (\epsilon \log \epsilon) = 0$$

we have

$$\lim_{\epsilon \to 0} \iint_C \left(u\frac{\partial v}{\partial \nu} - v\frac{\partial u}{\partial \nu}\right) = -2n\int_\zeta^{z_s} u(\xi, \eta, z)\, dz,$$

where (ξ, η, z) is on S and is in the part of S excluded by C. We thus obtain the formula

$$-2\pi\int_{z_s}^{\zeta} u(\xi, \eta, z)\, dz = \iint_S \left(u\frac{\partial v}{\partial \nu} - v\frac{\partial u}{\partial \nu}\right) dS + \iiint vf\, d\tau,$$

and the value of u at P may be derived from this formula by differentiating with respect to ζ. The result is

$$-2\pi u(\xi, \eta, \zeta) = \frac{\partial}{\partial \zeta}\left[\iint_S \left(u\frac{\partial v}{\partial \nu} - v\frac{\partial u}{\partial \nu}\right) dS + \iiint vf\, d\tau\right].$$

EXAMPLE

Prove that a solution of the equation

$$\frac{\partial^2 u}{\partial x^2} + \frac{\partial^2 u}{\partial y^2} = \frac{1}{c^2}\frac{\partial^2 u}{\partial t^2}$$

is given by the following generalisation of Kirchhoff's formula,

$$2\pi u(x, y, t) = \int_\sigma [c^2(t-t_1)^2 - r^2]^{-\frac{1}{2}} \{\cos \widehat{nt} - \cos \widehat{nr} \,.\, c(t-t_1)/r\}\, u(x_1, y_1, t_1)\, dS_1$$

$$+ \int_\sigma [c^2(t-t_1)^2 - r^2]^{-\frac{1}{2}} \left\{\frac{\partial u}{\partial t_1}\cos \widehat{nt} - \left(\frac{\partial u}{\partial x_1}\cos \widehat{nx} + \frac{\partial u}{\partial y_1}\cos \widehat{ny}\right)\right\} dS_1,$$

in which $r^2 = (x - x_1)^2 + (y - y_1)^2$ and the integration extends over the area σ cut out on a surface S in the (x_1, y_1, t_1) space by the characteristic cone

$$(x_1 - x)^2 + (y_1 - y)^2 = c^2(t_1 - t)^2,$$

the time t being chosen so as to satisfy the inequality $t < t_1$.

[V. Volterra.]

§ 2·71. *Integral equations of electromagnetism.* Let us consider a region of space in which for some range of values of t the components of the field-vectors E and H and their first derivatives are continuous functions of x, y, z and t.

Take a closed surface S in this region and assign a time t to each point of S and the enclosed space in accordance with some arbitrary law

$$t = f(x, y, z),$$

where f is a function with continuous first derivatives. We shall suppose that this function gives for the chosen region a value of t lying within the assigned range and shall use the symbol $\mathbf{T}$ to denote the vector with components $\frac{\partial f}{\partial x}, \frac{\partial f}{\partial y}, \frac{\partial f}{\partial z}$.

Writing $Q = H + iE$ as before we consider the integral

$$I = \iint [\mathbf{Q} - i(\mathbf{Q} \times \mathbf{T})]_n\, dS$$

taken over the closed surface S. The suffix n indicates the component along an outward-drawn normal of the vector $\mathbf{P}$ which is represented by the expression within square brackets. Transforming the surface integral into a volume integral we use the symbol Div $\mathbf{P}$ to denote the complete divergence when the fact is taken into consideration that $\mathbf{P}$ depends upon a time t which is itself a function of x, y and z. The symbol div $\mathbf{Q}$, on the other hand, is used to denote the partial divergence when the fact that $\mathbf{Q}$ depends upon x, y and z through its dependence on t is ignored. We then have the equation

$$I = \iiint \operatorname{div} \mathbf{P} \,.\, d\tau,$$

where $$\operatorname{div} \mathbf{P} = \operatorname{div} \mathbf{Q} + \mathbf{T} \,.\, \mathbf{R},$$

and $$\mathbf{R} = \frac{\partial \mathbf{Q}}{\partial t} - i \operatorname{curl} \mathbf{Q}.$$

Now div **Q** and **R** vanish on account of the electromagnetic equations and so these equations are expressed by the single equation $I = 0$. When f is constant $\mathbf{T} = 0$ and the equation $I = 0$ gives

$$\iint H_n dS = 0, \quad \iint E_n dS = 0,$$

which correspond to Gauss's theorem in magneto- and electrostatics. It may be recalled that Gauss's theorem is a direct consequence of the inverse square law for the radial electric or magnetic field strength due to an isolated pole. The contribution of a pole of strength e to an element $E\,dS$ of the second integral is, in fact, $e\,d\omega/4\pi$, where $d\omega$ is the elementary solid angle subtended by the surface element dS at this pole. On integrating over the surface it is seen that the contribution of the pole to the whole integral is e, $\frac{1}{2}e$ or zero according as the pole lies within the surface, on the surface or outside the surface.

This result is usually extended to the case of a volume distribution of electricity by a method of summation and in this case we have the equation

$$\iint E_n dS = \iiint \rho\, d\tau,$$

where ρ denotes the volume density of electricity.

Transforming the surface integral into a volume integral we have the equation

$$\iiint (\operatorname{div} \mathbf{E} - \rho)\, d\tau = 0,$$

which gives

$$\operatorname{div} \mathbf{E} = \rho.$$

Since $\mathbf{E} = -\nabla\phi$ the last equation is equivalent to Poisson's equation

$$\nabla^2\phi + \rho = 0,$$

in which the factor 4π is absent because the electromagnetic equations have been written in terms of rational units. Our aim is now to find a suitable generalisation of this equation. In order to generalise Gauss's theorem the natural method would be to start from the field of a moving electric pole and to look for some generalisation of the idea of solid angle. This method, however, is not easy, so instead we shall allow ourselves to be guided by the principle of the conservation of electricity. The integral which must be chosen to replace $\iiint \rho\, d\tau$ should be of such a nature that its different elements are associated with different electric charges when each element is different from zero. When the elements are associated with a series of different positions of the same group of charges which at one instant lie on a surface it may be called degenerate. In this case we can regard these charges as having a zero sum since a surface is of no thickness. Now it should be noticed that if we write

$$\theta = t - f(x, y, z),$$

the quantity

$$\frac{d\theta}{dt} = \frac{\partial\theta}{\partial t} + v_x \frac{\partial\theta}{\partial x} + v_y \frac{\partial\theta}{\partial y} + v_z \frac{\partial\theta}{\partial z} = 1 - (v \,.\, T)$$

vanishes when the particles of electricity move so as to keep $\theta = 0$, that is so as to maintain the relation $t = f(x, y, z)$, and in this case the integral

$$\iiint \rho \frac{d\theta}{dt} d\tau$$

is degenerate.

We shall try then the following generalisation of Gauss's theorem and examine its consequences:

$$I = i \iiint \rho \,[1 - (v \,.\, T)]\, d\tau.$$

Transforming the surface integral into a volume integral we have

$$0 = \iiint [\operatorname{div} \mathbf{Q} - i\rho + \mathbf{T} \,.\, (\mathbf{R} + i\rho\mathbf{v})]\, d\tau,$$

and since the function f is arbitrary this equation gives

$$\operatorname{div} \mathbf{Q} = i\rho, \quad \mathbf{R} = -\, i\rho\mathbf{v}.$$

Separating the real and imaginary parts we obtain the equations

$$\operatorname{curl} \mathbf{H} = \frac{\partial \mathbf{E}}{\partial t} + \rho\mathbf{v}, \quad \operatorname{div} \mathbf{E} = \rho,$$

$$\operatorname{curl} \mathbf{E} = -\frac{\partial \mathbf{H}}{\partial t}, \quad \operatorname{div} \mathbf{H} = 0,$$

which are the fundamental equations of the theory of electrons. The first two equations give

$$\frac{\partial\rho}{\partial t} + \operatorname{div}(\rho\mathbf{v}) = 0,$$

which is analogous to the equation of continuity in hydrodynamics. Our hypothesis is compatible, then, with the principle of the conservation of electricity. The integral equation

$$\iint [\mathbf{Q} - i(\mathbf{Q} \times \mathbf{T})]_n\, dS = i \iiint \rho\,[1 - (\mathbf{v} \,.\, \mathbf{T})]\, d\tau \qquad \text{......(A)}$$

will be regarded as more fundamental than the differential equations of the theory of electrons if the volume integral is interpreted as the total charge associated with the volume and is replaced by a summation when the charges are discrete. This fundamental equation may be used to obtain the boundary conditions to be satisfied at a moving surface of discontinuity which does not carry electric charges.

Let $t = f(x, y, z)$ be the equation of the moving surface and let the surface S be a thin biscuit-shaped surface surrounding a superficial cap S_0 at points of which t is assigned according to the law $t = f(x, y, z)$. At points of the surface S we shall suppose t to be assigned by a slightly different law $t = f_1(x, y, z)$ which is chosen in such a way that the points of S on one face have just not been reached by the moving surface $t = f(x, y, z)$, while the points on the other face have just been passed over by this surface. Taking the areas of these faces to be small and the thickness of the biscuit quite negligible the equation (A) gives

$$[\mathbf{Q}' - i(\mathbf{Q}' \times \mathbf{T})]_n = [\mathbf{Q}'' - i(\mathbf{Q}'' \times \mathbf{T})]_n,$$

where Q', Q'' are the values of Q on the two sides of the surface of discontinuity and the difference between f_1 and f has been ignored. Writing $q = Q' - Q''$ we have the equation

$$[\mathbf{q} - i(\mathbf{q} \times \mathbf{T})]_n = 0.$$

Now the direction of the cap S_0 is arbitrary and so q must satisfy the relation

$$\mathbf{q} = i(\mathbf{q} \times \mathbf{T}).$$

This gives $q^2 = 0$. Hence if $q = h + ie$ we have the relations

$$h^2 - e^2 = 0, \quad (h \, . \, e) = 0.$$

The equation also gives $\qquad (\mathbf{q} \, . \, \mathbf{T}) = 0,$

and

$$(\mathbf{q} \times \mathbf{T}) = i(\mathbf{q} \times \mathbf{T}) \times \mathbf{T} = i[\mathbf{T}(\mathbf{q} \, . \, \mathbf{T}) - \mathbf{q}\mathbf{T}^2]$$
$$= -iT^2\mathbf{q} \quad \text{or } T^2 = 1 \quad \text{if } \mathbf{q} \neq 0.$$

Hence the moving surface travels with the velocity of light.

A similar method may be used to find the boundary conditions at the surface of separation between two different media. We shall suppose that the media are dielectrics whose physical properties are in each case specified by a dielectric constant K and a magnetic permeability μ. For such a medium Maxwell's equations are

$$\operatorname{curl} \mathbf{H} = \frac{\partial \mathbf{D}}{\partial t}, \qquad \operatorname{div} \mathbf{D} = 0,$$

$$\operatorname{curl} \mathbf{E} = -\frac{\partial \mathbf{B}}{\partial t}, \quad \operatorname{div} \mathbf{B} = 0,$$

where

$$\mathbf{D} = K\mathbf{E}, \quad \mathbf{B} = \mu\mathbf{H}.$$

Instead of these equations we may adopt the more fundamental integral equations

$$\iint [\mathbf{D} - (\mathbf{H} \times \mathbf{T})]_n \, dS = 0,$$

$$\iint [\mathbf{B} + (\mathbf{E} \times \mathbf{T})]_n \, dS = 0,$$

which give the generalisations of Gauss's theorem. The boundary conditions derived from these equations by the foregoing method are

$$\mathbf{d} - (\mathbf{h} \times \mathbf{T}) = 0, \quad \mathbf{b} + (\mathbf{e} \times \mathbf{T}) = 0,$$

where **e**, **h**, **d**, **b** are the differences between the two values of the vectors **E**, **H**, **D**, **B** respectively on the two sides of the moving surface. These equations give

$$(\mathbf{d}\,.\,\mathbf{T}) = 0, \quad (\mathbf{b}\,.\,\mathbf{T}) = 0,$$

$$(\mathbf{d} \times \mathbf{T}) + T^2\mathbf{h} = (\mathbf{h}\,.\,\mathbf{T})\mathbf{T}; \quad (\mathbf{b} \times \mathbf{T}) - T^2\mathbf{e} = -(e\,.\,T)\,\mathbf{T}.$$

If the vector **v** represents the velocity along the normal of the moving surface we have

$$v^2\mathbf{T} = \mathbf{v}, \quad vT = 1,$$

hence the equations may be written in the form

$$d_\nu = 0, \quad b_\nu = 0, \quad h_\tau = (v \times d)_\tau, \quad e_\tau = -(v \times b)_\tau,$$

where d_ν, b_ν denote components of d and b normal to the moving surface and the suffix τ is used to denote a component in any direction tangential to the moving surface. When this surface is stationary the conditions take the simple form

$$d_\nu = 0, \quad b_\nu = 0, \quad h_\tau = 0, \quad e_\tau = 0$$

used by Maxwell, Rayleigh and Lorentz.

When a surface of discontinuity moves in a medium with the physical constants K and μ, we have Heaviside's equations (*Electrical Papers*, vol. II, p. 405)

$$K\,(\mathbf{e}\,.\,\mathbf{T}) = 0, \quad \mu\,(\mathbf{h}\,.\,\mathbf{T}) = 0,$$

$$K\,(\mathbf{e} \times \mathbf{T}) + T^2\,\mathbf{h} = 0, \quad \mu\,(\mathbf{h} \times \mathbf{T}) - T^2\,\mathbf{e} = 0,.$$

and so

$$K\mu\,[(\mathbf{h} \times \mathbf{T}) \times \mathbf{T}] = KT^2\,(\mathbf{e} \times \mathbf{T}) = -T^4\mathbf{h},$$

i.e.

$$K\mu = T^2$$

if

$$\mathbf{h} \neq 0.$$

The surface thus moves with a velocity v given by the equation

$$K\mu v^2 = 1.$$

§ **2·72**. *The retarded potentials of electromagnetic theory*. The electron equations

$$\text{curl}\,H = \frac{1}{c}\left(\frac{\partial E}{\partial t} + \rho v\right), \quad \text{div}\,E = \rho,$$

$$\text{curl}\,E = -\frac{1}{c}\frac{\partial H}{\partial t}, \qquad \text{div}\,H = 0$$

may be satisfied by writing

$$H = \text{curl}\,\mathbf{A}, \quad E = -\frac{1}{c}\frac{\partial \mathbf{A}}{\partial t} - \nabla\Phi,$$

where the potentials **A** and Φ satisfy the relations

$$\text{div}\,\mathbf{A} + \frac{1}{c}\frac{\partial \Phi}{\partial t} = 0, \qquad \text{......(A)}$$

$$\Box^2\mathbf{A} + \rho\frac{\mathbf{v}}{c} = 0, \quad \Box^2\Phi + \rho = 0.$$

The last equations are of the type to which Kirchhoff's formula is applicable and so we may write

$$\Phi = \iiint \frac{1}{r} [\rho] \, d\tau, \quad \mathbf{A} = \iiint \frac{1}{rc} [\rho \mathbf{v}] \, d\tau. \qquad \text{......(B)}$$

These are the retarded potentials of L. Lorenz.

The corresponding potentials for a moving electric pole were obtained by Liénard and Wiechert. They are similar to the above potentials except that the quantity $-c/M$ of § 1·93 takes the place of $1/r$. Let $\xi(t)$, $\eta(t)$, $\zeta(t)$ be the co-ordinates of the electric pole at time t and let a time τ be associated with the space-time point (x, y, z, t) by means of the relations

$$[x - \xi(\tau)]^2 + [y - \eta(\tau)]^2 + [z - \zeta(\tau)]^2 = c^2 (t - \tau)^2, \quad \tau < t, \quad \text{(C)}$$

then

$$M = [x - \xi(\tau)] \, \xi'(\tau) + [y - \eta(\tau)] \, \eta'(\tau) + [z - \zeta(\tau)] \, \zeta'(\tau) - c^2 (t - \tau),$$

and if e is the electric charge associated with the pole the expressions for the potentials are respectively

$$A_x = -\frac{e\xi'(\tau)}{4\pi M}, \quad A_y = -\frac{e\eta'(\tau)}{4\pi M}, \quad A_z = -\frac{e\zeta'(\tau)}{4\pi M}, \quad \Phi = -\frac{ec}{4\pi M}.$$

These satisfy the relation (A) and give the formulae of Hargreaves

$$E_x = \frac{e}{4\pi c} \frac{\partial(\sigma, \tau)}{\partial(x, t)}, \quad H_x = \frac{e}{4\pi} \frac{\partial(\sigma, \tau)}{\partial(y, z)},$$

where

$$\sigma = [x - \xi(\tau)] \, \xi''(\tau) + [y - \eta(\tau)] \, \eta''(\tau) + [z - \zeta(\tau)] \, \zeta''(\tau) + c^2 - [\xi'(\tau)]^2 - [\eta'(\tau)]^2 - [\zeta'(\tau)]^2.$$

It should be remarked that the retarded potentials (B) can be derived from the Liénard potentials by a process of integration analogous to that by which the potential function

$$V = \iiint \frac{\rho \, dx \, dy \, dz}{r}$$

is derived from the potential of an electric pole.

Instead of considering each electric pole within a small element of volume at its own retarded time τ we wish to consider all these electric poles at the same retarded time τ_0 belonging, say, to some particular pole $(\xi_0, \eta_0, \zeta_0, \tau_0)$. Writing

$$\xi(\sigma) = \xi_0(\sigma) + \alpha(\sigma), \quad \eta(\sigma) = \eta_0(\sigma) + \beta(\sigma), \quad \zeta(\sigma) = \zeta_0(\sigma) + \gamma(\sigma),$$

where $\alpha(\sigma)$, $\beta(\sigma)$, $\gamma(\sigma)$ are small quantities, we find that if τ is defined by (C),

$$(\tau - \tau_0) [(x - \xi_0) \xi_0' + (y - \eta_0) \eta_0' + (z - \zeta_0) \zeta_0' - c^2 (t - \tau_0)] + (x - \xi_0) \alpha + (y - \eta_0) \beta + (z - \zeta_0) \gamma = 0,$$

where ξ_0, η_0, ζ_0, ξ_0', η_0', ζ_0', α, β, γ are all calculated in this equation at time τ_0.

On account of the motion the pole (ξ, η, ζ) occupies at time τ the position given by the co-ordinates

$$\xi = \xi_0 + \alpha + (\tau - \tau_0)\,\xi_0',$$
$$\eta = \eta_0 + \beta + (\tau - \tau_0)\,\eta_0',$$
$$\zeta = \zeta_0 + \gamma + (\tau - \tau_0)\,\zeta_0'.$$

Hence
$$\frac{\partial(\xi, \eta, \zeta)}{\partial(\alpha, \beta, \gamma)} = 1 + \xi_0'\frac{\partial \tau}{\partial \alpha} + \eta_0'\frac{\partial \tau}{\partial \beta} + \zeta_0'\frac{\partial \tau}{\partial \gamma}$$
$$= -\frac{c^2(t - \tau_0)}{M_0}.$$

If ρ is the density of electricity when each particle in an element of volume is considered at the associated time τ and ρ_0 is the density when each particle is considered at time τ_0, we have

$$\rho d(\xi, \eta, \zeta) = \rho_0 d(\alpha, \beta, \gamma).$$

Therefore
$$\rho_0 = \rho\frac{cr}{cr - (r \cdot v)},$$

and so
$$\iiint \frac{[\rho_0]}{r}\,dx\,dy\,dz = \iiint \frac{\rho}{r - (r \cdot v)}\,dx\,dy\,dz.$$

Writing $\rho\,dx\,dy\,dz = de$ we obtain the Liénard potentials.

Similar analysis may be used to find the field of a dipole which moves in an arbitrary manner with a velocity less than c and at the same time changes its moment both in magnitude and direction.

Let us consider two electric poles which move along the two neighbouring curves

$$x = \xi(t), \quad y = \eta(t), \quad z = \zeta(t),$$
$$x = \xi(t) + \epsilon\alpha(t), \quad y = \eta(t) + \epsilon\beta(t), \quad z = \zeta(t) + \epsilon\gamma(t),$$

ϵ being a quantity whose square may be neglected. If τ_1 is defined in terms of x, y, z, t by the equation

$$[x - \xi(\tau_1) - \epsilon\alpha(\tau_1)]^2 + [y - \eta(\tau_1) - \epsilon\beta(\tau_1)]^2 + [z - \zeta(\tau_1) - \epsilon\gamma(\tau_1)]^2 = c^2(t - \tau_1)^2,$$

and $\tau_1 = \tau + \epsilon\theta$, we easily find that

$$M\theta + \alpha(\tau)[x - \xi(\tau)] + \beta(\tau)[y - \eta(\tau)] + \gamma(\tau)[z - \zeta(\tau)] = 0.$$

If M_1 is the quantity corresponding to M, we have

$$M_1 = M + \epsilon[\theta M\sigma + (x - \xi)\alpha' + (y - \eta)\beta' + (z - \zeta)\gamma' - \alpha\xi' - \beta\eta' - \gamma\zeta']$$
$$= M + \epsilon[\theta M\sigma + p],$$

say, where σ has the same meaning as before and primes denote differentiations with respect to τ. Now if

$$A_x' = -\frac{e\xi\xi'(\tau_1) + \epsilon\alpha'(\tau_1)\xi}{4\pi M_1}, \quad \Phi' = -\frac{ec}{4\pi M_1},$$
$$A_x = -\frac{e\xi'(\tau)}{4\pi M}, \quad \Phi = -\frac{ec}{4\pi M},$$

we have

$$a_x = \frac{1}{\epsilon}[A_x' - A_x] = -\frac{e}{4\pi M^2}[M\alpha' - p\xi' + M\theta\xi'' - M\theta\sigma\xi'],$$

$$\phi = \frac{1}{\epsilon}[\Phi' - \Phi] = \frac{ec}{4\pi M^2}[p + M\theta\sigma].$$

But $M\alpha' - p\xi' + M\theta\xi'' - M\theta\sigma\xi' \equiv (y-\eta)\,n' - (z-\zeta)\,m' - c^2(t-\tau)\,\alpha'$

$$+ c^2\alpha - n\eta' + m\zeta' + \sigma\{\alpha(t-\tau) - n(y-\eta) + m(z-\zeta)\},$$

where $\quad l = \beta\zeta' - \gamma\eta', \quad m = \gamma\xi' - \alpha\zeta', \quad n = \alpha\eta' - \beta\xi'.$

Hence we may write

$$a_x = -\frac{e}{4\pi}\left[\frac{\partial}{\partial y}\left(\frac{n}{M}\right) - \frac{\partial}{\partial z}\left(\frac{m}{M}\right) + \frac{\partial}{\partial t}\left(\frac{\alpha}{M}\right)\right],$$

$$\phi = \frac{ec}{4\pi}\left[\frac{\partial}{\partial x}\left(\frac{\alpha}{M}\right) + \frac{\partial}{\partial y}\left(\frac{\beta}{M}\right) + \frac{\partial}{\partial z}\left(\frac{\gamma}{M}\right)\right].$$

These results may be obtained also with the aid of the general theorem which gives the effect of an operation $\frac{d}{d\epsilon}$ analogous to differentiation,

$$\frac{d}{d\epsilon}\left[\frac{1}{M}\frac{df}{d\tau}\right] = \frac{\partial}{\partial x}\left[\frac{1}{M}\frac{\partial(f,\xi)}{\partial(\epsilon,\tau)}\right] + \frac{\partial}{\partial y}\left[\frac{1}{M}\frac{\partial(f,\eta)}{\partial(\epsilon,\tau)}\right] + \frac{\partial}{\partial z}\left[\frac{1}{M}\frac{\partial(f,\zeta)}{\partial(\epsilon,\tau)}\right] + \frac{\partial}{\partial t}\left[\frac{1}{M}\frac{\partial(f,\tau)}{\partial(\epsilon,\tau)}\right],$$

f being a function of τ and ϵ. Writing $f = \xi$, $\frac{\partial\xi}{\partial\epsilon} = \alpha$, $\frac{\partial\eta}{\partial\epsilon} = \beta$, $\frac{\partial\zeta}{\partial\epsilon} = \gamma$ the expression for a_x is at once obtained from that for A_x. Writing $f = \tau$ we obtain the expression for ϕ.

The formulae show that the field of the moving dipole may be derived from Hertzian vectors Π and Γ by means of the formulae

$$\mathbf{a} = \frac{\partial\mathbf{\Pi}}{c\,\partial t} + \operatorname{curl}\mathbf{\Gamma}, \quad \phi = -\operatorname{div}\mathbf{\Pi}, \quad \mathbf{\Pi} = -\frac{ec\mathbf{u}}{4\pi M}, \quad \Gamma = -\frac{e\mathbf{w}}{4\pi M}, \qquad \text{(D)}$$

where $\mathbf{u}$ and $\mathbf{w}$ are vector functions of τ with components (α, β, γ), (l, m, n) respectively. If $\mathbf{v}$ denotes the vector with components (ξ', η', ζ') we have the relations

$$(\mathbf{v}\,.\,\mathbf{w}) = 0, \quad (\mathbf{u}\,.\,\mathbf{w}) = 0,$$

consequently Hertzian vectors of types (D) do not specify the field of a moving electric dipole unless these relations are satisfied.

Since
$$\mathbf{A} = \frac{1}{c}\frac{\partial\mathbf{\Pi}}{\partial t} + \operatorname{curl}\mathbf{\Gamma}, \quad \mathbf{B} = \frac{1}{c}\frac{\partial\mathbf{\Gamma}}{\partial t} - \operatorname{curl}\mathbf{\Pi},$$
$$\Phi = -\operatorname{div}\mathbf{\Pi}, \qquad \Psi = -\operatorname{div}\mathbf{\Gamma},$$

where $\mathbf{B}$ and Φ are the electromagnetic potentials of magnetic type, we may write down the potentials for a moving magnetic dipole by analogy.

We simply replace $\mathbf{\Pi}$ by $\mathbf{\Gamma}$ and $\mathbf{\Gamma}$ by $-\mathbf{\Pi}$. Hence the potentials of a moving magnetic dipole are of type

$$a_x = \frac{m}{4\pi}\left[\frac{\partial}{\partial y}\left(\frac{\gamma}{M}\right) - \frac{\partial}{\partial z}\left(\frac{\beta}{M}\right) - \frac{\partial}{\partial t}\left(\frac{l}{M}\right)\right],$$

$$\phi = \frac{mc}{4\pi}\left[\frac{\partial}{\partial x}\left(\frac{l}{M}\right) + \frac{\partial}{\partial y}\left(\frac{m}{M}\right) + \frac{\partial}{\partial z}\left(\frac{n}{N}\right)\right].$$

Let us now calculate the rate of radiation from a stationary electric dipole whose moment varies periodically. Taking the origin at the centre of the dipole, we write

$$\Pi_x = \frac{1}{r} f(\tau), \quad \Pi_y = \frac{1}{r} g(\tau), \quad \Pi_z = \frac{1}{r} h(\tau), \quad \tau = t - r/c,$$

where f, g, h are periodic functions with period T. The vector is zero since there is no velocity.

In calculating the radiation we need only retain terms of order $1/r$ in the expressions for E and H. To this order of approximation we have

$$E_x = -\frac{1}{c^2 r} f'' + \frac{x}{c^2 r^3}(xf'' + yg'' + zh''),$$

$$H_x = \frac{1}{r}(yE_z - zE_y),$$

$$E_y H_z - E_z H_y = (x/r) E^2,$$

where $$E^2 = \frac{1}{c^4 r^2}(f''^2 + g''^2 + h''^2) - \frac{1}{c^4 r^4}(xf'' + yg'' + zh'')^2.$$

The rate of radiation is obtained by integrating cE^2 over a spherical surface $r = a$, where a is very large. With a suitable choice of the axis of z we may write

$$E^2 = \frac{1}{c^4 r^2}(f''^2 + g''^2 + h''^2)\sin^2\theta,$$

and the value of the integral over the sphere is

$$\frac{8\pi}{3c^3}(f''^2 + g''^2 + h''^2).$$

If now $$f''^2 + g''^2 + h''^2 = \left(\frac{A}{4\pi}\right)^2 \cos^2\frac{2\pi\tau}{T}\left(\frac{2\pi}{T}\right)^4,$$

the mean value over a period T is

$$\frac{A^2}{12\pi c^3}\left(\frac{2\pi}{T}\right)^4.$$

EXAMPLE

If
$$[l(s)]^2 + [m(s)]^2 + [n(s)]^2 = 1$$
and
$$L(x, y, z, t, s) = [x - \xi(s)]\, l(s) + [y - \eta(s)]\, m(s) + [z - \zeta(s)]\, n(s) - c(t - s),$$
prove that the potentials

$$A_x = \frac{1}{4\pi}\int_{-\infty}^{\tau} f'(s)\, \frac{l(s)\, ds}{L(x, y, z, t, s)} - f(\tau)\frac{\xi'(\tau)}{4\pi M},$$

$$A_y = \frac{1}{4\pi}\int_{-\infty}^{\tau} f'(s)\, \frac{m(s)\, ds}{L(x, y, z, t, s)} - f(\tau)\frac{\eta'(\tau)}{4\pi M},$$

$$A_z = \frac{1}{4\pi}\int_{-\infty}^{\tau} f'(s)\, \frac{n(s)\, ds}{L(x, y, z, t, s)} - f(\tau)\frac{\zeta'(\tau)}{4\pi M},$$

$$\Phi = \frac{1}{4\pi}\int_{-\infty}^{\tau} f'(s)\, \frac{ds}{L(x, y, z, t, s)} - f(\tau)\,\frac{c}{4\pi M}$$

are wave-functions satisfying the condition

$$\operatorname{div} A + \frac{1}{c}\frac{\partial \Phi}{\partial t} = 0.$$

Show also that in the field derived from these potentials the charge associated with the moving point $\xi(\tau)$, $\eta(\tau)$, $\zeta(\tau)$ is $f(\tau)$, the variation with the time being caused by the radiation of electric charges from the moving singularity in a varying direction specified by the direction cosines $l(\tau)$, $m(\tau)$, $n(\tau)$.

§ **2·73**. *The reciprocal theorem of wireless telegraphy.* If we multiply the electromagnetic equations

$$\dot{B}_1 = -\operatorname{curl}(cE_1), \quad C_1 = \operatorname{curl}(cH_1)$$

for a field (E_1, H_1) by H_2, $-E_2$ respectively, where (E_2, H_2) are the field vectors of a second field in the same medium, and multiply the field equations

$$\dot{B}_2 = -\operatorname{curl}(cE_2), \quad C_2 = \operatorname{curl}(cH_2)$$

for this second field by $-H_1$, E_1 respectively and then add all our equations together, we obtain an equation which may be written in the form

$$(H_2 \cdot \dot{B}_1) - (H_1 \cdot \dot{B}_2) + (E_1 \cdot C_2) - (E_2 \cdot C_1) = c \operatorname{div}(E_2 \times H_1) - c \operatorname{div}(E_1 \times H_2). \quad \text{......(A)}$$

We now assume that both fields are periodic and have the same frequency $\omega/2\pi$. Introducing the symbol T for the time factor $e^{-i\omega t}$ and assuming that it is understood that only the real part of any complex expression in an equation is retained, we may write

$$H_1 = Th_1, \quad H_2 = Th_2, \quad E_1 = Te_1, \quad E_2 = Te_2,$$
$$B_1 = Tb_1, \quad B_2 = Tb_2, \quad C_1 = Tc_1, \quad C_2 = Tc_2,$$

where the vectors e_1, h_1, c_1, etc. depend only on x, y and z.

Now let κ, μ and σ be the specific inductive capacity, permeability and conductivity of the medium at the point (x, y, z) and let a denote the quantity $(k\omega + i\sigma)/c$, then we have the equations

$$\dot{B}_1 = -i\omega\mu h_1 T, \quad \dot{B}_2 = -i\omega\mu h_2 T, \quad C_1 = -ice_1 Ta, \quad C_2 = -ice_2 Ta,$$

which indicate that the left-hand side of equation (A) vanishes. The equation thus reduces to the simple form*

$$\operatorname{div}(e_2 \times h_1) = \operatorname{div}(e_1 \times h_2).$$

This equation may be supposed to hold for the whole of the medium surrounding two antennae† if these sources of radiation are excluded by small spheres K_1 and K_2. An application of Green's theorem then gives

$$\int_{K_1} (e_2 \times h_1)_n \, dS + \int_{K_2} (e_2 \times h_1)_n \, dS = \int_{K_1} (e_1 \times h_2)_n \, dS + \int_{K_2} (e_1 \times h_2)_n \, dS,$$

an equation which may be written briefly in the form

$$J_{11} + J_{21} = J_{12} + J_{22}.$$

Let the first antenna be at the origin of co-ordinates and let us suppose for simplicity that it is an electrical antenna whose radiation may be represented approximately in the immediate neighbourhood of O by the field derived from a Hertzian vector ΠT with a single component $\Pi_z T$, where

$$4\pi\Pi_z = Me^{ipR} \quad (c^2p^2 = k\mu\omega^2 + i\sigma\mu\omega = \mu\omega ac),$$

and M is the moment of the dipole. In making this assumption we assume that the primary action of the source preponderates over the secondary actions arising from waves reflected or diffracted by the homogeneities of the surrounding medium.

Using a_1 to denote the value of a at O and writing Π' for $\partial\Pi/\partial r$, (ξ_2, η_2, ζ_2) for the components of e_2, we have

$$J_{11} = -ia_1 \int_{K_1} \{(x^2 + y^2)\,\zeta_2 - xz\xi_2 - yz\eta_2\}\, \Pi' R^{-2} dS.$$

For the integration over the surface K_1 the quantities R^2, Π' and the vector e_2 may be treated as constants, for K_1 is very small and e_2 varies continuously in the neighbourhood of O. We also have

$$\int yz\,dS = \int zx\,dS = \int xy\,dS = 0, \quad \int x^2 dS = \int y^2 dS = \int z^2 dS = 4\pi R^4/3,$$

$$\int x\,dS = 0, \quad \int x^3 dS = 0, \quad \int xy^2 dS = 0.$$

Therefore $\qquad J_{11} = -8\pi i a_1 R^2 \Pi' \zeta_2/3,$

and, since $\qquad \lim_{R\to 0} R^2\Pi' = -M_1/4\pi,$

we have finally $\qquad J_{11} = 2ia_1 M_1 \zeta_2/3.$

Now the rate at which the antenna at O radiates energy is

$$\bar{S} = k\omega^4 M^2/12\pi V^3, \qquad V = c\,(\epsilon\mu)^{-\frac{1}{2}}.$$

* H. A. Lorentz, *Amsterdam. Akad.* vol. IV, p. 176 (1895–6).

† A. Sommerfeld, *Jahrb. d. drahtl. Telegraphie*, Bd. XXVI, S. 93 (1925); W. Schottky, *ibid.* Bd. XXVII, S. 131 (1926).

Assuming that $\bar{S}$ is the same for both antennae we obtain the useful expression

$$J_{11} = 2ia_1 (\bar{S} V_1^3)^{\frac{1}{2}} k_1^{-\frac{1}{2}} (12\pi)^{\frac{1}{2}} \zeta_2/3.$$

The integral J_{21} is seen to be zero because it involves only terms which change sign when the signs of x, y and z are changed.

Evaluating J_{22} and J_{11} in a similar way we obtain the equation

$$(\kappa_1 + i\sigma_1/\omega)(V_1^3/\kappa_1)^{\frac{1}{2}} \zeta_2 = (\kappa_2 + i\sigma_2/\omega)(V_2^3/\kappa_2)^{\frac{1}{2}} \zeta_1,$$

where ξ_1, η_1, ζ_1 are the components, at the second antenna O_2, of the vector e_1. The amplitudes and phases of the field strength received at the two antennae are thus the same when both antennae are of the electric type and are situated at places where the medium has the same properties and emits energy at the same rate. When the two antennae are both of magnetic type the corresponding relation is

$$(\mu_2 V_2^3)^{\frac{1}{2}} \gamma_1 = (\mu_1 V_1^3)^{\frac{1}{2}} \gamma_2,$$

where $\alpha_1, \beta_1, \gamma_1$ are the components of h_1 and $\alpha_2, \beta_2, \gamma_2$ are the components of h_2. The antennae are again supposed to be directed along the axis of z but there is a more general theorem in which the two antennae have arbitrary directions.

The relation (A) and the associated reciprocal relation remind one of the very general extension of Green's theorem which was given by Volterra* for the case of a set of partial differential equations associated with a variational principle. This extension of Green's theorem is closely connected with a property of self-adjointness which has been shown by Hirsch, Kürschák, Davis and La Paz† to be characteristic of certain equations associated with a variational principle. In the case of the Eulerian equation $F = 0$ associated with a variational principle $\delta I = 0$, where

$$I = \int f[x_1, x_2, \dots x_n; u; u_1, u_2, \dots u_n] \, dx_1 dx_2 \dots dx_n,$$

$$u_s = \frac{\partial u}{\partial x_s}, \quad u_{rs} = \frac{\partial^2 u}{\partial x_r \partial x_s}, \qquad (r, s = 1, 2, \dots n),$$

the equation which is self-adjoint is the "equation of variation" for v

$$0 = v\frac{\partial F}{\partial u} + v_1 \frac{\partial F}{\partial u_1} + \dots v_n \frac{\partial F}{\partial u_n} + v_{11}\frac{\partial F}{\partial u_{11}} + v_{12}\frac{\partial F}{\partial u_{12}} + \dots .$$

* V. Volterra, *Rend. Lincei* (4), t. VI, p. 43 (1890).

† A. Hirsch, *Math. Ann.* Bd. XLIX, S. 49 (1897); J. Kürschák, *ibid.* Bd. LX, S. 157 (1905); D. R. Davis, *Trans. Amer. Math. Soc.* vol. XXX, p. 710 (1928); L. La Paz, *ibid.* vol. XXXII, p. 509 (1930).

CHAPTER III

TWO-DIMENSIONAL PROBLEMS

§ **3·11**. *Simple solutions and methods of generalisation of solutions.* A simple solution of a linear partial differential equation of the homogeneous type is one which can be expressed in the form of a product of a number of functions each of which has one of the independent variables as its argument. Thus Laplace's equation

$$\frac{\partial^2 V}{\partial x^2} + \frac{\partial^2 V}{\partial y^2} = 0$$

possesses a simple solution of type

$$V = e^{-my} \cos m (x - x'), \qquad \text{......(A)}$$

where m and x' are arbitrary constants; the equation

$$\frac{\partial V}{\partial t} = \kappa \frac{\partial^2 V}{\partial x^2}$$

possesses the simple solution

$$V = e^{-\kappa m^2 t} \cos m (x - x'), \qquad \text{......(B)}$$

and the wave-equation

$$\frac{\partial^2 V}{\partial x^2} = \frac{1}{c^2} \frac{\partial^2 V}{\partial t^2}$$

possesses the simple solution

$$V = \frac{1}{m} \sin mct \cos m (x - x'). \qquad \text{......(C)}$$

The last one is of great historical interest because it was used by Brook Taylor in a discussion of the transverse vibrations of a fine string. It should be noticed that the end conditions $V = 0$ when $x = \pm a/2$ are satisfied by a solution of this type only if $ma = 2n + 1$, where n is an integer. There are thus periodic solutions of period

$$T = 2\pi/mc = 2\pi a/(2n + 1).$$

If $M (m, x')$ denotes one of these simple solutions a more general solution may be obtained by multiplying by an arbitrary function of m and x' and then summing or integrating with respect to the parameters m and x'. This method of superposition is legitimate because the partial differential equations are linear. When infinite series and infinite ranges of integration are used it is not quite evident that the resulting expression will be a solution of the appropriate partial differential equation and some process of verification is necessary. If, for instance, we take as our generalisation the integral

$$V = \int_0^\infty M(m, x') f(m)\, dm \qquad (t > 0,\ y > 0),$$

and distinguish between solutions of the different equations by writing v for V when we are dealing with a solution of the second equation and y for V when we are dealing with a solution of the third equation, we easily find that when $f(m) = 1$ we have

$$V = \frac{y}{(x - x')^2 + y^2},$$

$$2v = (\pi/\kappa t)^{\frac{1}{2}} \exp[-(x - x')^2/4\kappa t],$$

$$y = \frac{\pi}{2}, \frac{\pi}{4} \text{ or } 0 \text{ according as } |x - x'| \lesseqqgtr ct \qquad (t > 0).$$

It is easily verified that these expressions are indeed solutions of their respective equations. These solutions are of fundamental importance because each one has a simple type of point of discontinuity. In the last case the points of discontinuity for y move with constant velocity c.

We may generalise each of these particular solutions by writing V, v or y equal to

$$\int_0^\infty \int_{-\infty}^\infty M(m, x')\, F(x')\, dx' dm,$$

where the integration with regard to x' precedes that with respect to m. When the order of integration can be changed without altering the value of the repeated integral the resulting expressions are respectively

$$V = \int_{-\infty}^\infty \frac{yF(x')\, dx'}{(x - x')^2 + y^2},$$

$$2v = (\pi/\kappa t)^{\frac{1}{2}} \int_{-\infty}^\infty \exp[-(x - x')^2/4\kappa t]\, F(x')\, dx',$$

$$y = \frac{\pi}{2} \int_{x-ct}^{x+ct} F(x')\, dx'.$$

The last expression evidently satisfies the differential equation when $F(x)$ is a function with a continuous derivative; y represents, moreover, a solution which satisfies the conditions

$$y = 0, \quad \frac{\partial y}{\partial t} = c\pi F(x),$$

when $\qquad t = 0.$

When the function $F(x)$ is of a suitable type the functions V and v also satisfy simple boundary conditions. This may be seen by writing

$$x' = x + y \tan(\theta/2)$$

in the first integral and $\qquad x' = x + 2u(\kappa t)^{\frac{1}{2}}$

in the second. The resulting classical formulae

$$V = \frac{1}{2}\int_{-\pi}^{\pi} F\,[x + y \tan(\theta/2)]\,d\theta,$$

$$v = \pi^{\frac{1}{2}}\int_{-\infty}^{\infty} F\,[x + 2u\,(\kappa t)^{\frac{1}{2}}]\,du$$

suggest that $V = \pi F(x)$ when $y = 0$ and $v = \pi F(x)$ when $t = 0$.

These results are certainly true when the function $F(x)$ is continuous and integrable over the infinite range but require careful proof. The theorems suggest that in many cases*

$$\pi F(x) = \int_0^{\infty} dm \int_{-\infty}^{\infty} \cos m\,(x - x')\,F(x')\,dx'.$$

This is a relation of very great importance which is known as Fourier's integral theorem. Much work has been done to determine the conditions under which the theorem is valid.

A useful equivalent formula is

$$2\pi F(x) = \int_{-\infty}^{\infty} dm \int_{-\infty}^{\infty} e^{im(x-x')}\,F(x')\,dx'.$$

When $F(x)$ is an even function of x Fourier's integral theorem may be replaced by the reciprocal formulae

$$F(x) = \int_0^{\infty} \cos mx\,G(m)\,dm,$$

$$G(m) = \frac{2}{\pi}\int_0^{\infty} \cos mx\,F(x)\,dx,$$

and when $F(x)$ is an odd function of x the theorem may be replaced by the reciprocal formulae

$$F(x) = \int_0^{\infty} \sin mx\,H(m)\,dm,$$

$$H(m) = \frac{2}{\pi}\int_0^{\infty} \sin mx\,F(x)\,dx.$$

The formulae require modification at a point x, where $F(x)$ is discontinuous. If $F(x)$ approaches different finite values from different sides of

* The theorem is usually established for a continuous function which is of bounded variation and is such that $\int_{-\infty}^{\infty} |F(x)|\,dx$ and $\int_{-\infty}^{0} |F(x)|\,dx$ exist. $F(x)$ may also have a finite number of points of discontinuity at which $F(x+0)$ and $F(x-0)$ exist but in this case the integral represents $\frac{\pi}{2}[F(x+0) + F(x-0)]$. Proofs of the theorem are given in Carslaw's *Fourier Series and Integrals*; in Whittaker and Watson's *Modern Analysis*; and in Hobson's *Functions of a Real Variable*.

the point x the integral is found to be equal to the mean of these values instead of one of them. Thus in the last pair of formulae we can have

$$F(x) = 1 \quad x < \phi, \quad H(m) = \frac{2}{\pi}[1 - \cos m],$$
$$= 0 \quad x > \phi,$$

but the integral gives $\qquad F(1) = \tfrac{1}{2}.$

EXAMPLE

If $S(x, t) = (\pi\kappa t)^{-\frac{1}{2}} \exp[-x^2/4\kappa t]$ and $f(x)$ is continuous bit by bit a solution of $\frac{\partial v}{\partial t} = \kappa \frac{\partial^2 v}{\partial x^2}$, which satisfies the condition $v = f(x)$ when $t = 0$ and $-\infty < x < \infty$, is given by the formula

$$v = \tfrac{1}{2}\int_{-\infty}^{\infty} S(x - x_0, t) f(x_0)\, dx_0 + \tfrac{1}{2}\sum_{n=1}^{m} [f(x_n) - \bar{f}(x_n)]$$
$$\times \int_0^{\infty} [S(x - x_n - \xi, t) - S(x - x_n + \xi, t)]\, d\xi,$$

where $2\bar{f}(x_n) = f(x_n + 0) + f(x_n - 0)$ and the summation extends over all the points of discontinuity of $f(x)$.

§ 3·12. *A study of Fourier's inversion formula.* The first step is to establish the Riemann-Lebesgue lemmas*.

Let $g(x)$ be integrable in the Riemann sense in the interval $a \leqslant x \leqslant b$ and when the integral is improper let $|g(x)|$ be integrable. We shall prove that in these circumstances

$$\lim_{k\to\infty} \int_a^b \sin(kx)\,.\,g(x)\,dx = 0.$$

Let us first consider the case when $g(x)$ is bounded in the range (a, b) and G is the upper bound of $|g(x)|$. We divide the range (a, b) into n parts by the points $x_1, x_2, \dots x_{n-1}$ and form the sums

$$S_n = U_1(x_1 - a) + U_2(x_2 - x_1) + \dots U_n(b - x_{n-1}),$$
$$s_n = L_1(x_1 - a) + L_2(x_2 - x_1) + \dots L_n(b - x_{n-1}),$$

where U_r, L_r are the bounds of $g(x)$ in the interval $x_{r-1} \leqslant x \leqslant x_r$, so that in this interval

$$g(x) = g_r(x_{r-1}) + \omega_r(x), \quad |\omega_r(x)| \leqslant U_r - L_r.$$

Since $g(x)$ is integrable we may choose n so large that $S_n - s_n < \epsilon$, where ϵ is any small positive quantity given in advance. Now

$$\left|\int_a^b g(x)\sin kx\,dx\right| = \left|\Sigma\, g_r(x_{r-1}) \int \sin kx\,.\,dx + \Sigma \int \omega_r(x) \sin kx \cdot dx\right|,$$

* The proof in the text is due to Prof. G. H. Hardy and is based upon that in Whittaker and Watson's *Modern Analysis*.

the summations on the right being from $r = 1$ to $r = n$ and the integrations from x_{r-1} to x_r. With the same convention

$$\left|\int_a^b g(x) \sin kx\, dx\right| < \Sigma \left| g_r(x_{r-1})\right| . \left|\int \sin kx\, .\, dx\right| + \Sigma \int \left|\omega_r(x)\right| dx$$

$$< 2nG/k + S_n - s_n < 2nG/k + \epsilon.$$

Keeping n fixed after ϵ has been chosen and making k sufficiently large we can make the last expression less than 2ϵ and so the theorem follows for the case in which $g(x)$ is bounded in (a, b). When $g(x)$ is unbounded and $|g(x)|$ integrable in (a, b) we may, by the definition of the improper integral, enclose the points at which $g(x)$ is unbounded in a finite number of intervals $i_1, i_2, \dots i_p$ such that

$$\sum_{r=1}^{p} \int_{i_r} |g(x)|\, dx < \epsilon.$$

Now let G denote the upper bound of $g(x)$ for values of x outside the intervals i_r and let $e_1, e_2, \dots e_{p+1}$ denote the portions of the interval (a, b) which do not belong to $i_1, i_2, \dots i_p$, then we may prove as before that

$$\left|\int_a^b g(x) \sin kx\, .\, dx\right| < \left|\sum_{r=1}^{p+1} \int_{e_r} g(x) \sin kx\, .\, dx\right| + \sum_{r=1}^{p} \int_{i_r} \left| g(x) \sin kx \right| dx$$

$$< 2nG/k + 2\epsilon.$$

Now the choice of ϵ fixes n and G, consequently the last expression may be made less than 3ϵ by taking a sufficiently large value of k. Hence the result follows also when $g(x)$ is unbounded, but subject to the above restriction.

Some restriction of this type is necessary because in the case when $g(x)$ is the unbounded function $x^{-1}(1 - x^2)^{-\frac{1}{2}}$ for which $|g(x)|$ is not integrable in the range $(-1, 1)$ we have

$$\int_{-1}^{1} \sin kx\, g(x)\, dx = \pi \int_0^k J_0(\tau)\, d\tau,$$

and as $k \to \infty$ $$\int_0^k J_0(\tau)\, d\tau = \int_0^\infty J_0(\tau)\, d\tau = 1.$$

The next step is to show that if x is an internal point of the interval $(-\alpha, \beta)$, where α and β are positive, and if $f(x)$ satisfies in $(-\alpha, \beta)$ the following conditions:

(1) $f(x)$ is continuous except at a finite number of points of discontinuity, and if $f(x)$ has an improper integral $|f(x)|$ is integrable;

(2) $f(x)$ is of bounded variation, then

$$\lim_{k\to\infty} \int_{-\alpha}^{\beta} \frac{\sin k(t - x)}{t - x} f(t)\, dt = \frac{\pi}{2}[f(x + 0) + f(x - 0)] = \pi \bar{f}(x), \text{ say.}$$

Let us write $$\int_{-\alpha}^{\beta} \frac{\sin k(t - x)}{t - x} f(t)\, dt = \int_{-\alpha}^{x} + \int_{x}^{\beta}$$

and transform the integrals by the substitutions $t = x - u$ and $t = x + u$ respectively, then

$$\int_{-\alpha}^{\beta} \frac{\sin k(t-x)}{t-x} f(t)\, dt$$

$$= \int_0^{\alpha+x} \frac{\sin ku}{u}[f(x-u) - f(x-0)]\, du + f(x-0)\int_0^{\alpha+x} \sin ku \,.\, du/u$$

$$+ \int_0^{\beta-x} \frac{\sin ku}{u}[f(x+u) - f(x+0)]\, du + f(x+0)\int_0^{\beta-x} \sin ku \,.\, du/u.$$

Now let c denote one of the two positive quantities $\alpha + x$, $\beta - x$, then

$$\int_0^c \sin ku \,.\, du/u = \int_0^{kc} \sin v \,.\, dv/v \to \frac{\pi}{2} \text{ as } k \to \infty.$$

Also, let $F(u)$ denote one of the two functions $f(x-u) - f(x-0)$, $f(x+u) - f(x+0)$, then $F(0) = 0$ and $F(u)$ is of bounded variation in the interval $(0 \leqslant u \leqslant c)$. We may therefore write

$$F(u) = H_1(u) - H_2(u),$$

where $H_1(u)$ and $H_2(u)$ are positive increasing functions such that

$$H_1(0) = H_2(0) = 0.$$

Given any small positive quantity ϵ we can now choose a positive number z such that

$$0 \leqslant H_1(u) < \epsilon, \quad 0 \leqslant H_2(u) < \epsilon,$$

whenever $0 \leqslant u \leqslant z$. We next write

$$\int_0^c \sin ku \,.\, F(u)\, du/u = \int_z^c \sin ku \,.\, F(u)\, du/u$$

$$+ \int_0^z \sin ku \,.\, H_1(u)\, du/u - \int_0^z \sin ku \,.\, H_2(u)\, du/u.$$

Let $H(u)$ denote either of the two functions $H_1(u)$, $H_2(u)$; since this function is a positive increasing function the second mean value theorem for integrals may be applied and this tells us that there is a number v between 0 and z for which

$$\left|\int_0^z \sin ku \,.\, H(u)\, du/u\right| = \left|H(z)\int_v^z \sin ku \,.\, du/u\right|$$

$$= H(z)\left|\int_{kv}^{kz} \sin s \,.\, ds/s\right|.$$

Since $\int_0^\infty \sin s \,.\, ds/s$ is a convergent integral, $\int_\tau^\infty \sin s \,.\, ds/s$ has an upper bound B which is independent of τ and it is then clear that

$$\left|\int_0^z \sin ku \,.\, H(u)\, du/u\right| \leqslant 2BH(z) < 2B\epsilon.$$

By the first lemma k may be chosen so large that

$$\left|\int_z^c \sin ku \,.\, F(u)\, du/u\right| < \epsilon,$$

and so we have the result that

$$\lim_{k\to\infty}\int_0^c \sin ku \,.\, F(u)\, du/u = 0.$$

It now follows that

$$\lim_{k\to\infty}\int_{-a}^{\beta} \frac{\sin k(t-x)}{t-x} f(t)\, dt = \frac{\pi}{2}[f(x+0)+f(x-0)].$$

To extend this result to the case in which the limits are $-\infty$ and ∞ we shall assume that for $x>\beta$

$$f(x) = P_1(x) - P_2(x),$$

where $P_1(x)$ and $P_2(x)$ are positive functions which decrease steadily to zero as x increases to ∞. A similar supposition will be made for the range $x < -a$, the positive functions now being such that they decrease steadily to zero as x decreases to $-\infty$. Since

$$\frac{P_1(t)}{t-x}, \quad (x<\beta<t<\gamma)$$

is a positive decreasing function of t for $t>\beta$ we may apply the second mean value theorem for integrals and this tells us that

$$\int_\beta^\gamma \frac{\sin k(t-x)}{t-x} P_1(t)\, dt = \frac{P_1(\beta)}{\beta-x}\int_\beta^\xi \sin k(t-x)\, dt + \frac{P_1(\gamma)}{\gamma-x}\int_\xi^\gamma \sin k(t-x)\, dt,$$

$$(\beta<\xi<\gamma).$$

Now let $|P_1(x)| < M$ for $x>\beta$, then

$$\left|\int_\beta^\gamma \frac{\sin k(t-x)}{t-x} P_1(t)\, dt\right| < \frac{M}{k(\beta-x)}\left\{\left|\int_{k(\beta-x)}^{k(\xi-x)} \sin u\, du\right| + \left|\int_{k(\xi-x)}^{k(\gamma-x)} \sin u\, du\right|\right\}$$

$$< 4M/k(\beta-x).$$

By making k large enough we can make $4M/k(\beta-x)$ as small as we please; moreover, this quantity is independent of γ, and so we can conclude that

$$\lim_{k\to\infty}\int_\beta^\infty \frac{\sin k(t-x)}{t-x} P_1(t)\, dt = 0.$$

Similar reasoning may be applied to the integral involving $P_2(t)$ and to the integrals arising from the range $t<-a$. It finally follows that

$$\lim_{k\to\infty}\int_{-\infty}^{\infty} \frac{\sin k(t-x)}{t-x} f(t)\, dt = \pi\bar{f}(x),$$

or

$$\pi\bar{f}(x) = \lim_{k\to\infty}\int_{-\infty}^{\infty} dt\int_0^k \cos s(t-x) f(t)\, ds.$$

To justify a change in the order of integration it will be sufficient to justify the change in the order of integration in the repeated integral

$$\int_q^\infty dt \int_0^k \cos s\,(t-x)\, P_1(t)\, ds,$$

where $q > \beta$, for the other integral with limit $-\infty$ may be treated in the same way and a change in the order of integration for the remaining integral between finite limits may be justified by the standard analysis.

Now let us assume that

$$\int_q^\infty P_1(t)\, dt \qquad \text{......(A)}$$

exists, then

$$\left| \int_q^\infty dt \int_0^k \cos s\,(t-x)\, P_1(t)\, ds - \int_0^k ds \int_q^\infty \cos s\,(t-x)\, P_1(t)\, dt \right| < 2k \int_q^\infty P_1(t)\, dt.$$

But, since the integral (A) exists we can choose q so large that

$$\int_q^\infty P_1(t)\, dt$$

is as small as we please. The order of integration can therefore be changed and so we have finally

$$\pi \bar{f}(x) = \int_0^\infty ds \int_{-\infty}^\infty \cos s\,(t-x) f(t)\, dt.$$

The assumptions which have been made are:

(1) For $x > \beta$, $f(x) = P_1(x) - P_2(x)$, where $P_1(x)$ and $P_2(x)$ are positive decreasing functions integrable in the range (β, ∞).

(2) A corresponding supposition for $x < -\alpha$.

(3) $f(x)$ of bounded variation in a range enclosing the point x.

(4) $f(x)$ discontinuous at only a finite number of points in $(-\alpha, \beta)$ and $|f(x)|$ integrable in $(-\alpha, \beta)$.

§ **3·13**. To illustrate the method of summation we shall try to find a potential which is zero when $x = 0$ and when $x = 1$. We shall be interested here in the case when the potential has a logarithmic singularity at the point $x = x'$, $y = 0$.

We first note that $M(m, x')$ is a simple combination of primary solutions and by an extension of the method of images used in the solution of physical problems by means of primary solutions we may satisfy the boundary condition at $x = 0$ by means of a simple potential of type $M(m, x') - M(m, -x')$. This can be written in the form

$$2e^{-my} \sin(mx) \,.\, \sin(mx'),$$

and it is readily seen that the boundary condition at $x = 1$ may be satisfied by writing $m = n\pi$, where n is an integer. We now multiply by a function of n and sum over integral values of n. To obtain a series which can be summed by means of logarithms we choose $f(n) = 1/n$ so that our series is

$$V = \sum_{n=1}^{\infty} \frac{1}{n} e^{-n\pi y} [\cos n\pi (x - x') - \cos n\pi (x + x')].$$

If $y > 0$ the sum of this series is*

$$V = \tfrac{1}{2} \log \frac{\cosh (\pi y) - \cos \pi (x + x')}{\cosh (\pi y) - \cos \pi (x - x')}.$$

To extend our solution to negative values of y we write it in the form

$$V = \sum_{n=1}^{\infty} \frac{2}{n} e^{-n\pi |y|} \sin (n\pi x) \sin (n\pi x').$$

The expression for V may be written in an alternative form

$$V = \tfrac{1}{2} \sum_{n=1}^{\infty} \log \frac{y^2 + (x + x' + 2n)^2}{y^2 + (x - x' + 2n)^2}$$

which shows that it may be derived from two infinite sets of line charges arranged at regular intervals.

This expression shows also that the potential V becomes infinite like $-\frac{1}{2} \log [(x - x')^2 + y^2]$ in the neighbourhood of $x = x'$, $y = 0$, it thus possesses the type of singularity characteristic of a Green's function and so we may adopt the following expression for the Green's function for the region between the lines $x = 0$, $x = 1$, when the function is to be zero on these lines

$$G(x, x'; y, y') = \sum_{n=1}^{\infty} \frac{1}{n} e^{-n\pi |y-y'|} \sin (n\pi x) \sin (n\pi x').$$

A corresponding solution of the equation

$$\frac{\partial^2 U}{\partial x^2} + \frac{\partial^2 U}{\partial y^2} + k^2 U = 0$$

is obtained by writing

$$\exp [- | y - y' | (n^2\pi^2 - k^2)^{\frac{1}{2}}]$$

in place of

$$\exp [- n\pi | y - y' |]$$

and

$$2n/(n^2 - k^2/\pi^2)$$

in place of the factor $2/n$.

§ **3·14**. As another illustration of the use of the simple solutions of Laplace's equation we shall consider the problem of the cooling of the fins of an air-cooled airplane engine when the fins are of the longitudinal type.

The problem will be treated for simplicity as two-dimensional.

A fin will be regarded as rectangular in section, of thickness 2τ, and of length a. Assuming that the end $x = 0$ is maintained at temperature θ_0 by the cylinder of the engine and that it is sufficient to assume a steady state, the problem is to find a solution of $\frac{\partial^2 \theta}{\partial x^2} + \frac{\partial^2 \theta}{\partial y^2} = 0$ and the boundary con-

* See, for instance, T. Boggio, *Rend, Lombardo* (2) 42:611–624 (1909).

ditions* $\frac{\partial\theta}{\partial y} = 0$ along $y = 0$, $k\frac{\partial\theta}{\partial y} = -q\theta$ along $y = \tau$, $k\frac{\partial\theta}{\partial x} = -q\theta$ along $x = a$.

The first two of these three conditions are satisfied by writing

$$\theta = \sum_{m=1}^{\infty} A_m \cosh[s_m(x - c_m)] \cos(s_m y), \quad ks_m \tau \tan(s_m \tau) = q.$$

This equation gives ∞^1 values of s_m and when s_m has been chosen the corresponding value of c_m is given uniquely by the equation

$$ks_m \tanh[s_m(c_m - a)] = q$$

which will ensure that the third condition is satisfied.

To make $\theta = \theta_0$ when $x = 0$ we have finally to determine the constant coefficients A_m in such a way that

$$\theta_0 = \sum_{m=1}^{\infty} A_m \cosh(s_m c_m) \cos(s_m y).$$

This may be done with the aid of the orthogonal relations

$$\int_0^\tau \cos(ys_m)\cos(ys_n)\,dy = 0, \quad m \neq n$$

$$= \frac{\tau}{2}\left[1 + \frac{\sin(2s_m\tau)}{2s_m\tau}\right], \quad m = n.$$

Therefore $$A_m = 4\theta_0 \operatorname{sech}(s_m c_m) \frac{\sin(s_m\tau)}{2s_m\tau + \sin(2s_m\tau)},$$

$$\theta = 4\theta_0 \sum_{m=1}^{\infty} \frac{\kappa s_m \cosh s_m(x-a) - q\sinh s_m(x-a)}{[\kappa s_m \cosh s_m a + q \sinh s_m a][2s_m\tau + \sin 2s_m\tau]} \cosh(s_m y)\sin(s_m\tau).$$

Harper and Brown derive from this expression a formula for the effectiveness of the fin, which they define as the ratio H/H_0, where

$$H_0 = 2q(a+\tau)\theta_0, \quad H = q\int \theta\, dS.$$

For numerical computations it is convenient to adopt an approximate method in which the variation of θ in the y direction is not taken into consideration. Results can then be obtained for a tapered fin.

The approximate method has been used by Binnie† in his discussion of the problem for the fins of annular shape which run round a cylinder barrel.

§ **3·15**. For some purposes it is useful to consider simple solutions of a complex type. Thus the equation

$$\frac{\partial v}{\partial t} = \nu \frac{\partial^2 v}{\partial x^2}$$

* The formal solution is obtained by D. R. Harper and W. B. Brown (*N.A.C.A. Report*, No. 158, Washington, 1923), but is not used in their computations.

† *Phil. Mag.* (7), vol. II, p. 449 (1926).

is satisfied by $$v = Ae^{\iota\sigma t \pm (1+i)\beta x},$$
if $2\nu\beta^2 = \sigma$. Retaining only the real part we have*
$$v = Ae^{-\beta x}\cos(\sigma t - \beta x). \qquad \text{......(A)}$$

This solution is readily interpreted by considering a viscous liquid which is set in motion by the periodic motion of the plane $x = 0$, the quantity v being velocity in one direction parallel to this plane (§ 2·56). The prescribed motion of the plane $x = 0$ is
$$v = A\cos\sigma t = V, \text{ say.}$$

The vibrations are propagated with velocity σ/β in the direction perpendicular to the plane but are rapidly damped, for the amplitude diminishes in the ratio $e^{-2\pi}$ as the wave travels a distance of one wave-length $2\pi/\beta$. For an assigned value of σ this wave-length is very small when ν is very small, when ν is assigned the wave-length is very small if σ is very large.

The equation (A) has been used by G. I. Taylor† to represent the range of potential temperature at a height x in the atmosphere, the potential temperature being defined as usual, as the temperature which a mass of air would have if it were brought isentropically (i.e. without gain or loss of heat and in a reversible manner) to a standard pressure.

The following examples to illustrate the use of the solution (A) are given by G. Green‡.

Suppose that two different media are in contact, the boundary surface being $x = a$ and the boundary conditions
$$v_1 = v_2, \quad K_1\frac{\partial v_1}{\partial x_1} = K_2\frac{\partial v_2}{\partial x_2}, \quad \text{for } x = a.$$

Let there be a periodic source of "plane-waves" on the side x, then the solution is of type
$$v_1 = \theta e^{-\beta_1 x}\cos(\sigma t - \beta_1 x) + A\theta e^{\beta_1(x-2a)}\cos[\sigma t + \beta_1(x-2a)], \quad x < a,$$
$$v_2 = B\theta e^{-\beta_2(x-c)}\cos[\sigma t - \beta_2(x-c)], \quad x > a,$$
where $$c = a[1 - (\nu_2/\nu_1)^{\frac{1}{2}}], \quad \beta_1 = (\sigma/2\nu_1)^{\frac{1}{2}}, \quad \beta_2 = (\sigma/2\nu_2)^{\frac{1}{2}},$$
$$pA = K_1\sqrt{\nu_2} - K_2\sqrt{\nu_1}, \quad pB = 2K_1\sqrt{\nu_2}, \quad p = K_1\sqrt{\nu_2} + K_2\sqrt{\nu_1}.$$
There is, of course, the physical difficulty that the expression for the incident waves becomes infinite when $x = -\infty$.

If we take the associated problem in which the incident waves correspond to a periodic supply of heat $q\cos\sigma t$ at the origin, the solution is
$$v_1 = (q/2K_1)(\nu_1/\sigma)^{\frac{1}{2}}[e^{-\beta_1 x}\cos(kt - \beta_1 x - \pi/4) + Ae^{\beta_1(x-2a)}\cos(kt + \beta_1 x - 2a\beta_1 - \pi/4)],$$
$$v_2 = (q/2K_2)(\nu_2/\sigma)^{\frac{1}{2}}Be^{-\beta_2(x-c)}\cos(kt - \beta_2 x + \beta_2 c - \pi/4),$$
where A and B have the same values as before. It is noteworthy that A

* The theory is due to Stokes, *Papers*, vol. III, p. 1. See Lamb's *Hydrodynamics*, p. 586.

† *Proc. Roy. Soc. London*, A, vol. XCIV, p. 137 (1918).

‡ G. Green, *Phil. Mag.* (7), vol. III, p. 784 (1927).

and B are independent of σ and that when the expressions in these solutions are integrated with respect to σ from 0 to ∞ the physically correct solution for the case of the instantaneous generation of a quantity of heat q at the origin at time $t = 0$ is obtained in the form

$$v_1 = (q/2K_1)(\nu_1/\pi t)^{\frac{1}{2}}\left[e^{-\frac{x^2}{4\nu_1 t}} + Ae^{-\frac{(2a-x)^2}{4\nu_2 t}}\right],$$

$$v_2 = (q/2K_2)(\nu_2/\pi t)^{\frac{1}{2}}\, e^{-\frac{(x-c)^2}{4\nu_2 t}}.$$

EXAMPLES

1. In the problem of the oscillating plane the viscous drag exerted by the fluid is, per unit area,

$$-\left(\tfrac{1}{2}\sigma\rho\mu\right)^{\frac{1}{2}}\left(V + \frac{1}{\sigma}\frac{dV}{dt}\right) = \mu\beta(\sin\sigma t - \cos\sigma t).$$

[Rayleigh.]

2. Discuss the equations

$$\frac{\partial u}{\partial t} - 2\omega v = \nu\frac{\partial^2 u}{\partial z^2},$$

$$\frac{\partial v}{\partial t} + 2\omega u = \nu\frac{\partial^2 v}{\partial z^2},$$

$$\omega = \Omega\sin\phi, \quad (\phi \text{ a constant}),$$

where Ω is a constant representing the angular velocity of the earth, and ϕ is the latitude.

[V. W. Ekman.]

§ **3·16**. The solution (A of § 3·15) may be generalised by regarding A as a function of β and then integrating with respect to β.

A solution of a very general character is thus given by

$$v = \int_0^\infty e^{-\beta x}\cos[\beta x - 2\nu\beta^2 t]\,\phi(\beta)\,d\beta + \int_0^\infty e^{-\beta x}\sin[\beta x - 2\nu\beta^2 t]\,\psi(\beta)\,d\beta,$$

where $\phi(\beta)$ and $\psi(\beta)$ are arbitrary functions of a suitable character. Solutions of this type have been used by Rayleigh and by G. Green.

Some useful identities may be obtained by comparing solutions of problems in the conduction of heat that are obtained by two different methods when the solution is known to be unique.

For instance, if we use the method of simple solutions we can construct a solution

$$v = \frac{1}{\pi}\sum_{n=-\infty}^{n}\int_0^{2\pi} e^{i(x-\xi)n - \nu t n^2} f(\xi)\,d\xi$$

which is periodic in x with the period 2π.

When $t = 0$ the series is simply the Fourier series of the function $f(x)$ and the inference is that with a suitable type of function $f(x)$ our solution is one which satisfies the initial condition $v = f(x)$ when $t = 0$. Now such a solution can also be expressed by means of Laplace's integral

$$v = (\nu t\pi)^{-\frac{1}{2}} \int_{-\infty}^{\infty} e^{-\frac{(x-\xi)}{4\nu t}} f(\xi)\, d\xi,$$

and this may be written in the form

$$v = (\nu t\pi)^{-\frac{1}{2}} \sum_{n=-\infty}^{\infty} \int_{0}^{2\pi} e^{-\frac{(x-\xi+2n\pi)^2}{4\nu t}} f(\xi)\, d\xi.$$

When the order of integration and summation can be changed, a comparison of the two solutions indicates that

$$\sum_{n=-\infty}^{\infty} e^{in(x-\xi)-n^2\nu t} = (\pi/\nu t)^{\frac{1}{2}} \sum_{n=-\infty}^{\infty} e^{-\frac{(x-\xi+2n\pi)^2}{4\nu t}}.$$

This identity, which is due to Poisson, has recently, in the hands of Ewald, become of great importance in the mathematical theory of electromagnetic waves in crystals. The identity can be established rigorously in several ways:

(1) With the aid of Fourier series.

(2) By the calculus of residues.

(3) By the theory of elliptic functions (theta functions).

(4) By means of the functional relation for the ζ-function, Riemann's method of deriving this functional relation being performed backwards.

An elementary proof based on the equations

$$\left(1 + \frac{x_n}{n}\right)^n \to e^x \quad \text{as } n \to \infty \quad \text{if } x_n \to x, \qquad \text{......(1)}$$

$$2^{-2n} n^{\frac{1}{2}} \binom{2n}{n+r} \to \pi^{-\frac{1}{2}} e^{-x_2} \quad \text{as } n \to \infty \quad \text{if } rn^{-\frac{1}{2}} \to x \qquad \text{......(2)}$$

has been given recently by Pólya*.

We have $2^{n-1} \leqslant n!$ for $n = 1, 2, 3, \ldots$ and so, for $0 \leqslant x \leqslant 1$,

$$e^{2x} = 1 + \frac{2x}{1!} + \ldots \leqslant 1 + 2x + 2x^2 + \ldots = \frac{1+x}{1-x}.$$

Also, for $0 \leqslant x \leqslant \frac{1}{7}$,

$$\frac{1+x}{1-x} = 1 + 2x + 2x^2 + \frac{2x^3}{1-x} \leqslant 1 + 2x + 2x^2 + \frac{7x^3}{3} < e^{2x+x^3}.$$

Therefore

$$e^{-2x-x^3} \leqslant \frac{1-x}{1+x} \leqslant e^{-2x}. \qquad \text{......(B)}$$

* G. Pólya, *Berlin, Akad. Wiss. Ber.* p. 158 (1927).

On account of the symmetry of the binomial coefficients it is sufficient to prove (2) for $r \geqslant 0$. In this case

$$n^{\frac{1}{2}} 2^{-2n} \binom{2n}{n+r} = n^{\frac{1}{2}} 2^{-2n} \binom{2n}{n} \frac{n}{n+r} \frac{\left(1-\frac{1}{n}\right)\left(1-\frac{2}{n}\right)\cdots\left(1-\frac{r-1}{n}\right)}{\left(1+\frac{1}{n}\right)\left(1+\frac{2}{n}\right)\cdots\left(1+\frac{r-1}{n}\right)}$$

$$\leqslant n^{\frac{1}{2}} 2^{-2n} \binom{2n}{n} e^{-r(r-1)/n} \to \pi^{-\frac{1}{2}} e^{-x^2},$$

the upper estimate in (B) having been applied. A use of the lower estimate gives an analogous result, which, on account of the fact that $r^4 n^{-3} \to 0$, completes the proof of (2).

Putting $x = z\omega^\nu = ze^{2\pi i\nu/l}$ in the identity

$$(\sqrt{x} + 1/\sqrt{x})^{2m} = \sum_{\nu=-m}^{m} \binom{2m}{m+\nu} x^\nu,$$

we obtain $$\sum_{-l<2\nu<l} [(\sqrt{(z\omega^\nu)} + 1/\sqrt{(z\omega^\nu)})]^{2m} = l \sum_{\nu=-k}^{k} \binom{2m}{m+l\nu} z^{l\nu},$$

where $k = [m/l]$ is the integral part of m/l.

Now let s be an arbitrary fixed complex number and t a fixed real positive number. Putting $l = \sqrt{[(mt)]}$, $z = e^{s/l}$, and dividing the series by 2^{2m}, we obtain the relation

$$\left.\begin{aligned} \sum_{-l<2\nu<l} \cosh^{2m}\left(\frac{s+2\pi i\nu}{2l}\right) &= \sum_{-l<2\nu<l} \left\{1 + \frac{(s+2\pi i\nu)^2}{8l^2} + \cdots\right\}^{8l^2\frac{m}{4l^2}} \\ &= \sum_{\nu=-k}^{k} \frac{[\sqrt{tm}]}{2^{2m}} \binom{2m}{m+[\nu\sqrt{(tm)}]} e^{s\nu}. \end{aligned}\right\} \quad \ldots\ldots\text{(C)}$$

Applying the limit (1) on the left and (2) on the right we finally obtain

$$\sum_{\nu=-\infty}^{\infty} e^{\frac{s+2\pi i\nu}{4l}} = \sum_{-\infty}^{\infty} \sqrt{(t/\pi)}\, e^{-t\nu^2+s\nu},$$

which is a form of Poisson's formula.

To justify the limiting process which has just been performed in which the limit is taken for each term separately, it is sufficient to find a quantity independent of m which dominates each term in each series.

There is little difficulty in finding a suitable dominating quantity for the terms on the right-hand side, but to find a suitable quantity for the terms on the left-hand side of the equation Pólya finds it necessary to prove the following lemma. Given two constants a and b for which $a > 0$, $0 < b < \pi$, we can find two other constants A and B such that $A > 0$, $B > 0$ and

$$|\cosh z| \leqslant e^{Ax^2 - By^2},$$

when $-a \leqslant x \leqslant a$, $-b \leqslant y \leqslant b$ and $z = x + iy$. We have, in fact,

$$|\cosh z|^2 = \tfrac{1}{2}(1 + \cosh 2x) - \sin^2 y,$$

but, for $-a \leqslant x \leqslant a$, we have

$$\tfrac{1}{2}(1 + \cosh 2x) \leqslant 1 + \tfrac{1}{2} \sum_{n=1}^{\infty} \frac{4^n a^{2n-2} x^2}{(2n)!} = 1 + 2Ax^2, \text{ say.}$$

On the other hand, since $\sin y/y$ decreases as y increases from 0 to π, we have for $-b \leqslant y \leqslant b$,

$$\frac{\sin y}{y} \geqslant \frac{\sin b}{b} = \sqrt{2B}, \text{ say, } \sqrt{2B} > 0.$$

It follows from the inequalities that have just been established that

$$|\cosh z|^2 \leqslant 1 + 2Ax^2 - 2By^2 \leqslant e^{2(Ax^2 - By^2)},$$

and this proves the lemma.

To apply the lemma to the series (C) we note that in the first member

$$|\pi\nu/l| \leqslant \pi/2,$$

we therefore take $$b = \pi/2.$$

If s is real, the sum in the first member of (C) is dominated by the series

$$\sum_{\nu=-\infty}^{\infty} e^{\frac{As^2m}{2l^2} - \frac{2B\pi^2\nu^2m}{l^2}}.$$

It is easy to dominate this series by one free from m. The case in which s is not real can also be treated in a similar manner.

EXAMPLES

1. If
$$P = \int_0^\infty e^{-x\theta}[\cos(x\theta) - \sin(x\theta)]\cos(2\kappa t\theta^2)\,d\theta,$$
$$Q = \int_0^\infty e^{-x\theta}[\cos(x\theta) + \sin(x\theta)]\sin(2\kappa t\theta^2)\,d\theta,$$
show by partial integration that
$$xP = -2\kappa t\frac{\partial Q}{\partial x}, \quad xQ = -2\kappa t\frac{\partial P}{\partial x}.$$
Show also that, as $t \to 0$,
$$4P \to 4Q \to \pi^{\frac{1}{2}}(\kappa t)^{-\frac{1}{2}},$$
and that consequently
$$4P = 4Q = \pi^{\frac{1}{2}}(\kappa t)^{-\frac{1}{2}}e^{-x^2/4\kappa t}.$$

2. If
$$C = \int_{-a}^\infty e^{-2ay}\cos(y^2 - a^2)\,dy,$$
$$S = \int_{-a}^\infty e^{-2ay}\sin(y^2 - a^2)\,dy,$$
prove that
$$C + S = \sqrt{(\pi/2)}, \quad C - S = 2\int_0^a e^{2\theta^2}\,d\theta.$$
[G. Green.]

§ 3·17. *Conduction of heat in a moving medium.* When the temperature depends on only one co-ordinate, the height above a fixed horizontal plane, and the vertical velocity of the medium is w, the equation of conduction is

$$\frac{\partial\theta}{\partial t} + v\frac{\partial\theta}{\partial y} = \kappa\frac{\partial^2\theta}{\partial y^2}, \qquad \text{......(A)}$$

where κ is the diffusivity. When v is constant the equation possesses a simple solution of type

$$\theta = Me^{\lambda y + \mu t}, \quad \mu + v\lambda = \kappa\lambda^2,$$

which may be generalised by summation or integration over a suitable set of values of μ. In particular, if we regard θ as made up of periodic terms and generalise by integration over all possible periods, we obtain a solution

$$\theta = \int_0^\infty e^{ay}\left[f(\alpha)\sin(by+\alpha t) + g(\alpha)\cos(by+\alpha t)\right] d\alpha,$$

where
$$2\kappa a = v - w, \qquad bw = -\alpha,$$
$$2w^2 = v^2 + (v^2 + 16\kappa\alpha^2)^{\frac{1}{2}},$$

and $f(\alpha)$, $g(\alpha)$ are suitable arbitrary functions. The integral may be used in the Stieltjes sense so that it can include the sum of a number of terms corresponding to discrete values of α.

When v varies periodically in such a way that $v = u(1 + r\cos\alpha t)$, where u, r and α are constants, a particular solution may be obtained by writing

$$\theta = Me^{\lambda y + f(t)}, \qquad \text{......(B)}$$

where $f(t)$ is a function which is easily determined with the aid of the differential equation. When v is an arbitrary function of t the equation (A) has a simple solution of type

$$\theta = e^{is[y - \int v dt] - \kappa s^2 t}$$

which may be generalized into

$$\theta = \int_{-\infty}^{\infty} e^{is[y - \int v dt] - \kappa s^2 t} F(s) ds$$

where $F(s)$ is a suitable arbitrary function of s.

The solution (B) has been used by McEwen* for a comparison of the results computed from theory with the results of a series of temperature observations made off Coronado Island about 20 miles from San Diego in California. The coefficient κ is to be interpreted as an "eddy conductivity" in the sense in which this term is used by G. I. Taylor. This is explained by McEwen as follows:

At a depth exceeding 40 metres the direct heating of sea water by the absorption of solar radiation is less than 1 per cent. of that at the surface. Also, the temperature range at that depth would bear the same proportion to that at the surface if the variation in rate of gain of heat were due only to the variation in this rate of absorption. The direct absorption of solar radiation cannot then be the cause of the observed seasonal variation of temperature, which amounts to 5° C. at a depth of 40 metres and exceeds 1° at a depth of 100 metres. Laboratory experiments show, moreover,

* *Ocean Temperatures, their relation to solar radiation and oceanic circulation* (University of California Semicentennial Publications, 1919).

that the ordinary process of heat conduction in still water is wholly inadequate to produce a transfer of heat with sufficient rapidity to account for the whole phenomenon. It is now generally recognised that a much more rapid transfer of heat results from an alternating vertical circulation of the water in which, at any given instant, certain portions of the water are moving upward while others are moving downward. The resultant flow of a given column of water may be either upward or downward, or may be zero. The motion may be described as turbulent and a vivid picture of the process may be obtained by supposing that heat is conveyed from one layer to another by means of eddies. This complicated process produces a transfer of heat from level to level which, when analysed statistically, will be assumed to be governed by the same law as conduction except that the "eddy conductivity" or "Mischungsintensität" will depend mainly on the intensity of the circulation or mixing process.

An equation which is more general than (A) has been obtained by S. P. Owen* in a study of the distribution of temperature in a column of liquid flowing through a tube.

Assuming, as an inference from Nettleton's experiments, that the shape of the isothermals is independent of the character of the flow, Owen considers an element of length δy fixed in space and estimates the amounts of heat entering and leaving the element across its two faces perpendicular to the y-axis to be

$$A\left\{-k\frac{\partial\theta}{\partial y}+\rho s v\theta\right\},$$

and
$$A\left\{-k\frac{\partial}{\partial y}\left(\theta+\frac{\partial\theta}{\partial y}\delta y\right)+\rho s v\left(\theta+\frac{\partial\theta}{\partial y}\delta y\right)\right\}+Ep(\theta-\theta_0)\,\delta y$$

respectively, where A is the area, p the perimeter of the cross-section of the tube, θ the temperature of the element, E the emissivity, k, ρ and s the thermal conductivity, density, and specific heat of the liquid respectively, and where θ_0 is the temperature of the enclosure which surrounds the tube.

Owen thus obtains the equation

$$A\left\{k\frac{\partial^2\theta}{\partial y^2}-\rho s v\frac{\partial\theta}{\partial y}\right\}\delta y-Ep(\theta-\theta_0)\,\delta y=A\rho s\frac{\partial\theta}{\partial t}\,\delta y,$$

or
$$a^2\frac{\partial^2\theta}{\partial y^2}-v\frac{\partial\theta}{\partial y}-\frac{Ep}{A\rho s}(\theta-\theta_0)=\frac{\partial\theta}{\partial t},$$

where
$$a^2=k/\rho s.$$

EXAMPLES

1. Prove that a temperature θ which satisfies the equation

$$\frac{\partial\theta}{\partial t}+v\frac{\partial\theta}{\partial y}=\kappa\frac{\partial^2\theta}{\partial y^2},$$

and the conditions

$$\theta=0 \text{ when } y=0,\quad \theta=\theta_1 \text{ when } y=b,\quad \theta=0 \text{ when } t=0,$$

* *Proc. London Math. Soc.* vol. XXIII, p. 238 (1925).

is given by the formula

$$\theta = \theta_1\left[\frac{e^{vy/\kappa}-1}{e^{vb/\kappa}-1} + \frac{2}{b^2}\sum_{n=1}^{\infty}(-)^n \frac{n\pi}{(n\pi/b)^2+(v/2\kappa)^2} \times e^{v(y-b)/2\kappa}\sin(n\pi y/b)\, e^{-\{(n^2\pi^2\kappa/b^2)+(v^2/4\kappa)\}t}\right].$$

[Somers, *Proc. Phys. Soc. London*, vol. XXV, p. 74 (1912); Owen, *loc. cit.*]

2. If in the last example the receiver is maintained at a temperature which is a periodic function of the time, so that the condition $\theta = \theta_1$ when $y = b$ is replaced by $\theta = \Theta\cos\omega t$ when $y = b$, the solution is

$$\theta = \Theta e^{v(y-b)/2\kappa}(\cosh 2nb - \cos 2mb)^{-1}[(\cos m\xi\cosh n\eta - \cos m\eta\cosh n\xi)\cos\omega t - (\sin m\xi\sinh n\eta - \sin m\eta\sinh n\xi)\sin\omega t]$$
$$+ 2\Theta b^{-2}\sum_{p=1}^{\infty}(-)^p e^{v(y-b)/2\kappa}\{p\pi a_n^2/[\Sigma a_p^4 + (\omega/\kappa)^2]\}\, e^{-\varkappa a_p^2 t},$$

where

$$\xi = y-b,\quad \eta = y+b,\quad a_p^2 = (p\pi/b)^2 + (v^2/4\kappa^2),$$

$$\begin{matrix} m \\ n \end{matrix} = \{(v/2\kappa)^4 + (\omega/\kappa)^2\}\begin{matrix}\cos\\ \sin\end{matrix}\{\tfrac{1}{2}\tan^{-1}(4\kappa\omega/v^2)\}.$$

§ **3·18**. *Theory of the unloaded cable.* Consider a cable in the form of a loop (Fig. 13) having an alternator A at the sending end and a receiving instrument B at the receiving end. We shall suppose that the alternator is generating a simple periodic electromotive force which may be represented as the real part of the expression Ee^{int}, where E and n are constants. Naturally, we are interested only in the real part of any complex quantity which is used to represent a physical entity.

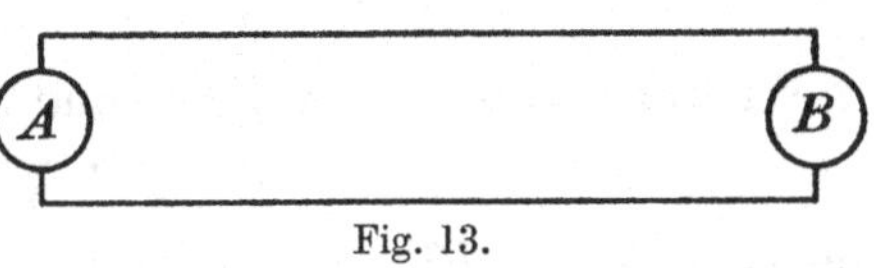

Fig. 13.

Now, if $C\delta x$ is the capacity of an element of length δx with regard to the earth, the capacity of a length δx with regard to a similar element in the return cable must be $\frac{1}{2}C\delta x$. Hence, if I_0 is the current in the alternator and V_0 the potential difference of the two sides of the cable at the sending end,

$$\tfrac{1}{2}C\frac{\partial V_0}{\partial t} = -\frac{\partial I_0}{\partial x}.$$

Now V_0 is the difference between the generated electromotive force Ee^{int} and the drop in voltage down the alternator circuit and a capacity C_0 in series with it, consequently we have the equation

$$L_0\frac{\partial I_0}{\partial t} + R_0 I_0 + \frac{1}{C_0}\int I_0\,dt + V_0 = Ee^{int}.$$

Assuming that I can be expressed as the real part of $X(x)\,e^{int}$ and that $I = I_0$ at the receiving end, we find on differentiating the last equation with respect to t and multiplying by C_0,

$$(1 - C_0L_0n^2 + inC_0R_0)\,I_0 + C_0\frac{\partial V_0}{\partial t} = inC_0Ee^{int}.$$

Hence the boundary condition at the sending end is

$$-2\frac{\partial I_0}{\partial x}=inCE\,e^{int}-h_0I_0,$$

where
$$h_0C_0=C\,(1-C_0L_0n^2+inC_0R_0).$$

Similarly, if I_1 is the current at the receiving end and if the receiving apparatus is equivalent to an inductive resistance (L_1, R_1) in series with a capacity C_1, we have the boundary condition

$$2\frac{\partial I_1}{\partial x}=-h_1I_1,$$

where
$$h_1C_1=C\,(1-C_1L_1n^2+inC_1R_1).$$

Assuming that there is no leakage, the differential equation for I is*

$$RC\frac{\partial I}{\partial t}+LC\frac{\partial^2 I}{\partial t^2}=\frac{\partial^2 I}{\partial x^2}, \qquad \text{......(A)}$$

and if
$$X=K_1\cos\mu\,(l-x)+K_2\sin\mu\,(l-x),$$

where l is the distance between the alternator and receiving instrument, and K_1, K_2 are constants to be determined, we have

$$\mu^2=C\,(n^2L-inR).$$

Writing $\mu=\alpha+i\beta$, where α and β are real, we have

$$\alpha^2-\beta^2=LCn^2,\quad 2\alpha\beta=-CRn,$$

and so, if $R^2+n^2L^2=G^2$, we have

$$2\alpha^2=Cn\,(G+nL),\qquad 2\beta^2=Cn\,(G-nL).$$

When nL is large in comparison with R we may write

$$G-nL=R^2/2nL,$$

and we have

$$\alpha=n\sqrt{(CL)},\qquad \beta=\tfrac{1}{2}R\sqrt{(C/L)},$$

the wave-velocity being $(CL)^{-\frac{1}{2}}$. In this case the wave-velocity and attenuation constant are approximately independent of the frequency, consequently a wave-form built up from waves of high frequency travels with very little distortion.

The constants K_1 and K_2 are easily determined from the boundary conditions and we find that

$$FK_1=2i\mu nCE,\quad FK_2=ih_1nCE,$$

where
$$F=(h_0h_1-4\mu^2)\sin\mu l+2\mu\,(h_0+h_1)\cos\mu l.$$

When $E=0$ the differential equation possesses a finite solution only when $F=0$ and this, then, is the condition for free oscillations. The roots of the equation $F=0$, regarded as an equation for n, are generally complex.

* Our presentation is based upon that of J. A. Fleming in his book, *The propagation of electric currents in telephone and telegraph conductors.*

This may be seen by considering the special case when $C_0 = C_1 = \infty$. This means that there are short circuits in place of the transmitting and receiving apparatus.

We now have $h_0 = h_1 = 0$, $\mu^2 \sin \mu l = 0$, and if we satisfy this equation by writing $\mu l = s\pi$, where s is an integer, the equation

$$s^2\pi^2 = \mu^2 l^2 = l^2 C\,(n^2 L - inR)$$

gives complex values for n.

When $R_0 = R_1 = R$ the roots of the equation for n are all real. This may be proved with the aid of the following theorem due to Koshliakov*.

Let $$\phi_0 + i\psi_0 = \sum_{s=1}^{m} m_s \log (z - z_s) - \sum_{s=1}^{n} k_s \log (z - \zeta_s)$$

be the complex potential of the two-dimensional flow produced by a number of sources and sinks, the sources being all above the axis of x and the sinks all on or below the axis of x.

Writing $$z_s = a_s + ib_s, \quad \zeta_s = \xi_s - i\eta_s,$$

where a_s, b_s, ξ_s, η_s are all real, we shall suppose that

$$b_s > 0, \quad \eta_s \geqslant 0, \quad m_s \geqslant 0, \quad k_s \geqslant 0.$$

Now suppose that when x is real and complex

$$\prod_{s=1}^{n} \frac{(x - z_s)^{m_s}}{(x - \zeta_s)^{k_s}} = f(x) + ig(x),$$

where $f(x)$ and $g(x)$ are real when x is real. If we superpose on the flow produced by the sources and sinks a rectilinear flow specified by the stream-function $\psi_1 = x - y \tan \omega$, the stream-function of the total flow is $\psi = \psi_0 + \psi_1$ and the points in which a stream-line $\psi = \theta$ cuts the axis of x are given by the transcendental equation

$$\tan^{-1}\frac{g(x)}{f(x)} + x = \theta,$$

or $$g(x) \cos (x - \theta) + f(x) \sin (x - \theta) = 0.$$

We wish to show in the first place that the roots of this equation are all real. Writing

$$G(x) + iF(x) \equiv e^{i(x-\theta)} \prod_{s=1}^{n} \frac{(x - a_s - ib_s)^{m_s}}{(x - \xi_s + i\eta_s)^{k_s}},$$

$$G(x) - iF(x) \equiv e^{i(\theta-x)} \prod_{s=1}^{n} \frac{(x - a_s + ib_s)^{m_s}}{(x - \xi_s - i\eta_s)^{k_s}},$$

we have
$$G(x) + iF(x) = e^{i(x-\theta)}\,[f(x) + ig(x)],$$
$$G(x) - iF(x) = e^{i(\theta-x)}\,[f(x) - ig(x)],$$
$$F(x) = f(x) \sin (x - \theta) + g(x) \cos (x - \theta),$$
$$G(x) = f(x) \cos (x - \theta) - g(x) \sin (x - \theta).$$

* *Mess. of Math.* vol. LV, p. 132 (1926). Koshliakov considers only the case $m_s = 1$, $k_s = 0$ without any hydrodynamical interpretation of the result.

Hence, if $z = x + iy$ is a root of the equation $F(z) = 0$, we have for this root

$$e^{i(z-\theta)} \prod_{s=1}^{n} \frac{(z - a_s - ib_s)^{m_s}}{(z - \xi_s + i\eta_s)^{k_s}} = e^{i(\theta - z)} \prod_{s=1}^{n} \frac{(z - a_s + ib_s)^{m_s}}{(z - \xi_s - i\eta_s)^{k_s}}.$$

Now let M_1 and M_2 be the moduli of the expressions on the two sides of the equation, then the equation tells us that $M_1^2 = M_2^2$, but

$$M_1^2 = e^{-2y} \prod_{s=1}^{n} \frac{[(x - a_s)^2 + (y - b_s)^2]^{m_s}}{[(x - \xi_s)^2 + (y + \eta_s)^2]^{k_s}},$$

$$M_2^2 = e^{2y} \prod_{s=1}^{n} \frac{[(x - a_s)^2 + (y + b_s)^2]^{m_s}}{[(x - \xi_s)^2 + (y - \eta_s)^2]^{k_s}},$$

and from these equations it appears that $y > 0$, $M_1^2 < M_2^2$, while if $y < 0$, $M_1^2 > M_2^2$. Hence we must have $y = 0$, and so the roots of the equation $F(z) = 0$ are all real.

Let us next determine the effect on the roots of varying the value of θ. If x is a real root of the equation $F(x) = 0$, we have

$$(dx/d\theta)\,[f'(x) \sin(x - \theta) + g'(x) \cos(x - \theta) + G(x)] = G(x),$$

but

$$\frac{\cos(x - \theta)}{f(x)} = \frac{\sin(x - \theta)}{-g(x)} = \frac{1}{G(x)},$$

therefore $(dx/d\theta)\,[f(x)\,g'(x) - f'(x)\,g(x) + \{G(x)\}^2] = [G(x)]^2$.

Now

$$\frac{f'(x) + ig'(x)}{f(x) + ig(x)} = \sum_{s=1}^{n} \left(\frac{m_s}{x - a_s - ib_s} - \frac{k_s}{x - \xi_s + i\eta_s} \right),$$

$$\frac{f'(x) - ig'(x)}{f(x) - ig(x)} = \sum_{s=1}^{n} \left(\frac{m_s}{x - a_s + ib_s} - \frac{k_s}{x - \xi_s - i\eta_s} \right),$$

therefore

$$\frac{f(x)\,g'(x) - f'(x)\,g(x)}{[f(x)]^2 + [g(x)]^2} = \sum_{s=1}^{n} \left[\frac{m_s b_s}{(x - a_s)^2 + b_s^2} + \frac{k_s \eta_s}{(x - \xi_s)^2 + \eta_s^2} \right].$$

The right-hand side is clearly positive and so $dx/d\theta$ is positive for all real values of θ. This means that when x increases, the point in which the stream-line meets the axis of x moves to the right (i.e. the direction in which x increases).

If we increase θ by $\frac{1}{2}\pi$, $F(x)$ is transformed into $-G(x)$, and if we add another $\frac{1}{2}\pi$ to θ, the function $-G(x)$ is transformed into $-F(x)$, consequently we surmise that the roots of $F(x) = 0$ are separated by those of $G(x) = 0$. To prove this we adopt Koshliakov's method of proof and calculate the derivative

$$\frac{d}{dx}\frac{F(x)}{G(x)} = \frac{f(x)\,g'(x) - f'(x)\,g(x) + [F(x)]^2 + [G(x)]^2}{[G(x)]^2}.$$

This is clearly positive for all values of x and infinite, perhaps, at the roots of $G(x)=0$. It is clear from a graph that the roots of $F(x)$ are separated by those of $G(x)=0$, for the curve

$$y=\frac{F(x)}{G(x)}$$

consists of a number of branches each of which has a positive shape.

In Koshliakov's case when $k_s=0$, $m_s=1$ the functions $f(x)$ and $g(x)$ are polynomials such that the roots of the equation $f(x)+ig(x)=0$ are of type $z_s=a_s+ib_s$, where $b_s>0$. The associated equation $F(x)=0$ is now of a type which frequently occurs in applied mathematics. In particular, if

$$f(x)=\beta_1\beta_2-x^2,\quad g(x)=(\beta_1+\beta_2)x,$$

the roots of the equation $f(x)+ig(x)=0$ are $i\beta_1$ and $i\beta_2$ and so we have the result that if β_1 and β_2 are both positive, the roots of the equation

$$(\beta_1+\beta_2)x\cos(x-\theta)+(\beta_1\beta_2-x^2)\sin(x-\theta)=0 \quad \text{......(B)}$$

are all real and increase with θ.

The theorem may be applied to the cable equation by writing this in the form

$$\tan\mu l=-\frac{2\mu\{\gamma_0+\gamma_1-\mu^2(\lambda_0+\lambda_1)\}}{(\gamma_0-\lambda_0\mu^2)(\gamma_1-\lambda_1\mu^2)-4\mu^2},$$

where $\gamma_0 C_0=C,\quad \gamma_1 C_1=C,\quad L_0=\lambda_0 L,\quad L_1=\lambda_1 L.$

Now

$$(\gamma_0-\lambda_0\mu^2)(\gamma_1-\lambda_1\mu^2)-4\mu^2+2i\mu\{\gamma_0+\gamma_1-\mu^2(\lambda_0+\lambda_1)\}$$
$$=(\gamma_0+2i\mu-\lambda_0\mu^2)(\gamma_1+2i\mu-\lambda_1\mu^2),$$

and when the expression on the right is equated to zero, the resulting algebraic equation for μ has roots of type $a+ib$, where b is positive, hence Koshliakov's theorem may be applied and the conclusion drawn that μl is real. Since in the present case $\mu^2=CLn^2$, the corresponding value of n is also real.

When $\theta=0$, $x=\omega l$, $\beta_1=\beta_2=lh$, the equation (B) becomes identical with the equation

$$2h\omega\cos\omega l=(\omega^2-h^2)\sin\omega l, \quad \text{......(C)}$$

which occurs in the theory of the conduction of heat in a finite rod, when there is radiation at the ends, into a medium at zero temperature.

The equations of this problem are in fact

$$\frac{\partial v}{\partial t}=\kappa\frac{\partial^2 v}{\partial x^2},\quad 0<x<l,$$
$$v=f(x),\quad \text{for } t=0,$$
$$-\frac{\partial v}{\partial x}+hv=0 \text{ at } x=0,\quad \frac{\partial v}{\partial x}+hv=0 \text{ at } x=l,$$

and are satisfied by

$$v=e^{-\kappa\omega^2 t}[A\cos\omega x+B\sin\omega x]$$

if

$$-\omega B+hA=0,$$
$$\omega(B\cos\omega l-A\sin\omega l)+h(B\sin\omega l+A\cos\omega l)=0.$$

Eliminating A/B the equation (C) is obtained. The problem is finally solved by a summation over the roots of this equation, the root $\omega = 0$ being excluded.

Equations similar to (A) occur in other branches of physics and many useful analogies may be drawn. In the theory of the transverse vibrations of a string we may suppose that the motion of each element of the string is resisted by a force proportional to its velocity*. The partial differential equation then becomes

$$\frac{\partial^2 y}{\partial t^2} + \kappa \frac{\partial y}{\partial t} = c^2 \frac{\partial^2 y}{\partial x^2},$$

which is of the same form as (A) if $\kappa = R/L$, $c^2 = 1/LC$.

An equation of the same type occurs also in Rayleigh's theory of the propagation of sound in a narrow tube, taking into consideration the influence of the viscosity of the medium†.

Let X denote the total transfer of fluid across the section of the tube at the point x. The force, due to hydrostatic pressure, acting on the slice between x and $x + dx$, is

$$- S \frac{\partial p}{\partial x} dx = a^2 \rho \, dx \frac{\partial^2 X}{\partial x^2},$$

where S is the area of the cross-section, p is the pressure in the fluid, ρ is the density and a is the velocity of propagation of sound waves in an unlimited medium of the same material.

The force due to viscosity may be inferred from the investigation for a vibrating plane (§ 3·15), provided that the thickness of the layer of air adhering to the walls of the tube be small in comparison with the diameter. Thus, if P be the perimeter of the inner section of the tube and V the velocity of the current at a distance from the walls of the tube, the tangential force on a slice of volume Sdx is, by the result of (§ 3·15, Ex. 1), equal to

$$- P dx \sqrt{(\tfrac{1}{2} n \rho \mu)} \left(V + \frac{1}{n} \frac{\partial V}{\partial t} \right),$$

where $n/2\pi$ is the frequency of vibration.

Replacing VS by $\dfrac{\partial X}{\partial t}$ we can say that the equation of motion of the fluid for disturbances of this particular frequency is

$$\rho \, dx \frac{\partial^2 X}{\partial t^2} + \sqrt{(\tfrac{1}{2} n \rho \mu)} \, . \, P dx \left(\frac{\partial X}{\partial t} + \frac{1}{n} \frac{\partial^2 X}{\partial t^2} \right) \Big/ S$$
$$= a^2 \rho \frac{\partial^2 X}{\partial x^2} dx,$$

or

$$\frac{\partial^2 X}{\partial t^2} \left\{ 1 + \frac{P}{S} \sqrt{\left(\frac{\mu}{2 n \rho} \right)} \right\} + \frac{P}{S} \left(\frac{n \mu}{2 \rho} \right)^{\frac{1}{2}} \frac{\partial X}{\partial t}$$
$$= a^2 \frac{\partial^2 X}{\partial x^2}.$$

* Rayleigh, *Theory of Sound*, vol. I, p. 232. † *Ibid.* vol. II, p. 318.

This equation has been used as a basis for some interesting analogies between acoustic and electrical problems *. We shall write it in the abbreviated form

$$H\frac{\partial^2 X}{\partial t^2} + K\frac{\partial X}{\partial t} = a^2\frac{\partial^2 X}{\partial x^2}.$$

Rayleigh's equation has been used recently by L. F. G. Simmons and F. C. Johansen in a discussion of their experiments on the transmission of air waves through pipes†.

At the end $x = 0$ the boundary condition is taken to be

$$X = X_0 \sin(nt), \qquad \text{......(D)}$$

and a solution is built up from elementary solutions of type

$$X = Ae^{int \pm mx},$$

where $$a^2m^2 = -Hn^2 + iKn.$$

Since m is complex, we write $m = \alpha + i\beta$. A solution appropriate for a pipe of length l with a free end ($x = l$) at which $\frac{\partial X}{\partial x} = 0$ is

$$X = A\{e^{-\alpha x}\sin(nt - \beta x) + e^{-\alpha x'}\sin(nt - \beta x')\} + C\{e^{-\alpha x}\cos(nt - \beta x) + e^{-\alpha x'}\cos(nt - \beta x')\},$$

where $x' = 2l - x$, and where the constants A, C are chosen so that

$$X_0 = A\{1 + e^{-2\alpha l}\cos 2\beta l\} + Ce^{-2\alpha l}\sin 2\beta l,$$

$$0 = -Ae^{-2\alpha l}\sin 2\beta l + C\{1 + e^{-2\alpha l}\cos 2\beta l\}.$$

These equations give

$$A\Gamma = (1 + e^{-2\alpha l}\cos 2\beta l)\,X_0, \qquad C = e^{-2\alpha l}\sin 2\beta l\,.\,X_0,$$

where $$\Gamma = 1 + 2e^{-2\alpha l}\cos 2\beta l + e^{-4\alpha l}.$$

In the case of a pipe with a fixed end the boundary condition is $X = 0$ at $x = l$, and we write

$$X = A[e^{-\alpha x}\sin(nt - \beta x) - e^{-\alpha x'}\sin(nt - \beta x')] + C[e^{-\alpha x}\cos(nt - \beta x) - e^{-\alpha x'}\cos(nt - \beta x')].$$

The boundary condition (D) is now satisfied if

$$X_0 = A\{1 - e^{-2\alpha l}\cos 2\beta l\} - Ce^{-2\alpha l}\cos 2\beta l,$$

$$0 = Ae^{-2\alpha l}\sin 2\beta l + C\{1 - e^{-2\alpha l}\cos 2\beta l\}.$$

Therefore

$$GA = (1 - e^{-2\alpha l}\cos 2\beta l)\,X_0, \qquad GC = -X_0e^{-2\alpha l}\sin 2\beta l,$$

where $$G = 1 - 2e^{-2\alpha l}\cos 2\beta l + e^{-4\alpha l}.$$

* See a recent discussion by W. P. Mason in the *Bell System Technical Journal*, vol. VI, p. 258 (1927).

† *Advisory Committee for Aeronautics*, vol. II, p. 661 (1924–5) (R.-M. 957, Ae. 176).

If γ denotes the ratio of the specific heats for air, the pressure at any point exceeds the normal pressure p_0 by the quantity

$$p - p_0 = -p_0 \gamma \frac{\partial \xi}{\partial x},$$

where $X = \xi S$. The excess pressure at the fixed end is consequently

$$p - p_0 = 2\xi_0 e^{-\alpha l} \gamma p_0 (\alpha^2 + \beta^2)^{\frac{1}{2}} G^{-\frac{1}{2}} \sin(nt - \beta l + \phi),$$

where
$$\tan \phi = \frac{\beta A + \alpha C}{\alpha A + \beta C}.$$

The following conclusion is derived from a comparison of theory with experiment:

"Marked divergence between observed and calculated results shows that existing formulae relating to the transmission of sound waves through pipes cannot be successfully employed for correcting air pulsations of low frequency and finite amplitude."

§ **3·21**. *Vibration of a light string loaded at equal intervals.* In recent years much work has been done on methods of approximation to solutions of partial differential equations by means of a method in which the partial differential equation is replaced initially by a partial difference equation or an equation in which both differences and differential coefficients appear. Such a method is really very old and its first use may be in the well-known problem of the light string loaded at equal intervals. This problem was discussed by Bernoulli* and later in greater detail by Lagrange†.

Let the string be initially along the axis of x and let the loading masses, which we assume to be all equal, be concentrated at the points

$$x = na, \quad n = 0, \pm 1, \pm 2, \ldots.$$

Let y_n be the transverse displacement in a direction parallel to the y-axis of the mass originally at the point na, then if the tension P is regarded as constant, we have for the motion of the nth particle

$$am\ddot{y}_n = P(y_{n+1} - y_n) + P(y_{n-1} - y_n).$$

Writing $k^2 am = P$, the equation becomes

$$\ddot{y}_n = k^2 (y_{n+1} + y_{n-1} - 2y_n). \qquad \ldots\ldots\text{(A)}$$

Let us now put $\quad u_{2n} = \dot{y}_n, \; u_{2n+1} = k(y_n - y_{n+1}),$

then

$$\dot{u}_{2n} = k(u_{2n-1} - u_{2n+1}),$$

$$\dot{u}_{2n+1} = k(u_{2n} - u_{2n+2}),$$

or, if s is any integer,

$$\dot{u}_s = k(u_{s-1} - u_{s+1}).$$

* Johann Bernoulli, *Petrop. Comm.* t. III, p. 13 (1728); *Collected Works*, vol. III, p. 198.

† J. L. Lagrange, *Mécanique Analytique*, t. I, p. 390.

This is a difference equation satisfied by the Bessel functions and a particular solution which will be found useful is given by*

$$u_s = AJ_{s-\sigma}(2kt),$$

where A and σ are arbitrary constants and

$$J_n(z) = \sum_{s=0}^{\infty} \frac{(-)^s}{s!} \frac{(\tfrac{1}{2}z)^{m+2s}}{\Gamma(m+s+1)}. \qquad \text{......(B)}$$

Let us first consider the ideal case of an endless string and suppose that initially all the masses except one are in their proper positions on the axis of x and have no velocity, while the particle which should be at $x = na$ has a displacement $y_n = \eta$ and a velocity $\dot{y}_n = v$, then the initial conditions are

$$u_{2n} = v, \quad u_{2n+1} = k\eta, \quad u_{2n-1} = -k\eta,$$

while u_s is initially zero if s does not have one of the three values $2n - 1$, $2n$, $2n + 1$. A solution which satisfies these conditions is

$$u_s = v\, J_{s-2n}(2kt) + k\eta\, [J_{s-2n-1}(2kt) - J_{s-2n+1}(2kt)],$$

for, when $t = 0$, $J_\tau(2kt)$ is zero except when $\tau = 0$ and then the value is unity.

When all the masses have initial velocities and displacements the solution obtained by superposition is

$$u_s = \Sigma v_n J_{s-2n}(2kt) + k\Sigma\eta_n [J_{s-2n-1}(2kt) - J_{s-2n+1}(2kt)]. \qquad \text{......(C)}$$

If $v_n = 0$ we find by integration that

$$y_s = \Sigma\eta_n J_{2s-2n}(2kt). \qquad \text{......(D)}$$

Let us now discuss the case when this series reduces to one term, namely, the one corresponding to $n = 0$. Referring to the known graph of the function $J_{2s}(2kt)$, to known theorems relating to the real zeros and to the asymptotic representation†

$$J_{2s}(2kt) = (\pi kt)^{-\frac{1}{2}} \cos\left(2\,kt - \frac{4s+1}{4}\pi\right), \qquad \text{......(E)}$$

we obtain the following picture of the motion:

The disturbed mass swings back into its stationary position, passes this and returns after reaching an extreme position for which $|\,y\,| < \eta_0$. Its motion always approaches more and more to an ordinary simple harmonic motion with frequency initially greater than k/π, but which is very close to this value after a few oscillations. The amplitude gradually decreases, the law of decrease being eventually $(\pi kt)^{-\frac{1}{2}}$. This diminution

* T. H. Havelock, *Phil. Mag.* (6), vol. XIX, p. 191 (1910); E. Schrödinger, *Ann. d. Phys.* Bd. XLIV, S. 916 (1914); M. Koppe, *Pr.* (No. 96) *Andreas-Realgymn. Berlin* (1899), reviewed in *Fortschritte der Math.* (1899).

† Whittaker and Watson, *Modern Analysis*, p. 368. The formula is due to Poisson. An extension of the formula is obtained and used by Koppe in his investigation. The complete asymptotic expansion is given in *Modern Analysis*.

depends on the fact that the vibrational energy of the mass is gradually transferred to its neighbours, which part with it gradually themselves and so on along the string in both directions. After a long time, when $2kt$ is so large that the asymptotic representation (E) can be used for the Bessel functions of low order, the masses in the neighbourhood of the origin vibrate approximately in the manner specified by the "limiting vibration" of our arrangement, neighbouring points being in opposite phase. The amplitudes, however, decrease according to the law mentioned above and the range over which this approximate description of the vibration is valid gets larger and larger.

According to the formula (D) all the masses are set in motion at the outset, and all, except the one originally displaced, begin to move in the positive direction if $\eta_0 > 0$.

Let us consider the way in which the mass originally at $x = na$ begins its motion. The larger n is, the slower is the beginning of the motion and the longer does it continue in one direction. This is because $J_{2n}(2kt)$ vanishes like $A_n t^{2n}$, as t approaches zero, A_n being the constant multiplier in the expansion (B). Also because the first value ot $2kt$ for which the function vanishes lies between $\sqrt{\{(2n)(2n+2)\}}$ and $\sqrt{\{(2)(2n+1)(2n+3)\}}$.

It is interesting to note that in this elementary disturbance there is no question of a propagation with a definite velocity c as we might expect from the analogous case of the stretched string. Let us, however, examine the case in which all the particles are set in motion initially and in such a way that the resulting motion is periodic.

Writing $y_n = Y_n e^{2ik\omega t}$ we have the difference equation

$$Y_{n+1} + Y_{n-1} = 2(1 - 2\omega^2) Y_n.$$

If $\omega = \sin\phi$ this equation is satisfied by

$$Y_n = A \sin 2n\phi + B \cos 2n\phi. \qquad \text{......(F)}$$

Choosing the particular solution

$$Y_n = Ce^{-2in\phi},$$

we have

$$y_n = Ce^{2i(kt \sin\phi - n\phi)}.$$

Making $kt \sin\phi - n\phi$ constant we see that the phase velocity is

$$c = \frac{ak \sin\phi}{\phi}.$$

The period T is given by the equation

$$T = \frac{\pi}{k \sin\phi},$$

and the wave-length λ by the equation

$$\lambda = cT = \pi a/\phi.$$

The phase velocity thus depends on the wave-length and so there is a phenomenon analogous to dispersion. Introducing the idea of a group velocity U such that

$$\frac{\partial\lambda}{\partial t} + U\frac{\partial\lambda}{\partial x} = 0,$$

that is, such that λ does not vary in the neighbourhood of a geometrical point travelling with velocity U, we next consider a geometrical point which travels with the waves. For this point λ varies in a manner given by the equation*

$$\frac{\partial\lambda}{\partial t} + c\frac{\partial\lambda}{\partial x} = \lambda\frac{\partial c}{\partial x} = \lambda\frac{dc}{d\lambda}\frac{\partial c}{\partial x},$$

the second member expressing the rate at which two consecutive wave-crests are separating from one another. Eliminating the derivatives of λ we obtain the formula of Stokes and Rayleigh,

$$U = c - \lambda\frac{dc}{d\lambda} = U(\lambda), \text{ say.}$$

In the present case

$$U = ak\cos(\pi a/\lambda) = ak\cos\phi.$$

When $\lambda \to \infty$, $U \to ak = U(\infty) = c(\infty)$.

Hence for long waves the group velocity is approximately the same as the wave velocity. For the shortest waves $\phi = \frac{1}{2}\pi$, we have $U = 0$.

When there are only n masses the two extreme ones being at distance a from a fixed end of the string, the equations of motion are

$$\ddot{y}_1 + k^2(2y_1 - 0 - y_2) = 0,$$
$$\ddot{y}_2 + k^2(2y_2 - y_1 - y_3) = 0,$$
$$\cdots\cdots\cdots\cdots\cdots\cdots$$
$$\ddot{y}_n + k^2(2y_n - y_{n-1} - 0) = 0.$$

Assuming $y_s = Y_s e^{2ik\omega t}$ as before and eliminating the quantities Y_s from the resulting equations we obtain the following condition for free oscillations:

$$D_n = \begin{vmatrix} 2\cos 2\phi & -1 & 0 & 0 \\ -1 & 2\cos 2\phi & -1 & 0 \\ 0 & -1 & 2\cos 2\phi & -1 \\ \cdots & \cdots & \cdots & \cdots \end{vmatrix} = 0,$$

where there are n rows and columns in the determinant. Since

$$D_n = 2\cos 2\phi \,.\, D_{n-1} - D_{n-2},$$

it is readily shown by induction that

$$D_n = \frac{\sin 2(n+1)\phi}{\sin 2\phi}.$$

* See Lamb's *Hydrodynamics*, p. 359.

This is zero if $2(n+1)\phi = r\pi, r = 1, 2, \dots n$; we thus obtain n different natural frequencies of vibration. When the motion corresponding to any one of these natural frequencies is desired we use an expression of type (F) for Y_n and the end condition $Y_n = 0$ will be satisfied by writing $B = 0$. Hence one of the natural vibrations is given by

$$y_m = A \sin 2m\phi e^{2ikt \sin \phi},$$

where
$$2(n+1)\phi = r\pi \quad (r = 1, 2, \dots n).$$

If the velocity is initially zero we write

$$y_s = A \sin 2s\phi \,.\, \cos(2kt \sin \phi).$$

Let us examine more fully the case in which $n = 2$. The possible values of ϕ are $\frac{\pi}{6}$ and $\frac{\pi}{3}$, consequently in the first case

$$y_1 = A \sin \frac{\pi}{3} \cos\left(2kt \sin \frac{\pi}{6}\right), \quad y_2 = A \sin \frac{2\pi}{3} \cos\left(2kt \sin \frac{\pi}{6}\right);$$

y_1 and y_2 have the same sign and the string does not cross the axis of x. In the second case

$$y_1 = A \sin \frac{2\pi}{3} \cos\left(2kt \sin \frac{\pi}{3}\right), \quad y_2 = A \sin \frac{4\pi}{3} \cos\left(2kt \sin \frac{\pi}{3}\right);$$

y_1 and y_2 have opposite signs, the string crosses the axis of z at its middle point which is a node of the vibration.

When $n = 3$ we find in a similar way that there is one vibration without a node, one with a node and one with two nodes.

The extension to the case in which n has any integral value is clear. The general vibration, moreover, is built up by superposition from the elementary vibrations which have respectively $0, 1, 2, \dots n-1$ nodes, the nodes of one elementary vibration such as the sth being separated by those of the $(s-1)$th.

If we regard this solution as valid for all integral values of s, we may apply it to the infinite string. The initial value of y_s is now $A \sin 2s\phi$ and so by applying the general formula we are led to the surmise that there is a relation

$$\sin 2s\phi \,.\, \cos(2kt \sin \phi) = \sum_{p=-\infty}^{\infty} \sin 2p\phi \,.\, J_{2s-2p}(2kt),$$

which is true for all real values of ϕ. This relation is easily proved with the aid of well-known formulae.

An equation similar to (A) occurs in the theory of the vibrations of a row of similar simple pendulums (a, m) whose bobs are in a horizontal line and equally spaced, consecutive bobs being connected by springs as shown in Fig. 14. Using y_n to denote the horizontal deflection of the nth bob along the line of bobs and supposing that the constants of the springs are all equal, the equations of motion are of type

$$m\ddot{y}_n = k\,(y_{n+1} - y_n) - k\,(y_n - y_{n-1}) - \frac{mg}{a}\,y_n. \qquad \text{......(G)}$$

The periodic solutions of this equation give a good illustration of the filter properties of chains of electric circuits that were discovered by G. A. Campbell*. The mechanical system may, in fact, be regarded as an analogue of the following electrical system consisting of a chain of electrical circuits each of which contains elements with inductance and capacitance (Fig. 15).

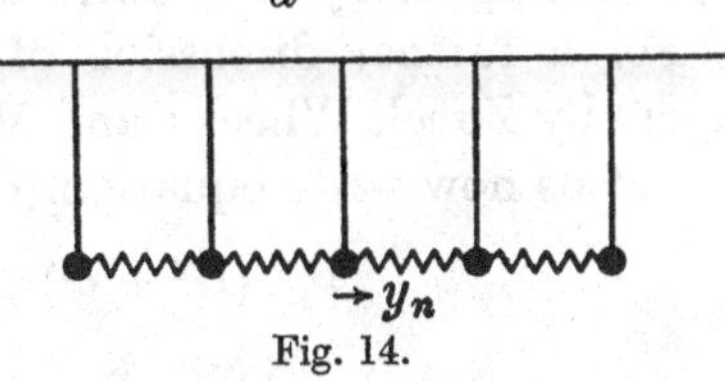

Fig. 14.

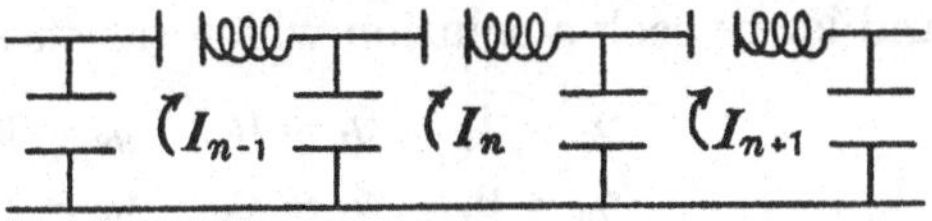

Fig. 15.

The following discussion is based largely upon that of T. B. Brown†. When the chain is of infinite length and the motion is periodic an appropriate solution is obtained by writing

$$y_n = Ar^{-n}\sin\,(pt - n\phi).$$

If $mg - ap^2 = 2akQ$ the equations for r and ϕ are

$$(1 - r^2)\sin\phi = 0, \qquad (r^2 + 1)\cos\phi = 2r\,(1 + Q).$$

These are satisfied by $\phi = 0$, $r \neq 1$ and by $r = 1$, $\cos\phi = 1 + Q$. In the latter case there is transmission without attenuation but with a change of phase from section to section, the phase velocity corresponding to a frequency $f = p/2\pi$ being

$$V = \frac{px}{\phi} = \frac{2\pi f x}{\phi},$$

where x is the length of each section. This type of transmission is possible only when Q lies between 0 and $-$ 2, that is when f lies between f_1 and f_2, where

$$2\pi f_1 = \sqrt{\left(\frac{g}{a}\right)}, \quad 2\pi f_2 = \sqrt{\left(\frac{4k}{m} + \frac{g}{a}\right)}.$$

This range of frequencies gives a pass band or transmission band. On the other hand, when $\phi = 0$ we have $r = 1 + Q \pm [(1 + Q)^2 - 1]^{\frac{1}{2}}$, and it is clear that $r > 0$ when $f < f_1$, $r < 0$ when $f > f_2$. The negative value of r indicates that adjacent sections are moving in opposite directions with amplitudes decreasing from section to section as we proceed in one direction down the line. We may use a positive value of r if we take $\phi = \pi$ instead of $\phi = 0$.

It should be noticed that r is real only when f lies outside the pass

* U.S. Patent No. 1,227,113 (1917); *Bell System Tech. Journ.* p. 1 (Nov. 1922).

† *Journ. Opt. Soc. America*, vol. VIII, p. 343 (1924).

band. There are two regions in which f may lie and these are called stop bands or suppression bands; one of these is direct and the other reverse. The stopping efficiency of each section is represented by $\log_e |r|$ and this is plotted against f in Brown's diagram.

For a further discussion of wave-filters reference may be made to papers by Zobel, Wheeler and Murnaghan.

Let us now write equation (G) in the form

$$(D^2 + c^2)\, y_n = b^2 (y_{n+1} + y_{n-1}),$$

where
$$b^2 = \frac{k}{m}, \quad c^2 = \frac{2k}{m} + \frac{g}{a}, \quad D \equiv \frac{d}{dt},$$

and let us seek a solution which satisfies the initial conditions

$$\begin{matrix} y_0 = 1, & y_1 = 0, & y_2 = 0, \dots \\ \dot{y}_0 = 0, & \dot{y}_1 = 0, & \dot{y}_2 = 0, \dots \end{matrix} \quad \text{when } t = 0.$$

One way of finding the desired solution is to expand y_n in ascending powers of b^2. Writing

$$y_n = (n, n)\, b^{2n} + (n, n+2)\, b^{2n+4} + \dots,$$

it is found by substitution in the equation that

$$y_n = \frac{1}{2^n}\left[\left(\frac{2b^2}{D^2 + c^2}\right)^n + \frac{(n+1)(n+2)}{2(2n+2)}\left(\frac{2b^2}{D^2 + c^2}\right)^{n+2} + \frac{(n+1)(n+2)(n+3)(n+4)}{2 \, . \, 4(2n+2)(2n+4)}\left(\frac{2b^2}{D^2 + c^2}\right)^{n+4} + \dots\right] \cos ct,$$

the law of the coefficients being easily verified. The meaning which must be given to

$$\left(\frac{2b^2}{D^2 + c^2}\right)^m \cos ct$$

is one in which the Taylor expansion of the expression in powers of t starts with $t^{2m} \, . \, (2b^2)^m/(2m)!$

An expression which seems to be suitable for our purpose is obtained by writing

$$\cos ct = \frac{1}{2\pi i}\int_C e^{zt} \frac{z\,dz}{z^2 + c^2},$$

where C is a circle with its centre at the origin and with a radius greater than c. The result of the operation is then

$$\left(\frac{2b^2}{D^2 + c^2}\right)^m \cos ct = \frac{1}{2\pi i}\int_C e^{zt} \frac{z\,dz}{z^2 + c^2}\left(\frac{2b^2}{z^2 + c^2}\right)^m,$$

and we obtain the formal expansion

$$y_n = \frac{1}{2^n} \cdot \frac{1}{2\pi i}\int_C e^{zt} \frac{z\,dz}{z^2 + c^2}\left[\left(\frac{2b^2}{z^2 + c^2}\right)^n + \frac{(n+1)(n+2)}{2(2n+2)}\left(\frac{2b^2}{z^2 + c^2}\right)^{n+2} + \dots\right].$$

In particular,

$$y_0 = \frac{1}{2\pi i}\int_C \frac{e^{zt} z\,dz}{[(z^2+c^2)^2 - 4b^4]^{\frac{1}{2}}},$$

$$y_1 = \frac{1}{2\pi i}\int_C \frac{e^{zt} z\,dz}{[(z^2+c^2)^2 - 4b^4]^{\frac{1}{2}}} \frac{2b^2}{z^2+c^2+[(z^2+c^2)^2-4b^4]^{\frac{1}{2}}},$$

and generally

$$y_n = \frac{1}{2\pi i}\int_C \frac{e^{zt} z\,dz}{[(z^2+c^2)^2 - 4b^4]^{\frac{1}{2}}} \left\{\frac{2b^2}{z^2+c^2+[(z^2+c^2)^2-4b^4]^{\frac{1}{2}}}\right\}^n.$$

It is easy to verify that this expression satisfies equation (G) and the prescribed initial conditions.

When $c^2 = 2b^2$ the formula reduces to

$$y_n = \frac{1}{2\pi i}\int_C \frac{e^{zt}}{(z^2+4b^2)^{\frac{1}{2}}} \left[\frac{b}{(z^2+4b^2)^{\frac{1}{2}}+z}\right]^{2n},$$

which must be equivalent to $J_{2n}(2bt)$.

The solution of the equation

$$(D + c^2)\, y_n = b^2 (y_{n+1} + y_{n-1}),$$

which satisfies the initial conditions

$$y_0 = 1, \quad y_1 = 0, \quad y_2 = 0, \dots \qquad \text{when } t = 0,$$

is given by the formula

$$y_n = \frac{1}{2\pi i}\int_C \frac{e^{z^2t} z\,dz}{[(z^2+c^2)^2 - 4b^4]^{\frac{1}{2}}} \left\{\frac{2b^2}{z^2+c^2+[(z^2+c^2)^2-4b^4]^{\frac{1}{2}}}\right\}^n.$$

An equation which is slightly more general than (A) occurs in the theory of the torsional vibrations of a shaft with several rotating masses*. This theory can be regarded as an extension of that of § 1·54 and as a preliminary study leading up to the more general case of a shaft whose sectional properties vary longitudinally in an arbitrary manner.

Let $I_1, I_2, I_3, \dots I_N$ be the moments of inertia of the rotating masses about the axis of the shaft, $\theta_1, \theta_2, \theta_3, \dots \theta_N$ the angles of rotation of these masses during vibration, $k_1, k_2, k_3, \dots k_N$ the spring constants of the shaft for the successive intervals between the rotating masses. Then

$$k_1(\theta_1 - \theta_2),\ k_2(\theta_2 - \theta_3),\ \dots\ k_{N-1}(\theta_{N-1} - \theta_N)$$

are torque moments for these intervals. Neglecting the moments of inertia of these intervening portions of the shaft in comparison with $I_1, I_2, \dots I_N$ the kinetic energy T and the potential energy V of the vibrating system are given by the equations

$$2T = I_1\dot\theta_1^2 + I_2\dot\theta_2^2 + \dots I_N\dot\theta_N^2,$$

$$2V = k_1(\theta_1-\theta_2)^2 + k_2(\theta_2-\theta_3)^2 + \dots k_{N-1}(\theta_{N-1}-\theta_N)^2,$$

* See S. Timoshenko, *Vibration Problems in Engineering*, p. 138; J. Morris, *The Strength of Shafts in Vibration*, ch. x (Crosby Lockwood, London, 1929).

and Lagrange's equations give

$$I_n \ddot{\theta}_n + k_n (\theta_n - \theta_{n+1}) - k_{n-1} (\theta_{n-1} - \theta_n) = 0,$$
$$n = 1, 2, \dots N, \quad \theta_0 \equiv \theta_1, \quad \theta_{N+1} \equiv \theta_N.$$

Except for the two end equations, for which $n = 1, N$, respectively, these equations are the same as those of a light string loaded at unequal intervals. When $k_1 = k_2 = k_3 = \dots = k_N$ the foregoing analysis can be used with slight modifications. A second case of some mathematical interest arises when

$$k_1 = k_3 = k_5 = \dots k$$
$$k_2 = k_4 = k_6 = \dots h.$$

EXAMPLES.

1. By considering special solutions of the equation of the loaded string prove that the following relations are indicated:

$$(n - x)^2 = \sum_{m=-\infty}^{\infty} m^2 J_{n-m}(x),$$

$$n^2 + k^2 t^2 = \sum_{m=-\infty}^{\infty} m^2 J_{2(n-m)}(2kt).$$

2. Prove that the equation

$$\frac{dF_n}{dx} = \tfrac{1}{2} [F_{n-1}(x) + F_{n+1}(x) - 2F_n(x)]$$

is satisfied by $\quad F_n(x) = e^{-(x-a)} \sum_{m=-\infty}^{\infty} I_{n-m}(x-a) F_m(a),$

where $\quad I_n(x) = i^{-n} J_n(ix),$

and obtain the solution in the form of a contour integral.

3. Prove that the equation

$$\dot{y}_n = b^4 [y_{n+2} + y_{n-2} + 2y_n] - 2b^2c^2 [y_{n+1} + y_{n-1}] + c^4 y_n$$

is satisfied by

$$y_n = \frac{1}{2\pi i} \int_C \frac{e^{z^4 t} z\,dz}{[(z^2 + c^2)^2 - 4b^4]^{\frac{1}{2}}} \left\{ \frac{2b^2}{z^2 + c^2 + [(z^2 + c^2)^2 - 4b^4]^{\frac{1}{2}}} \right\}^n.$$

4. Each mass in a system is connected with its immediate neighbours on the two sides by elastic rods capable of bending but without inertia. Assuming that the potential energy of bending is

$$V = \dots \tfrac{1}{2} b (2y_{r-1} - y_{r-2} - y_r)^2 + \tfrac{1}{2} b (2y_r - y_{r-1} - y_{r+1})^2 + \dots,$$

prove that the oscillations of the system are given by an equation of type

$$4\tau^2 \ddot{y}_r = -y_{r-2} + 4y_{r-1} - 6y_r + 4y_{r+1} - y_{r+2}$$

when $r > 1$ and obtain the two end equations. [Lord Rayleigh, *Phil. Mag.* (5), vol. XLIV, p. 356 (1897); *Scientific Papers*, vol. IV, p. 342.]

5. Prove that a solution of the last equation is given by

$$y_r = J_r\left(\frac{t}{\tau}\right) \cos\left(\frac{r\pi}{2} - \frac{t}{\tau}\right). \qquad \text{[Havelock.]}$$

§ 3·31. *Potential function with assigned values on a circle.* Let the origin and scale of measurement be chosen so that the circle is the unit circle $|z| = 1$ and let $z' = e^{i\theta'}$ be the complex number for a point P' on this circle. Our problem is to find a potential function V which satisfies the condition

$$V \to f(\theta') \text{ as } z \to z'.$$

To make the problem more precise the way in which the point z approaches z' ought to be specified and something must be said about the restrictions, if any, which must be laid on the function $f(\theta')$. These points will be considered later; for the present it will be supposed simply that $f(\theta')$ is real and uniquely defined for each value of θ' when θ' is a real angle between $-\pi$ and π. The mode of approach which will be considered now is one in which z moves towards z' along a radius of the unit circle. In other words, if $z = re^{i\theta}$, where r and θ are real, we shall suppose that θ remains equal to θ' and that $r \to 1$.

Now let $\bar{z} = r \,.\, e^{-i\theta}$ be the complex quantity conjugate to z; an attempt will be made first of all to represent V by means of a finite or infinite series

$$V_0 = c_0 + \sum_{n=1}^{\infty} (c_n z^n + c_{-n} \bar{z}^n), \qquad \text{......(A)}$$

where c_0 is a real constant and c_n, c_{-n} are conjugate complex constants. When the series contains only a finite number of terms it evidently represents a potential function and in the limiting process $z \to z'$ it tends to the value $\Sigma c_n z'^n$, where the summation extends over all integral values of n for which $c_n \neq 0$. Negative indices are included because $\bar{z}^n \to z'^{-n}$.

Supposing now that the finite series represents the function $f(\theta')$, the coefficient c_n is evidently given by the formula

$$c_n = \frac{1}{2\pi} \int_{-\pi}^{\pi} f(\theta')\, e^{-in\theta'} d\theta', \qquad \text{......(B)}$$

for the integral of $e^{im\theta'}$ between $-\pi$ and π is zero unless $m = 0$, consequently the term $c_n z'^n$ in the series for $f(\theta')$ is the only one which contributes to the value of the integral.

A function $f(\theta')$ which can be represented as the sum of a finite number of terms of type $c_n e^{-in\theta'}$ is evidently of a special nature and the natural thing to do is to endeavour to extend the solution which has just been found by considering the case in which an infinite number of the constants c_n defined by the formula (B) are different from zero. The series (A) formed from these constants then contains an infinite number of terms.

Let us now assume that the function $f(\theta')$ is integrable in the interval $-\pi \leqslant \theta' \leqslant \pi$. Since the series

$$1 + \sum_{1}^{\infty} r^n \left[e^{in(\theta-\theta')} + e^{-in(\theta-\theta')}\right] = \kappa(\theta - \theta')$$

is uniformly convergent for all points of this interval if $|r| < 1$, it may be integrated term by term after it has been multiplied by $f(\theta')$. The potential function V may, consequently, be expressed in the form

$$V = \frac{1}{2\pi}\int_{-\pi}^{\pi} \kappa(\theta - \theta') f(\theta')\, d\theta', \qquad \text{......(C)}$$

where $\kappa(\omega) = 1 + 2\sum_{n-1}^{\infty} r^n \cos n\omega = \dfrac{1 - r^2}{1 - 2r\cos\omega + r^2}.$

The integral representing our potential function V is generally called Poisson's integral and will be denoted here by the symbol $P(r, \theta)$ to indicate that it depends on both r and θ. This integral is of great importance in the theory of Fourier series as well as in the theory of potential functions.

The formula (C) may be obtained in another way by using the Green's function for the circle. If $P(r, \theta)$ is the pole of the Green's function, $Q(r^{-1}, \theta)$ the inverse point and $P'(r', \theta')$ an arbitrary point which is inside the circle (or on the circle) when P is inside the circle and outside (or on) the circle when P is outside the circle, an appropriate expression for the Green's function is

$$G = \log\frac{P'Q}{P'P} - \log\frac{AQ}{AP},$$

where A is an arbitrary point on the circle. This expression is evidently zero when P' is on the circle, it becomes infinite in the desired manner when P' approaches P and it is evidently a potential function which is regular except at P.

The formula

$$V = \frac{1}{2\pi}\int_0^{2\pi} \frac{|1 - r^2|\, f(\theta')\, d\theta'}{1 - 2r\cos(\theta - \theta') + r^2}$$

represents a potential function which takes the value $f(\theta)$ on the circle and is regular both inside and outside the circle.

When $r = 0$ the formula gives the relation

$$V_0 = \frac{1}{2\pi}\int_0^{2\pi} f(\theta')\, d\theta',$$

where V_0 is the value of V at the centre of the circle. This is the two-dimensional form of Gauss's mean value theorem.

When $f(\theta)$ is real for real values of θ the formula (C) may be written in the form

$$V = \frac{1}{4\pi}\int_0^{2\pi} \frac{e^{i\theta'} + z}{e^{i\theta'} - z} f(\theta')\, d\theta' + \frac{1}{4\pi}\int_0^{2\pi} \frac{e^{i\theta'} + \bar{z}}{e^{i\theta'} - \bar{z}} f(\theta')\, d\theta'$$

of the sum of two conjugate complex quantities each of which takes the value $\frac{1}{2}f(\theta)$ at the point $z = e^{i\theta}$ on the unit circle, and we deduce Schwarz's more general expression

$$F(z) = ib_0 + \frac{1}{2\pi}\int_0^{2\pi} \frac{e^{i\theta'} + z}{e^{i\theta'} - z} f(\theta')\, d\theta' \qquad \text{......(D)}$$

for a function $F(z)$ whose real part on the unit circle is $f(\theta)$. The imaginary part of $F(z)$ is a potential function iW which is given by the formula

$$W = b_0 + \frac{1}{\pi}\int_0^{2\pi} \frac{r \sin(\theta - \theta') f(\theta')\, d\theta'}{1 - 2r\cos(\theta - \theta') + r^2}.$$

If in the formula (D) we have $f(2\pi - \theta) = f(\theta)$ we obtain Boggio's formula for a function $F(z)$ whose real part takes an assigned value $f(\theta)$ on the semicircle $z = e^{i\theta}$, $0 \leqslant \theta \leqslant \pi$, and whose imaginary part vanishes on the line $z = \cos\alpha$, $0 \leqslant \alpha \leqslant \pi$, i.e. the diameter of the semicircle,

$$F(z) = \frac{1}{\pi}\int_0^{\pi} \frac{(1 - z^2) f(\theta')\, d\theta'}{1 - 2z\cos\theta' + z^2}.$$

When $V = f(z)$, where $f(z)$ is a function which is analytic in the unit circle $|z' - z| = 1$, Gauss's mean value theorem may be written in the form

$$f(z) = \frac{1}{2\pi}\int_0^{2\pi} f(z + e^{i\theta})\, d\theta,$$

and is then a particular case of Cauchy's integral theorem. By means of the substitution $z' = \rho z$, $F(\rho z) = f(z)$ the theorem may be extended to a circle of radius ρ.

If on the circle we have $|f(z')| \leqslant M$, the formula shows that $|f(z)| \leqslant M$.

More generally we can say that if $f(z)$ is a function which is regular and analytic in a closed region G and is free from zeros in G, then the greatest value of $|f(z)|$ is attained at some point of the boundary of G and the least value of $|f(z)|$ is also attained at some point on the boundary of G. In this statement values of $f(z)$ for points outside G are not taken into consideration at all.

If $f(z)$ is constant the theorem is trivial. If $f(z)$ is not constant and has its greatest value M at some point z_0 inside G we can find a small neighbourhood of z_0 entirely within G for which $|f(z)| < M$, and if C is a small circle in this neighbourhood and with z_0 as centre this inequality holds for each point of C and so

$$M = |f(z_0)| = \frac{1}{2\pi}\left|\int_0^{2\pi} f(z_0 + \rho e^{i\theta})\, d\theta\right|$$
$$\leqslant \frac{1}{2\pi}\int_0^{2\pi} |f(z_0 + \rho e^{i\theta})|\, d\theta < \frac{1}{2\pi}\int_0^{2\pi} M\, d\theta < M,$$

which leads to a contradiction. The theorem relating to the minimum value of $|f(z)|$ may be derived from the foregoing by considering the analytic function $1/f(z)$.

§ **3·32.** *Elementary treatment of Poisson's integral**. To find the limit

$$\lim_{r \to 1} P(r, \theta)$$

it will be assumed in the first place that $f(\theta)$ is integrable according to

* This treatment is based upon that given in Carslaw's *Fourier Series and Integrals*.

Riemann's definition and that if it is not bounded it is of such a nature that the integral

$$\int_{-\pi}^{\pi} f(\theta')\, d\theta'$$

is absolutely convergent.

Let us now suppose that θ is a point of the interval $-\pi \leqslant \theta \leqslant \pi$ which does not coincide with one of the end points. We shall suppose further that the limit

$$\lim_{\tau \to 0} [f(\theta+\tau) + f(\theta-\tau)] \qquad \text{......(E)}$$

exists and is equal to $2F(\theta)$, where $F(\theta)$ is simply a symbol for a quantity which is defined by this limit when θ is chosen in advance. No knowledge of the properties of the function $F(\theta)$ will be required.

Now let a function $\Phi(\theta')$ be defined for all values of θ' in the interval $(-\pi \leqslant \theta' \leqslant \pi)$ by the equation

$$\Phi(\theta') = f(\theta') - F(\theta).$$

Then

$$P(r, \theta) - F(\theta) = \frac{1}{2\pi} \int_{-\pi}^{\pi} \kappa(\theta - \theta') [f(\theta') - F(\theta)]\, d\theta'$$

$$= \frac{1}{2\pi} \int_{-\pi}^{\pi} \kappa(\theta - \theta')\, \Phi(\theta')\, d\theta.$$

Since, by hypothesis, the limit (E) exists, a positive number η can be found so as to satisfy the conditions

$$| f(\theta + \alpha) + f(\theta - \alpha) - 2F(\theta) | < \epsilon,$$

when $\qquad 0 < \alpha \leqslant \eta, \quad \theta - \eta > -\pi, \quad \theta + \eta < \pi,$

ϵ being an arbitrary small positive quantity chosen in advance. Then

$$\int_{\theta-\eta}^{\theta+\eta} \kappa(\theta - \theta')\, \Phi(\theta')\, d\theta' = \int_0^{\eta} \kappa(\alpha) [\Phi(\theta + \alpha) + \Phi(\theta - \alpha)]\, d\alpha$$

$$= \int_0^{\eta} \kappa(\alpha) [f(\theta + \alpha) + f(\theta - \alpha) - 2F(\theta)]\, d\alpha,$$

and so

$$\left| \int_{\theta-\eta}^{\theta+\eta} \kappa(\theta - \theta')\, \Phi(\theta')\, d\theta' \right| < \epsilon \int_0^{\eta} \kappa(\alpha)\, d\alpha < \epsilon \int_{-\pi}^{\pi} \kappa(\alpha)\, d\alpha = 2\pi\epsilon.$$

Also, when $0 < r < 1$,

$$\left| \int_{-\pi}^{\theta-\eta} \kappa(\theta - \theta')\, \Phi(\theta')\, d\theta' + \int_{\theta+\eta}^{\pi} \kappa(\theta - \theta')\, \Phi(\theta')\, d\theta' \right|$$

$$< \kappa(\eta) \int_{-\pi}^{\pi} | \Phi(\theta') |\, d\theta' < \kappa(\eta) \left[\int_{-\pi}^{\pi} | f(\theta') |\, d\theta' + 2\pi | F(\theta) | \right]$$

$$< 2\pi A \kappa(\eta), \text{ say,}$$

where A is a positive quantity.

But, when $0 < r < 1$,

$$\kappa(\eta) < \frac{2(1-r)}{(1-r)^2 + 4r \sin^2 \eta/2} < \frac{1-r}{2r \sin^2 \eta/2}.$$

Hence $2\pi A\kappa(\eta) < 2\pi\epsilon$ if r is so chosen that

$$\frac{1-r}{2r\sin^2\eta/2} < \frac{\epsilon}{A},$$

and this inequality is satisfied if

$$r > \frac{1}{1+\dfrac{2\epsilon}{A}\sin^2\dfrac{\eta}{2}}.$$

Combining the two results we find that

$$|P(r, \theta) - F(\theta)| < 2\epsilon,$$

if

$$1 > r > \left[1+\frac{2\epsilon}{A}\sin^2\frac{\eta}{2}\right]^{-1}.$$

Hence when the limit (E) exists,

$$P(r, \theta) \to F(\theta) \quad \text{as } r \to 1.$$

When θ is a point of continuity of the function $f(\theta)$ we have, of course, $F(\theta) = f(\theta)$ and so V tends to the assigned value. To prove that V is a potential function when $r < 1$ it is sufficient to remark that the series

$$\Sigma n c_n z^{n-1}$$

obtained by differentiating (A) term by term with respect to z is uniformly convergent for $r < s < 1$, where s is independent of r and θ, hence $\frac{\partial V}{\partial z}$ exists and is a function of z only. The equation $\frac{\partial^2 V}{\partial \bar{z}\,\partial z} = 0$ then follows immediately.

The behaviour of Poisson's integral in the neighbourhood of a point on the circle at which $f(\theta)$ is discontinuous is quite interesting. Let us suppose that $f(\theta)$ has different values $f_1(\theta)$ and $f_2(\theta)$ when the point θ is approached along the circle from different sides, then if the point θ is approached along a chord in a direction making an angle $\alpha\pi$ with the direction of the curve for which θ increases the definite integral tends to the value

$$(1-\alpha) f_1(\theta) + \alpha f_2(\theta).$$

A proof of this theorem is given by W. Gross, *Zeits. f. Math.* Bd. II, S. 273 (1918). When $f(\theta)$ is continuous round the circle we have the result that $V \to f(\theta)$ as any point on the circle is approached along an arbitrary chord through the point. This theorem has also been proved by P. Painlevé, *Comptes Rendus*, t. CXII, p. 653 (1891) and by L. Lichtenstein, *Journ. f. Math.* Bd. CXL, S. 100 (1911).

EXAMPLES

1 Show by means of Poisson's formula that if

$$\begin{aligned} f(\theta) &= -1 \qquad (-\pi < \theta < 0) \\ &= 1 \qquad (0 < \theta < \pi), \end{aligned}$$

the potential V is given by the equation

$$V = 1 + \frac{2}{\pi} \tan^{-1} \frac{r^2 - a^2}{2ar \sin \theta} \qquad (r^2 \leqslant a^2).$$

2. Let the unit circle $z = e^{i\theta}$ be divided into n arcs by points of division $\theta_1, \theta_2, \ldots \theta_n$, where $0 < \theta_1 < \theta_2 < \ldots < \theta_n = 2\pi$. Let $\phi + i\psi = f(z)$ be analytic for $|z| < 1$ and let ϕ satisfy the following conditions on the circle

$$\phi = c_m, \quad \theta_{m-1} < \theta < \theta_m, \quad c_{m+1} = c_1,$$

c_m being an arbitrary constant, then

$$2\pi f(z) = -2\pi c_1 + \sum_{s=1}^{n} (c_{s+1} - c_s)[\theta_s + 2i \log (e^{i\theta_s} - z)].$$

[H. Villat, *Bull. de la Soc. Math. de France*, t. XXXIX, p. 443 (1911); "Aperçus théoriques sur la résistance des fluides," *Scientia* (1920).]

§ 3·33. *Fourier series which are conjugate.* When r is put equal to 1 in the series (A) the resulting series may be written in the form

$$\sum_{n=-\infty}^{\infty} c_n e^{in\theta},$$

and is the "Fourier series" associated with the function $f(\theta)$.

Separating the real and imaginary parts, the series may be written in the form

$$a_0 + \sum_{\nu=1}^{\infty} (a_\nu \cos \nu\theta + b_\nu \sin \nu\theta), \qquad \text{......(A')}$$

where

$$a_0 = \frac{1}{2\pi} \int_{-\pi}^{\pi} f(\theta')\, d\theta',$$

$$a_\nu = \frac{1}{\pi} \int_{-\pi}^{\pi} f(\theta') \cos \nu\theta'\, d\theta',$$

$$b_\nu = \frac{1}{\pi} \int_{-\pi}^{\pi} f(\theta') \sin \nu\theta' d\theta'.$$

The constants a_ν, b_ν are called the "Fourier constants" associated with the function $f(\theta')$. In terms of these constants the series for V is

$$V = a_0 + \sum_{\nu=1}^{\infty} r^\nu (a_\nu \cos \nu\theta + b_\nu \sin \nu\theta). \qquad \text{......(B')}$$

When all the coefficients are real the series for the conjugate potential W is

$$b_0 + r(a_1 \sin \theta - b_1 \cos \theta) + r^2 (a_2 \sin \theta - b_2 \cos \theta) + \ldots, \qquad \text{......(C')}$$

and this is associated with the series

$$b_0 + (a_1 \sin \theta - b_1 \cos \theta) + (a_2 \sin \theta - b_2 \cos \theta) + \ldots, \qquad \text{......(D')}$$

which, when $b_0 = 0$, is called the conjugate* of the Fourier series (A'). There is now a considerable amount of knowledge relating to the conjugate series. One question of importance in potential theory is that of the

* Sometimes it is this series with the sign changed which is called the conjugate series. See L. Fejér, *Crelle*, vol. CXLII, p. 165 (1913); G. H. Hardy and J. E. Littlewood, *Proc. London Math. Soc.* (2), vol. XXIV, p. 211 (1926).

existence of a function $g(\theta)$ of which the foregoing series is the Fourier series. In this connection we may mention a theorem, due to Fatou*, which states that if $f(\theta)$ is everywhere continuous and the potential W is expressed in the form $W = W(r, \theta)$, the necessary and sufficient condition for the existence of the limit

$$\lim_{r \to 1} W(r, \theta) = g(\theta) \qquad \text{......(E')}$$

for any assigned value of θ is that the limit

$$\lim_{\epsilon \to 0} \int_{\epsilon}^{\pi} [f(\theta + \tau) - f(\theta - \tau)] \cot \frac{\tau}{2}\, d\tau = -2\pi g(\theta) \qquad \text{......(F')}$$

should exist. Fatou has also shown that if $f(\theta)$ has a finite lower bound and is such that $f(\theta)$ is integrable in the sense of Lebesgue then the limit (F') exists almost everywhere.

Lichtenstein† has recently added to this theorem by showing that the integral

$$\int_{-\pi}^{\pi} [g(\theta)]^2\, d\theta$$

exists when

$$\int_{-\pi}^{\pi} [f(\theta)]^2\, d\theta$$

exists.

For further properties of Poisson's integral and conjugate Fourier series reference may be made to the book of G. C. Evans on the logarithmic potential‡ and to Fichtenholz's paper in *Fundamenta Mathematicae* (1929).

Fatou's expression for $g(\theta)$, when $f(\theta)$ is given, is

$$\begin{aligned} g(\theta) &= \frac{1}{2\pi} \int_0^{\pi} \{f(\theta - \tau) - f(\theta + \tau)\} \cot \frac{\tau}{2}\, d\tau \\ &= \frac{1}{2\pi} P \int_{-\pi}^{\pi} f(\xi) \cot \frac{\theta - \xi}{2}\, d\xi. \end{aligned}$$

In the last integral the symbol P denotes that the integral has its principal value. Villat has deduced this expression by a limiting process with the aid of the result of Ex. 2, § 3·32. The formula is quite useful in the hydrodynamical theory of thin aerofoils.

An alternative expression, obtained by an integration by parts, is

$$g(\theta) = \frac{1}{2\pi} \int_{-\pi}^{\pi} f'(\xi) \log \sin^2 \left(\frac{\theta - \xi}{2}\right) d\xi.$$

§ **3·34**. *Abel's theorem for power series.* When for any fixed value of θ the Fourier series converges to a sum which may be denoted for the moment by $g(\theta)$, it may be shown with the aid of a property of power series discovered by N. H. Abel that $V \to g(\theta)$ as $r \to 1$. But since

* *Acta Math.* vol. xxx, p. 335 (1906).

† *Crelle's Journ.* vol. cxli, p. 12 (1912).

‡ *Amer. Math. Soc. Colloquium Publications*, vol. vi (1927).

$V \to F(\theta)$ we must have $g(\theta) = F(\theta)$ and so the Fourier series represents $F(\theta)$ whenever it is convergent.

The series for V may be written in the form

$$V = u_0 + ru_1 + r^2u_2 + \ldots, \qquad \ldots\ldots(\mathrm{G}')$$

and it should be noted that the coefficients $r, r^2, \ldots$ occurring in the different terms are all positive and form a decreasing sequence. The theorem to be proved is applicable to the more general series

$$V = v_0u_0 + v_1u_1 + v_2u_2 + \ldots,$$

where the factors v_0, v_1, v_2 are all positive and such that

$$v_{n+1} \leqslant v_n, \quad v_0 = 1.$$

Let us write

$$s_0 = u_0, \quad s_1 = u_0 + u_1, \quad s_2 = u_0 + u_1 + u_2, \ldots,$$

and suppose that the quantities $s_0, s_1, s_2, \ldots$ possess an upper limit H and a lower limit h, then

$$h \leqslant s_n \leqslant H, \qquad \text{for } n = 0, 1, 2, \ldots.$$

Now if V_n denotes the sum of the first $n + 1$ terms of the series V,

$$\begin{aligned} V &= v_0u_0 + v_1u_1 + \ldots v_nu_n \\ &= (v_0 - v_1)s_0 + (v_1 - v_2)s_1 + \ldots (v_{n-1} - v_n)s_{n-1} + v_ns_n, \end{aligned}$$

and in this series not one of the partial sums s_m has a negative coefficient. Hence

$$V_n \leqslant (v_0 - v_1)H + (v_1 - v_2)H + \ldots (v_{n-1} - v_n)H + v_nH,$$

and $\quad V_n \geqslant (v_0 - v_1)h + (v_1 - v_2)h + \ldots (v_{n-1} - v_n)h + v_nh.$

Summing the two series we obtain the inequality

$$v_0h \leqslant V_n \leqslant v_0H,$$

which shows that $|V_n| < v_0k$, where k is a fixed quantity greater than either $|h|$ or $|H|$.

Similarly, if

$$R_n{}^m = v_mu_m + v_{m+1}u_{m+1} + \ldots v_{m+n}u_{m+n},$$

we have the inequality

$$|R_n{}^m| < v_mk_m,$$

where k_m is a positive quantity greater than any one of the quantities

$$|u_m|, \quad |u_m + u_{m+1}|, \quad |u_m + u_{m+1} + u_{m+2}|, \ldots.$$

If now the series $u_0 + u_1 + u_2 + \ldots$ is convergent and ϵ is any arbitrarily chosen small positive quantity, a number $m(\epsilon)$ can be found such that

$$|u_m + u_{m+1} + \ldots u_{m+n}| < \epsilon,$$

for $n = 0, 1, 2, \ldots$ and $m > m(\epsilon)$. When m is chosen in this way we may take $k_m = \epsilon$ and since $v_m \leqslant v_0 \leqslant 1$ we have the inequality

$$|R_n{}^m| < \epsilon.$$

When the quantity v_n is a function of a variable r which lies in the unit interval $0 \leqslant r \leqslant 1$ the foregoing inequality shows that the series (G′) is uniformly convergent for all values of r in this interval and so represents a continuous function of r. In the case under consideration we have $v_n = r^n$ and the conditions imposed on v_n are satisfied if $0 < r \leqslant 1$. The function V is consequently continuous at $r = 1$ and so

$$u_0 + u_1 + u_2 + \ldots = \lim_{r \to 1} P(r, \theta) = F(\theta).$$

§ 3·41. *The analytical character of a regular logarithmic potential**. Poisson's integral may be used to prove that a logarithmic potential V which is regular in a region D is an analytic function of x and y.

We may, without loss of generality, take the origin at an arbitrary point within D. Let C denote the circle $x^2 + y^2 = a^2$ which lies entirely within D, then for a point $x = a\cos\alpha$, $y = a\sin\alpha$ on this circle, $V = f(\alpha)$, where $f(\alpha)$ is a continuous function of α and so by Poisson's formula

$$V = \frac{1}{2\pi}\int_0^{2\pi} \frac{(a^2 - r^2) f(\alpha)\, d\alpha}{a^2 + r^2 - 2ar\cos(\theta - \alpha)},$$

where $x = r\cos\theta$, $y = r\sin\theta$ and $r < a$.

Now the series

$$\frac{a^2 - r^2}{a^2 + r^2 - 2ar\cos(\theta - \alpha)} = 1 + 2\sum_{n=1}^{\infty}\left(\frac{r}{a}\right)^n \cos n(\theta - \alpha)$$

is absolutely and uniformly convergent and so can be integrated term by term after being multiplied by $f(\alpha)\, d\alpha/2\pi$. Therefore

$$V = a_0 + \sum_{n=1}^{\infty}\left(\frac{r}{a}\right)^n (a_n \cos n\theta + b_n \sin n\theta).$$

Now if in the polynomial

$$\left(\frac{r}{a}\right)^n (a_n \cos n\theta + b_n \sin n\theta)$$
$$= \tfrac{1}{2} a^{-n} [a_n (x + iy)^n + a_n (x - iy)^n - ib_n (x + iy)^n + ib_n (x - iy)^n]$$

we replace each term of type $c_{pq} x^p y^q$ by its modulus, the resulting expression will be less than the corresponding expression obtained by doing the same thing to each term of type $e_{pq} x^p y^q$ in the expansion of each of the four binomials and adding the results. Now this last expression is less than

$$2\,[|x| + |y|]^n M a^{-n},$$

where M is the upper bound of a_n and b_n. Now let

$$|x| < s, \quad |y| < s,$$

where $s < a/2$, then

$$2\,[|x| + |y|]\, M a^{-n} < 2\,(2s/a)^n M,$$

and the series of moduli is convergent. The series for V is thus a power

* E. Picard, *Cours d'Analyse*, t. II, p. 18.

series in x and y which is absolutely convergent for $|x| < s$, $|y| < s$, it thus represents an analytic function.

Since, moreover, the origin was chosen at an arbitrary point in D it follows that V is analytic at each point within D.

For the parabolic equation $\frac{\partial z}{\partial y} = \frac{\partial^2 z}{\partial x^2}$ there is a theorem given by Holmgren* which indicates that z is an analytic function of x in the neighbourhood of a point (x_0, y_0) in a region R within which z is regular.

If through the point (x_0, y_0) there is a segment $a \leqslant y \leqslant b$ of the line $x = x_0$ which lies entirely within R, there is a number c such that for $|x - x_0| < c$, $a \leqslant y \leqslant b$ there is an expansion

$$z(x, y) = \Sigma \frac{(x - x_0)^{2n}}{(2n)!} \phi^{(n)}(y) + \Sigma \frac{(x - x_0)^{2n+1}}{(2n+1)!} \psi^{(n)}(y),$$

where
$$\phi(y) = z(x_0, y), \quad \psi(y) = \frac{\partial}{\partial x_0} z(x_0, y).$$

These functions $\phi(y)$, $\psi(y)$ are continuous (D, ∞) in $a \leqslant y \leqslant b$ and their derivatives satisfy inequalities of type

$$|\phi^{(n)}(y)| < Mc^{-2n}(2n)!, \quad |\psi^{(n)}(y)| < Mc^{-2n}(2n)!.$$

§ 3·42. *Harnack's theorem*†. Let W_s, for each positive integral value of s, be a potential function which is continuous $(D, 2)$ (i.e. regular) in a closed region R and let the infinite series

$$w_1 + w_2 + w_3 + \dots \qquad \text{......(A)}$$

converge uniformly on the boundary B of R, then the series converges uniformly throughout R and represents a potential function which is regular and analytic in R. The sum $w_n + w_{n+1} + \dots w_{n+p}$ is a potential function regular in R. If it is not a constant it assumes its extreme value on B and if N_p is the numerically greatest of these we shall have

$$|w_n + w_{n+1} + \dots w_{n+p}| \leqslant |N_p|.$$

Since, however, the series converges uniformly on B we can choose a number $m(\epsilon)$ such that when $n > m(\epsilon)$ we have

$$|N_p| < \epsilon,$$

for all positive integral values of p. This inequality, combined with the previous one, proves that the series (A) converges uniformly in R and so represents a continuous function w. Now let C be any circle which lies entirely within R and let Poisson's formula be used to obtain expressions for potential functions W, W_1, W_2, W_3, ... regular within C and having respectively the same boundary values as the functions w, w_1, w_2, w_3, ...

* E. Holmgren, *Arkiv for Mat., Astr. och Fysik*, Bd. I (1904); Bd. III (1906); Bd. IV (1907); *Comptes Rendus*, t. CXLV, p. 1401 (1907).

† Kellogg calls this Harnack's first theorem. See *Potential Theory*, p. 248. The theorem was given by Harnack in his book. It has been extended to other equations of elliptic type by L. Lichtenstein, *Crelle's Journal*, Bd. CXLII, S. 1 (1913).

Since a potential function with assigned boundary values on C is unique if it is required also to be regular within C we have $W_s = w_s$ $(s = 1, 2, \ldots)$. Furthermore, since the series (A) is uniformly convergent it may be integrated term by term after multiplication by the appropriate Poisson factor. Therefore at any point within C

$$\begin{aligned} W &= W_1 + W_2 + W_3 + \ldots \\ &= w_1 + w_2 + w_3 + \ldots = w. \end{aligned}$$

Hence within C the function w is identical with the regular potential function which has the same values as w at points on C. Since C is an arbitrary circle within R it follows that w is a regular potential function at all points of R and is consequently analytic at each point of R.

For recent work relating to the analytical character of the solutions of elliptic partial differential equations reference may be made to L. Lichtenstein, *Enzyklopädie der Math. Wiss.*, II C. 12; T. Radó, *Math. Zeits.* Bd. XXV, S. 514 (1926); S. Bernstein, *ibid.* Bd. XXVIII, S. 330 (1928); H. Lewy, *Gött. Nachr.* (1927), *Math. Ann.* Bd. CI, S. 609 (1929).

§ **3·51**. *Schwarz's alternating process.* H. A. Schwarz* has used an alternating process, somewhat similar to that used by R. Murphy† in the treatment of the electrical problem of two conducting spheres, to solve the first boundary problem of potential theory for the case of a region bounded by a contour made up of a finite number of analytic arcs meeting at angles different from zero.

To indicate the process we consider the simple case of two contours $a\alpha$, $b\beta$ bounding two areas A, B which have a common part C bounded by α and β, while a and b bound a region D represented by $A + B - C$.

We shall use the symbols a, b, α, β to denote also the parameters by means of which the points on these curves may be expressed in a uniform continuous manner and shall use the symbols m and n to denote the points common to the curves a and b. We shall suppose, moreover, that the choice of parameters is made in such a way that m and n are represented by the parameters m and n whether they are regarded as points on a, b, α or β. This can always be done by subjecting parameters chosen for each curve to suitable linear transformations.

Our problem now is to find a potential function V which is regular within D and which satisfies the boundary conditions $V = f(a)$ on a, $V = g(b)$ on b, where $f(m) = g(m)$, $f(n) = g(n)$.

We shall suppose that $f(a)$ is continuous on a and that $g(b)$ is continuous on b. We shall suppose also that a function $h(\alpha)$, which is continuous on α, is chosen so as to satisfy the conditions

$$h(m) = f(m), \quad h(n) = f(n).$$

* *Berlin Monatsberichte* (1870); *Gesammelte Werke*, Bd. II, S. 133.

† *Electricity*, p. 93, Cambridge (1833).

We now form a sequence of logarithmic potentials $u_1, u_2, \ldots$ regular in A, and a sequence of logarithmic potentials $v_1, v_2, \ldots$ regular in B; these potentials being chosen so as to satisfy the following boundary conditions in which $u_s(\beta)$ denotes the value of u_s on β, and $v_s(\alpha)$ denotes the value of v_s on α, $(s = 1, 2, \ldots)$:

$$\begin{array}{ll} u_1 = f(a) \text{ on } a, & u_1 = h(\alpha) \text{ on } \alpha, \\ u_2 = f(a) \text{ on } a, & u_2 = v_1(\alpha) \text{ on } \alpha, \\ u_3 = f(a) \text{ on } a, & u_3 = v_2(\alpha) \text{ on } \alpha, \\ \ldots\ldots\ldots & \ldots\ldots\ldots \\ v_1 = g(b) \text{ on } b, & v_1 = u_1(\beta) \text{ on } \beta, \\ v_2 = g(b) \text{ on } b, & v_2 = u_2(\beta) \text{ on } \beta, \\ v_3 = g(b) \text{ on } b, & v_3 = u_3(\beta) \text{ on } \beta, \\ \ldots\ldots\ldots & \ldots\ldots\ldots \end{array}$$

Writing $\quad u_s = u_1 + (u_2 - u_1) + (u_3 - u_2) + \ldots (u_s - u_{s-1}),$

$$v_s = v_1 + (v_2 - v_1) + (v_3 - v_2) + \ldots (v_s - v_{s-1}),$$

our object now is to show that as $s \to \infty$ the series for u_s and v_s converge and represent potentials which are exactly the same in C. To establish the convergence of the series we shall make use of the following lemma.

We note that $w_s = u_s - u_{s-1}$ is a logarithmic potential which is regular in A and which is zero on a. Let $\delta_s(\alpha)$ be its value at a point on α and let δ_s be the maximum value of $|\delta_s(\alpha)|$.

Now let ϕ be the logarithmic potential which is regular in A and which satisfies the boundary conditions $\phi = 1$ on α, $\phi = 0$ on a. As the point (x, y) approaches one of the points of discontinuity m, ϕ tends to a value θ such that $0 < \theta < 1$. Now a regular potential function attains its greatest value in a region on the boundary of the region, theretore $\phi < 0$ for all points of A and so there is a positive number ϵ between 0 and 1 such that, on β, $\phi < \epsilon < 1$.

Now $w_s + \delta_s\phi$ is zero on a and positive on α and is a logarithmic potential regular in A. Its least value is therefore attained on the boundary of A and so $w_s + \delta_s\phi > 0$ within A. This inequality may be written in the form

$$\delta_s(\phi - \epsilon) + w_s + \epsilon\delta_s > 0,$$

and since $\phi < \epsilon$ on β it follows that $w_s + \epsilon\delta_s > 0$ on β.

In a similar way we can show that $w_s - \delta_s\phi < 0$ in A and so we may conclude that $w_s - \epsilon\delta_s < 0$ on β. Combining the inequalities we may write

$$|w_s| < \epsilon\delta_s.$$

The number ϵ was derived from the function ϕ associated with A. In a similar way there is a number η associated with the region B and the curve α. Let κ be the greater of these two numbers if the two numbers are not equal.

Writing $t_s = v_s - v_{s-1}$ and using the symbol $\tau_s(\beta)$ to denote the value of t_s on β we use τ_s to denote the maximum value of $|\tau_s(\beta)|$ on β. We then find in a similar way that

$$|t_s| < \eta\tau_s,$$

and so we may write

$$|w_s| < \kappa\delta_s, \quad |t_s| < \kappa\tau_s.$$

We thus obtain the successive inequalities

$$|v_2 - v_1| = |u_2 - u_1| \text{ on } \beta.$$

Therefore $$\tau_2 < \kappa\delta_2,$$

$$|u_3 - u_2| = |v_2 - v_1| \text{ on } \alpha.$$

Therefore $$\delta_3 < \kappa\tau_2 < \kappa^2\delta_2, \ldots \quad \delta_{s+2} < \kappa^{2s}\delta_2,$$

$$\tau_3 < \kappa\delta_3 < \kappa^2\tau_2, \ldots \quad \tau_{s+2} < \kappa^{2s}\tau_2.$$

The series for u_s and v_s thus converge uniformly at all points of the boundary of C and so by Harnack's theorem represent regular logarithmic potentials which we may denote by u and v respectively. Since $u_s = v_{s-1}$ on α and $u_s = v_s$ on β it follows that $u = v$ on the boundary of C and so $u = v$ throughout C. Since, moreover, the series for u converges uniformly on the boundary of A and the series for v converges uniformly on the boundary of B these series may be used to continue the potential function $u = v$ beyond the boundary of C into the regions A and B, and the potential function thus defined will have the desired values on a and b.

§ **3·61**. *Flow round a circular cylinder.* To illustrate the use of the complex potential in hydrodynamics we shall consider the flow represented by a complex potential χ which is the sum of a number of terms

$$\chi_1 = U(z + a^2/z), \quad \chi_2 = ik \log z, \quad \chi_3 = ic \log \frac{z - z_0}{z - z_1},$$

$$\chi_4 = -ic' \log \frac{z - z_2}{z - z_3}.$$

Writing $z = re^{i\theta}$, $\chi = \phi + i\psi$ we consider first the case in which $\chi = \chi_1$ and U is real. We then have

$$u - iv = d\chi/dz = U(1 - a^2/z^2),$$

$$\psi = U \sin\theta \, (r - a^2/r).$$

The stream-function ψ is zero on the circle $r^2 = a^2$ and also on the line $y = 0$. There is thus one stream-line which divides into two parts at a point S where it meets the circle; these two portions reunite at a second point S' on the circle and the stream-line leaves the circle along the line $y = 0$. Since $z^2 = a^2$ at the points S and S' these points are points of stagnation ($u = v = 0$). It will be noticed that the stream-line $y = 0$ cuts the boundary $r^2 = a^2$ orthogonally. This is in accordance with the general theorem of § 1·72.

At a great distance from the circle we have $u - iv = U$, $\psi = Uy$, and so the stream-lines are approximately straight lines parallel to the axis of x. Our function ψ is thus the stream-function for a type of steady flow past a circular cylinder. This flow is not actually possible in nature, the observed flow being more or less turbulent while for a certain range of speed depending upon the viscosity of the fluid and the size of the cylinder, eddies form behind the cylinder and escape downstream periodically* in such a way as to form a vortex street in which a vortex of one sign is almost equidistant from two successive vortices of the opposite sign and each vortex of this sign is almost equidistant from two successive vortices of the other sign. Vortices of one sign lie approximately on a line parallel to the axis of x and vortices of the other sign on a parallel line.

Some light on the formation of this asymmetric arrangement of vortices is furnished by a study of the equilibrium and stability of a pair of vortices of opposite signs which happen to be present in the flow round the circular cylinder.

The flow may be represented approximately by writing

$$\chi = \chi_1 + \chi_3 + \chi_4,$$

and choosing z_0, z_1, z_2, z_3 so that the circle $r^2 = a^2$ is a stream-line. This condition may be satisfied by writing

$$z_0 = r_0 e^{i\theta_0}, \quad z_1 = a^2 r_0^{-1} e^{i\theta_0},$$

$$z_2 = r_2 e^{i\theta_2}, \quad z_3 = a^2 r_2^{-1} e^{i\theta_2}.$$

If A, B, C, D are the points specified by the complex numbers z_0, z_1, z_2, z_3, respectively, these equations mean that B is the inverse of A and C the inverse of D.

In the theory of Helmholtz and Kelvin vortices move with the fluid. When the vortices are isolated line vortices this result is generally replaced by the hypothesis that the velocity of any rectilinear vortex ω is equal to the resultant of the velocities produced at its location by all the other vortices which together with ω produce the resultant flow at an arbitrary point. In using this hypothesis the uniform flow U is supposed to be produced by a double vortex at infinity and the complex potential Ua^2/z is interpreted as that of a double vortex at the origin of co-ordinates O.

The vortex at A will be stationary when

$$0 = U\left(1 - \frac{a^2}{z_0^2}\right) - \frac{ic}{z_0 - z_1} - \frac{ic'}{z_0 - z_2} + \frac{ic'}{z_0 - z_3}.$$

Taking for simplicity the case when $r_2 = r_0$, $\theta_2 = -\theta_0$, $c' = c$, and

* Th. v. Kármán, *Gött. Nachr.* p. 547 (1912); *Phys. Zeits.* p. 13 (1912). The vortices have been observed experimentally by Mallock, *Proc. Roy. Soc. London*, vol. IX, p. 262 (1907); and by Bénard, *Comptes Rendus*, vol. CXLVII, pp. 839–970 (1908); vol. CLVI, pp. 1003–1225 (1913); vol. CLXXXII, pp. 1375–1823 (1926); vol. CLXXXIII, pp. 20–184 (1926).

separating the real and imaginary parts of the expression on the right, after multiplying it by z_0, we obtain the equations

$$0 = U(r_0 - a^2 r_0^{-1}) \cos\theta_0 - \tfrac{1}{2}c \cot\theta_0 + a^2 c r_0^2 \Omega^{-1} \sin 2\theta_0,$$

$$0 = U(r_0 + a^2 r_0^{-1}) \sin\theta_0 - c r_0^2/(r_0^2 - a^2) - \tfrac{1}{2}c + c r_0^2 (r_0^2 - a^2 \cos 2\theta_0)\,\Omega^{-1},$$

where $$\Omega = r_0^4 - 2a^2 r_0^2 \cos 2\theta_0 + a^4.$$

The first equation gives

$$2U\Omega \sin\theta_0 = -c r_0 (a^2 - r_0^2),$$

and when this value of U is substituted in the second equation it is found that

$$r_0^2 - a^2 = \pm 2r_0^2 \sin\theta_0.$$

This result was obtained by Föppl*, who also studied the stability of the vortices. The result tells us that the vortex can be in equilibrium if $AB = AD$. To confirm this result by geometrical reasoning we complete the parallelogram $BADE$ and determine a point N on the axis of y such that $ON = AN$. Let M be the point of intersection of BC and AN, G the point of intersection of AC and BD.

On the understanding that all lines used to represent velocities are to be turned through a right angle in the clockwise direction the velocities at A due to the different vortices may be represented as follows:

Those due to the vortices at B and D by c/AB and c/AD respectively. Since $AB = AD$ these two velocities together may be represented by $c.AE/AB^2$ along AE.

The velocity due to the vortex at C may be represented by c/CA along CA and equally well by cGA/AB^2 along GA. The resultant velocity at A due to the vortices at B, C and D may thus be represented by $c.GE/AB^2$ along GE.

On the other hand, the velocity U is represented by U along ON, and the velocity due to the double vortex at O by $U.NM/ON$ along NM. The velocity in the flow round the cylinder in the absence of the vortices is thus represented by $U.OM/ON$.

Now $M\hat{A}B = M\hat{B}A = O\hat{C}M$, therefore O, M, A, C are concyclic and so $O\hat{M}C = O\hat{A}C = O\hat{E}G$. This means that OM and EG are parallel. By choosing c so that $c.EG/AB^2 = U.OM/ON$ the resultant velocity at A will be zero. Since the triangles ONA, OAD are similar, the equation for c becomes simply

$$c = U\frac{OM}{ON}\frac{AB^2}{EG} = U\frac{OM}{ON}\frac{AB^2}{AG} = U\frac{OM}{ON}AC = AC^2.U/a$$
$$= U(r_0^2 - a^2)(1 - a^4/r_0^4)/a$$

and implies that the strength of the vortex at A is greater the greater the distance of A from the origin.

* L. Föppl, *München Sitzungsber.* (1913). See also Howland, *Journ. Roy. Aeron. Soc.* (1925); M. Dupont, *La Technique Aéronautique*, Dec. 15 (1926) and Jan. 15 (1927); W. G. Bickley, *Proc. Roy. Soc. Lond.* A, vol. CXIX, p. 146 (1928).

The stream-lines in the flow studied by Föppl are quite interesting and have been carefully drawn by W. Müller*. There are four points of stagnation on the circle, two of these, S and S', lie on the line $y = 0$, while the other two, S_0, S_0', are images of each other in the line $y = 0$. Stream-lines orthogonal to the circle start at S_0 and S_0' and unite at a point T on the line y where they cut this line orthogonally. This point T is also a point of stagnation. Outside these stream-lines the flow is very similar to that round a contour formed from arcs of two circles which cut one another orthogonally; within the region bounded by these stream-lines there is a circulation of fluid and the flow between T and the circle is opposite in direction to that of the main stream. The stream-lines are, indeed, very similar to those which have been frequently observed or photographed in the case of the slow motion round a cylinder†.

Let us now consider the case when there is only one vortex outside the cylinder and a circulation round the cylinder. We now put

$$\chi = \chi_1 + \chi_2 + \chi_3.$$

In this case

$$u - iv = U(1 - a^2/z^2) + ik/z + ic\,[(z - r_0 e^{i\theta_0})^{-1} - (z - r_0^{-1} a^2 e^{i\theta_0})^{-1}],$$

and the component velocities of the vortex A are given by

$$u_0 - iv_0 = U(1 - a^2 r_0^{-2} e^{-2i\theta_0}) + ikr_0^{-1} e^{-i\theta_0} - icr_0 e^{-i\theta_0}(r_0^2 - a^2)^{-1},$$

while for its image B

$$u_1 + iv_1 = -a^2 (u_0 - iv_0)\, r_0^{-2} e^{2i\theta_0}.$$

If X, Y are the components of the resultant force on the cylinder per unit length, we have

$$X + iY = -\tfrac{1}{2}\rho a \int_0^{2\pi} \left(u^2 + v^2 + 2\frac{\partial\phi}{\partial t}\right) e^{i\theta} d\theta = (X_q + iY_q) + (X_\phi + iY_\phi), \text{ say.}$$

Now when $z = ae^{i\theta}$,

$$u^2 + v^2 = 4U^2 \sin^2\theta + k^2/a^2 + c^2 (r_0^2 - a^2)^2/a^2R^4$$
$$+ 4U \sin\theta . [k/a - c(r_0^2 - a^2)/aR^2] - 2kc(r_0^2 - a^2)/a^2R^2,$$

where $$R^2 = a^2 + r_0^2 - 2ar_0 \cos(\theta - \theta_0).$$

Therefore $$X_q + iY_q = 2\pi i\rho\, \{kU - c(u_0 + iv_0)\}.$$

We have also for $r = a$

$$\frac{\partial\phi}{\partial t} + i\frac{\partial\psi}{\partial t} = -ic\frac{u_0 + iv_0}{ae^{i\theta} - r_0 e^{i\theta_0}} + ic\frac{u_1 + iv_1}{ae^{i\theta} - (a^2/r_0) e^{i\theta_0}},$$

* *Zeits. für technische Physik*, Bd. VIII, S. 62 (1927); *Mathematische Strömungslehre* (Springer, Berlin, 1928), p. 124.

† See especially the photographs published by Camichel in *La Technique Aéronautique*, Nov. 15 (1925) and Dec. 15 (1925).

therefore

$$\int_0^{2\pi} \left(\frac{\partial \phi}{\partial t} + i\,\frac{\partial \psi}{\partial t}\right) ae^{i\theta}\, d\theta = 2\pi i c\,(u_1 + iv_1),$$

$$\int_0^{2\pi} \left(\frac{\partial \phi}{\partial t} + i\,\frac{\partial \psi}{\partial t}\right) ae^{-i\theta}\, d\theta = 2\pi i c a^2 r_0^{-2}\,(u_0 + iv_0)\, e^{-2i\theta_0},$$

$$\int_0^{2\pi} \frac{\partial \phi}{\partial t}\, ae^{i\theta}\, d\theta = \pi i c\,(u_1 + iv_1) - \pi i c a^2 r_0^{-2}\,(u_0 - iv_0)\, e^{2i\theta_0} = 2\pi i c\,(u_1 + iv_1).$$

Combining these results we have

$$X + iY = 2\pi i\rho\,\{kU - c\,(u_0 + iv_0) + c\,(u_1 + iv_1)\}.$$

This result may be extended to the case in which there are any number of vortices outside the cylinder*, the general result being

$$X + iY = 2\pi i\rho \left\{kU - \sum_{s=0}^{\infty} c_s\,(u_{2s} + iv_{2s} - u_{2s+1} - iv_{2s+1})\right\}.$$

In the special case when there is only one vortex and $k = 0$, $\theta_0 = 0$, we have $u_1 + iv_1 = -\,a^2 r_0^{-2}\,(u_0 - iv_0)$,

$$u_0 - iv_0 = U\,(1 - a^2 r_0^{-2}) - icr_0\,(r_0^2 - a^2)^{-1},$$

$$X + iY = 2\pi\rho c\,[c/r_0 - iU\,(1 - a^4 r_0^{-4})].$$

Introducing the coefficients of lift and drag, defined by $X = \rho SU^2 . C_D$, $Y = \rho SU^2 . C_L$, S being the projected area, we find

$$C_D = (c/aU)^2\, \pi a/r_0, \quad C_L = -\,\pi\,(c/aU)\,(1 - a^4 r_0^{-4}).$$

These results were obtained by W. G. Bickley† who plots the lift-drag curves for $r = 2a$, $4a$ and $6a$, and compares them with the published curves for Flettner rotors (rotating cylinders with end plates). With the last two values the agreement is fair except for low values of the lift.

The stream-lines for the case of a single vortex outside the cylinder have been drawn by W. Müller‡.

EXAMPLES

1. If in a type of flow similar to that considered by Föppl the vortices at z_0 and z_2 are not images of each other in the line $y = 0$, one of the conditions that the vortices may be stationary in the flow round the cylinder is

$$(r_0 - a^2 r_0^{-1}) \cos\theta_0 = (r_2 - a^2 r_2^{-1}) \cos\theta_2.$$

2. If in Föppl's flow the vortices move so that they are always images of each other in the line $y = 0$ the resultant force on the cylinder is a drag if

$$4r_0^4 \sin^2\theta_0 > (r_0^2 - a^2)^2.$$

[Bickley.]

* H. Bateman, *Bull. Amer. Math. Soc.* vol. XXV, p. 358 (1919); D. Riabouchinsky, *Comptes Rendus*, t. CLXXV, p. 442 (1922); M. Lagally, *Zeits. f. angew. Math. u. Mech.* Bd. II, S. 409 (1922). In this formula the even suffixes refer to the vortices outside the cylinder and the odd suffixes to the image vortices inside the cylinder.

† *Loc. cit. ante*, p. 251.

‡ *Loc. cit. ante*, p. 252.

3. A plate of width $2a$ is placed normal to a steady stream of velocity U and vortices form behind the plate at the points

$$z_1 = x_0 + iy_0, \quad z_2 = x_0 - iy_0.$$

Prove that the conditions are satisfied by

$$X = U(z^2 + a^2)^{\frac{1}{2}} + i\kappa \log \frac{(z^2 + a^2)^{\frac{1}{2}} - (z_1^2 + a^2)^{\frac{1}{2}}}{(z^2 + a^2)^{\frac{1}{2}} - (z_2^2 + a^2)^{\frac{1}{2}}}.$$

Prove also that when

$$y_0 \sqrt{3} = -x_0 + 2(x_0^2 + a^2)^{\frac{1}{2}}, \quad \kappa^2 \sqrt{3} = 4U^2 x_0 y_0,$$

the velocity does not take infinite values at the edges of the plate and the vortices are stationary.

[D. Riabouchinsky.]

§ 3·71. *Elliptic co-ordinates.* Problems relating to an ellipse or an elliptic cylinder may be conveniently solved with the aid of the substitution

$$x + iy = c \cosh(\xi + i\eta) = c \cosh \zeta,$$

which gives

$$x = c \cosh \xi \cos \eta,$$

$$y = c \sinh \xi \sin \eta.$$

The curves ξ = constant are confocal ellipses,

$$\frac{x^2}{c^2 \cosh^2 \xi} + \frac{y^2}{c^2 \sinh^2 \xi} = 1,$$

the semi-axes of the typical ellipse being $a = c \cosh \xi$ and $b = c \sinh \xi$. The angle η can be regarded as the excentric angle of a point on the ellipse.

The curves η = constant are confocal hyperbolas, the semi-axes of the typical hyperbola being $a' = c \cos \eta$ and $b' = c \sin \eta$.

The first problem we shall consider is that of the determination of the viscous drag on a long elliptic cylinder which moves parallel to its length through the fluid in a wide tube whose internal surface is a confocal elliptic cylinder*.

Considering a cylindrical element of fluid bounded by planes parallel to the plane of xy and a curved surface generated by lines perpendicular to this plane; the viscous drag per unit length on the curved surface of the cylinder is

$$\int \mu \frac{\partial w}{\partial n} ds,$$

taken round the contour of the cross-section, w being the velocity parallel to a generator and μ being the coefficient of viscosity.

If the fluid is not being forced through the tube under pressure the pressure may be assumed to be constant along the tube and so in steady motion the total viscous drag on the cylindrical element must be zero.

* C. H. Lees, *Proc. Roy. Soc.* A, vol. XCII, p. 144 (1916).

Transforming the line integral into an integral over the enclosed area, we obtain the equation

$$\frac{\partial^2 w}{\partial x^2} + \frac{\partial^2 w}{\partial y^2} = 0.$$

The boundary conditions are $w = 0$ when $\xi = \xi_1$, and $w = v$ when $\xi = \xi_2$, we therefore write

$$w(\xi_1 - \xi_2) = v(\xi_1 - \xi),$$

$$(\xi_1 - \xi_2)\,\mu \frac{\partial w}{\partial n} = -\,v\mu \frac{\partial \xi}{\partial n} = v\mu \frac{\partial \eta}{\partial s}.$$

Since η varies from 0 to 2π in a complete circuit round the contour of the cross-section, the total viscous force per unit length of the cylinder is

$$\frac{2\pi\mu v}{\xi_1 - \xi_2} = \frac{2\pi\mu v}{\log(a_1 + b_1) - \log(a_2 + b_2)}.$$

If the inner ellipse reduces to a straight line of length $2c$, the total drag on the plane is D per unit length, where

$$D[\log(a_1 + b_1) - \log(2c)] = 2\pi\mu v,$$

and the resistance per unit area at the point x is

$$(D/2\pi)(c^2 - x^2)^{-\frac{1}{2}}.$$

It is clear from this expression that the resistance per unit area, i.e. the shearing stress, is much greater near the edges of the strip than near its centre line.

The foregoing analysis may be used with a slight modification to determine the natural charges on two confocal elliptic cylinders regarded as conductors at different potentials. If V is the potential at (ξ, η) and $V = 0$ for $\xi = \xi_1$, $V = v$ for $\xi = \xi_2$, we have

$$V(\xi_1 - \xi_2) = v(\xi_1 - \xi),$$

and the density of charge on the cylinder $\xi = \xi_1$ is

$$-\frac{1}{4\pi}\frac{\partial V}{\partial \xi}\frac{\partial \xi}{\partial n} = \frac{1}{4\pi}\frac{\partial V}{\partial \xi}\frac{\partial \eta}{\partial s} = \frac{v}{4\pi(\xi_1 - \xi_2)\,c}(\sinh^2 \xi_1 - \sin^2 \eta_1)^{-\frac{1}{2}}.$$

When the inner cylinder reduces to the strip whose cross-section is S_1S_2 we have, when $v = 2(\xi_1 - \xi_2)$,

$$\sigma_1 = (1/2\pi c)\operatorname{cosec}\eta_1,$$

and if, moreover, the outer cylinder is of infinite size σ_1 becomes the natural charge on the strip when the total charge per unit length is equal to unity; this is the charge density on each side of the strip.

To find the stream-function for steady irrotational flow round an elliptic cylinder when there is no circulation round the cylinder, we write $\psi = \psi_1 + \psi_2$, where

$$\psi_1 = Uy - Vx = c(U \sinh \xi \sin \eta - V \cosh \xi \cos \eta)$$

is the stream-function for the steady flow at a great distance from the cylinder and ψ_2 is the stream-function for a superposed disturbance in this flow produced by the cylinder. To satisfy the boundary condition $\psi = 0$ at the surface of the cylinder, and the condition that the component velocities derived from ψ_2 are negligible at infinity, we write

$$\psi_2 = e^{-\xi} (A \cos \eta + B \sin \eta).$$

Choosing the constants so that $\psi = 0$ on the cylinder, we have

$$e^{-\xi_1} A = cV \cosh \xi_1, \quad e^{-\xi_1} B = -cU \sinh \xi_1$$
$$= a_1 V \qquad\qquad = -b_1 U$$

where a_1, b_1 are the semi-axes of the ellipse $\xi = \xi_1$. We have also

$$a_1 + b_1 = c_1 e^{-\xi_1}, \quad a_1 - b_1 = c_1 e^{-\xi_1}$$

Therefore $\psi_2 = (a_1 + b_1)^{\frac{1}{2}} (a_1 - b_1)^{-\frac{1}{2}} e^{-\xi} (Va_1 \cos \eta - Ub_1 \sin \eta)$,

$$\phi + i\psi = (U - iV) z + (Ub_1 + iVa_1)(a_1 + b_1)^{\frac{1}{2}} (a_1 - b_1)^{-\frac{1}{2}} e^{-\zeta}.$$

To find the electrical potential of a conducting elliptic cylinder which is under the influence of a line charge parallel to its generators, we need an expression for the logarithm

$$\log (z_0 - z) = \log [c (\cosh \zeta_0 - \cosh \zeta)]$$
$$= \zeta_0 + \log \tfrac{1}{2}c + \log (1 - e^{\zeta - \zeta_0})(1 - e^{-\zeta - \zeta_0})$$
$$= \zeta_0 + \log \tfrac{1}{2}c - 2 \sum_1^{\infty} n^{-1} e^{-n\zeta_0} \cosh n\zeta, \quad |\xi| < |\xi_0|.$$

Writing this equal to $\phi_0 + i\psi_0$ we have

$$\phi_0 = \xi_0 + \log \tfrac{1}{2}c - 2 \sum_{n=1}^{\infty} n^{-1} e^{-n\xi_0} (\cosh n\xi \cos n\eta \cos n\eta_0 + \sinh n\xi \sin n\eta \sin n\eta_0).$$

To obtain a potential which is constant over the elliptic cylinder $\xi = \xi_1$, we write $\phi = \phi_0 + \phi_1$, where

$$\phi_1 = \sum_{n=1}^{\infty} (A_n e^{-n\xi} \cos n\eta + B_n e^{-n\xi} \sin n\eta).$$

Each term of this series is indeed a potential function which vanishes at infinity. Choosing the constants A_n, B_n, so that the boundary condition $\phi = 0$ on $\xi = \xi_1$ is satisfied by $\phi = \phi_0 + \phi_1$, we have

$$nA_n e^{-n\xi_1} = 2e^{-n\xi_0} \cosh n\xi_1 \cos n\eta_0,$$
$$nB_n e^{-n\xi_1} = 2e^{-n\xi_0} \sinh n\xi_1 \sin n\eta_0.$$

Hence when $\xi_1 < \xi < \xi_0$,

$$\phi = \xi_0 + \log \tfrac{1}{2}c + \sum_{n=1}^{\infty} n^{-1} e^{n(\xi_1 - \xi_0)} \sinh n(\xi_1 - \xi) \cos n(\eta_0 - \eta).$$

Summing the series we find that

$$\phi = \xi + \xi_0 - \xi_1 + \log \tfrac{1}{2}c + \tfrac{1}{2} \log \frac{\cosh (\xi_0 - \xi) - \cos (\eta_0 - \eta)}{\cosh (\xi + \xi_0 - 2\xi_1) - \cos (\eta_0 - \eta)}.$$

The corresponding stream-function is

$$\psi = \eta + \tan^{-1}\frac{\sin(\eta_0 - \eta)}{1 - e^{\xi-\xi_0}\cos(\eta_0 - \eta)} + \tan^{-1}\frac{\sin(\eta_0 - \eta)}{1 - e^{2\xi_1-\xi_0-\xi}\cos(\eta_0 - \eta)},$$

and when $\xi_1 = 0$ the value of $\frac{\partial \phi}{\partial \xi}$ for $\xi = 0$ is

$$\left(\frac{\partial \phi}{\partial \xi}\right)_0 = 1 - \frac{\sinh \xi_0}{\cosh \xi_0 - \cos(\eta_0 - \eta)}.$$

The surface density of the charge on the plate $\xi = 0$ is thus

$$\frac{1}{4\pi}\frac{d\eta}{ds}\left[1 - \frac{\sinh \xi_0}{\cosh \xi_0 - \cos(\eta_0 - \eta)}\right],$$

and the total charge is zero. When the total charge per unit length is 1, and the total charge per unit length of the line is -1, the surface density of the charge on the cylinder is

$$\frac{1}{2\pi}\frac{d\eta}{ds}\frac{\sinh \xi_0}{\cosh \xi_0 - \cos(\eta_0 - \eta)}.$$

This is what C. Neumann* calls the induced charge density or the induced loading; it represents, of course, the charge on one side of the plate $\xi = 0$. We shall write this expression in the form

$$\sigma(0, \eta; \xi_0, \eta_0) = \frac{1}{2\pi}\frac{d\eta}{ds} S(0, \eta; \xi_0, \eta_0),$$

and shall use a corresponding expression

$$\sigma(\xi_1, \eta_1; \xi_0, \eta_0) = \frac{1}{2\pi}\frac{d\eta}{ds} S(\xi_1, \eta_1; \xi_0, \eta_0),$$

in which

$$S(\xi_1, \eta_1; \xi_0, \eta_0) = \frac{\sinh(\xi_0 - \xi_1)}{\cosh(\xi_0 - \xi_1) - \cos(\eta_0 - \eta_1)}, \qquad \ldots\ldots(A)$$

and $\sigma(\xi_1, \eta_1; \xi_0, \eta_0)$ is the density of the induced charge for the elliptic cylinder $\xi = \xi_1$.

Let us now consider the problem in which a function V is required to satisfy the condition $V = f(\eta_1)$ on the cylinder $\xi = \xi_1$, while V is a regular potential function outside the cylinder $\xi = \xi_1$ but not necessarily vanishing at infinity. Some idea of the nature of the solution may be obtained by first considering the two cases

$$f(\eta_1) = \cos m\eta, \quad V = e^{-m(\xi-\xi_1)}\cos m\eta,$$

$$f(\eta_1) = \sin m\eta, \quad V = e^{-m(\xi-\xi_1)}\sin m\eta.$$

Since

$$S(\xi_1, \eta_1; \xi, \eta) = 1 + 2\sum_{m=1}^{\infty} e^{-m(\xi-\xi_1)}\cos m(\eta - \eta_1),$$

* *Leipzig. Ber.* Bd. LXII, S. 87 (1910).

the solution is given in these cases by the formula

$$V = \frac{1}{2\pi}\int_0^{2\pi} S(\xi_1, \eta_1; \xi, \eta) f(\eta_1) \, d\eta_1 \qquad \text{......(B)}$$
$$= \int \sigma(\xi_1, \eta_1; \xi, \eta) f(\eta_1) \, ds_1,$$

and we may write

$$\sigma(\xi_1, \eta_1; \xi, \eta) = \sigma(\xi_1, \eta_1) S(\xi_1, \eta_1; \xi, \eta), \qquad \text{......(C)}$$

where $\sigma(\xi_1, \eta_1)$ is the natural density per unit length when the total charge per unit length on the cylinder $\xi = \xi_1$ is unity. This is a particular case of a general theorem due to C. Neumann*, which tells us that the density of the induced charge for a cylinder whose cross-section is a closed curve can be found when the natural density on the cylinder and the corresponding potential is known. The expression for the induced charge is then of the form (C), where ξ and η are conjugate functions such that ξ = constant are the equipotentials and η = constant, the lines of force associated with the natural charge. The undetermined constant factor occurring in the expressions for functions ξ and η which satisfy the last condition should be chosen so that η increases by 2π in one circuit round the cross-section of the cylinder.

The formula (A) gives a potential which satisfies the conditions of the problem for a wide class of functions and for this class of functions we have the interesting relation

$$2\pi f(\eta) = \lim_{\xi \to \xi_1} \int_0^{2\pi} \frac{\sinh(\xi - \xi_1) f(\eta_1) \, d\eta_1}{\cosh(\xi - \xi_1) - \cos(\eta - \eta_1)} \quad (\xi > \xi_1).$$

The question naturally arises whether the function V given by (B) is the only function which fulfils the conditions of the problem. To discuss this question we shall consider the case when the ellipse $\xi = \xi_1$ reduces to the line $\xi = 0$, i.e. the line S_1S_2.

It will be noticed that when $f(\eta) = 1$ the formula (B) gives $V = 1$. Now the potential ϕ which is the real part of the expression

$$\phi + i\psi = z(z^2 - c^2)^{-\frac{1}{2}} = \coth \zeta,$$

satisfies the condition that $\phi = 0$ on the line $\xi = 0$ and $\phi = 1$ at infinity. Furthermore, the function ϕ_1, which is the real part of

$$\phi_1 + i\psi_1 = c(z^2 - c^2)^{-\frac{1}{2}} = \operatorname{cosech} \zeta,$$

satisfies the conditions

$$\phi_1 = 0 \text{ when } \xi = 0, \quad \phi_1 = 0 \text{ when } \xi = \infty.$$

Hence a more general potential which satisfies the same conditions as V is

$$V + A\phi + B\phi_1,$$

* *Leipzig. Ber.* Bd. LXII, S. 278 (1910).

where A and B are arbitrary constants. Now

$$\frac{\sinh \xi}{\cosh \xi - \cos (\eta - \eta_0)} = R \frac{1 + e^{-\zeta + i\eta_0}}{1 - e^{-\zeta + i\eta_0}}$$

$$= R \frac{\sinh \zeta + i \sin \eta_0}{\cosh \zeta - \cos \eta_0}.$$

Hence, if

$$v + iu = \frac{1}{2\pi} \int_{-\pi}^{\pi} \left[\frac{\sinh \zeta + i \sin \eta_0}{\cosh \zeta - \cos \eta_0} - \frac{\cosh \zeta - A}{\sinh \zeta} \right] f(\eta_0)\, d\eta_0 + V + iU, \quad \text{......(D)}$$

where U and V are constants, the potentials u and v are conjugate functions which can be regarded as component velocities in a two-dimensional flow of an incompressible inviscid fluid. These component velocities satisfy the conditions

$$u = U, \quad v = V \text{ at infinity}, \quad v = f(\eta) \text{ on the line } S_1 S_2.$$

This result is of some interest in connection with Munk's theory of thin aerofoils. In this theory an element ds of a thin aerofoil in a steady stream of velocity U parallel to Ox is supposed to deflect the air so as to give it a small component velocity $v = u \dfrac{dy_0}{dx_0}$ in a direction parallel to the axis of y. Assuming that $u = U + \epsilon$, where ϵ is a small quantity of the same order of smallness as y_0 and dy_0/dx_0, we neglect $\epsilon \dfrac{dy_0}{dx_0}$, as it is of order ϵ^2, and write $v = U \dfrac{dy_0}{dx_0}$. This is now taken to be the y-component of velocity at points of the line S_1S_2 and the corresponding component velocities (u, v) for the region outside the aerofoil may be supposed, with a sufficient approximation, to be given by an expression of type (D). In this expression, however, the coefficient A is given the value 1 so that the velocity at the trailing edge will not be infinite.

Now when $|\zeta|$ is large we may write

$$(\cosh \zeta - \cos \eta_0)^{-1} = \operatorname{sech} \zeta + \cos \eta_0 \operatorname{sech}^2 \zeta + \ldots,$$

$$\sinh \zeta = \cosh \zeta - \tfrac{1}{2} \operatorname{sech} \zeta - \tfrac{1}{8} \operatorname{sech}^3 \zeta + \ldots,$$

$$\operatorname{cosech} \zeta = \operatorname{sech} \zeta + \tfrac{1}{2} \operatorname{sech}^3 \zeta + \ldots.$$

Hence, when $V = 0$ the flow at a great distance from the origin is of type

$$v + iu = iU + \beta_1/z + \beta_2 z^2 + \ldots,$$

where

$$\beta_1 = \frac{c}{2\pi} \int_{-\pi}^{\pi} (1 + \cos \eta_0 + i \sin \eta_0) f(\eta_0)\, d\eta_0,$$

$$\beta_2 = \frac{c^2}{2\pi} \int_{-\pi}^{\pi} (i \sin \eta_0 \cos \eta_0 - \sin^2 \eta_0) f(\eta_0)\, d\eta_0,$$

and by Kutta's theorem the lift, drag and moment per unit length of the aerofoil are given by the expressions of § 4·71

$$L + iD = \tfrac{1}{2}\rho \int (v + iu)^2\, dz = -\, 2\pi\rho U \beta_1,$$

$$M = \tfrac{1}{2}\rho R \int (v + iu)^2\, z dz = -\, 2\pi\rho U R \beta_2.$$

Therefore
$$L = -\, \rho c U^2 \int_{-\pi}^{\pi} (1 + \cos\eta_0) \frac{dy_0}{dx_0}\, d\eta_0,$$

$$D = -\, \rho c U^2 \int_{-\pi}^{\pi} \sin\eta_0 \frac{dy_0}{dx_0}\, d\eta_0 = 0,$$

$$M = \rho c^2 U^2 \int_{-\pi}^{\pi} \sin^2\eta_0 \frac{dy_0}{dx_0}\, d\eta_0.$$

These are the expressions obtained by Munk* by a slightly different form of analysis. A more satisfactory theory of thin aerofoils in which the thickness is taken into consideration, has been given by Jeffreys and is sketched in § 4·73.

Since
$$dx_0 = -\, c \sin\eta_0 d\eta_0, \quad x_0 = c\cos\eta_0,$$

we may write
$$L = -\, 2\rho U^2 \int_{-c}^{c} (c + x_0)(c^2 - x_0^2)^{-\frac{1}{2}} \frac{dy_0}{dx_0}\, dx_0$$

$$= 2\rho c U^2 \int_{-c}^{c} (c + x_0)^{-\frac{1}{2}} (c - x_0)^{-\frac{3}{2}}\, y_0 dx_0,$$

$$M = 2\rho U^2 \int_{-c}^{c} (c^2 - x_0^2)^{-\frac{1}{2}}\, x y_0 dx_0.$$

§ **3·81**. *Bipolar co-ordinates.* Problems relating to two circles which intersect at two points S_1 and S_2 with rectangular co-ordinates $(c, 0)$, $(-\,c, 0)$ respectively may be treated with the aid of the conformal transformation

$$z = ic \cot \tfrac{1}{2}\zeta, \qquad \text{......(A)}$$

where $z = x + iy$, $\zeta = \xi + i\eta$ and (x, y), (ξ, η) are the rectangular co-ordinates of two corresponding points P and π. We shall say that the point P is in the z-plane and the point π in the ζ-plane. The transformation may be said to map one plane on the other.

It is easily seen that

$$\frac{z + c}{z - c} = e^{-i\zeta} = e^{\eta - i\xi},$$

$$e^\eta = r_2/r_1, \quad S_2\hat{P}S_1 = \xi,$$

$$r_1^2 = (x - c)^2 + y^2 = |\, z - c\,|^2 = 2cMe^{-\eta},$$

$$r_2^2 = (x + c)^2 + y^2 = |\, z + c\,|^2 = 2cMe^{\eta},$$

where
$$M = \frac{c}{\cosh\eta - \cos\xi}.$$

* National Advisory Committee for Aeronautics, *Report*, p. 191 (1924); see also J. S. Ames, *Report*, p. 213 (1925) and C. A. Shook, *Amer. Journ. Math.* vol. XLVIII, p. 183 (1926).

The curves ξ = constant are clearly circles through the points S_1 and S_2, while the curves η = constant are circles having these points as inverse points. The two sets of curves form in fact two orthogonal systems of circles, as is to be expected since the transformation (A) is conformal, and the corresponding sets of lines are perpendicular.

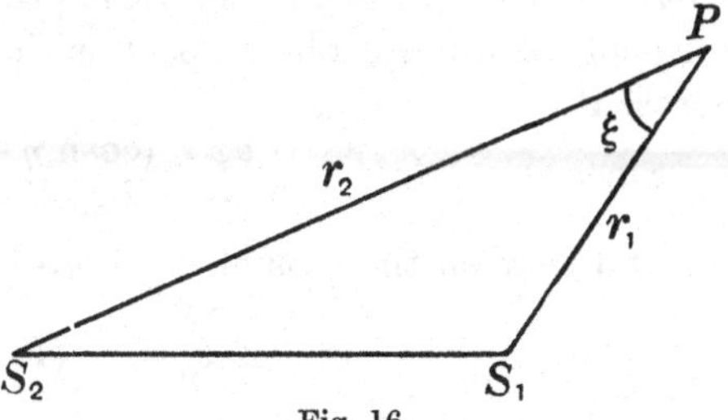

Fig. 16.

The expressions for x and y in terms of ξ and η are $x = M \sinh \eta$, $y = M \sin \xi$. At a point P_0 of the line S_1S_2 we have $\xi = \pi$, therefore

$$x_0 = c \tanh(\eta_0/2), \quad y = 0,$$

and the natural loading for this line is

$$\sigma_0 = (1/\pi c) \cosh(\eta_0/2).$$

The loading induced by a charge -1 at the point P (ξ, η) is, on the other hand,

$$\sigma_0{}^* = \frac{2\cosh^2(\eta_0/2)\cos(\xi/2)\cosh[(\eta - \eta_0)/2]}{\pi c[\cosh(\eta - \eta_0) + \cos\xi]}$$

$$= \sigma_0 \frac{\cosh(\eta/2) + \cosh[(\eta/2) - \eta_0]}{\cosh(\eta - \eta_0) + \cos\xi} \cos\frac{\xi}{2}.$$

EXAMPLE

A potential function v is regular in the semicircle $y > 0$, $x^2 + y^2 < a^2$ and satisfies the boundary conditions $v = A$ when $y = 0$, $\frac{\partial v}{\partial r} = B$ when $x^2 + y^2 = a^2$, prove that

$$v = A - A' \int_0^\infty \cos m\eta \frac{\sinh m(\xi - \pi)}{m} \frac{dm}{\cosh^2 \frac{m\pi}{2}},$$

where $x + iy = ia \cot \frac{1}{2}(\xi + i\eta)$, $A' = aB$.

§ **3·82**. *Effect of a mound or ditch on the electric potential.* Let us now consider the complex potential

$$\chi = \phi + i\psi = \frac{2c}{\kappa} \cot \frac{\zeta}{\kappa}, \qquad \text{......(B)}$$

where κ is a real constant at our disposal. The potential ϕ is zero when $\xi = 0$, for in this case χ becomes $\frac{2c}{\kappa} \cot \frac{i\eta}{\kappa}$, and is a purely imaginary quantity. It is also zero when $\xi = \frac{1}{2}\kappa\pi$, for then $\chi = -\frac{2c}{\kappa} \tan \frac{i\eta}{\kappa}$, and is again a purely imaginary quantity. The potential ϕ is thus zero on a continuous line made up of the portion of the line $y = 0$ outside the segment S_1S_2 and of the circular arc through S_1S_2 at points of which

S_1S_2 subtends the angle $\frac{1}{2}\kappa\pi$. Thus the complex potential χ provides us with the solution of an electrical problem relating to a conductor in the form of an infinite plane sheet with a circular mound or ditch running across it.

Since
$$c\frac{\partial\phi}{\partial y} = \frac{(\cosh\eta - \cos\xi)^2}{\cosh\eta\cos\xi - 1}\frac{\partial\phi}{\partial\xi},$$
we find that on the axis of y, where $\eta = 0$,
$$c\frac{\partial\phi}{\partial y} = -(1 - \cos\xi)\frac{\partial\phi}{\partial\xi}.$$

Also
$$\frac{\partial\phi}{\partial\xi} = -\frac{2c}{\kappa^2}\operatorname{cosec}^2\frac{\xi}{\kappa},$$
consequently the potential gradient on the axis $x = 0$ is
$$\frac{2}{\kappa^2}\operatorname{cosec}^2\frac{\xi}{\kappa}(1 - \cos\xi).$$

At the vertex where $\xi = \frac{1}{2}\kappa\pi$ it is
$$\frac{2}{\kappa^2}(1 - \cos\tfrac{1}{2}\kappa\pi).$$

On the plane $y = 0$, we have $\xi = 0$, and the gradient is
$$\frac{2}{\kappa^2}\operatorname{cosech}^2\frac{\eta}{\kappa}.(\cosh\eta - 1).$$

As $\eta \to 0$, $x \to \infty$ and the gradient tends to the value 1 which will be regarded as the normal value.

As $\eta \to \pm\infty$, $x \to \pm c$ and the gradient tends to become zero or infinite according as $\kappa \gtrless 2$.

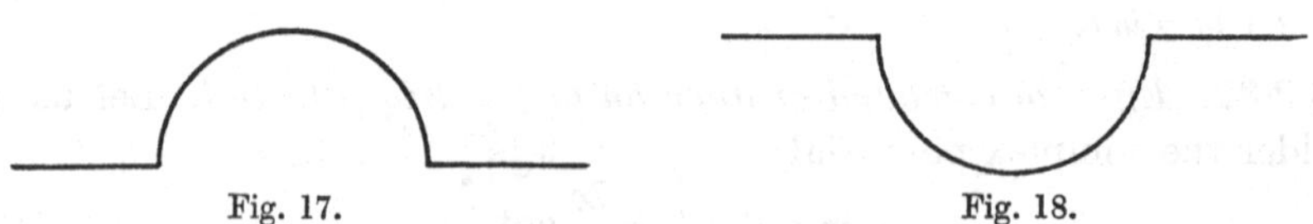

Fig. 17. Fig. 18.

When $\kappa = 1$ we have a semicircular mound.

The gradient on the line $x = 0$ is everywhere greater than the normal value, at the vertex it is 2, and at a point at distance $2c$ above the vertex it is 10/9.

When $\kappa = 3$ we have a semicircular ditch.

The gradient on the line $x = 0$ is everywhere less than the normal value, at the bottom of the ditch it is 2/9 and at a point $(0, c)$, at distance $2c$ above the bottom, it is 8/9. By making $\kappa \to 0$ we obtain values of the gradient for the case of a cylinder standing on an infinite plane. We must naturally make $c \to 0$ at the same time, in order to obtain a cylinder of finite radius a. The appropriate complex potential is

$$\chi = \phi + i\psi = a\pi \cot \frac{a\pi}{y - ix}. \qquad \text{......(C)}$$

On the line $x = 0$ the potential gradient is

$$\frac{a^2\pi^2}{y^2} \operatorname{cosec}^2 \left(\frac{a\pi}{y}\right),$$

and tends to the normal value as $y \to \pm\infty$.

When $y = 2a$ the gradient is $\frac{\pi^2}{4}$, which is nearly 2·5. At a distance $2a$ above the summit, $y = 4a$ and the gradient is $\frac{\pi^2}{8} = 1{\cdot}2337$. On the axis of x the gradient is

$$\frac{a^2\pi^2}{x^2} \operatorname{cosech}^2 \left(\frac{a\pi}{x}\right).$$

As $x \to 0$ the gradient diminishes rapidly to zero, consequently the surface density of electricity is very small in the neighbourhood of the point of contact.

EXAMPLE

Fluid of constant density moves above the infinite plane $y = 0$ with uniform velocity U. A cylinder of radius a is placed in contact with the plane with its generators perpendicular to the flow. Prove that the stream function is derived from a complex potential of type (C) multiplied by U and calculate the upward thrust on the cylinder.

[H. Jeffreys, *Proc. Camb. Phil. Soc.* vol. XXV, p. 272 (1929).]

§ **3·83**. *The effect of a vertical wall on the electric potential.* Let h be the height of the wall, $\chi = \phi + i\psi$ the complex potential. If a is a constant, the substitution

$$az = ih\,(a^2 + \chi^2)^{\frac{1}{2}} \qquad \text{......(D)}$$

makes the point $z = ih$ correspond to $\chi = 0$, the points on the axis of x correspond to the points on $\phi = 0$ for which $\psi^2 > a^2$, while the points on the axis of y for which $y < h$ correspond to the points on $\phi = 0$ for which $\psi^2 < a^2$. Hence, if ϕ be regarded as the electric potential, a conducting surface consisting of the plane $y = 0$ and the conducting wall $(x = 0, y \leqslant h)$ will be at zero potential*.

If (r, θ), (r', θ') are the bipolar co-ordinates of a point P relative to S, the top of the wall, and to S', the image of this point in the plane $y = 0$, we have

$$h^2\chi^2 = -a^2(z^2 + h^2) = -a^2 rr' e^{i(\theta+\theta')}.$$

Therefore
$$h\phi = a\,(rr')^{\frac{1}{2}} \sin \tfrac{1}{2}(\theta + \theta'),$$
$$h\psi = -a\,(rr')^{\frac{1}{2}} \cos \tfrac{1}{2}(\theta + \theta').$$

Therefore $$2h^2\phi^2 = a^2\{[(x^2 - y^2 + h^2)^2 + 4x^2y^2]^{\frac{1}{2}} - (x^2 - y^2 + h^2)\}.$$

* C. H. Lees, *Proc. Roy. Soc. London*, A, vol. XCI, p. 440 (1915).

The equipotentials have been drawn by Lees from the equation

$$y^2 (1 + a^2x^2/h^2\phi^2) = h^2 + x^2 + h^2\phi^2/a^2.$$

To determine the surface density of electricity we differentiate equation (D), then

$$h^2\chi \frac{\partial \chi}{\partial x} = - a^2 z, \quad h^2 \chi \frac{\partial \chi}{\partial y} = - ia^2 z.$$

When $x = 0$, $z = iy$, $h\chi = - ia (h^2 - y^2)^{\frac{1}{2}}$, and so

$$\frac{\partial \phi}{\partial x} = ay/h (h^2 - y^2)^{\frac{1}{2}}.$$

The surface density is thus zero at the base and infinite at the top of the wall.

When $y = 0$, $z = x$, $h\chi = - ia (h^2 + x^2)^{\frac{1}{2}}$, and so

$$\frac{\partial \phi}{\partial y} = ax/h (h^2 + x^2)^{\frac{1}{2}}.$$

As $x \to \infty$ this tends to the value a/h which may be regarded as the normal value of the gradient. At a distance from the foot equal to h the vertical gradient is 0·707 times the normal gradient.

The curve along which the electric field strength has the constant value F is given by

$$a^2 (x^2 + y^2) = h^2F^2 [(x^2 - y^2 + h^2)^2 + 4x^2y^2]^{\frac{1}{2}},$$

that is, by

$$a^2R^2 = h^2F^2rr',$$

where (R, Θ) are the polar co-ordinates of P with respect to the origin. The curves $F =$ constant may be obtained by inversion from the family of Cassinian ovals with S and S' as poles, they are the equipotential curves for two unit line charges at S and S', and a line charge of strength -2 at the origin O. The rectangular hyperbola

$$y^2 - x^2 = \tfrac{1}{2}h^2$$

is a particular curve of the family. This hyperbola meets the axis of y at a point where the horizontal gradient is equal to the normal gradient. The force is equal in magnitude to the normal gradient at all points of this hyperbola. At points above the hyperbola the force is greater than a/h, at points below the hyperbola it is less than a/h.

The curves along which the force has a fixed direction are the lemniscates defined by the equation

$$\theta + \theta' - 2\Theta = \text{constant}.$$

Each lemniscate passes through S and S' and has a double point at O.

It should be noticed that the transformation

$$az = ih (a^2 - \zeta^2)^{\frac{1}{2}}$$

enables us to map the upper half of the ζ-plane on the region of the upper

z-plane bounded by the line $y = 0$ and the vertical wall $x = 0$, $y < h$. This transformation makes the points at infinity in the two planes corresponding points. It may be observed also that if $a^2 - \zeta^2 = r^2 e^{2i\theta}$, the angle 2θ ranges from $-\pi$ to π. Hence, since $az = hr \sin\theta + ihr\cos\theta$, $hr\cos\theta$ is never negative and so it is the upper portion of the cut z-plane which corresponds to the upper portion of the ζ-plane. If we invert the z-plane from a point on the negative portion of the axis of y we obtain a region inside a circle which is cut along a radius from a point on the circumference to a point not on the circumference. The upper half of the ζ-plane maps into the interior of this region.

If, on the other hand, we invert from the origin of the z-plane, the cut upper half plane inverts into a half plane with a cut along the y-axis from infinity to a point some distance above the origin. The point at infinity in the ζ-plane now maps into the origin in the z-plane.

CHAPTER IV

CONFORMAL REPRESENTATION

§ 4·11. Many potential problems in two dimensions may be solved with the aid of a transformation of co-ordinates which leaves $\nabla^2 V = 0$ unaltered in form. It is easily seen that the transformation

$$\xi = f(x, y), \quad \eta = g(x, y)$$

furnished by the equation

$$\zeta \equiv \xi + i\eta = F(x + iy) \equiv F(z)$$

possesses this property when the function F is analytic, because a function of ζ which is analytic in some region Γ of the ζ-plane is also analytic in the corresponding region G of the z-plane when regarded as a function of z. In using a transformation of this kind it is necessary, however, to be cautious because singularities of a potential function may be introduced by the transformation, and the transformation may not always be one-to-one, i.e. a point P in the ζ-plane may not always correspond to a single point Q in the z-plane and vice versa.

Let V be a function of ξ and η which is continuous $(D, 1)$, then

$$\frac{\partial V}{\partial x} = \frac{\partial V}{\partial \xi}\frac{\partial \xi}{\partial x} + \frac{\partial V}{\partial \eta}\frac{\partial \eta}{\partial x}, \qquad \frac{\partial V}{\partial y} = \frac{\partial V}{\partial \xi}\frac{\partial \xi}{\partial y} + \frac{\partial V}{\partial \eta}\frac{\partial \eta}{\partial y}.$$

These equations show that if the derivatives of ξ and η are not all finite at a point (x, y) in the z-plane, the derivatives $\frac{\partial V}{\partial x}, \frac{\partial V}{\partial y}$ may be infinite even though $\frac{\partial V}{\partial \xi}$ and $\frac{\partial V}{\partial \eta}$ are finite. A possible exception occurs when $\frac{\partial V}{\partial \xi}$ and $\frac{\partial V}{\partial \eta}$ both vanish, i.e. at a point of equilibrium or stagnation.

At any point (x_0, y_0) in the neighbourhood of which the function $F(z)$ can be expanded in a Taylor series which converges for $|z - z_0| < c$, we have

$$F(z) = \sum_{n=0}^{\infty} a_n (z - z_0)^n,$$

where $$z = x + iy, \quad z_0 = x_0 + iy_0,$$

and if $$\zeta = \xi + i\eta = F(z), \quad \zeta_0 = \xi_0 + i\eta_0 = F(z_0),$$

we may write

$$d\zeta = \zeta - \zeta_0 = F(z) - F(z_0) = dz\,[F'(z) + \epsilon],$$

where $dz = z - z_0$ and $\epsilon \to 0$ as $dz \to 0$.

Hence $$d\zeta = dz.F'(x) \text{ approximately.}$$

This relation shows that the (x, y) plane is mapped conformally on the (ξ, η) plane for all points at which $|F'(z)|$ is neither zero nor infinite. We have in fact the approximate relations

$$d\sigma = ds\,|F'(z)|,$$
$$\phi = \theta + \alpha,$$

where
$$dz = ds\,.\,e^{i\theta}, \quad d\zeta = d\sigma\,.\,e^{i\phi},$$
$$F'(z) = |F'(z)|\,e^{i\alpha}.$$

These relations show that the ratio of the lengths of two corresponding linear elements is independent of the direction of either and that the angle between two linear elements dz, δz at the point (x, y) is equal to the angle between the two corresponding linear elements at (ξ, η). The first angle is, in fact, $\theta - \theta'$, while the second angle is

$$\phi - \phi' = (\theta + \alpha) - (\theta' + \alpha) = \theta - \theta'.$$

These theorems break down if some of the first coefficients in the expansion

$$\zeta - \zeta_0 = \sum_{n=1}^{\infty} a_n (z - z_0)^n \qquad \text{......(A)}$$

are zero. If, for instance, $a_1 = a_2 = \ldots a_{m-1} = 0$, we have for small values of $|z - z_0|$

$$\zeta - \zeta_0 = a_m (z - z_0)^m,$$

and the relation between the angles is

$$\phi = m\theta + \alpha_m, \text{ where } a_m = |a_m|\,e^{i\alpha_m}.$$

This gives

$$\phi - \phi' = m(\theta - \theta').$$

More generally if there is an expansion of type (A) in which the lowest index m is not an integer a similar relation holds.

§ **4·12**. The way in which conformal representation may be used to solve electrical and hydrodynamical problems is best illustrated by means of examples. One point to be noticed is that frequently the transformation does not alter the essential physical character of the problem because an electric charge concentrated at a point (line charge) corresponds to an equal electric charge concentrated at the corresponding point, a point source in a two-dimensional hydrodynamical problem corresponds to a point source and so on.

These results follow at once from the fact that if $\phi + i\psi$ is the complex potential we may write

$$\phi + i\psi = f(x + iy) = g(\xi + i\eta),$$

and if ϕ is the electric potential, the integral $\int d\psi$ taken round a closed curve is $\pm 4\pi$ times the total charge within the curve. Now the interior of a closed curve is generally mapped into the interior of a corresponding closed curve and a simple circuit generally corresponds to a simple circuit, moreover ψ is the same in both cases and so the theorem is easily proved. It should be noted that a simple circuit may fail to correspond to a simple circuit when the closed curve contains a point at which the conformal character of the transformation breaks down. Another apparent exception arises when a point (x, y) corresponds to points at infinity in the (ξ, η) plane, but there is no great difficulty if these points at infinity are imagined to possess a certain unity. In fact mathematicians are accustomed to speak of the point at infinity when discussing problems of conformal representation. This convention is at once suggested by the results obtained by inversion and is found to be very useful. There is no ambiguity then in talking of a point charge or source at infinity.

We have seen that certain angles are unaltered by a conformal transformation and can consequently be regarded as invariants of the transformation. Certain other quantities are easily seen to be invariants. Writing

$$d(x, y) = \frac{\partial(x, y)}{\partial(\xi, \eta)} d(\xi, \eta) \equiv J d(\xi, \eta),$$

where $d(x, y)$, $d(\xi, \eta)$ are elements of area in the two planes, we have

$$J\left(\frac{\partial^2 \phi}{\partial x^2} + \frac{\partial^2 \phi}{\partial y^2}\right) = \frac{\partial^2 \phi}{\partial \xi^2} + \frac{\partial^2 \phi}{\partial \eta^2},$$

$$J\left(\frac{\partial^2 \psi}{\partial x^2} + \frac{\partial^2 \psi}{\partial y^2}\right) = \frac{\partial^2 \psi}{\partial \xi^2} + \frac{\partial^2 \psi}{\partial \eta^2},$$

$$J\left[\left(\frac{\partial \phi}{\partial x}\right)^2 + \left(\frac{\partial \phi}{\partial y}\right)^2\right] = \left(\frac{\partial \phi}{\partial \xi}\right)^2 + \left(\frac{\partial \phi}{\partial \eta}\right)^2.$$

The quantities ϕ and ψ are usually taken to be invariants in a conformal transformation and the foregoing relations indicate that

$$\iint\left(\frac{\partial^2 \phi}{\partial x^2} + \frac{\partial^2 \phi}{\partial y^2}\right) dx\,dy, \quad \iint\left(\frac{\partial^2 \psi}{\partial x^2} + \frac{\partial^2 \psi}{\partial y^2}\right) dx\,dy \quad \text{and} \quad \iint\left[\left(\frac{\partial \psi}{\partial x}\right)^2 + \left(\frac{\partial \psi}{\partial y}\right)^2\right] dx\,dy$$

are invariants. In the theory of electricity the first integral is proportional to the total charge associated with the area over which the integration takes place. In hydrodynamics the second integral represents the total vorticity associated with the area and the third integral is proportional to the kinetic energy when the density of the fluid is constant. The invariant character of the integrals $\int(u\,dx + v\,dy)$ and $\int(u\,dy - v\,dx)$ is easily recognised because these represent $\int d\phi$ and $\int d\psi$ respectively.

§ 4·21. *The transformation* $w = z^n$. When n is a positive integer the transformation $w = z^n$ does not give a (1, 1) correspondence between the w-plane and the z-plane but it is convenient to consider an n-sheeted surface instead of a single plane as the domain of w. For a given value of w the equation $z^n = w$ has n roots. If one of these is Z_1 the others are respectively $Z_2 = Z_1\omega$, $Z_3 = Z_1\omega^2$, ... $Z_n = Z_1\omega^{n-1}$, where $\omega = e^{2\pi i/n}$.

If $z = re^{i\theta}$ we may adopt the convention that for

$$Z_1, \quad 0 \leqslant n\theta \leqslant 2\pi,$$

$$Z_2, \quad 2\pi \leqslant n\theta \leqslant 4\pi,$$

$$\cdots\cdots\cdots\cdots\cdots\cdots\cdots\cdots\cdots$$

$$Z_n, \quad (2n-2)\,\pi \leqslant n\theta \leqslant 2n\pi.$$

Defining the sheet (m) to be that for which $W_m = Z_m{}^n$ we can say that W_1 is in the first sheet, W_2 in the second sheet, and so on. The n sheets together form a "Riemann surface" and we can say that there is a (1, 1) correspondence between the z-plane and the Riemann surface composed of the sheets (1), (2), ... (n). If $w = Re^{i\Theta}$ we have $\Theta = n\theta$, and so when $w = W_m$, we have $(2m-2)\,\pi < \Theta < 2m\pi$.

The z-plane is divided into n parts by the lines joining the origin to the corners of a regular polygon, one of whose corners is on the axis of x. These n portions of the z-plane are in a (1, 1) correspondence with the n sheets of the Riemann surface. The n lines just mentioned each belong to two portions and so correspond to lines common to two sheets. It is by crossing these lines that a point passes from one sheet to another as the angle Θ steadily increases. The point O in the w-plane is a winding point of the Riemann surface, its order is defined as the number $n - 1$.

A circle $| w - W | = a^n$ corresponds to a curve $| z^n - Z^n | = a^n$, which belongs to the class of lemniscates*

$$r_1 r_2 \ldots r_n = a^n,$$

where $r_1, r_2, \ldots r_n$ are the distances of the point z from the points $Z_1, Z_2, \ldots Z_n$ which correspond to W. In the present case the poles of the lemniscate are at the corners of a regular polygon and the equation of the lemniscate can be expressed in the form

$$r^{2n} - 2r^n R^n \cos n\,(\theta - \Theta) + R^{2n} = a^{2n} \quad (re^{i\theta} = z,\ Re^{i\Theta} = Z).$$

When $n = 2$ a circle in the z-plane corresponds to a limaçon. To see this we write $w = u + iv$, $z = x + iy$, then

$$u = x^2 - y^2, \quad v = 2xy.$$

* This is the name used by D. Hilbert, *Gött. Nachr.* S. 63 (1897). The name cassinoid is used by C. J. de la Vallée Poussin, *Mathesis* (3), t. II, p. 289 (1902), Appendix. The geometrical properties and types of curves of this kind are discussed by H. Hilton, *Mess. of Math.* vol. XLVIII, p. 184 (1919), reference being made to the earlier work of Serret, La Goupillière and Darboux.

Hence, if

$$(x+a)^2 + y^2 = c^2,$$

we have $$[u^2 + v^2 - 2a^2u + (a^2 - c^2)^2]^2 = 4c^4 (u^2 + v^2),$$

or, if $$U = u + c^2 - a^2, \quad V = v, \quad U = R\cos\Theta, \quad V = R\sin\Theta,$$

$$(U^2 + V^2 - 2c^2U)^2 = 4a^2c^2 (U^2 + V^2),$$

$$R = 2ac + 2c^2 \cos\Theta.$$

EXAMPLES

1. The curve $r^{2n} - 2r^nc^n \cos n\theta + c^nd^n = 0$ has n ovals each of which is its own inverse with respect to a circle centre O and radius $\sqrt{(cd)}$. The ordinary foci $B_1, B_2, \ldots B_n$ invert into the singular foci $A_1, A_2, \ldots A_n$, the polar co-ordinates of B_s being given by $r = d$, $n\theta = 2s\pi$.

2. Line charges of strength $+1$ are placed at the corners of a regular polygon of n corners and centre O, while line charges of strength -1 are placed at the corners of another regular polygon of n corners and centre O. Prove that the equipotentials are n-poled lemniscates. [Darboux and Hilton.]

3. Prove also that the lines of force are n-poled lemniscates passing through the vertices of the regular polygons.

§ **4·22.** *The bilinear transformation.* The transformation

$$\zeta = \frac{az+b}{\alpha z + \beta}, \qquad \ldots\ldots(\text{A})$$

in which a, b, α, β are complex constants, is of special interest because it is the only type of transformation which transforms the whole of the z-plane in a one-to-one manner into the whole of the ζ-plane and gives a conformal mapping of the neighbourhood of each point.

If $\alpha \neq 0$ there are generally two points in the z-plane for which $\zeta = z$. These are given by the quadratic equation

$$\alpha z^2 + (\beta - a)\, z - b = 0.$$

Let us choose our origin in the z-plane so that it is midway between these points, then $\beta = a$ and if we write $b = \alpha c^2$ the self-corresponding points are given by $z = \pm c$. The transformation may now be written in the form

$$\frac{\zeta + c}{\zeta - c} = \frac{a + c\alpha}{a - c\alpha}\,\frac{z + c}{z - c}.$$

From this relation a geometrical construction for the transformation is easily derived. Writing
$$a + c\alpha = \rho e^{i\omega},$$

$$\zeta + c = R_1 e^{i\Theta_1}, \quad z + c = r_1 e^{i\theta_1},$$

$$\zeta - c = R_2 e^{i\Theta_2}, \quad z - c = r_2 e^{i\theta_2},$$

we have the relations $$\frac{R_1}{R_2} = \rho\,\frac{r_1}{r_2},$$

$$\Theta_1 - \Theta_2 = \omega + (\theta_1 - \theta_2).$$

If S_1 and S_2 are the self-corresponding points these relations tell us that a circle through S_1 and S_2 generally corresponds to a circle through S_1 and S_2, but in an exceptional case it may correspond to a straight line, namely the line S_1S_2.

Again, a circle which has S_1 and S_2 as inverse points corresponds to a circle which has S_1 and S_2 as inverse points.

By a suitable displacement of the z and ζ-planes we can make any given pair of points the self-corresponding points provided the self-corresponding points are distinct, for if the displacements are specified by the complex quantities u and v respectively, the transformation may be written in the form

$$\zeta + v = \frac{a(z+u)+b}{\alpha(z+u)+\beta},$$

and we can choose u and v so that the equation $\zeta = z$ has assigned roots z_1 and z_2.

We may conclude from this that the transformation maps any circle into either a straight line or a circle; a result which may be proved in many ways. One proof depends upon the theorem that in a bilinear transformation of type (A) the cross-ratio of four values of z is equal to the cross-ratio of the four values of ζ; i.e.

$$\frac{(z-z_1)(z_2-z_3)}{(z-z_2)(z_3-z_1)} = \frac{(\zeta-\zeta_1)(\zeta_2-\zeta_3)}{(\zeta-\zeta_2)(\zeta_3-\zeta_1)}.$$

Now the cross-ratio is real when the four points lie on either a straight line or a circle, hence four points on a circle must map into four points which are either collinear or concyclic. If in the transformation (A) we choose u so that $\alpha u + \beta = 0$, and v so that $\alpha v = a$, the transformation takes the form

$$\zeta z = k^2, \qquad \text{......(B)}$$

where

$$\alpha^2 k^2 = \alpha b - a\beta.$$

By a suitable rotation of the axes of reference we can reduce the transformation to the case in which k is real, and this is the case which will now be discussed.

The transformation evidently consists of an inversion in a circle of radius k with centre at the origin followed by a reflection in the axis of x. The points $z = \pm k$ are self-corresponding points and if these are denoted by S_1 and S_2 it is easily seen that two corresponding points P and Q lie on a circle through S_1 and S_2. The figure has a number of interesting properties which will be enumerated.

1. Since $z(\zeta + k) = k(z + k)$ the angles S_1PO, QS_1O are equal, and so the angles S_1PO, S_2PQ are also equal.

2. The triangles S_1PO, QPS_2 are similar, and so

$$PS_1 . PS_2 = PO . PQ.$$

3. If C is the middle point of PQ we have $CS_1.CS_2 = CP^2$, also PQ bisects the angle S_1CS_2.

The four points S_1, P, S_2, Q on the circle form a harmonic set. This follows from the relation

$$\frac{1}{z-k}+\frac{1}{z+k}=\frac{2}{z-\zeta},$$

which is easily derived from (B).

The angle PS_1C is equal to the angle S_1PO. The lines S_1P, PO, CS_1 thus form an isosceles triangle.

In the case when the self-corresponding points coincide we have

$$\alpha - \beta = 2ac, \quad b = - ac^2,$$

where c is the self-corresponding point. The transformation may now be written in the form

$$\frac{1}{\zeta - c}=\frac{a}{\beta + ac}+\frac{1}{z-c}, \quad \beta + ac \neq 0.$$

It may be built up from displacements and transformations of the type just considered and so needs no further discussion. The only other interesting special case is that in which the transformation then consists of a displacement followed by a magnification and rotation.

§ **4·23**. *Poisson's formula and the mean value theorem.* Bôcher has shown by inversion that Poisson's formula may be derived from Gauss's theorem relating to the mean value of a potential function round a circle.

Let C and C' be inverse points with respect to the circle Γ of radius a, and let $CC' = c$. Inverting with respect to a circle whose centre is C' and radius c, the point S on the circle transforms into a point S'. We shall suppose that C' is outside the circle Γ, then S is inside the circle Γ. Let ds, ds' be corresponding arcs at S and S' respectively and let the polar co-ordinates of C and C' be (r, θ), (r', θ) respectively, where $rr' = a^2$. The circle Γ inverts into a circle with centre C and radius given by the formula $aa' = cr$, for

$$CS' = c\frac{CS}{C'S} = c\frac{CA}{C'A}$$

$$= c\frac{a-r}{r'-a} = \frac{cr}{a} = a'.$$

Also $\quad r^2C'S^2 = a^2.CS^2 = a^2[r^2 + a^2 - 2ar\cos(\sigma - \theta)],$

where (a, σ) are the polar co-ordinates of S.

Writing $ds' = a'd\sigma'$, we have

$$d\sigma' = \frac{c^2 ds}{a'.C'S^2} = \frac{ca^2 d\sigma}{r.C'S^2} = \frac{rc d\sigma}{a^2 - 2ar\cos(\sigma-\theta) + r^2},$$

and $rc = a^2 - r^2$, consequently the formula of Poisson becomes

$$V = \frac{1}{2\pi}\int_0^{2\pi} V' d\sigma'.$$

This formula states that the mean value of a potential function round the circumference of a circle is equal to the value of the function at the centre of the circle. Hence Poisson's formula may be derived from this mean value theorem and is true under the same conditions as the mean value theorem.

§ **4·24.** *The conformal representation of a circle on a half plane**. If two plane areas A and A_1 can be mapped on a third area A_0 they can be mapped on one another, consequently the problem of mapping A on A_1 reduces to that of mapping A and A_1 on some standard area A_0.

This standard area A_0 is generally taken to be either a circle of unit radius or a half plane. The transition from the circle $x^2 + y^2 \leqslant 1$ in the z-plane to the half plane $v \geqslant 0$ in the w-plane is made by means of the substitutions

$$z = x + iy, \quad w = u + iv, \quad z(i + w) = i - w,$$

$$Dx = 1 - u^2 - v^2, \quad Dy = 2u, \qquad \qquad \text{......(I)}$$

where

$$D = u^2 + (1 + v)^2,$$

$$4/D = (1 + x)^2 + y^2.$$

When $v = 0$, the substitution $u = \tan\theta$ gives

$$x = \cos 2\theta, \quad y = \sin 2\theta.$$

As 2θ varies from $-\pi$ to π, the variable u varies from $-\infty$ to ∞ and so the real axis in the w-plane is mapped in a uniform manner on the unit circle $x^2 + y^2 = 1$ in the z-plane.

Since, moreover,

$$D(1 - x^2 - y^2) = 4v, \quad Dy = 2u,$$

we have $v \geqslant 0$ when $x^2 + y^2 \leqslant 1$, consequently the interior of the circle is mapped on the upper half of the w-plane.

When u and v are both infinite or when either of them is infinite, we have $x = -1$, $y = 0$; hence the point at infinity in the w-plane corresponds to a single point in the z-plane and this point is on the unit circle.

The transformation (I) may be applied to the whole of the z-plane; it maps the region outside the circle $x^2 + y^2 = 1$ on the lower half $(v < 0)$ of the w-plane. A line $y = mx$ drawn through the centre of the circle corresponds to a circle $m(1 - u^2 - v^2) = 2u$ which passes through the point $(0, 1)$ which corresponds to the centre of the circle, and through the point $(0, -1)$ which corresponds to the point at infinity in the z-plane. This circle cuts the line $v = 0$ orthogonally.

Two points which are inverse points with respect to the circle $x^2 + y^2 = 1$ map into points which are images of each other in the line $v = 0$.

* This presentation in §§ 4·24, 4·61 and 4·62 follows closely that given in Forsyth's *Theory of Functions* and the one given in Darboux's *Théorie générale des surfaces*, t. I, pp. 170–180.

The upper half of the w-plane may be mapped on itself in an infinite number of ways. To see this, let us consider the transformation

$$\zeta = \frac{aw+b}{cw+d}, \quad w = \frac{d\zeta - b}{a - c\zeta}, \qquad \text{......(II)}$$

in which a, b, c and d are real constants and $\zeta = \xi + i\eta$.

When w is real ζ is also real and vice versa, hence the real axes correspond. Furthermore,

$$\eta\,[(cu+d)^2 + c^2v^2] = (ad - bc)\,v,$$

hence if $ad - bc$ is positive, η is positive when v is positive. There are three effective constants in this transformation, namely, the ratios of a, b and c to d, hence by a suitable choice of these constants any three points on the axis of u may be mapped into any three points on the axis of ξ. In fact, if u_1, u_2, u_3 are the values of u corresponding to the values ξ_1, ξ_2, ξ_3 of ξ, we can say from the invariance of the cross-ratio that

$$\frac{(\zeta - \xi_1)(\xi_2 - \xi_3)}{(\zeta - \xi_2)(\xi_3 - \xi_1)} = \frac{(w - u_1)(u_2 - u_3)}{(w - u_2)(u_3 - u_1)},$$

and so the equation of the transformation may be written down in the previous form, the coefficients being

$$\begin{aligned} a &= \xi_2\xi_3(u_2 - u_3) + \xi_3\xi_1(u_3 - u_1) + \xi_1\xi_2(u_1 - u_2), \\ b &= u_2u_3\xi_1(\xi_2 - \xi_3) + u_3u_1\xi_2(\xi_3 - \xi_1) + u_1u_2\xi_3(\xi_1 - \xi_2), \\ c &= u_1(\xi_2 - \xi_3) + u_2(\xi_3 - \xi_1) + u_3(\xi_1 - \xi_2), \\ d &= u_2u_3(\xi_2 - \xi_3) + u_3u_1(\xi_3 - \xi_1) + u_1u_2(\xi_1 - \xi_2). \end{aligned}$$

The quantity $ad - bc$ is given by the formula

$$ad - bc = (\xi_2 - \xi_3)(\xi_3 - \xi_1)(\xi_1 - \xi_2)(u_2 - u_3)(u_3 - u_1)(u_1 - u_2).$$

If u_1, u_2, u_3 are all different the coefficients c and d cannot vanish simultaneously, for the equations $c = 0$, $d = 0$ give

$$\frac{\xi_2 - \xi_3}{u_1(u_2^2 - u_3^2)} = \frac{\xi_3 - \xi_1}{u_2(u_3^2 - u_1^2)} = \frac{\xi_1 - \xi_2}{u_3(u_1^2 - u_2^2)},$$

and these equations imply that

$$u_1(u_2^2 - u_3^2) + u_2(u_3^2 - u_1^2) + u_3(u_1^2 - u_2^2) = 0,$$

or

$$(u_2 - u_3)(u_3 - u_1)(u_1 - u_2) = 0,$$

if the quantities ξ_1, ξ_2, ξ_3 are also all different. In a similar way it can be shown that a and c cannot vanish simultaneously and that a and b cannot vanish simultaneously. Poincaré has remarked that the transformation (II) can usually be determined uniquely so as to satisfy the requirement that an assigned point ζ and an assigned direction through this point should correspond to an assigned point w and an assigned direction through this point. The proof of the theorem may be left to the reader, who should examine also the special case in which one or both of the points is on the real axis in the plane in which it lies.

EXAMPLES

1. Prove that Poisson's formula

$$V = \frac{1}{2\pi}\int_{-\pi}^{\pi} \frac{(1-r^2) f(\theta')\, d\theta'}{1 - 2r\cos(\theta - \theta') + r^2} \qquad |r| \leqslant 1$$

maps into the formula of § 3·11,

$$V = \frac{1}{\pi}\int_{-\infty}^{\infty} \frac{yF(x')\, dx'}{(x-x')^2 + y^2},$$

where $f(\theta') = F(\tan \frac{1}{2}\theta')$.

2. Prove that the transformation

$$w = \frac{z - z_0}{z - \bar{z}_0}$$

maps the half plane $y \geqslant 0$ on the unit circle $|w| \leqslant 1$ in such a way that the point z_0 maps into the centre of the circle.

§ **4·31.** *Riemann's problem.* The standard problem of conformal representation will be taken to be that of mapping the area A in the z-plane on the upper half of the w-plane in such a way that three selected points on the boundary of A map into three selected points on the axis of u. This is the problem considered by B. Riemann in his dissertation. The problem may be made more precise by specifying that the function $f(w)$ which gives the desired relation

$$z = f(w)$$

should possess the following properties:

(1) $f(w)$ should be uniform and continuous for all values of w for which $v > 0$. If w_0 is any one of these values $f(w)$ should be capable of expansion in a Taylor series of ascending powers of $w - w_0$ which has a radius of convergence different from zero.

(2) The derivative $f'(w)$ should exist and not vanish for $v > 0$; indeed, if $f'(w) = 0$ for $w = w_0$ there will be at least two points in the neighbourhood of w_0 for which z has the same value. This is contrary to the requirement that the representation should be biuniform.

(3) $f(w)$ should be continuous also for all real values of w, but it is not required that in the neighbourhood of one of these values, w_0, the function $f(w)$ can be expanded in a Taylor series of ascending powers of $w - w_0$, for, as far as the mapping is concerned, $f(w)$ is defined only for $v \geqslant 0$.

(4) Considered as a function of z, the variable w should satisfy the same conditions as $f(w)$. If $f(w)$ satisfies all these requirements it will give the solution of the problem. The solution is, moreover, unique because if two functions

$$z = f(w), \quad z = g(w)$$

give different solutions of the problem, the transformation

$$f(w) = g(W)$$

will map the upper half plane into itself in such a way that the points u_1, u_2, u_3 map into themselves. Now it can be proved that a transformation which maps the upper half plane into itself is bilinear and so the relation between w and W is

$$\frac{(w-u_1)(u_2-u_3)}{(w-u_2)(u_3-u_1)}=\frac{(W-u_1)(u_2-u_3)}{(W-u_2)(u_3-u_1)},$$

or

$$\frac{w-u_1}{w-u_2}=\frac{W-u_1}{W-u_2}.$$

This reduces to

$$(w-W)(u_1-u_2)=0,$$

and so $W=w$.

§ 4·32. *The general problem of conformal representation.* The general type of region which is considered in the theory of conformal representation may be regarded as a carpet which is laid down on the z-plane. This carpet is supposed to have a boundary the exact nature of which requires careful specification because with the aim of obtaining the greatest possible generality, different writers use different definitions of the boundary curve. There may, indeed, be more than one boundary curve, for a carpet may, for instance, have a hole in its centre. For simplicity we shall suppose that each boundary curve is a simple closed curve composed of a finite number of pieces, each piece having a definite direction at each of its points. At a point where two pieces meet, however, the directions of the two tangents need not be the same; a carpet may, for instance, have a corner. The tangent may actually turn through an angle 2π as we pass from one piece of a boundary curve to another and in this case the boundary has a sharp point which may point either inwards or outwards. It turns out that the former case presents a greater difficulty than the latter.

In special investigations other restrictions may be laid on each piece of a boundary curve and from the numerous restrictions which have been used we shall select the following for special mention.

(1) The direction of the tangent is required to vary continuously as a point moves along the curve (smooth curve*).

(2) The curvature is required to vary continuously as a point moves along the curve.

(3) The curve should be rectifiable, i.e. it should be possible to define the length of any portion of the curve with the aid of a definite integral which has a precise meaning specified beforehand, such as the meaning given to an integral by Riemann, Stieltjes or Lebesgue.

A simple curve which possesses the first property may be called a curve (CT), one which possesses the properties 1 and 2 a curve (CTC), a curve which possesses the properties 1 and 3 may be called a curve (RCT).

* A curve which is made up of pieces of smooth curves joined together may be called smooth bit by bit ("Stückweise glatte Kurve"; see Hurwitz-Courant, Berlin (1925), *Funktionentheorie*).

The carpet will be said to be simply connected when a cross cut starting from any point of the boundary and ending at any other divides the carpet into two pieces. A carpet shaped like a ring is not simply connected because a cut starting from a point on one boundary and ending at a point on the other does not divide the carpet into two pieces. Such a carpet may, however, be made simply connected by making a cut of this type. When we consider a carpet with n boundaries which are simple closed curves we shall suppose that the boundaries can be converted into one by a suitable number of cuts which will at the same time render the carpet simply connected. It will be supposed, in fact, that the carpet is not twisted like a Möbius' strip when the cuts have been made.

Any closed curve on a simply connected carpet can be continuously deformed until it becomes an infinitely small circle. This cannot always be done on a ring-shaped carpet as may be seen by considering a circle concentric with the boundary circles of a ring, and it cannot be done in the case of a curve which runs parallel to the edge of a singly twisted Möbius' strip formed by joining the ends of a thin rectangular strip of paper after the strip has been given a single twist through 180°. Such a closed curve is said to be irreducible and the connectivity of a carpet may be defined with the aid of the number of different types of irreducible closed curves that can be drawn on it. Two closed curves are said to be of different types when one cannot be deformed into the other without any break or crossing of the boundary. It is not allowed, for instance, to cut the curve into pieces and join these together later or in any way to make the curve into one which does not close.

A simply connected carpet may cover the plane more than once; it may, for instance, be folded over, or it may be double, triple, etc. In the latter case it is called a Riemann surface, i.e. a surface consisting of several sheets which connect with one another at certain branch lines in such a way as to give a simply connected surface. When there are only two sheets it is often convenient to regard them as the upper and lower surfaces of a single carpet with a cut or branch line through which passage may be made from one surface to the other. In the case of a ring-shaped carpet we generally consider only the upper surface, but if the lower surface is also considered and a passage is allowed from one surface to the other across either one or both of the edges of the ring a surface with two sheets is obtained, but this doubly sheeted surface is not simply connected because a curve concentric with the two edges is still irreducible. In a more general theory of conformal representation the mapping of multiply connected surfaces is considered, but these will be excluded from the present considerations.

A carpet may also have an infinite number of boundaries or an infinite number of sheets, but these cases will also be excluded. When we speak of an area A we shall mean the right side of a carpet which is bounded by a simple closed curve formed of pieces of type ($RCTC$) and is not folded over in any way. The carpet will be supposed, in fact, to be simply connected and smooth, the word smooth being used here as equivalent to the German word "schlicht," which means that the carpet is not folded or wrinkled in any way. The function $F(z)$ maps the circular area $|z| < 1$ into a smooth region if

$$\frac{F(z_1) - F(z_2)}{z_1 - z_2} \neq 0,$$

whenever $|z_1| < 1$ and $|z_2| < 1$.

We shall be occupied in general with the conformal representation of one simple area on another, and for brevity we shall speak of this as a mapping. In advanced works on the theory of functions the problem of conformal representation is considered also for the case of Riemann surfaces and the more general theory of the conformal representation of multiply connected surfaces is treated in books on the differential geometry of surfaces.

For many purposes it will be sufficient to consider the problem of conformal representation for the case of boundaries made up of pieces of curves having the property that the co-ordinates of their points can be expressed parametrically in the form

$$x = f(t), \quad y = g(t),$$

where the functions $f(t)$ and $g(t)$ can be expanded in power series of type

$$\sum_{n=0}^{\infty} a_n (t - t_0)^n, \qquad \text{......(III)}$$

which are absolutely and uniformly convergent for all values of the parameter t that are needed for the specification of points on the arc under consideration. In such a case the boundary is said to be composed of analytic curves and this is the type of boundary that was considered in the pioneer work of H. A. Schwarz, but the restriction of the theory to boundaries composed of analytic curves is not necessary* and a method of removing this restriction was found by W. F. Osgood†. His work has been followed up by that of many other investigators‡.

In the power series (III) the quantities t_0, a_n are constants which, of course, may be different for different pieces of the boundary.

There are really two problems of conformal representation. In one problem the aim is simply to map the open region A bounded by a curve

* In the modern work the boundary considered is a Jordan curve, that is, a curve whose points may be placed in a continuous (1, 1) correspondence with the points of a circle.

† *Trans. Amer. Math. Soc.* vol. I, p. 310 (1900).

‡ Particularly E. Study, C. Carathéodory, P. Koebe and L. Bieberbach.

a on the open region B bounded by a curve b, there being no specified requirement about the correspondence of points on the two boundaries. In the second problem the aim is to map the closed realm* A on the closed realm B in such a way that each point P on a corresponds to only one point Q which is on b and so that each point Q on b corresponds to only one point P which is on a. It is this second problem which is of most interest in applied mathematics. If, moreover, in the first problem the correspondence between the boundaries is not one-to-one the applied mathematician is anxious to know where the uniformity of the correspondence breaks down.

Existence theorems are more easily established for the first problem than for the second and fortunately it always happens in practice that a solution of the first problem is also a solution of the second; but this, of course, requires proof and such a proof must be added to an existence theorem that is adapted only for the first problem.

The methods of conformal representation are particularly useful because they frequently enable us to deduce the solution of a boundary problem for one closed region A from the solution of a corresponding boundary problem for another region B which is of a simpler type. When the function which effects the mapping is given by an explicit relation the process of solution is generally one of simple substitution of expressions in a formula, but when the relation is of an implicit nature or is expressed by an infinite series or a definite integral the direct method of substitution becomes difficult and a method of approximation may be preferable. A method of approximation which is admirable for the purpose of establishing the existence of a solution may not be the best for purposes of computation.

§ **4·33**. *Special and exceptional cases.* It is easy to see that it is not possible to map the whole of the complex z-plane on the interior of a circle. Indeed, if there were a mapping function $f(z)$ which gave the desired representation, $f(z)$ would be analytic over the whole plane and $|f(z)|$ would always lie below a certain positive value determined by the radius of the circle into which the z-plane maps, consequently by Liouville's theorem $f(z)$ would be a constant. A similar argument may be used for the case of the pierced z-plane with the point z_0 as boundary. By means of a transformation $z - z_0 = 1/z'$ the region outside z_0 can be mapped into the whole of the z'-plane when the point $z' = \infty$ is excluded. The mapping function is again an integral function for which $|f(z')| < M$, and is thus a constant. On account of this result a region considered in the mapping problem is supposed to have more than one external point, a point on the boundary being regarded as an external point.

* We use realm as equivalent to the German word "Bereich," and region as equivalent to "Gebiet."

The next case in order of simplicity is the simply connected region with at least two boundary points A and B. If these were isolated the region would not be simply connected. We shall therefore assume that there is a curve of boundary points joining A and B. This curve may contain all the boundary points (Case 1) or it may be part of a curve of boundary points which may either be closed or terminated by two other end-points C and F. The latter case is similar to the first, while the case of a closed curve is the one which we wish eventually to consider.

The simplest example of the first case is that in which the end-points are $z = 0$ and $z = \infty$, the boundary consisting of the positive x-axis. The region bounded by this line can be regarded as one sheet of a two-sheeted Riemann surface with the points 0 and ∞ as winding points of the second order, passage from one sheet to the other being made possible by a junction of the sheets along the positive x-axis. The whole of this Riemann surface is mapped on the z'-plane by means of the simple transformation $z' = \sqrt{z}$, which sends the one sheet in which we are interested into the half plane $0 \leqslant \theta' \leqslant \pi$, where $z' = r'e^{i\theta'}$.

In the case when the boundary consists of a curve joining the points $z = a$, $z = b$, these points are regarded as winding points of the second order for a two-sheeted Riemann surface whose sheets connect with each other along the boundary curve. This surface is mapped on the whole z'-plane by means of the transformation

$$z'\left(\frac{z-a}{z-b}\right)^{\frac{1}{2}} - cz' = 1,$$

and in this transformation one sheet goes into the interior, the other into the exterior of a certain closed curve C. The mapping problem is thus reduced to the mapping of the interior of C on a half plane or a unit circle.

Finally, by means of a transformation of type

$$z = Az' + B,$$

we can transform the region enclosed by C into a region which lies entirely within the unit circle $|z| \leqslant 1$ and our problem is to map this region on the interior of the unit circle $|\zeta| \leqslant 1$ by means of a transformation of type $\zeta = f(z)$.

§ **4·41**. *The mapping of the unit circle on itself.* If a and $\bar{a}$ are any two conjugate complex quantities and α is a real angle the quantities $e^{i\alpha} - a$ and $1 - \bar{a}e^{i\alpha}$ have the same modulus, consequently if β is another real angle the transformation

$$(1 - \bar{a}z)\,\zeta = e^{i\beta}(z - a) \qquad \text{......(A)}$$

maps $|z| = 1$ into $|\zeta| = 1$, and it is readily seen that the interior of one circle maps into the interior of the other. It should be noticed that this transformation maps the point $z = a$ into the centre of the circle $|\zeta| = 1$. If we put $a = 0$ the transformation reduces to the rotation

$$\zeta = ze^{i\beta},$$

which leaves the centre of the circle unaltered. If we can prove that this is the most general conformal transformation which maps the interior of the unit circle into itself in such a way that the centre maps into the centre it will follow that the formula (A) gives the most general transformation which maps the unit circle into itself.

The following proof is due to H. A. Schwarz.

Let $f(z)$ be an analytic function of z which is regular in the circle $|z| = 1$ and satisfies the conditions

$$|f(z)| < 1 \text{ for } |z| < 1, \quad f(0) = 0.$$

If $$\phi(z) = f(z)/z, \quad \phi(0) = f'(0),$$

the function $\phi(z)$ is also regular in the unit circle, and if $|z| = r$, where $r < 1$, we have

$$|\phi(z)| < 1/r.$$

But since $\phi(z)$ is analytic in the circle $|z| = r$ the maximum value of $|\phi(z)|$ occurs on the boundary of this region and not within it, hence for a point z_0 within the circle $|z| = r$, or on its circumference, we have the inequality $|\phi(z_0)| < 1/r$ (Schwarz's inequality*).

Passing to the limit $r \to 1$ we have the inequality

$$|\phi(z_0)| < 1 \text{ for } |z_0| \leqslant 1.$$

Now let $\zeta = f(z)$, $z = g(\zeta)$ be the mapping functions which map a circle on itself in such a way that the centre maps into the centre, then by Schwarz's inequality

$$|\zeta/z| \leqslant 1, \quad |z/\zeta| \leqslant 1.$$

Hence $|\zeta/z| = 1$, and so $|\phi(z)|$ is equal to unity within the unit circle. Now an analytic function whose modulus is constant within the unit circle is necessarily a constant, hence $\zeta = ze^{i\beta}$ where β is a constant real angle.

Since the unit circle is mapped on a half plane by a bilinear transformation, it follows that a transformation which maps a half plane into itself is necessarily a bilinear transformation.

§ **4·42**. *Normalisation of the mapping problem.* Let Γ be the unit circle $|\zeta| \leqslant 1$ in the ζ-plane and suppose that a smooth region G in the z-plane can be mapped in a (1, 1) manner on the interior of Γ. Since Γ can be mapped on itself by a bilinear transformation in such a way that two prescribed linear elements correspond, it is always possible to normalise

* This is often called Schwarz's lemma as another inequality is known as Schwarz's inequality. The lemma of § 4·61 is then called Schwarz's principle or continuation theorem. This second inequality is used in § 4·81.

the mapping so that a prescribed linear element in the region G corresponds to the centre of the unit circle and the direction of the positive real axis, that is to what we may call the "chief linear element." We can then, without loss of generality, imagine the axes in the z-plane to be chosen so that the origin lies in G on the prescribed linear element and so that this linear element is the chief linear element for the z-plane. This means that the normalised mapping function $\zeta = f(z)$ satisfies the conditions $f(0) = 0$, $f'(0) > 0$. Finally, by a suitable choice of the unit of length in the z-plane, or by a transformation of type $z' = kz$, we can make $f'(0) = 1$. The transformation is then fully normalised and $f(z)$ is a completely normalised mapping function. The power series which represents the function in the neighbourhood of $z = 0$ is of type

$$f(z) = z + a_2 z^2 + \dots.$$

The coefficients $a_2, a_3, \dots$ in this series are not entirely arbitrary, in fact it appears that a_2 is subject to the inequality* $|a_2| < 2$. To prove this we consider the function $g(z)$ defined by the equation $f(z)\,g(z) = 1$. We have

$$zg(z) = 1 + b_1 z + b_2 z^2 + \dots,$$
$$b_1 = -a_2, \quad b_2 = a_2^2 - a_3, \dots.$$

If $0 < c < 1$, the transformation $\gamma = g(z)$ maps the circular ring $c < z < 1$ on a region Λ in the γ-plane bounded by a curve C and a curve C_c which can be represented parametrically by the equation

$$c\gamma = e^{-i\alpha} + b_1 c + b_2 c^2 e^{i\alpha} + c^3 \omega(c, \alpha),$$

where α is the parameter and $\omega(c, \alpha)$ remains bounded as α varies between 0 and 2π. Writing $b_2 = be^{2i\beta}$, $cd = 1$, we remark that the equation

$$\gamma e^{-i\beta} = de^{-i(\alpha+\beta)} + bce^{i(\alpha+\beta)},$$

gives the parametric representation of an ellipse with semi-axes $d + bc$, $d - bc$ respectively, and so the area A_c of the curve C_c differs from πd^2 by cB, where $|B|$ remains bounded as $c \to 0$. Now the area of the region Λ is a quantity A_c' given by the equation

$$A_c' = \int_c^1 \int_0^{2\pi} |\gamma'|^2 \rho \, d\rho \, d\alpha = \pi \left(d^2 - 1 + \sum_{n=2}^{\infty} n |b_n|^2 - \sum_{n=2}^{\infty} n |b_n|^2 c^{2n} \right)$$

and $A_c > A_c'$; also as $c \to 0$ the difference $\pi d^2 - A_c'$ tends to the area A of the region enclosed by C. Since $A \geqslant 0$ we have the inequality

$$\sum_{n=2}^{\infty} n |b_n|^2 \leqslant 1.$$

Now the function

$$[g(z^2)]^{\frac{1}{2}} = \frac{1}{z} + \beta_1 z + \dots, \quad 2\beta_1 = b_1$$

likewise maps the unit circle $|z| < 1$ on a smooth region, and so by the last theorem

$$|\beta_1|^2 \leqslant 1.$$

* See, for instance, L. Bieberbach, *Berlin. Sitzungsber.* Bd. XXXVIII, S. 940 (1916).

Since $b_1 = -a_2$ the last inequality becomes simply

$$|a_2|^2 \leqslant 4 \quad \text{or} \quad |a_2| \leqslant 2.$$

We have $|\beta_1| = 1$ when and only when $\beta_2 = \beta_3 = \ldots = 0$, consequently $|a_2| = 2$ when and only when

$$[g(z^2)]^{\frac{1}{2}} = \frac{1}{z} + ze^{i\sigma}, \text{ where } \sigma \text{ is real,}$$

or

$$g(z) = (z^{-\frac{1}{2}} + e^{i\sigma} z^{\frac{1}{2}})^2,$$

that is, when

$$f(z) = \frac{z}{(1 + ze^{i\alpha})^2}.$$

EXAMPLES

1. A transformation which maps the unit circle into itself in a one-to-one manner and transforms the chief linear element into itself is necessarily the identical transformation. [Schwarz and Poincaré.]

2. A region enclosing the origin which can be mapped on itself with conservation of the chief linear element consists either of the whole plane or of the whole plane pierced at the origin. [T. Radó, *Szeged Acta*, t. I, p. 240 (1923).]

3. If the region $|z| < 1$ is mapped smoothly on a region W in the w-plane by the function $w = f(z) = z + a_1 z^2 + a_2 z^3 + \ldots$, prove that, when $|z| = r < 1$,

$$\frac{r}{(1+r)^2} \leqslant |f(z)| \leqslant \frac{r}{(1-r)^2}.$$

Hence show that if w_0 is a point not belonging to the region W

$$|w_0| < \tfrac{1}{4}.$$

The value $|w_0| = \frac{1}{4}$ is attained at the point $w_0 = \frac{1}{4}e^{-i\alpha}$ when

$$f(z) = \frac{z}{(1 + ze^{i\alpha})^2}.$$

4. If a region W of the w-plane is mapped smoothly on the circle $|z| < 1$ by the function $w = f(z)$ and if Z_1, Z_2 are any two points which do not lie within the circle, then

$$\left| \frac{[f(z_1) - f(z_2)] f'(0)}{[f(0) - f(z_1)][f(0) - f(z_2)]} \right| \leqslant 4.$$

[G. Pick, *Leipziger Berichte*, Bd. LXXXI, S. 3 (1929).]

§ 4·43. *The derivative of a normalised mapping function.* Now let $f(z)$ be regular in the unit circle $|z| < 1$, which we shall call K. We shall study the behaviour of $f'(z)$ in the neighbourhood of an interior point z_0 of K. Let $\bar{z}_0$ be the conjugate of z_0, then the transformation

$$z' = \frac{z - z_0}{1 - z\bar{z}_0}$$

maps the circle K into itself and sends the point z_0 into the point $z' = 0$. Thus

$$f(z) = f\left(\frac{z + z_0}{1 + \bar{z}_0 z'}\right).$$

Writing $$f^*(z) = f\left(\frac{z+z_0}{1+\bar{z}_0 z}\right) - f(z_0),$$

we see that the function $f^*(z)$ maps the circle K on a smooth region and leaves the origin fixed. If $z_0 = re^{i\theta}$ we have by Taylor's theorem

$$f^*(z) = z(1-r^2) f'(z_0) + \tfrac{1}{2}z^2(1-r^2)\left[(1-r^2) f''(z_0) - 2\bar{z}_0 f'(z_0)\right] + \dots.$$

The function†

$$F(z) = \frac{f^*(z)}{(1-r^2) f'(z_0)} = z + A_2 z^2 + \dots$$

is thus a normalised mapping function, and so by the theorem of § 4·42, $|A_2| \leqslant 2$, i.e.

$$\left|(1-r^2)\frac{f''(z_0)}{f'(z_0)} - 2\bar{z}_0\right| \leqslant 4.$$

Writing $$z_0 \frac{f''(z_0)}{f'(z_0)} = P + iQ = r\frac{\partial}{\partial r}(u+iv),$$

where $f'(z) = e^{u+iv}$ and u and v are real, we have the inequality

$$\left| r\frac{\partial}{\partial r}(u+iv) - \frac{2r^2}{1-r^2}\right| \leqslant \frac{4r}{1-r^2}.$$

Therefore $$-\frac{4-2r}{1-r^2} \leqslant \frac{\partial u}{\partial r} \leqslant \frac{4+2r}{1-r^2},$$

$$-\frac{4}{1-r^2} \leqslant \frac{\partial v}{\partial r} \leqslant \frac{4}{1-r^2}.$$

Integrating between 0 and r we obtain the two inequalities‡

$$\log\frac{1-r}{(1+r)^3} \leqslant u \leqslant \log\frac{1+r}{(1-r)^3},$$

$$|v| \leqslant 2\log\frac{1+r}{1-r}.$$

The first of these may be written in the more general form

$$\frac{1-|z|}{(1+|z|)^3} \leqslant \left|\frac{f'(z)}{f'(0)}\right| \leqslant \frac{1+|z|}{(1-|z|)^3},$$

where now $\zeta = f(z)$ is a function which maps K on a smooth region not

† This function was used by L. Bieberbach, *Math. Zeitschr.* Bd. IV, S. 295 (1919), and later by R. Nevanlinna, see Bieberbach, *Math. Zeitschr.* Bd. IX, S. 161 (1921). The following analysis which is due to Nevanlinna is derived from the account in Hurwitz-Courant, *Funktionentheorie*, S. 388 (1925).

‡ The first of these was given by T. H. Gronwall, *Comptes Rendus*, t. CLXII, p. 249 (1916), and by J. Plemelj and G. Pick, *Leipziger Ber.* Bd. LXVIII, S. 58 (1916). In the form

$$k(r) \leq |f'(z)| \leq h(r)$$

it is known as Koebe's Verzerrungssatz (distortion theorem). The precise forms for $k(r)$ and $h(r)$ were derived also by G. Faber, *Münchener Ber.* S. 39 (1916) and L. Bieberbach, *Berliner Ber.* Bd. XXXVIII, S. 940 (1916). The inequality satisfied by $|v|$ was discovered by Bieberbach, *Math. Zeitschr.* Bd. IV, S. 295 (1919).

containing the point at infinity but is not necessarily a normalised mapping function.

The second theorem is called the rotation theorem, as it indicates limits for the angle through which a small area is rotated in the conformal mapping. The other theorem gives limits for the ratio in which the area changes in size. This theorem has been much used by Koebe* in his investigations relating to the conformal representation of regions and has been used also in hydrodynamics† and aerodynamics.

When $f(z)$ maps K on a convex region it can be shown that $|f'(z)|$ lies within narrower limits‡. Study has shown, moreover, that in this case any circle within K and concentric with it also maps into a convex region§. Many other inequalities relating to conformal mapping are given in a paper by J. E. Littlewood, *Proc. London Math. Soc.* (2), vol. XXIII, p. 481 (1925).

§ **4·44**. *The mapping of a doubly carpeted circle with one interior branch point.* Let P be a point within the unit circle $|z'| < 1$ and let $r^2 e^{i\theta}$ $(0 < r < 1)$ be the value of z' at P. The transformation||

$$z' = z\frac{(1+r^2)z - 2re^{i\theta}}{2rz - (1+r^2)e^{i\theta}} = \phi(z) \qquad \text{......(A)}$$

satisfies the conditions $\phi(0) = 0$, $\phi'(0) > 0$, $|z'| = |z|$, when $|z| = 1$, and so represents a partially normalised transformation which maps the unit circle in the z-plane on a doubly carpeted unit circle in the z'-plane, the two sheets having a junction along a line extending from P to the boundary. We shall regard this line as a cut in that sheet which contains the chief element corresponding to the chief element in the z-plane.

It is evident that $|z'| < |z|$ whenever $|z| < 1$, and so $|z'| < 1$ when $|z| < 1$. This means that $|z'| < |z|$ whenever $|z'| < 1$.

From this we conclude that for all values of z' for which $|z'| < r^2$ there is a positive number $q(r)$ greater than unity for which $|z| > q(r)|z'|$. Indeed, if there were no such quantity $q(r)$ there would be at least one point in the circle $|z'| < r^2$, for which $|z| = |z'|$. An expression for $q(r)$ may be obtained by writing

$$s = \frac{2r}{1+r^2}, \quad z = \rho e^{i\alpha},$$

and considering the points s, $1/s$ on the real axis. If R_1, R_2 are the distances of the point z from these points respectively we have

$$|z'| = \rho\frac{R_1}{sR_2}.$$

* *Gött. Nachr.* (1909); *Crelle*, Bd. CXXXVIII, S. 248 (1910); *Math. Ann.* Bd. LXIX.

† Ph. Frank and K. Löwner, *Math. Zeitschr.* Bd. III, S. 78 (1919).

‡ T. H. Gronwall, *Comptes Rendus*, t. CLXII, p. 316 (1916).

§ E. Study, *Konforme Abbildung einfach zusammenhängender Bereiche*, p. 110 (Teubner, Leipzig, 1913). A simple proof depending on a use of Schwarz's inequality has been given recently by T. Radó, *Math. Ann.* Bd. CII, S. 428 (1929).

|| C. Carathéodory, *Math. Ann.* Bd. LXXII, S. 107 (1912).

The oval curve for which $\rho R_1/sR_2 = r^2$ lies entirely within a circle $R_2/R_1 =$ constant which touches it at a point $x_0 = -\rho$ on the real axis for which

$$\rho(\rho + s) = r^2(1 + \rho s),$$

$$\rho = \frac{r}{1 + r^2}[(2 + 2r^4)^{\frac{1}{2}} - 1 + r^2].$$

The constant is found to be

$$\frac{1 + \rho s}{\rho + s} = \frac{\rho}{r^2} = \frac{1}{r(1 + r^2)}[(2 + 2r^4)^{\frac{1}{2}} - 1 + r^2]$$

and we may take this as our value of $q(r)$. We can see that it is greater than 1 when $r < 1$ because

$$2 + 2r^4 - (1 + r - r^2 + r^3)^2 = (1 - r^4)(1 - r)^2.$$

It is clear from the inequality $|z| > q(r)|z'|$ that points corresponding to those which lie within the circle $|z'| < r^2$ in the z'-plane lie within the larger circle $|z| < r^2 q(r)$ in the z-plane. If r^2 is the minimum distance from the origin of points on a closed continuous curve C' which lies entirely within the unit circle $|z'| \leqslant 1$, the transformation (A) maps the interior of C' into the interior of a closed* curve C which lies entirely between the two circles $|z| \leqslant 1$, $|z| = r^2 q(r)$. The shortest distance from the origin to a point of C may be greater than $r^2 q(r)$ but it lies between this quantity and r, i.e. the value of $|z|$ corresponding to the branch point $z = e^{i\theta} r^2$. This second minimum distance may be used as the constant of type r^2 in a second transformation of type (A). Let us call it r_1^2 and use the symbol C_1 to denote the curve into which C is mapped by the new transformation. The minimum distance from the origin of a point of this curve is a quantity r_2^2 which is not less than a quantity $r_1^2 q(r_1)$ associated with the number r_1.

If we consider the worst possible case in which the minimum distance for a curve C_{n+1} derived from a curve C_n with minimum distance r_n^2 is always $r_n^2 q(r_n)$, we have a sequence of numbers $r_1, r_2, \dots r_n$, which are derived successively by means of the recurrence relation

$$r_{n+1}^2 = \frac{r_n}{1 + r_n^2}[(2 + 2r_n^4)^{\frac{1}{2}} - 1 + r_n^2].$$

Since $r_n < 1$ for all values of n and $r_{n+1} > r_n$, the sequence tends to a limit R which must be given by the equation

$$R^2 = \frac{R}{1 + R^2}[(2 + 2R^4)^{\frac{1}{2}} - 1 + R^2].$$

This equation gives the value $R = 1$. Hence as $n \to \infty$ the curve C_n lies between two circles which ultimately coincide.

* The curve C may in some cases close by crossing the line which corresponds to the cut in the z'-plane. This will not happen if the cut is drawn so that it does not intersect C' again.

The convergence to the limit is very slow, as may be seen by considering a few successive values of r_n^2:

r_1^2	r_2^2	r_3^2
·25	·283	·309

The best possible case from the point of view of convergence is that in which $r_1^2 = r$. This case occurs when the curve C' is shaped something like a cardioid with a cusp at P.

Though useful for establishing the existence of the conformal mapping of a region on the unit circle, the present transformation is not as useful as some others for the purpose of transforming a given curve into another curve which is nearly circular, unless the given curve happens to be shaped something like a cardioid or a limaçon with imaginary tangents at the double point. We shall not complete the proof of the existence of a mapping function for a region bounded by a Jordan curve. This is done in books on the theory of functions such as those of Bieberbach and Hurwitz-Courant. Reference may be made also to the tract on conformal transformation which is being written by Carathéodory*, to E. Goursat's *Cours d'Analyse Mathématique*, t. III, and to Picard's *Traité d'Analyse.*

§ **4·45.** *The selection theorem.* Let us suppose that the set or sequence of functions $u_1(x, y)$, $u_2(x, y)$, $u_3(x, y)$, ... possesses the following properties:

(1) It is *uniformly bounded.* This means that in the region of definition R the functions all satisfy an inequality of type

$$| u_s(x, y) | \leqslant M,$$

where M is a number independent of s and of the position of the point (x, y) of the region R.

(2) It is *equicontinuous*†. This means that for any small positive number ϵ there is an associated number δ independent of s, x and y but depending on ϵ in such a way that whenever

$$(x' - x)^2 + (y' - y)^2 \leqslant \delta^2$$

we have

$$| u_s(x', y') - u_s(x, y) | < \tfrac{1}{3}\epsilon.$$

We now suppose that the sequence contains an unlimited number of functions and that an infinite number of these functions forming a subsequence $u_{m1}(x, y)$, $u_{m2}(x, y)$, ... can be selected by a selection rule (m). Our aim now is to find a sequence (1), (2), (3), ... of selection rules such that the "diagonal sequence" $u_{11}(x, y)$, $u_{22}(x, y)$, ... converges uniformly in R.

* Carathéodory's proof is given in a paper in *Schwarz-Festschrift,* and in *Math. Ann.* Bd. LXXII, S. 107 (1912). Koebe's proof will be found in his papers in *Journ. für Math.* Bd. CXLV, S. 177 (1915); *Acta Math.* t. XL, p. 251 (1916).

† The idea of equal continuity was introduced by Ascoli, *Mem. d. R. Acc. dei Lincei,* t. XVIII (1883).

The first step is to construct a sequence of points $P_1, P_2, \ldots$ everywhere dense in R. This may be done by choosing our origin outside R and using for the co-ordinates of P_s expressions of type

$$x_s = p2^{-q}, \quad y_s = p'2^{-q}, \quad q \geqslant 0$$

where p, p' and q are integers, and where the index $s = I(p, p', q)$ is a positive integer with the following properties:

$$\begin{aligned} I(p, p', q) &> I(p_0, p_0', q_0), \quad \text{whenever} \quad q > q_0, \\ I(p, p', q) &> I(p_0, p_0', q), \quad \text{whenever} \quad p > p_0, \\ I(p, p', q) &> I(p, p_0', q), \quad \text{whenever} \quad p' > p_0'. \end{aligned}$$

Since the functions $u_s(x, y)$ are bounded, their values at P_1 have at least one limit point $U_1(x_1, y_1)$. We therefore choose the sequence $u_{1n}(x, y)$ so that it converges at P_1 to this limit $U_1(x_1, y_1)$. Since, moreover, the functions $u_{1n}(x, y)$ are uniformly bounded their values at P_2 have at least one limit point $U_2(x_2, y_2)$; we therefore select from the infinite sequence $u_{1n}(x, y)$ a second infinite sequence $u_{2n}(x, y)$ which converges at P_2 to $U_2(x_2, y_2)$. These functions $u_{2n}(x, y)$ are uniformly bounded and their values at P_3 have at least one limit point $U_3(x_3, y_3)$, we therefore select from the sequence $u_{2n}(x, y)$ an infinite subsequence $u_{3n}(x, y)$ which converges at P_3 to $U_3(x_3, y_3)$, and so on.

We now consider the sequence $u_{11}(x, y), u_{22}(x, y), u_{33}(x, y), \ldots$. Since the functions are all equicontinuous we have

$$| u_{mm}(x', y') - u_{mm}(x, y) | < \tfrac{1}{3}\epsilon$$

for any two points P and P' whose distance PP' is not greater than δ.

Next let 2^{-q} be less than δ and let r be such that the set $P_1, P_2, \ldots P_r$ contains all the points of R for which q has a selected value satisfying this inequality, then a number N can be chosen such that for $m > N$

$$|u_{ll}(x_s, y_s) - u_{mm}(x_s, y_s) | < \tfrac{1}{3}\epsilon$$

for all values of l greater than m and for all points (x_s, y_s) for which q has the selected value. This number N should, in fact, be chosen so that the sequences $u_{mm}(x, y)$ converge for all these points P_s. These points P_s form a portion of a lattice of side 2^{-q} and so there is at least one of these points P' within a distance δ from z. We thus have the additional inequalities

$$\begin{aligned} &| u_{ll}(x, y) \;\;\, - u_{ll} \;\,(x', y') \;| < \tfrac{1}{3}\epsilon, \\ &| u_{ll}(x', y') - u_{mm}(x', y') \,| < \tfrac{1}{3}\epsilon, \\ \therefore \quad &| u_{ll}(x, y) \;\;\, - u_{mm}(x, y) \;\;\, | < \epsilon. \end{aligned}$$

This inequality holds for all points P in R and proves that the diagonal sequence converges uniformly in R to a continuous limit function $u(x, y)$.

There is a similar theorem for sequences of functions of any number of variables, and for infinite sets of functions which are not denumerable.

In the case of a sequence of functions $f_1(z)$, $f_2(z)$, ... of a complex variable $z = x + iy$ there is equicontinuity when

$$|f_s(z') - f_s(z)| < \tfrac{1}{3}\epsilon$$

for any pair of values z, z' for which $|z' - z| < \delta$, δ, as before, being independent of s and of the position of z in the region R. A sufficient condition for equicontinuity, due to Arzelà*, is that

$$\left|\frac{f_s(z') - f_s(z)}{z' - z}\right| < M_1$$

for all functions $f_s(z)$ of the set and for all pairs of points z, z' of the domain, M_1 being a number independent of s, z and z'. We need in fact only take $\delta M_1 = \frac{1}{3}\epsilon$ to obtain the desired inequality.

In the particular case when each function $f_s(z)$ possesses a derivative it is sufficient for equicontinuity that $|f_s'(z)| < M_2$, where M_2 is independent of s and z. The result follows from the formula for the remainder in Taylor's theorem.

Montel† has shown that if a family of functions $f_s(z)$ is uniformly bounded in a region R it is equicontinuous in any region R' interior to R.

Suppose, in fact, that $|f_s(z)| < M$ for any point z in R and for any function $f_s(z)$ of the family, the suffix s being used simply as a distinguishing mark and not as a representative integer.

Let D be a domain bounded by a simple rectifiable curve C and such that R contains D while D contains R'. We then have for any point ζ within R'

$$f_s'(\zeta) = \frac{1}{2\pi i}\int_C \frac{f_s(z)\,dz}{(z-\zeta)^2}.$$

Therefore $$|f_s'(\zeta)| < \frac{Ml}{2\pi h^2},$$

where l is the length of C and h is the lower bound of the distance between a point of C and a point of R'. This inequality shows that the functions $f_s(z)$ are equicontinuous in R'.

Now if $f_s(z) = u_s(x, y) + iv_s(x, y)$ where u_s and v_s are real, the functions u_s, v_s are likewise equicontinuous and uniformly bounded in R'. We can then select from the set u_s a sequence $u_{11}, u_{22}, u_{33}, \ldots$ which converges uniformly to a function $u(x, y)$ which is continuous in R'. Also from the associated sequence $v_{11}, v_{22}, v_{33}, \ldots$ we can select an infinite subsequence $v_{aa}, v_{bb}, v_{cc}, \ldots$ which converges uniformly in R' to a continuous function $v(x, y)$. The series

$$f_{aa}(z) + [f_{bb}(z) - f_{aa}(z)] + [f_{cc}(z) - f_{bb}(z)] + \ldots$$

then converges uniformly to $u(x, y) + iv(x, y)$, which we shall denote by

* *Mem. della R. Acc. di Bologna* (5), t. VIII.

† *Annales de l'École Normale* (3), t. XXIV, p. 233 (1907).

the symbol $f(z)$. On account of the uniform convergence the function $f(z)$ is, by Weierstrass' theorem, an analytic function of z in R'. Indeed if $f_s^{(n)}(z)$ denotes the nth derivative of any function $f_s(z)$ of our set and C' is any rectifiable simple closed curve contained within R', we have by Cauchy's theorem and the property of uniform convergence

$$\begin{aligned} f_{aa}^{(n)}(z) &+ [f_{bb}^{(n)}(z) - f_{aa}^{(n)}(z)] + \dots \\ &= \frac{n!}{2\pi i}\int_{C'} \frac{f_{aa}(\zeta)\,d\zeta}{(\zeta - z)^{n+1}} + \frac{n!}{2\pi i}\int_{C'} \frac{f_{bb}(\zeta) - f_{aa}(\zeta)}{(\zeta - z)^{n+1}}\,d\zeta + \dots \\ &= \frac{n!}{2\pi i}\int_{C'} \frac{f(\zeta)\,d\zeta}{(\zeta - z)^{n+1}}. \end{aligned}$$

Since $f(\zeta)$ is continuous in R' and on C' the integral on the right represents an analytic function. When $n = 0$ this tells us that the sequence

$$f_{aa}(z), f_{bb}(z), \dots$$

converges to an analytic function which, of course, is $f(z)$. When $n \neq 0$ the relation tells us that the sequence $f_{aa}^{(n)}(z)$, $f_{bb}^{(n)}(z)$, ... converges to $f^{(n)}(z)$.

We may conclude from Cauchy's expression for $f_s^{(n)}(z)$ as a contour integral that $f_s^{(n)}(z)$ is uniformly bounded in any region containing R' and contained in R. It then follows that the set $f_s^{(n)}(z)$ is equicontinuous in R'. Hence from the sequence $f_{aa}(z), f_{bb}(z), f_{cc}(z)$, ... we can select an infinite sequence $f_{\alpha\alpha}(z), f_{\beta\beta}(z), f_{\gamma\gamma}(z)$, ... such that $f_{\alpha\alpha}'(z), f_{\beta\beta}'(z), f_{\gamma\gamma}'(z)$, ... converges uniformly in R' to an analytic function which can be no other than $f'(z)$. At the same time the sequence $f_{\alpha\alpha}(z), f_{\beta\beta}(z)$, ... converges uniformly to $f(z)$. This process may be repeated any number of times so as to give a partial sequence of functions converging uniformly to $f(z)$ and having the property that the associated sequences of derivatives up to an assigned order n converge uniformly to the corresponding derivatives of $f(z)$.

We now consider a sequence of contours C_1, C_2, ... C_n having for limit the contour C_0 which bounds R, the contours C_1, C_2, ... C_n bounding domains D_1, D_2, ..., each of which contains the preceding and has R as limit. From our set of functions $f_s(z)$ we can select a sequence $f_{s1}(z)$ which converges uniformly in D_1 towards a limit function, from the sequence $f_{s1}(z)$ we can cull a new sequence $f_{s2}(z)$ which converges uniformly in D_2 to a limit function and so on. The diagonal sequence $f_{11}(z), f_{22}(z)$, ... then converges uniformly throughout the open region R to a limit function.

Hence we have Montel's theorem that an infinite set of uniformly bounded analytic functions admits at least one continuous limit function, both boundedness and continuity being understood to refer to the open region R in which the functions are defined to be analytic.

For further developments relating to this important theorem reference must be made to Montel's paper. For the case of functions of a real variable

A. Roussel* has recently invented a new method. The selection theorem has been extended by Montel to functions of bounded variation.

§ **4·46.** *Mapping of an open region.* Let R be a simply connected bounded region which contains the origin O and has at least two boundary points. Let S be the set of analytic functions $f_s(z)$ which are uniform, regular, smooth and bounded in R and for which

$$f_s(0) = 0, \quad f_s'(0) = 1, \quad |f_s(z)| < M.$$

Let U_s be the upper limit of $|f_s(z)|$ in R and let ρ be the lower limit of all the quantities U_s. There is then a sequence $f_r(z)$ of the functions $f_s(z)$ for which $U_r \to \rho$. Since, moreover, this sequence is uniformly bounded, we can apply the selection theorem and construct an infinite subsequence which converges uniformly to a limit function $f(z)$ in any closed partial region R' or R. This function $f(z)$ is a regular analytic function in R and satisfies the conditions $f(0) = 0, f'(0) = 1$. Being a uniform limit function of a sequence of smooth mapping functions it is smooth in R and its U is ρ.

The function $f(z)$ thus maps R on a region T which lies in the circle with centre O and radius ρ. If T does not completely fill the circle, there will be a value $re^{i\phi}$, with $r < \rho$, which is not assumed by our function $f(z)$ in R. We shall now show that this is impossible and that consequently T does fill the circle.

Let $r = a^2\rho$, then $a < 1$ and if we write

$$f_0(z) = \frac{2a}{1+a^2}\rho e^{i\phi}\frac{v(z)-a}{av(z)-1},$$

where
$$[v(z)]^2 = \frac{f(z) - a^2\rho e^{i\phi}}{a^2 f(z) - \rho e^{i\phi}},$$
$$v(0) = a,$$

we have $f_0(0) = 0$, $f_0'(0) = 1$ and the function $f_0(z)$ is uniform, regular, smooth and bounded in R.

Now let U_0 be the upper limit of $f_0(z)$ in R. We may find an inequality satisfied by this quantity by observing that

$$a\left|\frac{v(z)-a}{av(z)-1}\right|$$

is of the form r_1/r_2, where r_1, r_2 are the distances of the point $v(z)$ from the points a, $1/a$ respectively which are inverse points with respect to the circle $|z| = 1$.

On the other hand,

$$a^2|v(z)|^2 = \rho_1/\rho_2,$$

where ρ_1, ρ_2 are the distances of the point $f(z)$ from the points $a^2\rho e^{i\phi}$, $a^{-2}\rho e^{i\phi}$ which are inverse points with respect to the circle $|z| = \rho$. Now

* *Journ. de Math.* (9), t. v, p. 395 (1926). See also *Bull. des Sciences Math.* t. LII, p. 232 (1928).

the point $f(z)$ lies either on this circle or within it and so ρ_1/ρ_2 has a value which is constant either on $|z| = \rho$ or on a circle within $|z| = \rho$ and with the same pair of inverse points. This constant for a circle with this pair of inverse points has its greatest value for the circle $|z| = \rho$ if circles lying outside this circle are excluded. This greatest value is, moreover,

$$\frac{\rho - a^2\rho}{\rho - a^{-2}\rho} = a^2.$$

Hence we have the inequality $|v(z)|^2 \leqslant 1$. By a similar argument we conclude that r_1/r_2 has its greatest value when the point $v(z)$ is at some place on the circle $|z| = 1$ and this value is a. Hence

$$\left|\frac{v(z) - a}{av(z) - 1}\right| \leqslant 1,$$

and so $U_0 < \rho$. We have thus found a function for which $U_0 < \rho$, and this is incompatible with the definition of ρ as the lower limit of the quantities U_s. The region T must then completely fill the circle of radius ρ and so the function $f(z)$ maps R on this circle. The radius ρ is consequently called the radius of the region R.

This analysis, which is due to L. Fejér and F. Riesz, is taken from a paper by T. Radó*. The analysis has been carried further by G. Julia† who first selects from the functions $f_s(z)$ the polynomials $p_s^{(n)}(z)$ of degree n. Among these polynomials there is one polynomial $p^{(n)}(z)$ whose maximum modulus has a minimum value m_n. It is clear that $m_n \geqslant \rho$. Julia specifies a type of region R for which the sequence $p^{(n)}(z)$ possesses a limit function $f(z)$ mapping the region R on the circle of radius ρ.

§ **4·51**. *Conformal representation and the Green's function.* Consider in the xy-plane a region A which is simply connected and which contains the origin of co-ordinates. We shall assume that the boundary of A is smooth bit by bit. We write

$$G(x, y) = \log(1/r) - H(x, y)$$

for the Green's function associated with the origin as view-point, r being short for $(x^2 + y^2)^{\frac{1}{2}}$. Let us write $H(0, 0) = \log \tau \log(1/\rho)$, then τ is the capacity constant or constant of Robin‡. Now let

$$Z = \phi(z) = z + c_2 z + c_3 z^3 + \dots$$

be the uniquely determined function which maps the interior of a circle $|Z| < \rho$ on A in such a manner that

$$\phi(0) = 0, \quad \phi'(0) = 1.$$

* *Szeged Acta*, t. I, p. 240 (1923).

† *Comptes Rendus*, t. CLXXXIII, p. 10 (1926).

‡ The boundary of A may also be taken to be a closed Jordan curve, in which case τ is the transfinite diameter.

It will be shown that the Green's function $G(x, y)$ is

$$G(x, y) = \log(\rho/r), \quad r = |\phi(z)|.$$

Bieberbach* has proved a theorem relating to the area of the region A which is expressed by the inequality area $\geqslant \pi\rho^2$. This means that among all regions A for which $H(0, 0)$ has a prescribed value the circle possesses the smallest area.

For the theorem relating to the Green's function we may, with advantage, adopt a more general standpoint. Let us suppose that the transformation $w = f(z)$ maps the area A on the interior of a unit circle in the w-plane in such a way that to each point of the circle there corresponds only one point of the area A and vice versa. Let the centre of the circle correspond to the point z_0 of the area A, then z_0 is a simple root of the equation $f(z_0) = 0$ and $f(z) = 0$ has no other root in the interior of A. This is true also for the boundary if it is known that there is a (1, 1) correspondence between the points of the unit circle and the points of the boundary of A. We may therefore write

$$f(z) = (z - z_0)\, e^{p(z)},$$

where the function $p(z)$ is analytic in A.

Putting $p(z) = P + iQ$, $z - z_0 = re^{i\theta}$, where P, Q, r and θ are all real, we have

$$w = f(z) = \exp\{\log r + P + i(Q + \theta)\}.$$

Now, by hypothesis, the boundary of A maps into the boundary of the unit circle, therefore $\log r + P$ must be zero on the boundary of A. This means that $\log r + P$ is a potential function which is infinite like $\log r$ at the point (x_0, y_0), is zero on the boundary of A and is regular inside A except at (x_0, y_0). This potential has just the properties of the function $G(x, y; x_0, y_0)$, where $G(x, y, x_0, y_0)$ is the Green's function for the area A when (x_0, y_0) is taken as view-point, consequently the problem of the conformal mapping of A on the unit circle is closely related to that of finding the function G.

Writing $\sigma = Q + \theta$, $0 \leqslant \sigma \leqslant 2\pi$, we have on the boundary of the circle

$$dw = ie^{i\sigma}\, d\sigma,$$

while on the boundary of A

$$dz = |dz|\, e^{i\psi},$$

where ψ is the angle which the tangent makes with the real axis. Since dz/dw is neither zero nor infinite, the function

$$-i \log\left(i \frac{dz}{dw}\right)$$

* L. Bieberbach, *Rend. Palermo*, vol. XXXVIII, p. 98 (1914). This theorem is discussed in § 4·91.

is analytic within the circle and its real part takes the value $\psi - \sigma$ on the boundary of the circle. On the other hand, the function

$$F(w) = -i \log[-i(1-w)^2 dz/dw]$$

is analytic within the circle and its real part takes the value ψ on the boundary.

If ψ is a known function of σ on the boundary of A, Schwarz's formula gives

$$F(w) = -i \log k + \frac{1}{2\pi}\int_0^{2\pi} \psi(\sigma) \frac{e^{i\sigma}+w}{e^{i\sigma}-w} d\sigma,$$

where k is an arbitrary constant. The preceding formula then gives z by means of the equation

$$z - z' = i\int_{w'}^{w} \frac{e^{iF(w)}\,dw}{(1-w)^2}.$$

The relation between ψ and σ is partly known when the boundary of A is made up of segments of straight lines but in the general case ψ is an unknown function of σ and the present analysis gives only a functional equation for the determination of ψ.

To see this we suppose that on the circumference of the circle

$$F = \psi + i\phi$$

where ϕ and ψ are real, then

$$|dz| = \tfrac{1}{4}\operatorname{cosec}^2\frac{\sigma}{2}\,|d\sigma|\,e^{-\phi},$$

and the curvature of the boundary of A is

$$C = \frac{d\psi}{|dz|} = 4\sin^2\frac{\sigma}{2}\frac{d\psi}{|d\sigma|}e^{\phi},$$

and may be regarded as a known function of ψ, say $C(\psi)$. Making use of the relation between ϕ and ψ of § 3·33

$$\phi(\sigma) = b - \frac{1}{2\pi}\int_0^{2\pi} \psi'(\sigma_0)\log\left[\tfrac{1}{4}\operatorname{cosec}^2\frac{\sigma_0-\sigma}{2}\right]d\sigma_0,$$

where b is a constant, we obtain the functional equation*

$$\psi'(\sigma) = \frac{1}{4C(\psi)\sin^2\dfrac{\sigma}{2}}e^{-\phi(\sigma)},$$

where $\phi(\sigma)$ is defined by the foregoing equation.

EXAMPLE

Prove that

$$|\operatorname{grad} H|_{x=0,y=0} < 2/\rho.$$

[K. Löwner.]

§ **4·61**. *Schwarz's lemma*. It was remarked that a Taylor expansion for $f(w)$ in powers of $w - w_0$ is not required for points w_0 on the real axis,

* T. Levi Civita, *Rend. Palermo*, vol. XXIII, p. 33 (1907); H. Villat, *Annales de l'École Normale*, t. XXVIII, p. 284 (1911); U. Cisotti, *Idromeccanica piana*, Milan, p. 50 (1921).

but when $f(w)$ is real* for real values of w belonging to a finite interval, Schwarz has shown that it is possible to make an analytical continuation of $f(w)$ into a region for which v is negative. Let us consider an area S bounded by a curve ACB of which the portion AB is on the line $v = 0$ within the interval just mentioned.

Let S' be the image of S in the line $v = 0$ and let the value of $f(w)$ for a point w' of S' be defined as follows. We write

$$w = u + iv, \quad w' = u - iv,$$

$$f(w) = \xi + i\eta, \quad f(w') = \xi - i\eta,$$

where u, v, ξ, η are all real. The function $f(w)$ being now defined within the region $S + S'$ we write

$$g(w, \zeta) = 1/2\pi i\,(w - \zeta)$$

and consider the two integrals

$$I = \int_S g(w, \zeta) f(w)\, dw, \quad I' = \int_{S'} g(w, \zeta) f(w)\, dw$$

taken round the boundaries of S and S' Since $f(w)$ is analytic within both S and S' we have

$$I = f(\zeta), \quad I' = 0 \qquad \text{when } \zeta \text{ lies within } S,$$

$$I = 0, \qquad I' = f(\zeta) \text{ when } \zeta \text{ lies within } S'.$$

Hence in either case $I + I' = f(\zeta)$ and so

$$f(\zeta) = \int_{S+S'} g(w, \zeta) f(w)\, dw,$$

for the two integrals along the line AB are taken in opposite directions and so cancel each other.

Now the integral in this equation can be expanded in a Taylor series of ascending powers of $\zeta - \zeta_0$ for any point ζ_0 within the area $S + S'$ whether ζ_0 is on the real axis or not. The integral in fact represents a function which is analytic within the area $S + S'$ and can be used to define $f(\zeta)$ within $S + S'$. In this case, when ζ_0 is on AB, $f(\zeta)$ can be expanded in a power series of the foregoing type and the coefficients in this series, being of type

$$\frac{1}{2\pi i} \int_{S+S'} (w - \zeta_0)^{n+1} f(w)\, dw,$$

are all real.

* It is assumed here that $f(w)$ has a definite finite real integrable value for these real values of w. In a recent paper, *Bull. des Sciences Math.* t. LII, p. 289 (1928), G. Valiron has given an extension of Schwarz's lemma in which it is simply assumed that the imaginary part $i\eta$ of $f(w)$ tends uniformly to zero as $v \to 0$. If, then, the function $f(w)$ is holomorphic in the semicircle $|w| < R$, $v > 0$, it is holomorphic in the whole of the circle $|w| < R$.

Let us now use z_0 to denote the value of z corresponding to this value ζ_0 of w. The equation

$$z - z_0 = f(w) - f(\zeta_0) = a(w - \zeta_0) + b(w - \zeta_0)^2 + c(w - \zeta_0)^3 + \dots$$

can be solved for $w - \zeta_0$ by the reversion of series if $a \neq 0$, and the series thus obtained is of type

$$w - \zeta_0 = A(z - z_0) + B(z - z_0)^2 + C(z - z_0)^3 + \dots,$$

where the coefficients A, B, C are all real. The exceptional case $a = 0$ occurs only when the correspondence between w and z at the point ζ_0 ceases to be uniform.

§ 4·62. *The mapping function for a polygon.* Let us now consider an area A in the z-plane which is bounded by a contour formed of straight portions L_1, L_2, ... L_n. Let z_0 denote a point on one of the lines L and let $h\pi$ be the angle which this line makes with the real axis, also let w_0 be the value of w corresponding to z.

It is easily seen that the function

$$f(w) = (z - z_0)\, e^{-ih\pi}$$

has the properties of a mapping function for points z within A, and consequently also for the corresponding region in the w-plane; it is real when the point z is on the line L in the neighbourhood of z_0 and changes sign as z passes through the value z_0; consequently, when considered as a function of w it is real on the real axis and changes sign as w passes through the value w_0. Schwarz's lemma may, then, be applied to this function to define its continuation across the real axis and it is thus seen that we may write

$$e^{-ih\pi}(z - z_0) = (w - w_0)\, P(w - w_0),$$

where $P(w - w_0)$ denotes a power series of positive integral powers of $w - w_0$ including a constant term which is not zero. From this equation it follows that in the neighbourhood of the point w_0

$$\frac{dz}{dw} = e^{ih\pi} P_0(w - w_0),$$

where $P_0(w - w_0)$ is real when w and w_0 are real.

Taking logarithms and differentiating again, we see that the function

$$F(w) = \frac{d}{dw}\left(\log \frac{dz}{dw}\right)$$

is real and finite in the neighbourhood of $w = w_0$.

Next, let z_1 denote the point of intersection of two consecutive lines L, L', intersecting at an angle $\alpha\pi$; the argument of $z_1 - z$ varies from $h\pi$ to $h\pi - \alpha\pi$ as the point z passes from the line L to L' through the point of intersection (Fig. 19). Hence the function

$$J = [(z_1 - z)\, e^{-ih\pi}]^{\frac{1}{\alpha}}$$

is real and positive on L and negative on L'. Moreover, it has the required properties within A, and when considered as a function of w it has the required mapping properties in the region corresponding to A and is real on the real axis. By Schwarz's lemma we may continue this function across the real axis and may write for points w in the neighbourhood of w_1,

$$J = (w - w_1) P_1 (w - w_1),$$

where $P_1 (w - w_1)$ is a power series with real coefficients and with a constant term which is not zero. This equation gives

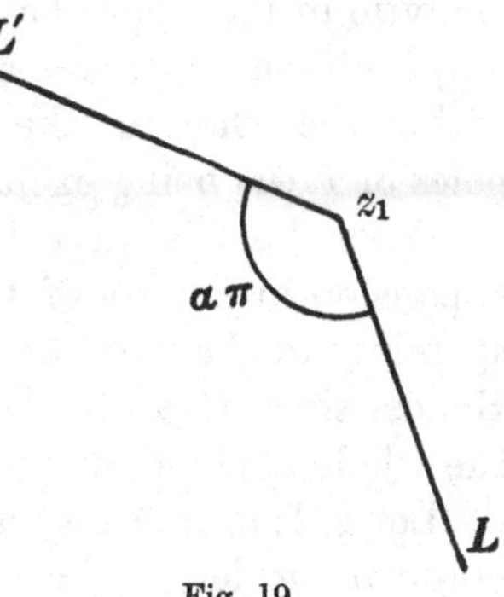

Fig. 19.

$$z - z_1 = e^{ih\pi} (w - w_1)^{\alpha} P_2 (w - w_1),$$

where $P_2 (w - w_1)$ is another power series with real coefficients. This equation indicates that for points in the neighbourhood of w_1

$$\frac{dz}{dw} = e^{ih\pi} (w - w_1)^{\alpha - 1} P_3 (w - w_1),$$

where $P_3 (w - w_1)$ is a power series with real coefficients. Taking logarithms and differentiating we find that

$$F (w) = \frac{d}{dw} \left(\log \frac{dz}{dw} \right) = \frac{\alpha - 1}{w - w_1} + T (w - w_1),$$

where $T (w - w_1)$ is a power series with real coefficients. The function

$$F (w) - \frac{\alpha - 1}{w - w_1}$$

is thus analytic in the neighbourhood of $w = w_1$.

For a point z_2 on the boundary of A which corresponds to w we have (if z_2 is not a corner of the polygon)

$$z - z_2 = \frac{A_1}{w} + \frac{A_2}{w^2} + \ldots.$$

Therefore $$\frac{dz}{dw} = - \frac{A_1}{w^2} p \left(\frac{1}{w} \right),$$

$$F (w) = \frac{d}{dw} \left(\log \frac{dz}{dw} \right) = - \frac{2}{w} + \frac{1}{w^2} p_1 \left(\frac{1}{w} \right),$$

where $p (1/w)$ is a power series. The expansion for $z - z_2$ may, indeed, be obtained by mapping the half plane w into itself by means of the substitution $w = - 1/w_1$, and by then using the result already obtained for an ordinary point z_0 on L.

The function $F (w)$ is real for all real values of w, as the foregoing investigation shows, is analytic in the whole of the upper half of the w-plane and is real on the real axis, the fact that it is analytic being a consequence of the supposition that the inverse function $z = g(w)$ is analytic in the upper half of the w-plane. Applying Schwarz's lemma we may continue this function $F(w)$ across the real axis and define it analytically within the whole of the w-plane, the points on the real axis which are poles of $F(w)$ being excluded.

When $|w|$ is large $|F(w)|$ is negligibly small, as is seen from the expansion in powers of $1/w$; moreover, $F(w)$ has only simple poles corresponding to the vertices of the polygon A and these are finite in number. Hence, since $F(w)$ outside these poles is a uniform analytic function for the whole w-plane, it must be a rational function.

Let $a, b, c, \ldots l$ be the values of w corresponding to the vertices of the polygon and let $\alpha\pi, \beta\pi, \ldots \lambda\pi$ be the interior angles at these vertices, then

$$F(w) = \Sigma \frac{\alpha - 1}{w - a} = \frac{d}{dw} \log \frac{dz}{dw},$$

and there is a condition

$$\Sigma (\alpha - 1) = -2,$$

which must be introduced because the sum of the interior angles of a closed polygon with n vertices is equal to $(n-2)\pi$.

Integrating the differential equation for z we obtain

$$z = C \int (w-a)^{\alpha-1} (w-b)^{\beta-1} \ldots (w-l)^{\lambda-1} \, dw + C',$$

where C and C' are arbitrary constants. By displacing the area A without changing its form or size but perhaps changing its orientation we can reduce the equation to the form

$$z = K \int (w-a)^{\alpha-1} (w-b)^{\beta-1} \ldots (w-l)^{\lambda-1} \, dw,$$

where K is a constant. This is the celebrated formula of Schwarz and Christoffel* If one of the angular points with interior angle $\mu\pi$ corresponds to an infinite value of w, the number of factors in the integrand is $n-1$ instead of n, and the equation

$$\Sigma (\alpha - 1) = -2$$

may be written in the form

$$\Sigma (\alpha - 1) = -1 - \mu,$$

where now the summation extends to the $n-1$ values of α which appear in the integral.

Since we can choose arbitrarily the values of w corresponding to three vertices of the polygon, there are still $n-3$ constants besides C and C' to be determined when the polygon is given. In the case of the triangle there is no difficulty. We can choose a, b and c arbitrarily; α, β and γ are known from the angles of the triangle and by varying K we can change the size of the triangle until the desired size is obtained.

* E. B. Christoffel, *Annali di Mat.* (2), t. I, p. 95 (1867); t. IV, p. 1 (1871); *Ges. Werke*, Bd. I, S. 255. H. A. Schwarz, *Journ. für Math.* Bd. LXX, S. 105 (1869); *Ges. Abh.* Bd. II, S. 65.

An interesting example of the conformal representation of a triangle with one corner at infinity is furnished by the equation

$$z - z_0 = \int_i^w f(s)\, ds, \text{ where } f(s) = (2a/\pi)(1 - s^2)^{\frac{1}{2}}/s, \quad w = u + iv.$$

When w lies between 1 and ∞ we have

$$z - z_0 = \int_i^1 f(s)\, ds + \int_1^w f(s)\, ds$$
$$= b + ic, \text{ say,}$$

where b is a constant and c varies from 0 to ∞. Thus, the portion $w > 1$ of the real axis corresponds to a line parallel to the axis of y.

Again, if $0 < w < 1$, we may write

$$z - z_0 = \int_i^1 f(s)\, ds - \int_w^1 f(s)\, ds$$
$$= b - d,$$

where d varies from 0 to ∞. The portion $0 < w < 1$ of the real axis corresponds, then, to a line parallel to the axis of x and extending from $z = z_0 + b$ to $-\infty$.

When $-1 < w < 0$ we may write

$$z - z_0 = \int_i^{-1} f(s)\, ds + \int_{-1}^w f(s)\, ds$$
$$= b' + d',$$

and so the corresponding line in the z-plane extends from $-\infty$ to $b' + z_0$ and is parallel to the axis of x. When $-\infty < w < 1$, we have

$$z - z_0 = \int_i^{-1} f(s)\, ds - \int_w^{-1} f(s)\, ds$$
$$= b' - ic',$$

where c' ranges from 0 to ∞, and so this part of the u-axis corresponds to a line from b' parallel to the y-axis. The two lines parallel to the y-axis can be shown to be portions of the same line separated by a gap. We have in fact

$$b - b' = \int_i^1 f(s)\, ds - \int_i^{-1} f(s)\, ds = \int_{-1}^1 f(s)\, ds,$$

where the integral is taken along the semicircle with the points $-1, +1$ as extremities of a diameter. On this semicircle we may put

$$s = \cos\theta + i\sin\theta,$$

and so

$$(\pi/2a)(b - b') = i\int_\pi^0 - d\theta\, [-2i\sin\theta\, e^{i\theta}]^{\frac{1}{2}}$$
$$= i(1 - i)\int_0^\pi \left(\cos\frac{\theta}{2} + i\sin\frac{\theta}{2}\right)(\sin\theta)^{\frac{1}{2}}\, d\theta$$
$$= 2i\int_0^\pi \cos\frac{\theta}{2}(\sin\theta)^{\frac{1}{2}}\, d\theta = 2i\sqrt{2}\,\Gamma(\tfrac{5}{4})\,\Gamma(\tfrac{3}{4}) = i.$$

The figure in the z-plane is thus of the type shown in Fig. 20. To solve an electrical problem with the aid of this transformation we put $w = ie^{\frac{1}{2}\chi}$, where χ is the complex potential $\psi + i\phi$. This transformation maps the half w-plane for which $v \geqslant 0$ on a strip of the χ-plane lying between the lines $\phi = \pm \pi$.

Performing the integration we find that

$$z - z_0 = \frac{a}{\pi}\left[2\tau - \log\frac{\tau + 1}{\tau - 1} - 2\sqrt{2} - 2\log(\sqrt{2} + 1)\right],$$

where $$\tau = (e^{\chi} + 1)^{\frac{1}{2}}.$$

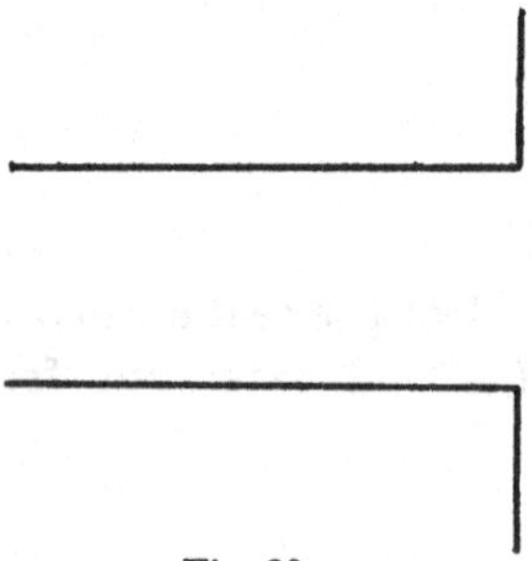

Fig. 20.

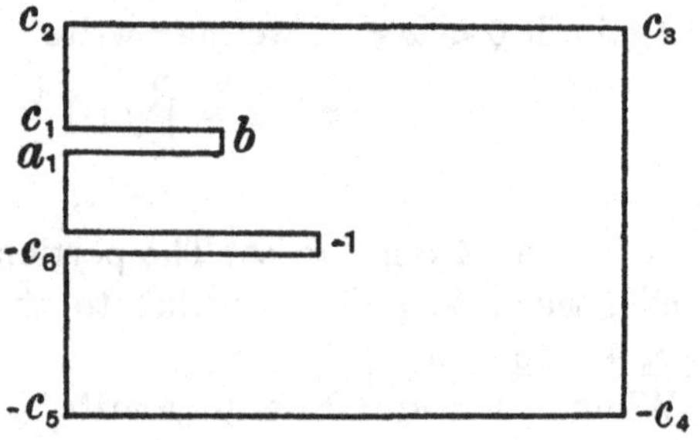

Fig. 21.

When the real part of w is large and negative the chief part of the expression for z is

$$-\frac{a}{\pi}\log(\tau - 1) = -\frac{a}{\pi}\log(1 + \tfrac{1}{2}e^{\chi} + \dots - 1) = \frac{a}{\pi}\log 2 - \frac{a}{\pi}\chi.$$

This gives a field that is approximately uniform. On the other hand, when the real part of w is large and positive, the chief part of the expression for z is

$$\frac{2a\tau}{\pi} = \frac{2a}{\pi}e^{\chi/2},$$

and we may thus get an idea of the nature of the field at a point outside the gap and at some distance from its surfaces. These results are of some interest in the theory of the dynamo. Another interesting example, in which the polygon is originally of the form shown in Fig. 21, gives edge corrections for condensers*.

Assigning values of w to the corners in the manner indicated, the transformation is of type

$$z = C(c_1c_2c_3c_4c_5c_6)^{-\frac{1}{2}}\int\frac{(w + 1)(w - b)\,dw}{(w - a_1)^{\frac{1}{2}}(w + a_2)^{\frac{1}{2}}}[(c_1 - w)\dots(c_6 + w)]^{\frac{1}{2}}.$$

* J. J. Thomson, *Recent Researches in Electricity and Magnetism*, 1893; Maxwell, *Electricity and Magnetism*, French translation by Potier, ch. II, Appendix; J. G. Coffin, *Proc. Amer. Acad. of Arts and Sciences*, vol. XXXIX, p. 415 (1903).

Making $a_s \to 0$, $c_s \to \infty$ we finally obtain

$$z = C\int (w+1)(w-b)\frac{dw}{w},$$

where C and b are constants to be determined.

Integrating, we find that

$$z = C\,[w + \tfrac{1}{2}(w-b)^2 - b\log(-w) + \Gamma],$$

where Γ is a constant of integration. Since $z = 0$ when $w = -1$, we have

$$\Gamma = 1 - \tfrac{1}{2}(1+b)^2.$$

When w is small and positive, the imaginary part of z must be ih, and the real part must be negative. Since the argument of $-w$ in both conditions are satisfied by taking $C = -h/b\pi$, therefore

$$z = -\frac{h}{2b\pi}[(w+1)^2 - 2b(w+1) - 2b\log(-w)].$$

Assuming that the potential ϕ is zero on AA' and equal to V on BB', we may write

$$\chi = \psi - i\phi = \frac{V}{\pi}(\log w - i\pi).$$

Fig. 22.

The charge per unit length on BB' from the edge $(w = b)$ to a point $P\ (w = s)$ so far from B that the surface density is uniform is

$$q = -\frac{1}{4\pi}(\psi_P - \psi_B) = -\frac{V}{4\pi^2}\log(s/b).$$

Now when s is very small and positive, $z = x + ih$, and so

$$x + ih = -\frac{h}{2b\pi}(1 - 2b - 2ib\pi - 2b\log s).$$

Therefore $$\log(s/b) = \pi x/h + 1/2b - 1 - \log b,$$

and so $$q = -\frac{V}{4\pi^2}\left[\frac{\pi x}{h} + \frac{1-2b}{2b} - \log b\right].$$

When $b = 1$ we have the well-known result

$$q = \frac{V}{4\pi h}\left(\frac{h}{2\pi} - x\right),$$

in which it must be remembered that x is negative.

When w is very small and negative, $z = x$, and so

$$x = -\frac{h}{2b\pi}[1 - 2b - 2b\log(-s)],$$

$$q = \frac{V}{4\pi^2}\log(-s)$$

$$= -\frac{V}{4\pi h}\left[\frac{h}{\pi}\left(1 - \frac{1}{2b}\right) - x\right].$$

When $b = 1$ $\quad q = -\dfrac{V}{4\pi h}\left(\dfrac{h}{2\pi} - x\right),$

$b = \infty$ $\quad q = -\dfrac{V}{4\pi h}\left(\dfrac{h}{\pi} - x\right).$

Since $w = b$ at the point B, the value of z for this point is

$$z = -\frac{h}{2b\pi}[1 - b^2 - 2b\log b - 2bi\pi],$$

and so the upper plate projects a distance d beyond the lower one, where

$$d = \frac{h}{\pi}\left[\log b + \tfrac{1}{2}\left(b - \frac{1}{b}\right)\right].$$

Many important electrostatic problems relating to condensers are solved by means of conformal representation in an admirable paper by A. E. H. Love*. The problem of the parallel plate condenser is treated for planes of unequal breadth and for planes of equal breadth arranged asymmetrically. The formulae involve elliptic functions. The hydrodynamical problems relating to two parallel planes, when the motion is discontinuous, are treated in a paper by E. G. C. Poole†.

Some applications of conformal representation to problems relating to gratings are given in a paper by H. W. Richmond‡. The general problem of the conformal mapping of a plane with two rectilinear or two circular slits has been discussed recently by J. Hodgkinson and E. G. C. Poole§.

§ **4·63**. *The mapping function for a rectangle.* When $n = 4$ and $\alpha = \beta = \gamma = \delta = \frac{1}{2}$, the polygon is a rectangle and z is represented by an elliptic integral which can be reduced to the normal form

$$z = H\int_0^t dt\,[(1 - t^2)(1 - k^2t^2)]^{-\frac{1}{2}}$$

by a transformation of type

$$w(Ct + D) = At + B. \qquad \text{......(A)}$$

If, in fact, the integral is

$$\int dw\,[(w - p)(w - q)(w - r)(w - s)]^{-\frac{1}{2}},$$

we have $\quad (Ct + D)(w - p) = (A - Cp)\,t + B - Dp,$

$$(Ct + D^2)\,dw = (AD - BC)\,dt,$$

* *Proc. London Math. Soc.* (2), vol. XXII, p. 337 (1924). † *Ibid.* p. 425.
‡ *Ibid.* p. 389. § *Ibid.* vol. XXIII, p. 396 (1925).

and so the transformation reduces the integral to the normal form if

$$\begin{aligned} A - Cp &= B - Dp, \\ A - Cq &= Dq - B, \\ A - Cr &= k(B - Dr), \\ A - Cs &= k(Ds - B). \end{aligned}$$

These equations give

$$\begin{aligned} C(q-p) &= 2B - D(p+q), \\ C(s-r) &= 2kB - kD(r+s), \\ 2A - C(p+q) &= D(q-p), \\ 2A - C(r+s) &= kD(s-r), \\ C[s-r-k(q-p)] &= kD[p+q-r-s], \\ C[r+s-(p+q)] &= D[q-p+k(r-s)], \end{aligned}$$

$$k^2(q-p)(r-s) + k[(q-p)^2 + (r-s)^2 - (p+q-r-s)^2] + (q-p)(r-s) = 0.$$

This equation gives two values of k which are both real if

$$[(q-p)^2 + (r-s)^2 - (p+q-r-s)^2]^2 > 4(q-p)^2(r-s)^2,$$

that is, if

$$[(q-p+r-s)^2 - (p+q-r-s)^2][(q-p-r+s)^2 - (p+q-r-s)^2] > 0,$$

or, if $$4(q-s)(r-p)(q-r)(s-p) > 0.$$

If $r \not> p \not> q \not> s$ this is evidently true and since the product of the two values of k is unity, we may conclude that one value of k is greater than 1, the other less than 1. This latter value should be chosen for the transformation. With this value

$$\frac{AD - BC}{D^2} = \frac{(1-k^2)(q-p)(s-r)}{s-r-k(q-p)} > 0,$$

consequently, the transformation (A) transforms the upper half of the w-plane into the upper half of the t-plane.

When the normal form of the integral is used the lengths of the sides of the rectangle are a and b respectively, where

$$a = H\int_{-1}^{1} dt\,[(1-t^2)(1-k^2t^2)]^{-\frac{1}{2}} = 2HK,$$

$$b = H\int_{1}^{1/k} dt\,[(1-t^2)(1-k^2t^2)]^{-\frac{1}{2}} = HK',$$

and where $4K$ and $2iK$ are the periods of the elliptic function sn u defined by the equation $x = \text{sn}\, u$, where

$$u = \int_0^x dt\,[(1-t^2)(1-k^2t^2)]^{-\frac{1}{2}}.$$

With the aid of this function t can be expressed in the form

$$t = \text{sn}\,(z/H).$$

The modulus k may be calculated with the aid of Jacobi's well-known formula

$$k = 4\sqrt{q}\left[\frac{(1+q^2)(1+q^4)(1+q^6)\dots}{(1+q)(1+q^3)(1+q^5)\dots}\right]^4$$

in which $$q = \exp[-\pi K'/K] = \exp[-2\pi b/a].$$

When the region is of the type shown in Fig. 23 the internal angles of the polygon are $3\pi/2$ at four corners and $\pi/2$ at the other eight. The transformation is thus of the type

$$z = A\,[(w-c_1)(w-c_2)(w-c_3)(w-c_4)]^{\frac{1}{2}}\,[(w-p_1)\dots(w-p_8)]^{-\frac{1}{2}}dw + B.$$

A particular transformation of this type is obtained by assigning positive values of w to corners of the polygon which lie above the axis of x and negative values of w to corners which lie below the axis of x, points which are images of each other in the axis of x being given parameters whose sum is zero. The transformation is now

$$z = Cb_1b_2\int\frac{(w^2-1)^{\frac{1}{2}}(w^2-c^2)^{\frac{1}{2}}\,dw}{[(w^2-a_1{}^2)(w^2-a_2{}^2)(w^2-b_1{}^2)(w^2-b_2{}^2)]^{\frac{1}{2}}} + B.$$

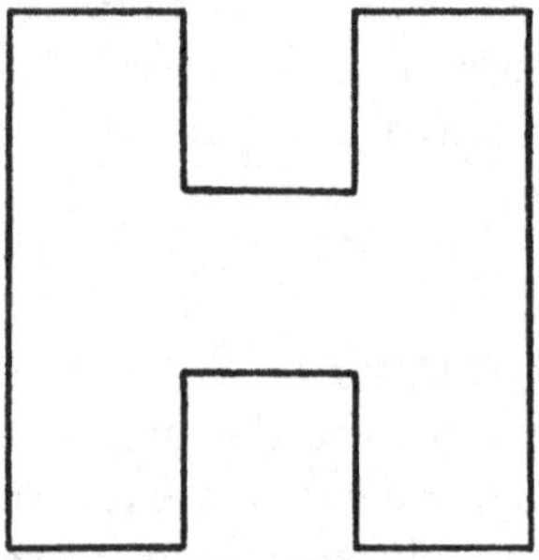

Fig. 23.

Fig. 24.

Making the parameters a_1, a_2 tend to zero and the parameters b_1, b_2 tend to infinity, the transformation becomes

$$z = C\int\frac{dw}{w^2}(w^2-1)^{\frac{1}{2}}(w^2-c^2)^{\frac{1}{2}} + B,$$

and the interior of the polygon becomes a region which extends to infinity.

To use this transformation for the solution of an electrical problem in which the two pole pieces in Fig. 24 are maintained at different potentials, we write*

$$w = ic^{\frac{1}{2}}e^{\frac{1}{2}\chi},\quad \chi = \psi + i\phi,$$

so as to map the half of the w-plane for which $v \geqslant 0$ on the strip $-\pi \leqslant \phi \leqslant \pi$. This will make $w = 0$ correspond to $z = 0$ if $B = 0$, and the lower limit of the integral is $i\sqrt{c}$.

Writing $ck = 1$ we find that the lengths a and b in the figure are given by the equations

$$2b = Cc\int_1^c \frac{ds}{s^2}f(s),$$

$$-2ia = Cc\int_{i\sqrt{c}}^1 \frac{ds}{s^2}f(s) - Cc\int_{i\sqrt{c}}^{-1}\frac{ds}{s^2}f(s),$$

* Riemann-Weber, *Differentialgleichungen der Physik*, Bd. II, S. 304.

where $f(s) = [(1-s^2)(1-k^2s^2)]^{\frac{1}{2}}$. These integrals are easily reduced to standard forms of elliptic integrals, thus

$$\int_1^c \frac{ds}{s^2} f(s) = \int_1^c \frac{ds}{s^2 f(s)} [f(s)]^2$$

$$= -\int_1^c \frac{ds}{f(s)}(1-k^2s^2) - k^2\int_1^c \frac{ds}{f(s)} + \int_1^c \frac{ds}{s^2 f(s)}.$$

Now if we put $ksr = 1$, the last integral becomes

$$\int_1^c \frac{k^2r^2dr}{f(r)} = -\int_1^c \frac{(1-k^2s^2)\,ds}{f(s)} + \int_1^c \frac{ds}{f(s)},$$

and we eventually find that

$$b = Cc\,[2E' - (1-k^2)K'],$$

$$a = 2Cc\,[2E - (1-k^2)K].$$

Therefore $$\frac{2b}{a} = \frac{2E' - (1-k^2)K'}{2E - (1-k^2)K}.$$

EXAMPLE

Prove that if $OABC$ is the rectangle with sides $x = 0$, $x = K$, $y = 0$, $y = K'$ and

$$\phi + i\psi = \log(\text{sn } z),$$

we have $\psi = 0$ on OA, AB, BC; $\psi = \pi/2$ on CO. Prove also that if

$$\phi + i\psi = \log(\text{cn } z),$$

where $(\text{cn } z)^2 + (\text{sn } z)^2 = 1$, we have $\psi = 0$ on OA, OC; $\psi = -\pi/2$ on BA, BC.

See Greenhill's *Elliptic Functions*, ch. IX.

§ **4·64**. *Conformal mapping of the region outside a polygon.* In order to map the region outside a polygon on the upper half of the w-plane, we may proceed in much the same way as before, but we must now use the external angles of the polygon and must consider the point in the w-plane which corresponds to points at infinity in the z-plane. Let us suppose that the w-plane is chosen so that this point is given by $w = i$, then there should be an equation of the form

$$z = \frac{C-1}{w-i} + C_0 + C_1(w-i) + \dots,$$

where the coefficients C_m are constants. This gives

$$\frac{dz}{dw} = -\frac{C-1}{(w-i)^2} + C_1 + \dots,$$

$$\frac{d}{dw}\log\frac{dz}{dw} = -\frac{2}{w-i} + P(w-i),$$

where $P(w-i)$ is a power series in $w-i$.

Since $\frac{d}{dw}\log\frac{dz}{dw}$ is to be real it must be of the form

$$\frac{d}{dw}\log\frac{dz}{dw} = \Sigma\frac{\alpha-1}{w-a} - \frac{2}{w-i} - \frac{2}{w+i}.$$

Therefore

$$z = C\int (w-a)^{\alpha-1}(w-b)^{\beta-1}\dots(w-l)^{\lambda-1}(1+w^2)^{-2}\,dw + C', \quad\dots\text{(I)}$$

where C and C' are arbitrary constants of integration. The relation between the indices α is now

$$\Sigma(\alpha-1)=2,$$

for the sum of the exterior angles of a polygon with n vertices is $(n+2)\pi$.

The region outside a polygon can be mapped on the exterior of a unit circle with the aid of a transformation of type

$$z = H\int (w-a)^{\alpha-1}(w-b)^{\beta-1}\dots w^{-2}dw, \quad |a| = |b| = \dots = 1,$$

where, as before,

$$\Sigma(\alpha-1)=2.$$

When the integrand is expanded in ascending powers of w^{-1} there will be a term of type w^{-1} which will, on integration, give rise to a logarithmic term unless the condition

$$\Sigma a(\alpha-1)=0$$

is satisfied.

When the polygon has only two vertices and reduces to a rectilinear cut of finite length in the z-plane, we have $\alpha=\beta=2$. The second condition may be satisfied by assigning the values $w=\pm 1$ to the ends of the cut. The transformation is now

$$z = H\int (w^2-1)\,w^{-2}dw,$$

and the length of the cut evidently depends on the value of H. Taking $H=\frac{1}{2}$ for simplicity, the transformation becomes

$$2z = w + w^{-1}.$$

This is the transformation discussed in § 4·73.

The general theorem (I) indicates that the region outside a straight cut may be mapped on the upper half of the w-plane by means of the transformation

$$z = 2\int \frac{1-w^2}{(1+w^2)^2}\,dw = \frac{2w}{1+w^2}.$$

The region outside a cut in the form of a circular arc may be obtained from the region outside a straight cut by inversion. If the arc is taken to be that part of the circle $z=-ie^{2i\theta}$, for which $-\alpha<\theta<\alpha$, the transformation

$$z = -i\,\frac{w^2+1+2iw\tan\alpha}{w^2+1-2iw\tan\alpha}$$

maps the region outside the arc on a half plane.

Suppose that in the w-plane there is an electric charge at the point $w=i(\sec\alpha+\tan\alpha)=is$, say, and that the real axis is a conductor.

The corresponding charge in the z-plane will be at infinity and the circular arc will be a conductor which must be charged with a charge of the same amount but of opposite sign. The solution of the potential problem in the w-plane is evidently

$$\chi = \phi + i\psi = \log \frac{w - is}{w + is}.$$

This gives $\quad w = -is \coth(\tfrac{1}{2}\chi), \quad z = -i\,\dfrac{1 + e^{-\chi}\sin\alpha}{1 + e^{\chi}\sin\alpha},$

and finally

$$\chi = -\log\left[\frac{i}{2}\operatorname{cosec}\alpha\,\{z + 1 + (z^2 + 2iz\cos 2\alpha - 1)^{\frac{1}{2}}\}\right].$$

The two-valued function $(z^2 + 2iz\cos 2\alpha - 1)^{\frac{1}{2}}$ may be regarded as one-valued in the region outside the cut and must be defined so that it is equal to i when $z = 0$ and is of the form $-z - i\cos 2\alpha$ when $|z|$ is very large. Changing the signs of ϕ and χ we have

$$x = K\,[2\sin\alpha\cosh\phi\sin\psi + \sin^2\alpha\sin 2\psi],$$

$$y = -K\,[1 + 2\sin\alpha\cosh\phi\cos\psi + \sin^2\alpha\cos 2\psi],$$

where $\quad K^{-1} = 1 + 2e^{-\phi}\sin\alpha\cos\psi + e^{-2\phi}\sin^2\alpha.$

With the aid of these equations Bickley has drawn the equipotentials and lines of force for the case of a semicircular arc. The charge resides for the most part on the outer face, the surface density becoming infinite at the edges. The field appears to be approximately uniform on the axis just above the centre of the circle.

The field at a great distance from the circular arc is roughly that due to an equal charge at the point $z = -i\cos^2\alpha$, for when χ is large, the equation

$$z = -i\,\frac{1 + e^{\chi}\sin\alpha}{1 + e^{-\chi}\sin\alpha} = -i\,[1 + e^{\chi}\sin\alpha]\,[1 - e^{-\chi}\sin\alpha]\ldots$$

may be written in the form

$$z + i\cos^2\alpha = -ie^{\chi}\sin\alpha + \text{negligible terms}.$$

This point, which may be called the "centre of charge," is the middle point of that portion of the central radius cut off by the chord and the arc.

On the circular arc

$$\chi = i\psi \quad\text{and}\quad z = -ie^{2i\theta}.$$

Therefore $\quad \sin(\psi - \theta)\sin\alpha = \sin\theta.$

The surface density is thus proportional to $S\cos\theta + 1$ on the convex face and to $S\cos\theta - 1$ on the concave face, S denoting the quantity

$$S = (\sin^2\alpha - \sin^2\theta)^{-\frac{1}{2}}.$$

If E is the charge per unit length of a cylindrical conductor whose cross-section is the circular arc and d is the diameter of the circle, the surface density σ is given by Love's formula

$$2\pi\sigma d = E \mid \sec v \mid (\operatorname{cosec} \alpha - \cos v),$$

where $\sin v = -\tan\theta \cot\alpha$.

The solution of the electrical problem of a conducting plate under the influence of a line charge parallel to the plate but not in its plane may be derived from the preceding analysis by inversion from a point O on the unoccupied part of the circle. Let AB be the cross-section of the plate, D the foot of the perpendicular from O on AB, OC' the bisector of the angle AOB, then the surface density σ is given by Love's formula

$$\sigma = \frac{E}{2\pi}\frac{OD}{OP^2}\frac{\operatorname{cosec}\alpha - \cos v}{\mid \cos v \mid},$$

where now $\quad \sin v = \cot \alpha \tan (P'OC'),$

$$\cos v = \pm \operatorname{cosec}\alpha \frac{(A'P'.B'P')^{\frac{1}{2}}}{OA'},$$

$A'P'C'B'$ being perpendicular to OC' (Fig. 25). Thus

$$\sigma = \frac{E}{2\pi}\frac{OD}{OP^2}\frac{OA' \mp (A'P'.B'P')^{\frac{1}{2}}}{(A'P'.B'P')^{\frac{1}{2}}}.$$

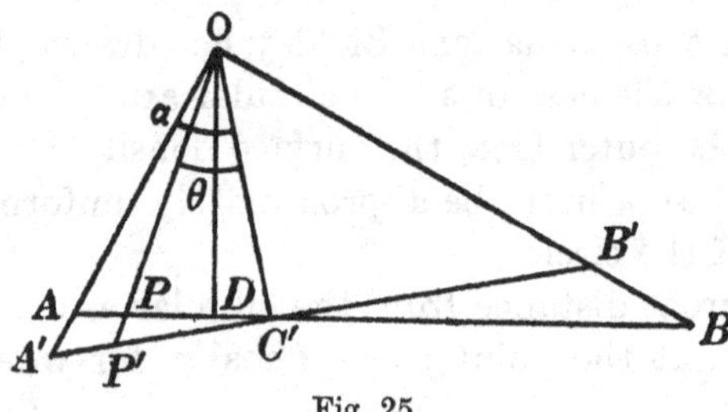

Fig. 25.

This is easily converted into the expression given in § 3·81.

The region outside a rectangle may be mapped on the interior of the unit circle in the ζ-plane with the aid of the transformation

$$z = \int_1^\zeta ds\,(1 - 2s^2 \cos 2\alpha + s^4)^{\frac{1}{2}}/s^2,$$

while a transformation which maps the region outside the rectangle into the region outside the circle is obtained by using a minus sign in front of the integral.

Let us use this transformation to determine the drag on a long thin rod of rectangular section which is moved slowly parallel to its length through a viscous liquid contained in a wide pipe of nearly circular section. We write $\log \zeta = iw = i(u + iv)$, where v is the velocity at any point in the z-plane, then

$$z = 2^{\frac{1}{2}} \int_0^w (\cos 2\alpha - \cos 2s)^{\frac{1}{2}}\,ds = 2\int_0^w (\sin^2\alpha - \sin^2 s)^{\frac{1}{2}}\,ds.$$

Putting $\sin s = \sin \alpha \sin \theta$, this becomes

$$z = 2\int_0^\theta \frac{\sin^2 \alpha\,(1 - \sin^2 \beta)\,d\beta}{(1 - \sin^2 \alpha \sin^2 \beta)^{\frac{1}{2}}} = 2\int_0^\theta (1 - \sin^2 \alpha \sin^2 \beta)^{\frac{1}{2}}\,d\beta$$

$$- 2\int_0^\theta \cos^2 \alpha\,(1 - \sin^2 \alpha \sin^2 \beta)^{-\frac{1}{2}}\,d\beta.$$

At the corner A immediately to the right of the origin O in the z-plane, we have $\theta = \frac{1}{2}\pi$, and

$$x_A = 2E\,(k) - 2k'^2\,K\,(k),$$

where $k = \sin \alpha$ and $E\,(k)$, $K\,(k)$ are the complete elliptic integrals to modulus k. The drag on the half side OA of the rectangle is proportional to w_A, and since

$$\sin w_A = \sin \alpha \sin (\tfrac{1}{2}\pi) = \sin \alpha,$$

we have $w_A = \alpha$. The drag on the side OA is thus equal to $(\alpha/2\pi)$ times the drag of the whole rectangle. [C. H. Lees, *Proc. Roy. Soc.* A, vol. XCII, p. 144 (1916).]

EXAMPLE

A line charge Q at the origin is partly shielded by a cylindrical shell of no radial thickness having the line charge for its axis, the trace of the shell on the xy-plane being that part of the circular arc $z = ae^{2i\theta}$ for which $-\pi < -2\omega < 2\theta < 2\omega < \pi$. Prove that the potential ϕ is given by the formula

$$\phi + i\psi = Q \log \frac{(z - a)\cos^2 \omega + (z + a)\sin^2 \omega + R}{(z + a)\sin^2 \omega - (z - a)\cos^2 \omega + R},$$

where R denotes that branch of the radical $[z^2 - 2az \cos 2\omega + a^2]^{\frac{1}{2}}$, whose real part is positive when the point z is external to the circle.

The surface density σ of the induced charge at a point θ on the charged arc is

$$\sigma = -Q\,\{\sec \omega\,(\tan^2 \omega - \tan^2 \theta)^{-\frac{1}{2}} \pm 1\}/4\pi a,$$

the upper sign corresponding to the density on the concave side, the lower sign to the density on the convex side. The latter is zero when $2\omega = \pi$, that is, when the circle closes.

[Chester Snow, *Scientific Papers of the Bureau of Standards*, No. 542 (1926).]

§ **4·71**. *Applications of conformal representation in hydrodynamics.* Consider the two-dimensional flow round an airplane wing whose span is so great that the hypothesis of two-dimensional flow is useful. Let u, v be the component velocities, p the pressure, ρ the density, L the lift per unit length of span, D the drag per unit length and M the moment about the origin of co-ordinates, this moment being also per unit length. These quantities may be calculated from the flux of momentum across a very large contour C which completely surrounds the aerofoil. In fact, if l, m are the direction cosines of the normal to the element ds, we have

$$L + iD = -\rho \int (v + iu)\,(ul + vm)\,ds - \int p\,(m + il)\,ds,$$

$$M = -\rho \int (xv - yu)\,(ul + vm)\,ds - \int p\,(xm - yl)\,ds;$$

the sign of M is such that a diving couple is regarded as positive. The equations may be rewritten in the form

$$L + iD = \tfrac{1}{2}\rho \int (v + iu)^2 \, dz - \int [p + \tfrac{1}{2}\rho (u^2 + v^2)] (m + il) \, ds,$$

$$M = \tfrac{1}{2}\rho \int [(u^2 - v^2)(mx + ly) + 2uv (my - lx)] \, ds + \int (ly - mx) [p + \tfrac{1}{2}\rho (u^2 + v^2)] \, ds,$$

where $z = x + iy$. Now when the motion is irrotational outside the aerofoil the quantity $p + \tfrac{1}{2}\rho (u^2 + v^2)$ is constant, also

$$\int (m + il) \, ds = 0, \quad \int (ly - mx) \, ds = 0,$$

hence

$$L + iD = \tfrac{1}{2}\rho \int (v + iu)^2 \, dz.$$

Taking the contour to be a circle of radius r, we have

$$M = \tfrac{1}{2}\rho \int [(u^2 - v^2)\, 2xy + 2uv (y^2 - x^2)] \, ds/r$$

$$= \tfrac{1}{2}R\rho \int [u^2 - v^2 - 2iuv] [x^2 - y^2 + 2ixy] \, ds/ir$$

$$= \tfrac{1}{2}R\rho \int (v + iu)^2 z \, dz,$$

where the symbol R is used to denote the real part of the expression which follows it. These are the formulae of Blasius* but the analysis is merely a development of that given by Kutta and Joukowsky.

The integrals may be evaluated with the aid of Cauchy's theory of residues by expanding $v + iu$ in the form

$$v + iu = iU + \frac{a_1}{z} + \frac{a_2}{z^2} + \ldots.$$

When the region outside the aerofoil is mapped on the region outside the circle $|z'| = a$ by a transformation of type

$$z' = c_0 + n\left(z + \frac{c_1}{z} + \frac{c_2}{z^2} + \ldots\right),$$

the flow round the aerofoil may be made to correspond to a flow round the circle by using the same complex potential χ in each case. Now for our flow round the circle we may write

$$\frac{d\chi}{dz'} = U'e^{-i\alpha} - U'e^{i\alpha} a^2/z'^2 + \frac{i\kappa}{2\pi z'},$$

* *Zeits. f. Math. u. Phys.* Bd. LVIII, S. 90 (1909), Bd. LIX, S. 43 (1910).

where U', α and κ are constants, therefore

$$v + iu = i\frac{d\chi}{dz} = i\frac{d\chi}{dz'}\frac{dz'}{dz}$$

$$= in\left[1 - \frac{c_1}{z^2} - \frac{2c_2}{z^3}\dots\right]\left[U'e^{-i\alpha} - U'e^{i\alpha}\frac{a^2}{n^2z^2} - \frac{i\kappa c_0}{2\pi n^2z^2} + \frac{i\kappa}{2\pi nz} + \dots\right]$$

$$= inU'e^{-i\alpha} - \frac{i\kappa}{2\pi z} + \frac{\kappa c_0}{2\pi nz^2} - \frac{inc_1U'}{z^2}e^{-i\alpha} - inU'e^{i\alpha}\frac{a^2}{n^2z^2} + \dots.$$

If $z' = ae^{i\beta}$ is the point of stagnation on the circle which maps into the trailing edge of the aerofoil, we have

$$\kappa = 2\pi aU' \sin(\alpha - \beta),$$

$$U = nU'e^{-i\alpha},$$

$$L + iD = \kappa\rho U'ne^{-i\alpha} = \kappa\rho U = 2\pi a\rho UU' \sin(\alpha - \beta).$$

§ 4·72. *The mapping of a wing profile on a nearly circular curve.* For the study of the flow of an inviscid incompressible fluid round an aerofoil of infinite span, it is useful to find a transformation which will map the region outside the aerofoil on the region outside a curve which is nearly circular.

If the profile has a sharp point at the trailing edge at which the tangents to the upper and lower parts of the curve meet at an angle σ, it is convenient to make use of a transformation of type

$$\frac{z - \kappa c}{z + \kappa c} = \left(\frac{\zeta - c}{\zeta + c}\right)^{\kappa}, \qquad \text{......(A)}$$

where $\sigma = (2 - \kappa)\pi$.

If a circle is drawn through the point $-c$ in the ζ-plane so that it just encloses the point c, cutting the line $(-c, c)$ in a point $c + d$, say, where d is small, this circle will be mapped by the transformation into a wing-shaped curve in the z-plane. This curve closely surrounds the lune formed from two circular arcs meeting at an angle σ at each of their points of junction, $z = \kappa c$, $z = -\kappa c$. The curve actually passes through the point $-\kappa c$ and has the same tangents there as the lune derived from a circle in the ζ-plane which passes through the points $(-c, c)$ and touches the former circle at the point $-c$.

If we start with the profile in the z-plane and wish to derive from it a nearly circular curve with the aid of a transformation of this type, the rule is to place the point $-\kappa c$ at the trailing edge and the point κc inside the contour very close to the place where the curvature is greatest*. This rule works well for thin aerofoils, but it has been found by experience that by increasing the magnitude of d an aerofoil with a thick head may be obtained from a circle, and that the thickness of the middle portion of the aerofoil is governed partly by the value of σ. Hence in endeavouring to

* F. Höhndorf, *Zeits. f. ang. Math. u. Mech.* Bd. VI, S. 265 (1926).

map a thick aerofoil on a nearly circular curve the point κc may be taken at an appreciable distance from the boundary. Another point to be noticed is that when a circle is transformed into an aerofoil by means of the transformation (A) the smaller the distance of the centre from the line $(-c, c)$ the smaller is the camber of the corresponding aerofoil and the more symmetric is the head. The point κc, moreover, lies very nearly on the line of symmetry.

The actual transformation may be carried out graphically with the aid of two corresponding systems of circles indicated by the use of bipolar co-ordinates. The circles in one plane are the two mutually orthogonal coaxial systems having the points $(-c, c)$ as common points and limiting points respectively; the corresponding circles in the other plane for two mutually orthogonal systems having the points $(-c, c)$ as common points and limiting points respectively. This is the method recommended by Kármán and Trefftz. Another construction recommended by Höhndorf depends upon the substitutions

$$t = \frac{z - \kappa c}{z + \kappa c}, \quad \tau = \frac{\zeta - c}{\zeta + c},$$

by which the transformation may be written in the form

$$\tau = t^{\frac{1}{\kappa}} = t^{\frac{1}{2}} \sqrt[\eta]{(t^{\frac{1}{2}})},$$

where

$$\eta = \frac{2\pi - \sigma}{\sigma}.$$

The plan is to first consider the transformation from z to ζ, given by the equation

$$\tau = t^{\frac{1}{2}} \text{ or } \frac{\zeta - c}{\zeta + c} = \left(\frac{z - \kappa c}{z + \kappa c}\right)^{\frac{1}{2}}.$$

This transformation may be performed graphically* by writing

$$z_0 = z + \kappa c, \quad \zeta_0 = \zeta + c,$$

when the relation becomes

$$\left(1 - \frac{2\kappa c}{z_0}\right)^{\frac{1}{2}} = 1 - \frac{2c}{\zeta_0}.$$

When $t^{\frac{1}{2}}$ has been found its ηth root may be determined graphically and when this is multiplied by $t^{\frac{1}{2}}$ the value of τ is obtained, and from this ζ is easily derived.

Höhndorf gives a table of values of η corresponding to different angles σ. When $\sigma = 4°$, $\eta = 89$, when $\sigma = 8°$, $\eta = 44$, and when $\sigma = 10°$,

* The transformation may also be performed graphically by writing it in the form

$$z = \frac{\kappa}{2}\left(\zeta + \frac{c^2}{\zeta}\right)$$

when it is desired to pass from a figure in the ζ-plane to a corresponding figure in the z-plane.

$\eta = 35$. When the transformation (A) is expressed by means of series, the results are

$$z = \zeta + \frac{\kappa^2 - 1}{3}\frac{c^2}{\zeta} - \frac{(\kappa^4 - 5\kappa^2 + 4)\,c^4}{45\zeta^3} + \dots,$$

$$\zeta = z + \frac{1 - \kappa^2}{3}\frac{c^2}{z} - \frac{(4\kappa^4 - 5\kappa^2 + 1)\,c^4}{45z^3} + \dots.$$

§ **4·73.** *Aerofoil of small thickness**. We have seen that the transformation

$$z' = z + a^2/z$$

maps the circle C given by $|z| = a$ into a flat plate P' extending from $z' = 2a$ to $z' = -2a$ and back. On the other hand, if the A's are small quantities the transformation

$$\zeta = z\{1 + \sum_{n=0}^{\infty} A_n (a/z)^n\}$$

maps C into a curve Γ differing slightly from a circle, and if we then put

$$\zeta' = \zeta + a^2/\zeta,$$

Γ maps into a curve Π' differing slightly from a flat plate. Now for a point on Γ

$$\zeta = a(1 + r)\,e^{i\theta},$$

where θ is a real angle and r a real quantity which is small; therefore to the first order in r

$$\zeta' = 2a(\cos\theta + ir\sin\theta),$$

and so
$$\xi' = 2a\cos\theta, \quad \eta' = 2ar\sin\theta.$$

For points on C and P' we may use a real angle ω and write

$$z = ae^{i\omega}, \quad x' = 2a\cos\omega, \quad y' = 0,$$

then
$$(1 + r)\,e^{i(\theta-\omega)} = 1 + \sum_{n=0}^{\infty} A_n e^{-in\omega},$$

and since $\theta - \omega$ is small we have to the first order, with $A_n = B_n + iC_n$,

$$r = \Sigma\,(B_n \cos n\omega + C_n \sin n\omega),$$
$$\theta - \phi = \Sigma\,(C_n \cos n\omega - B_n \sin n\omega).$$

Hence by Fourier's theorem

$$\pi B_n = \int_{-\pi}^{\pi} r\cos n\theta\,d\theta = \int_{-\pi}^{\pi} \frac{\eta'}{2a}\frac{\cos n\theta}{\sin\theta}\,d\theta,$$

$$\pi C_n = \int_{-\pi}^{\pi} r\sin n\theta\,d\theta = \int_{-\pi}^{\pi} \frac{\eta'}{2a}\frac{\sin n\theta}{\sin\theta}\,d\theta,$$

$$2\pi B_0 = \int_{-\pi}^{\pi} r\,d\theta = \int_{-\pi}^{\pi} \frac{\eta'}{2a}\frac{d\theta}{\sin\theta}.$$

* H. Jeffreys, *Proc. Roy. Soc.* A, vol. CXXI, p. 22 (1928).

Since $\sin\theta$ and $\sin n\theta$ are odd functions of θ, whilst $\cos n\theta$ is an even function of θ, it appears that C_n depends on the sums and B_n on the differences of the values of η' corresponding to angles $\pm\theta$; thus the C_n's depend on the camber of the aerofoil, the B_n's on its thickness.

When θ is small, that is, for points near the trailing edge of the aerofoil, we have approximately

$$\frac{\eta'}{2a-\xi'} = \frac{r\sin\theta}{1-\cos\theta} = \frac{2r}{\theta},$$

and when $\pi - \theta$ is a small quantity ω we have

$$\frac{\eta'}{\xi'+2a} = \frac{2r}{\omega}.$$

Thus r vanishes at $\theta = 0$ because the slope of the section is finite there; but at $\theta = \pi$ the section and the axis meet at right angles at a point which may be called the leading edge. If the curvature at this point is $1/R$ we have to a close approximation

$$2R = \frac{\eta'^2}{\xi+2a} = \frac{4a^2r^2\sin^2\theta}{2a(1+\cos\theta)} = 2ar^2(1-\cos\theta) = 4ar^2,$$

consequently r is finite and equal to $(R/2a)^{\frac{1}{2}}$ at the leading edge. If at a great distance from the circle C the flow in the z-plane is represented approximately by a velocity U making an angle α with the axis of x, we have

$$\chi = \phi + i\psi = Uze^{-i\alpha} + Ue^{i\alpha}a^2/z + \frac{i\kappa}{2\pi}\log(z/a),$$

where κ is the circulation round the cylinder. Taking χ to be the complex potential for a corresponding flow in the ζ'-plane the component velocities (u, v) in this plane are given by the equation

$$u - iv = \frac{d\chi}{d\zeta'} = \frac{d\chi}{dz}\bigg/\frac{d\zeta}{dz}\frac{d\zeta'}{d\zeta}.$$

To determine κ we make the velocity finite at the trailing edge where ξ' is a maximum and $\theta = 0$, $r = 0$, $\zeta = a$, $\dfrac{d\zeta'}{d\zeta} = 0$.

Hence $d\chi/dz$ is zero and so

$$Ue^{-i\alpha} - Ue^{i\alpha}\frac{a^2}{z^2} + \frac{i\kappa}{2\pi z} = 0.$$

But when $\theta = 0$

$$a/z = e^{i(\theta-\phi)} = e^{i\beta},$$

say, where

$$\beta = \Sigma C_n.$$

Therefore

$$\kappa = 4\pi aU\sin(\alpha+\beta).$$

Now

$$u - iv = \frac{Ue^{-i\alpha} - Ue^{i\alpha}\, a^2/z^2 + i\kappa/2\pi z}{\{1 - \Sigma\,(n-1)\, A_n\,(a/z)^2\}\,(1 - a^2/\zeta^2)}$$

$$= \frac{Ue^{-i\alpha} - Ue^{i\alpha} a^2\,(1+B_0)^2/\zeta'^2 + i\kappa\,(1 + B_0 + A_1 a/\zeta')/2\pi\zeta'}{\{1 + B_0 - A_2\,(a/\zeta')^2\}\,(1 - a^2/\zeta'^2)} + O(\zeta'^{-3})$$

$$= \frac{Ue^{-i\alpha}}{1+B_0} + \frac{i\kappa}{2\pi\zeta'}$$

$$+ \zeta'^{-2}\left[\frac{A_2 a^2 U e^{-i\alpha}}{(1+B_0)^2} + \frac{Ua^2e^{-i\alpha}}{1+B_0} - Ua^2e^{i\alpha}\,(1+B_0) + \frac{i\kappa a A_1}{2\pi\,(1+B_0)}\right] + \ldots.$$

Therefore by the Kutta-Joukowsky theorem the lift per unit span of the aerofoil is

$$L = \frac{\rho\kappa U}{1+B_0} = 4\pi\rho a\,(1+B_0)\sin(\alpha+\beta)\,V^2, \quad V\,(1+B_0) = U,$$

and the lift coefficient is

$$K_L = (L/4\rho a V^2) = \pi\,(1+B_0)\sin(\alpha+\beta).$$

The thickness thus affects the lift through B_0, which is a positive constant for a given wing.

The moment about O, that is a point midway between the leading and trailing edges of the aerofoil, is equal to $\pi\rho$ times the real part of the coefficient of $+\,i\zeta'^{-2}$ in the expansion of $\left(\frac{d\chi}{d\zeta'}\right)^2$. This coefficient is

$$\frac{i\kappa^2}{4\pi^2} - 2iV^2a^2\left\{(1+A_2)\,e^{-2i\alpha} - (1+B_0)^2 + \frac{i\kappa A_1 e^{-i\alpha}}{2\pi a V}\right\},$$

and so to the first order in the B's and C's the moment is M, where

$$M = 2\pi\rho V^2a^2\{C_2\cos 2\alpha - (1+B_2)\sin 2\alpha + 2B_1\cos\alpha\sin(\alpha+\beta) + 2C_1\sin\alpha\sin(\alpha+\beta)\}.$$

The moment about the leading edge is

$$M_0 = M + 2aL = 2\pi\rho V^2a^2\{2\alpha\,(1 + 2B_0 + B_1 - B_2) + C_2 + 2\beta\,(2 + 2B_0 + B_1)\},$$

where terms of orders α^2, αB_n, αC_n have been retained, but terms of orders α^3, $\alpha^2 B_n$, $\alpha^2 C_n$, dropped. When squares and products of the B's and C's are neglected, the moment coefficient K_M is

$$K_M = \frac{M + 2aL}{(4a)^2\rho V^2} = \tfrac{1}{4}\pi\alpha\,(1 + 2B_0 + B_1 - B_2) + \tfrac{1}{8}\pi C_2 + \tfrac{1}{2}\pi\beta$$

$$= \tfrac{1}{4}K_L\,(1 + B_0 + B_1 - B_2) + \tfrac{1}{8}\pi C_2 + \tfrac{1}{4}\pi\beta.$$

The moment coefficient at zero lift is thus

$$\tfrac{1}{8}\pi C_2 + \tfrac{1}{4}\pi\beta,$$

and is independent of the thickness to this order of approximation. The thickness, however, affects the coefficient of K_L.

For further applications of conformal representation in hydrodynamics the reader is referred to H. Glauert's *Aerofoil and Airscrew Theory* (Cambridge, 1926) and to H. Villat's *Leçons sur l'Hydrodynamique* (Gauthier-Villars, Paris, 1929).

§ **4·81**. *Orthogonal polynomials associated with a given curve**. Let $f(z)$ be a function which is defined for points of the z-plane which lie on a closed continuous rectifiable curve C which is free from double points. If $ds = |dz|$, the integral

$$\int_C f(z)\, ds$$

denotes as usual the limiting value

$$\lim_{m\to\infty} \sum_{\nu=1}^{m} f(\zeta_\nu)\,|z_\nu - z_{\nu-1}|,$$

where $z_0, z_1, z_2, \ldots$ represent successive points on C for which

$$\lim_{m\to\infty} [\text{Maximum value of } |z_\nu - z_{\nu-1}| \text{ for } 0 \leqslant \nu \leqslant m] = 0, \quad (z_1 = z_m),$$

and ζ_ν denotes an arbitrary point of C which lies between $z_{\nu-1}$ and z_ν. We have in particular

$$\int_C ds = l,$$

where l denotes the length of the curve C. We shall suppose now that the unit of length is chosen so that $l = 1$.

Using $\bar{z}$ to denote the complex quantity conjugate to z, we write

$$h_{pq} = \int_C z^p \bar{z}^q\, ds = \bar{h}_{qp},$$

$$D_0 = h_{00} = 1, \quad D_1 = \begin{vmatrix} h_{00} & h_{01} \\ h_{10} & h_{11} \end{vmatrix},$$

$$D_n = \begin{vmatrix} h_{00} & h_{01} \ldots h_{0n} \\ h_{10} & \ldots\ldots\ h_{1n} \\ \ldots & \ldots\ldots\ldots \\ h_{n0} & \ldots\ldots\ h_{nn} \end{vmatrix}.$$

Let H_{mn} denote the co-factor of h_{mn} in the determinant D_n, and let α_n be a constant whose value will be determined later; then, if

$$P_n(z) = \alpha_n (H_{0n} + zH_{1n} + \ldots z^n H_{nn}),$$

it is easily seen that

$$\begin{aligned} \int_C P_n(z)\, \bar{z}^\nu\, ds &= \alpha_n (h_{0\nu} H_{0n} + h_{1\nu} H_{1n} + \ldots h_{n\nu} H_{nn}) \\ &= 0 \text{ for } 0 \leqslant \nu \leqslant n-1 \\ &= \alpha_n D_n \text{ for } \nu = n, \end{aligned}$$

$$\int_C |P_n(z)|^2\, dz = \alpha_n \bar{\alpha}_n H_{nn} D_n = \alpha_n \bar{\alpha}_n D_{n-1} D_n.$$

* See a remarkable paper by Szegö, *Math. Zeits.* Bd. IX, S. 218 (1921).

The polynomials thus form an orthogonal system which is normalised by choosing α_n, so that $\alpha_n = \bar{\alpha}_n = (D_n D_{n-1})^{-\frac{1}{2}}$.

If $\overline{P_m(z)}$ denotes the complex quantity conjugate to $P_m(z)$, the orthogonal relations may be written in the form

$$\int_C P_n(z)\,\overline{P_m(z)}\,ds = 0, \quad m \neq n$$
$$= 1, \quad m = n.$$

Let us now suppose that C is an analytic curve and that $\zeta = \gamma(z)$ is the function which maps the interior of C smoothly on the region $|\zeta| < 1$ of the ζ-plane in such a way that $\gamma(a) = 0$, $\gamma'(a) > 0$. Since C is an analytic curve $\gamma(z)$ is also regular and smooth in a region enclosing the curve C. It is known, moreover, that there is one and only one function, $z = g(\zeta)$, which is regular and smooth for $|\zeta| \leqslant 1$ and maps the interior of $|\zeta| = 1$ on the interior of the curve C. The derivatives of the functions $\gamma(z)$, $g(\zeta)$ are connected by the relation $\gamma'(z)\,g'(\zeta) = 1$, where z and ζ are associated points of the two planes.

Our aim now is to show that

$$\gamma(z) = \lim_{n \to \infty} \frac{2\pi}{K_n(a, a)} \int_a^z [K_n(a, w)]^2\,dw,$$

where $$K_n(a, z) = \overline{P_0(a)}\,P_0(z) + \ldots \overline{P_n(a)}\,P_n(z).$$

We shall first of all prove an important property of the polynomial $K_n(a, z)$.

Let a be arbitrary and $G_n(z)$ a polynomial of the nth degree with the property

$$\int_C |G_n(z)|^2\,ds = 1,$$

then the maximum value of $|G_n(a)|^2$ is $K_n(a, a)$, and this value is attained when $G_n(z) = \epsilon K_n(a, z)\,[K_n(a, a)]^{-\frac{1}{2}}$, where ϵ is an arbitrary constant such that $|\epsilon| = 1$.

Let us write

$$G_n(z) = t_0 P_0(z) + t_1 P_1(z) + \ldots t_n P_n(z),$$

where the coefficient t_ν is determined by Fourier's rule and is

$$t_\nu = \int_C G_n(z)\,\overline{P_\nu(z)}\,ds.$$

We then have

$$1 = \int_C |G_n(z)|^2\,ds = |t_0|^2 + |t_1|^2 + \ldots |t_n|^2,$$
$$G(a) = t_0 P_0(a) + t_1 P_1(a) + \ldots t_n P_n(a),$$

and by Schwarz's inequality

$$|G_n(a)|^2 \leqslant K_n(a, a).$$

The sign of equality can be used when

$$t_\nu = \epsilon \overline{P_\nu (a)} \, [K_n (a, a)]^{-\frac{1}{2}};$$

that is, when $$G_n (z) = \epsilon K_n (a, z) \, [K_n (a, a)]^{-\frac{1}{2}}.$$

When the point a is within the region bounded by C, and $F(z)$ is any function which is regular and analytic in the closed inner realm of C, we have the inequality

$$| F (a) | \leqslant \frac{1}{2\pi\delta} \int_C | F (z) | \, ds,$$

where δ is the least distance of the point a from the curve C. To prove this we remark that Cauchy's theorem gives

$$F (a) = \frac{1}{2\pi i} \int_C \frac{F (z) \, dz}{z - a},$$

and so $$| F (a) | \leqslant \frac{1}{2\pi} \int_C \left| \frac{F (z)}{z - a} \right| ds \leqslant \frac{1}{2\pi\delta} \int_C | F (z) | \, ds.$$

In the special case when $F (z) = [G_n (z)]^2$ the inequality gives

$$[G_n (a)]^2 \leqslant \frac{1}{2\pi\delta}.$$

This is true for all polynomials $G_n (z)$, and so, in particular,

$$K_n (a, a) \leqslant \frac{1}{2\pi\delta}.$$

Since δ is independent of n, this inequality establishes the convergence of the series

$$K (a, a) = | P_0 (a) |^2 + | P_1 (a) |^2 + \dots$$

for the case in which the point a lies in the region bounded by C. Again, we have the inequality

$$|K_n (a, z) |^2 \leqslant [|P_0 (a) || P_0 (z) | + | P_1 (a) || P_1 (z) | + \dots]^2 \leqslant K_n (a, a) \, K_n (z, z),$$

and if $R_n{}^m (a, z)$ denotes the remainder

$$R_n{}^m (a, z) = \overline{P_{n+1} (a)} \, P_{n+1} (z) + \dots \overline{P_{n+m} (a)} \, P_{n+m} (z),$$

we have $$| R_n{}^m (a, z) |^2 \leqslant R_n{}^m (a, a) \, R_n{}^m (z, z).$$

Since the series $K (a, a)$ is convergent when a lies within C we can find a number $N (a)$ such that if $n > N (a)$ we have for all values of m

$$| R_n{}^m (a, a) | < \epsilon,$$

where ϵ is a small positive quantity given in advance, hence if N is the greater of the two quantities $N (a)$, $N (z)$, we have for $n > N$

$$| R_n{}^m (a, z) |^2 \leqslant \epsilon^2.$$

This establishes the convergence of the series

$$K (a, z) = \overline{P_0 (a)} \, P_0 (z) + \overline{P_1 (a)} \, P_1 (z) + \dots.$$

To prove that the series is uniformly convergent in any closed realm R lying entirely within C we note that the quantities $K_n(a, z)$ are uniformly bounded in the sense that

$$|K_n(a, z)|^2 \leqslant K_n(a, a) K_n(z, z) \leqslant \left(\frac{1}{2\pi\delta}\right)^2.$$

The general selection theorem of § 4·45 now tells us that from the sequence $K_n(a, z)$ we may select a partial sequence of functions which converges uniformly in R towards a limit function $f(z)$. Since, however, the sequence converges to $K(a, z)$ this limit function $f(z)$ must be identical with $K(a, z)$ and so the series which represents $K(a, z)$ converges uniformly in R, a and z being points within R.

We now consider the integral

$$J_n = \int_C |K_n(a, z) - \lambda \{\gamma'(z)\}^{\frac{1}{2}}|^2 ds,$$

where λ is a constant which is at our disposal. We have

$$\int_C |K_n(a, z)|^2 ds = K_n(a, a),$$

$$\int_C |\gamma'(z)| ds = \int_0^{2\pi} d\theta = 2\pi,$$

$$\int_C K_n(a, z) \overline{\{\gamma'(z)\}^{\frac{1}{2}}} ds = \int_0^{2\pi} K_n\{a, g(\zeta)\} \frac{|g'(\zeta)|}{\{g'(\zeta)\}^{\frac{1}{2}}} d\theta$$

$$= \int_0^{2\pi} K_n\{a, g(\zeta)\} \sqrt{g'(\zeta)}\, d\theta,$$

where $\zeta = e^{i\theta}$.

Now the function $K_n\{a, g(\zeta)\} \sqrt{g'(\zeta)}$ is regular and analytic for $|\zeta| \leqslant 1$, and so the last integral is equal to

$$2\pi K_n\{a, g(0)\} \sqrt{g'(0)} = 2\pi K_n(a, a) [\gamma'(a)]^{-\frac{1}{2}}.$$

Choosing $$\lambda = \frac{1}{2\pi} \overline{[\gamma'(a)]^{\frac{1}{2}}},$$

we have finally $$J_n = \frac{1}{2\pi} |\gamma'(a)| - K_n(a, a).$$

Our object now is to show that

$$\lim_{n\to\infty} J_n = 0.$$

Let us write $$L(\zeta) = [g'(\zeta)]^{\frac{1}{2}}.$$

Since $g'(\zeta) \neq 0$ for $|\zeta| \leqslant 1$, a branch of $L(\zeta)$ is a regular analytic function for $|\zeta| \leqslant 1$.

We now consider the set of analytic functions $E(\zeta)$ regular in $|\zeta| \leqslant 1$ and such that

$$\int_0^{2\pi} |L(\zeta) E(\zeta)|^2 d\theta = 1.$$

Let a be a fixed number whose modulus is less than unity, then the maximum value of $|E(a)|^2$ is

$$[1-|a|^2]^{-1}|L(a)|^{-2}.$$

To see this we put

$$L(\zeta)E(\zeta)=t_0+t_1\zeta+\ldots+t_n\zeta^n+\ldots,$$

then on the above supposition

$$|t_0|^2+|t_1|^2+\ldots+|t_n|^2+\ldots=\frac{1}{2\pi},$$

and Schwarz's inequality gives

$$|E(a)|^2|L(a)|^2<\sum_{n=0}^{\infty}|t_n|^2\sum_{n=0}^{\infty}|a|^{2n}=\frac{1/2\pi}{1-|a|^2}.$$

The sign of equality holds when, and only when,

$$t_n=c\bar{a}^n, \qquad n=0,1,2,\ldots,$$

that is, when
$$L(\zeta)E(\zeta)=\frac{c}{1-\zeta\bar{a}}.$$

Now
$$\int_0^{2\pi}\frac{d\theta}{|1-\bar{a}\zeta|^2}=\frac{2\pi}{1-|a|^2},$$

therefore
$$2\pi|c|^2=1-|a|^2,$$

and so
$$E(\zeta)=\epsilon\frac{(1-|a|^2)^{\frac{1}{2}}}{1-\zeta\bar{a}}\frac{1}{(2\pi)^{\frac{1}{2}}L(\zeta)}.$$

Now let
$$E(\zeta)=E\{\gamma(z)\}=G(z)=G\{g(\zeta)\},$$

then $E(\zeta)$ and $G(z)$ are simultaneously regular, and

$$1=\int_0^{2\pi}|L(\zeta)E(\zeta)|^2d\theta=\int_0^{2\pi}|g'(\zeta)||E(\zeta)|^2d\theta=\int_C|G(z)|^2ds.$$

Finally,
$$E(0)=G(a),$$

so that
$$\max|E(0)|^2=\max|G(a)|^2,$$

therefore

$$K(a,a)=\max|G(a)|^2=\max|E(0)|^2=\frac{1/2\pi}{|L(0)|^2}=\frac{1}{2\pi|g'(0)|},$$

and so

$$\lim_{n\to\infty}J_n=\frac{1}{2\pi}|\gamma'(a)|-K(a,a)=\frac{1}{2\pi}\{|\gamma'(a)|-|g'(0)|\}=0.$$

Since
$$[K_n(a,z)-\lambda\{\gamma'(z)\}^{\frac{1}{2}}]^2=F_n(z)$$

is a regular analytic function in the closed inner realm of C, we have for any point z_0 within C whose least distance from C is δ,

$$|F_n(z_0)|^2<\frac{1}{2\pi\delta}\int_C|F_n(z)|^2dz=\frac{J_n}{2\pi\delta},$$

and so as $n\to\infty$,
$$\lim_{n\to\infty}|F_n(z_0)|=0.$$

Hence $$K(a, z_0) = \lim_{n\to\infty} K_n(a, z_0) = \lambda\{\gamma'(z_0)\}^{\frac{1}{2}}.$$

Furthermore, since

$$K(a, a) = \frac{1}{2\pi} \mid \gamma'(a) \mid = 2\pi\lambda^2,$$

we have $$\gamma'(z_0) = 2\pi \frac{[K(a, z_0)]^2}{K(a, a)},$$

and so $$\gamma(z) = \frac{2\pi}{K(a, a)} \int_a^z [K(a, z_0)]^2 \, dz_0.$$

If the curve C instead of being of unit length is of length l the orthogonal polynomials $P_n(z)$ are defined so that

$$\frac{1}{l} \int_C \mid P_n(z) \mid^2 ds = 1,$$

and the general formula for the mapping function becomes

$$\gamma(z) = \frac{2\pi\epsilon}{l} \cdot \frac{1}{K(a, a)} \int_a^z [K(a, z_0)]^2 \, dz_0,$$

where ϵ is a number with unit modulus and is equal to unity when the mapping function is required to be such that $\gamma'(a) > 0$.

A study of the expansion of functions in series of the orthogonal polynomials $P_n(z)$ has been made recently by Szegö and by V. Smirnoff, *Comptes Rendus*, t. CLXXXVI, p. 21 (1928).

§ **4·82**. *The mapping of the region outside C'.* If we write

$$z'(z - a) = 1, \quad ww' = 1,$$

the interior of C maps into the region outside a closed curve C' in such a way that the point $z = a$ maps into the point at infinity in the z'-plane. The interior of the unit circle $\mid w \mid < 1$ is likewise mapped into the region $\mid w' \mid > 1$, the point $w = 0$ corresponding to $w' = \infty$.

Hence the region $\mid w' \mid > 1$ is mapped on the region outside C' in such a way that $w' = \infty$ corresponds to $z' = \infty$, the relation between the variables being of type

$$z' = \tau w' + \tau_0 + \tau_1 (w')^{-1} + \dots + \tau_n (w')^{-n} + \dots.$$

Since $w = \gamma(z)$, the function which gives the conformal representation is

$$w' = [\gamma\{a + (z')^{-1}\}]^{-1} = \psi(z'), \text{ say.}$$

Szegö has shown that the function $\psi(z')$ may also be obtained directly with the aid of an orthogonal system of polynomials $\Pi_n(z')$ associated with the curve C'.

If $\tau = \mid \tau \mid e^{i\alpha}$, we have in fact the formula

$$\psi(z') = e^{-i\alpha} \lim_{n\to\infty} \frac{\Pi_{n+1}(z')}{\Pi_n(z')}.$$

EXAMPLES

1. If z_0 is a root of the equation $P_n(z) = 0$, prove that

$$z_0 \int_C \left| \frac{P_n(z)}{z - z_0} \right|^2 ds = \int_C \left| \frac{P_n(z)}{z - z_0} \right|^2 z\, ds.$$

Hence show that z_0 lies within the smallest convex closed realm R which contains the curve C.

2. If C is a circle of unit radius,

$$P_n(z) = z^n,$$
$$K(a, z) = \frac{1}{1 - z\bar{a}},$$
$$\gamma(z) = \frac{z - a}{1 - z\bar{a}}.$$

3. If the curve C is a double line joining the points $-1, 1$, the polynomial $P(z)$ becomes proportional to the Legendre polynomial. Note that in this case the series $K(a, z)$ fails to converge, but this does not contradict the general convergence theorem because now the points a and z do not lie within C.

4. If $R_n(z)$ is any polynomial and a any point within the curve C, prove that

$$\frac{1}{l} \int_C \overline{K_n(a, z)}\, R_n(z)\, ds = R_n(a).$$

§ **4·91**. *Approximation to the mapping function by means of polynomials**. Let a circle of radius R be drawn round the origin in the z-plane and let

$$w = f(z) = a_0 + a_1 z + a_2 z^2 + \dots$$

be a power series converging uniformly in its whole interior. This maps the circle on a region of the complex w-plane. For the area of this region we easily find the expressions

$$A = \int_0^R \int_0^{2\pi} | f'(z) |^2 r\, dr\, d\theta \qquad (z = re^{i\theta})$$
$$= 2\pi \int_0^R r\, dr \sum_{n=1}^{\infty} | a_n |^2 n^2 r^{2n-2}$$
$$= \pi R^2 | a_1 |^2 + \pi \sum_{n=2}^{\infty} n | a_n |^2 R^{2n} + \dots.$$

The area of the image region is always greater than $\pi R^2 | a_1 |^2$ when $a_1 \neq 0$ and is always greater than $\pi n | a_n |^2 R^{2n}$ when $a_n \neq 0$. When the mapping function $f(z)$ is such that $a_1 = 1$ the result is that the area of the picture is greater than that of the original region unless the picture happens to be a circle of radius R.

Suppose now that we are given a simply connected smooth limited region B of the z-plane. Let $d\tau$ be an element of area of this region, we then look for a function $f(z)$ regular in B which makes the integral

$$I = \iint | f'(z) |^2 d\tau$$

* L. Bieberbach, *Rend. Palermo*, vol. XXXVIII, p. 98 (1914).

as small as possible. To make the problem definite we add the restrictions that $f(0) = 0, f'(0) = 1$, and that $f(z)$ is a polynomial of the nth degree. These conditions are satisfied by writing

$$f(z) = z + a_2 z^2 + \ldots + a_n z^n. \qquad \ldots\ldots(A)$$

If $f(z) + \epsilon g(z)$ is a comparison function we have to formulate the conditions that the integral

$$J(\epsilon) = \iint |f'(z) + \epsilon g'(z)|^2 d\tau = \iint [f'(z) + \epsilon g'(z)] [\overline{f'(z)} + \bar{\epsilon} \overline{g'(z)}] d\tau$$

may be a minimum for $\epsilon = 0$. These conditions are

$$\frac{\partial J}{\partial \epsilon} = \iint g'(z) \overline{f'(z)} d\tau = 0,$$

$$\frac{\partial J}{\partial \bar{\epsilon}} = \iint f'(z) \overline{g'(z)} d\tau = 0,$$

$$\frac{\partial^2 J}{\partial \epsilon \partial \bar{\epsilon}} = \iint g'(z) \overline{g'(z)} d\tau = \iint |g'(z)|^2 d\tau.$$

The inequality is always satisfied, but the two equations are satisfied for all forms of the polynomial $g(z)$ only when the coefficients a_s satisfy certain linear equations. If

$$\iint z^p \bar{z}^q d\tau = z_{pq},$$

where $\bar{z}$ is the conjugate of z, these equations are

$$2z_{0,1} + 4z_{1,1}a_2 + 6z_{2,1}a_3 + \ldots 2nz_{n-1,1}a_n = 0,$$

$$\cdots\cdots\cdots\cdots\cdots\cdots\cdots\cdots\cdots\cdots\cdots\cdots$$

$$nz_{0,n-1} + (n.2)\, z_{1,n-1}a_2 + (n.3)\, z_{2,n-1}a_3 + \ldots (n.n)\, z_{n-1,n-1}a_n = 0,$$

and

$$2z_{1,0} + (2.2)\, z_{1,1}\bar{a}_2 + (2.3)\, z_{1,2}\bar{a}_3 + \ldots (2.n)\, z_{1,n-1}\bar{a}_n = 0,$$

$$\cdots\cdots\cdots\cdots\cdots\cdots\cdots\cdots\cdots\cdots\cdots\cdots$$

$$nz_{n-1,0} + (n.2)\, z_{n-1,1}\bar{a}_2 + (n.3)\, z_{n-1,2}\bar{a}_3 + \ldots (n.n)\, z_{n-1,n-1}\bar{a}_n = 0.$$

These linear equations are associated with the Hermitian form

$$\sum_{p=0}^{n-1} \sum_{q=0}^{n-1} (p+1)(q+1)\, z_{pq} a_p \bar{a}_q,$$

and possess a single set of solutions for which I is a minimum. By giving different values to n we obtain a sequence of polynomials which in many cases converges towards a limit function $F(z)$. The question to be settled is whether this function $F(z)$, among all mapping functions with the properties $F(0) = 0$, $F'(0) = 1$, gives the smallest possible area to the picture into which B is mapped. The following simple example tells us that this is not always the case. Consider the region B which arises from a circle when the outer half of one of its radii is added to the boundary (Fig. 26). There is no polynomial which maps this region B on a region of smaller area. For

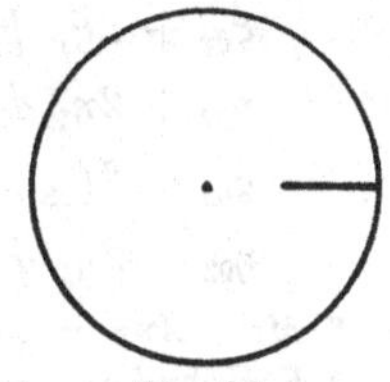

Fig. 26.

by means of a polynomial the region B is mapped on another region which has the same area as the region on which the complete circle is mapped and, unless the polynomial is simply z, this region has an area which is greater than that of the circle. Hence in this case all minimal polynomials are equal to z and $F(z)$ is also equal to z.

Bieberbach has investigated the convergence of the sequence of polynomials to the desired mapping function for the type of region discovered by Carathéodory*. For such a region the boundary is contained in the boundary of another region which has no point in common with the first. The interior of a polygon is a particular region of this type and so also is the interior of a Jordan curve.

Bieberbach's method of approximation has been used recently in aerofoil theory for the mapping of a circle on a region which is nearly circular†.

Introducing polar co-ordinates, $z = re^{i\theta}$, and supposing that on the boundary $r = 1 + \gamma$, where γ is small, we may write

$$z_{pq} = \int_0^{2\pi} d\theta \int_0^{1+\gamma} r^{p+q+1} e^{(p-q)i\theta} dr = \frac{1}{p+q+2}\int_0^{2\pi} (1+\gamma)^{p+q+2} e^{(p-q)i\theta} d\theta.$$

Hence retaining only terms up to the second order in the binomial expansion of $(1+\gamma)^{p+q+2}$,

$$z_{pq} = \int_0^{2\pi} \gamma . \cos(p-q)\,\theta . d\theta + \frac{p+q+1}{2}\int_0^{2\pi} \gamma^2 . \cos(p-q)\,\theta . d\theta$$

$$+ i\left[\int_0^{2\pi} \gamma . \sin(p-q)\,\theta . d\theta + \frac{p+q+1}{2}\int_0^{2\pi} \gamma^2 . \sin(p-q)\,\theta . d\theta\right], \quad p \neq q,$$

$$z_{pp} = \frac{\pi}{p+1} + \int_0^{2\pi} p\,d\theta + (p+\tfrac{1}{2})\int_0^{2\pi} \gamma^2 d\theta.$$

These quantities may be determined from the profile of the nearly circular curve when this is given.

Now writing $\quad z_{pq} = \xi_{pq} + i\eta_{pq}, \quad a_p = \phi_p + i\psi_p,$

where ξ_{pq}, η_{pq}, ϕ_p and ψ_p are all real, and neglecting all the coefficients after a_4 in the expansion (A), we obtain the following equations for the determination of $\phi_2, \phi_3, \phi_4, \psi_2, \psi_3, \psi_4$:

$$\begin{aligned}
&\xi_{01} + 2\xi_{11}\phi_2 + 3\xi_{12}\phi_3 + 3\eta_{12}\psi_3 + 4\xi_{13}\phi_4 + 4\eta_{13}\psi_4 = 0,\\
&\eta_{01} + 2\xi_{11}\psi_2 - 3\eta_{12}\phi_3 + 3\xi_{12}\psi_3 - 4\eta_{13}\phi_4 + 4\xi_{13}\psi_4 = 0,\\
&\xi_{02} + 2\xi_{12}\phi_2 - 2\eta_{12}\psi_2 + 3\xi_{22}\phi_3 + 4\xi_{23}\phi_4 + 4\eta_{23}\psi_4 = 0,\\
&\eta_{02} + 2\eta_{12}\phi_2 + 2\xi_{12}\psi_2 + 3\xi_{22}\psi_3 - 4\eta_{23}\phi_4 + 4\xi_{23}\psi_4 = 0,\\
&\xi_{03} + 2\xi_{13}\phi_2 - 2\eta_{13}\psi_2 + 3\xi_{23}\phi_3 - 3\eta_{23}\psi_3 + 4\xi_{33}\phi_4 = 0,\\
&\eta_{03} + 2\eta_{13}\phi_2 + 2\xi_{13}\psi_2 + 3\eta_{23}\phi_3 + 3\xi_{23}\psi_3 + 4\xi_{33}\psi_4 = 0.
\end{aligned}$$

* *Math. Ann.* vol. LXXII, p. 107 (1912).

† F. Höhndorf, *Zeits. f. ang. Math. u. Mech.* Bd. VI, S. 265 (1926). The conformal representation of a region which is nearly circular is discussed in a very general way by L. Bieberbach, *Sitzungsber. der preussischen Akademie der Wissenschaften*, S. 181 (1924).

Eliminating ϕ_4 and ψ_4 we obtain the equations

$$\begin{aligned}
&(0133) + 2\,(1133)\,\phi_2 + 3\,(1233)\,\phi_3 + 3\,[1233]\,\psi_3 = 0,\\
&[0133] + 2\,(1133)\,\psi_2 - 3\,[1233]\,\phi_3 + 3\,(1233)\,\psi_3 = 0,\\
&(0233) + 2\,(1233)\,\phi_2 - 2\,[1233]\,\psi_2 + 3\,(2233)\,\phi_3 = 0,\\
&[0233] + 2\,[1233]\,\phi_2 + 2\,(1233)\,\psi_2 + 3\,(2233)\,\psi_3 = 0,
\end{aligned}$$

where

$$(pqrs) = \xi_{pq}\xi_{rs} - \xi_{pr}\xi_{qs} - \eta_{pr}\eta_{qs}, \quad [pqrs] = \eta_{pq}\eta_{rs} + \xi_{pr}\eta_{qs} - \eta_{pr}\xi_{qs},$$

and these finally give the values

$$\nu N\phi_\nu = Z_{2\nu-3}, \quad \nu N\psi_\nu = Z_{2\nu-2}, \qquad \nu = 2, 3, 4,$$

where

$$\begin{aligned}
N &= (1233)^2 + [1233]^2 - (1133)\,(2233),\\
Z_1 &= (0133)\,(2233) - (0233)\,(1233) - [0233]\,[1233],\\
Z_2 &= [0133]\,(2233) + (0233)\,[1233] - [0233]\,(1233),\\
Z_3 &= [0133]\,[1233] + (0233)\,(1133) - (0133)\,(1233),\\
Z_4 &= [0233]\,(1133) - (0133)\,[1233] - [0133]\,(1233).
\end{aligned}$$

§ **4·92.** *Daniell's orthogonal potentials.* Consider a set of polynomials $p_0(z), p_1(z), \ldots$ defined by the equations*

$$p_n(z) = \frac{1}{\sqrt{D_{n-1}D_n}}\begin{vmatrix} (0,0) & (1,0) & (n,0)\\ (0,1) & (1,1) & (n,1)\\ \cdots & \cdots & \cdots\\ (0,n-1) & (1,n-1) & (n,n-1)\\ 1 & z & z^n \end{vmatrix},$$

where

$$D_n = \begin{vmatrix} (0,0) & (0,1) & (0,n)\\ (1,0) & (1,1) & (1,n)\\ \cdots & \cdots & \cdots\\ (n,0) & (n,1) & (n,n) \end{vmatrix}, \quad D_0 = 1,$$

and

$$(m, n) = \frac{1}{A}\iint z^m \bar{z}^n d\tau,$$

the integral being taken over the region to be mapped on a unit circle. A denotes here the area of the region and $d\tau$ an element of area enclosing the point z. These polynomials satisfy the orthogonal conditions

$$\begin{aligned}\frac{1}{A}\iint p_m(z)\,\overline{p_n(z)}\,d\tau &= 0, \quad m \neq n\\ &= 1, \quad m = n.\end{aligned}$$

A mapping function $f(z)$ which satisfies the conditions $f(a) = 0$, $f'(a) = 1$, is given formally† by the expansion

$$Sf(x) = \overline{p_0(a)}\int_a^z p_0(z')\,dz' + \overline{p_1(a)}\int_a^z p_1(z')\,dz' + \ldots,$$

where

$$S = \overline{p_0(a)}\,p_0(a) + \overline{p_1(a)}\,p_1(a) + \ldots.$$

* These equations are analogous to those used by Szegö.

† This series does not always represent an appropriate mapping function as may be seen from a consideration of the circular region with a cut extending half-way along a radius as in § 4·91.

To see this we write

$$f'(z) = a_0 p_0(z) + a_1 p_1(z) + a_2 p_2(z) + \dots,$$
$$\overline{f'(z)} = \bar{a}_0 \overline{p_0(z)} + \bar{a}_1 \overline{p_1(z)} + \bar{a}_2 \overline{p_2(z)} + \dots,$$

where $a_0, a_1, \dots; \bar{a}_0, \bar{a}_1, \dots$ are coefficients to be determined, so that the integral

$$\frac{1}{A}\iint f'(z)\,\overline{f'(z)}\,.\,d\tau = a_0\bar{a}_0 + a_1\bar{a}_1 + \dots$$

may be a minimum subject to the conditions

$$f'(a) = 1, \quad \overline{f'(a)} = 1. \qquad \dots\dots(A)$$

Differentiating with respect to $a_0, a_1, \dots; \bar{a}_0, \bar{a}_1, \dots$ in turn we find that

$$\bar{a}_n = k p_n(a), \quad n = 0, 1,$$
$$a_n = \overline{k p_n(a)}, \quad n = 0, 1, \dots$$

where k and $\bar{k}$ are Lagrangian multipliers to be determined by means of the equations (A). We easily find that

$$1 = kS, \quad \bar{k}S = 1,$$

and so

$$Sf'(z) = \overline{p_0(a)}\,p_0(z) + \overline{p_1(a)}\,p_1(z) + \dots.$$

If $p_n(z) = u_n - iv_n$, where u_n and v_n are real potentials which can be derived from a potential function ϕ_n by means of the equations

$$u_n = \frac{\partial \phi_n}{\partial x}, \quad v_n = \frac{\partial \phi_n}{\partial y},$$

we have

$$\frac{1}{A}\iint\left(\frac{\partial \phi_m}{\partial x}\frac{\partial \phi_n}{\partial x} + \frac{\partial \phi_m}{\partial y}\frac{\partial \phi_n}{\partial y}\right)d\tau = 0, \quad m \neq n$$
$$= 1, \quad m = n.$$

The potentials $\phi_0, \phi_1, \dots$ thus form an orthogonal system of the type considered by P. J. Daniell*. This definition of orthogonal potentials is easily extended by using a type of integral suggested by the appropriate problem in the Calculus of Variations.

For the unit circle itself the orthogonal polynomials are

$$p_n(z) = z^n.(n+1)^{\frac{1}{2}},$$

and the mapping function is consequently given by the equations

$$f'(z) = \left(\frac{1 - a\bar{a}}{1 - z\bar{a}}\right)^2,$$

$$f(z) = \frac{(1 - a\bar{a})(z - a)}{1 - z\bar{a}}.$$

§ **4·93**. *Fejér's theorem.* Let

$$Z = f(z) = a_0 + a_1 z + a_2 z^2 + \dots \qquad \dots\dots(1)$$

be the function mapping a region D in the Z-plane on the unit circle d with

* *Phil. Mag.* (7), vol. II, p. 247 (1926).

equation $|z| < 1$ in the z-plane. We shall suppose that D is bounded by a Jordan curve C and that $Z = H(\theta)$ is the point on C which corresponds to the point $z = e^{i\theta}$ on the unit circle c which bounds the region d. At this point, if the series converges

$$Z = a_0 + a_1 e^{i\theta} + a_2 e^{2i\theta} + \dots a_n e^{in\theta} + \dots$$
$$= u_0 + u_1 + u_2 + \dots \quad \text{say.} \qquad \dots\dots(2)$$

Now by Cauchy's form of Taylor's theorem

$$\frac{1}{2\pi i}\int \zeta^{-n-1} f(\zeta)\, d\zeta = a_n, \qquad n \geqslant 0,$$
$$= 0, \qquad n < 0,$$

where n is an integer and the contour is a simple one enclosing the origin and lying within the circle of convergence of the power series. On account of the continuity of $f(z)$ in d we may deform the contour until it becomes the same as c without altering the values of the integrals. Hence, writing $\zeta = e^{i\alpha}$ we get

$$2\pi a_n = \int_0^{2\pi} H(\alpha)\, e^{-in\alpha} \qquad n \geqslant 0, \qquad 0 = \int_0^{\infty} H(\alpha)\, e^{in\alpha}\, d\alpha.$$

These equations show that the series (2) is the Fourier series of the continuous function $H(\theta)$ and is consequently summable $(C, 1)$ (§ 1·16).

Now consider a circle $|z| = \rho$ where $\rho < 1$. The function $f(z)$ maps the interior of this circle on the interior of a region R whose area A is, by § 4·91, equal to the convergent series

$$\pi[|a_1|^2 \rho^2 + 2|a_2|^2 \rho^4 + \dots].$$

This area A is bounded for all values of ρ and is less than B, say,

$$\therefore\ \pi[|a_1|^2 \rho^2 + 2|a_2|^2 \rho^4 + \dots n|a_n|^2 \rho^{2n}] < B$$

for $0 < \rho \leqslant 1$ and so

$$\pi[|a_1|^2 + 2|a_2|^2 + \dots n|a_n|^2] < B.$$

This inequality shows that the series $\Sigma n|a_n|^2$ is convergent. Now this property combined with the fact that (2) is summable $(C, 1)$ is sufficient to show that the series (2) converges. Writing

$$s_n = u_0 + u_1 + \dots u_n, \quad (n+1) S_n = s_0 + s_1 + \dots s_n$$

we have $S_n = s_n - \sigma_n$ where $(n+1)\sigma_n = u_1 + 2u_2 + \dots nu_n$. It is sufficient then to show that $\sigma_n \to 0$ as $n \to \infty$. With the notation $v_n = |u_n|$ we have the inequality

$$[(m+1) v_{m+1} + \dots (m+p) v_{m+p}]^2$$
$$< [(m+1) + \dots (m+p)][(m+1) v^2_{m+1} + \dots (m+p) v^2_{m+p}].$$

The first factor on the right is less than $1 + 2 + \dots (m+p)$ which is less than $(m+p+1)^2$. Also, since the series $\Sigma n v_n^2$ converges we can choose

m so large that the second factor is less than ϵ^2 whatever p may be. We may now write $n = m + p$,

$$\sigma_n^* = \frac{v_1 + 2v_2 + \ldots mv_m}{m+p+1} + \frac{(m+1)\,v_{m+1} + \ldots\,(m+p)\,v_{m+p}}{m+p+1},$$

where the second term on the right is less than ϵ and since p is at our disposal we may choose it so large that the first term on the right is less than ϵ. Hence we can choose n so large that $|\,\sigma_n\,| < \sigma_n^* < 2\epsilon$ and so $|\,\sigma_n\,| \to 0$ as $n \to \infty$.

It follows then that the series (2) converges and that the co-ordinates (X, Y) of a point on a simple closed Jordan curve can be expressed as Fourier series with θ as parameter. The theorem implies that the series (1) converges uniformly throughout d and that the mapping by means of the function $f(z)$ may be extended to regions which are slightly larger than d and D.

Reference is made to Fejér's papers, *Münchener Sitzungsber.* (1910), *Comptes Rendus*, t. CLVI, p. 46 (1913) for further developments. Also to the book by P. Montel and J. Barbotte, *Leçons sur les familles normales de fonctions analytiques* (Gauthier-Villars, Paris, 1927), p. 118.

CHAPTER V

EQUATIONS IN THREE VARIABLES

§ 5·11. *Simple solutions and their generalisation.* Commencing as before with some applications of the simple solutions we consider the equation

$$\rho \frac{\partial^2 v}{\partial t^2} = \mu \left(\frac{\partial^2 v}{\partial x^2} + \frac{\partial^2 v}{\partial z^2} \right)$$

of the propagation of Love-waves in the direction of the x-axis. If now ρ and μ have the values ρ_0, μ_0 respectively for $z > 0$, and the values ρ_1, μ_1 respectively for $z < 0$, it is useful to consider a solution of type

$$v = v_0 = (A \cos sz + B \sin sz) \sin \kappa (x - qt), \quad z > 0,$$

$$v = v_1 = Ce^{hz} \sin \kappa (x - qt), \quad z < 0,$$

where the constants are connected by the relations

$$\rho_0 \kappa^2 q^2 = \mu_0 (s^2 + \kappa^2), \quad \rho_1 \kappa^2 q^2 = \mu_1 (\kappa^2 - h^2).$$

In the expression for v_2 we take $h > 0$ so that there is no deep penetration of the waves.

The boundary conditions are

$$\mu_0 \frac{\partial v_0}{\partial z} = 0, \text{ when } z = a,$$

$$\left. \begin{aligned} \mu_0 \frac{\partial v_0}{\partial z} &= \mu_1 \frac{\partial v_1}{\partial z} \\ v_0 &= v_1 \end{aligned} \right\}, \text{ when } z = 0.$$

These equations give

$$A \sin sa = B \cos sa,$$

$$A = C,$$

$$\mu_0 sB = \mu_1 hC.$$

Putting $\mu_0 = c_0^2 \rho_0$, $\mu_1 = c_1^2 \rho_1$ we have with $s = \kappa \operatorname{cosech} \omega_0$, $h = \kappa \operatorname{sech} \omega_1$,

$$c_0 = q \tanh \omega_0, \quad c_1 = q \coth \omega_1,$$

$$\cosh \omega_1 = \frac{\mu_1}{\mu_0} \sinh \omega_0 \cot (a\kappa \operatorname{cosech} \omega_0) = \left(1 - \frac{c_0^2}{c_1^2} \tanh^2 \omega_0\right)^{-\frac{1}{2}},$$

and it is readily seen that there are no waves of the present type unless $c_0 < c_1$.

Matuzawa has examined the case of three media arranged so that in his notation

$$v = v_1 = A_1 (e^{s_1 z} + e^{-s_1 z}) \cos (pt + fx), \quad \rho = \rho_1, \quad \mu = \mu_1, \quad 0 > z > -h,$$

$$v = v_2 = (A_2 e^{s_2 z} + Be^{-s_2 z}) \cos (pt + fx), \quad \rho = \rho_2, \quad \mu = \mu_2, \quad -h > z > -H,$$

$$v = v_3 = A_3 e^{s_3 z} \cos (pt + fx), \quad \rho = \rho_3, \quad \mu = \mu_3, \quad -H > z.$$

The boundary conditions

$$v_1 = v_2, \quad \mu_1 \frac{\partial v_1}{\partial z} = \mu_2 \frac{\partial v_2}{\partial z} \text{ at } z = -h,$$

$$v_2 = v_3, \quad \mu_2 \frac{\partial v_2}{\partial z} = \mu_3 \frac{\partial v_3}{\partial z} \text{ at } z = -H,$$

give

$$A_1 (e^{-s_1 h} + e^{s_1 h}) = A_2 e^{-s_2 h} + B_2 e^{s_2 h},$$

$$A_1 \mu_1 s_1 (e^{-s_1 h} - e^{s_1 h}) = A_2 \mu_2 s_2 e^{-s_2 h} - B_2 \mu_2 s_2 e^{s_2 h},$$

$$A_2 e^{-s_2 H} + B_2 e^{s_2 H} = A_3 e^{-s_3 H},$$

$$A_2 \mu_2 s_2 e^{-s_2 H} - B_2 \mu_2 s_2 e^{s_2 H} = A_3 \mu_3 s_3 e^{-s_3 H}.$$

Eliminating A_1, A_2, A_3, B_2 and writing $\tau_1 = \tanh s_1 h$, $\tau_2 = \tanh s_2 h$, $T_2 = \tanh s_2 H$, we obtain the equation

$$\frac{\mu_1 s_1 \tau_1}{\mu_2 s_2} = - \frac{T_2 - \tau_2 + \dfrac{\mu_3 s_3}{\mu_2 s_2}(1 - \tau_2 T_2)}{1 - \tau_2 T_2 + \dfrac{\mu_3 s_3}{\mu_2 s_2}(T_2 - \tau_2)}.$$

The cases s_1, s_2, s_3 all real and s_2 imaginary, s_1, s_3 real, are not compatible with an equation of this type. When s_2 is real it appears that there is only one value of s_1 and this is an imaginary quantity; when s_2 is an imaginary quantity it appears that there are two possible values of s_1 and these are both imaginary.

Matuzawa has examined the six possible cases

A	B	C	D	E	F
$c_1 < c_2 < c_3$	$c_1 < c_3 < c_2$	$c_2 < c_1 < c_3$	$c_3 < c_1 < c_2$	$c_2 < c_3 < c_1$	$c_3 < c_2 < c_1$

and concludes that in cases B, D and F there is no solution.

§ 5·12. The simple solutions considered so far correspond to the case of travelling waves. We shall next consider a case of standing waves and shall take the equation of a vibrating membrane

$$\frac{\partial^2 w}{\partial t^2} = c^2 \left(\frac{\partial^2 w}{\partial x^2} + \frac{\partial^2 w}{\partial y^2} \right).$$

Let the boundary of the membrane consist of the axes of co-ordinates and the lines $x = a$, $y = b$. The expression

$$w = \sin \frac{m\pi x}{a} \sin \frac{n\pi y}{b} \{A_{mn} \cos pt + B_{mn} \sin pt\}$$

satisfies the condition $w = 0$ on the boundary and is a simple solution of the differential equation if

$$p^2 = c^2 \pi^2 \left(\frac{m^2}{a^2} + \frac{n^2}{b^2} \right).$$

This equation gives the possible frequencies of vibration, m and n being integers. A more general type of vibration may be obtained by summing with respect to m and n from $m = 1$ to ∞ and $n = 1$ to ∞.

The resulting double Fourier series is usually a solution which is sufficiently general to make it possible to satisfy assigned initial conditions

$$w = w_0, \quad \frac{\partial w}{\partial t} = \dot{w}_0 \text{ for } t=0,$$

by using coefficients A_{mn}, B_{mn} determined by Fourier's rule

$$A_{mn} = \frac{4}{ab}\int_0^a\int_0^b w_0 \sin\frac{m\pi x}{a}\sin\frac{n\pi y}{b}\,dx\,dy,$$

$$B_{mn} = \frac{4}{abp}\int_0^a\int_0^b \dot{w}_0 \sin\frac{m\pi x}{a}\sin\frac{n\pi y}{b}\,dx\,dy.$$

EXAMPLE

Find the nodal lines of the solutions

$$w = A\sin\frac{\pi x}{a}\sin\frac{\pi y}{a}\left(\cos\frac{\pi x}{a}+\cos\frac{\pi y}{a}\right)\cos pt,$$

$$w = A\sin\frac{2\pi x}{a}\sin\frac{2\pi y}{a}\cos pt,$$

$$w = C\left\{\sin\frac{3\pi x}{a}\sin\frac{\pi y}{a} - \sin\frac{\pi x}{a}\sin\frac{3\pi y}{a}\right\}\cos pt,$$

which are suitable for the representation of the vibration of a square if p has an appropriate value in each case.

§ **5·13**. *Reflection and refraction of electromagnetic waves.* In a non-conducting medium the equations of the electromagnetic field are*

$$\operatorname{curl} H = \dot{D}/c, \quad \operatorname{div} D = 0,$$
$$\operatorname{curl} E = -\dot{B}/c, \quad \operatorname{div} B = 0,$$

and the constitutive relations are

$$D = \kappa E, \quad B = \mu H,$$

where the coefficients κ and μ can be regarded as constants if the material is homogeneous and the frequency of the waves is not too high.

If all the field vectors are independent of z, their components satisfy the two-dimensional wave-equation

$$\frac{\partial^2\theta}{\partial x^2}+\frac{\partial^2\theta}{\partial y^2} = \frac{1}{V^2}\frac{\partial^2\theta}{\partial t^2},$$

where $$V^2 = c^2/\kappa\mu = 1/s^2, \text{ say.}$$

The permeability of all substances is practically unity for frequencies as great as that of light. Hence for light waves it is permissible to write

$$V = c/\sqrt{\kappa},$$

and in this case we may also write $\kappa = n^2$, where n is the index of refraction of the medium.

Let us now suppose that the medium with the constants (κ_1, μ_1) is on

* For convenience we denote a partial differentiation with respect to the time t by a dot.

the side $x < 0$ of the plane $x = 0$, and that on the other side of this plane there is a medium with constants (κ_2, μ_2).

We shall suppose that when $x < 0$ there is an incident and a reflected wave, but that for $x > 0$ there is only a transmitted wave. We shall suppose further that the electric vector in all the waves is parallel to the axis of z, then with a view of being able to satisfy the boundary conditions we assume

$$E_z = A_1 e_1 + A_1' e_1' \quad (x < 0), \quad E_z = A_2 e_2 \quad (x > 0),$$

where e_1, e_1', e_2 denote respectively the exponentials

$$e_1 = e^{i\omega[s_1(x\cos\phi_1 + y\sin\phi_1) - t]}, \quad e_1' = e^{i\omega[s_1(y\sin\phi_1 - x\cos\phi_1) - t]},$$

$$e_2 = e^{i\omega[s_2(x\cos\phi_2 + y\sin\phi_2) - t]}.$$

The corresponding expressions for the components of H are

$$H_x = (cs_1/\mu_1)(A_1 e_1 + A_1' e_1')\sin\phi_1, \quad x < 0,$$
$$H_x = (cs_2/\mu_2) A_2 e_2 \sin\phi_2, \quad x > 0,$$
$$H_y = (cs_1/\mu_1)(A_1' e_1' - A_1 e_1)\cos\phi_1, \quad x < 0,$$
$$H_y = -(cs_2/\mu_2) A_2 e_2 \cos\phi_2, \quad x > 0.$$

The boundary conditions are that the tangential components of E and H are to be continuous. These conditions give

$$A_1 + A_1' = A_2,$$
$$\sin\phi_1 . \mu_2 s_1 (A_1 + A_1') = \mu_1 s_2 A_2 \sin\phi_2,$$
$$\cos\phi_1 . \mu_2 s_1 (A_1 - A_1') = \mu_1 s_2 A_2 \cos\phi_2.$$

Hence
$$\frac{\sin\phi_1}{\sin\phi_2} = \frac{\mu_1 s_2}{\mu_2 s_1} = \frac{n_2}{n_1} \quad \text{when } \mu_1 = \mu_2.$$

This is the familiar relation of Snell. Writing

$$A' = RA, \quad A_2 = TA,$$

where R and T are the coefficients of reflection and transmission respectively, we have

$$\frac{1-R}{1+R} = \frac{\sin\phi_1\cos\phi_2}{\sin\phi_2\cos\phi_1}, \quad R = -\frac{\sin(\phi_1 - \phi_2)}{\sin(\phi_1 + \phi_2)},$$

$$\frac{2-T}{T} = \frac{\sin\phi_1\cos\phi_2}{\sin\phi_2\cos\phi_1}, \quad T = \frac{2\sin\phi_2\cos\phi_1}{\sin(\phi_1 + \phi_2)}.$$

In the case when the electric vector is in the plane of incidence we write

$$E_x = -(C_1 e_1 + C_1' e_1')\sin\phi_1, \qquad E_x = -C_2 e_2 \sin\phi_2,$$
$$E_y = (C_1 e_1 - C_1' e_1')\cos\phi_1 \quad (x < 0), \qquad E_y = C_2 e_2 \cos\phi_2 \quad (x > 0),$$
$$H_z = (cs_1/\mu_1)(C_1 e_1 + C_1' e_1'), \qquad H_z = (cs_2/\mu_2) C_2 e_2,$$

and the boundary conditions give

$$(s_1/\mu_1)(C_1 + C_1') = s_2 C_2/\mu_2,$$
$$(C_1 - C_1')\cos\phi_1 = C_2\cos\phi_2,$$
$$\kappa_1 (C_1 + C_1')\sin\phi_1 = \kappa_2 C_2 \sin\phi_2.$$

The third equation implies that the x-component of D is continuous at $x = 0$. These equations give

$$\frac{\sin\phi_1}{\sin\phi_2} = \frac{\kappa_2\mu_2 s_1}{\kappa_1\mu_1 s_2} = \frac{n_2}{n_1}, \quad \text{when } \mu_1 = \mu_2.$$

Thus Snell's law holds as before. If we further write

$$C_1' = \rho C_1, \quad C_2 = \tau C_2,$$

so that ρ, τ are the coefficients of reflection and transmission respectively, we have

$$\rho = \frac{\tan(\phi_1 - \phi_2)}{\tan(\phi_1 + \phi_2)},$$

$$\tau = \frac{2\sin\phi_2\cos\phi_1}{\sin(\phi_1 + \phi_2)\cos(\phi_1 - \phi_2)}.$$

Of the four quantities R, T, ρ, τ only one can vanish, viz. the polarizing angle Φ_1 is defined as the angle of incidence for which $\rho = 0$. This angle is given by the equation $\tan(\phi_1 + \phi_2) = \infty$, and so

$$\tan\Phi_1 = n_2/n_1.$$

When the incident light is unpolarized it consists of a mixture of waves in some of which E is parallel to the axis of z and in the others H is parallel to the axis of z. When such light strikes the surface $x = 0$ at the polarizing angle the waves of the second kind are transmitted *in toto*, and so the reflected light consists merely of waves of the first kind and is thus linearly polarized.

Reflection and refraction of plane waves of sound. Consider a homogeneous medium whose natural density is ρ_0. When waves of sound traverse the medium the density ρ and pressure p at an arbitrary point $Q(x, y, z)$ have at time t new values which may be expressed in the forms

$$\rho = \rho_0(1 + s), \quad p = p_0(1 + As),$$

where p_0 is the undisturbed pressure and A is a coefficient depending on the compressibility. The quantity s is called the condensation and will be assumed to be so small that its square may be neglected.

We now suppose that the velocity components (u, v, w) of the medium at the point Q can be derived from a velocity potential ϕ which depends on the time. Bernoulli's integral

$$\int\frac{dp}{\rho} + \frac{\partial\phi}{\partial t} = \text{constant}$$

then gives the approximate equation

$$\frac{\partial\phi}{\partial t} + c^2 s = 0,$$

where $c^2 = p_0 A/\rho_0 = (dp/d\rho)_0$ is the local velocity of sound and is constant since A and ρ_0 are constants. The equation of continuity

$$\frac{\partial\rho}{\partial t} + \frac{\partial}{\partial x}(\rho u) + \frac{\partial}{\partial y}(\rho v) + \frac{\partial}{\partial z}(\rho w) = 0,$$

and the equations $u = \dfrac{\partial \phi}{\partial x}$, $v = \dfrac{\partial \phi}{\partial y}$, $w = \dfrac{\partial \phi}{\partial z}$, when u, v, w are small, give the wave-equation

$$\frac{\partial^2 \phi}{\partial t^2} = c^2 \nabla^2 \phi.$$

The conditions to be satisfied at the surface separating two media are that the pressure and the normal component of velocity must be continuous. On account of Bernoulli's equation the continuity of pressure implies that $\rho \dfrac{\partial \phi}{\partial t}$ is continuous.

Let us now consider the case in which two media are separated by the plane $x = 0$. We shall suppose that in the medium on the left there is an initial train of plane waves represented by the velocity potential

$$\phi_0 = a_0 e^{in(t - \xi x - \eta y)},$$

and that these waves are partly reflected and partly transmitted. We therefore assume that

$$\phi_1 = a_0 e^{in(t-\xi x - \eta y)} + a_1 e^{in(t+\xi x - \eta y)}, \quad x < 0,$$
$$\phi_2 = a_2 e^{in(t-\zeta x - \eta y)}, \quad x > 0.$$

The boundary conditions give

$$\rho_1 (a_0 + a_1) = \rho_2 a_2,$$
$$\xi (a_0 - a_1) = \zeta a_2,$$

where ρ_1 and ρ_2 are the values of the natural density for $x < 0$ and $x > 0$ respectively. If c_1 and c_2 are the two associated velocities of sound, we must have

$$c_1 \xi = \cos \alpha_1, \quad c_1 \eta = \sin \alpha_1,$$
$$c_2 \zeta = \cos \alpha_2, \quad c_2 \eta = \sin \alpha_2.$$

Therefore

$$a_1 = a_0 \frac{c_2 \rho_2 \cos \alpha_1 - c_1 \rho_1 \cos \alpha_2}{c_2 \rho_2 \cos \alpha_1 + c_1 \rho_1 \cos \alpha_2} = a_0 \frac{\rho_2 \cot \alpha_1 - \rho_1 \cot \alpha_2}{\rho_2 \cot \alpha_1 + \rho_1 \cot \alpha_2},$$

$$a_2 = a_0 \frac{2 c_2 \rho_1 \cos \alpha_1}{c_2 \rho_2 \cos \alpha_1 + c_1 \rho_1 \cos \alpha_2} = a_0 \frac{2 \rho_1 \cot \alpha_1}{\rho_2 \cot \alpha_1 + \rho_1 \cot \alpha_2}.$$

The equation $\qquad c_2 \sin \alpha_1 = c_1 \sin \alpha_2$

gives a law of refraction analogous to Snell's law.

When the second medium ends at $x = b$, where $b > 0$, and for $x > b$ the medium is the same as the first, there are three forms for the velocity potential:

$$\phi_1 = a_0 e^{in(t-\xi x-\eta y)} + a_1 e^{in(t+\xi x-\eta y)}, \quad x < 0,$$
$$\phi_2 = a_2 e^{in(t-\zeta x-\eta y)} + a_3 e^{in(t+\zeta x-\eta y)}, \quad b > x > 0,$$
$$\phi_3 = a_4 e^{in(t-\xi x-\eta y)},$$

and the boundary conditions give

$$\rho_1 (a_0 + a_1) = \rho_2 (a_2 + a_3), \quad \xi (a_0 - a_1) = \zeta (a_2 - a_3),$$
$$\rho_2 a_2 e^{-in\zeta b} + \rho_2 a_3 e^{in\zeta b} = \rho_1 a_4 e^{-in\xi b},$$
$$\zeta a_2 e^{-in\zeta b} - \zeta a_3 e^{in\zeta b} = \xi a_4 e^{-in\xi b}.$$

Therefore

$$a_2 + a_3 = e^{-in\xi b}\left[\frac{\rho_1 a_4}{\rho_2}\cos(nb\zeta) + \frac{i\xi a_4}{\zeta}\sin(nb\zeta)\right],$$

$$a_2 - a_3 = e^{-in\xi b}\left[\frac{\rho_1 a_4}{\rho_2}\,i\sin(nb\zeta) + \frac{\xi a_4}{\zeta}\cos(nb\zeta)\right],$$

$$a_0 e^{in\xi b} = a_4\left[\cos(nb\zeta) + \tfrac{1}{2}i\left\{\frac{\xi\rho_2}{\zeta\rho_1} + \frac{\zeta\rho_1}{\xi\rho_2}\right\}\sin(nb\zeta)\right],$$

$$a_1 e^{in\xi b} = \tfrac{1}{2}ia_4\left\{\frac{\xi\rho_2}{\zeta\rho_1} - \frac{\zeta\rho_1}{\xi\rho_2}\right\}\sin(nb\zeta).$$

It should be noticed that these equations give

$$|a_0|^2 - |a_1|^2 = |a_4|^2,$$

$$\rho_2|a_2|^2 - \rho_2|a_3|^2 = \rho_1|a_4|^2\,\frac{\cot\alpha_1}{\cot\alpha_2},$$

and the first of these equations indicates that the sum of the energies per wave-length of the reflected and transmitted waves in the first medium is equal to the energy per wave-length in the incident wave.

It should be noticed that if $\sin(nb\zeta) \neq 0$, the condition for no reflected wave ($a_1 = 0$) is $\xi\rho_2 = \zeta\rho_1$, and is independent of the thickness of the second medium.

We have assumed so far that there is a real angle α_2 which satisfies the equation $c_1 \sin\alpha_2 = c_2 \sin\alpha_1$, but if $c_2 > c_1$ it may happen that there is no such angle. If the value of $\sin\alpha_2$ given by this equation is greater than unity, $\cos\alpha_2$ will be imaginary and the solution appropriate for a single surface of separation ($x = 0$) will be of type

$$\phi_1 = a_0 e^{in(t-\xi x-\eta y)} + a_1 e^{in(t+\xi x-\eta y)}, \quad x < 0,$$

$$\phi_2 = a_2 e^{in(t-\eta y)-\theta x}, \quad x > 0,\ \theta > 0.$$

In this case there is no proper wave in the second medium, and on account of the exponential factor $e^{-\theta x}$ the intensity of the disturbance falls off very rapidly as x increases. The corresponding solution of the problem for the case in which the second medium is of thickness b is obtained from the formulae already given by replacing ζ by $-i\theta$. It is thus found that

$$2a_2 = e^{nb(\theta-i\xi)}\left[\frac{\rho_1}{\rho_2} + i\,\frac{\xi}{\theta}\right]a_4,$$

$$2a_3 = e^{-nb(\theta+i\xi)}\left[\frac{\rho_1}{\rho_2} - i\,\frac{\xi}{\theta}\right]a_4,$$

$$a_0 e^{in\xi b} = a_4\left[\cosh(nb\theta) + \tfrac{1}{2}i\left(\frac{\xi\rho_2}{\theta\rho_1} - \frac{\theta\rho_1}{\xi\rho_2}\right)\sinh(nb\theta)\right],$$

$$a_1 e^{in\xi b} = \tfrac{1}{2}ia_4\left\{\frac{\xi\rho_2}{\theta\rho_1} + \frac{\theta\rho_1}{\xi\rho_2}\right\}\sinh(nb\theta).$$

The coefficient a_3 of the disturbance of type $e^{in(t-\eta y)+\theta x}$ which increases in intensity with x is seen to be very small so that this disturbance is small even when $x = b$.

In the present case the reflection is not quite total, for some sound reaches the medium $x > b$. The change of phase on reflection is easily calculated by expressing a_1/a_0 in the form $Re^{i\alpha}$.

Let us now consider briefly the case when $nb\zeta = k\pi$, where k is an integer. In this case $\sin(nb\zeta) = 0$; there is no reflected wave and the formulae become simply

$$\phi_1 = a_0 e^{in(t-\xi x-\eta y)}, \qquad x < 0,$$
$$\phi_2 = a_0 (\rho_1/\rho_2)\, e^{in(t-\zeta x-\eta y)} + a_3 e^{in(t-\eta y)} \sin(\zeta n x), \quad b > x > 0,$$
$$\phi_3 = a_0 e^{in(t-\xi x-\eta y)}, \qquad x > b.$$

It will be noticed that the value of n is precisely one for which there is a potential ϕ fulfilling the conditions $\frac{\partial \phi}{\partial t} = 0$ for $x = 0$ and $x = b$.

The slab of material between $x = 0$ and $x = b$ can be regarded as in a state of free vibration of such an intensity that there is no interference with the travelling waves.

The absorption of plane waves of sound by a slab of soft material has been treated by Rayleigh* by an ingenious approximate method in which the material is regarded as perforated by a large number of cylindrical holes with axes parallel to the axis of x and the velocity potential within these holes is supposed to satisfy an equation of type

$$c^2 \nabla^2 \phi = \frac{\partial^2 \phi}{\partial t^2} + h \frac{\partial \phi}{\partial t},$$

where h is a positive constant. The new term is supposed to take into consideration the effect of dissipation.

At a very short distance from the mouth ($x = 0$) of a channel it is assumed that the terms $\frac{\partial^2 \phi}{\partial y^2}$ and $\frac{\partial^2 \phi}{\partial z^2}$ may be neglected and that the solution is effectively of type

$$\phi = e^{int} \{a' \cos k'x + b' \sin k'x\},$$

where

$$c^2 k'^2 = n^2 - inh.$$

If the channel is closed at $x = b$, we have $\frac{\partial \phi}{\partial x} = 0$ there, and so we may write

$$\phi = A' e^{int} \cos k'(x - b).$$

When x is very small

$$u = \frac{\partial \phi}{\partial x} = k' A' e^{int} \sin(k'b),$$

$$c^2 s = -\frac{\partial \phi}{\partial t} = -inA' e^{int} \cos(k'b),$$

$$\frac{u}{c^2 s} = \frac{ik'}{n} \tan(k'b).$$

* *Phil. Mag.* (6), vol. XXXIX, p. 225 (1920); *Papers*, vol. VI, p. 662.

If, for $x < 0$, we adopt the same expression as before, viz.

$$\phi_1 = a_0 e^{in(t-\xi x-\eta y)} + a_1 e^{in(t+\xi x-\eta y)},$$

we have
$$\left(\frac{u_1}{c^2 s_1}\right)_{x=0} = \frac{\xi (a_0 - a_1)}{a_0 + a_1}.$$

Now let σ be the perforated area of the slab and σ' the area free from holes. The transition from one state of motion on the side $x < 0$ to the other state on the side $x > 0$ is assumed to be of such a nature that

$$(\sigma + \sigma') u_1 = \sigma u,$$

$$s_1 = s.$$

These equations give the relation

$$\frac{a_0 - a_1}{a_0 + a_1} = \frac{ik'}{n\xi} \frac{\sigma}{\sigma + \sigma'} \tan (k'b)$$

for the determination of the intensity of the reflected wave. When $h = 0$, we have $| a_1 | = | a_0 |$ and the reflection is total, as it should be. When $\sigma = 0$, $a_1 = a_0$, and there is again total reflection. On the other hand, if $\sigma' = 0$, the partitions between the channels being infinitely thin, we have, when $h = 0$,

$$a_1 = a_0 \frac{n\xi \cos (k'b) - ik' \sin (k'b)}{n\xi \cos (k'b) + ik' \sin (k'b)} = a_0 \frac{\cos \alpha_1 \cos (k'b) - i \sin (k'b)}{\cos \alpha_1 \cos (k'b) + i \sin (k'b)}.$$

In the case of normal incidence $\alpha_1 = 0$, $a_1 = a_0 e^{-2ik'b}$, and the effect is the same as if the wall were transferred to $x = b$. When h is very small but the term k_2 in the complex expression $k' = k_1 + ik_2$ is so large that the vibrations in the channels are sensibly extinguished before the stopped end is reached, we may write

$$\cos (ik_2 b) = \tfrac{1}{2} e^{k_2 b}, \quad \sin (ik_2 b) = \tfrac{1}{2} i e^{k_2 b}, \quad \tan (k'b) = - i,$$

and the formula becomes

$$\frac{a_0 - a_1}{a_0 + a_1} = \frac{\sigma}{(\sigma + \sigma') \cos \alpha_1}.$$

EXAMPLES

1. In the reflection of plane waves of sound at a plane interface between two media the velocity of the trace of a wave-front on the plane interface is the same in the two media. [Rayleigh.]

2. When the velocity of sound at altitude z is c and the wind velocity has components $(u, v, 0)$, the axis of z being vertical, the laws of refraction are expressed by the equations

$$\phi = \phi_0, \quad c \operatorname{cosec} \theta + u \cos \phi + v \sin \phi = c_0 \operatorname{cosec} \theta_0 + u_0 \cos \phi_0 + v_0 \sin \phi_0 = \lambda, \text{ say,}$$

where (θ, ϕ) are the spherical polar co-ordinates of the wave-normal relative to the vertical polar axis and the suffix 0 is used to indicate values of quantities at the level of the ground.

3. Prove that the ray-velocity (the rays being defined as the bicharacteristics as in § 1·93) is obtained by compounding the wind velocity with a velocity c directed along the wave-normal. See also Ex. 1, § 12·1.

4. The range and time of passage of sound which travels up into the air and down again are given by the equations

$$x = 2\int_0^Z (c^2 \cos\phi + us)\, dz/\Gamma,$$

$$y = 2\int_0^Z (c^2 \sin\phi + vs)\, dz/\Gamma,$$

$$t = 2\int_0^Z s\, dz/\Gamma,$$

where $$s = \lambda - u\cos\phi - v\sin\phi,\quad \Gamma = s\,(s^2 - c^2)^{\frac{1}{2}},$$

and Z is defined by the equation $s = c$.

§ 5·21. *Some problems in the conduction of heat.* Our first problem is to find a solution of the equation

$$\frac{\partial\theta}{\partial t} = \kappa\left(\frac{\partial^2\theta}{\partial x^2} + \frac{\partial^2\theta}{\partial y^2}\right), \qquad \text{......(A)}$$

which will satisfy the conditions

$$\theta = \exp\,[ip\,(t - x/c)] \text{ when } y = 0,\quad \theta = 0 \text{ when } y = \infty.$$

Assuming as a trial solution

$$\exp\,[ip\,(t - x/c - y/b) - ay],$$

we find that $$ip = \kappa\left[\left(a + \frac{ip}{b}\right)^2 - \frac{p^2}{c^2}\right].$$

Therefore $$b = 2a\kappa,\quad p^2\left(\frac{1}{b^2} + \frac{1}{c^2}\right) = a^2.$$

The result tells us that if the temperature at the ground ($y = 0$) varies in a manner corresponding to a travelling periodic disturbance, the variation of temperature at depth y will also correspond to a periodic disturbance travelling with the same velocity but this disturbance lags behind the other in phase and has a smaller amplitude.

The solution may be generalised by writing

$$b = c\tan\phi,\quad a = (c/2\kappa)\tan\phi,\quad p = (c^2/2\kappa)\tan\phi\sin\phi,$$

$$\theta = \int_0^{\frac{\pi}{2}} f(\phi)\, d\phi . \exp\,[(ic/2\kappa)(ct - x)\tan\phi\sin\phi - (cy/2\kappa)(\tan\phi + i\sin\phi)],$$

where c is regarded as a constant independent of ϕ and $f(\phi)$ is a suitable arbitrary function.

If we wish this solution to satisfy the conditions

$$\theta = g\,(ct - x) \text{ when } y = 0,\quad \theta = 0 \text{ when } y = \infty,$$

the function $f(\phi)$ must be derived from the integral equation

$$g\,(u) = \int_0^{\frac{\pi}{2}} f(\phi)\, d\phi . \exp\,[(icu/2\kappa)\tan\phi\sin\phi],\quad (-\infty < u < \infty).$$

When the function $g(u)$ is of a suitable type, Fourier's inversion formula gives

$$f(\phi) = (c/4\pi\kappa)\int_{-\infty}^{\infty}(1+\sec^2\phi)\sin\phi . g(u)\,du . \exp[-(icu/2\kappa)\tan\phi\sin\phi],$$
$$0 < \phi < \pi/2.$$

In particular, if

$$g(u) = (2\kappa/cu)\sin[\tan\alpha\sin\alpha\,(cu/2\kappa)],$$

where α is a constant, we have

$$\theta = \int_0^\alpha \sin\phi\,(1+\sec^2\phi)\,d\phi . e^{-(cy/2\kappa)\tan\phi}$$
$$\times \sin[(c/2\kappa)\{(ct-x)\tan\phi\sin\phi - y\sin\phi\}].$$

Another solution may be obtained by making c a function of ϕ and then integrating; for instance, if $c = 2\kappa\cos\phi$ we obtain the solution

$$\theta = \int_0^{\frac{\pi}{2}} f(\phi)\,d\phi . \exp[i\sin^2\phi\,(2\kappa t\cos\phi - x) - y(\sin\phi + i\sin\phi\cos\phi)].$$

It should be noticed that the definite integral

$$\Theta(x, y, z, t) = \int_0^{\frac{\pi}{2}} f(\phi)\,d\phi . \exp[i\sin^2\phi\,(2\kappa t\cos\phi - x) - z\sin\phi - iy\sin\phi\cos\phi]$$

is a solution of the two partial differential equations

$$\frac{\partial^2\Theta}{\partial x^2} + \frac{\partial^2\Theta}{\partial y^2} + \frac{\partial^2\Theta}{\partial z^2} = 0, \quad \frac{\partial\Theta}{\partial t} = 2\kappa\frac{\partial^2\Theta}{\partial y\partial z}, \qquad \text{......(B)}$$

and is of such a nature that the function

$$\theta(x, y, t) = \Theta(x, y, y, t)$$

is a solution of equation (A). It is easy to verify, in fact, that if $\Theta(x, y, z, t)$ is any solution of equations (B) the associated function $\theta(x, y, t)$ is a solution of (A), for we have

$$\frac{\partial^2\theta}{\partial x^2} + \frac{\partial^2\theta}{\partial y^2} = \frac{\partial^2\Theta}{\partial x^2} + \frac{\partial^2\Theta}{\partial y^2} + \frac{\partial^2\Theta}{\partial z^2} + 2\frac{\partial^2\Theta}{\partial y\partial z}$$
$$= 2\frac{\partial^2\Theta}{\partial y\partial z} = \frac{1}{\kappa}\frac{\partial\Theta}{\partial t} = \frac{1}{\kappa}\frac{\partial\theta}{\partial t}.$$

Again, if we take $c = 2\kappa\cot\phi$, we obtain an integral

$$\theta(x, y, t) = \int_0^{\frac{\pi}{2}} f(\phi)\,d\phi . \exp[i(2\kappa t\cos\phi - x\sin\phi) - y(1 + i\cos\phi)],$$

which is a solution of (A), and the associated integral

$$\Theta(x, y, z, t) = \int_0^{\frac{\pi}{2}} f(\phi)\,d\phi . \exp[i(2\kappa t\cos\phi - x\sin\phi - y\cos\phi) - z] \qquad \text{......(C)}$$

is likewise a solution of the equations (B). Indeed, if c is any suitable function of ϕ the integral

$$\Theta(x, y, z, t) = \int_0^{\frac{\pi}{2}} f(\phi)\, d\phi . \exp [(ic/2\kappa)(ct - x) \tan \phi \sin \phi - (c/2\kappa)(z \tan \phi + iy \sin \phi)]$$

is a solution of the equations (B).

It should be noticed that the particular solution (C) is of type

$$\Theta(x, y, z, t) = e^{-z} F(x, y - 2\kappa t),$$

where $F(u, v)$ is a solution of the equation

$$\frac{\partial^2 F}{\partial u^2} + \frac{\partial^2 F}{\partial v^2} + F = 0. \qquad \text{......(D)}$$

This indicates that if F is any solution of this equation, then the function

$$\theta(x, y, t) = e^{-y} F(x, y - 2\kappa t)$$

is a solution of the equation (A). This is easily verified by differentiation. Since there is also a solution $\theta = e^{-\kappa t} F(x, y)$, we have two different ways of deriving a particular solution of the equation (A) from a particular solution of the equation (D).

Since $F(u, v) = J_0 \sqrt{[u^2 + v^2]}$ is a particular solution of equation (D) there is a certain surface distribution of temperature

$$\theta = J_0 \sqrt{[x^2 + 4\kappa^2 t^2]}, \quad \text{when } y = 0,$$

which is propagated downwards as a travelling disturbance gradually damped on the way, the velocity of propagation being 2κ.

If, on the other hand, we take $F(u, v) = \cos mu . \exp v [m^2 - 1]^{\frac{1}{2}}$, we obtain a distribution of temperature

$$\theta(x, y, t) = e^{-y} \cos mx . \exp \{(y - 2\kappa t)[m^2 - 1]^{\frac{1}{2}}\}, \quad m^2 > 1 \qquad \text{......(E)}$$

in which a periodic surface distribution is decaying at the same proportional rate at every point of the surface. If $m^2 < 2$ the foregoing distribution gives $\theta = 0$ when $y = \infty$. The periodic distribution now travels upwards with constant velocity

$$c = 2\kappa (m^2 - 1)^{\frac{1}{2}}/[1 - (m^2 - 1)^{\frac{1}{2}}],$$

and the rate of damping at depth y is the same as that at the surface, but at any instant the temperature at this depth is a fraction

$$\exp [-1 + (m^2 - 1)^{\frac{1}{2}}]$$

of that at the surface. When $m^2 = 2$ there is a distribution of temperature

$$\theta = e^{-2\kappa t} \cos (x\sqrt{2}),$$

which is independent of the depth but does not satisfy the condition* $\theta = 0$ when $y = \infty$. When $m^2 > 2$ the distribution (E) gives $\theta = 0$ when

* In this case there is no solution of type $\theta = e^{-2\kappa t}\, Y(y) \cos (x\sqrt{2})$ which gives the foregoing surface value of t and a value $\theta = 0$ when $y = \infty$ for $Y''(y) = 0$.

$y = -\infty$, and the material into which conduction takes place may be supposed to be on the side $y < 0$. In this case the velocity of propagation is

$$c = 2\kappa\,(m^2 - 1)^{\frac{1}{2}}/[(m^2 - 1)^{\frac{1}{2}} - 1],$$

and the temperature at depth $|\,y\,|$ is at any instant a fraction

$$\exp - [(m^2 - 1)^{\frac{1}{2}} - 1]$$

of that at the surface.

We have seen in § 2·432 that if $\theta\,(x, y, t)$ is a solution of equation (A) then the function

$$\phi\,(x, y, t) = t^{-1} e^{-\frac{x^2+y^2}{4\kappa t}}\,\theta\left(\frac{ax}{t}, \frac{ay}{t}, -\frac{a^2}{t}\right)$$

is a second solution. If, in particular, we take the function

$$\theta\,(x, y, t) = e^{-\kappa t}\,F\,(x, y),$$

where $F\,(u, v)$ satisfies (D), we obtain the solution

$$\phi\,(x, y, t) = t^{-1} e^{-\frac{x^2+y^2-4a^2\kappa^2}{4\kappa t}}\,F\left(\frac{ax}{t}, \frac{ay}{t}\right). \qquad \text{......(F)}$$

If $r^2 = x^2 + y^2$ there is a solution

$$\phi = t^{-1} e^{-\frac{r^2-4a^2\kappa^2}{4\kappa t}}\,J_0\,(ar/t) \qquad \text{......(G)}$$

depending only on r and t which at time $t = 0$ is zero at all points outside the circle $r = 2a\kappa$. When $t > 0$ the temperature at points of the circle is given by $\phi = t^{-1}J_0\,(2a^2\kappa/t)$. The circle can thus be regarded as a source of fluctuations in temperature which are transmitted by conduction to the external space. The total flow of heat from this circular source in the interval $t = 0$ to $t = \infty$ may be obtained by calculating the integral

$$- K\,.\,2\pi\,(2a\kappa)\int_0^\infty \left(\frac{\partial\phi}{\partial r}\right)_{r=2a\kappa} dt.$$

Now
$$\frac{\partial\phi}{\partial r} = \left[-\frac{r}{2\kappa t^2}J_0\,(ar/t) + \frac{a}{t^2}J_0'\,(ar/t)\right] e^{-\frac{r^2-4a^2\kappa^2}{4\kappa t}},$$

$$\left(\frac{\partial\phi}{\partial r}\right)_{r=2a\kappa} = -\,(a/t^2)\,J_0\,(2a^2\kappa/t) + (a/t^2)\,J_0'\,(2a^2\kappa/t).$$

Also
$$\int_0^\infty dt\,(a/t^2)\,J_0'\,(2a^2\kappa/t) = -\,1/2\kappa a,$$

$$\int_0^\infty dt\,(a/t^2)\,J_0\,(2a^2\kappa/t) = 1/2\kappa a.$$

Hence
$$\int_0^\infty dt\left(\frac{\partial\phi}{\partial r}\right)_{r=2a\kappa} = -\,1/\kappa a,$$

and so the total flow of heat from the circle is

$$4\pi K.$$

This is independent of a and so our formula holds also for a point source. The temperature function of a point source of "strength" Q is thus

$$\phi = (Q/4\pi\kappa t)^{-1} e^{-r^2/4\kappa t}, \qquad \text{......(H)}$$

while that of a circular source is

$$\phi = (Q/4\pi\kappa t)^{-1} e^{-\frac{r^2-4a^2\kappa^2}{4\kappa t}} J_0(ar/t). \qquad \text{......(I)}$$

This result is easily extended to a space of n dimensions, thus in three-dimensional space the temperature function for a spherical source of strength Q is

$$\phi = (Q/4\pi\kappa t)^{-\frac{3}{2}} e^{-\frac{r^2-4a^2\kappa^2}{4\kappa t}} \sin(ar/t)/(ar/t). \qquad \text{......(J)}$$

The solution for an instantaneous source uniformly distributed over a circular cylinder has been obtained by Lord Rayleigh* by integrating the solution for an instantaneous line source. The result is

$$v = \frac{\sigma a}{4\pi\kappa t}\int_0^{2\pi} e^{-\frac{r^2+a^2-2ar\cos\theta}{4\kappa t}}\, d\theta = \frac{\sigma a}{2\kappa t} e^{-\frac{r^2+a^2}{4\kappa t}} I_0\left(\frac{ra}{2\kappa t}\right). \qquad \text{......(K)}$$

A more general solution is

$$v = \frac{\sigma a}{2\kappa t}\cos n\phi \,.\, e^{-\frac{r^2+a^2}{4\kappa t}} I_n\left(\frac{ra}{2\kappa t}\right). \qquad \text{......(L)}$$

Integration with respect to t from 0 to ∞ gives a corresponding solution of Laplace's equation and we have the identity

$$\left.\begin{aligned} \int_0^\infty \frac{dt}{t} I_n\left(\frac{ra}{2\kappa t}\right) e^{-\frac{a^2+r^2}{4\kappa t}} &= \frac{1}{n}\left(\frac{r}{a}\right)^n \quad r < a, \\ &= \frac{1}{n}\left(\frac{a}{r}\right)^n \quad r > a. \end{aligned}\right\} \qquad \text{......(M)}$$

The temperature θ due to an instantaneous line doublet† of strength q may be derived by differentiating with respect to y the temperature ϕ due to an instantaneous line source of strength q. Since the latter is

$$\phi = (q/4\pi\kappa t)\, e^{-r^2/4\kappa t},$$

we have

$$\theta = (qy/8\pi\kappa^2 t^2)\, e^{-r^2/4\kappa t}. \qquad \text{......(N)}$$

The temperature due to a continuous line doublet of constant strength Q is obtained by integrating with respect to t between 0 and t. Denoting this temperature by Θ we have

$$\Theta = \int_0^t \theta\, dt = (qy/2\pi\kappa r^2)\int_0^t \frac{d}{dt}\left[e^{-r^2/4\kappa t}\right] dt = (qy/2\pi\kappa r^2)\, e^{-r^2/4\kappa t}.$$

* *Phil. Mag.* vol. XXII, p. 381 (1911); *Papers*, vol. VI, p. 51.

† See Carslaw's *Fourier Series and Integrals*, p. 345 (1906). The direction of the doublet is that of the axis of y. The doublet is supposed to be "located" at the origin.

This solution may be used to find a solution of (A) which takes the value $F(x)$ when $y=0$ and is zero when $y=\infty$ and when $t=0$. If θ is to be such that θ, $\frac{\partial\theta}{\partial x}$ and $\frac{\partial\theta}{\partial y}$ are continuous for $y>0$, an appropriate expression for θ is

$$\left.\begin{aligned}\theta &= \frac{1}{\pi}\int_{-\infty}^{\infty}\frac{y\,dx'}{(x-x')^2+y^2}F(x')\,e^{-\frac{(x-x')^2+y^2}{4\kappa t}}\\ &= \frac{1}{\pi}\int_{-\frac{\pi}{2}}^{\frac{\pi}{2}} d\alpha\, F(x+y\tan\alpha)\,e^{-\frac{y^2\sec^2\alpha}{4\kappa t}}.\end{aligned}\right\} \quad \text{......(O)}$$

The first integral evidently satisfies (A) if $y>0$, and the second integral tends to $F(x)$ as $y\to 0$ if $F(x)$ is a continuous function of x.

In the special case when $F(x)=1$ the expression for θ takes the form

$$\theta = \frac{1}{\pi}\int_{-\frac{\pi}{2}}^{\frac{\pi}{2}} e^{-\frac{y^2\sec^2\alpha}{4\kappa t}}\,d\alpha,$$

and can be expressed in the well-known form*

$$2\pi^{-\frac{1}{2}}\int_u^{\infty} e^{-v^2}\,dv, \quad \text{......(P)}$$

where $u^2 = y^2/4\kappa t$ and $u>0$.

If the boundary $y=0$ is maintained at the temperature $F(x,t)$ the solution which is zero when $y=\infty$ and when $t=0$ is given by the formula

$$\theta = \frac{y}{4\pi\kappa}\int_{-\infty}^{\infty}dx'\int_0^t \frac{F(x',t')}{(t-t')^2}\,e^{-\frac{(x-x')^2+y^2}{4\kappa(t-t')}}\,dt'. \quad \text{......(Q)}$$

There is a similar formula for a space of three dimensions.

If $\theta=F(x,y,t)$ when $z=0$ and $\theta=0$ when $z=\infty$ and when $t=0$, the appropriate solution is

$$v = \frac{z}{8(\pi\kappa)^{\frac{3}{2}}}\int_0^t\int_{-\infty}^{\infty}\int_{-\infty}^{\infty}\frac{F(x',y',t')}{(t-t')^{\frac{5}{2}}}\,e^{-\frac{(x-x')^2+(y-y')^2+z^2}{4\kappa(t-t')}}\,dt'\,dx'\,dy'. \quad \text{......(R)}$$

In this case an element of the integrand corresponds to an instantaneous doublet whose direction is that of the axis of z.

Let us next consider a case of steady heat conduction in a fluid moving vertically with constant velocity w. The fundamental equation is

$$w\frac{\partial\theta}{\partial z} = \kappa\left(\frac{\partial^2\theta}{\partial x^2}+\frac{\partial^2\theta}{\partial y^2}+\frac{\partial^2\theta}{\partial z^2}\right), \quad \text{......(S)}$$

where κ is the diffusivity. Writing $\theta = \Theta e^{wz/2\kappa}$ the equation satisfied by Θ is

$$\nabla^2\Theta = \lambda^2\Theta,$$

* The transformation from one integral to the other can be made by successive differentiation and integration with respect to u of the firſt integral.

where $\lambda = w/2\kappa$. A fundamental solution of this equation is given by

$$\Theta = AR^{-1}e^{-\lambda R},$$

where $R^2 = (x-\xi)^2 + (y-\eta)^2 + (z-\zeta)^2$, $(\xi, \eta, \zeta$ constant).

In particular, if $\xi = \eta = \zeta = 0$, we have the solution $\Theta = Ar^{-1}e^{-\lambda r}$, where r is the distance from the origin, and this corresponds to the solution

$$\theta = Ar^{-1}e^{\lambda(z-r)}. \qquad \text{......(T)}$$

This solution has been used by H. A. Wilson* and H. Mache† to account for the following phenomenon.

If a bead of easily fusible glass (O) be placed a few millimetres above the tip of the inner cone (K) of the flame of a Bunsen burner, a sharply defined yellow space (SS') of luminous sodium vapour is formed in the current of gas which is ascending vertically with considerable velocity. This space envelops the bead and broadens out in the higher part of the flame, as shown in Fig. 27. Provided the gas-pressure is not too high, the critical velocity of Osborne Reynolds, at which turbulence sets in, will not be exceeded even in these parts of the flame, so that the flow remains laminar, and the sodium vapour developed from the bead is driven into the hot gas solely under the influence of diffusion.

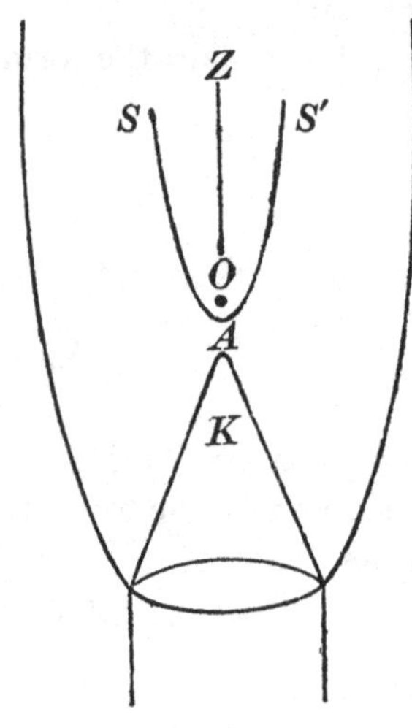

Fig. 27.

The fact that the vapour extends beneath the bead in the direction OA is proof of the high values of the coefficient of diffusion assumed at high temperatures, and at this point diffusion must be able to more than counteract the upward flow. Since an isothermal surface corresponds in the theory of diffusion to a surface of equal partial pressure, it is supposed that for suitable constant values of A and θ the equation (T) represents the surface enclosing the sodium vapour developed from the glass bead. When κ is small and w large, this surface approximates to the form of a paraboloid of revolution with the origin as focus.

Mache obtains the solution by integrating the effect of an instantaneous source which is successively at the different positions of a point moving relative to the medium with velocity w. In fact

$$(\kappa\pi)^{-\frac{1}{2}} \int_0^\infty t^{-\frac{3}{2}}\, dt . e^{-\frac{x^2+y^2+(z-wt)^2}{4\kappa t}} = \frac{2}{r} e^{\lambda(z-r)},$$

where $\lambda = w/2\kappa$.

* *Phil. Mag.* (6), vol. XXIV, p. 118 (1912); *Proc. Camb. Phil. Soc.* vol. XII, p. 406 (1904).

† *Phil. Mag.* (6), vol. XLVII, p. 724 (1924).

A similar solution has been used by O. F. T. Roberts* to give the distribution of density in a smoke cloud when the smoke is produced continuously at one point, and at a constant rate. The case in which the smoke is produced continuously along a horizontal line at right angles to the direction of the wind is solved by integrating the solution for the previous case.

§ **5·31**. *Two-dimensional motion of a viscous fluid.* If (u, v) are the component velocities at the point (x, y) at time t, p the pressure at this point, the equations of motion, when the fluid is incompressible and of uniform density ρ, are

$$\frac{\partial u}{\partial t} + u\frac{\partial u}{\partial x} + v\frac{\partial u}{\partial y} = -\frac{1}{\rho}\frac{\partial p}{\partial x} + \nu\nabla^2 u,$$

$$\frac{\partial v}{\partial t} + u\frac{\partial v}{\partial x} + v\frac{\partial v}{\partial y} = -\frac{1}{\rho}\frac{\partial p}{\partial y} + \nu\nabla^2 v,$$

while the equation of continuity is

$$\frac{\partial u}{\partial x} + \frac{\partial v}{\partial y} = 0.$$

This last equation may be satisfied by writing

$$v = \frac{\partial \psi}{\partial x}, \quad u = -\frac{\partial \psi}{\partial y},$$

where ψ is the stream-function, and if

$$\zeta = \frac{\partial v}{\partial x} - \frac{\partial u}{\partial y} = \nabla^2\psi$$

is the vorticity at the point (x, y) at time t, we have

$$\frac{\partial \zeta}{\partial t} + \frac{\partial(\psi, \zeta)}{\partial(x, y)} = \nu\nabla^2\zeta,$$

or

$$\frac{d\zeta}{dt} \equiv \frac{\partial \zeta}{\partial t} + u\frac{\partial \zeta}{\partial x} + v\frac{\partial \zeta}{\partial y} = \nu\nabla^2\zeta.$$

If $s = xv - yu$, we have

$$\frac{\partial s}{\partial x} = x\frac{\partial v}{\partial x} - y\frac{\partial u}{\partial x} + v, \qquad \frac{\partial s}{\partial y} = x\frac{\partial v}{\partial y} - y\frac{\partial u}{\partial y} - u,$$

$$\frac{\partial^2 s}{\partial x^2} = x\frac{\partial^2 v}{\partial x^2} - y\frac{\partial^2 u}{\partial x^2} + 2\frac{\partial v}{\partial x}, \quad \frac{\partial^2 s}{\partial y^2} = x\frac{\partial^2 v}{\partial y^2} - y\frac{\partial^2 u}{\partial y^2} - 2\frac{\partial u}{\partial y}.$$

Hence $$\frac{\partial s}{\partial t} + u\frac{\partial s}{\partial x} + v\frac{\partial s}{\partial y} = -\frac{1}{\rho}\left(x\frac{\partial p}{\partial y} - y\frac{\partial p}{\partial x}\right) + \nu\nabla^2 s - 2\nu\zeta.$$

If $x^2 + y^2 = r^2$ we may write

$$\frac{\partial s}{\partial x} = r\frac{\partial v}{\partial r} + v, \quad \frac{\partial s}{\partial y} = -r\frac{\partial u}{\partial r} - u,$$

$$u\frac{\partial s}{\partial x} + v\frac{\partial s}{\partial y} = r\left(u\frac{\partial v}{\partial r} - v\frac{\partial u}{\partial r}\right).$$

* *Proc. Roy. Soc. London*, vol. CIV, p. 640 (1923).

If the flow is of such a nature that p depends only on r and v/u is independent of r, we have

$$\frac{\partial s}{\partial t} = \nu\,(\nabla^2 s - 2\zeta).$$

Since $s = r\dfrac{\partial \psi}{\partial r}$, we have

$$\frac{\partial s}{\partial r} = r\frac{\partial^2\psi}{\partial r^2} + \frac{\partial \psi}{\partial r} = r\nabla^2\psi - \frac{1}{r}\frac{\partial^2\psi}{\partial\theta^2}.$$

Hence in the special case when ψ depends only on r, and the velocity is everywhere perpendicular to the radius from the origin, we have the differential equation

$$\frac{\partial s}{\partial t} = \nu\left(\nabla^2 s - \frac{2}{r}\frac{\partial s}{\partial r}\right) = \nu\left(\frac{\partial^2 s}{\partial r^2} - \frac{1}{r}\frac{\partial s}{\partial r}\right).$$

This indicates that the velocity $V = s/r$ satisfies the equation

$$\frac{\partial V}{\partial t} = \nu\left[\frac{\partial^2 V}{\partial r^2} + \frac{1}{r}\frac{\partial V}{\partial r} - \frac{V}{r^2}\right],$$

which is of the same form as the equation of the conduction of heat when the temperature Θ is of the form $\Theta = V\cos\theta$.

In the present case ψ and ζ are related since they both depend on r and so the equation for ζ is

$$\frac{\partial \zeta}{\partial t} = \nu\nabla^2\zeta.$$

The equation satisfied by ψ is

$$\frac{\partial \psi}{\partial t} = \nu\nabla^2\psi + f(t),$$

where $f(t)$ is an arbitrary function of t.

In the particular case when

$$\psi = t^{-1}e^{-r^2/4\nu t},$$

we have $\quad s = -(r^2/2\nu t^2)\,e^{-r^2/4\nu t}, \quad V = -(r/2\nu t^2)\,e^{-r^2/4\nu t}.$

The total angular momentum is in this case

$$2\pi\rho\int_0^\infty sr\,dr = -8\nu\rho\pi,$$

and is constant. The kinetic energy is on the other hand

$$\pi\rho\int_0^\infty V^2 r\,dr = \pi\rho/2t^2.$$

This type of vortex motion has been discussed by G. I. Taylor* in connection with the decay of eddies. The corresponding type of vortex motion in which

$$\zeta = t^{-1}e^{-r^2/4\nu t},$$

has been discussed by Oseen†, Terazawa‡ and Levy§.

* *Technical Report, Advisory Committee for Aeronautics*, vol. I, 1918–19, p. 73.

† C. W. Oseen, *Arkiv f. Mat., Astr. o. Fys.* Bd. VII (1911).

‡ K. Terazawa, *Report Aer. Res. Inst., Tokyo Imp. Univ.* (1922).

§ H. Levy, *Phil. Mag.* (7), vol. II, p. 844 (1926).

§ **5·32.** *Solutions of the form* $\psi = X(x, t) + Y(y, t)$. The condition to be satisfied is

$$\frac{\partial^3 X}{\partial x^2 \partial t} + \frac{\partial^3 Y}{\partial y^2 \partial t} + \frac{\partial X}{\partial x}\frac{\partial^3 Y}{\partial y^3} - \frac{\partial Y}{\partial y}\frac{\partial^3 X}{\partial x^3} = \nu\left(\frac{\partial^4 X}{\partial x^4} + \frac{\partial^4 Y}{\partial y^4}\right).$$

Differentiating successively with respect to x and y we get

$$\frac{\partial^2 X}{\partial x^2}\frac{\partial^4 Y}{\partial y^4} - \frac{\partial^2 Y}{\partial y^2}\frac{\partial^4 X}{\partial x^4} = 0.$$

We can satisfy this equation either by writing

$$\frac{\partial^2 X}{\partial x^2} = 0, \quad \frac{\partial^2 Y}{\partial y^2} = 0, \qquad \text{......(A)}$$

$$X = xa'(t) + b(t), \quad Y = yA'(t) + B(t), \qquad \text{......(B)}$$

or by writing

$$\frac{\partial^4 X}{\partial x^4} = [\mu(t)]^2 \frac{\partial^2 X}{\partial x^2}, \quad \frac{\partial^4 Y}{\partial y^4} = [\mu(t)]^2 \frac{\partial^2 Y}{\partial y^2}. \qquad \text{......(C)}$$

The supposition

$$\frac{\partial^4 X}{\partial x^4} = [\lambda(t)]^2 \frac{\partial^4 Y}{\partial y^4}, \quad \frac{\partial^2 X}{\partial x^2} = [\lambda(t)]^2 \frac{\partial^2 Y}{\partial y^2} \qquad \text{......(D)}$$

leads to

$$\frac{\partial^4 X}{\partial x^4} = 0, \quad \frac{\partial^4 Y}{\partial y^4} = 0.$$

These equations follow from (C) if we put $\mu(t) = 0$.

Solving equations (C) for X and Y we get

$$X = a(t)\, e^{x\mu(t)} + b(t)\, e^{-x\mu(t)} + xc(t) + d(t),$$

$$Y = A(t)\, e^{y\mu(t)} + B(t)\, e^{-y\mu(t)} + yC(t) + D(t).$$

Substituting in the original equation and assuming that $a(t)$, $b(t)$, $A(t)$, $B(t)$ are not zero, we find that $\mu(t)$ must be a constant μ and that the functions a, b, c, A, B, C must satisfy the equations

$$\mu^2 a' - Ca\mu^3 = \nu a\mu^4, \quad \mu^2 b' + Cb\mu^3 = \nu b\mu^4,$$

$$\mu^2 A' + cA\mu^3 = \nu A\mu^4, \quad \mu^2 B' - cB\mu^3 = \nu B\mu^4,$$

primes denoting differentiations with respect to t.

If the functions $c(t)$ and $C(t)$ are chosen arbitrarily, $a(t)$, $b(t)$, $A(t)$ and $B(t)$ may be determined by means of these equations when their initial values are given. In particular, if $a = A = 0$, $c = C = 0$, we can have

$$b = Pe^{\nu\mu^2 t}, \quad B = Qe^{\nu\mu^2 t},$$

$$\psi = Pe^{\nu\mu^2 t - \mu x} + Qe^{\nu\mu^2 t - \mu y},$$

$$u = \mu Q e^{\nu\mu^2 t - \mu y}, \quad v = -\mu P e^{\nu\mu^2 t - \mu x}.$$

This represents a growing disturbance in which each velocity component is propagated like a plane wave. The pressure is given by the equation

$$V + \int \frac{dp}{\rho} = C + \mu^2 PQ\, e^{2\nu\mu^2 t - \mu(x+y)}.$$

The fluid may be supposed to occupy the region $x > 0$, $y > 0$. If so, fluid enters this region across the plane $x = 0$ $(Q > 0)$ and leaves it at the plane $y = 0$ $(P > 0)$. The amount entering the region is equal to the amount leaving the region if $P = Q$, the density ρ being assumed constant.

If $V = 0$ and p_∞ is the pressure at infinity ($x = \infty$, $y = \infty$), we have

$$p - p_\infty = \rho\mu^2 P^2 e^{2\mu^2\nu t - \mu(x+y)}.$$

The pressure is generally greater than p_∞ and is propagated like a plane wave with velocity

$$c = \mu\nu\sqrt{2} = \nu \frac{\zeta}{\dfrac{u-v}{\sqrt{2}}}.$$

Thus the velocity of a plane pressure wave in an incompressible fluid is equal to ν times the ratio of the vorticity and the transverse component of velocity.

When the motion is steady the equation to be satisfied is

$$ae^{\mu x}(\nu\mu + C) + be^{-\mu x}(\nu\mu - C) + Ae^{\mu y}(\nu\mu - c) + Be^{-\mu y}(\nu\mu + c) = 0,$$

and we have four typical solutions:

$$\psi = px^2 + cx + qy^2 + Cy + D,$$
$$\psi = \nu\mu y + be^{-\mu x} + cx + d,$$
$$\psi = \nu\mu(x + y) + Ae^{\mu y} + be^{-\mu x} + d,$$
$$\psi = Ae^{\mu y} + \nu\mu x + Cy + D.$$

Returning to the first case we note that when $\dfrac{\partial^2 X}{\partial x^2} = 0$ the equations (B) do not give all possible solutions, for if

$$X = xa'(t) + b(t),$$

the original equation becomes

$$\frac{\partial^3 Y}{\partial y^2 \partial t} + a'(t)\frac{\partial^3 Y}{\partial y^3} = \nu \frac{\partial^4 Y}{\partial y^4}.$$

Writing $U = \dfrac{\partial^2 Y}{\partial y^2}$ we have the simpler equation

$$\frac{\partial U}{\partial t} + a'(t)\frac{\partial U}{\partial y} = \nu \frac{\partial^2 U}{\partial y^2},$$

which possesses a solution of type

$$\zeta \equiv U = \int_0^\infty e^{-\nu\lambda^2 t} \cos\lambda[y - a(t)]\,\omega(\lambda)\,d\lambda,$$

where $\omega(\lambda)$ is a suitable arbitrary function. For the corresponding motion

$$\psi = xa'(t) + yc(t) + b(t) + \int_0^\infty e^{-\nu\lambda^2 t}\cos\lambda[y - a(t)]\,\omega(\lambda)\frac{d\lambda}{\lambda^2},$$
$$u = c(t) - \int_0^\infty e^{-\nu\lambda^2 t}\sin\lambda[y - a(t)]\,\omega(\lambda)\frac{d\lambda}{\lambda},$$
$$v = a'(t).$$

This solution may be used to study laminar motion. The corresponding solution for the case in which the motion is steady is

$$\psi = Kx + Pe^{\frac{Ky}{\nu}} + Qy^2 + Ry + S,$$

where P, Q, R, S, K are arbitrary constants. If $K \to 0$ while the coefficients P, Q, R, S become infinite in a suitable manner, a limiting form of the solution gives the well-known solution

$$\psi = Ay^3 + Qy^2 + Ry + S.$$

It may be mentioned here that an attempt to find a stream-function ψ depending on a parameter s but not on t, and such that

$$\zeta = \frac{\partial \psi}{\partial s},$$

led to the equation
$$\frac{\partial^2 \psi}{\partial x^2} - \frac{\partial^2 \psi}{\partial y^2} = f(x, y) \frac{\partial^2 \psi}{\partial x \, \partial y}.$$

The conditions for the compatibility of this equation and

$$\nabla^2 \psi = \frac{\partial \psi}{\partial s}$$

seem to require $f(x, y)$ to be a constant. By a suitable choice of axes the former equation may then be reduced to the form

$$\frac{\partial^2 \psi}{\partial x \, \partial y} = 0,$$

and so
$$\psi = X(x, s) + Y(y, s).$$

EXAMPLES

1. In the case when there is a radial velocity U and a transverse velocity V, both of which depend only on r and t and when the pressure p depends only on r and t, the equations for U and V are

$$\frac{\partial V}{\partial t} + U \frac{\partial V}{\partial r} + \frac{UV}{r} = \nu \left\{ \frac{\partial^2 V}{\partial r^2} + \frac{1}{r} \frac{\partial V}{\partial r} - \frac{1}{r^2} V \right\} \frac{\partial}{\partial r}(rU) = 0.$$

Hence show that V satisfies the equation

$$\frac{\partial V}{\partial t} = \left(\nu \frac{\partial}{\partial r} - \frac{K}{r} \right) \left(\frac{\partial V}{\partial r} + \frac{V}{r} \right),$$

where K is a constant. If $\sigma = K/2\nu$, prove that there is a solution of type

$$V = r^{2\sigma+1} t^{-\sigma-2} e^{-r^2/4\nu t},$$

and verify that the total angular momentum about the origin remains constant.

2. Prove that the equation for V is satisfied by a series of type

$$V = r^n t^{-m} \left\{ 1 - \frac{m}{(1 + n - 2\sigma)(n + 3)} (r^2/\nu t) + \frac{m(m+1)}{(1 + n - 2\sigma)(3 + n - 2\sigma)(n + 3)(n + 5)} (r^2/\nu t)^2 - \dots \right\},$$

and verify that when $n = 1$, $m = 2$,

$$V = rt^{-2}\left\{1 - \frac{1}{1-\sigma}(r^2/4\nu t) + \frac{1}{(1-\sigma)(2-\sigma)}(r^2/4\nu t)^2 - \dots\right\}$$

$$= rt^{-2}e^{-(r^2/4\nu t)}\left\{1 + \frac{\sigma}{\sigma-1}(r^2/4\nu t) + \frac{\sigma}{\sigma-2}\cdot\frac{1}{2!}(r^2/4\nu t)^2 + \dots\right\}.$$

This is a particular case of Kummer's identity $F(\alpha; \gamma; x)\, e^{-x} = F(\gamma - \alpha; \gamma; -x)$, where $F(\alpha; \gamma; x)$ is the confluent hypergeometric function (Ch. IX).

3. Prove that there is a type of two-dimensional flow in which

$$\zeta = k^2\psi,$$

and ψ is consequently of the form

$$\psi = e^{-\nu k^2 t} F(x, y),$$

where $F(x, y)$ satisfies the differential equation

$$\frac{d^2F}{dx^2} + \frac{d^2F}{dy^2} + k^2F = 0.$$

Discuss the cases in which

$$F = \cos \alpha x \cos \beta y, \qquad \alpha^2 + \beta^2 = k^2,$$
$$F = e^{-ay} \cos bx, \qquad b^2 - a^2 = k^2.$$

Prove that in the latter case if $a^2 > b^2$ there is a growing disturbance which is propagated with velocity $\nu k^2/a$, and show that

$$\frac{\nu k^2}{a} = -\nu\frac{\zeta}{u}.$$

CHAPTER VI

POLAR CO-ORDINATES

§ **6·11.** *The elementary solutions.* If we make the transformation

$$x = r \sin\theta \cos\phi, \quad y = r \sin\theta \sin\phi, \quad z = r\cos\theta,$$

the wave-equation becomes

$$\frac{\partial^2 W}{\partial r^2} + \frac{2}{r}\frac{\partial W}{\partial r} + \frac{1}{r^2 \sin\theta}\frac{\partial}{\partial\theta}\left(\sin\theta \frac{\partial W}{\partial\theta}\right) + \frac{1}{r^2\sin^2\theta}\frac{\partial^2 W}{\partial\phi^2} - \frac{1}{c^2}\frac{\partial^2 W}{\partial t^2} = 0.$$

This is satisfied by a product of type

$$W = R(r)\,\Theta(\theta)\,\Phi(\phi)\,T(t), \qquad \text{......(I)}$$

if

$$\frac{d^2T}{dt^2} + k^2c^2T = 0,$$

$$\frac{d^2\Phi}{d\phi^2} + m^2\Phi = 0,$$

$$\frac{1}{\sin\theta}\frac{d}{d\theta}\left(\sin\theta\frac{d\Theta}{d\theta}\right) + \left[n(n+1) - \frac{m^2}{\sin^2\theta}\right]\Theta = 0,$$

$$\frac{d^2R}{dr^2} + \frac{2}{r}\frac{dR}{dr} + \left[k^2 - \frac{n(n+1)}{r^2}\right]R = 0,$$

where k, m and n are constants.

The first equation is satisfied by

$$T = a\cos(kct) + b\sin(kct),$$

where a and b are arbitrary constants; the second equation is satisfied by

$$\Phi = A\cos m\phi + B\sin m\phi,$$

where A and B are arbitrary constants. The third equation is reduced by the substitution $\cos\theta = \mu$ to the form

$$\frac{d}{d\mu}\left[(1-\mu^2)\frac{d\Theta}{d\mu}\right] + \left[n(n+1) - \frac{m^2}{1-\mu^2}\right]\Theta = 0. \qquad \text{......(II)}$$

Its solution can be expressed in terms of the associated Legendre functions $P_n^m(\mu)$ and $Q_n^m(\mu)$ which will be defined presently.

When $k = 0$ the fourth equation has the two independent solutions r^n and $r^{-(n+1)}$, except in the special case when $n = -(n+1)$, i.e. when $n = -\frac{1}{2}$. Making the substitution $w = r^{\frac{1}{2}}R$ in this case we obtain the equation

$$\frac{d^2w}{dr^2} + \frac{1}{r}\frac{dw}{dr} = 0,$$

which is satisfied by $w = C + D\log r$, where C and D are arbitrary constants.

The fact that r^n and r^{-n-1} are solutions of the equation for R furnishes us with an illustration of Kelvin's theorem that if $f(x, y, z)$ is a solution of Laplace's equation, then

$$\frac{1}{r} f\left(\frac{x}{r^2}, \frac{y}{r^2}, \frac{z}{r^2}\right)$$

is also a solution. The transformation in fact transforms $r^n \Theta\Phi$ into $r^{-n-1}\Theta\Phi$; it also transforms $r^{-\frac{1}{2}}(C + D \log r)\Theta\Phi$ into $r^{-\frac{1}{2}}(C - D\log r)\Theta\Phi$.

When $m = 0$ and $n = 0$ the differential equation for Θ is satisfied by

$$\Theta = 1 \text{ and } \Theta = \int \frac{d\mu}{1-\mu^2} = \frac{1}{2}\log\frac{1+\mu}{1-\mu}.$$

Thus, in addition to the potential functions 1 and $\dfrac{1}{r}$, we have the potential functions

$$\frac{1}{2}\log\frac{r+z}{r-z} \quad\text{and}\quad \frac{1}{2r}\log\frac{r+z}{r-z}.$$

It should be noticed that

$$\frac{\partial}{\partial z}\left(\frac{1}{2}\log\frac{r+z}{r-z}\right) = \frac{1}{r}.$$

In fact we have

$$\frac{\partial}{\partial z}\log(r+z) = \frac{1}{r},$$

$$\frac{\partial}{\partial z}\log(r-z) = -\frac{1}{r},$$

and it is easily verified that $\log(r+z)$ and $\log(r-z)$ are solutions of Laplace's equation. These formulae are all illustrations of the theorem that if W is a solution of Laplace's equation (or of the wave-equation), then

$$V = \frac{\partial W}{\partial z}$$

is also a solution of Laplace's equation (or of the wave-equation).

§ 6·12. In the case of the wave-equation the solution corresponding to $1/r$ is e^{ikr}/r, and there are associated wave-functions

$$\frac{1}{r}\cos k(r - ct), \quad \frac{1}{r}\sin k(r - ct),$$

which are, of course, particular cases of the wave-function

$$\frac{1}{r} f\left(t - \frac{r}{c}\right),$$

in which $f(\tau)$ is an arbitrary function which is continuous (D, 2).

§ 6·13. In the case of the conduction of heat the fundamental equation possesses solutions of the form (I) where R, Θ, Φ satisfy the same differential equations as before but T is of type

$$a \exp(-k^2h^2t),$$

where a is a constant and h^2 is the diffusivity. Thus there are solutions of type

$$\frac{1}{r}\cos kr.e^{-k^2h^2t},\quad \frac{1}{r}\sin kr.e^{-k^2h^2t},$$

which depend only on r and t. The second of these is the one suitable for the solution of problems relating to a solid sphere. If, in particular, there is heat generated at a uniform rate in the interior of the sphere the differential equation for the temperature θ is

$$\frac{\partial\theta}{\partial t} = h^2\nabla^2\theta + b^2,$$

where b is a constant. There is now a particular integral $-b^2r^2/6h^2$ which must be added to a solution of $\frac{\partial\theta}{\partial t} = h^2\nabla^2\theta$.

If initially $\theta = \theta_0$ throughout the sphere, θ_0 being a constant, and the boundary $r = a$ is suddenly maintained at temperature θ_1 from the time $t = 0$ to a sufficiently great time T, the condition at the surface is satisfied by writing

$$\theta = \theta_1 + \frac{b^2}{6h^2}(a^2 - r^2) + \frac{1}{r}\sum_{m=1}^{\infty} D_m \sin\frac{m\pi r}{a}\, e^{-\frac{m^2\pi^2h^2t}{a^2}},$$

while the initial condition is satisfied by writing*

$$D_m = (-)^m\left[\frac{2a^3b^2}{h^2m^3\pi^3} + \frac{2a(\theta_1 - \theta_0)}{m\pi}\right].$$

As $t \to \infty$, θ tends to the value $\theta_1 + \frac{b^2(a^2 - r^2)}{6h^2}$ and $\frac{\partial\theta}{\partial r}$ to the value $-\frac{b^2r}{3h^2}$, so that the flow of heat across the surface is, per second,

$$-4\pi a^2K\left(\frac{d\theta}{dr}\right)_{r=a} = \frac{4}{3}\frac{\pi b^2a^3K}{h^2}.$$

Writing $b^2 = \frac{Q}{\rho\sigma}$ and $h^2 = \frac{K}{\rho\sigma}$, where ρ is the density and σ the specific heat of the substance, we have the result that the rate of flow of heat across the surface is $4Q\pi a^3/3$, a result to be anticipated.

If, on the other hand, the initial temperature is $\theta_1 + \frac{b^2(a^2 - r^2)}{6h^2}$ and the surface of the sphere radiates heat to a surrounding medium at temperature θ_2 at a rate $E(\theta_a - \theta_2)$ per square centimetre, where θ_a is the (variable) surface temperature of the surface of the sphere, the solution is

$$\theta = B - \frac{b^2r^2}{6h^2} + \frac{1}{r}\Sigma D_n e^{-n^2h^2t}\sin nr.$$

* The constant D_m is obtained by Fourier's rule from the expansion of $\theta_0 - \theta_1 - \frac{b^2}{6h^2}(a^2 - r^2)$ in a sine series.

The surface condition is satisfied by writing

$$B = \theta_2 + \frac{ab^2(Ea + 2K)}{6Eh^2},$$
$$an = \phi_m,$$

where ϕ_m is the mth root of the transcendental equation

$$\tan\phi = \frac{K\phi}{K - Ea}.$$

The initial condition gives

$$Fr = \sum_{m=1}^{\infty} D_n \sin nr,$$

where
$$F = \theta_1 - \theta_2 - \frac{Kab^2}{3Eh^2},$$

and the extended form of Fourier's rule gives

$$D_n = \frac{2F}{a}\,\frac{K^2\phi_m{}^2 + (Ea - K)^2}{K^2\phi_m{}^2 + Ea(Ea - K)}\int_0^a r.\sin\left(\frac{r\phi_m}{a}\right) dr$$
$$= \frac{2a^2EF}{\phi_m}\,\frac{[\phi_m{}^2K^2 + (Ea - K)^2]^{\frac{1}{2}}}{[\phi_m{}^2K^2 + Ea(Ea - K)]}.$$

These results have been used by J. H. Awbery* in a discussion of the cooling of apples when in cold storage.

§ **6·21**. *Legendre functions.* The method of differentiation will now be used to derive new solutions of Laplace's equation from the fundamental solutions $\frac{1}{r}$ and $\frac{1}{2r}\log\frac{r+z}{r-z}$.

After differentiating n times with respect to z the new functions are of form $r^{-n-1}\Theta$, consequently we write

$$\frac{1}{r^{n+1}} P_n(\mu) = \frac{(-)^n}{n!}\frac{\partial^n}{\partial z^n}\left(\frac{1}{r}\right),$$
$$\frac{1}{r^{n+1}} Q_n(\mu) = \frac{(-)^n}{n!}\frac{\partial^n}{\partial z^n}\left(\frac{1}{2r}\log\frac{r+z}{r-z}\right),$$

and we shall adopt these equations as definitions of the functions $P_n(\mu)$ and $Q_n(\mu)$ for the case when n is a positive integer and θ is a real angle. The first equation indicates that there is an expansion of type

$$(r^2 - 2ar\mu + a^2)^{-\frac{1}{2}} = \sum_{n=0}^{\infty}\frac{a^n}{r^{n+1}} P_n(\mu), \quad |a| < |r|$$

and this equation may be used to obtain various expansions for $P_n(\mu)$. Thus

$$P_n(\mu) = \frac{1.3\ldots(2n-1)}{1.2\ldots n}$$
$$\times\left[\mu^n - \frac{n(n-1)}{2(2n-1)}\mu^{n-2} + \frac{n(n-1)(n-2)(n-3)}{2.4(2n-1)(2n-3)}\mu^{n-4} + \ldots\right]$$
$$= F\left(-n, n+1; 1; \frac{1-\mu}{2}\right) = (-)^n F\left(-n, n+1; 1; \frac{1+\mu}{2}\right),$$

* *Phil. Mag.* (7), vol. IV, p. 629 (1927).

where $F(a, b; c; x)$ denotes the hypergeometric series

$$1 + \frac{a.b}{1.c}x + \frac{a(a+1)b(b+1)}{1.2.c(c+1)}x^2 + \ldots.$$

§ **6·22.** *Hobson's theorem.* The first expansion for $P_n(\mu)$ is a particular case of a general expansion given by E. W. Hobson*. If $f(x, y, z)$ is a homogeneous polynomial of the nth degree in x, y, z,

$$f\left(\frac{\partial}{\partial x}, \frac{\partial}{\partial y}, \frac{\partial}{\partial z}\right)\frac{1}{r} = (-)^n.1.3\ldots(2n-1)\,r^{-2n-1}$$
$$\times\left[1 - \frac{r^2\nabla^2}{2(2n-1)} + \frac{r^4\nabla^4}{2.4(2n-1)(2n-3)} - \ldots\right]f(x, y, z).$$

When $f(x, y, z) = z^n$ this becomes

$$\frac{\partial^n}{\partial z^n}\left(\frac{1}{r}\right) = (-)^n r^{-2n-1}.1.3\ldots(2n-1)$$
$$\times\left[z^n - \frac{n(n-1)}{2(2n-1)}r^2z^{n-2} + \frac{n(n-1)(n-2)(n-3)}{2.4(2n-1)(2n-3)}r^4z^{n-4} - \ldots\right],$$

which is equivalent to the expansion for $n!\,r^{-n-1}P_n(\mu)$.

Assuming that the theorem is true for $f(x, y, z) = z^n$ it is easy to see that the theorem must also be true for $f(x, y, z) = (\xi x + \eta y + \zeta z)^n$, where $\xi x + \eta y + \zeta z$ is derived from z by a transformation of rectangular axes, for such a transformation transforms $\frac{\partial}{\partial z}$ into $\xi\frac{\partial}{\partial x} + \eta\frac{\partial}{\partial y} + \zeta\frac{\partial}{\partial z}$ and leaves ∇^2 unaltered.

To prove that the theorem is true in general it is only necessary to show that $f(x, y, z)$ can be expressed in the form

$$f(x, y, z) = \sum_{s=1}^{k} A_s(\xi_s x + \eta_s y + \zeta_s z)^n,$$

where the coefficients A_s are constants.

To determine such a relation we choose k points such that they do not all lie on a curve of degree n and such that a curve of degree n can be drawn through the remaining $k-1$ points when any one of the group of k points is omitted. Let ξ_s, η_s, ζ_s be proportional to the homogeneous co-ordinates of the sth point and let $\psi_s(x, y, z) = 0$ be the equation of the curve of degree n which passes through the remaining $k-1$ points.

Assuming that a relation of the desired type exists we operate on both sides of the equation with the operator $\psi_s\left(\frac{\partial}{\partial x}, \frac{\partial}{\partial y}, \frac{\partial}{\partial z}\right)$. The result is

$$\psi_s\left(\frac{\partial}{\partial x}, \frac{\partial}{\partial y}, \frac{\partial}{\partial z}\right)f(x, y, z) = n!\,A_s\psi_s(\xi_s, \eta_s, \zeta_s).$$

Giving s the values $1, 2, \ldots k$ all the coefficients are determined. Since a curve of the nth degree can be drawn through $\frac{1}{2}n(n+3)$ arbitrary points,

* *Proc. Lond. Math. Soc.* (1), vol. XXIV, p. 55 (1892–3).

the number k should be taken to be $\frac{1}{2}(n+1)(n+2)$, which is exactly the number of terms in the general homogeneous polynomial $f(x, y, z)$ of degree n. The coefficients A_s could, of course, be obtained by equating coefficients of the different products $x^a y^b z^c$ and solving the resulting linear equations, but it is not evident *a priori* that the determinant of this system of linear equations is different from zero. The foregoing argument shows that with our special choice of the quantities ξ_s, η_s, ζ_s the determinant is indeed different from zero because with a special choice of f, say

$$f = \sum_{s=1}^{k} B_s (\xi_s x + \eta_s y + \zeta_s z)^n,$$

the equations can be solved.

The solution is, moreover, unique because if there were an identical relation

$$0 \equiv \sum_{s=1}^{k} C_s (\xi_s x + \eta_s y + \zeta_s z)^n,$$

the foregoing argument would give

$$0 = n!\, C_s \psi_s (\xi_s, \eta_s, \zeta_s).$$

Hobson's theorem has been generalised so as to be applicable to Laplace's equation for a Euclidean space of m dimensions. Writing

$$\nabla_m^2 \equiv \frac{\partial^2}{\partial x_1^2} + \frac{\partial^2}{\partial x_2^2} + \dots + \frac{\partial^2}{\partial x_m^2},$$

$$r^2 = x_1^2 + x_2^2 + \dots + x_m^2,$$

and using $f(x_1, x_2, \dots x_m)$ to denote a homogeneous polynomial of degree n, the general relation is

$$f\left(\frac{\partial}{\partial x_1}, \frac{\partial}{\partial x_2}, \dots \frac{\partial}{\partial x_m}\right) r^{2-m} = (-)^n (m-2)(m) \dots (m+2n-4) . r^{2-m-2n}$$

$$\times \left[1 - \frac{r^2 \nabla_m^2}{2(m+2n-4)} + \frac{r^4 \nabla_m^4}{2.4(m+2n-4)(m+2n-6)} - \dots\right] f(x_1, x_2, \dots x_m).$$

§ **6·23.** *Potential functions of degree zero.* When $n = 0$ the differential equation satisfied by the product $U = \Theta\Phi$ may be written in the form

$$\frac{\partial^2 U}{\partial s^2} + \frac{\partial^2 U}{\partial \phi^2} = 0,$$

where $$s = -\int \frac{d\mu}{1-\mu^2} = \int \frac{d\theta}{\sin\theta} = \log \tan \frac{\theta}{2}.$$

It follows that there are solutions of type

$$U = f(s + i\phi) = F\left(\tan \frac{\theta}{2} e^{i\phi}\right),$$

where f is an arbitrary function and $f(u) = F(e^u)$.

This solution may be written in the form

$$U = F\left(\frac{x+iy}{r+z}\right),$$

where F is an arbitrary function. The general solution of Laplace's equation of degree zero may thus be written in the form

$$U = F\left(\frac{x+iy}{r+z}\right) + G\left(\frac{x-iy}{r+z}\right),$$

where F and G are arbitrary functions*. The general solution of degree -1 may be obtained from this by inversion and is

$$V = \frac{1}{r} F\left(\frac{x+iy}{r+z}\right) + \frac{1}{r} G\left(\frac{x-iy}{r+z}\right).$$

Solutions of degree $-(n+1)$ may be obtained from the last solution by differentiation. In particular, there is a potential function of type

$$V = \frac{\partial^n}{\partial z^n}\left[\frac{1}{r}\left(\frac{x+iy}{z+r}\right)^m\right],$$

which is of the form $r^{-n-1}\chi(\theta)\,e^{im\phi}$. The function must consequently be expressible in terms of Legendre functions. When m is a positive integer equal to or less than n we have in fact the formula of Hobson

$$\frac{\partial^n}{\partial z^n}\left[\frac{1}{r}\left(\frac{x+iy}{z+r}\right)^m\right] = (-)^n (n-m)!\, r^{-n-1} P_n^m(\mu)\, e^{im\phi}.$$

When m is a positive integer greater than or equal to n we have the expansion

$$\frac{\partial^n}{\partial z^n}\left[\frac{1}{r(z+r)^m}\right] = (-)^n\left[\frac{1.3\ldots(2n-1)}{r^{2n+1}(z+r)^{m-n}} + \frac{1.3\ldots(2n-3)}{r^{2n}(z+r)^{m-n+1}}\frac{2n-1}{1}(m-n)\right.$$
$$+ \frac{1.3\ldots(2n-5)}{r^{2n-1}(z+r)^{m-n+2}}\frac{(2n-2)(2n-3)}{1.2}(m-n)(m-n+1) + \ldots$$
$$\left.\ldots + \frac{1}{r^{n+1}}\frac{1}{(z+r)^m}(m-n)(m-n+1)\ldots(m-1)\right],$$

which may be used to define the function $\chi(\theta)$ in this case. In particular, we have the relation

$$\frac{\partial^n}{\partial z^n}\left[\frac{1}{r(z+r)^n}\right] = (-)^n\frac{1.3\ldots(2n-1)}{r^{2n+1}}.$$

When this is used to transform the expression for $r^{-n-1}P_n^m(\mu)\,e^{im\phi}$, we find that†

$$\frac{1}{r^{n+m+1}} P_n^m(\mu) = (-)^{n-m}\frac{1.3\ldots(2m-1)}{(n-m)!}\sin^m\theta\,\frac{\partial^{n-m}}{\partial z^{n-m}}\left(\frac{1}{r^{2m+1}}\right).$$

* W. F. Donkin, *Phil. Trans.* (1857).

† This formula is given substantially by E. W. Hobson, *Proc. London Math. Soc.* (1), vol. XXII, p. 442 (1891). Some other expressions for the Legendre functions are given by Hobson in the article on "Spherical Harmonics" in the *Encyclopædia Britannica*, 11th edition.

EXAMPLE

Prove that if m is a positive integer

$$\frac{i^m}{2\pi}\int_0^{2\pi}\frac{e^{ima}da}{z+ix\cos a+iy\sin a}=\frac{1}{r}\left(\frac{x+iy}{z+r}\right)^m,$$

$$\frac{i^m}{2\pi}\int_0^{2\pi}\log(z+ix\cos a+iy\sin a)\,e^{ima}\,da=-\frac{1}{m}\left(\frac{x+iy}{z+r}\right)^m,$$

$$\frac{1}{2\pi}\int_0^{2\pi}\tan^{-1}\left(\frac{x\cos a+y\sin a}{z}\right)e^{(2m+1)ia}=\frac{(-)^m}{2m+1}\left(\frac{x+iy}{z+r}\right)^{2m+1}.$$

§ **6·24.** *Upper and lower bounds for the function $P_n(\mu)$.* We shall now show that when $-1 \leqslant \mu \leqslant 1$ the function $P_n(\mu)$ lies between -1 and $+1$.

This may be proved with the aid of the expansion

$$P_n(\mu)=2\left[\frac{1.3\ldots(2n-1)}{2.4\ldots 2n}\cos n\theta+\frac{1}{2}\cdot\frac{1.3\ldots(2n-3)}{2.4\ldots(2n-2)}\cos(n-2)\theta\right.$$
$$\left.+\frac{1.3}{2.4}\cdot\frac{1.3\ldots(2n-5)}{2.4\ldots(2n-4)}\cos(n-4)\theta+\ldots\right],$$

which is obtained by writing

$$(1-2x\cos\theta+x^2)^{-\frac{1}{2}}=(1-xe^{i\theta})^{-\frac{1}{2}}(1-xe^{-i\theta})^{-\frac{1}{2}},$$

and expanding each factor in ascending powers of x by the binomial theorem, assuming that $|x|<1$.

It should be observed that each coefficient in the expansion is positive, consequently $P_n(\mu)$ has its greatest value when $\theta=0$ and $\mu=1$, for then each cosine is unity.

If, on the other hand, we replace each cosine by -1, we obtain a quantity which is certainly not greater than $P_n(\mu)$. Hence we have the inequality

$$-1 \leqslant P_n(\mu) \leqslant 1, \text{ for } -1 \leqslant \mu \leqslant 1.$$

When n is an odd integer $P_n(\mu)$ takes all values between -1 and $+1$, but when n is an even integer $P_n(\mu)$ has a minimum value which is not equal to -1. This minimum value is $-\frac{1}{2}$ for $P_2(\mu)$ and $-\frac{3}{7}$ for $P_4(\mu)$.

§ **6·25.** *Expressions for the Legendre polynomials as nth derivatives.* Lagrange's expansion theorem tells us that if

$$z=\mu+a\phi(z),$$

the Taylor expansion of $f'(z)\dfrac{dz}{d\mu}$ in powers of a is of type

$$\sum_{n=0}^{\infty}\frac{a^n}{n!}\frac{d^n}{d\mu^n}[f'(\mu)\{\phi(\mu)\}^n].$$

Writing

$$2\phi(z)=z^2-1,\quad f'(z)=1,$$

$$az=1-(1-2\mu a+a^2)^{\frac{1}{2}},\quad \frac{dz}{d\mu}=(1-2\mu a+a^2)^{-\frac{1}{2}};$$

a comparison of coefficients in this expansion and the expansion

$$(1 - 2\mu a + a^2)^{-\frac{1}{2}} = \sum_0^\infty a^n P_n(\mu)$$

gives us the formula of Rodrigues,

$$P_n(\mu) = \frac{1}{2^n n!} \frac{d^n}{d\mu^n} [(\mu^2 - 1)^n].$$

If, on the other hand, we write

$$\phi(z) = 2(\sqrt{z} - t),$$

we have
$$z = \mu + 2a(\sqrt{z} - t),$$

$$z - 2a\sqrt{z} + a^2 = a^2 - 2at + \mu,$$

$$\sqrt{z} = a \pm \sqrt{a^2 - 2at + \mu},$$

$$\frac{1}{\sqrt{z}} \frac{dz}{d\mu} = (a^2 - 2at + \mu)^{-\frac{1}{2}} = \mu^{-\frac{1}{2}} + \sum_1^\infty a^n \mu^{-\frac{n+1}{2}} P_n\left(\frac{t}{\sqrt{\mu}}\right).$$

Hence
$$\mu^{-\frac{n+1}{2}} P_n\left(\frac{t}{\sqrt{\mu}}\right) = \frac{2^n}{n!} \frac{\partial^n}{\partial \mu^n} \frac{(\sqrt{\mu} - t)^n}{\sqrt{\mu}},$$

or
$$\frac{1}{r^{n+1}} P_n\left(\frac{t}{r}\right) = \frac{1}{n!} \frac{\partial^n}{(r\,dr)^n} \frac{(r - t)^n}{r}$$

This formula is due to A. W. Conway, the previous one to E. Laguerre.

Replacing t by z we have the following expression for a zonal harmonic

$$\frac{1}{r^{n+1}} P_n(\mu) = \frac{1}{n!} \frac{\partial^n}{(r\,dr)^n} \frac{(r - z)^n}{r},$$

z and r being regarded as independent.

§ **6·26.** *The associated Legendre functions.* The differential equation (II) of § 6·11 is transformed by the substitution $\Theta = (1 - \mu^2)^{\frac{m}{2}} P$ to the form

$$(1 - \mu^2) \frac{d^2 P}{d\mu^2} - 2(m + 1)\mu \frac{dP}{d\mu} + [n(n + 1) - m(m + 1)] P = 0,$$

but this equation is satisfied by $P = \dfrac{d^m v}{d\mu^m}$, where v is a solution of Legendre's equation

$$(1 - \mu^2) \frac{d^2 v}{d\mu^2} - 2\mu \frac{dv}{d\mu} + n(n + 1) v = 0,$$

particular solutions of which are $P_n(\mu)$ and $Q_n(\mu)$.

Hence we adopt as our definitions of the functions $P_n^m(\mu)$ and $Q_n^m(\mu)$ for positive integral values of n and m

$$P_n^m(\mu) = (1-\mu^2)^{\frac{m}{2}} \frac{d^m}{d\mu^m} P_n(\mu),$$

$$Q_n^m(\mu) = (1-\mu^2)^{\frac{m}{2}} \frac{d^m}{d\mu^m} Q_n(\mu),$$

$$-1 \leqslant \mu \leqslant 1.$$

With the aid of these equations we may obtain the difference equations satisfied by $P_n^m(\mu)$ and $Q_n^m(\mu)$:

$$(n-m+1)\, P^m{}_{n+1} - (2n+1)\,\mu P_n{}^m + (n+m)\, P^m{}_{n-1} = 0,$$

$$\sqrt{1-\mu^2}\, P_n{}^{m+1} = 2m\mu P_n{}^m - (n+m)(n-m+1)\sqrt{1-\mu^2}\, P_n{}^{m-1},$$

$$P^m{}_{n-1} = \mu P_n{}^m - (n-m+1)\sqrt{1-\mu^2}\, P_n{}^{m-1},$$

$$P^m{}_{n+1} = \mu P_n{}^m + (n+m)\sqrt{1-\mu^2}\, P_n{}^{m-1},$$

$$\sqrt{1-\mu^2}\, P_n{}^{m+1} = (n+m+1)\,\mu P_n{}^m - (n-m+1)\, P^m{}_{n+1},$$

and the following expressions for the derivative

$$(1-\mu^2)\frac{d}{d\mu} P_n{}^m(\mu) = (n+1)\,\mu P_n{}^m(\mu) - (n-m+1)\, P^m{}_{n+1}(\mu)$$

$$= (n+m)\, P^m{}_{n-1}(\mu) - n\mu P_n{}^m(\mu).$$

Similar expressions hold for the derivative of $Q_n^m(\mu)$.

Expressions for the Legendre functions of different order and degree n are easily obtained from the difference equations or from the original definitions. In particular

$P_0{}^0 = 1.$

$P_1{}^0 = \cos\theta, \quad P_1{}^1 = \sin\theta.$

$P_2{}^0 = \frac{1}{2}(3\cos^2\theta - 1), \quad P_2{}^1 = 3\sin\theta\cos\theta, \quad P_2{}^2 = 3\sin^2\theta.$

$P_3{}^0 = \frac{1}{2}(5\cos^3\theta - 3\cos\theta), \quad P_3{}^1 = \frac{1}{2}\sin\theta\,(15\cos^2\theta - 3),$

$P_3{}^2 = 15\sin^2\theta\cos\theta, \quad P_3{}^3 = 15\sin^3\theta.$

$P_4{}^0 = \frac{1}{8}(35\cos^4\theta - 30\cos^2\theta + 3), \quad P_4{}^1 = \frac{1}{2}\sin\theta\,(35\cos^3\theta - 15\cos\theta),$

$P_4{}^2 = \frac{1}{2}\sin^2\theta\,(105\cos^3\theta - 15), \quad P_4{}^3 = 105\sin^3\theta\cos\theta, \quad P_4{}^4 = 105\sin^4\theta.$

$P_5{}^0 = \frac{1}{8}(63\cos^5\theta - 70\cos^3\theta + 15\cos\theta),$

$P_5{}^1 = \frac{1}{8}\sin\theta\,(315\cos^4\theta - 210\cos^2\theta + 15),$

$P_5{}^2 = \frac{1}{2}\sin^2\theta\,(315\cos^3\theta - 105\cos\theta),$

$P_5{}^3 = \frac{1}{2}\sin^3\theta\,(945\cos^2\theta - 105),$

$P_5{}^4 = 945\sin^4\theta\cos\theta, \quad P_5{}^5 = 945\sin^5\theta.$

EXAMPLES

1. Prove that if m and n are positive integers

$$\left(\frac{\partial}{\partial x}-i\frac{\partial}{\partial y}\right)^n\left[\frac{1}{r}\left(\frac{x+iy}{r+z}\right)^{n+m}\right]=(n-m)!\,r^{-n-1}P_n{}^m(\mu)\,e^{im\phi},$$

$$\left(\frac{\partial}{\partial x}-i\frac{\partial}{\partial y}\right)^n\left[\frac{1}{r}\left(\frac{x+iy}{r+z}\right)^{n-m}\right]=(n+m)!\,r^{-n-1}P_n{}^{-m}(\mu)\,e^{-im\phi}.$$

2. Prove that

$$\left(\frac{\partial}{\partial x}+i\frac{\partial}{\partial y}\right)[r^nP_n{}^m(\mu)\,e^{im\phi}] = -r^{n-1}P_{n-1}{}^{m+1}(\mu)\,e^{i(m+1)\phi},$$

$$\left(\frac{\partial}{\partial x}-i\frac{\partial}{\partial y}\right)[r^nP_n{}^m(\mu)\,e^{im\phi}] = (n+m)(n+m-1)\,r^{n-1}P_{n-1}{}^{m-1}(\mu)\,e^{i(m-1)\phi},$$

$$\left(\frac{\partial}{\partial x}+i\frac{\partial}{\partial y}\right)[r^{-n-1}P_n{}^m(\mu)\,e^{im\phi}] = -r^{-n-2}P_{n+1}{}^{m+1}(\mu)\,e^{i(m+1)\phi},$$

$$\left(\frac{\partial}{\partial x}-i\frac{\partial}{\partial y}\right)[r^{-n-1}P_n{}^m(\mu)\,e^{im\phi}] = (n-m+1)(n-m+2)\,r^{-n-2}P_{n+1}{}^{m-1}\,e^{i(m-1)\phi}.$$

§ 6·27. *Extensions of the formulae of Rodrigues and Conway.* By differentiating the formula of Rodrigues m times with respect to μ we obtain the formula

$$P_n{}^m(\mu)=\frac{(1-\mu^2)^{\frac{m}{2}}}{2^n.n!}\frac{d^{n+m}}{d\mu^{n+m}}(\mu^2-1)^n. \qquad \text{......(A)}$$

We shall use a similar definition for negative integral values of m and shall write

$$P_n{}^{-m}(\mu)=\frac{(1-\mu^2)^{-\frac{m}{2}}}{2^n.n!}\frac{d^{n-m}}{d\mu^{n-m}}(\mu^2-1)^n. \qquad \text{......(B)}$$

Expanding by Leibnitz's theorem we obtain

$$\frac{d^{n+m}(\mu+1)^n(\mu-1)^n}{d\mu^{n+m}}$$
$$=\sum_{s=0}^{n-m}\frac{(n+m)!}{(m+s)!\,(n-s)!}\frac{n!}{(n-m-s)!}\frac{n!}{s!}(\mu+1)^{n-m-s}(\mu-1)^s,$$

$$\frac{d^{n-m}(\mu+1)^n(\mu-1)^n}{d\mu^{n-m}}$$
$$=\sum_{s=0}^{n-m}\frac{(n-m)!}{s!\,(n-m-s)!}\frac{n!}{(n-s)!}\frac{n!}{(m+s)!}(\mu+1)^{n-s}(\mu-1)^{m+s}.$$

Comparing the two series, we obtain the relation of Rodrigues

$$P_n{}^{-m}(\mu)=(-)^m\frac{(n-m)!}{(n+m)!}P_n{}^m(\mu). \qquad \text{......(C)}$$

This may be derived also from the equations of Schendel

$$P_n^m(\mu) = \frac{1}{2^n(n-m)!}\left(\frac{1-\mu}{1+\mu}\right)^{\frac{m}{2}} \frac{d^n}{d\mu^n}[(\mu-1)^{n-m}(\mu+1)^{n+m}] \quad \text{(D)}$$

$$= \frac{(-)^m}{2^n(n-m)!}\left(\frac{1+\mu}{1-\mu}\right)^{\frac{m}{2}} \frac{d^n}{d\mu^n}[(\mu-1)^{n+m}(\mu+1)^{n-m}], \quad \text{(E)}$$

which may likewise be proved with the aid of Leibnitz's theorem. We have in fact

$$\frac{d^n}{d\mu^n}[(\mu-1)^{n-m}(\mu+1)^{n+m}]$$
$$= \Sigma \frac{n!}{s!(n-s)!}\frac{(n-m)!}{(n-m-s)!}\frac{(n+m)!}{(m+s)!}(\mu-1)^{n-m-s}(\mu+1)^{m+s}.$$

By differentiating Conway's formula m times with respect to t and multiplying by $(r^2-t^2)^{\frac{m}{2}}$ we obtain the formula

$$\frac{1}{r^{n+1}}P_n^m\left(\frac{t}{r}\right) = (-)^m(r^2-t^2)^{\frac{m}{2}}\frac{1}{(n-m)!}\left(\frac{\partial}{r\partial r}\right)^n\frac{(r-t)^{n-m}}{r}, \quad m \geqslant 0.$$

Making use of the formula (C) we may also write

$$\frac{1}{r^{n+1}}P_n^{-m}\left(\frac{t}{r}\right) = (r^2-t^2)^{\frac{m}{2}}\frac{1}{(n+m)!}\left(\frac{\partial}{r\partial r}\right)^n\frac{(r-t)^{n-m}}{r}, \quad m \geqslant 0.$$

Changing the sign of m we have

$$\frac{1}{r^{n+1}}P_n^m\left(\frac{t}{r}\right) = (r^2-t^2)^{-\frac{m}{2}}\frac{(n-m)!}{1}\left(\frac{\partial}{r\partial r}\right)^n\frac{(r-t)^{n+m}}{r}, \quad m \leqslant 0.$$

This formula also holds for $m > 0$.

§ 6·28. *Integral relations.* The Legendre functions satisfy some interesting integral relations which may be found as follows:

Writing down the differential equations satisfied by $P_n^m(\mu)$ and $P_l^k(\mu)$

$$\frac{d}{d\mu}\left[(1-\mu^2)\frac{dP_n^m}{d\mu}\right] + \left[n(n+1) - \frac{m^2}{1-\mu^2}\right]P_n^m = 0,$$

$$\frac{d}{d\mu}\left[(1-\mu^2)\frac{dP_l^k}{d\mu}\right] + \left[l(l+1) - \frac{k^2}{1-\mu^2}\right]P_l^k = 0,$$

let us first put $k = m$ and multiply these equations respectively by P_l^k and P_n^m and subtract, we then find that

$$\frac{d}{d\mu}\left[(1-\mu^2)\left(P_l^m\frac{dP_n^m}{d\mu} - P_n^m\frac{dP_l^m}{d\mu}\right)\right]$$
$$+ (n-l)(n+l+1)P_n^mP_l^m = 0.$$

Integrating between -1 and $+1$ the first term vanishes on account of the factor $1-\mu^2$ and so we find that if $l \neq n$

$$\int_{-1}^{1} P_n^m(\mu)\, P_l^m(\mu)\, d\mu = 0.$$

Next, if we put $l = n$ and multiply by P_n^k, P_n^m respectively and subtract we find in a similar way that if $m^2 \neq k^2$

$$\int_{-1}^{1} P_n^m(\mu)\, P_n^k(\mu)\, \frac{d\mu}{1-\mu^2} = 0.$$

To find the values of the integrals in the cases $l = n$, $k = m$ we may proceed as follows:

If we multiply the first difference equation by $P^m_{n+1}(\mu)$ and integrate between -1 and $+1$ we obtain the relation

$$(n-m+1)\int_{-1}^{1} [P^m_{n+1}(\mu)]^2\, d\mu = (2n+1)\int_{-1}^{1} \mu P^m_{n+1} P_n^m\, d\mu,$$

while if we multiply it by $P^m_{n-1}(\mu)$ and integrate we obtain the relation

$$(n+m)\int_{-1}^{1} [P^m_{n-1}(\mu)]^2\, d\mu = (2n+1)\int_{-1}^{1} \mu P_n^m P^m_{n-1}\, d\mu.$$

Changing n into $n-1$ in the previous relation we find that

$$(2n+1)(n-m)\int_{-1}^{1} [P_n^m(\mu)]^2\, d\mu = (2n-1)(n+m)\int_{-1}^{1} [P^m_{n-1}(\mu)]^2\, d\mu.$$

But

$$P_m^m(\mu) = (1-\mu^2)^{\frac{m}{2}} \frac{d^m}{d\mu^m} P_m(\mu) = 1.3 \ldots (2m-1)\,(1-\mu^2)^{\frac{m}{2}}.$$

Therefore

$$\int_{-1}^{1} [P_m^m(\mu)]^2\, d\mu = 1^2.3^2 \ldots (2m-1)^2 \int_{-1}^{1} (1-\mu^2)^m\, d\mu = \frac{2}{2m+1}(2m)!,$$

and so
$$\int_{-1}^{1} [P_n^m(\mu)]^2\, d\mu = \frac{2}{2n+1}\,\frac{(n+m)!}{(n-m)!}.$$

Let us next multiply the difference equations

$$(1-\mu^2)\frac{dP_n^m}{d\mu} = (n+m)\, P^m_{n-1} - n\mu P_n^m,$$

$$(1-\mu^2)\frac{dP^m_{n-1}}{d\mu} = n\mu P^m_{n-1} - (n-m)\, P_n^m$$

by $(1-\mu^2)^{-1} P^m_{n-1}$ and $(1-\mu^2)^{-1} P_n^m$ respectively and add.

Integrating between -1 and $+1$ we obtain the relation

$$(n+m)\int_{-1}^{1} [P^m_{n-1}]^2 \frac{d\mu}{1-\mu^2} - (n-m)\int_{-1}^{1} [P_n^m]^2 \frac{d\mu}{1-\mu^2}$$

$$= \int_{-1}^{1} \frac{d}{d\mu}[P_n^m P^m_{n-1}]\, d\mu = 0 \text{ if } m > 0.$$

Now

$$\int_{-1}^{1} [P_n^m(\mu)]^2 \frac{d\mu}{1-\mu^2} = 1^2.3^2 \ldots (2m-1)^2 \int_{-1}^{1} (1-\mu^2)^{m-1} d\mu$$

$$= 2.(2m-1)!.$$

Therefore $$\int_{-1}^{1} [P_n^m(\mu)]^2 \frac{d\mu}{1-\mu^2} = \frac{1}{m} \cdot \frac{(n+m)!}{(n-m)!}.$$

These relations are of great importance in the theory of expansions in series of Legendre functions. See Appendix, Note III.

§ 6·29. *Properties of the Legendre coefficients.* If the function $f(x)$ is integrable in the interval $-1 \leqslant x \leqslant 1$, which we shall denote by the symbol I, the quantities

$$C_n = (n + \tfrac{1}{2}) \int_{-1}^{1} f(x) P_n(x) \, dx \qquad \text{......(I)}$$

are called the *Legendre constants.* If these constants are known for all the above specified values of n and certain restrictions are laid on the function $f(x)$ this function is determined uniquely by its constants. An important case in which the function is unique is that in which the function $(1-x^2)^{\frac{1}{2}} f(x)$ is continuous throughout I. To prove this we shall show that if $\phi(x) = (1-x^2)^{-\frac{1}{2}} \psi(x)$, where $\psi(x)$ is continuous in I, then the equations

$$\int_{-1}^{1} \phi(x) P_n(x) \, dx = 0 \quad (n = 0, 1, 2, \ldots) \qquad \text{......(II)}$$

imply that $\phi(x) = 0$.

The first step is to deduce from the relations (II) that

$$\int_{-1}^{1} \phi(x) x^n \, dx = 0 \quad (n = 0, 1, 2, \ldots).$$

This step is simple because x^n can be represented as a linear combination of the polynomials $P_0(x)$, $P_1(x)$, ... $P_n(x)$.

The theorem to be proved is now very similar to one first proved by Lerch*. The following proof is due to M. H. Stone†.

If $\phi(x) \neq 0$ for a value $x = \xi$ in I we may, without loss of generality, assume that $\phi(\xi) > 0$, and we may determine a neighbourhood of ξ throughout which $\phi(x) \geqslant m > 0$. Now if $A > 0$ the polynomial

$$p(x) = A - \tfrac{1}{8} A (x - \xi)^2 (x^2 + 1)$$

is not negative in I and has a single maximum at $x = \xi$. We choose the constant A so that in the above-mentioned neighbourhood of ξ there are

* *Acta Math.* vol. XXVII (1903).

† *Annals of Math.* vol. XXVII, p. 315 (1926).

two distinct roots of the equation $p(x) = 1$ which we denote by x_1, x_2, the latter root being the greater. We thus have the inequalities

$$\begin{array}{ll} 0 \leqslant p(x) \leqslant 1, & -1 \leqslant x \leqslant x_1, \\ & x_2 \leqslant x \leqslant 1, \\ p(x) \geqslant 1, \quad \phi(x) \geqslant m, & x_1 \leqslant x \leqslant x_2, \\ \psi(x) \geqslant -M, & -1 \leqslant x \leqslant 1, \end{array}$$

where M is a positive quantity such that $-M$ is a lower bound for the continuous function $\psi(x)$.

Writing $p_n(x) = [p(x)]^n$, we have

$$\int_{-1}^{1} \phi(x)\, p_n(x)\, dx = 0, \quad n = 1, 2. \qquad \ldots\ldots(A)$$

On the other hand

$$\int_{x_1}^{x_2} \phi(x)\, p_n(x)\, dx \geqslant m \int_{x_1}^{x_2} p_n(x)\, dx,$$

$$\int_{-1}^{x_1} \phi(x)\, p_n(x)\, dx \geqslant -M \int_{-1}^{x_1} (1 - x^2)^{-\frac{1}{2}}\, dx,$$

$$\int_{x_2}^{1} \phi(x)\, p_n(x)\, dx \geqslant -M \int_{x_2}^{1} (1 - x^2)^{-\frac{1}{2}}\, dx,$$

$$\int_{-1}^{1} \phi(x)\, p_n(x)\, dx \geqslant m \int_{x_1}^{x_2} p_n(x)\, dx - M \int_{-1}^{1} (1 - x^2)^{-\frac{1}{2}}\, dx$$

$$\geqslant m \int_{x_1}^{x_2} p_n(x)\, dx - \pi M.$$

Since $\int_{x_1}^{x_2} p_n(x)\, dx \to \infty$ as $n \to \infty$ we can choose a number N such that the right-hand side is positive for $n \geqslant N$. This contradicts (A) and so we must conclude that $\phi(x) = 0$ throughout I.

Lerch's theorem is that if $\psi(x)$ is a real continuous function and $\int_0^1 x^n \psi(x)\, dx = 0$ for $n = 0, 1, 2, \ldots$ to ∞, then $\psi(x) = 0$.

By Weierstrass's theorem the function $\psi(x)$ may be approximated uniformly throughout the interval $(0, 1)$ by a polynomial $G(x)$. In other words, a polynomial $G(x)$ can be chosen so that $\psi(x) = G(x) + \delta\theta(x)$, where $|\theta(x)| < 1$, and δ is any small positive number chosen in advance. Now if $\psi(x)$ is not zero throughout the interval $(0, 1)$ we can choose our number δ so that

$$0 < \delta \int_0^1 |\psi(x)|\, dx < \int_0^1 [\psi(x)]^2\, dx. \qquad \ldots\ldots(B)$$

But, since $G(x)$ is a polynomial, we have

$$\int_0^1 G(x)\, \psi(x)\, dx = 0.$$

Therefore $$\int_0^1 \psi(x)\,[\psi(x) - \delta\theta(x)]\,dx = 0$$

or $$\int_0^1 [\psi(x)]^2\,dx = \delta\int_0^1 \theta(x)\,\psi(x)\,dx$$

$$\leqslant \delta\int_0^1 |\theta(x)|\,|\psi(x)|\,dx \leqslant \delta\int_0^1 |\psi(x)|\,dx.$$

This contradicts (B) and so we must have $\psi(x) = 0$. Putting $x = e^{-t}$ we deduce that if $\int_0^\infty e^{-zt}\phi(t)\,dt = 0$ for $z > 0$, and $\phi(t)$ is continuous for $t \geqslant 0$, then $\phi(t) \equiv 0$.

EXAMPLES

1. When m and n have positive real parts

$$A\int_0^1 Q_m(z)\,Q_n(z)\,dz = \psi(n+1) - \psi(m+1) - \tfrac{1}{2}\pi\,[(B-C)\sin(\tfrac{1}{2}m + \tfrac{1}{2}n)\,\pi - (B+C)\sin(\tfrac{1}{2}m - \tfrac{1}{2}n)\pi],$$

where $\psi(z) = \dfrac{d}{dz}\log\Gamma(z)$,

$$A = (m-n)(m+n+1)$$

$$B = \frac{B(\tfrac{1}{2}n + \tfrac{1}{2}, \tfrac{1}{2}m + 1)}{B(\tfrac{1}{2}m + \tfrac{1}{2}, \tfrac{1}{2}n + 1)} \text{ and } BC = 1$$

[S. C. Dhar and N. G. Shabde, *Bull. Calcutta Math. Soc.* v. 24, 177–186 (1932).]

Show also that with the same notation

$$A\int_1^\infty Q_m(z)\,Q_n(z)\,dz = \psi(m+1) - \psi(n+1)$$

[Ganesh Prasad, *Proc. Benares Math. Soc.* v. 12, pp. 33–42, 19.]

2. Show by means of the relation

$$\int_{-1}^1 P_m(\mu)\,P_n(\mu)\,d\mu = 0,$$

that when n is a positive integer the equation $P_n(\mu) = 0$ has n distinct roots which all lie in the interval $-1 \leqslant \mu \leqslant 1$.

3. Prove that when m and n are positive integers

$$\int_{-1}^1 (1+z)^{m+n}\,P_m(z)\,P_n(z)\,dz = \frac{2^{m+n+1}\,\{(m+n)!\}^4}{(m!\,n!)^2\,(2m+2n+1)!}.$$

[E. C. Titchmarsh.

An elementary proof of this formula is given by R. G. Cooke, *Proc. London Math. Soc.* (2), vol. XXIII (1925); *Records of Proceedings*, p. xix.

§ **6·31**. *Potential function with assigned values on a spherical surface S.*

Let P, P' be two inverse points with respect to a sphere of radius a. If O is the centre of the sphere we have then

$$OP.OP' = a^2,$$

and O, P, P' lie on a line. The point O is sometimes called the *centre of inversion.*

If P lies inside the sphere, P' lies outside; if P is outside the sphere, P' is inside. If P is on the sphere, P' coincides with P. If P describes a curve or surface P' will describe the inverse curve or surface and it is clear that a curve or surface will intersect the sphere at points where it meets its inverse. If a curve or surface inverts into itself it must intersect the sphere S orthogonally at the points where it meets it because at these points two consecutive inverse points lie on the surface and on a line through O. This line is then a tangent to the surface and a normal to S at the same point. If M_s is any point on S the triangles OPM_s, OM_sP' are similar, and we have

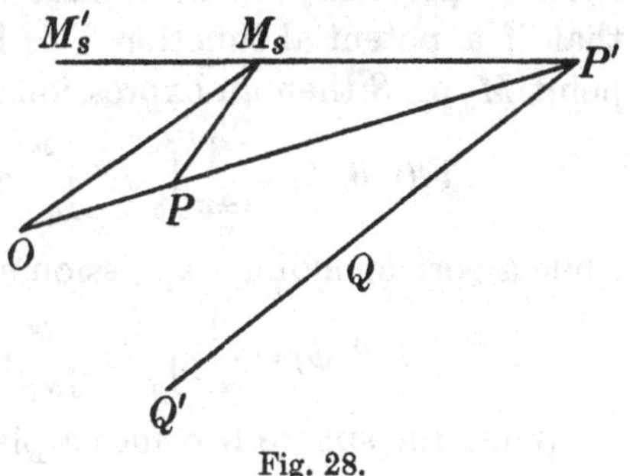

Fig. 28.

$$\frac{OP}{PM_s} = \frac{OM_s}{P'M_s}.$$

If charges proportional to OP and $-OM_s$ are placed at P and P' respectively, the sum of their potentials at any point M_s on S will be zero. Writing $OP = r$, $PM = R$, $P'M = R'$, where M is any point, we see that the function

$$G_{PM} = \frac{1}{R} - \frac{a}{r} \cdot \frac{1}{R'}$$

is zero when M is on S and is infinite like $\frac{1}{R}$ at the point P. We shall call this function the *Green's function* for the sphere. G_{PM} is easily seen to be a symmetric function of the co-ordinates of P and M, for if M' is the inverse of M we have

$$\frac{OM}{P'M} = \frac{OP}{PM'}.$$

The point P' is called the *electrical image* of P and G_{PM} represents the potential at M when the sphere S, regarded as a conducting surface at zero potential, is influenced by a unit charge at P. When a becomes infinite and O recedes to infinity the sphere becomes a plane, P' is then the optical image of P in this plane, and the virtual charge at P' is equal and opposite to that at P.

Now let $OM = r'$ and $P\hat{O}M = \omega$, then

$$R^2 = r^2 + r'^2 - 2rr' \cos \omega,$$

$$R'^2 = r'^2 + \frac{a^4}{r^2} - 2r' \frac{a^2}{r} \cos \omega,$$

$$\left(\frac{\partial G}{\partial r'}\right)_{r'=a} = \frac{r^2 - a^2}{a\,(a^2 + r^2 - 2ar \cos \omega)^{\frac{3}{2}}}.$$

Let (r, θ, ϕ), (r, θ', ϕ') be the spherical polar co-ordinates of the point P and a point M_s on the surface of S, then the theorem of § 2·32 tells us that if a potential function V is known to have the value $F(\theta', \phi')$ at a point M_s on S then an expression for V suitable for the space outside S is

$$V(r, \theta, \phi) = \frac{1}{4\pi}\int_0^{\pi} d\theta' \int_0^{2\pi} d\phi' \frac{a\,(r^2 - a^2)\,F(\theta', \phi') \sin \theta'}{(a^2 + r^2 - 2ar \cos \omega)^{\frac{3}{2}}},$$

while a corresponding expression suitable for the space inside S is*

$$V(r, \theta, \phi) = \frac{1}{4\pi}\int_0^{\pi} d\theta' \int_0^{2\pi} d\phi' \frac{a\,(a^2 - r^2)\,F(\theta', \phi') \sin \theta'}{(a^2 + r^2 - 2ar \cos \omega)^{\frac{3}{2}}}.$$

When the sphere becomes a plane the corresponding expression is

$$V(x, y, z) = \frac{1}{2\pi}\int_{-\infty}^{\infty} dx' \int_{-\infty}^{\infty} dy' \frac{\pm f(x', y')}{[(x - x')^2 + (y - y')^2 + z^2]^{\frac{1}{2}}},$$

the upper or lower sign being taken according as $z \gtrless 0$. In this case $f(x', y') = V(x', y', 0)$ is the value of V on the plane $z = 0$.

§ **6·32**. *Derivation of Poisson's formula from Gauss's mean value theorem.* Poisson's formula may also be obtained by inversion, using the method of Bôcher.

Let us take P' as centre of inversion and invert the sphere S into itself. The radius of inversion is then $c = (r_0^2 - a^2)^{\frac{1}{2}} = \frac{a}{r}(a^2 - r^2)^{\frac{1}{2}}$, where c is the length of the tangent from P' to the sphere, it is real when P' is outside the sphere and imaginary when P' is within the sphere. (In Fig. 28 $OP' = r_0$.)

Let Q, Q' be two corresponding points on S, then the relation between corresponding elements of area is

$$\frac{dS'}{dS} = \left(\frac{P'Q'}{P'Q}\right)^2 = \left(\frac{c}{P'Q}\right)^4 = \left(\frac{cr}{a.PQ}\right)^4.$$

Writing $dS' = a^2 d\Omega'$, $dS = a^2 d\Omega$, where $d\Omega'$ and $d\Omega$ are elementary solid angles, we have

$$d\Omega' = \left(\frac{cr}{a.PQ}\right)^4 d\Omega = (a^2 - r^2)^2 (r^2 + a^2 - 2ar \cos \omega)^{-2}\, d\Omega,$$

where ω is the angle between OQ and OP.

* This is generally called "Poisson's integral," both formulae having been proved by S. D. Poisson, *Journ. École Polyt.* vol. XIX (1823). The formula for the interior of the sphere had, however, been given previously by J. L. Lagrange, *ibid.* vol. XV (1809).

Now if $V'_{Q'}$ is a potential function when expressed in terms of the co-ordinates of Q', the function

$$\frac{c}{P'Q} V'_{Q'} \equiv \frac{cr}{a.PQ} V'_{Q'} = V_Q, \text{ say,}$$

is a potential function when expressed in terms of the co-ordinates of Q, consequently the mean value theorem

$$4\pi V_0' = \int V'_{Q'} d\Omega',$$

gives
$$4\pi V_0' = \frac{a}{cr}(a^2 - r^2)^2 \int V_Q d\Omega \, [r^2 + a^2 - 2ar \cos \omega]^{\frac{3}{2}},$$

and since $c.V_0' = P'P.V_P$, we have $crV_0' = (a^2 \sim r^2) V_P$, and our formula is the same as that derived from the theory of the Green's function.

This method is easily extended to the case of hyperspheres in a space of n dimensions. The relation between the contents of corresponding elements of the hyperspheres is now

$$\frac{dS'}{dS} = \left(\frac{P'Q'}{P'Q}\right)^{n-1} = \left(\frac{c}{P'Q}\right)^{2n-2} = \left(\frac{cr}{a.PQ}\right)^{2n-2},$$

while the relation between corresponding potentials is

$$\left(\frac{c}{P'Q}\right)^{n-2} V'_{Q'} \equiv \left(\frac{cr}{a.PQ}\right)^{n-2} V_Q' = V_Q, \text{ say.}$$

Writing the mean value theorem in the form

$$V_0' \int dS' = \int V'_{Q'} dS',$$

the generalised formula of Poisson is

$$\left(\frac{a^2 \sim r^2}{cr}\right)^{n-2} V_P \int dS' = \left(\frac{cr}{a}\right)^n \int V_Q \, dS \, [r^2 + a^2 - 2ar \cos \omega]^{-\frac{n}{2}}$$

or
$$V_P \int dS' = \pm \, a^{n-2} (a^2 - r^2) \int V_Q \, dS \, [r^2 + a^2 - 2ar \cos \omega]^{-\frac{n}{2}}.$$

§ 6·33. *Some applications of Gauss's mean value theorem.* The mean value theorem may be used to obtain some interesting properties of potential functions.

In the first place, if a function V is harmonic in a region R it can have neither a maximum nor a minimum in R.

If the contrary were true and V did have a maximum or minimum value at a point P of R the mean value of V over a small sphere with centre at P would not be equal to the value at P. If now the sphere is made so small that it lies entirely within R, Gauss's theorem may be applied and we arrive at a contradiction. Since a function which is continuous over a region consisting of a closed set of points has finite upper and lower bounds which are actually attained, we have the theorem:

If a function is harmonic in a region R with boundary B and is continuous in the domain $R + B$, the greatest and least values of V in the domain $R + B$ are attained on the boundary B.

One immediate consequence of the last theorem is that if the function V is harmonic in R, continuous in $R + B$ and constant on B it is constant on $R + B$. This theorem is important in electrostatics because it tells us that the potential is constant throughout the interior of a closed hollow conductor if it is known to be constant on the interior surface of the conductor. Another interesting consequence of the theorem is that if the function V is harmonic in R, continuous in $R + B$ and positive on B it is positive in $R + B$. For if it were zero or negative at some point of R the least value of V in R would not be attained on the boundary*.

This theorem may be restated as follows:

If V_1 and V_2 be functions harmonic in R and continuous in $R + B$, and if V_1 is greater than (equal to or less than) V_2 at every point of B, then V_1 is greater than (equal to or less than) V_2 at every point of $R + B$.

A converse of Gauss's theorem, due to Koebe, is given in Kellogg's *Foundations of Potential Theory*, p. 224.

§ **6·34**. *The expansion of a potential function in a series of spherical harmonics.* If $V(x, y, z)$ is a potential function which is continuous throughout the interior of a sphere S and on its boundary, and whose first derivatives $\frac{\partial V}{\partial x}, \frac{\partial V}{\partial y}, \frac{\partial V}{\partial z}$ are likewise continuous and the second derivatives finite and integrable (for simplicity we shall suppose them to be continuous) then V admits of a representation by means of Poisson's formula and it will be shown that V can be expanded in a convergent power series in the co-ordinates x, y, z relative to the centre of S. Writing

$$\cos\omega = \cos\theta\cos\theta' + \sin\theta\sin\theta'\cos(\phi - \phi') = \mu,$$

we have

$$\frac{a(r^2 - a^2)}{(a^2 + r^2 - 2ar\cos\omega)^{\frac{3}{2}}} = \sum_{n=0}^{\infty} (2n + 1)\left(\frac{a}{r}\right)^{n+1} P_n(\mu) \quad |r| > a,$$

$$\frac{a(a^2 - r^2)}{(a^2 + r^2 - 2ar\cos\omega)^{\frac{3}{2}}} = \sum_{n=0}^{\infty} (2n + 1)\left(\frac{r}{a}\right)^{n} P_n(\mu) \quad |r| < a.$$

Substituting in the expressions for V we may integrate term by term because the series are absolutely and uniformly convergent on account of the inequality $|P_n(\mu)| \leqslant 1$.

* See a paper by G. E. Raynor, *Annals of Math.* (2), vol. XXIII, p. 183 (1923).

We thus obtain the expansions

$$V = \sum_{n=0}^{\infty} (2n+1) \left(\frac{a}{r}\right)^{n+1} S_n(\theta, \phi) \quad |r| > a,$$

$$V = \sum_{n=0}^{\infty} (2n+1) \left(\frac{r}{a}\right)^{n} S_n(\theta, \phi) \quad |r| < a,$$

where in each case

$$S_n(\theta, \phi) = \frac{1}{4\pi} \int_0^{\pi} \int_0^{2\pi} F(\theta', \phi') P_n(\mu) \sin\theta' d\theta' d\phi'.$$

The function $r^n S_n(\theta, \phi)$ is called a spherical harmonic or solid harmonic of degree n, it is a polynomial of the nth degree in x, y, z and is a solution of Laplace's equation because $r^n P_n(\mu)$ is a solution.

The function $S_n(\theta, \phi)$ is called a surface harmonic, it may be expressed in terms of elementary products of type $P_n^m(\cos\theta) e^{im\phi}$ by expanding $P_n(\mu)$ in a Fourier series of type

$$P_n(\mu) = \sum_{m=-n}^{n} F_n^m(\theta, \theta') e^{im(\phi-\phi')}.$$

By expanding $r^n P_n(\mu)$ in a series of this form and substituting in Laplace's equation (in polar co-ordinates) we get a series of type $\Sigma C_m e^{im(\phi-\phi')}$ each term of which must be separately zero, consequently each term in our expansion of $r^n P_n(\mu)$ is a solution of Laplace's equation and is a polynomial of degree n in x, y and z. Similarly, if r', θ', ϕ' are regarded as polar co-ordinates of a point (x', y', z'), $r'^n P_n(\mu)$ is a solution of Laplace's equation relative to the co-ordinates of this point. We infer then that

$$F_n^m(\theta, \theta') = A_n^m P_n^m(\cos\theta) P_n^{-m}(\cos\theta'),$$

where A_n^m is a constant to be determined.

We thus have the result that

$$S_n(\theta, \phi) = \sum_{m=-n}^{n} B_n^m P_n^m(\cos\theta) e^{im\phi},$$

where $B_n^m = \dfrac{1}{4\pi} A_n^m \displaystyle\int_0^{\pi} \int_0^{2\pi} F(\theta', \phi') P_n^{-m}(\cos\theta') e^{-im\phi'} \sin\theta' d\theta' d\phi'.$

To determine the constant A_n^m we consider the particular case when

$$V = r^n P_n^m(\cos\theta) e^{im\phi},$$

$$F(\theta', \phi') = a^n P_n^m(\cos\theta') e^{im\phi'},$$

then

$$\begin{aligned}(2\nu+1) S_\nu(\theta, \phi) &= a^n P_n^m(\cos\theta) e^{im\phi} \quad \nu = n \\ &= 0 \quad \nu \neq n,\end{aligned}$$

and consequently

$$\begin{aligned}1 &= \frac{2n+1}{4\pi} A_n^m \int_0^{\pi} \int_0^{2\pi} P_n^m(\cos\theta') P_n^{-m}(\cos\theta') \sin\theta' d\theta' d\phi' \\ &= (-)^m A_n^m,\end{aligned}$$

or $\quad A_n^m = (-)^m.$

Hence we have the expansion

$$P_n(\mu) = \sum_{m=-n}^{n} (-)^m P_n^m(\cos\theta)\, P_n^{-m}(\cos\theta')\, e^{im(\phi-\phi')}.$$

§ 6·35. *Legendre's expansion.* Transforming the last equation with the aid of the relation of Rodrigues,

$$P_n^{-m}(\mu) = (-)^m \frac{(n-m)!}{(n+m)!} P_n^m(\mu),$$

we obtain Legendre's expansion*

$$P_n(\mu) = 2\sum_{m=1}^{\infty} \frac{(n-m)!}{(n+m)!} P_n^m(\cos\theta)\, P_n^m(\cos\theta') \cos m(\phi-\phi') + P_n(\cos\theta)\, P_n(\cos\theta'),$$

and the expression for B_n^m may be written in the alternative form

$$B_n^m = \frac{1}{4\pi}\frac{(n-m)!}{(n+m)!}\int_0^{\pi}\int_0^{2\pi} S_n(\theta',\phi')\, P_n^m(\cos\theta')\, e^{-im\phi'} \sin\theta' d\theta' d\phi'.$$

One simple deduction from the expansion for $S_n(\theta, \phi)$ is that a simple expression can be obtained for the mean value of $S_n(\theta, \phi)$ round a circle on the sphere. Let the circle in fact be $\theta = \alpha$, then the mean value in question is obtained by integrating our series for $S_n(\theta, \phi)$ between $\phi = 0$ and $\phi = 2\pi$ and afterwards dividing by 2π. The result is that

$$\overline{S_n(\theta, \phi)} = B_n^0 P_n(\cos\alpha).$$

Now when $\theta = 0$, $P_n^m(\cos\theta) = 0$ except when $m = 0$, and then the value is unity, hence

$$S_n(0, \phi) = B_n^0,$$

and so

$$\overline{S_n(\theta, \phi)} = S_n(0, \phi)\, P_n(\cos\alpha),$$

where the coefficient $S_n(0, \phi)$ is the value of $S_n(\theta, \phi)$ at the pole of the circle. This theorem may be extended so as to give the mean value of a function $f(\theta, \phi)$, which can be expanded in a series of type

$$f(\theta, \phi) = \sum_{n=0}^{\kappa} c_n S_n(\theta, \phi).$$

The result is

$$\overline{f(\theta, \phi)} = \sum_{n=0}^{\kappa} c_n P_n(\cos\alpha)\, S_n(0, \phi).$$

If the analytical form of the function f is not given, but various graphs are available, the present result may sometimes be used to find the coefficients in the expansion

$$f(\theta, \phi) = \sum_{n=0}^{\infty} C_n P_n(\cos\theta) + \sum_{n=1}^{\infty}\sum_{m=0}^{n} P_n^m(\cos\theta)\,[A_n^m \cos m\phi + B_n^m \sin m\phi].$$

To use the method in practice it is convenient to have a series of curves in which f is plotted against ϕ for different values of θ and a series of curves in which f is plotted against θ for different values of ϕ. The two

* Legendre, *Hist. Acad. Sci. Paris*, t. II, p. 432 (1789).

meridians $\phi = \beta$ and $\phi = \beta + \pi$ may be regarded as one great circle with a pole $\theta = \frac{\pi}{2}$, $\phi = \beta + \frac{\pi}{2}$.

Mean values round "parallels of latitude" for which θ has various constant values will give linear equations involving only the coefficients C_n.

Since $P_n(0) = 0$ when n is odd and $P_n^m(0) = 0$ when n is even and m is odd, the mean values round meridian circles will give equations involving only the coefficients A_n^m and B_n^m in which both m and n are even, but terms of type C_n will also occur. To illustrate the method we shall suppose that the function $f(\theta, \phi)$ is of such a nature that spherical harmonics of odd order or degree do not occur in the expansion and that a good approximation to the function may be obtained by taking terms of orders and degrees up to $n = 4$ and $m = 4$. We have then to determine the nine coefficients C_0, C_2, C_4, A_2^2, B_2^2, A_4^4, B_4^4, A_4^2, B_4^2. Three of these may be determined from the mean values of f round parallels of latitude, say $\theta = \frac{\pi}{2}$, $\theta = \frac{\pi}{3}$, $\theta = \frac{\pi}{6}$. Two equations connecting A_2^2, A_4^4, A_4^2 may be obtained from the mean values of f round the meridians $\phi = 0$, π and $\phi = \frac{\pi}{2}, \frac{3\pi}{2}$, while two equations involving B_2^2, B_4^2, A_4^4 may be obtained from the mean values round the circles $\phi = \frac{\pi}{4}, \frac{5\pi}{4}$; $\phi = \frac{3\pi}{4}$, $\phi = \frac{7\pi}{4}$. Further equations may be obtained from the mean values round the circles $\phi = \frac{\pi}{6}, \frac{7\pi}{6}$; $\phi = \frac{\pi}{3}, \frac{4\pi}{3}$; $\phi = \frac{2\pi}{3}, \frac{5\pi}{3}$; $\phi = \frac{5\pi}{6}, \frac{11\pi}{6}$.

Having found C_0, C_2, C_4 from the first three equations and having expressed A_2^2, A_4^2, B_2^2, B_4^2 in terms of A_4^4 with the aid of the next four, two of the last set of equations can be transformed into equations for A_4^4 and B_4^4.

When the two sets of curves have been drawn the mean values of f round the different circles may be found with the aid of a planimeter.

§ 6·36. *Expansion of a polynomial in a series of surface harmonics.* When $r^n F(\theta, \phi)$ is a polynomial of the nth degree in x, y, z the expansion of $F(\theta, \phi)$ in a series of surface harmonics may be obtained in an elementary way by using the operator ∇^2. Let us write $r^n F(\theta, \phi) = f_n(x, y, z)$. The first step is to determine a polynomial $f_{n-2}(x, y, z)$ such that

$$f_n(x, y, z) - r^2 f_{n-2}(x, y, z)$$

is a solution of Laplace's equation of type $r^n S_n(\theta, \phi)$. The equation

$$\nabla^2 [f_n - r^2 f_{n-2}] = 0$$

gives just enough equations to determine the coefficients in f_{n-2}. To show that the determinant of this system of linear equations does not vanish we must show that

$$\nabla^2 [r^2 f_{n-2}] \not\equiv 0.$$

If, however, $r^2 f_{n-2}$ were a spherical harmonic of degree n we should have $\int (r^2 f_{n-2}) f_{n-2} dS = 0$ when integrated over the spherical surface, because f_{n-2} can be expressed in terms of surface harmonics $S_m(\theta, \phi)$ of degree less than n. But this equation is impossible unless f_{n-2} vanishes identically.

Having found f_{n-2} we repeat the process with f_{n-2} in place of f_n and so on. We thus obtain a series of equations

$$f_n - r^2 f_{n-2} = r^2 S_n,$$
$$f_{n-2} - r^2 f_{n-4} = r^2 S_{n-2},$$
$$\cdots\cdots\cdots\cdots$$

from which we find that

$$f_n = r^n [S_n + S_{n-2} + \ldots] = r^n F(\theta, \phi).$$

When $n = 2m$, where m is an integer and $f_n = f$, the spherical harmonics are determined by the system of equations*

$$\nabla^{2m} f = (2m, 2)(2m+1, 3) S_0,$$
$$\nabla^{2m-2} f = (2m, 4)(2m+1, 5) r^2 S_0 + (2m-2, 2)(2m+3, 7) r^2 S_2,$$
$$\nabla^{2m-4} f = (2m, 6)(2m+1, 7) r^4 S_0 + (2m-2, 4)(2m+3, 9) r^4 S_2 + (2m-4, 2)(2m+5, 11) r^4 S_4,$$
$$\cdots\cdots\cdots\cdots\cdots\cdots$$

where $$(a, b) = a(a-2)(a-4)\ldots b.$$

Solving these linear equations we find that†

$$r^{2k} S_{2k} (2m-2k, 2)(2m+2k+1, 4k+3)$$
$$= \left[1 - \frac{r^2\nabla^2}{2(4k-1)} + \frac{r^4\nabla^4}{2.4(4k-1)(4k-3)} - \ldots\right] \nabla^{2m-2k} f(x, y, z)$$
$$= \frac{r^{4k+1}}{(4k-1, 1)} \psi_{2k}\left(\frac{\partial}{\partial x}, \frac{\partial}{\partial y}, \frac{\partial}{\partial z}\right)\frac{1}{r},$$

where $$\psi_{2k}(x, y, z) = \nabla^{2m-2k} f(x, y, z).$$

The equivalence of the two expressions for S is a consequence of Hobson's theorem (§ 6·22).

There is a corresponding theorem for a space of n dimensions. The fundamental formula for the effect of the operator

$$\nabla_n^2 \equiv \frac{\partial^2}{\partial x_1^2} + \frac{\partial^2}{\partial x_2^2} + \ldots \frac{\partial^2}{\partial x_n^2}$$

is $$\nabla_n^2 (r^{2p} v_q) = 2p(2p + 2q + n - 2) r^{2p-2} v_q + r^{2p} \nabla_n^2 v_q,$$

where $$r^2 = x_1^2 + x_2^2 + \ldots x_n^2, \quad v_q = v_q(x_1, x_2, \ldots x_n).$$

* If $v_q(x, y, z)$ is a rational integral homogeneous function of degree m, we have
$$\nabla^2 (r^{2p} v_q) = 2p(2p + 2q + 1) r^{2p-2} v_q + r^{2p} \nabla^2 v_q.$$
Hence if $\nabla^2 v_q = 0$ the effect of successive operations with ∇^2 is easily determined.

† G. Prasad, *Math. Ann.* vol. LXXII, p. 435 (1912).

The equations are now

$$\nabla_n^{2m} f = (2m, 2)(2m+n-2, n) S_0,$$

$$\nabla_n^{2m-2} f = (2m, 4)(2m+n-2, n+2) r^2 S_0 + (2m-2, 2)(2m+n, n+4) r^2 S_2,$$

....

$$r^{2k} S_{2k} (2m-2k, 2)(2m+2k+n-2, 4k+n)$$

$$= \left[1 - \frac{r^2 \nabla_n^2}{2(4k+n-4)} + \frac{r^4 \nabla_n^4}{2.4(4k+n-4)(4k+n-6)} - \dots\right] \nabla^{2m-2k} f$$

$$= \frac{r^{4k+n-2}}{(4k-4+n, n-2)} \psi_{2k} \left(\frac{\partial}{\partial x_1}, \frac{\partial}{\partial x_2}, \dots \frac{\partial}{\partial x_n}\right) r^{2-n},$$

where $\psi_{2k}(x_1, x_2, \dots x_n) = \nabla_n^{2m-2k} f(x_1, x_2, \dots x_n)$,

and f is a homogeneous polynomial of degree $2m$.

§ **6·41**. *Legendre functions and associated functions.* It should be observed that Laplace's equation possesses solutions of type

$$r^n P_n^m(\mu) e^{\pm im\phi}, \quad r^n Q_n^m(\mu) e^{\pm im\phi},$$

when n and m are any numbers. It is useful, therefore, to have definitions of the functions $P_n^m(\mu)$ and $Q_n^m(\mu)$ which will be applicable in such cases and also when μ is not restricted to the real interval $-1 \leqslant \mu \leqslant 1$.

The need for such definitions will appear later, but one reason why they are needed may be mentioned here.

In an attempt to generalise the method of inversion for transforming solutions of Laplace's equation* it was found that if

$$X = \frac{r^2 - a^2}{2(x-iy)}, \quad Y = -i\frac{r^2+a^2}{2(x-iy)}, \quad Z = \frac{az}{x-iy}, \quad \dots\dots(\text{I})$$

and if $f(X, Y, Z)$ is a solution of Laplace's equation in the co-ordinates X, Y, Z, then

$$(x-iy)^{-\frac{1}{2}} f(X, Y, Z)$$

is a solution of Laplace's equation in the variables x, y, z. Introducing polar co-ordinates, we find that

$$R = iae^{i\phi}, \quad r = iae^{i\Phi}, \quad \sin\Theta = \operatorname{cosec}\theta.$$

The standard simple solutions of Laplace's equation give rise, then, to new simple solutions of type

$$(\sin\theta)^{-\frac{1}{2}} P_n^m(i\cot\theta) e^{i(n+\frac{1}{2})\phi} r^{-\frac{1}{2}+m},$$

and we are led to infer the existence of reciprocal relations between associated Legendre functions with real and imaginary arguments and of more general relations when the arguments are complex quantities or real quantities not restricted to the interval $-1 \leqslant z \leqslant 1$.

Definitions of the associated Legendre functions $Q_n^m(z)$ for all values

* *Proc. London Math. Soc.* (2), vol. VII, p. 70 (1908).

of n, m and z have been given by E. W. Hobson* and by E. W. Barnes†. The definitions adopted by Barnes are as follows:

Let
$$z = x + iy, \quad w = \log\frac{z+1}{z-1},$$
$$\gamma(m, n, s) = \frac{i}{2}\frac{2^s\,\Gamma(1-m-s)}{\Gamma(s-n)\,\Gamma(n+1+s)\,\Gamma(-s)}e^{-\frac{1}{2}mw},$$
then, if
$$|\arg(z-1)| < \pi,$$
$$P_n^m(z) = -\sin n\pi\int\gamma(m, n, s)(z-1)^s\,ds,$$
where the integral is taken along a path parallel to the imaginary axis with loops if necessary to ensure that positive sequences of poles of the integrand lie to the right of the contour, and negative sequences to the left. Also
$$Q_n^m(z) = \frac{1}{2\pi}e^{-mw}\int\gamma(m, n, s)(-z-1)^s\,ds + e^{\mp n\pi i}I_m,$$
where $I_m = \pi\operatorname{cosec} n\pi . P_n^m(z)$, and the upper or lower sign is taken in the exponential factor multiplying I_m according as $y \gtrless 0$.

The functions $P_n^m(z)$, $Q_n^m(z)$ are not generally one-valued. To render their values unique a barrier is introduced from $-\infty$ to 1. When m is not a positive integer and z is not on the cross-cut, $P_n^m(z)$ is expressible in the form
$$P_n^m(z) = \frac{1}{\Gamma(1-m)}e^{\frac{1}{2}mw}F\{-n, n+1; 1-m; \tfrac{1}{2}(1-z)\},$$
where $F(a, b; c; x)$ denotes the hypergeometric function or its analytical continuation. This formula, which gives a convergent series when $|1-z| < 2$, shows that $z = 1$ is a singular point in the neighbourhood of which $P_n^m(z)$ has the form
$$(z-1)^{-\frac{1}{2}m}\{C_0 + C_1(z-1) + \ldots\}.$$
Under like conditions
$$\begin{aligned}-2Q_n^m(z) . \sin n\pi . \Gamma(-m-n) \\ = \frac{\pi\Gamma(m)}{\Gamma(1+m+n)}e^{\frac{1}{2}mw}F\{-n, n+1; 1-m; \tfrac{1}{2}(1-z)\} \\ + \frac{\pi\Gamma(-m)}{\Gamma(1+n-m)}e^{-\frac{1}{2}mw}F\{-n, n+1; 1+m; \tfrac{1}{2}(1-z)\},\end{aligned}$$
where, as before,
$$e^w = \frac{z+1}{z-1}.$$

The definition of $Q_n^m(z)$ given by Barnes differs from that given by Hobson, the relation between the two definitions being given by the formula
$$\sin n\pi\,[Q_n^m(z)]_B = e^{-im\pi}\sin(n+m)\pi . [Q_n^m(z)]_H.$$

* *Phil. Trans.* A, vol. CLXXXVII, p. 443 (1896).

† *Quart. Journ.* vol. XXXIX, p. 97 (1908).

It follows from the definitions that

$$P_n^m(-z) = e^{\mp n\pi i} P_n^m(z) - 2\frac{\sin n\pi}{\pi} Q_n^m(z),$$

$$Q_n^m(-z) = -Q_n^m(z) . e^{\pm n\pi i},$$

$$P^m_{-n-1}(z) = P_n^m(z),$$

$$Q^m_{-n-1}(z) = Q_n^m(z) - \pi \cot n\pi . P_n^m(z),$$

$$\frac{P_n^{-m}(z)}{\Gamma(1-m+n)} - \frac{P_n^m(z)}{\Gamma(1+m+n)} = 2Q_n^m(z)\,\Gamma(-m-n)\frac{\sin m\pi \sin n\pi}{\pi^2},$$

$$Q_n^{-m}(z)\,\Gamma(m-n) = Q_n^m(z)\,\Gamma(-m-n).$$

When $m = 0$, or when m is an integer, $P_n^m(z)$ has no singularity at $z = 1$. This is evident from the expression for $P_n^m(z)$ in terms of the hypergeometric function in the cases when m is negative or zero and may be derived from the formula

$$P_n^{-m}(z) = P_n^m(z)\frac{\Gamma(1-m+n)}{\Gamma(1+m+n)}$$

in the case when m is positive. We add some theorems without proofs.

1°. The nature of the singularity of $P_n^m(z)$ at $z = -1$ may be inferred from the formula

$$P_n^m(z)\,e^{-\frac{1}{2}m\varpi} = \frac{\Gamma(-m)}{\Gamma(1-m+n)\,\Gamma(-m-n)} F\{-n, n+1; 1+m; \theta\}$$

$$+\,\theta^{-m}\frac{\Gamma(m)}{\Gamma(-n)\,\Gamma(1+n)} F\{1-m+n, -m-n; 1-m; \theta\},$$

where $\theta = \frac{1}{2}(1+z)$. When m is a positive integer,

$$P_n^m(z) = (z^2-1)^{\frac{1}{2}m}\frac{\Gamma(m+n+1)}{2^m\Gamma(1+n-m)\,\Gamma(m+1)}$$

$$\times F\{m-n, m+n+1; m+1; \tfrac{1}{2}(1-z)\}. \quad \text{......(A)}$$

2°. When in addition n is an integer there are three cases:

(1) $0 < n < m$. In this case $P_n^m(z) = 0$ but $\Gamma(1+n-m)\,P_n^m(z)$ is a solution of the differential equation.

(2) $n \geqslant m$. In this case the formula (A) is valid.

(3) $n < 0$. In this case, if $-n > m$,

$$P_n^m(z) = (z^2-1)^{\frac{1}{2}m}\frac{\Gamma(m-n)}{2^m\Gamma(-m-n)\,\Gamma(m+1)}$$

$$\times F\{m-n, m+n+1; m+1; \tfrac{1}{2}(1-z)\}.$$

3°. If $-n \leqslant m$, $P_n^m(z) = 0$, but $\Gamma(-m-n)\,P_n^m(z)$ is a solution of the differential equation.

4°. When $m = 0$ and n is not an integer and is not zero,

$$(\pi \operatorname{cosec} n\pi)^2 P_n(z) = -\frac{1}{2\pi i}\int \{\Gamma(-s)\}^2\,\Gamma(-n+s)\,\Gamma(n+1+s)\,\theta^s\,ds$$

$$= \sum_{t=0}^{\infty} \frac{\Gamma(t-n)\,\Gamma(n+1+t)}{(t\,!)^2}\,\theta^t$$

$$\times \{\log\theta - 2\psi(1+t) + \psi(t-n) + \psi(n+1+t)\},$$

where
$$\theta = \tfrac{1}{2}(1+z) \text{ and } \psi(u) = \frac{d}{du}\log\Gamma(u).$$

Hence $P_n(z)$ has a logarithmic singularity at $x = -1$, at which it becomes infinite like

$$\pi^{-1}\sin n\pi\,.\,F\{-n, n+1; 1; \theta\}\log\theta + \text{a power series in } \theta.$$

5°. When $m = 0$, we have seen that $P_n(z)$ has a cross-cut from $-\infty$ to -1; when, however, $\tfrac{1}{2}m$ is not an integer and not zero, $P_n{}^m(z)$ has a cross-cut from $-\infty$ to 1, and is therefore not defined by the preceding formulae when $-1 < z < 1$. It is convenient to have a single value of the function in this interval, and one which is real when m and n are real. It is therefore assumed that as $\epsilon \to 0$ and $-1 < x < 1$,

$$P_n{}^m(x) = \lim e^{\frac{1}{2}m\pi i} P_n{}^m(x+\epsilon i) = \lim e^{-\frac{1}{2}\pi m i} P_n{}^m(x-\epsilon i)$$

$$= \frac{1}{\Gamma(1-m)}\left(\frac{1+x}{1-x}\right)^{\frac{1}{2}m} F\{-n, n+1; 1-m; \tfrac{1}{2}-\tfrac{1}{2}x\},$$

$$Q_n{}^m(x) = \tfrac{1}{2}\lim\{Q_n{}^m(x+\epsilon i) + Q_n{}^m(x-\epsilon i)\}$$

$$= -\frac{\pi\cos\frac{1}{2}m\pi}{2\sin n\pi\,\Gamma(-m-n)}\,[\Phi^{(m)}(x) + \Phi^{(-m)}(x)],$$

where

$$\Phi^{(m)}(x) = \frac{\Gamma(m)}{\Gamma(1+m+n)}\left(\frac{1+x}{1-x}\right)^{\frac{1}{2}m} F(-n, n+1; 1-m; \tfrac{1}{2}-\tfrac{1}{2}x).$$

6°. The function $Q_n{}^m(x)$ has a cross-cut between -1 and 1. For values of z for which $|z| > 1$ the function can be expanded in a convergent power series in $1/z$. If $|\arg(z \pm 1)| < \pi$ and $(z^2-1)^{\frac{1}{2}m} = (z-1)^{\frac{1}{2}m}(z+1)^{\frac{1}{2}m}$,

$$Q_n{}^m(z) = \frac{\sin(n+m)\pi}{\sin n\pi}\;\frac{\Gamma(n+m+1)\,\Gamma(\frac{1}{2})}{2^{n+1}\,\Gamma(n+\frac{3}{2})}\;\frac{(z^2-1)^{\frac{1}{2}m}}{z^{n+m+1}}$$

$$\times F(\tfrac{1}{2}n+\tfrac{1}{2}m+1, \tfrac{1}{2}n+\tfrac{1}{2}m+\tfrac{1}{2}; n+\tfrac{3}{2}; z^{-2}).$$

The values for cases in which $|z| < 1$ may be deduced by analytical continuation of the hypergeometric function and use of the foregoing definition when z is real.

§ **6·42**. *Reciprocal relations**. Barnes has shown that the power series in $1/z$ can, under the foregoing conditions, be expressed in the form

$$Q_n^m(z) = C\frac{(z^2-1)^{\frac{1}{2}m}}{z^{n+m+1}}\left(1-\frac{1}{z^2}\right)^{-\frac{1}{2}(n+m+1)}$$
$$\times F\left\{\tfrac{1}{2}(n+m+1), \tfrac{1}{2}(n-m+1); n+\tfrac{3}{2}; \frac{1}{1-z^2}\right\},$$

where
$$C = \frac{\sin(n+m)\pi}{\sin n\pi}\frac{\Gamma(n+m+1)\Gamma(\frac{1}{2})}{2^{n+1}\Gamma(n+\frac{3}{2})}.$$

Putting $z = i\cot\theta$, we have

$$Q_n^m(i\cot\theta) = Ci^{-n-1}\sin^{n+1}\theta$$
$$\times F\{\tfrac{1}{2}(n+m+1), \tfrac{1}{2}(n-m+1); n+\tfrac{3}{2}; \sin^2\theta\}.$$

Now

$$F\{\tfrac{1}{2}(n+m+1), \tfrac{1}{2}(n-m+1); n+\tfrac{3}{2}; \sin^2\theta\}$$
$$= F\{n+m+1, n-m+1; n+\tfrac{3}{2}; \sin^2\tfrac{1}{2}\theta\}$$
$$= (\cos\tfrac{1}{2}\theta)^{-2n-1}F[m+\tfrac{1}{2}, \tfrac{1}{2}-m; n+\tfrac{3}{2}; \sin^2\tfrac{1}{2}\theta].$$

Therefore

$$Q_n^m(i\cot\theta) = Ci^{-n-1}2^{n+\frac{1}{2}}(\sin\theta)^{\frac{1}{2}}(\cot\tfrac{1}{2}\theta)^{-n-\frac{1}{2}}$$
$$\times F[m+\tfrac{1}{2}, \tfrac{1}{2}-m; n+\tfrac{3}{2}; \sin\tfrac{1}{2}\theta].$$

But

$$P_{-m-\frac{1}{2}}^{-n-\frac{1}{2}}(\cos\theta) = \frac{1}{\Gamma(n+\frac{3}{2})}(\cot\tfrac{1}{2}\theta)^{-n-\frac{1}{2}}F\{m+\tfrac{1}{2}, \tfrac{1}{2}-m; n+\tfrac{3}{2}; \sin^2\tfrac{1}{2}\theta\}.$$

Therefore

$$Q_n^m(i\cot\theta) = -\frac{\pi^{\frac{3}{2}}i^{-n-1}}{\sin n\pi.\Gamma(-m-n)}(\tfrac{1}{2}\sin\theta)^{\frac{1}{2}}P_{-m-\frac{1}{2}}^{-n-\frac{1}{2}}(\cos\theta).$$

Writing $-m-\frac{1}{2}$ in place of n and $-n-\frac{1}{2}$ in place of m, the formula becomes

$$Q_{-m-\frac{1}{2}}^{-n-\frac{1}{2}}(i\cot\theta) = \frac{\pi^{\frac{3}{2}}i^{m-\frac{1}{2}}}{\cos m\pi.\Gamma(m+n+1)}(\tfrac{1}{2}\sin\theta)^{\frac{1}{2}}P_n^m(\cos\theta).$$

Again, Barnes has shown that when $|1-z^2| > 1$,

$$P_n^m(z) = C_1(z^2-1)^{-\frac{1}{2}(n+1)}F\left\{\tfrac{1}{2}(n+1-m), \tfrac{1}{2}(n+m+1); \tfrac{3}{2}+n; \frac{1}{1-z^2}\right\}$$
$$+ C_2(z^2-1)^{\frac{1}{2}n}F\left\{\tfrac{1}{2}(m-n), \tfrac{1}{2}(-m-n); \tfrac{1}{2}-n; \frac{1}{1-z^2}\right\},$$

where

$$C_1 = \frac{2^{-n-1}\Gamma(-n-\frac{1}{2})}{\pi^{\frac{1}{2}}\Gamma(-m-n)}, \quad C_2 = \frac{2^n\Gamma(n+\frac{1}{2})}{\pi^{\frac{1}{2}}\Gamma(n-m+1)};$$

* Judging from a conversation with Dr Barnes in 1908 he had at that time noted at least one explicit reciprocal relation between the functions $P_n^m(z)$ and $Q_n^m(z)$.

consequently, using again the transformations of the hypergeometric series, we find that

$$P_n^m(i\cot\theta) = (2\pi\operatorname{cosec}\theta)^{-\frac{1}{2}}\left[\frac{\Gamma(-n-\frac{1}{2})}{\Gamma(-m-n)}e^{-\frac{1}{2}(n+1)i\pi}(\tan\tfrac{1}{2}\theta)^{n+\frac{1}{2}}\right.$$
$$\times F\{\tfrac{1}{2}+m, \tfrac{1}{2}-m;\ n+\tfrac{3}{2}; \sin^2\tfrac{1}{2}\theta\}$$
$$\left.+\frac{\Gamma(n+\frac{1}{2})}{\Gamma(n-m+1)}e^{\frac{1}{2}ni\pi}(\cot\tfrac{1}{2}\theta)^{n+\frac{1}{2}}F\{\tfrac{1}{2}-m, \tfrac{1}{2}+m; \tfrac{1}{2}-n; \sin^2\tfrac{1}{2}\theta\}\right]$$
$$= \frac{2}{\pi}\sin(m+\tfrac{1}{2})\pi.\Gamma(m+n+1)(2\pi\operatorname{cosec}\theta)^{-\frac{1}{2}}\lim_{\epsilon\to 0}Q_{-m-\frac{1}{2}}^{-n-\frac{1}{2}}(\cos\theta - i\epsilon),$$

and so

$$P_{-m-\frac{1}{2}}^{-n-\frac{1}{2}}(i\cot\theta) = -\frac{2}{\pi}\sin n\pi.\Gamma(-m-n)(2\pi\operatorname{cosec}\theta)^{-\frac{1}{2}}\lim_{\epsilon\to 0}Q_n^m(\cos\theta - i\epsilon)$$
$$= -\frac{2}{\Gamma(1+m+n)}\frac{\sin n\pi}{\sin(m+n)\pi}(2\pi\operatorname{cosec}\theta)^{-\frac{1}{2}}\lim_{\epsilon\to 0}Q_n^m(\cos\theta - i\epsilon).$$

This is very similar to the reciprocal formula obtained by F. J. W. Whipple*, which may be written in the form

$$\frac{\sin n\pi}{\sin(n+m)\pi}Q_n^m(\cosh\alpha) = \frac{\pi\Gamma(1+m+n)}{(2\pi\sinh\alpha)^{\frac{1}{2}}}P_{-m-\frac{1}{2}}^{-n-\frac{1}{2}}(\coth\alpha).$$

EXAMPLES

1. Prove that when n is a positive integer

$$\frac{d}{dn}P_n(\mu) = \frac{2}{2^n.n!}\left(\frac{d}{d\mu}\right)^n\{(\mu^2-1)^n\log\tfrac{1}{2}(1+\mu)\} - P_n(\mu).\log\tfrac{1}{2}(1+\mu).$$

[A. E. Jolliffe, *Mess. of Math.* vol. XLIX, p. 125 (1919).]

2. Prove that if $2x = t^{\frac{1}{2}} + t^{-\frac{1}{2}}$,

$$P_{-\frac{1}{2}}(x) = F\left(\tfrac{1}{2}, \tfrac{1}{2}; 1; \frac{1-x}{2}\right)$$
$$= t^{\frac{1}{4}}F(\tfrac{1}{2}, \tfrac{1}{2}, 1, 1-t)$$
$$= \frac{2}{\pi}t^{\frac{1}{4}}\left[(2\log 2 - \tfrac{1}{2}\log t)F(\tfrac{1}{2}, \tfrac{1}{2}; 1; t)\right.$$
$$\left.-\left\{\left(\frac{1}{2}\right)^2 t + \left(\frac{1.3}{2.4}\right)^2 1 + \left(\frac{2}{3.4}\right)t^2 + \left(\frac{1.3.5}{2.4.6}\right)^2\left(1+\frac{2}{3.4}+\frac{2}{5.6}\right)t^3 + \ldots\right\}\right].$$

Prove also that

$$P'_{-\frac{1}{2}}(x) = -\tfrac{1}{8}F\left(\tfrac{3}{2}, \tfrac{3}{2}; 2; \frac{1-x}{2}\right)$$
$$= -\tfrac{1}{8}t^{\frac{3}{4}}F(\tfrac{3}{2}, \tfrac{3}{2}; 3; 1-t)$$
$$= -\frac{1}{\pi}t^{\frac{3}{4}}[(4\log 2 - 4 - \log t)F(\tfrac{3}{2}, \tfrac{3}{2}; 1; t) - 4\{(\tfrac{3}{2})^2(\tfrac{1}{3}-\tfrac{1}{2})t$$
$$+ \tfrac{1}{2}[(\tfrac{3}{2})^2 - (\tfrac{5}{2})^2](\tfrac{1}{3}+\tfrac{1}{5}-\tfrac{1}{2}-\tfrac{1}{4})t^2 + \ldots\}].$$

[H. V. Lowry, *Phil. Mag.* (7), vol. II, p. 1184 (1926).]

* *Proc. London Math. Soc.* (2), vol. XVI, p. 301 (1917).

3. If K is the quarter period of elliptic functions with modulus k and complementary modulus $k' = (1-k^2)^{\frac{1}{2}}$, prove that

$$\begin{aligned} K &= \frac{\pi}{2} F(\tfrac{1}{2}, \tfrac{1}{2}; 1; k^2) \\ &= \frac{\pi}{2} P_{-\frac{1}{2}}(1-2k^2) \\ &= \frac{\pi}{2k'} F\left(\tfrac{1}{2}, \tfrac{1}{2}; 1; \frac{k^2}{k^2-1}\right) \\ &= \frac{\pi}{2k'} P_{-\frac{1}{2}}\left(\frac{1+k^2}{1-k^2}\right) \\ &= \frac{\pi}{2(1-k)} F\left(\tfrac{1}{2}, \tfrac{1}{2}; 1; \frac{4k}{(1+k)^2}\right) \\ &= \frac{1}{1-k}\left[\left\{2\log 2 - \log\left(\frac{1-k}{1+k}\right)\right\} F\left\{\tfrac{1}{2}, \tfrac{1}{2}; 1; \frac{1-k^2}{1+k^2}\right\}\right. \\ &\qquad \left. - \left\{\left(\frac{1}{2}\right)^2\left(\frac{1-k}{1+k}\right)^2 + \left(\frac{1.3}{2.4}\right)^2\left(1+\frac{2}{3.4}\right)\left(\frac{1-k}{1+k}\right)^4 + \ldots\right\}\right]. \end{aligned}$$

The last series is recommended by Lowry for the calculation of K when k is nearly 1.

4. Prove that if n is a positive integer the equation $P_{n-\frac{1}{2}}(z) = 0$ has no root which lies in the range $1 < z < 3$.

5. Prove that if $2n+1 \neq 0$

$$\frac{1}{2}\int_0^{\pi} P_n(1-2\sin^2 a \sin^2\theta)\sin a \, da = \frac{\sin(2n+1)\theta}{(2n+1)\sin\theta}.$$

6. Prove that if n is a positive integer

$$P_n(1-2\sin^2 a\sin^2\theta) = [P_n(\cos\theta)]^2 + 2\Sigma\frac{(n-m)!}{(n+m)!}[P_n^m(\cos\theta)]^2\cos 2ma$$

and deduce that for $0 < \theta < \pi$

$$[P_n(\cos\theta)]^2 > \frac{\sin(2n+1)\theta}{(2n+1)\sin\theta}.$$

Hence show that the roots of the equation $P_n(\cos\theta) = 0$ can only lie within certain intervals in which $\sin(2n+1)\theta$ is negative.

§ **6·43**. *Potential functions of degree* $n+\frac{1}{2}$ *where* n *is an integer.* If we apply the imaginary transformation (I) to the potential functions of degree zero and -1 we find that if $\rho^2 = x^2+y^2$ and $F(\ )$, $G(\ )$ are arbitrary functions of their arguments, the functions

$$V = (x-iy)^{-\frac{1}{2}}\left[F(\rho - iz) + G\left(\frac{\rho - iz}{r^2}\right)\right],$$

$$V = (x+iy)^{-\frac{1}{2}}\left[F(\rho - iz) + G\left(\frac{\rho - iz}{r^2}\right)\right]$$

are solutions of Laplace's equation. In particular,

$$(x-iy)^{-\frac{1}{2}}(\rho - iz)^{n+1}$$

is a homogeneous solution of degree $n+\frac{1}{2}$. Differentiating this k times with respect to x we obtain a solution which may be written in the form

$$\sum_{s=0}^{k}\binom{k}{s}\frac{\partial^s}{\partial x^s}(x-iy)^{-\frac{1}{2}}\frac{\partial^{k-s}}{\partial x^{k-s}}(\rho - iz)^{n+1}.$$

The typical term of this series is a constant multiple of

$$\rho^{-s-\frac{1}{2}} e^{i(s+\frac{1}{2})\phi} \frac{\partial^{k-s}}{\partial x^{k-s}} (\rho - iz)^{n+1},$$

and when the derivative in this expression is expanded in powers of x, the coefficient of x^{k-s} is

$$\left(\frac{1}{\rho}\frac{\partial}{\partial \rho}\right)^{k-s} (\rho - iz)^{n+1}.$$

Now $$x = \tfrac{1}{2}\rho\,[e^{i\phi} + e^{-i\phi}],$$

consequently the term involving x^{k-s} gives rise to a term in our series with the exponential factor $e^{i(k+\frac{1}{2})\phi}$ and a number of terms with exponential factors of lower order. Taking all the terms with the exponential factor $e^{i(k+\frac{1}{2})\phi}$ we must get a solution of Laplace's equation and this solution is represented by the series

$$V = e^{i(k+\frac{1}{2})\phi} \sum_{s=0}^{k} C_s \rho^{k-2s-\frac{1}{2}} \left(\frac{1}{\rho}\frac{\partial}{\partial \rho}\right)^{k-s} (\rho - iz)^{n+1},$$

where $$C_s = (-)^s \frac{k!\,(2s)!}{2^{k+s}\,(s!)^2\,(k-s)!}.$$

We may conclude that if $f(a)$ is an arbitrary function with a suitable number of continuous derivatives, the series

$$V = e^{i(k+\frac{1}{2})\phi} \sum_{s=0}^{k} C_s \rho^{k-2s-\frac{1}{2}} \left(\frac{1}{\rho}\frac{\partial}{\partial \rho}\right)^{k-s} f(\rho - iz)$$

is a solution of Laplace's equation.

§ **6·44**. *Conical harmonics**. When V is independent of ϕ a set of solutions suitable for the treatment of problems relating to a cone may be obtained by writing $r = ce^s$, $V = (cr)^{-\frac{1}{2}} u$. The equation for u is then

$$\frac{\partial}{\partial \mu}\left\{(1-\mu^2)\frac{\partial u}{\partial \mu}\right\} + \frac{\partial^2 u}{\partial s^2} - \tfrac{1}{4}u = 0,$$

and there are solutions of type $u = \cos(ks)\,K^{(k)}(\mu)$, where $K^{(k)}(\mu)$ is a solution of the differential equation

$$\frac{d}{d\mu}\left\{(1-\mu^2)\frac{dK}{d\mu}\right\} - (k^2 + \tfrac{1}{4})\,K = 0.$$

Mehler writes

$$K^{(k)}(\mu) = \frac{2}{\pi}\cosh(k\pi)\int_0^\infty \frac{\cos k\alpha\,.\,d\alpha}{[2(\cosh\alpha + \mu)]^{\frac{1}{2}}},$$

and remarks that $K^{(k)}(\mu)$, $K^{(k)}(-\mu)$ are two essentially different solutions

* F. G. Mehler, *Math. Ann.* Bd. XVIII, S. 161 (1881); C. Neumann, *ibid.* S. 195; E. Heine, *Kugelfunktionen*, Bd. II, S. 217–250.

of the differential equation. The function $K^{(k)}(\mu)$ can be expanded in a power series

$$K^{(k)}(\mu) = F\left(\tfrac{1}{2}+ki, \tfrac{1}{2}-ki; 1; \frac{1-\mu}{2}\right)$$

$$= 1 + \frac{4k^2+1^2}{2^2}\left(\frac{1-\mu}{2}\right) + \frac{(4k^2+1^2)(4k^2+3^2)}{2^2.4^2}\left(\frac{1-\mu}{2}\right)^2 + \ldots,$$

and its relation to the Legendre function becomes clear.

There is also an expansion

$$K^{(k)}(\mu) = \frac{L\sqrt{\pi}}{\Gamma\left(\frac{3}{4}+\frac{ki}{2}\right)\Gamma\left(\frac{3}{4}-\frac{ki}{2}\right)} - \frac{2M\sqrt{\pi}}{\Gamma\left(\frac{1}{4}+\frac{ki}{2}\right)\Gamma\left(\frac{1}{4}-\frac{ki}{2}\right)},$$

where
$$L = 1 + \frac{4k^2+1^2}{2.4}\mu^2 + \frac{(4k^2+1^2)(4k^2+5^2)}{2.4.6.8}\mu^5 + \ldots,$$

$$M = \mu + \frac{4k^2+3^2}{4.6}\mu^3 + \frac{(4k^2+3^2)(4k^2+7^2)}{4.6.8.(10)}\mu^5 + \ldots.$$

Problems relating to a cone have also been treated with the aid of the ordinary Legendre functions*.

If, for instance, there is a charge q on the axis of z at a distance a from the origin, and the cone $\theta = \alpha$ is at zero potential, the potential V is given by the series

$$V = -2(q/a)\,\Sigma\,(r/a)^n \frac{P_n(\mu)}{\sin^2\alpha\left(\frac{\partial P_n}{\partial n}\frac{\partial P_n}{\partial \mu}\right)_{\theta=a}} \qquad r < a$$

$$= -2(q/r)\,\Sigma\,(a/r)^n \frac{P_n(\mu)}{\sin^2\alpha\left(\frac{\partial P_n}{\partial n}\frac{\partial P_n}{\partial \mu}\right)_{\theta=a}} \qquad r > a,$$

where the summations extend over all the positive values of n which make $P_n(\cos\alpha) = 0$.

EXAMPLES

1. Prove that when $0 < \theta < \frac{1}{2}\pi$

$$K^{(k)}(\cos\theta) = F\left(\frac{1}{4}+\frac{ki}{2}, \frac{1}{4}-\frac{ki}{2}; 1; \sin^2\theta\right)$$

$$= 1 + \frac{4\mu^2+1^2}{4^2}\sin^2\theta + \frac{(4\mu^2+1^2)(4\mu^2+5^2)}{4^2.8^2}\sin^4\theta + \ldots.$$

When $\frac{1}{2}\pi < \theta < \pi$ the series represents $K^{(k)}(-\cos\theta)$.

2. Prove that

$$K^{(k)}(\cos\theta) = \frac{2}{\pi}\int_0^\theta \frac{\cosh(\mu\beta)\,d\beta}{[2(\cos\beta-\cos\theta)]^{\frac{1}{2}}}.$$

3. Prove that

$$(r^2+c^2-2rc\cos\theta)^{-\frac{1}{2}} = (rc)^{-\frac{1}{2}}\int_0^\infty \frac{\cos(ks)}{\cosh(k\pi)} K^{(k)}(-\cos\theta)\,dk.$$

* See, for instance, H. M. Macdonald, *Camb. Phil. Trans.* vol XVIII, p. 292 (1899).

4. Prove that when k is large and positive

$$K^{(k)}(\cos\theta) \sim e^{k\theta}(1+\delta)(2\pi k\sin\theta)^{-\frac{1}{2}},$$

where $\delta \to 0$ as $k \to \infty$. The point $\theta = \pi$ must be excluded, for $K(\cos\theta)$ has no meaning for $\theta = \pi$.

5. Prove that

$$K^{(k)}(\mu) = \frac{\cosh(k\pi)}{\pi}\int_0^\infty \frac{x^{ki-\frac{1}{2}}dx}{\sqrt{(1+2\mu x+x^2)}},$$

$$K^{(k)}(-\mu) = \frac{\cosh k\pi}{\pi}\int_1^\infty \frac{K^{(k)}(\nu)\,d\nu}{\nu-\mu},$$

$$\int_0^{2\pi} K^{(k)}(\cosh u\cosh v - \sinh u\sinh v\cos\phi)\,d\phi = 2\pi K^{(k)}(\cosh u)\,K^{(k)}(\cosh v).$$

6. If $$f(x) = \int_0^\infty K^{(k)}(x)\tanh(k\pi)\,k\phi(k)\,dk,$$

show that under suitable conditions

$$\phi(k) = \int_1^\infty K^{(k)}(x)\,f(x)\,dx.$$

The results of examples 1–6 are all due to Mehler.

§ 6·51. *Solutions of the wave equation.* These may be found with the aid of the Green's substitution

$$x = s\sin\alpha\sin\beta\cos\phi, \quad y = s\sin\alpha\sin\beta\sin\phi,$$

$$z = s\sin\alpha\cos\beta, \quad ict = s\cos\alpha,$$

$$dx^2+dy^2+dz^2-c^2dt^2 = ds^2+s^2d\alpha^2+s^2\sin^2\alpha\,d\beta^2+s^2\sin^2\alpha\sin^2\beta\,.\,d\phi^2,$$

$$d(x, y, z, t) = s^2\sin^2\alpha\sin\beta\,d(s, \alpha, \beta, \phi).$$

The wave equation now becomes

$$\frac{\partial^2u}{\partial s^2}+\frac{3}{s}\frac{\partial u}{\partial s}+\frac{2}{s^2}\cot\alpha\frac{\partial u}{\partial\alpha}+\frac{1}{s^2}\frac{\partial^2u}{\partial\alpha^2}$$

$$+\frac{1}{s^2\sin^2\alpha\sin^2\beta}\frac{\partial}{\partial\beta}\left(\sin\beta\frac{\partial u}{\partial\beta}\right)+\frac{1}{s^2\sin^2\alpha\sin^2\beta}\frac{\partial^2u}{\partial\phi^2} = 0,$$

and possesses elementary solutions of the form

$$u = s^nA(\alpha)\,B(\beta)\cos m(\phi-\phi_0)$$

if $$\operatorname{cosec}\beta\frac{d}{d\beta}\left(\sin\beta\frac{dB}{d\beta}\right)+\left[\nu(\nu+1)-\frac{m^2}{\sin^2\beta}\right]B = 0,$$

$$\frac{d^2A}{d\alpha^2}+2\cot\alpha\frac{dA}{d\alpha}+\left[n(n+2)-\frac{\nu(\nu+1)}{\sin^2\alpha}\right]A = 0.$$

We may thus write

$$B = c_1P_m^{\nu}(\cos\beta)+c_2Q_m^{\nu}(\cos\beta),$$

$$A = \sqrt{\operatorname{cosec}\alpha}\,[b_1P_{n+\frac{1}{2}}^{\nu+\frac{1}{2}}(\cos\alpha)+b_2Q_{n+\frac{1}{2}}^{\nu+\frac{1}{2}}(\cos\alpha)],$$

where c_1, c_2, b_1, b_2 are arbitrary constants. On account of the reciprocal relation between the Legendre functions we may also write

$$A = \operatorname{cosec}\alpha\,[a_1P_{-\nu-1}^{-n-1}(i\cot\alpha)+a_2Q_{-\nu-1}^{-n-1}(i\cot\alpha)],$$

where a_1 and a_2 are new arbitrary constants.

The analysis is easily extended to Laplace's equation in $n+2$ variables. The appropriate substitution is

$$x_1 = r\cos\theta_1, \quad x_2 = r\sin\theta_1\cos\theta_2, \quad x_3 = r\sin\theta_1\sin\theta_2\cos\theta_3,$$

$$x_{n+1} = r\sin\theta_1\sin\theta_2\ldots\sin\theta_n\cos\phi, \quad x_{n+2} = r\sin\theta_1\sin\theta_2\ldots\sin\theta_n\sin\phi,$$

and Laplace's equation becomes

$$\frac{\partial}{\partial r}\left[r^{n+1}\sin^n\theta_1\sin^{n-1}\theta_2\ldots\sin\theta_n\frac{\partial u}{\partial r}\right]+\frac{\partial}{\partial\theta_1}\left[r^{n-1}\sin^n\theta_1\sin^{n-1}\theta_2\ldots\sin\theta_n\frac{\partial u}{\partial\theta_1}\right]$$
$$+\frac{\partial}{\partial\theta_2}\left[r^{n-1}\sin^{n-2}\theta_1\sin^{n-1}\theta_2\ldots\sin\theta_n\frac{\partial u}{\partial\theta_2}\right]+\ldots$$
$$+\frac{\partial}{\partial\phi}\left[r^{n-1}\sin^{n-2}\theta_1\sin^{n-3}\theta_2\ldots\operatorname{cosec}\theta_n\frac{\partial u}{\partial\phi}\right]=0.$$

Assuming that there is an elementary solution of type

$$u = R(r)\,\Theta_1(\theta_1)\,\Theta_2(\theta_2)\ldots\Theta_n(\theta_n)\cos(m\phi+\epsilon),$$

we obtain the equations

$$\frac{d^2\Theta_n}{d\theta_n^{\,2}}+\cot\theta_n\frac{d\Theta_n}{d\theta_n}+[\nu_n(\nu_n+1)-m^2\operatorname{cosec}^2\theta_n]\,\Theta_n=0,$$

$$\frac{d^2\Theta_s}{d\theta_s^{\,2}}+(n-s+1)\cot\theta_s\frac{d\Theta_s}{d\theta_s}+[\nu_s(\nu_s+n-s+1)$$
$$-\nu_{s+1}(\nu_{s+1}+n-s)\operatorname{cosec}^2\theta_s]\,\Theta_s=0,$$

$$\frac{d^2R}{dr^2}+\frac{n+1}{r}\frac{dR}{dr}-\frac{\nu_1(\nu_1+n)}{r^2}R=0,$$

and so we may write

$$\Theta_n = A_nP_{\nu_n}^{\ m}(\cos\theta_n)+B_nQ_{\nu_n}^{\ m}(\cos\theta_n),$$
$$\Theta_s = (\operatorname{cosec}\theta_s)^{\frac{1}{2}(n-s)}[A_sP_{\omega_s}^{\ \sigma_s}(\cos\theta_s)+B_sQ_{\omega_s}^{\ \sigma_s}(\cos\theta_s)],$$
$$R = Ar^{\nu_1}+Br^{-n-\nu_1},$$

where $\omega_s=\nu_s+\frac{1}{2}(n-s), \quad \sigma_s=\nu_{s+1}+\frac{1}{2}(n-s).$

If in place of Laplace's equation we consider the equation*

$$\sum_{s=1}^{n+2}\frac{\partial^2V}{\partial x_s^{\,2}}+k^2V=0,$$

the analysis is the same as before except that now the equation for R is

$$\frac{d^2R}{dr^2}+\frac{n+1}{r}\frac{dR}{dr}+\left[k^2-\frac{\nu_1(\nu_1+n)}{r^2}\right]R=0,$$

and the solution is

$$R=[AJ_{\nu_1+\frac{1}{2}n}(kr)+BJ_{-\nu_1-\frac{1}{2}n}(kr)]\,r^{-\frac{n}{2}}.$$

If $\nu_1 = s-\frac{1}{2}n$, where s is an integer, one of these Bessel functions must be replaced by $Y_s(kr)$, the second solution of Bessel's equation.

* E. W. Hobson, *Proc. London Math. Soc.* vol. xxv, p. 49 (1894).

It should be remarked that the elementary solid angle in the generalised space for which $x_1, x_2, \dots x_{n+2}$ are rectangular co-ordinates is

$$d\omega = \sin^n \theta_1 . \sin^{n-1} \theta_2 \dots \sin \theta_n \, d\,(\theta_1, \theta_2, \dots \theta_n . \phi).$$

When this is integrated over all angular space the result is

$$\frac{s\,[\Gamma(\tfrac{1}{2})]^s}{\Gamma\left(\dfrac{s}{2}+1\right)},$$

where $s = n + 2$. See, for instance, P. H. Schoute, *Mehrdimensionale Geometrie*, Bd. II, S. 289.

For ordinary space $n = 1$, and the foregoing result tells us that the equation $\nabla^2 V + k^2 V = 0$ is satisfied by

$$V = r^{-\frac{1}{2}} J_{s+\frac{1}{2}}(kr)\, P_s^m(\cos\theta) \cos m\phi.$$

In particular, when $s = 0$ we have the solution

$$V = r^{-\frac{1}{2}} J_{\frac{1}{2}}(kr) = (2/k\pi)^{\frac{1}{2}} \frac{\sin kr}{r},$$

and when $s = -1$ there is a corresponding solution

$$V = r^{-\frac{1}{2}} J_{-\frac{1}{2}}(kr) = (2/k\pi)^{\frac{1}{2}} \frac{\cos kr}{r}.$$

These may be combined so as to give the solution

$$V = \frac{e^{-ikr}}{r},$$

which arises naturally from the wave-function

$$V = \frac{1}{r} e^{ik(ct-r)}$$

suitable for the representation of waves diverging in three-dimensional space. This type of wave-function is fundamental in the theory of Hertzian waves and also in the theory of sound. The general function $r^{-\frac{1}{2}} J_{s+\frac{1}{2}}(kr)$ can be expressed in the form

$$\sigma^{-\frac{1}{2}} J_{s+\frac{1}{2}}(\sigma) = (-)^s (2\sigma)^s (2/\pi)^{\frac{1}{2}} \frac{d^s}{d(\sigma^2)^s}\left(\frac{\sin\sigma}{\sigma}\right),$$

and is easily seen to be of the form

$$P_s(\sigma) \sin\sigma + Q_s(\sigma)\cos\sigma,$$

where P and Q are polynomials in σ^{-1}.

For some purposes it is convenient to use the notation

$$\psi_n(x) = (\tfrac{1}{2}\pi x)^{\frac{1}{2}} J_{n+\frac{1}{2}}(x),$$
$$\chi_n(x) = (-)^n (\tfrac{1}{2}\pi x)^{\frac{1}{2}} J_{-n-\frac{1}{2}}(x) = -(\tfrac{1}{2}\pi x)^{\frac{1}{2}} Y_{n+\frac{1}{2}}(x),$$
$$\eta_n(x) = \psi_n(x) - i\chi_n(x),$$
$$\zeta_n(x) = \psi_n(x) + i\chi_n(x).$$

These new functions η_n, ζ_n are connected with Hankel's cylindrical functions by the relations

$$\eta_n(x) = (\tfrac{1}{2}\pi x)^{\frac{1}{2}} H_1^{\,n+\frac{1}{2}}(x), \quad \zeta_n(x) = (\tfrac{1}{2}\pi x)^{\frac{1}{2}} H_2^{\,n+\frac{1}{2}}(x).$$

When $|x|$ is large and $|\arg x| < \pi$ we have the asymptotic expansion*

$$\eta_n(x) = (-i)^{n+1} e^{ix}\left[1 + \frac{i}{2x}\frac{n(n+1)}{1!} + \left(\frac{i}{2x}\right)^2 \frac{(n-1)n(n+1)(n+2)}{2!} + \dots\right].$$

The series terminates when n is an integer and gives an exact representation of the function. In this case we may write

$$\zeta_n(kr) = i^{n+1} e^{-ikr}\left[1 - \frac{i}{2kr}\frac{n(n+1)}{1} + \dots \left(-\frac{i}{2kr}\right)^n \frac{(2n!)}{n!}\right],$$

and when kr is real a series for $|\zeta_n(kr)|^2$ may be obtained from the linear differential equation of the third order satisfied by $[\psi_n(kr)]^2$, $[\chi_n(kr)]^2$ and $\psi_n(kr)\,\chi_n(kr)$. This series

$$|\zeta_n(kr)|^2 = 1 + \tfrac{1}{2}n(n+1)\frac{1}{(kr)^2} + \frac{1.3}{2.4}(n-1)n(n+1)(n+2)\frac{1}{(kr)^4} + \dots \frac{1.3\dots(2n-1)}{2.4\dots 2n}(2n!)\frac{1}{(kr)^{2n}},$$

which contains only positive terms, shows that $|\zeta_n(kr)|$ decreases as kr increases, hence, if a series of the form

$$\sum_{n=0}^{\infty} \zeta_n(kr) f_n(\theta, \phi)$$

converges absolutely for any value of kr greater than zero, it converges absolutely for all greater values of kr.

For small values of $|r|$ the function $\psi_n(kr)$ is represented by the series

$$\psi_n(kr) = \frac{(kr)^{n+1}}{1.3\dots(2n+1)}\left[1 - \frac{(kr)^2}{2(2n+3)} + \frac{(kr)^4}{2.4(2n+3)(2n+5)} - \dots\right],$$

which converges for all values of r. For large values of $|r|$ we have approximately

$$\psi_n(kr) = \tfrac{1}{2}\left[i^{n+1} e^{-ikr} + (-i)^{n+1} e^{ikr}\right].$$

This formula is exact when $n = 0$ and $\psi_n(kr) = \sin kr$, and when $n = -1$ and $\psi_n(kr) = \cos kr$. In other cases it represents simply the first term of an expansion which terminates when n is an integer. It should be remarked that

$$\psi_1(kr) = \frac{\sin kr}{kr} - \cos kr,$$

$$\psi_2(kr) = \left[\frac{3}{(kr)^2} - 1\right]\sin kr - \frac{3}{kr}\cos kr.$$

* Whittaker and Watson, *Modern Analysis*, p. 368.

The functions $\psi_n(x)$ may be calculated successively with the aid of the difference equation

$$(2n+1)\,\psi_n(x) = x\,[\psi_{n-1}(x) + \psi_{n+1}(x)],$$

while their derivatives may be calculated with the aid of the relation

$$(2n+1)\,\frac{d\psi_n(x)}{dx} = (n+1)\,\psi_{n-1}(x) - n\psi_{n+1}(x).$$

Similar relations may be used to calculate the functions and their derivatives.

These functions are particularly useful for the solution of problems relating to the diffraction of waves by a sphere*.

EXAMPLES

1. Prove that if $z = r\mu$

$$e^{ikz} = \sqrt{\frac{\pi}{2kr}}\sum_{n=0}^{\infty} i^n (2n+1)\,P_n(\mu)\,J_{n+\frac{1}{2}}(kr)$$

and deduce that

$$\sqrt{\frac{\pi}{2kr}}\,J_{n+\frac{1}{2}}(kr) = \frac{1}{2i^n}\int_{-1}^{1} e^{ikr\mu}\,P_n(\mu)\,d\mu.$$ [Bauer.]

2. Show by means of the result of Example 1, that

$$\int_{-\infty}^{\infty} J_{n+\frac{1}{2}}(u)\,J_{m+\frac{1}{2}}(u)\,\frac{du}{u} = 0 \quad m \neq n$$
$$= \frac{2}{2n+1} \quad m = n.$$

3. Prove that

$$\frac{\sin k\sqrt{r^2+a^2-2ar\mu}}{\sqrt{r^2+a^2-2ar\mu}} = \sum_{n=0}^{\infty} \frac{\pi}{\sqrt{ra}}\,(2n+1)\,J_{n+\frac{1}{2}}(kr)\,J_{n+\frac{1}{2}}(ka)\,P_n(\mu).$$

[Clebsch.]

4. If $R^2 = r^2 + a^2 - 2ra\mu$ and $\Pi = e^{-ikR}/R$, prove that

$$\Pi = -\frac{i}{kra}\sum_{n=0}^{\infty}(2n+1)\,f_n(kr)\,P_n(\mu),$$

where $f_n(kr) = \zeta_n(ka)\,\psi_n(kr)$ or $\zeta_n(kr)\,\psi_n(ka)$ according as r is less than or greater than a. [Macdonald.]

5. Prove that $\psi_n(ka)\,\zeta_n'(ka) - \zeta_n(ka)\,\psi_n'(ka) = -i$.

6. Prove that if $R^2 = r^2 + a^2 - 2ar\mu$

$$\tfrac{1}{2}\int_{-1}^{1}\frac{\sin(kR)}{R}\,d\mu = \frac{\sin(kr)}{r}\,\frac{\sin(ka)}{ka},$$

$$\tfrac{1}{2}\int_{-1}^{1}\frac{\cos(kR)}{R}\,d\mu = \frac{\sin(kr)}{kr}\,\frac{\cos(ka)}{a} \quad r < a$$
$$= \frac{\cos(kr)}{r}\,\frac{\sin(ka)}{ka} \quad r > a.$$

[Rayleigh.]

* See for instance, Lamb's *Hydrodynamics*, 5th ed. p. 495; H. M. Macdonald, *Proc. Roy. Soc.* A, vol. xc, p. 50 (1914); G. N. Watson, *ibid.* vol. xcv, p. 83 (1918); Lord Rayleigh, *Phil. Trans.* A, vol. CCIII, p. 87 (1904); *Papers*, vol. v, p. 149; A. E. H. Love, *Proc. London Math. Soc.* vol. xxx, p. 308 (1899).

§ **6·52.** There is a second method of dealing with the homogeneous wave-functions which is due to Stieltjes*.

If we write

$$x = u\cos\phi,\quad y = u\sin\phi,\quad z = v\cos\chi,\quad w = v\sin\chi,$$

the equation $\Box W = 0$ becomes

$$\frac{\partial^2 W}{\partial u^2} + \frac{1}{u}\frac{\partial W}{\partial u} + \frac{1}{u^2}\frac{\partial^2 W}{\partial \phi^2} + \frac{\partial^2 W}{\partial v^2} + \frac{1}{v}\frac{\partial W}{\partial v} + \frac{1}{v^2}\frac{\partial^2 W}{\partial \chi^2} = 0.$$

This equation possesses solutions of type

$$W = u^m v^k e^{i(m\phi+k\chi)} f(u, v)$$

if

$$\frac{\partial^2 f}{\partial u^2} + \frac{2m+1}{u}\frac{\partial f}{\partial u} + \frac{\partial^2 f}{\partial v^2} + \frac{2k+1}{v}\frac{\partial f}{\partial v} = 0. \qquad \text{......(A)}$$

This equation belongs to the class of "harmonic equations" studied by Euler and Poisson, it retains the harmonic form in which the variables are separated when the variables u and v are subjected to a number of transformations of type

$$u + iv = F(\xi + i\eta).$$

The general theory of these transformations has been discussed in Ch. IV. At present we are interested only in the particular substitution

$$u + iv = se^{i\theta},$$

which transforms the equation into

$$\frac{\partial^2 W}{\partial s^2} + \frac{3}{s}\frac{\partial W}{\partial s} + \frac{1}{s^2}\frac{\partial^2 W}{\partial \theta^2} + \frac{1}{s^2\cos^2\theta}\frac{\partial^2 W}{\partial \phi^2} + \frac{1}{s^2\sin^2\theta}\frac{\partial^2 W}{\partial \chi^2} + \frac{\cot\theta - \tan\theta}{s^2}\frac{\partial W}{\partial \theta} = 0.$$

Putting $\cos 2\theta = \mu$ we find that there are elementary solutions of the form

$$W = s^{2n}\Theta(\mu)\, e^{im\phi + ik\chi},$$

where Θ satisfies the differential equation

$$\frac{d}{d\mu}\left\{(1-\mu^2)\frac{d\Theta}{d\mu}\right\} + \Theta\left\{n(n+1) - \frac{m^2}{2(1+\mu)} - \frac{k^2}{2(1-\mu)}\right\} = 0. \qquad \text{......(B)}$$

This equation is satisfied by

$$\Theta = (1+\mu)^{\frac{m}{2}}(1-\mu)^{\frac{k}{2}} F\left(n+1+p, p-n; k+1; \frac{1-\mu}{2}\right),$$

where $2p = m + k$ and F as usual denotes the hypergeometric series. If $n - p$ is a positive integer the series terminates, and if n is also a positive integer we obtain a solution in the form of a polynomial.

* *Comptes Rendus*, t. xcv, p. 901 (1882).

Writing $\nu = n - p$ we have a solution of equation (A) which may be written in the equivalent forms

$$f = (u^2 + v^2)^\nu F(\nu + m + k + 1, -\nu; k + 1; \tau)$$

$$= (-)^\nu (u^2 + v^2)^\nu \frac{\Gamma(\nu + m + 1)\Gamma(k + 1)}{\Gamma(\nu + k + 1)\Gamma(m + 1)} F(\nu + m + k + 1, -\nu; m + 1; 1 - \tau)$$

$$= u^{2\nu} F\left(-\nu, -m - \nu; k + 1; -\frac{v^2}{u^2}\right)$$

$$= (-)^\nu v^{2\nu} \frac{\Gamma(\nu + m + 1)\Gamma(k + 1)}{\Gamma(\nu + k + 1)\Gamma(m + 1)} F\left(-\nu, -k - \nu; m + 1; -\frac{u^2}{v^2}\right),$$

where
$$\tau = \frac{v^2}{u^2 + v^2}, \quad 1 - \tau = \frac{u^2}{u^2 + v^2}.$$

To express a given wave-function in a series of elementary wave-functions of the present type we need an expansion theorem relating to series of hypergeometric functions. A formula for the coefficients in such a series was given long ago by Jacobi and is derived in § 6·53. The conditions under which the series represents the function were investigated by Darboux and have been studied more recently by other writers. Corresponding studies have been made of other series of hypergeometric functions. Some references to the literature are given in Note III, Appendix.

An interesting reciprocal relation between solutions may be obtained by making use of the fact that if

$$X = \frac{s^2 - a^2}{2(x - iy)}, \quad Y = -i\frac{s^2 + a^2}{2(x - iy)}, \quad Z = \frac{az}{x - iy}, \quad W = \frac{aw}{x - iy},$$

and if $f(X, Y, Z, W)$ is a solution of

$$\frac{\partial^2 V}{\partial X^2} + \frac{\partial^2 V}{\partial Y^2} + \frac{\partial^2 V}{\partial Z^2} + \frac{\partial^2 V}{\partial W^2} = 0,$$

then
$$v = (x - iy)^{-1} f(X, Y, Z, W)$$

is a solution of
$$\frac{\partial^2 v}{\partial x^2} + \frac{\partial^2 v}{\partial y^2} + \frac{\partial^2 v}{\partial z^2} + \frac{\partial^2 v}{\partial w^2} = 0,$$

when considered as a function of x, y, z, w. Now, if

$$x = s\cos\theta\cos\phi, \quad y = s\cos\theta\sin\phi, \quad z = s\sin\theta\cos\chi, \quad w = s\sin\theta\sin\chi,$$
$$X = S\cos\Theta\cos\Phi, \quad Y = S\cos\Theta\sin\Phi, \quad Z = S\sin\Theta\cos\mathrm{X}, \quad W = S\sin\Theta\sin\mathrm{X},$$

we have the relations

$$S^2 = -a^2 e^{2i\phi}, \quad s^2 = -a^2 e^{2i\Phi}, \quad \cos\Theta = \sec\theta, \quad \mathrm{X} = \chi,$$

and so a solution

$$S^n \cos^m \Theta \sin^k \Theta e^{i(m\Phi + k\mathrm{X})} G(n, m, k, \cos\Theta)$$

corresponds to a solution of type

$$-(S\cos\theta\, e^{-i\phi})^{-1} . \sec^m\theta\, (i\tan\theta)^k . (s/ia)^m . e^{ik\chi} G(n, m, k, \sec\theta)\,(ia)^n e^{ni\phi}$$

i.e.
$$s^{m-1}(\cos\theta)^{-1-n-k}(\sin\theta)^k e^{(n+1)i\phi + ik\chi} G(n, m, k, \sec\theta).$$

EXAMPLES

1. Prove that if

$$(n, m, k) = s^n \cos^m \theta \sin^k \theta . e^{i(m\phi+k\chi)} F(\tfrac{1}{2}n + 1 + \tfrac{1}{2}m + \tfrac{1}{2}k, \tfrac{1}{2}m + \tfrac{1}{2}k - \tfrac{1}{2}n; k+1; \sin^2\theta)$$

$$\times \frac{\Gamma(\tfrac{1}{2}n + \tfrac{1}{2}m + \tfrac{1}{2}k + 1)}{\Gamma(k+1)\,\Gamma(\tfrac{1}{2}n - \tfrac{1}{2}m - \tfrac{1}{2}k + 1)},$$

then

$$\left(\frac{\partial}{\partial x} + i\frac{\partial}{\partial y}\right)(n, m, k) = 2(n-1, m+1, k),$$

$$\left(\frac{\partial}{\partial x} - i\frac{\partial}{\partial y}\right)(n, m, k) = \tfrac{1}{2}(n+m-k)(n+m+k)(n-1, m-1, k),$$

$$\left(\frac{\partial}{\partial z} + i\frac{\partial}{\partial w}\right)(n, m, k) = (k-n-m)(n-1, m, k+1),$$

$$\left(\frac{\partial}{\partial z} - i\frac{\partial}{\partial w}\right)(n, m, k) = (n+m+k)(n-1, m, k-1).$$

2. Prove that if $u = y - z$, $v = z - x$, $w = x - y$, the differential equation

$$(y-z)\frac{\partial^2 V}{\partial y\,\partial z} + (z-x)\frac{\partial^2 V}{\partial z\,\partial x} + (x-y)\frac{\partial^2 V}{\partial x\,\partial y} = (\beta-\gamma)\frac{\partial V}{\partial x} + (\gamma-\alpha)\frac{\partial V}{\partial y} + (\alpha-\beta)\frac{\partial V}{\partial z}$$

possesses a solution of type

$$V = u^m F\left[-m, \frac{1+2\alpha-\beta-\gamma-m}{3}; \frac{2+\gamma+\alpha-2\beta-2m}{3}; -\frac{v}{u}\right].$$

When this solution is a homogeneous polynomial of degree m in x, y and z it can be expressed in six different ways in terms of the hypergeometric function, the arguments being respectively

$$-\frac{v}{u}, -\frac{w}{u}, -\frac{w}{v}, -\frac{u}{v}, -\frac{u}{w}, -\frac{v}{w}.$$

3. If

$$u + iv = a\cosh(\alpha + i\beta)$$

and

$$x = u\cos\phi,\quad y = u\sin\phi,\quad z = v\cos\chi,\quad ict = v\sin\chi,$$

the wave-equation becomes

$$\frac{\partial^2 W}{\partial \alpha^2} + \frac{\partial^2 W}{\partial \beta^2} + 2\coth 2\alpha\frac{\partial W}{\partial \alpha} + 2\cot 2\beta\frac{\partial W}{\partial \beta} + (\operatorname{cosech}^2\alpha + \operatorname{cosec}^2\beta)\frac{\partial^2 W}{\partial \chi^2}$$
$$+ (\operatorname{sech}^2\alpha - \sec^2\beta)\frac{\partial^2 W}{\partial \phi^2} = 0.$$

Hence show that there are simple solutions of type

$$W = A\Theta(\cosh 2\alpha)\Theta(\cos 2\beta)e^{im\phi+ik\chi},$$

where k, m and A are arbitrary constants and $\Theta(\mu)$ satisfies the differential equation (B).

4. Prove that the differential equation

$$\frac{\partial^2 V}{\partial x^2} + \frac{\partial^2 V}{\partial y^2} + \frac{\partial^2 V}{\partial z^2} - \frac{1}{c^2}\frac{\partial^2 V}{\partial t^2} + h^2 V = 0$$

possesses simple solutions of the types

$$V = s^{-1}J_{2n+1}(hs)\Theta(\mu)e^{im\phi+ik\chi},$$

$$V = J_m(hu\cos\alpha)J_k(hv\sin\alpha)e^{im\phi+ik\chi},$$

and deduce the expansion of the second solution in a series of solutions of the first type.

§ **6·53.** *Jacobi's polynomial.* The function

$$H_n(\tau) = F(n+m+k+1, -n; k+1; \tau)$$

may be called Jacobi's polynomial*. It may be expressed in the form

$$(k+1)(k+2)\ldots(k+n)\,\tau^k(1-\tau)^m H_n(\tau) = \frac{d^n}{d\tau^n}\left[\tau^{k+n}(1-\tau)^{m+n}\right]. \qquad \ldots\ldots(C)$$

With the aid of this formula and integration by parts it is easily seen that

$$\int_0^1 H_n(\tau) H_{n'}(\tau)\,\tau^k(1-\tau)^m\,d\tau = 0 \quad n' \neq n$$

$$= \frac{\Gamma(n+1)}{m+k+2n+1}\,\frac{[\Gamma(k+1)]^2\,\Gamma(m+n+1)}{\Gamma(m+n+k+1)\,\Gamma(k+n+1)},$$

$$n' = n, \quad k > -1, \quad m > -1, \quad n' = n > -1.$$

When $m + k = 2p$, $m - k = 2q$ where p and q are positive integers, the polynomial can be expressed in terms of the Legendre polynomial with the aid of the formula

$$H_n\left(\frac{1+\mu}{2}\right) = (-)^n\,\frac{2^p(p-q)!\,n!}{(n+2p)!}\,\frac{d^p}{d\mu^p}\left[(1+\mu)^q\,\frac{d^q}{d\mu^q}P_{n+p}(\mu)\right]$$

$$p \geqslant q.$$

Many interesting expansions may be obtained by expanding special solutions of the equation (A) in series of Jacobi polynomials. A few of these will now be mentioned.

In the first place we have the two associated expansions

$$(1-\xi-\eta)^p H_p\left(\frac{\xi\eta}{\xi+\eta-1}\right) = \sum_{n=0}^{p} A_n H_n(\xi)\,H_n(\eta),$$

$$H_s(\xi)\,H_s(\eta) = \sum_{p=0}^{s} B_p (1-\xi-\eta)^p H_p\left(\frac{\xi\eta}{\xi+\eta-1}\right),$$

where

$$A_n = (m+k+2n+1)$$
$$\times\frac{\Gamma(n+k+1)\,\Gamma(p+1)\,\Gamma(p+m+1)\,\Gamma(n+m+k+1)}{\Gamma(n+m+1)\,\Gamma(p-n+1)\,\Gamma(n+1)\,\Gamma(k+1)\,\Gamma(p+n+m+k+2)},$$

$$B_p = (-)^{s+p}$$
$$\times\frac{\Gamma(s+1)\,\Gamma(m+k+s+p+1)\,\Gamma(s+m+1)\,\Gamma(k+1)}{\Gamma(p+1)\,\Gamma(s-p+1)\,\Gamma(s+m+k+1)\,\Gamma(p+m+1)\,\Gamma(s+k+1)}.$$

A proof of these relations may be based upon the fact that if we make the transformation

$$u^2 = (1-\xi)(1-\eta), \quad v^2 = -\xi\eta,$$

* C. G. J. Jacobi, *Crelle's Journal*, Bd. LVI, S. 156 (1859); *Werke*, Bd. VI, S. 191.

and take ξ and η as new independent variables the equation becomes

$$\xi(1-\xi)\frac{\partial^2 f}{\partial \xi^2} - [(m+1)\xi + (k+1)(\xi-1)]\frac{\partial f}{\partial \xi}$$
$$= \eta(1-\eta)\frac{\partial^2 f}{\partial \eta^2} - [(m+1)\eta + (k+1)(\eta-1)]\frac{\partial f}{\partial \eta},$$

and is consequently satisfied by $f = H_s(\xi) H_s(\eta)$.

To determine the coefficients B_p in the expansion of this solution in terms of the solutions already found it is sufficient to put $\eta = 0$ and to use the expansion

$$H_s(\xi) = \sum_{p=0}^{s} B_p (1-\xi)^p,$$

which is already known. To find the coefficients A_n we put $\eta = 0$ in the other expansion; we have then to find the coefficients in the expansion

$$(1-\xi)^p = \sum_{n=0}^{p} A_n H_n(\xi).$$

This may be done by evaluating the integral

$$\int_0^1 H_n(\xi)\, \xi^k (1-\xi)^{p+m} d\xi$$
$$= \frac{\Gamma(k+1)}{\Gamma(k+n+1)} \int_0^1 (1-\xi)^p \frac{d^n}{d\xi^n}[\xi^{k+n}(1-\xi)^{m+n}]\, d\xi$$
$$= \frac{\Gamma(k+1)\Gamma(p+1)}{\Gamma(k+n+1)\Gamma(p-n+1)} \int_0^1 \xi^{k+n}(1-\xi)^{m+p} d\xi$$
$$= \frac{\Gamma(k+1)\Gamma(p+1)\Gamma(m+p+1)}{\Gamma(p-n+1)\Gamma(m+k+n+p+2)}.$$

In the particular case when $m = k = 0$ we obtain an expansion which may be written in the form

$$\left(\frac{\mu+\mu'}{2}\right)^n P_n\left(\frac{1+\mu\mu'}{\mu+\mu'}\right)$$
$$= \sum_{s=0}^{n} (2s+1) \frac{[\Gamma(n+1)]^2}{\Gamma(n-s+1)\Gamma(n+s+2)} P_s(\mu) P_s(\mu').$$

When $\mu' = 1$ this gives the well-known expansion

$$\left(\frac{1+\mu}{2}\right)^n = \sum_{s=0}^{n} (2s+1) \frac{[\Gamma(n+1)]^2}{\Gamma(n-s+1)\Gamma(n+s+2)} P_s(\mu).$$

The second expansion gives

$$P_s(\mu) P_s(\mu') = \sum_{n=0}^{s} (-)^{s+n} \frac{\Gamma(n+s+1)}{[\Gamma(n+1)]^2 \Gamma(s-n+1)} \left(\frac{\mu+\mu'}{2}\right)^n P_n\left(\frac{1+\mu\mu'}{\mu+\mu'}\right).$$

If we write

$$2x = \mu + \mu', \quad y = \frac{1+\mu\mu'}{\mu+\mu'},$$

the quantities μ and μ' are the roots of the equation

$$\theta^2 - 2\theta x + 2xy - 1 = 0.$$

In particular, if $y = 1$, we have $\mu' = 1$, $\mu = 2x - 1$.

§ **6·54**. Some further interesting results may be deduced from the fact that if $V(x, y, z)$ is a solution of Laplace's equation, then

$$W = (x + iy)^{-\frac{1}{2}} V(\sqrt{x^2 + y^2}, z, ict)$$

is a solution of the equation.

In particular, if we take the solid harmonic

$$V = r'^n P_n^m (\cos\theta') e^{im\phi'},$$

where $$P_n^m(\mu) = \frac{\Gamma(n + m + 1)}{2^m \Gamma(m + 1)\Gamma(n - m + 1)} (1 - \mu^2)^{\frac{m}{2}} C_n^m(\mu),$$

$$C_n^m(\mu) = F\left(m + n + 1, m - n; m + 1; \frac{1 - \mu}{2}\right),$$

we obtain the wave-function

$$W = u^{-\frac{1}{2}} (u^2 + v^2)^n e^{im\chi - \frac{1}{2}i\phi} P_n^m\left(\frac{u}{\sqrt{u^2 + v^2}}\right).$$

Comparing this with the type of wave-function already obtained we find that

$$C_n^m(\mu) = F\left(\frac{m + n + 1}{2}, \frac{m - n}{2}; m + 1; 1 - \mu^2\right) \quad n - m \text{ even},$$

$$= \mu F\left(\frac{m + n + 2}{2}, \frac{m - n + 1}{2}; m + 1; 1 - \mu^2\right) \quad n - m \text{ odd},$$

the conditions under which the series terminates being given on the right.

The expression (C) for Jacobi's polynomial now gives the interesting formulae*

$$P_{2n}(\mu) = \frac{1}{n!}\mu \frac{d^n}{d\xi^n}[\mu^{2n-1}\xi^n],$$

$$P_{2n+1}(\mu) = \frac{1}{n!}\frac{d^n}{d\xi^n}[\mu^{2n+1}\xi^n],$$

where $$\xi = 1 - \mu^2.$$

Writing

$$\xi = 1 - \mu^2, \quad \eta = 1 - \nu^2, \quad \sigma^2 = 1 - \frac{\xi\eta}{\xi + \eta - 1} = \frac{\mu^2\nu^2}{\mu^2 + \nu^2 - 1},$$

* Given explicitly by A. Wangerin, *Jahresbericht deutsch. Math. Verein*, Bd. XXII, S. 385 (1914). The formulae are both included in the general formula given on p. 122 of the author's paper, *Proc. London Math. Soc.* (2), vol. III, p. 111 (1905).

the two expansion theorems give

$$(\mu^2 + \nu^2 - 1)^p P_{2p}(\sigma) = \sum_{s=0}^{p} (2s + \tfrac{1}{2}) \frac{\Gamma(p+1)\,\Gamma(p+\tfrac{1}{2})}{\Gamma(p-s+1)\,\Gamma(p+s+\tfrac{3}{2})} P_{2s}(\mu)\,P_{2s}(\nu),$$

$$(\mu^2 + \nu^2 - 1)^{p+\frac{1}{2}} P_{2p+1}(\sigma)$$

$$= \sum_{s=0}^{p} (2s + \tfrac{3}{2}) \frac{\Gamma(p+1)\,\Gamma(p+\tfrac{3}{2})}{\Gamma(p-s+1)\,\Gamma(p+s+\tfrac{5}{2})} P_{2s+1}(\mu)\,P_{2s+1}(\nu),$$

$$P_{2n}(\mu)\,P_{2n}(\nu)$$

$$= \sum_{p=0}^{n} (-)^{n+p} \frac{\Gamma(n+p+\tfrac{1}{2})}{\Gamma(p+\tfrac{1}{2})\,\Gamma(p+1)\,\Gamma(n-p+1)} (\mu^2+\nu^2-1)^p P_{2p}(\sigma),$$

$$P_{2n+1}(\mu)\,P_{2n+1}(\nu)$$

$$= \sum_{p=0}^{n} (-)^{n+p} \frac{\Gamma(n+p+\tfrac{3}{2})}{\Gamma(p+\tfrac{3}{2})\,\Gamma(p+1)\,\Gamma(n-p+1)} (\mu^2+\nu^2-1)^{p+\frac{1}{2}} P_{2p+1}(\sigma).$$

EXAMPLES

1. Prove that the differential equation

$$\frac{d^2y}{dx^2} = y\left[\frac{m(m-1)}{\sin^2 x} + \frac{n(n-1)}{\cos^2 x} - k^2\right]$$

is satisfied by

$$y = \sin^m x \cos^n x\, F\left(\frac{m+n+k}{2}, \frac{m+n-k}{2}; m+\tfrac{1}{2}; \sin^2 x\right).$$

[G. Darboux, *Théorie des Surfaces*, t. II, p. 199 (1889).]

Show also that

$$\frac{d^2y}{dx^2} = \left[k^2 - \frac{n(n+1)}{\cosh^2 x}\right] y$$

is satisfied by
$$y = e^{kx} F\left[-n, n+1; 1-k; \frac{1-\tanh x}{2}\right].$$

2. Prove that

$$\left(\frac{1+\mu}{2}\right)^m F[-n, n+m+1; m+1; \tfrac{1}{2}(1+\mu)]$$

$$= (-)^n \sum_{n}^{\infty} \frac{[\Gamma(m+1)]^2\,\Gamma(n+m+1)\,(2s+1)\,P_s(\mu)}{\Gamma(s-n+1)\,\Gamma(s+m+n+2)\,\Gamma(m+n+1-s)}, \qquad |1+\mu| < 2,$$

$$P_n(\cos\theta)\,P_n(\sin\theta) = \frac{[\Gamma(2n+1)]^2}{2^{2n}[\Gamma(n+1)]^4}\left[\cos^n\theta\sin^n\theta - \frac{n^2(n-1)^2}{2(2n-1)^2(2n-3)}\cos^{n-2}\theta\sin^{n-2}\theta\right.$$

$$\left. + \frac{n^2(n-1)^2(n-2)^2(n-3)^2}{2.4\,(2n-1)^2(2n-3)^2(2n-5)(2n-7)}\cos^{n-4}\theta\sin^{n-4}\theta - \ldots\right].$$

3. If $u = x^{-2}$ prove that

$$x^{2m+1} P_{2m}(x) = \frac{1}{m!}\frac{d^m}{du^m}[x(1-x^2)^m],$$

$$x^{2m+2} P_{2m+1}(x) = \frac{1}{m!}\frac{d^m}{du^m}[x^3(1-x^2)^m].$$

[L. Koschmieder, *Rev. Mat. Hispano-Amer.* (1924).]

§ 6·61. *Definite integrals for the Legendre functions.* Some useful definite integrals for the representation of Legendre functions may be obtained by deriving Newtonian potentials from four-dimensional potentials by integration with respect to one parameter.

Starting with the four-dimensional potential

$$W' = f(x + iy, z + iw),$$

we derive a second potential W from it by inversion.

If $s^2 = x^2 + y^2 + z^2 + w^2$ we have

$$W = \frac{1}{s^2} f\left(\frac{x+iy}{s^2}, \frac{z+iw}{s^2}\right),$$

and a Newtonian potential is derived from this by integrating with respect to w either between $-\infty$ and ∞ or round a closed contour in the complex w-plane. In particular, we may obtain in this way a Newtonian potential

$$V = \frac{1}{\pi}\int_{-\infty}^{\infty} \frac{(x+iy)^m (z+iw)^{n-m} dw}{(x^2+y^2+z^2+w^2)^{n+1}},$$

which may be expected to be a constant multiple of $r^{-n-1} P_n^m(\cos\theta) e^{im\phi}$. We thus obtain the formula

$$P_n^m(\mu) = \frac{2^n}{\pi} \frac{\Gamma(n+1)}{\Gamma(n-m+1)} \sin^m\theta \int_{-\infty}^{\infty} \frac{(\mu+it)^{n-m} dt}{(1+t^2)^{n+1}} \quad n \geqslant m,$$

which is certainly valid when n and m are positive integers as a simple expansion shows. The corresponding formula for $P_n(\mu)$ is

$$P_n(\mu) = \frac{2^n}{\pi}\int_{-\infty}^{\infty} \frac{(\mu+it)^n dt}{(1+t^2)^{n+1}} \quad n > -1,$$

and the formula for $P_n^m(\mu)$ may be derived from this by differentiation.

The last result may be obtained directly by expanding both sides of the equation

$$[x^2 + y^2 + (z-a)^2]^{-\frac{1}{2}} = \frac{1}{\pi}\int_{-\infty}^{\infty} \frac{dw}{x^2+y^2+(z-a)^2+(w-ia)^2}$$

in ascending powers of a and equating coefficients. The general formula may likewise be obtained for positive integral values of m and n by expanding the two sides of the equation

$$[(x-b)^2 + (y-ib)^2 + (z-a)^2]^{-\frac{1}{2}}$$
$$= \frac{1}{\pi}\int_{-\infty}^{\infty} \frac{dw}{(x-b)^2+(y-ib)^2+(z-a)^2+(w-ia)^2}$$

in ascending powers of a and b. We thus obtain the expansion

$$[(x-b)^2 + (y-ib)^2 + (z-a)^2]^{-\frac{1}{2}} = \sum_{n=0}^{\infty}\sum_{m=0}^{n} \frac{1}{m!}\frac{a^{n-m}b^m}{r^{n+1}} P_n^m(\mu) e^{im\phi},$$

which is easily obtained from the Taylor expansion

$$P_n(\cos\theta + k\sin\theta) = \sum_{m=0}^{n} \frac{k^m}{m!} P_n^m(\cos\theta),$$

by writing $k = \frac{b}{a} e^{i\phi}$.

A formula for $Q_n(\mu)$ due to Heine* may be derived from the fact that if $t^2 = x^2 + y^2 + w^2$, the function

$$W = t^{-1} f(z + it)$$

is a four-dimensional potential function.

Writing $w = \rho \sinh u$, where $\rho^2 = x^2 + y^2$, the integral

$$V = \frac{1}{\pi}\int_{-\infty}^{\infty} W dw$$

takes the form

$$V = \frac{1}{\pi}\int_{-\infty}^{\infty} f(z + i\rho \cosh u)\, du.$$

With suitable restrictions on the function f this integral represents a solution of Laplace's equation.

In particular, if x, y and z are not all real quantities

$$r^{-n-1} Q_n\left(\frac{z}{r}\right) = \tfrac{1}{2}\int_{-\infty}^{\infty} (z + i\rho \cosh u)^{-n-1} du$$

or
$$Q_n(s) = \tfrac{1}{2}\int_{-\infty}^{\infty} [s + (s^2 - 1)^{\frac{1}{2}} \cosh u]^{-n-1} du.$$

This equation may be deduced from the well-known formula

$$Q_n(s) = \frac{1}{2^{n+1}}\int_{-1}^{1} (1 - t^2)^n (s - t)^{-n-1} dt,$$

with the aid of the substitution†

$$t = \frac{e^u (s+1)^{\frac{1}{2}} - (s-1)^{\frac{1}{2}}}{e^u (s+1)^{\frac{1}{2}} + (s-1)^{\frac{1}{2}}}.$$

When x, y and z are real the corresponding formula is

$$\int_0^{\infty} (z + i\rho \cosh u)^{-n-1} du = r^{-n-1} Q_n(\mu) - \tfrac{1}{2} i\pi r^{-n-1} P_n(\mu),$$

where $z = \mu r$.

EXAMPLE

Prove that, if $a > 0$ and $|2z| < 1$,

$$\int_{-\infty}^{\infty} \frac{(1 - 2z + it)^{-b}\, dt}{(1 + it)^{c-b} (1 - it)^{a-c+1}} = \frac{2\pi\, \Gamma(a)}{2^a\, \Gamma(c)\, \Gamma(a + 1 - c)} F(a, b; c; z).$$

Show also that in the analytical continuation of the integral the line $2z = 1$ is generally a barrier.

* *Kugelfunktionen*, p. 147.

† Whittaker and Watson, *Modern Analysis*, p. 319.

CHAPTER VII

CYLINDRICAL CO-ORDINATES

§ 7·11. *The diffusion equation in two dimensions.* When cylindrical co-ordinates ρ, ϕ, z are used we have

$$x = \rho \cos \phi, \quad y = \rho \sin \phi, \quad z = z,$$

and the equation of the conduction of heat becomes

$$\frac{\partial v}{\partial t} = \kappa \left[\frac{\partial^2 v}{\partial \rho^2} + \frac{1}{\rho} \frac{\partial v}{\partial \rho} + \frac{1}{\rho^2} \frac{\partial^2 v}{\partial \phi^2} + \frac{\partial^2 v}{\partial z^2} \right].$$

Let us first consider solutions which are independent of z. Writing

$$v = e^{-\kappa \lambda^2 t} R(\rho) e^{im\phi}, \qquad \text{......(I)}$$

the equation for R is

$$\frac{d^2 R}{d\rho^2} + \frac{1}{\rho} \frac{dR}{d\rho} + \left(\lambda^2 - \frac{m^2}{\rho^2} \right) R = 0,$$

and is satisfied by

$$R = A_m J_m(\lambda\rho) + B_m Y_m(\lambda\rho),$$

where $J_m(\lambda\rho)$ and $Y_m(\lambda\rho)$ are the standard solutions of Bessel's equation, the definitions of which are given in § 7·21. In particular, when v is independent of ϕ the solution is of type

$$v = e^{-\kappa\lambda^2 t} [A_0 J_0(\lambda\rho) + B_0 Y_0(\lambda\rho)],$$

if $\lambda \neq 0$, but when $\lambda = 0$ the solution is of type

$$v = A + B \log \rho,$$

where A and B are arbitrary constants.

In the case of diffusion from a cylindrical rod $r = b$ to a coaxial cylinder which collects the diffusing substance we may use boundary conditions such as

$$v = 0 \text{ when } \rho = a, \quad \kappa \frac{\partial v}{\partial \rho} = -Q \text{ when } \rho = b.$$

If Q is constant there is a steady state given by

$$v = \frac{Qb}{\kappa} \log \frac{a}{\rho} \quad b < \rho < a.$$

If there is no rod inside the cylinder but an initial distribution of concentration, say $v = f(\rho)$ when $t = 0$, we may try to satisfy the conditions by a series of type

$$v = \sum_{n=1}^{\infty} c_n e^{-\kappa s_n^2 t} J_0(\rho s_n),$$

where the quantities $s_1, s_2, \ldots$ are the different values of s for which

$$J_0(a s_n) = 0.$$

The solution of this problem is facilitated by means of the formula

$$\int_0^a rJ_0(rs_m)\,J_0(rs_n)\,dr = 0 \quad s_m \neq s_n,$$
$$= \frac{a^2}{2}[J_1(as_n)]^2 \quad s_m = s_n.$$

The solution (I) may be generalised by making A and B functions of λ and integrating with respect to λ between 0 and ∞. Many interesting solutions of the equation may be expressed by means of definite integrals in this way as the following examples will show.

EXAMPLES

1. Prove that if $S = \rho^2/4\kappa t$

$$\int_0^\infty e^{-\kappa\lambda^2 t} J_0(\lambda\rho)\,\lambda d\lambda = \frac{1}{2\kappa t} e^{-S},$$

$$\int_0^\infty e^{-\kappa\lambda^2 t} \cos(\lambda\rho)(\lambda\rho)^{-\frac{1}{2}} \lambda^{\frac{1}{2}}\,d\lambda = (\pi/\kappa\rho t)^{\frac{1}{2}} e^{-S}.$$

These are particular cases of Sonine's general formula (Ex. 9, § 7·31).

2. Prove that

$$\int_0^\infty e^{-\kappa\lambda^2 t} Y_0(\lambda\rho)\,\lambda\,d\lambda = \frac{1}{2\pi\kappa t}\left[C + \log S + \int_0^S (e^s - 1)\,ds/s\right],$$

where C is Euler's constant.

3. Prove that
$$\int_0^\infty e^{-\kappa\lambda^2 t} J_0(\lambda\rho)\,d\lambda = \tfrac{1}{2}\sqrt{(\pi/\kappa t)}\,e^{-\frac{1}{2}S} I_0(\tfrac{1}{2}S),$$

$$\int_0^\infty e^{-\kappa\lambda^2 t} Y_0(\lambda\rho)\,d\lambda = -\tfrac{1}{2}(\pi\kappa t)^{-\frac{1}{2}} e^{-\frac{1}{2}S} K_0(\tfrac{1}{2}S).$$

[O. Heaviside, *Electromagnetic Theory*, vol. III, p. 271.]

§ **7·12**. *Motion of an incompressible viscous fluid in an infinite right circular cylinder rotating about its axis.* Let $\omega = \omega(r, t)$ denote the angular velocity of the fluid about the axis of the cylinder at a distance r from this axis, then the equations of motion of a viscous fluid take the form

$$\frac{\partial p}{\partial r} = \rho r\omega^2,$$

$$\frac{\partial^2\omega}{\partial r^2} + \frac{3}{r}\frac{\partial\omega}{\partial r} = \frac{1}{\nu}\frac{\partial\omega}{\partial t}.$$

If the boundary conditions are

$$\omega = F(r) \text{ for } t = 0, \quad \omega = G(t) \text{ for } r = a,$$

the solution given by McLeod* is

$$r\omega = \sum_{n=1}^{\infty} 2aJ_1(Nr)[j(N)]^2 e^{-\nu N^2 t}\int_0^1 \xi^2 F(a\xi)\,J_1(Na\xi)\,d\xi$$
$$- \sum_{n=1}^{\infty} 2\nu N J_1(Nr)\,j(N)\,e^{-\nu N^2 t}\int_0^t G(\tau)\,e^{\nu N^2\tau}\,d\tau,$$

* A. R. McLeod, *Phil. Mag.* (6), vol. XLIV, p. 1 (1922). Particular cases of the formula have been obtained by other writers to whom reference is made in McLeod's paper. A paper by K. Aichi, *Tokyo Math. Phys. Soc.* (2), vol. IV, p. 2200 (1922) also deals with this problem.

where $j(N) J_0(Na) = 1$ and $J_1(Na) = 0$, aN being the nth root of the Bessel function. When $G(t) = \Omega =$ constant and $F(r) = 0$, we have the case in which the fluid is initially at rest and the cylinder suddenly starts to rotate with angular velocity Ω.

When $G(t) = 0$ and $F(r) = \Omega =$ constant we have the case in which the fluid is initially rotating with angular velocity Ω and the cylinder is suddenly stopped.

The case in which the cylinder is of finite height $2h$ may be treated with the aid of the equation

$$\frac{\partial^2 \omega}{\partial r^2} + \frac{3}{r}\frac{\partial \omega}{\partial r} + \frac{\partial^2 \omega}{\partial z^2} = \frac{1}{\nu}\frac{\partial \omega}{\partial t}.$$

This equation was solved by Meyer* by means of an infinite series of type

$$\omega = \Sigma A_n e^{-\nu m^2 t + ilz} J_1(kr),$$

where l, m, k are functions of n connected by the relation

$$k^2 = m^2 - l^2.$$

A. F. Crossley† has given a solution of the case in which the boundary conditions are

$$\omega = \Omega(t) \text{ when } z = 0 \text{ and } \omega = \Omega(t) \text{ when } r = a,$$

$\Omega(t)$ being an assigned function of t. This is the case of the semi-infinite cylinder. He has also considered the case when a constant couple of magnitude G per unit length acts on the cylinder. The corresponding problem for an infinite cylinder has been treated by Havelock‡.

Many years ago Meyer§ applied similar analysis to the problem of the damping of the vibrations of an oscillating disc and obtained the following relation between the coefficient of damping k and the coefficient of viscosity

$$k = \frac{\pi a^4}{2I}(\tfrac{1}{2}\rho\mu n)^{\frac{1}{2}} + \frac{\pi a^3 \mu}{I},$$

where I is the moment of inertia of the disc and $n/2$ is the frequency of oscillation. Kobayashi ‖ has recently made a more exact calculation and has obtained a formula

$$k = \frac{\pi a^4}{2I}(\tfrac{1}{2}\rho\mu n)^{\frac{1}{2}} + \frac{\pi S a^3 \mu}{2I} - \frac{\pi^2 a^8 \rho \mu}{4I^2},$$

in which S is estimated by means of some approximations to be 2·11. Kobayashi's formula, however, does not agree with experiments as well as that of Meyer, and the reason for the discrepancy is yet to be found. One possible explanation is that the component velocities u and w have been ignored. A fuller treatment of the problem has been commenced.

* O. E. Meyer, *Wied. Ann.* Bd. XLIII, S. 1 (1891).

† *Proc. Camb. Phil. Soc.* vol. XXIV, pp. 234, 480 (1928).

‡ *Phil. Mag.* vol. XLII, p. 620 (1921).

§ O. E. Meyer, *Pogg. Ann.* Bd. CXIII, S. 55 (1861); *Wied. Ann.* Bd. XXXII, S. 642 (1887).

‖ I. Kobayashi, *Zeits. f. Phys.* Bd. XLII, S. 448 (1927).

EXAMPLES

1. If v satisfies the differential equation

$$\frac{\partial^2 v}{\partial r^2} + \frac{1}{r}\frac{\partial v}{\partial r} - \frac{v}{r^2} + \frac{\partial^2 v}{\partial z^2} = 0,$$

and the boundary conditions $v = 0$ for $r = a$, $v = Vr/a$ for $z = h$, $v = 0$ for $z = -h$, we have

$$v = V \sum_{n=1}^{\infty} \frac{2}{\rho_n J_2(\rho_n)} \frac{\sinh \dfrac{\rho_n (h+z)}{a}}{\sinh \dfrac{2h\rho_n}{a}} J_1\left(\frac{r\rho_n}{a}\right), \quad J_1(\rho_n) = 0.$$

2. If $v = V$ for $r = a$, $v = 0$ for $z = \pm h$, we have

$$v = \frac{4V}{\pi} \sum_{n=1}^{\infty} \frac{(-)^{m+1}}{2m-1} \cos \frac{(2m-1)\pi z}{2h} \frac{I_1[(m-\frac{1}{2})\pi r/h]}{I_1[(m-\frac{1}{2})\pi a/h]}.$$

[W. Hort.]

§ **7·13**. *The vibration of a circular membrane.* The equation of vibration in polar co-ordinates is

$$\frac{\partial^2 w}{\partial t^2} = c^2 \left[\frac{\partial^2 w}{\partial r^2} + \frac{1}{r}\frac{\partial w}{\partial r} + \frac{1}{r^2}\frac{\partial^2 w}{\partial \phi^2}\right],$$

and is satisfied by

$$w = (A_n \cos kct + B_n \sin kct) \cos n(\phi + \alpha_n) J_n(kr).$$

The boundary condition $w = 0$ for $r = a$ is satisfied if $J_n(ka) = 0$. The roots of this equation may be calculated by means of the following formula given by McMahon*

$$k_s a = \beta - \frac{4n^2 - 1}{8\beta} - \frac{4(4n^2-1)(28n^2-31)}{3.(8\beta)^3} - \dots$$

where

$$\beta = \tfrac{1}{4}\pi(2n + 4s - 1),$$

and $k_s a$ is the root corresponding to the suffix s in the series $k_1 a, k_2 a, k_3 a, \dots$ where the roots are arranged in ascending order of magnitude.

For the fundamental mode of vibration ($n = 0$) there is no nodal line. For the other modes there are nodal lines which may be concentric circles or diameters. The nodal lines for the simple cases are shown in Rayleigh's *Sound*, vol. I, p. 331.

The solution may be generalised by summation so as to be suitable for the representation of a solution which satisfies prescribed initial conditions

$$w = w_0, \quad \frac{\partial w}{\partial t} = \dot{w}_0 \text{ for } t = 0.$$

For the determination of the coefficients in the series the following formulae are particularly useful. If k and k' are different roots of the equation $J_n(ka) = 0$,

$$\int_0^a J_n(kr) J_n(k'r)\, r dr = 0,$$

$$2\int_0^a [J_n(kr)]^2 r dr = a^2 [J_n'(ka)]^2.$$

* *Annals of Math.* vol. IX, p. 23 (1894). See also Watson's *Bessel Functions*, p. 505.

§ **7·21.** *The simple solutions of the wave-equation.* In cylindrical co-ordinates the wave-equation is

$$\frac{\partial^2 u}{\partial \rho^2} + \frac{1}{\rho}\frac{\partial u}{\partial \rho} + \frac{1}{\rho^2}\frac{\partial^2 u}{\partial \phi^2} + \frac{\partial^2 u}{\partial z^2} = \frac{1}{c^2}\frac{\partial^2 u}{\partial t^2},$$

and possesses simple solutions of type

$$u = R(\rho)\, Z(z)\, \Phi(\phi)\, e^{ikct} = Ve^{ikct}, \text{ say}$$

if

$$\frac{d^2\Phi}{d\phi^2} + m^2\Phi = 0,$$

$$\frac{d^2 Z}{dz^2} \pm l^2 Z = 0,$$

$$\frac{d^2 R}{d\rho^2} + \frac{1}{\rho}\frac{dR}{d\rho} + \left(k^2 \mp l^2 - \frac{m^2}{\rho^2}\right) R = 0,$$

k, l and m being arbitrary constants. The last equation is satisfied by

$$R = aJ_n[\rho\sqrt{(k^2 \mp l^2)}] + bY_m[\rho\sqrt{(k^2 \mp l^2)}],$$

where a and b are arbitrary constants. When the lower sign is chosen it may be more advantageous to write the solution in the form

$$R = \alpha I_m[\rho\sqrt{(l^2 - k^2)}] + \beta K_m[\rho\sqrt{(l^2 - k^2)}],$$

where α and β are arbitrary constants.

For convenience the definitions and a few properties of the Bessel functions are listed below; for a full development of the properties of the functions reference may be made to Whittaker and Watson's *Modern Analysis*, to Watson's *Bessel Functions* and to Gray and Mathews' *Bessel Functions*. The notation used here is the same as that of Watson.

The function $J_m(x)$ is defined by the infinite series

$$J_m(x) = \sum_{s=0}^{\infty} \frac{(-)^s}{s!}\frac{(\frac{1}{2}x)^{n+2s}}{\Gamma(m+s+1)},$$

which converges for all finite values of x.

When m is an integer we have the relation

$$J_{-m}(x) = (-)^m J_m(x),$$

and it is necessary to define a second solution of the differential equation, because in this case the two solutions $J_m(x)$ and $J_{-m}(x)$ are not linearly independent.

The function $Y_m(x)$ which gives the second solution is defined by the equation

$$Y_m(x) = \lim_{\nu \to m} \frac{J_\nu(x)\cos\nu\pi - J_{-\nu}(x)}{\sin\nu\pi}$$

$$= \frac{1}{\pi} e^{-im\pi}\cos m\pi \left[\sum_{s=0}^{\infty} \frac{(-)^s(\frac{1}{2}x)^{m+2s}}{s!\,(m+s)!}\left\{2\log(\tfrac{1}{2}x) + 2\gamma - \sum_{n=1}^{m+s} n^{-1} - \sum_{n=1}^{s} n^{-1}\right\} - \sum_{s=0}^{m-1}\frac{(\frac{1}{2}x)^{2s-m}\,\Gamma(m-s)}{s!}\right],$$

where γ is Euler's constant.

When x is imaginary it is convenient to use the functions

$$I_m(x) = i^{-m} J_m(ix),$$

$$K_m(x) = \tfrac{1}{2}\pi \{I_{-m}(x) - I_m(x)\} \operatorname{cosec} m\pi = K_{-m}(x).$$

When m is an integer $K_m(x)$ is defined by the equation

$$K_m(x) = \lim_{\nu \to m} K_\nu(x).$$

When $|\arg x| < \frac{1}{2}\pi$ and $R(m + \frac{1}{2}) > 0$ the function $K_m(x)$ may be represented by the definite integrals

$$K_m(x) = \int_0^\infty e^{-x\cosh a} \cosh ma . da$$

$$= \frac{x^m \Gamma(\frac{1}{2})}{2^m \Gamma(m + \frac{1}{2})} \int_0^\infty e^{-x\cosh\phi} \sinh^{2m}\phi \, d\phi$$

$$= (2x)^m \pi^{-\frac{1}{2}} \Gamma(m + \tfrac{1}{2}) \int_0^\infty (u^2 + x^2)^{-m-\frac{1}{2}} \cos u . du.$$

In the first integral m is unrestricted.

The functions $J_m(x)$ and $Y_m(x)$ satisfy the difference equations

$$J_{m-1}(x) + J_{m+1}(x) = \frac{2m}{x} J_m(x),$$

$$\frac{d}{dx} J_m(x) = \frac{m}{x} J_m(x) - J_{m+1}(x)$$

$$= J_{m-1}(x) - \frac{m}{x} J_m(x),$$

$$x^{-m-s} J_{m+s}(x) = (-)^s \frac{d^s}{(xdx)^s} \{x^{-m} J_m(x)\}.$$

EXAMPLES

1. Prove that

$$\left(\frac{\partial}{\partial x} + i\frac{\partial}{\partial y}\right)^s [\rho^{m-n} J_n(\rho) e^{im\phi}] = (-)^s \rho^{m-n} J_{n+s}(\rho) e^{i(m+s)\phi},$$

$$\left(\frac{\partial}{\partial x} - i\frac{\partial}{\partial y}\right)^s [\rho^{n-m} J_n(\rho) e^{im\phi}] = \rho^{n-m} J_{n-s}(\rho) e^{i(m-s)\phi}.$$

2. Show that when m is a positive integer the relation

$$e^{im\phi} \int_0^\infty e^{-kz} J_m(k\rho) k^m dk = 1.3 \ldots (2m-1) \rho^m e^{im\phi} . r^{-(2m+1)}$$

may be deduced from Poisson's relation

$$\int_0^\infty e^{-kz} J_0(k\rho)\, dk = 1/r$$

by differentiation.

3. Prove that

$$\left(\frac{\partial}{\partial x}+i\frac{\partial}{\partial y}\right)^s [J_m(\rho) J_n(\rho) e^{(m+n)i\phi}] = (-)^s e^{(m+n+s)i\phi} [J_{m+s}(\rho) J_n(\rho)$$

$$+ sJ_{m+s-1}(\rho) J_{n+1}(\rho) + \frac{s(s-1)}{1.2} J_{m+s-2}(\rho) J_{n+2}(\rho) + \dots J_m(\rho) J_{n+s}(\rho)],$$

$$\left(\frac{\partial}{\partial x}-i\frac{\partial}{\partial y}\right)^s [J_m(\rho) J_n(\rho) e^{(m+n)i\phi}] = e^{(m+n-s)i\phi} [J_{m-s}(\rho) J_n(\rho)$$

$$+ sJ_{m-s+1}(\rho) J_{n-1}(\rho) + \frac{s(s-1)}{1.2} J_{m-s+2}(\rho) J_{n-2}(\rho) + \dots J_m(\rho) J_{n-s}(\rho)].$$

4. Prove that

$$(-)^s \left(\frac{\partial^2}{\partial x^2}+\frac{\partial^2}{\partial y^2}\right)^s [J_m(\rho) J_n(\rho) e^{(m+n)i\phi}] = \sum_{p=0}^{2s} \binom{2s}{p} J_{m+s-p}(\rho) J_{n-s+p}(\rho) e^{(m+n)i\phi},$$

$$2^n (\pi\rho)^{-\frac{1}{2}} J_{n-\frac{1}{2}}(2\rho) = \sum_{m=0}^{n} \binom{n}{m} J_{m-\frac{1}{2}}(\rho) J_{n-m-\frac{1}{2}}(\rho).$$

§ 7·22. The elementary solutions involving the function $K_m(l\rho)$ are useful for the representation of potentials of distribution of charge located near the axis of z. In particular, for a point charge at the origin, we have the formula of Basset

$$\frac{1}{r} = \frac{1}{\sqrt{(z^2+\rho^2)}} = \frac{2}{\pi}\int_0^\infty K_0(l\rho) \cos lz \, . \, dl,$$

and from this we may deduce the potential of a line charge of density $f(z)$

$$V = \frac{2}{\pi}\int_0^\infty K_0(l\rho)\, dl \int_{-\infty}^\infty \cos l(z-\zeta) f(\zeta)\, d\zeta.$$

In the neighbourhood of the axis $\rho = 0$, this function V becomes infinite like

$$-\frac{2}{\pi}\log\rho \int_0^\infty dl \int_{-\infty}^\infty \cos l(z-\zeta) f(\zeta)\, d\zeta.$$

If $f(z)$ is a function which can be expressed by means of a Fourier double integral, the foregoing expression becomes

$$-2f(z)\log\rho$$

at a place where $f(z)$ is continuous and

$$-[f(z-0)+f(z+0)]\log\rho$$

at a place where $f(z)$ has a finite discontinuity. This theorem relating to the behaviour of the potential of a line charge may be made more precise by means of modern results relating to Fourier integrals. The theorem has been generalised by Poincaré*, Levi-Civita† and Tonolo‡ so as to be applicable to a charge on a curved line. When the curve is plane the quantity ρ in the foregoing formula is simply the normal distance of a point

* *Acta Math.* vol. XXII, p. 89 (1899).

† *Rend. Lincei* (5), t. XVII (1908); *Rend. Palermo*, t. XXXIII, p. 354 (1912).

‡ *Math. Ann.* vol. LXXII, p. 78 (1912); see also A. Viterbi, *Rend. Lombardo* (2), t. XLII (1908).

from the curve, but when the curve is twisted the expression for ρ is more complicated and the conditions for the validity of the formula harder to find. Levi-Civita attributes the asymptotic formula for V to Betti, *Teorica delle forze Newtoniane* (Pisa, Nistri, 1879).

EXAMPLES

1. The potential of a row of equal unit point charges with co-ordinates

$$x = y = 0, \quad z = 2\pi n \quad (n = 0, \pm 1, \pm 2, \ldots)$$

can be expressed in a simple form by adding a compensating uniform line charge on the axis of z. The total potential is then

$$V = \frac{1}{\pi}\int_{-\infty}^{\infty} W dw,$$

where

$$W = \frac{1}{2u}\left[\frac{\sinh u}{\cosh u - \cos z} - 1\right]$$

$$= \frac{1}{u}\left[e^{-u}\cos z + e^{-2u}\cos 2z + \ldots\right],$$

and $u = (\rho^2 + w^2)^{\frac{1}{2}}$. Prove also that

$$V = \frac{2}{\pi}\sum_{n=1}^{\infty} K_0(nr)\cos nz,$$

and obtain Appell's formula (used in crystal theory by E. Madelung*)

$$V = C + \frac{1}{\pi}\log\left(\frac{4\pi}{r}\right) + \frac{2}{\pi}\sum_{n=1}^{\infty} K_0(nr)\cos nz,$$

for the potential of the point charges, C being an infinite constant. This result is allied to the general theorem of Lerch [*Ann. de Toulouse* (1), t. III (1889)], which states that if $s > 0$,

$$\pi^{-\frac{1}{2}}\Gamma(s+\tfrac{1}{2})\sum_{m=-\infty}^{\infty}[(x-m)^2+u^2]^{-s-\frac{1}{2}} = \Gamma(s)u^{-2s} + 2\sum_{n=1}^{\infty} A_n\cos 2\pi nx,$$

where

$$A_n = \int_0^{\infty} e^{-u^2 z - \frac{\pi^2 n^2}{z}} z^{s-1}\,dz.$$

2. Prove that a particular solution of the equation

$$\frac{\partial^2 V}{\partial\rho^2} + \frac{2s+1}{\rho}\frac{\partial V}{\partial\rho} + \frac{\partial^2 V}{\partial z^2} + k^2 V = 0, \quad (ak < 2\pi)$$

is

$$V = \rho^{-s}\sum_{n=1}^{\infty} K_s[(\rho/a)(4\pi^2n^2 - a^2k^2)^{\frac{1}{2}}]\int_0^a \cos\frac{2\pi n}{a}(z-\zeta)f(\zeta)\,d\zeta.$$

Examine the behaviour of this solution in the neighbourhood of the axis of z.

§ **7·31**. *Laplace's expression for a potential function which is symmetrical about an axis and finite on the axis.* Let us suppose that the potential function V is continuous $(D, 2)$ within a sphere S whose centre is at a point O on the axis of symmetry of the function, then by the theorem of § 6·34 V can be expanded in a power series of ascending powers of the co-ordinates x, y, z of a point relative to O. The fact that V is symmetrical about the

* *Phys. Zeit.* Bd. XIX, S. 524 (1918). Another method of calculating the potentials of periodic distributions of charge is given by C. N. Wall, *Phil. Mag.* (7), vol. III, p. 660 (1927).

axis, which we take as axis of z, means that this power series can be expressed in terms of ρ and z and so is of the form

$$V = \Sigma\, a_n r^n P_n(\mu),$$

where $\mu = \dfrac{z}{r}$. On the axis of z, $\mu = \pm 1$ and $P_n(\mu) = (\pm 1)^n$.

Therefore $$V = \Sigma\, a_n (\pm r)^n = \Sigma\, a_n z^n.$$

If the value of V is known on the axis and V can be expanded in a series of this form, the coefficients a_n are known and the function V is determined uniquely. Similarly, if we have on the axis

$$V = \Sigma\, b_n z^{-n-1},$$

the expression for V is given uniquely by

$$V = \Sigma\, b_n r^{-n-1} P_n(\mu).$$

Let us suppose that $V = f(z)$ when $\rho = 0$. Writing

$$V = f(z) + \rho^2 f_2(z) + \rho^4 f_4(z) + \dots,$$

we find, on substituting in $\nabla^2 V = 0$, that

$$4n^2 f_{2n}(z) + f_{2n-2}''(z) = 0,$$

where primes denote differentiations with respect to z. The formal expression for V is thus

$$V = f(z) - \frac{\rho^2}{2^2} f''(z) + \frac{\rho^4}{2^2 . 4^2} f^{\text{iv}}(z) - \dots \qquad \text{......(A)}$$

Since $$\frac{1}{\pi}\int_0^\pi \cos^{2n} \alpha . d\alpha = \frac{1}{2^{2n}} \binom{2n}{n},$$

$$\frac{1}{\pi}\int_0^\pi \cos^{2n+1} \alpha . d\alpha = 0,$$

we find that when $f(z+h)$ can be expanded in a Taylor series which is absolutely and uniformly convergent for $a > |h|$

$$V = \frac{1}{\pi}\int_0^\pi f(z + i\rho\cos\alpha)\, d\alpha \quad r < a.$$

This is Laplace's expression for the symmetrical potential function which reduces to $f(z)$ when $\rho = 0$. The formula may be deduced from Whittaker's general formula for a solution of $\nabla^2 V = 0$, namely

$$V = \frac{1}{2\pi}\int_0^{2\pi} f(z + ix\cos\omega + iy\sin\omega, \omega)\, d\omega. \qquad \text{......(B)}$$

The series (A) may be written in the symbolical form

$$V = J_0(\rho D) f(z),$$

where $D = \dfrac{\partial}{\partial z}$.

The foregoing method may be applied also to the wave-equation, and the formula thus obtained for a wave-function depending only on ρ, z and t, and reducing to $F(z, t)$ when $\rho = 0$, may be deduced at once from the generalisation of (B), namely

$$V = \frac{1}{2\pi}\int_0^{2\pi} f(z + ix\cos\omega + iy\sin\omega, ct - ix\sin\omega + iy\cos\omega, \omega)\,d\omega.$$

The appropriate formula is

$$V = \frac{1}{2\pi}\int_0^{2\pi} F\left[z + i\rho\cos\alpha, t - \frac{\rho}{c}\sin\alpha\right] d\alpha.$$

The following special cases of Laplace's formula and the formula just given are of special interest:

$$r^n P_n(\mu) = \frac{1}{\pi}\int_0^{\pi} (z + i\rho\cos\alpha)^n\, d\alpha,$$

$$r^{-n-1} P_n(\mu) = \frac{1}{\pi}\int_0^{\pi} (z + i\rho\cos\alpha)^{-n-1}\, d\alpha,$$

$$e^{-lz} J_0(l\rho) = \frac{1}{\pi}\int_0^{\pi} e^{-l(z+i\rho\cos\alpha)}\, d\alpha;$$

$$e^{-lz} J_0[\rho\sqrt{(k^2 + l^2)}] = \frac{1}{2\pi}\int_0^{2\pi} e^{-l(z+i\rho\cos\alpha)-ik\rho\sin\alpha}\, d\alpha;$$

the factor e^{ikct} has been omitted from both sides of the last equation.

The formula

$$\zeta^{-m} J_m(\zeta) = (-)^m \frac{d^m}{(\zeta d\zeta)^m} J_0(\zeta),$$

which holds for both types of Bessel functions, indicates that if U is a wave-function independent of ϕ, then

$$u = \rho^m \frac{\partial^m}{(\rho\partial\rho)^m}[Ue^{im\phi}]$$

is also a wave-function. The effect of this transformation in certain particular cases is indicated in the following table.

Table I

U	u
$[(z-a)^2+\rho^2]^{-\frac{1}{2}}$	$(-)^m\, 1.3\ldots(2m-1)[(z-a)^2+\rho^2]^{-m-\frac{1}{2}}\rho^m e^{im\phi}$
$r^n P_n(\mu)$	$(-)^m r^{n-m} P_{n-m}(\mu)\, e^{im\phi}$
$r^{-n-1} P_n(\mu)$	$(-)^m r^{-n-m-1} P_{n+m}{}^m(\mu)\, e^{im\phi}.$

Transforming the expansions of the first expression for U in series of Legendre functions and making use of the transformations of the Legendre functions indicated by the other two forms of U, we obtain the expansion

$$1.3\ldots(2m-1)\frac{a^{m+1}\rho^m}{[(z-a)^2+\rho^2]^{m+\frac{1}{2}}} = \sum_{n=0}^{\infty}\left(\frac{a}{r}\right)^{n+m+1} P_{n+m}{}^m(\mu) \quad r > a$$

$$= \sum_{n=2m}^{\infty}\left(\frac{r}{a}\right)^{n-m} P_{n-m}{}^m(\mu) \quad r < a.$$

On the other hand, if we calculate

$$\frac{1}{\pi}\frac{\partial^m}{(\rho\partial\rho)^m}\int_0^\pi (z + i\rho\cos\alpha)^n\, d\alpha,$$

by performing an integration by parts after each differentiation we obtain the formula

$$r^n P_n^m(\mu) = \frac{\rho^m}{1.3\ldots(2m-1)}\frac{(n+m)!}{(n-m)!}\int_0^\pi (z + i\rho\cos\alpha)^{n-m}\sin^{2m}\alpha\,.\,d\alpha.$$

The corresponding formula for $r^{-n-1}P_n^m(\mu)$ is obtained from this by replacing n by $-n-1$ in the integral.

EXAMPLES

1. A solution of the equation

$$\frac{\partial^2 V}{\partial s^2} + \frac{\partial^2 V}{\partial r^2} + \frac{n-2}{r}\frac{\partial V}{\partial r} = 0,$$

which reduces to $f(s)$ when $r = 0$, is given by the formula

$$V = \frac{1}{\pi}\int_0^\pi f(s + ir\cos\phi)(\sin\phi)^{n-3}\, d\phi,$$

where the function $f(z)$ is analytic in a rectangle $a \leqslant s \leqslant b$, $|r| < c$; z being equal to $s + ir$.

2. In the last example if $f(s) = (a^2 + s^2)^{1-\frac{1}{2}n}$, we have

$$V = \frac{1}{a^{n-2}} + \frac{1}{2}\frac{n-2}{n-1}\frac{r^2}{a^n}F\left(-1, -\frac{n-1}{2}; \frac{1}{2}; -\frac{s^2}{r^2}\right)$$
$$+ \frac{1.3}{2.4}\frac{(n-2)n}{(n-1)(n+1)}\frac{r^4}{a^{n+2}}F\left(-2, -\frac{n+1}{2}, \frac{1}{2}; -\frac{s^2}{r^2}\right)$$
$$+ \ldots.$$

3. What problem in potential theory suggests the inversion formula

$$\phi(z) = \frac{1}{\pi}\int_0^\pi f(z + ia\cos w)\, dw,$$

$$f(z) = \frac{1}{\pi}\int_0^\infty \frac{dk}{I_0(ka)}\int_{-\infty}^\infty \cos k(z-\zeta)\,\phi(\zeta)\, d\zeta?$$

4. If $t = s^2 + \sigma^2$, $r = 2s\sigma$, the equation

$$\frac{\partial^2 w}{\partial t^2} = \frac{\partial^2 w}{\partial r^2} + \frac{2m}{r}\frac{\partial w}{\partial r}$$

is transformed into

$$\frac{\partial^2 w}{\partial s^2} + \frac{2m}{s}\frac{\partial w}{\partial s} = \frac{\partial^2 w}{\partial \sigma^2} + \frac{2m}{\sigma}\frac{\partial w}{\partial \sigma}.$$

5. Prove that the differential equation

$$(1-x^2)\frac{\partial^2 V}{\partial x^2} - nx\frac{\partial V}{\partial x} = (1-y^2)\frac{\partial^2 V}{\partial y^2} - ny\frac{\partial V}{\partial y}$$

is transformed by the substitution

$$z = xy, \quad r = \sqrt{(1-x^2)(1-y^2)}$$

into

$$\frac{\partial^2 V}{\partial z^2} = \frac{\partial^2 V}{\partial r^2} + \frac{n-1}{r}\frac{\partial V}{\partial r}.$$

Hence show that this equation possesses a particular solution of type

$$V = \int_0^\pi f[xy + \cos\phi\sqrt{(1-x^2)(1-y^2)}]\sin^{n-2}\phi d\phi.$$

6. If the equation $\dfrac{\partial^2 W}{\partial x^2} + \dfrac{\partial^2 W}{\partial y^2} + \dfrac{\partial^2 W}{\partial z^2} + \dfrac{\partial^2 W}{\partial w^2} = 0$

is transformed by the substitution

$$x = uv\cos\phi, \quad y = uv\sin\phi, \quad z = \tfrac{1}{2}(u^2 - v^2),$$

it is satisfied by $V = \cos m\phi . \cos nw\, U(u, v)$ if

$$\frac{\partial^2 U}{\partial u^2} + \frac{\partial^2 U}{\partial v^2} + \frac{1}{u}\frac{\partial U}{\partial u} + \frac{1}{v}\frac{\partial U}{\partial v} - m^2\left(\frac{1}{u^2} + \frac{1}{v^2}\right) U - n^2(u^2 + v^2) U = 0.$$

[P. Humbert, *Comptes Rendus*, t. CLXX, p. 564 (1920).]

7. A solution of the equation

$$\frac{\partial w}{\partial t} = \kappa\left[\frac{\partial^2 w}{\partial r^2} + \frac{n-2}{r}\frac{\partial w}{\partial r}\right]$$

is given by

$$w = \int_{-\infty}^{\infty} F[r, 2u\sqrt{(\kappa t)}]\, e^{-u^2} du,$$

where $F(r, x)$ is an appropriate type of solution of the equation

$$\frac{\partial^2 F}{\partial x^2} = \frac{\partial^2 F}{\partial r^2} + \frac{n-2}{r}\frac{\partial F}{\partial r}.$$

8. The solution of $\dfrac{\partial^2 V}{\partial s^2} + \dfrac{\partial^2 V}{\partial r^2} + \dfrac{n-2}{r}\dfrac{\partial V}{\partial r} = 0,$

which reduces to s^p when $r = 0$, is given by

$$V = s^p - \frac{p(p-1)}{2(n-1)} s^{p-2} r^2 + \frac{p(p-1)(p-2)(p-3)}{2.4(n-1)(n+1)} s^{p-4} r^4 - \ldots.$$

9. Prove that if $n > -1$ and $t > 0$

$$\int_0^\infty e^{-k^2 t} J_n(kr)\, k^{n+1} dk = (2t)^{-n-1} r^n \exp\{-r^2/4t\}.$$

[H. Weber and N. Sonine.]

§ 7·32. *The use of definite integrals involving Bessel functions.* The potential function represented by the definite integral

$$V = \int_0^\infty e^{-lz} J_0(l\rho) F(l)\, dl, \qquad \ldots\ldots(A)$$

in which the function $F(l)$ is supposed to be one which will ensure uniform convergence and make the limit of V as $\rho \to 0$ equal to the result of making $\rho \to 0$ under the integral sign, will, when $z > 0$, take the value

$$f(z) = \int_0^\infty e^{-lz} F(l)\, dl$$

on the axis of z, and may often be identified immediately from the form of $f(z)$. If, for instance, $f(z) = z^{-1}$ the corresponding function V is r^{-1} and the analysis suggests that

$$\frac{1}{r} = (z^2 + \rho^2)^{-\frac{1}{2}} = \int_0^\infty e^{-lz} J_0(l\rho)\, dl.$$

This result is easily verified*. Again, if $F(l) = J_0(al)$ where a is an arbitrary real constant, we have by the preceding result

$$f(z) = (z^2 + a^2)^{-\frac{1}{2}}.$$

Now this $f(z)$ is the potential on the axis of a unit charge distributed uniformly round the circle $x^2 + y^2 = a^2$, $z = 0$. The function V in this case represents the potential of the ring at any point. It should be noticed that it is a symmetrical function of a and ρ.

In the case of a thin disc of electricity of uniform surface density σ on the circle $x^2 + y^2 \leqslant a^2$, $z = 0$, the potential may be derived from that of a ring by integrating with respect to a between 0 and c. The function $F(l)$ is consequently†

$$F(l) = 2\pi c\sigma \frac{J_1(cl)}{l}.$$

For a circular disc with dipoles normal to its plane and of strength m per unit area, the function F is obtained by differentiating with respect to z. It is consequently

$$F(l) = 2\pi mcJ_1(cl).$$

Similar formulae involving Bessel functions may be used for the representation of wave-functions. The natural generalisation of (A) is

$$V = \int_a^\infty e^{-lz} J_0[\rho\sqrt{(l^2 + k^2)}]\, F(l)\, dl,$$

where the lower limit a is at our disposal. Introducing a new variable $s = (l^2 + k^2)^{\frac{1}{2}}$ this becomes, with a suitable choice of a,

$$V = \int_0^\infty e^{\pm z(s^2-k^2)^{\frac{1}{2}}} J_0(s\rho)\, f(s)\,.\,sds\,.\,(s^2 - k^2)^{-\frac{1}{2}}.$$

When $f(s) = 1$ the formula gives Sommerfeld's representation‡ of the function $\frac{1}{r}e^{ikr}$, which has been used so much in studies relating to the propagation of Hertzian waves over the earth's surface. The upper or lower sign is chosen according as $z \gtrless 0$.

A wave potential may often be expressed in the form of a definite integral involving Bessel functions by making use of Hankel's inversion

* See Watson's *Bessel Functions*, p. 384.

† This result is obtained in a direct manner by A. Gray, *Phil. Mag.* (6), vol. XXXVIII, p. 201 (1919).

‡ *Ann. d. Phys.* Bd. XXVIII, S. 683 (1909). The formula was given, however, without proof in an examination question, Math. Tripos (1905). See Whittaker and Watson's *Modern Analysis*, Ewald, *Ann. d. Phys.* Bd. LXIV, S. 253 (1921). It was given by H. Lamb in a study of earthquake waves. *Roy. Soc. London, Trans.* v. 203A, pp. 1 42 (1904).

formula which holds for an extensive class of functions*. For continuous functions satisfying certain other conditions the inversion formula is

$$f(x) = \int_0^\infty J_m(xt)\, g(t)\, tdt, \quad g(t) = \int_0^\infty J_m(xt)\, f(x)\, xdx.$$

The inversion formula seems to be applicable in the case of the function which occurs under the integral sign in Sommerfeld's formula and gives the equation†

$$\int_0^\infty J_0(\lambda\rho)\, e^{ikr} \rho d\rho / r = (\lambda^2 - k^2)^{-\frac{1}{2}} e^{-|z|\sqrt{(\lambda^2-k^2)}} \quad \lambda^2 > k^2$$

$$= i(k^2 - \lambda^2)^{-\frac{1}{2}} e^{i|z|\sqrt{(k^2-\lambda^2)}} \quad \lambda^2 < k^2$$

which has been used by H. Lamb‡ in some of his physical investigations.

EXAMPLES

1. If
$$f(x) = \int_0^\infty J_m(xt)\, tg(t)\, dt$$
and
$$F(y) = y^{\frac{m+1}{2}} \int_0^\infty e^{-yx^2} x^{m+1} f(x)\, dx,$$
$$G(y) = y^{\frac{m+1}{2}} \int_0^\infty e^{-yx^2} x^{m+1} g(x)\, dx,$$
prove that under suitable conditions
$$F(y) = G(1/4y),$$
so that the relation between the functions f and g is a reciprocal one.

2. Prove that if the real part of $\nu + 1$ is positive the equation
$$f(x) = \int_0^\infty J_\nu(xt)\,(xt)^{\frac{1}{2}} f(t)\, dt$$
is satisfied by
$$f(x) = x^{\nu+\frac{1}{2}} e^{-\frac{1}{2}x^2}.$$
[S. Ramanujan.]

3. When ν is subject to the further restriction that its real part is less than 3/2 the equation of Ex. 2 possesses a second solution
$$f(x) = x^{\frac{1}{2}-\nu} e^{\frac{1}{2}x^2} \int_x^\infty t^{2\nu-1} e^{-\frac{1}{2}t^2}\, dt.$$
[W. N. Bailey, *Journ. London Math. Soc.* vol. v, p. 92 (1930).]

* Proofs of the formula are given in Nielsen's *Handbuch der Cylinderfunktionen*, Gray and Mathews' *Bessel Functions*, and Watson's *Bessel Functions*. New treatments of the relation have been given recently by E. C. Titchmarsh, *Proc. Camb. Phil. Soc.* vol. XXI, p. 463 (1923), and by M. Plancherel, *Proc. London Math. Soc.* (2), vol. XXIV, p. 62 (1926). An extension of the formula is given by G. H. Hardy, *Proc. London Math. Soc.* (2), vol. XXIII, p. lxi (1925), (Records for June 12, 1924). See also R. G. Cooke, *ibid.* vol. XXIV, p. 381 (1926).

† For this equation see N. Sonine, *Math. Ann.* vol. XVI (1880).

‡ *Proc. London Math. Soc.* (2), vol. VII, p. 140 (1909).

4. A potential which satisfies the conditions

$$\frac{\partial^2 V}{\partial t^2} = g\frac{\partial V}{\partial z} \quad \text{for } z = 0,$$

$$V = 0 \quad \text{for } z = \infty \text{ and for } t = 0,$$

$$\frac{\partial V}{\partial t} = gf(\rho)\cos m\phi \quad \text{for } t = 0,\ z = 0,$$

$$\nabla^2 V = 0,$$

is given by the formula

$$V = g\cos m\phi \int_0^\infty e^{-kz} J_m(k\rho)\sin(\sigma t)\, kF(k)\, dk/\sigma,$$

where
$$F(k) = \int_0^\infty f(a) J_m(ka)\, a\, da$$

and $\sigma^2 = gk$.

This result is useful for the study of waves caused by a local disturbance in deep sea water. [K. Terazawa.]

5. If the normal pressure on the infinite plane surface of a semi-infinite elastic solid is $f(\rho)$ when $\rho < a$ and is zero when $\rho > a$ the normal displacement w is given by the formula

$$2\mu w = -z\int_0^\infty e^{-kz} J_0(k\rho)\, F(k)\, dk - (1+\mu/\nu)\int_0^\infty e^{-kz} J_0(k\rho)\, F(k)\, dk/k,$$

where $\nu = \lambda + \mu$ and

$$F(k) = k\int_0^a J_0(k\rho) f(\rho)\, \rho\, d\rho.$$

If u and v are the lateral displacements

$$2\mu(ux + vy) = -\rho z\int_0^\infty e^{-kz} J_1(k\rho)\, F(k)\, dk + (\mu\rho/\nu)\int_0^\infty e^{-kz} J_1(k\rho)\, F(k)\, dk/k.$$

[H. Lamb, *Proc. London Math. Soc.* (1), vol. XXXIV, p. 276 (1902); K. Terazawa, *Phil. Trans.* A, vol. CCXVII, p. 35 (1916).]

§ 7·33. Another useful formula

$$2\pi f(r,\theta) = \int_0^\infty u\,du \int_{-\pi}^{\pi}\int_0^\infty f(\rho,\phi) J_0(uR)\, \rho\, d\rho\, d\phi, \quad R^2 = r^2 + \rho^2 - 2r\rho\cos(\theta - \phi)$$

was first given by Neumann. It is proved by Watson* under the following conditions:

(1) It is required that $f(r, \theta)$ should be a bounded function of the real variables r and θ whenever $-\pi < \theta < \pi$ and $0 \leqslant r$.

(2) The integral
$$\int_{-\pi}^{\pi}\int_0^\infty f(\rho,\phi)\, \rho^{\frac{1}{2}}\, d\rho\, d\phi$$

is supposed to exist and converge absolutely.

(3) $f(r, \theta)$ considered as a function of r, is required to be of bounded variation in the interval $(0, \infty)$ for every value of θ lying between $\pm\pi$, this variation being an integrable function of θ.

* *Bessel Functions*, p. 470. Less stringent conditions have been discovered recently by Fox, *Phil. Mag.* (7), vol. VI, p. 994 (1928).

(4) The total variation $F(r, \theta)$ of the function $f(r, \theta)$ in the interval (r, s) is required to tend uniformly to zero with respect to θ as $s \to r$ for all values of θ in the interval $(-\pi, \pi)$ save, perhaps, some exceptional values lying in intervals the sum of whose lengths is arbitrarily small.

(5) When $f(\rho, \phi)$ is discontinuous at a point (r, θ) the $f(r, \theta)$ outside the integral is to be interpreted to mean the mean value

$$\bar{f}(r, \theta) = \frac{1}{2\pi}\int_{-\pi}^{\pi} f(r', \theta')\, d\alpha, \qquad \begin{aligned} r' \cos\theta' &= r\cos\theta + a\cos\alpha, \\ r' \sin\theta' &= r\sin\theta + a\sin\alpha, \end{aligned}$$

where a is small. This mean value is supposed to be finite as $a \to 0$.

We have seen that if $u(z)$ is a solution of the equation

$$\frac{d^2u}{dz^2} = l^2 z,$$

the definite integral
$$V = \int_0^\infty u(z)\, J_0(l\rho)\, dl$$

is frequently a solution of Laplace's equation. We shall now consider the result obtained by taking u to be the Green's function for certain prescribed boundary conditions at the planes $z = \pm c$. If the conditions are $u = 0$ when $z = \pm c$, the appropriate function is

$$\begin{aligned} u = g(z, z') &= 2\,\frac{\sinh l(c-z)\,.\sinh l(c+z')}{\sinh 2lc} \qquad & z' \leqslant z \leqslant c \\ &= 2\,\frac{\sinh l(c-z')\,.\sinh l(c+z)}{\sinh 2lc} \qquad & -c \leqslant z \leqslant z', \end{aligned}$$

and if the boundary conditions are $\dfrac{\partial u}{\partial z} = 0$, when $z = \pm c$, the appropriate function is

$$\begin{aligned} u = \gamma(z, z') &= 2\,\frac{\cosh l(c-z)\,.\cosh l(c+z')}{\sinh 2lc} \qquad & z' \leqslant z \leqslant c \\ &= 2\,\frac{\cosh l(c-z')\,.\cosh l(c+z)}{\sinh 2lc} \qquad & -c \leqslant z \leqslant z'. \end{aligned}$$

The resulting integrals have been discussed by Fox* with the aid of the identities

$$g(z, z') = e^{-l|z-z'|} + h(z, z'),$$

$$\gamma(z, z') = e^{-l|z-z'|} + k(z, z'),$$

where
$$\sinh 2lc\,.\,h(z, z') = e^{-2lc}\cosh l(z-z') - \cosh l(z+z'),$$

$$\sinh 2lc\,.\,k(z, z') = e^{-2lc}\cosh l(z-z') + \cosh l(z+z').$$

Now the potentials

$$v_1 = \int_0^\infty h(z, z')\, J_0(l\rho)\, dl \quad \text{and} \quad v_2 = \int_0^l k(z, z')\, J_0(l\rho)\, dl$$

* *Loc. cit.*

are such that v_1, v_2, $\frac{\partial v_1}{\partial z}$ and $\frac{\partial v_2}{\partial z}$ are continuous at the plane $z = z'$, while

$$\int_0^\infty e^{-l|z-z'|} J_0(l\rho)\, dl = [\rho^2 + (z - z')^2]^{-\frac{1}{2}} = \frac{1}{r}$$

say, hence
$$U = \int_0^\infty g(z, z') J_0(l\rho)\, dl,$$

$$V = \int_0^\infty \gamma(z, z') J_0(l\rho)\, dl$$

are solutions of Laplace's equation which satisfy the boundary conditions*

$$U = 0 \text{ for } z = \pm c, \quad \frac{\partial V}{\partial z} = 0 \text{ for } z = \pm c,$$

and are such that $U - r^{-1}$, $V - r^{-1}$ are regular potential functions in the neighbourhood of the point $z = z'$, $x = 0$, $y = 0$. By shifting the axis of z to a new position the singularity of U and V may be made an arbitrary point (x', y', z') between the two planes, and U and V then become Green's functions for the space between the planes $z = \pm c$.

Another expression for U may be obtained by the method of images and by the formula of summation given in Example 1, § 7·22. The result is

$$U = \frac{1}{4c}\int_{-\infty}^{\infty} \frac{d\sigma}{s}\left[\frac{\sinh \dfrac{\pi s}{2c}}{\cosh \dfrac{\pi s}{2c} - \cos \dfrac{\pi (z - z')}{2c}} - \frac{\sin \dfrac{\pi s}{2c}}{\cosh \dfrac{\pi s}{2c} + \cos \dfrac{\pi (z + z')}{2c}}\right],$$

$$(s^2 = \sigma^2 + \rho^2).$$

Putting $z' = z$ in both expressions for U we obtain the relation

$$\int_0^\infty J_0(l\rho) \frac{dl}{\sinh 2lc} (\cosh 2lc - \cosh 2lz)$$

$$= \frac{1}{4c}\int_{-\infty}^{\infty} \frac{d\sigma}{s}\left[\coth \frac{\pi s}{4c} - \frac{\sinh \dfrac{\pi s}{2c}}{\cosh \dfrac{\pi s}{2c} + \cos \dfrac{\pi z}{c}}\right].$$

When $z = 0$ this becomes

$$\int_0^\infty J_0(l\rho) \tanh lc . dl = \frac{1}{2c}\int_{-\infty}^{\infty} \frac{d\sigma}{s \sinh \dfrac{\pi s}{2c}}$$

$$= \frac{1}{2c}\int_{-\infty}^{\infty} \frac{d\tau}{\sinh\left(\dfrac{\pi \rho}{2c} \cosh \tau\right)}.$$

* These results are given by Gray, Mathews and MacRobert in their treatise on Bessel Functions, and are proved by Fox, *loc. cit.*

Each integral is, of course, equal to the sum of the series

$$\frac{1}{\rho} - \frac{2}{(\rho^2 + 4c^2)^{\frac{1}{2}}} + \frac{2}{(\rho^2 + 16c^2)^{\frac{1}{2}}} - \frac{2}{(\rho^2 + 36c^2)^{\frac{1}{2}}} + \dots$$

obtained by the method of images. The second expression for U is given by T. Boggio, *Rend. Lombardo*, (2) vol. xlii., pp. 611–624 (1909).

EXAMPLES

1. Prove that when s is real

$$\cos sz . K_0 (s\rho) = \tfrac{1}{2}\int_{-\infty}^{\infty} V \cos s\zeta . d\zeta,$$

$$\sin sz . K_0 (s\rho) = \tfrac{1}{2}\int_{-\infty}^{\infty} V \sin s\zeta . d\zeta,$$

where
$$V = [(z - \zeta)^2 + w^2]^{-\frac{1}{2}}.$$

2. Prove that

$$\int_0^{\pi} \cos (sa \cos \theta)\, K_0 (sa \sin \theta) \sin \theta d\theta = 2\int_a^{\infty} \cos s\zeta . \frac{d\zeta}{\zeta} + \frac{2}{\zeta}\int_0^a \cos s\zeta . d\zeta,$$

and so obtain a verification of Gauss's theorem relating to the mean value of a potential function over a sphere whose centre is at the origin.

3. Prove that
$$\int_0^{\infty} e^{-\lambda z} \sin c\lambda . J_0 (\lambda\rho)\, \lambda^{-1} d\lambda = \sin^{-1} \frac{2c}{r_1 + r_2},$$

where
$$r_1^2 = z^2 + (\rho + c)^2, \quad r_2^2 = z^2 + (\rho - c)^2.$$

[A. B. Basset.]

4. Prove that the integral in Example 3 can also be expressed in the form

$$\tan^{-1} \frac{c + R \sin \Theta}{z + R \cos \Theta},$$

where
$$R^2 \cos 2\Theta = z^2 + \rho^2 - c^2, \quad R^2 \sin 2\Theta = 2cz.$$

5. Prove that when $\mu > 0$ and m and n are positive integers

$$r^{-n-1} P_n (\mu) = \frac{1}{\Gamma(n + 1)}\int_0^{\infty} e^{-\lambda z} J_0 (\lambda\rho)\, \lambda^n d\lambda,$$

$$r^{-n-1} P_n^m (\mu) = \frac{1}{\Gamma(n - m + 1)}\int_0^{\infty} e^{-\lambda z} J_m (\lambda\rho)\, \lambda^n d\lambda.$$

[E. W. Hobson, *Proc. London Math. Soc.* vol. xxv, p. 49 (1894). The first formula was given by Callandreau (see Whittaker and Watson's *Modern Analysis*, p. 364).]

6. Prove that
$$\int_{-\infty}^{\infty} \frac{ds}{z + is} e^{-\frac{a(r^2+s^2)}{2(z+is)}} = 2\pi e^{-az} J_0 (a\rho) \quad z > 0$$
$$= 0 \quad z < 0.$$

[N. Sonine, *Math. Ann.* vol. XVI, p. 25 (1880).]

7. A conducting cylinder $\rho = a$ is surrounded by a uniformly charged ring ($x^2 + y^2 = a^2$, $z = 0$). Prove that the potential outside the cylinder is given by

$$V = \int_0^{\infty} e^{-sz} J_0 (s\rho)\, J_0 (sb)\, ds - \frac{2}{\pi}\int_0^{\infty} \cos sz . K_0 (s\rho) \frac{I_0 (as)\, K_0 (bs)}{K_0 (as)} ds.$$

8. Prove that
$$\int_0^{\infty} e^{-\lambda z} J_m (\lambda\rho)\, e^{im\phi}\, d\lambda = \frac{1}{r}\left(\frac{x + iy}{r + z}\right)^m.$$

$$m > -1.$$

[H. Hankel.]

§ 7·41. *Potential of a thin circular ring.* We have already obtained one expression for the potential of a thin circular ring, but a simpler expression may be obtained by the method of inversion.

Consider first a point P in the plane of the ring. Let the origin be the centre of the ring, a the radius, and let $OP = r$. Let the mass (or charge) associated with a line element $ad\theta$ of the ring be $\sigma ad\theta$, then the potential at P is

$$V = \int_0^{2\pi} \frac{\sigma ad\theta}{R} = \int_0^{2\pi} \frac{\sigma d\phi}{\cos(\phi - \theta)} = \int_0^{2\pi} \frac{\sigma ad\phi}{R + r\cos\phi},$$

where θ is the angle ROP and ϕ the angle RPQ, while R denotes the distance of the point R from P.

Now $R + r\cos\phi = RN$ where N is the foot of the perpendicular from O on PR. We thus obtain the formula

$$V = \sigma \int_0^{2\pi} ad\phi\, (a^2 - r^2 \sin^2\phi)^{-\frac{1}{2}} = 4\sigma K(r/a), \quad (r < a)$$

where K is the complete elliptic integral of the first kind to modulus r/a.

When $r > a$ it is convenient to use another formula which will be obtained by inversion. This formula is included, however, in the general formula for the potential at an external point and this will now be obtained.

Let C be an external point, A and B the points on the ring which are respectively at the greatest and least distances from C. The plane CAB is perpendicular to the plane of the ring and passes through O. Let the circle CAB be drawn and let IJ be the diameter perpendicular to AB. Let CI meet AB in P, then CI bisects the angle ACB, and we have

$$\frac{AC}{BC} = \frac{AP}{PB}.$$

Hence, if $AC = r_1$, $BC = r_2$,

$$\frac{r}{a} = \frac{r_1 - r_2}{r_1 + r_2}.$$

Since $I\hat{B}A = I\hat{C}A = I\hat{C}B$, the triangles IPB, IBC are similar and so

$$IP.IC = IB^2.$$

This means that C is the inverse of P with respect to a sphere of radius IB. The theory of inversion now indicates that the potential at C is

$$V_C = \frac{IB}{IC} V_P.$$

Now the triangles ICB and ACP are similar. Therefore

$$\frac{IB}{IC} = \frac{AP}{AC} = \frac{BP}{BC} = \frac{r_1 + r_2}{2a}.$$

Therefore
$$V_C = \frac{2a}{r_1 + r_2} V_P$$
$$= \frac{4E}{\pi (r_1 + r_2)} K \left(\frac{r_1 - r_2}{r_1 + r_2}\right),$$
where E is the total mass (or charge) of the ring.

For a point Q in the plane but outside the circle we have
$$V_Q = 2\sigma \int_{-\alpha}^{\alpha} a d\psi \, (a^2 - r'^2 \sin^2 \psi)^{-\frac{1}{2}} = 4\sigma \, (a/r') \, K \, (a/r'),$$
where $r' = OQ$ and $r' \sin \alpha = a$. The expression for the integral is easily verified with the aic of the substitution $r' \sin \psi = a \sin \omega$.

Another expression is obtained by integrating the four-dimensional potential of a circular ring. This is
$$W = \int_0^{2\pi} \frac{\sigma a d\theta}{(x - a \cos \theta)^2 + (y - a \sin \theta)^2 + z^2 + w^2}$$
$$= \frac{2\pi a \sigma}{[(x^2 + y^2 + z^2 + w^2)^2 - 4a^2 (x^2 + y^2)]^{\frac{1}{2}}} = \frac{2\pi a \sigma}{[(w^2 + r_1^2)(w^2 + r_2^2)]^{\frac{1}{2}}}.$$
The corresponding Newtonian potential is thus
$$V = \frac{1}{\pi} \int_{-\infty}^{\infty} W dw = 2a\sigma \int_{-\infty}^{\infty} \frac{dw}{[(w^2 + r_1^2)(w^2 + r_2^2)]^{\frac{1}{2}}}.$$
Comparing with the previous result we obtain the equation
$$\int_{-\infty}^{\infty} [(w^2 + r_1^2)(w^2 + r_2^2)]^{-\frac{1}{2}} dw = \frac{4}{r_1 + r_2} K \left(\frac{r_1 - r_2}{r_1 + r_2}\right).$$
Still another expression for the potential has been obtained in the form of a definite integral and so we have the formula
$$\pi \int_0^{\infty} e^{-lz} J_0 (l\rho) \, J_0 (la) \, dl = \frac{4}{r_1 + r_2} K \left(\frac{r_1 - r_2}{r_1 + r_2}\right).$$

EXAMPLES

1. The stream-function for a thin circular ring is
$$S = \frac{\rho}{\pi} \int_0^{\infty} e^{-lz} J_1 (l\rho) \, J_0 (la) \, dl.$$

2. A disc carries a uniform charge distribution of total amount m. Prove that at a point in the plane of the disc the potential is
$$V = \frac{4m\rho}{\pi a^2} [E (a/\rho) - (1 - a^2/\rho^2) K (a/\rho)],$$
where a is the radius of the disc. Show also that the electric field strength is
$$F = \frac{4m}{\pi a^2} [K (a/\rho) - E (a/\rho)],$$
where $K (k)$, $E (k)$ are the complete elliptic integrals to modulus k.

§ 7·42. *The mean value of a potential function round a circle.* Let us first consider the four-dimensional potential

$$W = [(x - x_1)^2 + (y - y_1)^2 + (z - z_1)^2 + (w - w_1)^2]^{-1}.$$

The mean value round the circle $x^2 + y^2 = a$, $z = 0$, $w = 0$, is

$$\overline{W} = \frac{1}{2\pi}\int_0^{2\pi} [(x_1 - a\cos\theta)^2 + (y_1 - a\sin\theta)^2 + z_1^2 + w_1^2]^{-1}\, d\theta$$

$$= [(x_1^2 + y_1^2 + z_1^2 + w_1^2 + a^2)^2 - 4a^2(x_1^2 + y_1^2)]^{-\frac{1}{2}},$$

while the mean value round the circle $z^2 + w^2 = -a^2$, $x = 0$, $y = 0$, is

$$\overline{W} = \frac{1}{2\pi}\int_0^{2\pi} [x_1^2 + y_1^2 + (z_1 - ia\cos\psi)^2 + (w_1 - ia\sin\psi)^2]^{-1}\, d\psi$$

$$= [(x_1^2 + y_1^2 + z_1^2 + w_1^2 - a^2)^2 + 4a^2(z_1^2 + w_1^2)]^{-\frac{1}{2}}.$$

These two values are equal.

Now write $$V = \frac{1}{\pi}\int_{-\infty}^{\infty} W dw_1 = 1/R,$$

where $$R^2 = (x - x_1)^2 + (y - y_1)^2 + (z - z_1)^2.$$

The mean value of V round the first circle is

$$\overline{V} = \frac{1}{2\pi^2}\int_0^{2\pi} d\theta \int_{-\infty}^{\infty} W dw_1.$$

Changing the order of integration and making use of the previous result we are led to surmise that

$$\overline{V} = \frac{1}{2\pi^2}\int_{-\infty}^{\infty} dw_1 \int_0^{2\pi} \frac{d\psi}{x_1^2 + y_1^2 + (z_1 - ia\cos\psi)^2 + (w_1 - ia\sin\psi)^2}.$$

Again changing the order of integration and performing the integration with respect to w_1 we obtain

$$\overline{V} = \frac{1}{2\pi}\int_0^{2\pi} [x_1^2 + y_1^2 + (z_1 - ia\cos\psi)^2]^{-\frac{1}{2}}.$$

The equation thus indicated, viz.

$$\int_0^{2\pi} [(x_1 - a\cos\theta)^2 + (y_1 - a\sin\theta)^2 + z_1^2]^{-\frac{1}{2}}\, d\theta$$

$$= \int_0^{2\pi} [x_1^2 + y_1^2 + (z_1 - ia\cos\psi)^2]^{-\frac{1}{2}}\, d\psi,$$

may in some cases be established directly by writing down Laplace's integral (§ 7·31) for the potential of a uniform circular wire and recalling the fact already noticed in § 7·32 that this potential is symmetric in ρ and a. This equation tells us that if a potential function $V(x, y, z)$ arises from a finite (and perhaps an infinite) number of poles, its mean value round a circle $x^2 + y^2 = a$, $z = 0$, which does not pass through any of the poles, is

$$\overline{V} = \frac{1}{2\pi}\int_0^{2\pi} V(0, 0, ia\cos\psi)\, d\psi = \frac{1}{\pi}\int_0^{\pi} V(0, 0, ia\cos\psi)\, d\psi.$$

When the circle, round which the mean value of V is desired, lies on a given sphere of radius c, it is useful to consider the case in which V can be expanded in an absolutely and uniformly convergent series

$$V = \sum_{n=0}^{\infty} (c/r)^{n+1} S_n (\theta, \phi), \quad r > c.$$

Using our theorem to find the mean value of V round the circle

$$x^2 + y^2 = a^2, \quad z = b = \sqrt{(c^2 - a^2)},$$

we notice that $S_n (\theta, \phi)$ is constant on the axis of z, while the integral of type

$$\frac{1}{\pi} \int_0^{\pi} \left(\frac{c}{b + ia \cos \psi} \right)^{n+1}$$

is equal to $P_n (b/c) = P_n (\mu)$. Hence the mean value is

$$\overline{V} = \sum_{n=0}^{\infty} P_n (\mu) S_n (\theta_0, \phi_0),$$

where $S_n (\theta_0, \phi_0)$ is the value of $S_n (\theta, \phi)$ on the axis of the circle, and $\mu = \cos \alpha$, where α is the angle which a radius of the circle subtends at the centre of the sphere. This agrees with the result of § 6·35.

Since at points on the sphere

$$V = \sum_{n=0}^{\infty} S_n (\theta, \phi),$$

we have a means of finding the mean value of a function of the spherical polar co-ordinates θ, ϕ when this function can be expanded in an absolutely convergent series of spherical harmonics.

§ **7·51.** *An equation which changes from the elliptic to the hyperbolic type.* We shall find it interesting to discuss a simple boundary problem for an equation

$$\frac{\partial^2 V}{\partial r^2} + \frac{1}{r} \frac{\partial V}{\partial r} + \left(\frac{1}{r^2} - 1 \right) \frac{\partial^2 V}{\partial \theta^2} = 0,$$

which is elliptic when $r < 1$ and hyperbolic when $r > 1$. Writing $x = r \cos \theta$, $y = r \sin \theta$ we shall seek a solution which is such that $V = f(\theta)$ when $r = a$, and shall suppose that V, $\frac{\partial V}{\partial r}$ and $\frac{\partial V}{\partial \theta}$ are to be finite and continuous for $r < a$, while the second derivatives $\frac{\partial^2 V}{\partial r^2}$, $\frac{\partial^2 V}{\partial \theta^2}$ exist.

If $f(\theta) = \cos n\theta$ and $J_n (na) \neq 0$ there is a solution

$$V = \frac{J_n (nr)}{J_n (na)} \cos n\theta \qquad \text{......(A)}$$

which satisfies the foregoing requirements. Now if $z = z^{(n)}$ is the smallest positive root of the equation $J_n (z) = 0$, it is known* that when n is a positive integer $z^{(n)} > n$ and that as $n \to \infty$

$$z^{(n)}/n \to 1.$$

* Watson's *Bessel Functions*, p. 485.

Hence if $a < 1$ we certainly have $J_n(na) \neq 0$ and there is a single solution of type (A), but if $a > 1$ there is no solution of type (A) if a happens to have a value for which $J_n(na) = 0$.

In the more general case, when

$$f(\theta) = \sum_{n=0}^{\infty} (A_n \cos n\theta + B_n \sin n\theta),$$

a solution satisfying the condition $V = f(\theta)$, when $r = a$, may be given uniquely by the formula

$$V = \sum_{n=0}^{\infty} \frac{J_n(nr)}{J_n(na)} (A_n \cos n\theta + B_n \sin n\theta),$$

when $a < 1$, but when $a > 1$ there is considerable doubt with regard to the convergence of the series and it cannot be asserted that there is a solution of the boundary problem in this case until the matter of convergence has been settled.

In some cases the convergence may be discussed with the aid of the asymptotic expressions for the function $J_n(na)$ when n is large. The form of these is different according as $a - 1$ is positive or negative.

CHAPTER VIII

ELLIPSOIDAL CO-ORDINATES

§ **8·11.** *Confocal co-ordinates.* An important system of orthogonal co-ordinates is associated with a system of confocal quadrics in a space of n dimensions. Let $x_1, x_2, \dots x_n$ be rectangular co-ordinates relative to the principal axes of one of the quadrics, then the family of quadrics is represented by the equation

$$\sum_{s=1}^{n} \frac{x_s^2}{a_s^2 + \tau} = 1,$$

where τ is a variable parameter, and $a_1^2 + \tau$, $a_2^2 + \tau$, ... $a_n^2 + \tau$ are the squares of the semi-axes of a typical quadric of the family. It is supposed that each quadric possesses a centre; the case in which the quadrics are not central needs special treatment.

Let us write

$$F_\tau = 1 - \sum_{s=1}^{n} \frac{x_s^2}{a_s^2 + \tau} = \frac{P(\tau)}{Q(\tau)},$$

where

$$P(\tau) = (\tau - \xi_1)(\tau - \xi_2) \dots (\tau - \xi_n),$$
$$Q(\tau) = (\tau + a_1^2)(\tau + a_2^2) \dots (\tau + a_n^2).$$

We shall suppose that

$$a_1^2 > a_2^2 > a_3^2 > \dots > a_n^2.$$

Forming the product $Q(\tau) F_\tau$, and putting $\tau = -a_s^2$, we obtain the equations

$$-x_s^2 Q'(-a_s^2) = P(-a_s^2) \quad (s = 1, 2, \dots n) \qquad \text{......(A)}$$

which express the rectangular co-ordinates x_s in terms of the 'confocal' or 'elliptic' co-ordinates ξ_s.

The expression $Q'(-a_s^2)$, which is the value of the derivative $Q'(\tau)$ for $\tau = -a_s^2$, is the product of $n - 1$ factors, thus

$$Q'(-a_1^2) = (a_2^2 - a_1^2)(a_3^2 - a_1^2) \dots (a_n^2 - a_1^2),$$

each factor being in this case negative. In $Q'(-a_2^2)$ there is one positive factor, namely $a_1^2 - a_2^2$, in $Q'(-a_3^2)$ two positive factors, and so on.

Looking at the formula (A) we now see that $(-)^n P(-a_s^2)$ is positive or negative according as s is odd or even. Also $(-)^n P(-\infty)$ is positive, hence the roots of the equation $P(\tau) = 0$ may be arranged in order as follows:

$$-a_1^2 < \xi_1 < -a_2^2 < \xi_2 \dots < -a_n^2 < \xi_n.$$

The last root ξ_n is the parameter of that ellipsoidal quadric of the confocal family which passes through the point $(x_1, x_2, \dots x_n)$. The equation $F_\tau = 0$ shows that n quadrics of the family pass through this point, and the foregoing inequalities indicate that only one of these quadrics is ellipsoidal.

When a small element of length ds is expressed in terms of $\xi_1, \xi_2, \dots \xi_n$ it is found that

$$ds^2 = \sum_{m=1}^{n} dx_m^2 = \tfrac{1}{4}\sum_{m=1}^{n} \frac{P(-a_m^2)}{-Q'(-a_m^2)}\left[\sum_{p=1}^{n} \frac{d\xi_p}{a_m^2+\xi_p}\right]^2.$$

Now $$\sum_{m=1}^{n} \frac{P(-a_m^2)}{Q'(-a_m^2)}\frac{1}{a_m^2+\xi_p} = \frac{P(\xi_p)}{Q(\xi_p)} = 0.$$

Therefore $$\sum_{m=1}^{n} \frac{P(-a_m^2)}{Q'(-a_m^2)}\frac{1}{(a_m^2+\xi_p)(a_m^2+\xi_q)} = 0, \quad p \neq q,$$

$$\sum_{m=1}^{n} \frac{P(-a_m^2)}{Q'(-a_m^2)}\frac{1}{(a_m^2+\xi_p)^2} = -\frac{P'(\xi_p)}{Q(\xi_p)}.$$

Therefore $$ds^2 = \tfrac{1}{4}\sum_{p=1}^{n} \frac{P'(\xi_p)}{Q(\xi_p)}(d\xi_p)^2. \qquad \text{......(B)}$$

Laplace's equation in the co-ordinates $\xi_1, \xi_2, \dots \xi_n$ is thus

$$\nabla^2 V \equiv 4\sum_{p=1}^{n} \frac{\sqrt{Q(\xi_p)}}{P'(\xi_p)}\frac{\partial}{\partial \xi_p}\left(\sqrt{Q(\xi_p)}\frac{\partial V}{\partial \xi_p}\right) = 0. \qquad \text{......(C)}$$

This theorem, which was partially known to Green, is generally associated with the names of Lamé and Jacobi. The case of n variables is considered by Bôcher* who gives an extension suitable for the family of confocal cyclides.

Let us now denote ξ_m by λ. It is clear that Laplace's equation possesses a solution which is a function of λ only if

$$\sqrt{Q(\lambda)}\frac{\partial V}{\partial \lambda} = -C,$$

where C is a constant. Hence we have the solution

$$V = C\int_\lambda^\infty \frac{d\tau}{\sqrt{Q(\tau)}}.$$

This is a particular case of a more general solution, namely

$$V = C\int_\lambda^\infty \left\{\frac{P(\tau)}{Q(\tau)}\right\}^\kappa \frac{d\tau}{\sqrt{Q(\tau)}}, \quad \kappa \geqslant 0.$$

To verify that this is a solution we notice that if $\kappa > 1$

$$\frac{\partial V}{\partial \xi_p} = C\int_\lambda^\infty \left\{\frac{P(\tau)}{Q(\tau)}\right\}^\kappa \frac{d\tau}{\sqrt{Q(\tau)}}\frac{\kappa}{\xi_p - \tau},$$

$$\frac{\partial^2 V}{\partial \xi_p^2} = +C\int_\lambda^\infty \left\{\frac{P(\tau)}{Q(\tau)}\right\}^\kappa \frac{d\tau}{\sqrt{Q(\tau)}}\frac{\kappa(\kappa-1)}{(\xi_p-\tau)^2}.$$

* *Ueber die Reihenentwickelungen der Potentialtheorie*, Ch. III, Leipzig (1894).

Now $$\sum_{p=1}^{n} \frac{Q(\xi_p)}{P'(\xi_p)} \frac{1}{(\tau - \xi_p)^2} = -\frac{\partial}{\partial\tau} \frac{Q(\tau)}{P(\tau)} = \frac{QP'}{P^2} - \frac{Q'}{P},$$

$$\sum_{p=1}^{n} \frac{Q'(\xi_p)}{P'(\xi_p)} \frac{1}{\tau - \xi_p} = \frac{Q'}{P},$$

hence we have to show that when $\kappa > 1$

$$\kappa \int_\lambda^\infty \left[(\kappa - 1) \frac{P^{\kappa-2} P'}{Q^{\kappa-\frac{1}{2}}} - (\kappa - \tfrac{1}{2}) \frac{P^{\kappa-1} Q'}{Q^{\kappa+\frac{1}{2}}} \right] d\tau = 0,$$

that is, that $$\kappa \int_\lambda^\infty \frac{d}{d\tau} \left(\frac{P^{\kappa-1}}{Q^{\kappa-\frac{1}{2}}} \right) d\tau = 0,$$

and this is true.

When $\kappa \leqslant 1$ we cannot differentiate directly to form $\frac{\partial^2 V}{\partial \lambda^2}$, for $\frac{\partial V}{\partial \lambda}$ is of the form

$$\frac{\partial V}{\partial \lambda} = C\kappa \int_\lambda^\infty E(\tau) \frac{d\tau}{(\tau - \lambda)^{1-\kappa}} = -C \int_\lambda^\infty (\tau - \lambda)^\kappa E'(\tau)\, d\tau.$$

Therefore

$$\frac{\partial^2 V}{\partial \lambda^2} = C\kappa \int_\lambda^\infty (\tau - \lambda)^{\kappa-1} E'(\tau)\, d\tau$$

$$= C\kappa \int_\lambda^\infty \left[\frac{d}{d\tau} \{(\tau - \lambda)^{\kappa-1} E(\tau)\} - (\kappa - 1)(\tau - \lambda)^{\kappa-2} E(\tau) \right] d\tau$$

$$= + C \int_\lambda^\infty \left\{ \frac{P(\tau)}{Q(\tau)} \right\}^\kappa \frac{d\tau}{\sqrt{Q(\tau)}} \frac{\kappa(\kappa - 1)}{(\lambda - \tau)^2} - C \int_\lambda^\infty d\tau \frac{d}{d\tau} \left[\left\{ \frac{P(\tau)}{Q(\tau)} \right\}^\kappa \frac{1}{\sqrt{Q(\tau)}} \frac{\kappa}{\tau - \lambda} \right].$$

There is thus an extra term in $\frac{\partial^2 V}{\partial \lambda^2}$ and we have to prove now that

$$\int_\lambda^\infty \frac{d}{d\tau} \left[\frac{P^{\kappa-1}}{Q^{\kappa-\frac{1}{2}}} - \frac{P^\kappa}{Q^{\kappa+\frac{1}{2}}} \frac{Q(\lambda)}{P'(\lambda)(\tau - \lambda)} \right] d\tau = 0.$$

This is true because

$$Q(\tau) P'(\lambda)(\tau - \lambda) - P(\tau) Q(\lambda)$$

vanishes to the first order as $\tau \to \lambda$. It has thus been proved that the function

$$V = C \int_\lambda^\infty (F_\tau)^\kappa \frac{d\tau}{\sqrt{Q(\tau)}}, \quad \kappa \geqslant 0, \qquad \text{......(D)}$$

is a solution of Laplace's equation.

If we write $F_\tau = \omega$ we obtain an integral in which the limits for ω are 0 and 1. Since these are constants and occur to an arbitrary power κ in the integrand we may expect the integrand to be a solution of Laplace's equation for all values of the parameter ω. This is indeed true and the result may be stated as follows:

If τ is defined by the equation

$$\sum_{s=1}^{n} \frac{x_s^2}{a_s^2 + \tau} = 1 - \omega,$$

where ω is a constant parameter, the function

$$W = \frac{1}{\sqrt{Q(\tau)}} \frac{\partial \tau}{\partial \omega} = \frac{1}{\sqrt{Q(\tau)}\, \Sigma \dfrac{x_s^2}{(a_s^2 + \tau)^2}}$$

is a solution of Laplace's equation.

The potential function (D) may be used to solve some interesting problems. It may be used, in particular, to solve the hydrodynamical problem of the steady irrotational motion of an incompressible fluid past a stationary ellipsoid. Green's solution of this problem* was amplified by Clebsch† and extended to a space of n dimensions by C. A. Bjerknes‡ who also considered some additional types of motion of the fluid and called attention to the work of Dirichlet and Schering on the problem. The analysis is really a development of the formulae of Rodrigues§ for the gravitational potential of a solid homogeneous ellipsoid and of the early work of English and French writers on this subject. Historical references are given in Routh's *Analytical Statics*, vol. II (1902).

Let us consider a potential V defined by the equations

$$V = V_0 = \int_\lambda^\infty \frac{du}{\sqrt{Q(u)}} (F_u)^\kappa \quad \text{outside } F_0 = 0,$$

$$V = V_i = \int_0^\infty \frac{du}{\sqrt{Q(u)}} (F_u)^\kappa \quad \text{inside } F_0 = 0.$$

If $\kappa > 0$ we have at the boundary of $F_0 = 0$

$$V_0 = V_i, \quad \frac{\partial V_0}{\partial \lambda} = \frac{\partial V_i}{\partial \lambda},$$

while

$$\nabla^2 V_0 = 0.$$

If $\kappa \geqslant 1$ we have also

$$\nabla^2 V_i = 4\kappa \int_0^\infty du \frac{\partial}{\partial u} \frac{(F_u)^{\kappa-1}}{\sqrt{Q(u)}} = -4\kappa \frac{F_0^{\kappa-1}}{a_1 a_2 \dots a_n}.$$

Hence if the volume density ρ be defined by the equation

$$\nabla^2 V_i + 4\pi\rho h_n = 0,$$

where the constant h_n has the value

$$\frac{[\Gamma(\frac{1}{2})]^{n-2}}{\Gamma\left(\frac{n}{2} - 1\right)},$$

which is readily deducible from the solid angle determined in § 6·51, we

* *Edin. Trans.* (1833); *Papers*, p. 315.

† *Crelle*, Bd. LII, S. 103 (1856), Bd. LIII, S. 287 (1857).

‡ *Gött. Nachr.* pp. 439, 448, 829 (1873), p. 285 (1874); *Forh. Christiania*, p. 386 (1875).

§ *Correspondance sur l'École Polytechnique*, t. III (1815). See also G. Green, *Camb. Trans.* (1835); *Papers*, p. 187. He considered problems in a space of s dimensions for various distributions of density and different laws of attraction.

may regard V as the potential corresponding to a distribution of generalised matter (or electricity) of density

$$\rho = \frac{\kappa F_0^{\kappa-1}}{\pi a_1 a_2 \dots a_n h_n}.$$

In particular, if $\kappa = 1$ we have the potential of a homogeneous ellipsoid. The Newtonian potential of a solid homogeneous ellipsoid is thus

$$V = \pi\rho abc \int_\lambda^\infty \frac{du}{\sqrt{Q(u)}}\left\{1 - \frac{x^2}{a^2+u} - \frac{y^2}{b^2+u} - \frac{z^2}{c^2+u}\right\},$$

where a, b, c are the semi-axes, and

$$Q(u) = (a^2+u)(b^2+u)(c^2+u).$$

The component forces are represented by expressions of type

$$X = -\frac{\partial V}{\partial x} = 2\pi\rho abcx \int_\lambda^\infty \frac{du}{(a^2+u)\sqrt{Q(u)}},$$

and it may be concluded that these expressions represent solutions of Laplace's equation.

The quantity λ is defined for external points by the equation

$$\frac{x^2}{a^2+\lambda} + \frac{y^2}{b^2+\lambda} + \frac{z^2}{c^2+\lambda} = 1,$$

and the inequality $\lambda > 0$. For internal points the lower limit is zero instead of λ and we may write

$$V = \tfrac{1}{2}\rho\{D - Ax^2 - By^2 - Cz^2\},$$

where A, B, C, D are certain constants defined by the equations

$$A = \tfrac{1}{2}\pi abc \int_0^\infty \frac{du}{(a^2+u)\sqrt{Q(u)}}, \quad B = \tfrac{1}{2}\pi abc \int_0^\infty \frac{du}{(b^2+u)\sqrt{Q(u)}},$$

$$C = \tfrac{1}{2}\pi abc \int_0^\infty \frac{du}{(c^2+u)\sqrt{Q(u)}}, \quad D = \tfrac{1}{2}\pi abc \int_0^\infty \frac{du}{\sqrt{Q(u)}}.$$

The component forces at an internal point (x, y, z) are

$$X = -\frac{\partial V}{\partial x} = \rho Ax, \quad Y = -\frac{\partial V}{\partial y} = \rho By, \quad Z = -\frac{\partial V}{\partial z} = \rho Cz.$$

§ **8·12**. *Maclaurin's theorem.* The potential at an external point (x, y, z) may be written in the form

$$V = \pi\rho abc \int_0^\infty \frac{dv}{\sqrt{Q(\lambda+v)}}\left\{1 - \frac{x^2}{a^2+\lambda+v} - \frac{y^2}{b^2+\lambda+v} - \frac{z^2}{c^2+\lambda+v}\right\}$$

$$= \pi\rho abc \int_0^\infty \frac{dv}{\sqrt{(a_1^2+v)(b_1^2+v)(c_1^2+v)}}\left\{1 - \frac{x^2}{a_1^2+v} - \frac{y^2}{b_1^2+v} - \frac{z^2}{c_1^2+v}\right\},$$

where a_1, b_1, c_1 are the semi-axes of the confocal ellipsoid through the point (x, y, z). It is thus seen that the potentials at an external point of two homogeneous solid confocal ellipsoids are proportional to their masses.

§ 8·21. *Hypersphere.* When the quadric in S_n is a hypersphere

$$r^2 \equiv x_1^2 + x_2^2 + \dots x_n^2 = a^2,$$

we have

$$F_\tau = 1 - \frac{r^2}{a^2 + \tau},$$

and the potential corresponding to a density

$$\rho = \frac{C\kappa}{\pi a^n h_n}\left(1 - \frac{r^2}{a^2}\right)^{\kappa-1},$$

is

$$V = C\int_\lambda^\infty \frac{d\tau}{(a^2+\tau)^{n/2}}\left(1 - \frac{r^2}{a^2+\tau}\right)^\kappa, \quad r^2 > a^2$$

$$= C\int_0^\infty \frac{d\tau}{(a^2+\tau)^{n/2}}\left(1 - \frac{r^2}{a^2+\tau}\right)^\kappa, \quad r^2 < a^2,$$

where C is a constant and $\lambda = r^2 - a^2$.

In particular, if $\kappa = 1$, we have when $n > 2$,

$$V = \frac{4C}{n\,(n-2)}\,r^{2-n}, \quad (r^2 > a^2), \quad V = \frac{2C}{n-2}\,a^{2-n} - \frac{2C}{n}\,r^2 a^{-n}, \quad (r^2 < a^2).$$

The total mass associated with the hypersphere is in this case $\dfrac{4C}{n\,(n-2)}$ and so the volume of the hypersphere is

$$\frac{4\pi a^n h_n}{n\,(n-2)}.$$

Comparing this with the value

$$2\pi a^n \frac{[\Gamma(\frac{1}{2})]^{n-2}}{n\Gamma\left(\frac{n}{2}\right)}$$

already found, we find that

$$h_n = \frac{[\Gamma(\frac{1}{2})]^{n-2}}{\Gamma\left(\frac{n}{2} - 1\right)}.$$

The case in which $\rho = f(r)$ can be solved quite generally with the aid of the formulae

$$V = \frac{4\pi h_n}{n-2}\int_0^a \frac{f(s)\,s^{n-1}ds}{r^{n-2}}, \quad r > a,$$

$$V = \frac{4\pi h_n}{n-2}\left[\int_0^r \frac{f(s)\,s^{n-1}ds}{r^{n-2}} + \int_r^a f(s)\,s\,ds\right].$$

In particular, if $f(s) = s^n$

$$(n-2)\,V = \frac{4\pi h_n}{m+n}\,\frac{a^{m+n}}{r^{n-2}}, \quad r > a$$

$$= 4\pi h_n\left[\frac{r^{m+2}}{m+n} + \frac{a^{m+2} - r^{m+2}}{m+2}\right], \quad r < a.$$

A density
$$\rho = \sum_{m=0}^{\infty} C_m r^m$$
will give
$$V = \mu^2 \rho$$
if
$$\mu^2 C_{m+2} = -\frac{4\pi h_n}{(m+n)(m+2)} C_m,$$
$$4\pi h_n \sum_{m=0}^{\infty} \frac{a^{m+2}}{m+2} C_m = \mu^2 C_0.$$
Therefore
$$\rho = \Gamma\left(\frac{n}{2}\right) \sum_{m=0}^{\infty} (-)^m \frac{1}{\Gamma\left(\frac{n}{2}+m\right) m!} \left(\frac{r^2}{4s^2}\right)^m,$$
where
$$4s^2 = \mu^2/\pi h_n.$$
The quantity μ^2 must, moreover, be such that
$$\sum_{m=0}^{\infty} \frac{(-)^m}{\Gamma\left(\frac{n}{2}+m-1\right) m!} \left(\frac{a^2}{4s^2}\right)^m = 0.$$
In the notation of Bessel functions
$$\rho = \Gamma\left(\frac{n}{2}\right) \left(\frac{2s}{r}\right)^{\frac{n}{2}-1} J_{\frac{n}{2}-1}\left(\frac{r}{s}\right),$$
where
$$J_{\frac{n}{2}-2}\left(\frac{a}{s}\right) = 0.$$

§ 8·31. *Potential of a homoeoid and of an ellipsoidal conductor.* Many of the formulae relating to the attraction of ellipsoids and ellipsoidal shells may be obtained geometrically. The theorems will be proved for the ellipsoid in n dimensions and an extension will be made of the meaning of the word *homoeoid* introduced by Lord Kelvin and P. G. Tait in their *Natural Philosophy*. The analysis is an extension of that given by Poisson*.

A homoeoid is a shell bounded by two loci which are similar and similarly situated with regard to each other. If one locus is an n-dimensional ellipsoid and the centre of similitude is the centre of this locus, the second locus is an ellipsoid with the same principal axes.

Let $a_1, a_2, \ldots a_n$ be the semi-axes of the internal boundary of the shell, $a_1 + da_1, a_2 + da_2, \ldots a_n + da_n$ the corresponding semi-axes of the external boundary. Let OPQ be a line through the centre of the ellipsoids cutting them in P and Q respectively, and let $OP = r$, $OQ = r + dr$.

Let p and $p + dp$ be the distances of O from the tangent hyperplanes at P and Q which are, of course, parallel. Then dp is the thickness of the shell at P, and we have
$$\frac{da_1}{a_1} = \frac{da_2}{a_2} = \ldots \frac{da_n}{a_n} = \frac{dp}{p} = \frac{dr}{r} = d\epsilon, \text{ say.}$$

* *Mémoires de l'Institut de France* (1835).

Since the volume of a solid ellipsoid with semi-axes $a_1, a_2, \ldots a_n$ is

$$\frac{4\pi h_n a_1 a_2 \ldots a_n}{n(n-2)} = a_1 a_2 \ldots a_n \frac{\pi^{\frac{n}{2}}}{\Gamma\left(\frac{n}{2}+1\right)},$$

the volume of the shell is approximately

$$\frac{2\pi^{\frac{n}{2}}}{\Gamma\left(\frac{n}{2}\right)} a_1 a_2 \ldots a_n . d\epsilon.$$

Let us now imagine the shell to be filled with attracting matter of uniform density ρ, then the total mass of the shell is

$$M = \frac{2\pi^{\frac{n}{2}}}{\Gamma\left(\frac{n}{2}\right)} \rho a_1 a_2 \ldots a_n . d\epsilon.$$

The importance of the ellipsoidal homoeoid in potential theory arises from the fact that the attraction of a thin uniform shell of this type is zero at any internal point. This is a simple extension of the theorem established by Newton for spherical and spheroidal shells. It is important in electrostatics because it indicates at once the distribution over the surface of an ellipsoidal conductor of electricity which is in equilibrium.

Through any point I of the region enclosed by the internal boundary of the homoeoid let lines be drawn so as to generate a double cone of small solid angle $d\omega$ and to cut out from the ellipsoidal shell small pieces of contents dv, dv' respectively. Let a line $sSITt$ completely enclosed by this cone meet the boundaries of the shell in the points ST, st respectively, and let

$$IS = R, \quad Is = R + dR, \quad IT = R', \quad It = R' + dR'.$$

Since parallel chords of the two boundaries of the shell are bisected by the same diametral hyperplane, we have $dR' = dR$. Also, when $d\omega$ is very small, $dv = R^{n-1}\,dR\,d\omega$, $dv' = R'^{n-1}\,dR'd\omega$, hence

$$\frac{\rho\,dv}{R^{n-1}} = \frac{\rho\,dv'}{R'^{n-1}},$$

and so the attractions at I of the two small pieces balance. When $d\epsilon$ is very small we may write $dv = dp\,dS$, where dS is the area of a surface element; the mass of the element dv is thus $\rho p\,d\epsilon\,dS$ and so the surface density is $\sigma = \rho p\,d\epsilon$, thus

$$\sigma = \frac{Mp\Gamma\left(\frac{n}{2}\right)}{2\pi^{\frac{n}{2}} a_1 a_2 \ldots a_n}.$$

The potential of the homoeoid at an internal point is constant. When this constant value is known the potential at an external point may be found by means of Ivory's theorem as in Routh's *Analytical Statics*, vol. II, p. 102.

§ **8·32**. *Potential of a homogeneous elliptic cylinder.* This potential may be found by direct integration*. Let us take the focus S of the cross-section as origin, then the polar equation of the section is

$$r = \frac{b^2}{a + k\cos\theta} = f(\theta), \text{ say,}$$

where a and b are the semi-axes of the ellipse and $2k$ is the distance between the foci. If σ is the density of the line charges from which the cylinder is supposed to be built up, the potential of all these line charges is

$$V = -2\sigma\int_{-\pi}^{\pi} d\theta_0 \int_0^{f(\theta_0)} \log R \,.\, r_0\,dr_0,$$

where

$$R^2 = r^2 + r_0^2 - 2rr_0\cos(\theta - \theta_0),$$

the infinite constant in the potential of each line charge being omitted.

Now if $r > a + k$ we may write

$$\log R = \log r - \sum_1^\infty \frac{1}{n}\left(\frac{r_0}{r}\right)^n \cos n(\theta - \theta_0).$$

Therefore

$$V = -\sigma b^4 \int_{-\pi}^{\pi} \frac{d\theta_0 \,.\log r}{(a + k\cos\theta_0)^2} + 2\sigma \sum_1^\infty \frac{b^{2n+4}}{n(n+2)\,r^n}\int_{-\pi}^{\pi}\frac{d\theta_0\cos n(\theta - \theta_0)}{(a + k\cos\theta_0)^{n+2}}.$$

But

$$\int_{-\pi}^{\pi}\frac{d\theta_0\cos n(\theta - \theta_0)}{(a + k\cos\theta_0)^{n+2}} = (-)^n\binom{2n+1}{n}\frac{\pi k^n a}{2^{n-1}b^{2n+3}}\cos n\theta.$$

Therefore

$$V = -2\pi\sigma ab\left[\log r - \sum_1^\infty (-)^n \frac{2^{1-n}}{n(n+2)}\binom{2n+1}{n}\left(\frac{k}{r}\right)^n \cos n\theta\right]$$

$$r > a + k.$$

§ **8·33**. *Elliptic co-ordinates.* Potential problems relating to an elliptic cylinder may often be solved by using the elliptic co-ordinates of § 3·71. These are defined by the equations

$$x + iy = a\cosh(\xi + i\eta), \quad x = a\cosh\xi\cos\eta,$$

$$y = a\sinh\xi\sin\eta.$$

The same co-ordinates are also useful for the treatment of the vibrations of an elliptic membrane and the scattering of periodic electromagnetic waves by an obstacle having the form of either an elliptic or hyperbolic

* N. R. Sen, *Phil. Mag.* (6), vol. XXXVIII, p. 465 (1919); see also W. Burnside, *Mess. of Math.* vol. XVIII, p. 84 (1889).

cylinder. A screen containing a straight cut of constant width can be regarded as a limiting case of an obstacle whose surface is a hyperbolic cylinder. In terms of the co-ordinates ξ, η the equation

$$\frac{\partial^2 V}{\partial x^2} + \frac{\partial^2 V}{\partial y^2} + k^2 V = 0$$

becomes
$$\frac{\partial^2 V}{\partial \xi^2} + \frac{\partial^2 V}{\partial \eta^2} + a^2 k^2 V (\cosh^2 \xi - \cos^2 \eta) = 0,$$

and there are solutions of type $V = X(\xi)\, Y(\eta)$ if

$$\frac{d^2 X}{d\xi^2} + X (a^2 k^2 \cosh^2 \xi - A) = 0,$$

$$\frac{d^2 Y}{d\eta^2} + Y (A - a^2 k^2 \cos^2 \eta) = 0.$$

The equations to be solved are thus of type

$$\frac{d^2 y}{dz^2} + (a + 16b \cos 2z)\, y = 0. \qquad \text{......(A)}$$

This is known as Mathieu's equation or as the equation of the elliptic cylinder. The solutions of this equation have been studied by many writers. The best presentation of the results obtained is that in Whittaker and Watson's *Modern Analysis*, Ch. XIX.

A discussion of the zeros of solutions of this equation is given in a paper by Hille*. He calls any solution of the equation a Mathieu function, while Whittaker reserves this name for the solutions† with period 2π. Hille remarks that all solutions of the equation are entire functions of z of infinite genus.

The form of the solution when b is very large has been discussed by Jeffreys‡ who also considers the effect of a variation of b on the positions of the zeros.

Jeffreys has also discussed the equation for X, which he calls the "modified Mathieu's equation." The equations satisfied by X and Y are found to govern the free oscillations of water in an elliptic lake.

§ 8·34. *Mathieu functions.* When $b = 0$ and $a = n^2$ the differential equation (A) possesses two independent solutions $\cos nz$, $\sin nz$, which are periodic in z with period 2π, where n is an integer.

* E. Hille, *Proc. London Math. Soc.* (2), vol. XXIII, p. 185 (1925).

† E. L. Ince has shown that for no value of b, except $b = 0$, does Mathieu's equation possess two independent periodic solutions of period 2π. See *Proc. Camb. Phil. Soc.* vol. XXI, p. 117 (1922); *Proc. London Math. Soc.* (2), vol. XXIII, p. 56 (1925). See also E. Hille, *loc. cit.*; J. H. McDonald, *Trans. Amer. Math. Soc.* vol. XXIX, p. 647 (1927); Ž. Marković, *Proc. Camb. Phil. Soc.* vol. XXIII, p. 203 (1926–7). A more general type of equation, due to Hill, has been und by Ince to possess a similar property, *ibid.* p. 44.

‡ H. Jeffreys, *Proc. London Math. Soc.* (2), vol. XXIII, pp. 437, 455 (1925).

When $b \neq 0$ we can, for any fixed values of a and b, define an even solution $c(z)$ by the initial conditions $c(0) = 1$, $c'(0) = 0$ and an odd solution $s(z)$ by the initial conditions $s(0) = 0$, $s'(0) = 1$.

These two solutions of the equation form a fundamental system and are connected by the relation

$$c(z)\, s'(z) - s(z)\, c'(z) = 1.$$

When b is given there are certain values of a for which $c(z)$ is a periodic function of period 2π. These values are roots of a certain determinantal equation*

$$\begin{vmatrix} a - 1 + 8b & 8b & 0 & \dots \\ 8b & a - 9 & 8b & \dots \\ 0 & 8b & a - 25 & \dots \\ \dots & \dots & \dots & \dots \end{vmatrix} \qquad \text{......(B)}$$

There are also certain other values of a for which $s(z)$ is a periodic function of period 2π. The equation determining these values is obtained from the last equation by writing $a - 1 - 8b$ in place of $a - 1 + 8b$. These determinantal equations are obtained immediately by substituting Fourier series in the differential equation and writing down the conditions for the compatibility of the resulting difference equations.

There is a corresponding determinantal equation for the determination of the values of a for which $c(z)$ has a period π and also one for the determination of the values of a for which $s(z)$ has the period π.

Whittaker writes $ce_n(z)$ for the even periodic Mathieu function whose Fourier expansion has a unit coefficient for $\cos nz$, and writes $se_n(z)$ for the odd periodic Mathieu function whose Fourier expansion has a unit coefficient for $\sin nz$. The functions with even suffix have the period π, those with an odd suffix have the period 2π.

The analysis relating to these periodic functions has been much improved recently by S. Goldstein† who treats the difference equations by a method which has proved very successful in the theory of tides on a rotating globe‡. In the case of an even function with period 2π

$$ce_{2s+1}(z) = \sum_{r=0}^{\infty} A_{2r+1} \cos(2r+1) z, \quad A_{2s+1} = 1. \qquad \text{......(C)}$$

The difference equation which leads to (B) is

$$\{a - (2r+1)^2\} A_{2r+1} + 8b \{A_{2r-1} + A_{2r+3}\} = 0.$$

This shows that, as $r \to \infty$, the ratio

$$V_r = \frac{A_{2r+3}}{A_{2r+1}}$$

* See, for instance, E. L. Ince, *Proc. Camb. Phil. Soc.* vol. XXIII, p. 47 (1926–7); *Proc. Edinburgh Math. Soc.* vol. XLI, p. 94 (1923); *Ordinary Differential Equations*, p. 177 (1927).

† *Proc. Camb. Phil. Soc.* vol. XXIII, p. 223 (1928). Another method leading to useful results has been used by McDonald (*l.c.*).

‡ See Lamb's *Hydrodynamics*, 5th ed. p. 313.

tends to either zero or ∞. Now in order that the series (C) may converge the limit should be zero and not ∞.

To find the condition that this may be the case we write

$$C_r = (2r+1)^{-2},$$

then

$$V_{r-1} = \frac{8b\,C_r}{1 - a\,C_r - 8b\,C_r V_r},$$

and so when $V_r \to 0$ as $r \to \infty$ we have

$$V_{r-1} = \frac{8b\,C_r}{1 - a\,C_r -}\,\frac{64b^2\,C_r C_{r+1}}{1 - a\,C_{r+1}}\, - \,\frac{64b^2\,C_{r+1} C_{r+2}}{1 - a\,C_{r+2}}\, - \,\dots .$$

Now the difference equation

$$(a - 1 + 8b)\,A_1 + 8b\,A_3 = 0$$

gives

$$V_1 = -\frac{a - 1 + 8b}{8b}.$$

Hence we have the equation

$$1 - a = 8b + \frac{64b^2\,C_2}{1 - a\,C_2 +}\,\frac{64b^2\,C_2 C_3}{1 - aC_3}\, + \,\dots$$

for the determination of a.

For the asymptotic expansions of solutions of Mathieu's equation reference may be made to papers by W. Marshall* and J. Dougall†.

EXAMPLES

1. If

$$x = [(\rho-1)(\mu-1)(\nu-1)]^{\frac{1}{2}},\quad y = i(\rho\mu\nu)^{\frac{1}{2}}\cos\phi,\quad z = i(\rho\mu\nu)^{\frac{1}{2}}\sin\phi,\quad t = -\tfrac{1}{2}(\rho+\mu+\nu-1),$$

the equation

$$\frac{\partial^2 u}{\partial x^2} + \frac{\partial^2 u}{\partial y^2} + \frac{\partial^2 u}{\partial z^2} + \frac{\partial^2 u}{\partial t^2} = 0 \qquad \text{......(D)}$$

becomes

$$\sum_{\rho,\mu,\nu} \frac{\mu-\nu}{\sqrt{(\mu-1)(\nu-1)}}\frac{\partial}{\partial\rho}\left[\rho(\rho-1)^{\frac{1}{2}}\frac{\partial U}{\partial\rho}\right] + \frac{1}{4}\frac{(\rho-\mu)(\mu-\nu)(\nu-\rho)}{\rho\mu\nu\,[(\rho-1)(\mu-1)(\nu-1)]^{\frac{1}{2}}}\frac{\partial^2 U}{\partial\phi^2} = 0,$$

and possesses simple solutions of type

$$u = R(\rho)\,M(\mu)\,N(\nu)\cos m\phi,$$

if R, M, N satisfy equations of type

$$\rho^2(\rho-1)\,R'' + (\tfrac{3}{2}\rho^2 - \rho)\,R' + (\tfrac{1}{4}m^2 + h\rho + k\rho^2)\,R = 0.$$

The substitution $R(\rho) = \rho^{\frac{1}{2}m}\,S(\sin^2\theta)$ reduces this equation to the equation

$$\frac{d^2S}{d\theta^2} + (2m+1)\cot\theta\,\frac{dS}{d\theta} - 4\,[h + k + \tfrac{1}{4}m(m+1) - k\cos^2\theta]\,S = 0,$$

for the associated Mathieu functions‡. [P. Humbert§.]

* *Proc. Edinburgh Math. Soc.* vol. XL, p. 2 (1921).

† *Ibid.* vol. XLI, p. 26 (1923).

‡ E. L. Ince, *Proc. Roy. Soc. Edinburgh*, vol. XLII, p. 47 (1922); *Proc. Edinburgh Math. Soc.* vol. XLI, p. 94 (1923).

§ *Proc. Roy. Soc. Edinburgh*, vol. XLVI, p. 206 (1926); *Proc. Edinburgh Math. Soc.* vol. XL, p. 27 (1922); *Fonctions de Lamé et Fonctions de Mathieu*, Gauthier-Villars, Paris (1926).

2. The substitution

$$x = [(\rho - 1)(\mu - 1)(\nu - 1)]^{\frac{1}{2}}, \quad y = i(\rho\mu\nu)^{\frac{1}{2}}, \quad z = -\tfrac{1}{2}(\rho + \mu + \nu - 1),$$

transforms the equation (D) into

$$\Sigma \frac{\mu - \nu}{[\mu\nu(\mu - 1)(\nu - 1)]^{\frac{1}{2}}} \frac{\partial}{\partial \rho}\left[\rho(\rho - 1)^{\frac{1}{2}} \frac{\partial u}{\partial \rho}\right] + \frac{1}{4}\frac{\partial^2 u}{\partial t^2} \frac{(\rho - \mu)(\mu - \nu)(\nu - \rho)}{[\rho\mu\nu(\rho - 1)(\mu - 1)(\nu - 1)]^{\frac{1}{2}}} = 0,$$

and there are simple solutions of type

$$u = e^{\lambda t} R(\rho) M(\mu) N(\nu)$$

if

$$\rho(\rho - 1) R'' + (\rho - \tfrac{1}{2}) R' + (h + k\rho - \tfrac{1}{4}\lambda^2\rho^2) R = 0,$$

or

$$\frac{d^2R}{d\theta^2} + (\alpha + \beta\cos 2\theta + \gamma\cos 4\theta) R = 0,$$

where $\rho = \cos^2\theta$. This is an extension of Mathieu's equation considered by Whittaker* and Ince†.

§ 8·41. *Prolate spheroid.* When the ellipsoid has two equal axes the elliptic integrals of § 8·11 can be expressed in terms of circular functions‡.

In the case of the prolate spheroid

$$\frac{x^2 + y^2}{a^2} + \frac{z^2}{c^2} = 1, \quad c^2 > a^2,$$

the potential of the homoeoid is

$$V = \frac{M}{2}\int_\lambda^\infty \frac{du}{(a^2 + u)(c^2 + u)^{\frac{1}{2}}},$$

where

$$\frac{x^2 + y^2}{a^2 + \lambda} + \frac{z^2}{c^2 + \lambda} = 1, \quad \lambda \geqslant 0.$$

The integral may be evaluated with the aid of the substitutions

$$u = c^2\tan^2\beta, \quad \cos\beta = s,$$

and we find that

$$V = \frac{M}{2k}\log\frac{(c^2 + \lambda)^{\frac{1}{2}} + k}{(c^2 + \lambda)^{\frac{1}{2}} - k}, \quad k = (c^2 - a^2)^{\frac{1}{2}}.$$

Writing

$$x = \varpi\cos\phi, \quad y = \varpi\sin\phi,$$
$$z + i\varpi = k\cos(\xi + i\eta),$$

we have

$$z = k\cosh\eta\cos\xi = k\theta\mu, \text{ say},$$
$$\varpi = k\sinh\eta\sin\xi = k[(\theta^2 - 1)(1 - \mu^2)]^{\frac{1}{2}},$$
$$\frac{z^2}{k^2\cosh^2\eta} + \frac{\varpi^2}{k^2\sinh^2\eta} = 1.$$

* E. T. Whittaker, *Proc. Edinburgh Math. Soc.* vol. XXXIII, p. 75 (1914–15).

† E. L. Ince, *Proc. London Math. Soc.* vol. XXIII, p. 56 (1925).

‡ The results are all well known. Reference may be made to Heine's *Kugelfunktionen*, Bd. II, § 38; to Lamb's *Hydrodynamics*, Ch. V; and to Byerly's *Fourier Series and Spherical Harmonics*.

Comparing this with the equation

$$\frac{z^2}{c^2+\lambda}+\frac{\varpi^2}{a^2+\lambda}=1,$$

we see that we must have

$$k^2\cosh^2\eta = c^2+\lambda, \quad k^2\sinh^2\eta = a^2+\lambda,$$

$$V=\frac{M}{2k}\log\frac{\cosh\eta+1}{\cosh\eta-1}=\frac{M}{k}\log\coth\frac{\eta}{2}.$$

The potential may be expressed in another form by finding the distances R, R' of a point P from the real foci S, S' of the spheroid. Since

$$OS = OS' = k,$$

we have

$$\begin{aligned} R^2 &= (z-k)^2+\varpi^2 = k^2[(\cosh\eta\cos\xi-1)^2+\sinh^2\eta\sin^2\xi] \\ &= k^2[\cosh\eta-\cos\xi]^2, \end{aligned}$$

$$R = k(\cosh\eta-\cos\xi), \quad R' = k(\cosh\eta+\cos\xi),$$

$$\cosh\eta=\frac{R+R'}{2k}, \quad \cos\xi=\frac{R'-R}{2k},$$

$$V=\frac{M}{2k}\log\frac{R+R'+2k}{R+R'-2k}.$$

It is clear from this expression that V is constant on a prolate spheroid with S and S' as foci. On the surface of the conductor $R+R'=2c$, and V has the constant value V_0, where

$$V_0=\frac{M}{2k}\log\frac{c+k}{c-k}.$$

The capacity of the conductor is thus

$$C=\frac{2k}{\log\dfrac{c+k}{c-k}}.$$

The lines of force are given by $\xi=$ constant, $\phi=$ constant, and are confocal hyperbolas.

These results remain valid in the limiting case when the spheroid reduces to a thin rod SS', and in this case the potential must be capable of being expressed as an integral of the inverse distance along the rod. We have in fact

$$V=\frac{M}{2k}\int_{-k}^{k}\frac{ds}{[\varpi^2+(z-s)^2]^{\frac{1}{2}}}.$$

The line density is thus $M/2k$ and is uniform.

If we take as new co-ordinates the quantities θ, μ, ϕ, where $\theta=\cosh\eta$, $\mu=\cos\xi$, the square of the linear element ds is given by the equation

$$\begin{aligned} ds^2 &= dx^2+dy^2+dz^2 = dz^2+d\varpi^2+\varpi^2 d\phi^2 \\ &= k^2[(\cosh^2\eta-\cos^2\xi)(d\xi^2+d\eta^2)+\sinh^2\eta\sin^2\xi\,.\,d\phi^2] \\ &= k^2\left[\frac{\theta^2-\mu^2}{1-\mu^2}d\mu^2+\frac{\theta^2-\mu^2}{\theta^2-1}d\theta^2+(1-\mu^2)(\theta^2-1)\,d\phi^2\right]. \end{aligned}$$

Along the normal to a surface $\theta =$ constant, we have

$$dn = ds = k\left(\frac{\theta^2 - \mu^2}{\theta^2 - 1}\right)^{\frac{1}{2}} d\theta,$$

and so the potential gradient is

$$\frac{\partial V}{\partial n} = \frac{1}{k}\left(\frac{\theta^2 - 1}{\theta^2 - \mu^2}\right)^{\frac{1}{2}} \frac{\partial V}{\partial \theta} = -\frac{M}{k^2}\left[(\theta^2 - 1)(\theta^2 - \mu^2)\right]^{-\frac{1}{2}}.$$

At the vertex of the surface $\theta = \alpha$ the gradient is

$$-\frac{M}{k^2}(\alpha^2 - 1)^{-1} = -\frac{M}{k^2 \sinh^2 \eta},$$

while at the equator it is

$$-\frac{M}{\alpha k^2}(\alpha^2 - 1)^{-\frac{1}{2}} = -\frac{M}{k^2 \sinh \eta \cosh \eta}.$$

The ratio of the two gradients is $\coth \eta$, which is the ratio of the semi-axes of the spheroid. On the spheroid $\lambda = 0$, $\cosh \eta = c/k$, the surface density of electricity is

$$\sigma = -\frac{1}{4\pi}\frac{\partial V}{\partial n} = \frac{M}{4\pi a (c^2 - \mu^2 k^2)^{\frac{1}{2}}}.$$

§ **8·42**. *Oblate spheroid*. In the case of an oblate spheroid

$$\frac{x^2 + y^2}{a^2} + \frac{z^2}{c^2} = 1, \quad c^2 < a^2,$$

the potential of the spheroidal homoeoid is

$$V = \frac{M}{2}\int_\lambda^\infty \frac{du}{(a^2 + u)(c^2 + u)^{\frac{1}{2}}} = \frac{M}{k}\cot^{-1}\left[\frac{1}{k}(\lambda + c^2)^{\frac{1}{2}}\right],$$

where $k^2 = a^2 - c^2$ and λ is defined by

$$\frac{x^2 + y^2}{a^2 + \lambda} + \frac{z^2}{c^2 + \lambda} = 1, \quad \lambda \geqslant 0.$$

Making the substitutions

$$x = \varpi \cos \phi, \quad y = \varpi \sin \phi,$$

$$\varpi + iz = k \cos(\xi + i\eta),$$

which give

$$z = k \sinh \eta \sin \xi = k\theta\mu, \text{ say,}$$

$$\varpi = k \cosh \eta \cos \xi = k\left[(\theta^2 + 1)(1 - \mu^2)\right]^{\frac{1}{2}},$$

$$a^2 + \lambda = k^2 \cosh^2 \eta = k^2(\theta^2 + 1),$$

$$c^2 + \lambda = k^2 \sinh^2 \eta = k^2\theta^2,$$

we find that

$$V = \frac{M}{k}\cot^{-1}\theta.$$

At a point on the surface of the conductor $\lambda = 0$, $\theta = c/k$, and the potential has the constant value

$$V_0 = \frac{M}{k}\cot^{-1}\frac{c}{k}.$$

The capacity of the spheroidal conductor is thus

$$C = \frac{k}{\cot^{-1}\dfrac{c}{k}}.$$

As $c \to 0$ this approaches the value $\dfrac{2k}{\pi}$. This represents the capacity of an infinitely thin circular disc of radius k. If θ, μ, ϕ are taken as new co-ordinates the square of the linear element ds is given by the equation

$$ds^2 = k^2 (\theta^2 + \mu^2) \left[\frac{d\theta^2}{1+\theta^2} + \frac{d\mu^2}{1-\mu^2}\right] + k^2 (1+\theta^2)(1-\mu^2)\, d\phi^2.$$

Along the normal to the surface $\theta =$ constant, we have

$$dn = ds = kd\theta \left(\frac{\theta^2 - \mu^2}{\theta^2 + 1}\right)^{\frac{1}{2}}.$$

Therefore

$$\frac{\partial V}{\partial n} = \frac{1}{k}\left(\frac{\theta^2+1}{\theta^2+\mu^2}\right)^{\frac{1}{2}} \frac{\partial V}{\partial \theta} = -\frac{M}{k^2}\left[(\theta^2+1)(\theta^2+\mu^2)\right]^{-\frac{1}{2}}.$$

Thus at points of an equipotential surface $\theta =$ constant, the gradient varies like a constant multiple of $(\theta^2 + \mu^2)^{-\frac{1}{2}}$. The ratio of the potential gradients at the vertex and equator of this equipotential surface (which is likewise an oblate spheroid) is the ratio of the central radii which end at these places.

The surface density of electricity at a point of the spheroidal conductor is

$$\sigma = -\frac{1}{4\pi}\frac{\partial V}{\partial n} = \frac{M}{4\pi a}(c^2 + k^2\mu^2)^{-\frac{1}{2}}.$$

In the case of the disc this becomes simply

$$\frac{M}{4\pi k^2 \mu} = \frac{M}{4\pi k}(k^2 - \varpi^2)^{-\frac{1}{2}}.$$

§ **8·43**. *A conducting ellipsoidal column projecting above a flat conducting plane.* The electric potential for this case has been studied by Sir Joseph Larmor and J. S. B. Larmor* in connection with the theory of lightning conductors, and by Benndorf† in relation to the measurement of atmospheric potential gradients. It is clear that the potential

$$V = -z + Az \int_{\lambda}^{\infty} du \left[(a^2+u)(b^2+u)(c^2+u)^3\right]^{-\frac{1}{2}}$$

is zero over the plane $z = 0$, and also over the ellipsoid

$$\frac{x^2}{a^2} + \frac{y^2}{b^2} + \frac{z^2}{c^2} = 1,$$

* *Proc. Roy. Soc.* A, vol. xc, p. 312 (1914)

† *Wiener Berichte*, Bd. cix, S. 923 (1900); Bd. cxv, S. 425 (1906).

if A is defined by the equation

$$1 = A \int_0^\infty du \, [(a^2 + u)(b^2 + u)(c^2 + u)^3]^{-\frac{1}{2}},$$

and λ by the equation

$$\frac{x^2}{a^2 + \lambda} + \frac{y^2}{b^2 + \lambda} + \frac{z^2}{c^2 + \lambda} = 1, \quad \lambda \geqslant 0.$$

At a great distance from the ellipsoid V is approximately equal to $-z$ and the field is uniform.

When a and b are small compared with c so that the column is tall and slender, the Larmors remark that A is small and so the lateral effect of the column is on the whole small, though the gradient may be very high in the immediate neighbourhood of the vertex. When $a = b$ the gradient at the vertex is given by the formula

$$-\frac{\partial V}{\partial z} = \frac{2k^3}{a^2 c \log \dfrac{c + k}{c - k} - 2a^2 k},$$

where $k^2 = c^2 - a^2$. With $a = 7$, $c = 25$, $k = 24$, the value of this ratio is about 11·44, so that the gradient is more than eleven times the normal gradient.

In the case of a hemispherical projection of radius a the potential is simply

$$V = z\left(\frac{a^3}{r^3} - 1\right),$$

where r is the distance of the point (x, y, z) from the centre of the sphere. At points of the plane $z = 0$ we have

$$-\frac{\partial V}{\partial z} = 1 - \frac{a^3}{r^3},$$

while at points on the axis of z

$$-\frac{\partial V}{\partial z} = 1 - \frac{a^3}{z^3} + \frac{3a^3}{z^3}.$$

The potential gradient at the top of the mound has three times the normal value unity. At a point on the plane where $r = 2a$ the gradient is 7/8, while at a point on the axis where $z = 2a$ the gradient is 5/4. The effect of the projection on the force thus dies off more rapidly in a vertical than in a horizontal direction, but is always more marked at a given vertical distance from the sphere than at a given horizontal distance from the sphere.

§ **8·44**. *Point charge above a hemispherical boss.* Let the point charge be at P. Let Q be the image of P in the spherical surface of the boss, R the image of P in the plane, S the image of Q in the plane.

Let a be the radius of the sphere, h the distance OP, then the potential at T is

$$V = \frac{1}{TP} - \frac{1}{TR} + \frac{a}{h}\left(\frac{1}{TS} - \frac{1}{TQ}\right),$$

for this expression is zero on the plane and also zero on the hemisphere.

When OP is perpendicular to the plane the force on P due to the charges at the electrical images of P is

$$\frac{1}{4h^2} + \frac{a}{h}\frac{1}{\left(h - \frac{a^2}{h}\right)^2} - \frac{a}{h}\frac{1}{\left(h + \frac{a^2}{h}\right)^2} = \frac{ah}{(h^2 - a^2)^2} - \frac{ah}{(h^2 + a^2)^2} + \frac{1}{4h^2}.$$

When $h = 2a$ this expression becomes

$$\frac{1}{a^2}\left[\frac{2}{9} + \frac{1}{16} - \frac{2}{25}\right] = \frac{737}{3600a^2}.$$

This is greater than $\frac{5}{4}\left(\frac{1}{16a^2}\right)$, consequently the image force is increased by the presence of the hemispherical boss, and an electron would tend to return to the boss when acted upon by an external electric field $\frac{1}{16a^2}$ sufficiently large to just overcome the image force for a perfectly level surface.

§ 8·45. *Point charge in front of a plane conductor with a pit or projection facing the charge.* By inverting a spheroid with respect to a sphere whose centre I is at one vertex, we obtain an idea of the charge induced on a plane conductor by a point charge when there is a pit or projection facing the charge.

Let R be the radius of inversion, PP', QQ' pairs of inverse points. Let P be on the surface of the spheroid and let e be the charge associated with a surface element containing the point P, then

$$\Sigma\, IQ \frac{e}{QP} = \Sigma\, e \frac{IP'}{Q'P'}.$$

We shall regard $\frac{e}{R} IP'$ as the charge associated with the corresponding point P on the inverse surface S. Denoting this charge by e', and making use of the relation $IQ.IQ' = R^2$, we obtain

$$\frac{R}{IQ'}\Sigma \frac{e}{QP} = \Sigma \frac{e'}{Q'P'},$$

or

$$\frac{R}{IQ'} V_0 = V',$$

where V' is the potential at Q' due to the charges e' on S'.

Placing a charge $-RV_0$ at I the surface S' will be an equipotential for this charge and the charges e' on S'. These charges e' are in fact the charges induced on S' by the charge at I. The force exerted on this charge at I by the charges e' on S' is

$$\Sigma\, RV_C \frac{e'\cos\alpha}{IP'^2} = \Sigma\, V_0 \frac{e\cos\alpha}{IP'} = \frac{V_0}{R^2}\Sigma\, eIP\cos\alpha = \frac{V_0}{R^2}Mc,$$

where c is the distance of I from the centre of the charges e, which is in the present case the centre of the spheroid, and where M is the total charge on S. The force is thus

$$\frac{McV_0}{R^2}.$$

Except for the pit or projection facing the point I the surface S is very nearly plane. At infinity it can be regarded as identical with a plane whose distance from I is h, where

$$R^2 = 2\rho h,$$

ρ being the radius of curvature of the spheroid at I. Since $\rho = a^2/c$, where a and c are the semi-axes of the spheroid, we have

$$2a^2h = cR^2.$$

If now the point charge $-RV$ were simply in front of a perfectly level conducting plane at distance h from it, the force exerted on the charge by the induced charges on the plane would be equal to the force exerted by a charge rv at the optical image of I in the plane, and so would be

$$\frac{R^2V_0{}^2}{4h^2} = \frac{a^4V_0{}^2}{c^2R^2}.$$

Comparing this with the force on the charge when there is a pit or projection facing it, we find that the ratio of the two forces is ν, where for the case of a pit

$$\nu = \frac{a^4V_0}{Mc^3} = \frac{a^4}{c^3k}\cot^{-1}\frac{k}{c}, \qquad k = (a^2-c^2)^{\frac{1}{2}}.$$

The value of this ratio ν is given for various values of the ratio $\frac{a}{c}$. The table also includes corresponding values of the ratio $\frac{H}{h}$, where H is the distance of I from the bottom of the pit or the top of the projection as the case may be:

$\frac{a}{c}$	$\frac{1}{2}$	$\sqrt{\frac{1}{2}}$	1	$\sqrt{2}$	2	3
ν	—	—	1	3·14	9·67	—
$\frac{H}{h}$	$\frac{1}{4}$	$\frac{1}{2}$	1	2	4	9

§ **8·51**. *Laplace's equation in spheroidal co-ordinates.* The quantities θ, μ, ϕ are called spheroidal co-ordinates. In the case of the prolate spheroid, we have

$$\theta = \cosh \eta, \quad \mu = \cos \xi, \quad z = k \cos \xi \cosh \eta, \quad \varpi = k \sin \xi \sinh \eta,$$

and Laplace's equation is

$$0 = k^2 (\theta^2 - \mu^2) \nabla^2 V \equiv \frac{\partial}{\partial \theta}\left\{(\theta^2 - 1)\frac{\partial V}{\partial \theta}\right\} + \frac{\partial}{\partial \mu}\left\{(1 - \mu^2)\frac{\partial V}{\partial \mu}\right\} + \frac{\theta^2 - \mu^2}{(\theta^2 - 1)(1 - \mu^2)}\frac{\partial^2 V}{\partial \phi^2},$$

while in the case of the oblate spheroid

$$\theta = \sinh \eta, \quad \mu = \sin \xi, \quad z = k \sin \xi \sinh \eta, \quad \varpi = k \cos \xi \cosh \eta,$$

and Laplace's equation is

$$0 = k^2 (\theta^2 + \mu^2) \nabla^2 V \equiv \frac{\partial}{\partial \theta}\left\{(\theta^2 + 1)\frac{\partial V}{\partial \theta}\right\} + \frac{\partial}{\partial \mu}\left\{(1 - \mu^2)\frac{\partial V}{\partial \mu}\right\} + \frac{\theta^2 + \mu^2}{(\theta^2 + 1)(1 - \mu^2)}\frac{\partial^2 V}{\partial \phi^2}.$$

Some very simple solutions may be found by adopting as trial solutions

$$V = f(\theta \pm \mu) \quad \text{for the prolate spheroid,}$$
$$V = f(\theta \pm i\mu) \quad \text{for the oblate spheroid.}$$

It is found in each case that $f(u) = \frac{1}{u}$ satisfies requirements, and so we have the solutions

$$V = \frac{\theta}{\theta^2 - \mu^2}, \quad V = \frac{\mu}{\theta^2 - \mu^2}$$

for the prolate spheroid, and

$$V = \frac{\theta}{\theta^2 + \mu^2}, \quad V = \frac{\mu}{\theta^2 + \mu^2}$$

for the oblate spheroid.

§ **8·52**. *Lamé products for spheroidal co-ordinates.* The equation

$$\nabla^2 V + h^2 V = 0$$

is satisfied by a Lamé product of type

$$V = M(\mu)\, \Theta(\theta)\, \Phi(\phi)$$

if

$$\frac{d^2\Phi}{d\phi^2} + m^2\Phi = 0,$$

and

$$\left.\begin{aligned} &\frac{d}{d\mu}\left\{(1 - \mu^2)\frac{dM}{d\mu}\right\} + \left\{n(n + 1) - \frac{m^2}{1 - \mu^2} \pm h^2k^2(1 - \mu^2)\right\} M = 0, \\ &\frac{d}{d\theta}\left\{(\theta^2 \mp 1)\frac{d\Theta}{d\theta}\right\} - \left\{n(n + 1) - \frac{m^2}{1 \mp \theta^2} - h^2k^2(\theta^2 \mp 1)\right\} \Theta = 0, \end{aligned}\right\} \quad \text{(A)}$$

the upper or lower sign being taken according as the spheroidal co-ordinates μ, θ, ϕ are of the prolate or oblate type. When $h = 0$, we may write

$$\Phi = Ae^{im\phi} + Be^{-im\phi},$$
$$M = CP_n^m(\mu) + DQ_n^m(\mu),$$
$$\Theta = EP_n^m(\theta) + FQ_n^m(\theta)$$

for the case of the prolate spheroid and similar expressions for M and Φ, but the following expression

$$\Theta = EP_n^m(i\theta) + FQ_n^m(i\theta)$$

for the case of the oblate spheroid, A, B, C, D, E, F being arbitrary constants and $P_n^m(\mu)$, $Q_n^m(\mu)$ associated Legendre functions*. It should be noticed that we now require a knowledge of the properties of the associated Legendre functions $P_n^m(u)$ and $Q_n^m(u)$ for arguments u of types $u = \cosh\eta$ and $u = i\sinh\eta$.

When n and m are positive integers appropriate definitions are

$$P_n^m(u) = (u^2-1)^{\frac{1}{2}m}\frac{d^mP_n(u)}{du^m},$$

$$Q_n^m(u) = (u^2-1)^{\frac{1}{2}m}\frac{d^mQ_n(u)}{du^m}.$$

It has been found convenient to write $p_n(u)$ for $i^{-n}P_n(iu)$, and $q_n(u)$ for $i^{n+1}Q_n(iu)$, then when n is zero or a positive integer we have the expansions

$$p_n(u) = \frac{(2n)!}{2^n(n!)^2}\left\{u^n + \frac{n(n-1)}{2(2n-1)}u^{n-2} + \frac{n(n-1)(n-2)(n-3)}{2.4(2n-1)(2n-3)}u^{n-4} + \ldots\right\},$$

$$q_n(u) = \frac{2^n(n!)^2}{(2n+1)!}\left\{u^{-n-1} - \frac{(n+1)(n+2)}{2(2n+3)}u^{-n-3} + \frac{(n+1)(n+2)(n+3)(n+4)}{2.4(2n+3)(2n+5)}u^{-n-5} - \ldots\right\}.$$

If $\theta = \cot\beta$, it is found that

$$q_0(\theta) = \beta,\quad q_1(\theta) = 1 - \beta\cot\beta,$$
$$q_2(\theta) = \tfrac{1}{2}\beta(3\cot^2\beta + 1) - \tfrac{3}{2}\cot\beta.$$

The coefficient of β in the expression for $q_n(\theta)$ is equal to $p_n(\theta)$. We also have the expressions

$$p_n(\theta) = (-)^n\frac{1}{n!}(\sin\beta)^{n+1}\left(\frac{d}{\sin\beta . d\beta}\right)^n(\operatorname{cosec}\beta),$$

$$q_n(\theta) = \frac{1}{n!}(\sin\beta)^{n+1}\left(\frac{d}{\sin\beta . d\beta}\right)^n(\beta\operatorname{cosec}\beta),$$

which are readily verified with the aid of the differential equations satisfied by the functions $p_n(\theta)$ and $q_n(\theta)$.

* The fact that the Lamé products depend on the associated Legendre functions seems to have been first noticed by E. Heine, *Crelle*, Bd. XXVI, S. 185 (1843).

The integral of Laplace's type for the function $P_n^m(\theta)$ is

$$P_n^m(\theta) = \frac{\sinh^m \eta}{1.3 \ldots (2m-1)} \frac{(n+m)!}{(n-m)!} \int_0^\pi (\cosh \eta + \sinh \eta \cos \alpha)^{n-m} \sin^{2m} \alpha \, d\alpha.$$

It shows that $P_n^m(\cosh \eta)$ is positive when η is positive, for when the binomial in the integrand is expanded in powers of $\cos \alpha$ and integrated term by term the odd powers of $\cos \alpha$ do not contribute anything to the result, while the coefficients of the even powers are all positive. This process gives a finite series for $P_n^m(\cosh \eta)$ with the property that each term increases with η. Consequently, when η is positive $P_n^m(\cosh \eta)$ increases with η. When $m = 0$ and $\eta = 0$ the value of the function is unity, hence we have the result that

$$P_n(\cosh \eta) > 1, \quad \eta > 0.$$

§ **8·53**. *Spheroidal wave-functions.* When the spheroid θ = constant is of the prolate type the equations satisfied by M and Θ are both of the type

$$\frac{d}{dx}\left\{(1-x^2)\frac{dX}{dx}\right\} + \left\{\lambda^2 x^2 - \frac{m^2}{1-x^2} + \sigma\right\} X = 0,$$

where λ, σ and m are constants and $x = \mu$ when $X = M$ and $x = \theta$ when $X = \Theta$. If $X = (1-x^2)^{m/2} w$, we have the equation

$$(1-x^2)\frac{d^2w}{dx^2} - 2(m+1)x\frac{dw}{dx} + \{\lambda^2 x^2 + \sigma - m(m+1)\} w = 0.$$

This equation has been discussed by several writers*. The present investigation follows closely that of A. H. Wilson†.

Solutions are required which are finite for $-1 \leqslant x \leqslant 1$ so as to represent quantities of type M, and solutions are also required which are finite for $1 \leqslant x \leqslant a$, where a is some finite constant. These latter solutions are of interest for the representation of quantities of type Θ, and for this purpose a knowledge is also required of the behaviour of solutions of the equation for large values of $|x|$.

We commence with a study of a solution

$$w = \Sigma a_s x^s,$$

represented by a series containing even positive integral powers of the variable x. The recurrence relation

$$(s+1)(s+2)\,a_{s+2} = \{s(s-1) + 2s(m+1) - \sigma + m(m+1)\}\,a_s - \lambda^2 a_{s-2}$$

gives the following equation for the ratio

$$N_s = a_{s+2}/a_s,$$

$$N_s = \frac{(s+m+1)(s+m) - \sigma}{(s+1)(s+2)} - \frac{\lambda^2}{(s+1)(s+2)\,N_{s-2}}. \quad \ldots\ldots\text{(B)}$$

* Niven, *Phil. Trans.* A, vol. CLXXI, p. 231 (1892); R. C. Maclaurin, *Trans. Camb. Phil. Soc.* vol. XVII, p. 41 (1898); M. Abraham, *Math. Ann.* vol. LII, p. 81 (1899); J. W. Nicholson, *Proc. Roy. Soc. Lond.* A, vol. CVII, p. 43 (1925); E. G. C. Poole, *Quart. Journ.* vol. XLIX, p. 309 (1921).

† *Proc. Roy. Soc. Lond.* A, vol. CXVIII, p. 617 (1928).

When $N_s \to 1$ as $s \to \infty$ an approximate value of N_s for large values of s is obtained by writing $N = 1 - 2\tau/s$ in (B); it is then found that $\tau = 1 - m$. Therefore for large values of s the coefficients a_s approximate to those in the expansion of $(1 - x^2)^{-m}$ when $m \neq 0$, and of $\log(1 - x^2)$ when $m = 0$. In this case $X(x)$ is infinite for $x = \pm 1$.

If N_s^{-1} is unbounded the series represents an integral function and is therefore finite in the range $-1 \leqslant x \leqslant 1$. Also $N_s \sim \lambda^2/s^2$ and $w(x) \sim \cosh \lambda x$.

The condition that $\lim_{s \to \infty} N_s = 0$ gives a transcendental equation between σ and λ^2. When this condition is combined with the equation

$$N_{s-2} = \frac{\lambda^2}{(s+m+1)(s+m) - \sigma - (s+2)(s+1) N_s},$$

it is found that

$$N_0 = \frac{\lambda^2}{(m+3)(m+2) - \sigma -} \; \frac{3.4.\lambda^2}{(m+5)(m+4) - \sigma -} \; \frac{5.6.\lambda^2}{(m+7)(m+6) - \sigma -} \cdots,$$

the continued fraction being convergent for all values of λ and σ. Also $N_0 = a_2/a_0 = \{m(m+1) - \sigma\}/2$, and so the equation for σ becomes

$$m(m+1) - \sigma = \frac{1.2.\lambda^2}{(m+3)(m+2) - \sigma -} \; \frac{3.4.\lambda^2}{(m+5)(m+4) - \sigma} \; \frac{5.6.\lambda^2}{-\,(m+7)(m+6) - \sigma -} \cdots.$$

A better form is

$$m(m+1) - \sigma = \frac{1.2.\lambda^2/\{(m+2)(m+3)\}}{1 - \dfrac{\sigma}{(m+2)(m+3)}} \; \frac{3.4.\lambda^2/\{(m+2)(m+3)(m+4)(m+5)\}}{-\quad 1 - \dfrac{\sigma}{(m+4)(m+5)} \quad -} \cdots.$$

This continued fraction gives an infinite number of values of σ approximate expressions for which may be readily obtained. When $m = 0$ the first value of σ is given by the series

$$\sigma_1 = -\frac{1}{3}\lambda^2 - \frac{2}{135}\lambda^4 - \frac{4}{3^5.5.7}\lambda^6 + \frac{16}{3^7.5^3.7^2}\lambda^8 + \ldots,$$

which converges very rapidly. To find the second value of σ it is advantageous to write the relation between σ and λ^2 in the form

$$\frac{1.2.\lambda^2/\{(m+2)(m+3)\}}{m(m+1) - \sigma} = 1 - \frac{\sigma}{(m+2)(m+3)} \; \frac{3.4.\lambda^2/\{(m+2)(m+3)(m+4)(m+5)\}}{-\quad 1 - \dfrac{\sigma}{(m+4)(m+5)} \quad -} \cdots,$$

and it is readily found that

$$\sigma_3 = 6 - \frac{11}{21}\lambda^2 + \frac{94}{(21)^3}\lambda^4 + \ldots,$$

If the series for w contains only odd powers of x it is found that the recurrence relation is

$$(s+1)(s+2)\,a_{s+2} = \{(s+m)(s+m+1) - \sigma\}\,a_s - \lambda^2 a_{s-2},$$

and σ is given approximately by the continued fraction

$$(m+1)(m+2) - \sigma = \cfrac{2.3.\lambda^2/\{(m+3)(m+4)\}}{1 - \dfrac{\sigma}{(m+3)(m+4)}} - \cfrac{4.5.\lambda^2/\{(m+3)(m+4)(m+5)(m+6)\}}{1 - \dfrac{\sigma}{(m+5)(m+6)}} - \ldots.$$

Priestley* has discussed the solutions of equations (A) by the methods of integral equations. If the associated Legendre functions are defined by the equations

$$P_n^m(\mu) = \frac{1}{\Gamma(1-m)}\left(\frac{1+\mu}{1-\mu}\right)^{\frac{1}{2}m} F[n+1, -n; -m+1; \tfrac{1}{2}(1-\mu)],$$

$$Q_n^m(\mu) = \lim_{m \to \text{integer}} \tfrac{1}{2}\pi \operatorname{cosec} m\pi \left[\cos m\pi\, P_n^m(\mu) - \frac{\Gamma(n+m+1)}{\Gamma(n-m+1)} P_n^{-m}(\mu)\right],$$

we have

$$P_n^{+m}(-\mu) = \cos[(m+n)\pi]\, P_n^m(\mu) - (2/\pi)\sin[(m+n)\pi]\, Q_n^m(\mu),$$

and so we can construct an even function $\phi_n^m(\mu)$ and an odd function $\psi_n^m(\mu)$ by means of the relations

$$\phi_n^m(\mu) = (n, m)\cos\tfrac{1}{2}(m+n)\pi . P_n^{-m}(\mu) + (2/\pi)\, s . Q_n^m(\mu),$$
$$\psi_n^m(\mu) = (n, m)\sin\tfrac{1}{2}(m+n)\pi . P_n^{-m}(\mu) + (2/\pi)\, c . Q_n^m(\mu),$$

where

$$(n, m) = \frac{\Gamma(n+m+1)}{\Gamma(n-m+1)}, \quad s = \sin\tfrac{1}{2}(m-n), \quad c = \cos\tfrac{1}{2}(m-n).$$

Associated with these functions there is an even solution of the differential equation†

$$\frac{d}{d\mu}\left[(1-\mu^2)\frac{dM}{d\mu}\right] + \left[n(n+1) - \frac{m^2}{1-\mu^2}\right] M = h^2k^2(1-\mu^2) - M,$$

which may be derived by solving the integral equation

$$\phi_n^m(\mu) = U_n^m(\mu) - k^2h^2 \sec m\pi \int_0^\mu K_n^m(\mu, t)\, U_n^m(t)\, dt,$$

and an odd solution of the differential equation which may be obtained by solving the integral equation

$$\psi_n^m(\mu) = V_n^m(\mu) - k^2h^2 \sec m\pi \int_0^\mu K_n^m(\mu, t)\, V_n^m(t)\, dt.$$

* H. J. Priestley, *Proc. Lond. Math. Soc.* vol. xx, p. 37 (1922).

† This is the differential equation corresponding to the oblate spheroid.

In both of these integral equations the kernel $K_n^m(\mu, t)$ is

$$K_n^m(\mu, t) = (1 - t^2) \begin{vmatrix} Q_n^m(\mu) & P_n^{-m}(\mu) \\ Q_n^m(t) & P_n^{-m}(t) \end{vmatrix}.$$

The functions $U_n^m(\mu)$ and $V_n^m(\mu)$ are infinite for $\mu = \pm 1$, but

$$(1 - \mu^2)^{\frac{1}{2}m} U_n^m(\mu) \text{ and } (1 - \mu^2)^{\frac{1}{2}m} V_n^m(\mu)$$

are finite. This is not true when $m = 0$, the corresponding theorem is then that $U_n(\mu)/\log(1 - \mu)$ and $V_n(\mu)/\log(1 - \mu)$ are finite for $\mu = \pm 1$.

Solutions of the differential equation

$$\frac{d}{d\theta}\left\{(\theta^2 + 1)\frac{d\Theta}{d\theta}\right\} - \left\{n(n+1) - \frac{m^2}{1 + \theta^2} - h^2k^2(\theta^2 + 1)\right\}\Theta = 0,$$

which approximate to $\sin(kh\theta)/\theta$ and $\cos(kh\theta)/\theta$ respectively, as $\theta \to \infty$, are obtained as solutions of the integral equations

$$(1 + \theta^2)^{-\frac{1}{2}} \sin kh\theta = \Theta(\theta) - \int_\infty^\theta G_n^m(\theta, t)\,\Theta(t)\,dt,$$

$$(1 + \theta^2)^{-\frac{1}{2}} \cos kh\theta = \Theta(\theta) - \int_\infty^\theta G_n^m(\theta, t)\,\Theta(t)\,dt$$

respectively, where

$$G_n^m(\theta, t) = \left[n(n+1) - \frac{m^2 - 1}{1 + t^2}\right] \frac{\sin[kh(\theta - t)]}{kh[(1 + \theta^2)(1 + t^2)]^{\frac{1}{2}}}.$$

EXAMPLES

1. Prove that the integral

$$\phi = \int_0^1 dt \sinh(kh\mu t)\left\{\frac{\cosh kh(t + i\theta)}{t + i\theta} + \frac{\cosh kh(t - i\theta)}{t - i\theta}\right\}$$

represents a spheroidal wave-function of oblate type.

2. Prove also that if $F(\mu, \theta)$ is a solution of $\nabla^2 F + h^2 F = 0$, the integral

$$\Phi = \int_{-1}^1 \int_{-1}^1 F(t, -i\lambda)\,d\lambda\,dt\,e^{\pm kh(i\lambda\theta \pm \mu t)}$$

represents a second solution.

[J. W. Nicholson.]

§ **8·54**. *A relation between spheroidal harmonics of different types.* The four-dimensional potential of the spherical surface

$$x^2 + y^2 + w^2 = a^2, \quad z = 0,$$

when the surface density depends only on $x^2 + y^2$, is

$$W = a^2 \int_0^\pi f(\cos\theta) \sin\theta\,d\theta \int_0^{2\pi} \frac{d\phi}{R^2},$$

where

$$R^2 = (w - a\cos\theta)^2 + z^2 + (x - a\sin\theta\cos\phi)^2 + (y - a\sin\theta\sin\phi)^2,$$

and $f(\mu)$ is a function giving the law of density. Integrating with respect to ϕ and writing $\zeta = \cos\theta$, we obtain

$$W = \frac{\pi a}{(\rho^2 + w^2)^{\frac{1}{2}}} U(X, Y, Z),$$

where
$$U = \int_{-1}^{1} \frac{f(\zeta)\, d\zeta}{[(\zeta - Z)^2 + X^2 + Y^2]^{\frac{1}{2}}},$$

$$Z = \frac{ws^2}{2a(\rho^2 + w^2)},$$

$$X^2 + Y^2 + Z^2 = \frac{s^4 - 4a^2\rho^2}{4a^2(\rho^2 + w^2)},$$

$$s^2 = x^2 + y^2 + z^2 + w^2 + a^2, \quad \rho^2 = x^2 + y^2.$$

Now when (X, Y, Z) are regarded as rectangular co-ordinates in a three-dimensional space S, the function U is the Newtonian potential of a rod of varying density; we therefore introduce spheroidal co-ordinates ξ, η defined by the equations

$$\cos\xi \cosh\eta = Z = \frac{ws^2}{2a(\rho^2 + w^2)},$$

$$\sin\xi \sinh\eta = (X^2 + Y^2)^{\frac{1}{2}} = \frac{\rho[s^4 - 4a^2(\rho^2 + w^2)]^{\frac{1}{2}}}{2a(\rho^2 + w^2)},$$

and we find that

$$\cos\xi = \frac{w}{(\rho^2 + w^2)^{\frac{1}{2}}}, \quad \cosh\eta = \frac{s^2}{2a(\rho^2 + w^2)^{\frac{1}{2}}}.$$

Corresponding to the standard potential function

$$U = Q_n(\cosh\eta) P_n(\cos\xi),$$

we then have the four-dimensional potential

$$W = \frac{\pi a}{(\rho^2 + w^2)^{\frac{1}{2}}} Q_n\left[\frac{s^2}{2a(\rho^2 + w^2)^{\frac{1}{2}}}\right] P_n\left[\frac{w}{(\rho^2 + w^2)^{\frac{1}{2}}}\right],$$

and a corresponding Newtonian potential

$$V = \frac{1}{\pi}\int_{-\infty}^{\infty} W dw,$$

which reduces when $\rho = 0$ to the value

$$V_0 = a\int_{-\infty}^{\infty} \frac{dw}{w} Q_n\left[\frac{z^2 + a^2 + w^2}{2aw}\right].$$

To evaluate this integral we use the expansion

$$Q_n(u) = \frac{\pi^{\frac{1}{2}}\Gamma(n+1)}{2^{n+1}\Gamma(n+\frac{3}{2})} u^{-n-1} F\left(\frac{n+1}{2}, \frac{n+2}{2}; n+\tfrac{3}{2}, u^{-2}\right).$$

Now if m is zero or a positive integer,

$$\int_{-\infty}^{\infty} \frac{dw}{w}\left(\frac{2aw}{z^2+a^2+w^2}\right)^{n+2s+1}$$

$$= 2\pi\left(\frac{a^2}{z^2+a^2}\right)^{m+s+\frac{1}{2}} \frac{1.3\ldots(2m+2s-1)}{2.4\ldots(2m+2s)}, \quad n = 2m$$

$$= 0, \qquad n = 2m+1.$$

Hence, when $n = 2m$,

$$V_0 = 2\pi a \frac{\pi^{\frac{1}{2}}\Gamma(2m+1)}{2^{2m+1}\Gamma(2m+\frac{3}{2})} \frac{1.3\ldots(2m-1)}{2.4\ldots 2m}\left(\frac{a^2}{z^2+a^2}\right)^{m+\frac{1}{2}}$$

$$\times F\left(m+\tfrac{1}{2}, m+\tfrac{1}{2}; 2m+\tfrac{3}{2}; \frac{a^2}{z^2+a^2}\right)$$

$$= 2\pi a \frac{1.3\ldots(2m-1)}{2.4\ldots 2m} q_{2m}\left(\frac{|z|}{a}\right).$$

Introducing the spheroidal co-ordinates

$$z = a\mu\nu, \quad \rho = a\sqrt{\{(1-\mu^2)(\nu^2+1)\}},$$

we can say at once that

$$V = 2\pi a \frac{1.3\ldots(2m-1)}{2.4\ldots 2m} P_{2m}(\mu)\, q_{2m}(\nu)$$

is a potential function which takes the value V_0 when $\mu = 1$; consequently we have the formula

$$\int_{-\infty}^{\infty} \frac{dw}{(\rho^2+w^2)^{\frac{1}{2}}} Q_{2m}\left[\frac{s^2}{2a(\rho^2+w^2)^{\frac{1}{2}}}\right] P_{2m}\left[\frac{w}{(\rho^2+w^2)^{\frac{1}{2}}}\right]$$

$$= 2\pi \frac{1.3\ldots(2m-1)}{2.4\ldots 2m} P_{2m}(\mu)\, q_{2m}(\nu).$$

Since this potential function is obtained by projection from the four-dimensional potential of a spherical surface it represents the potential of a circular disc whose density is a function of the distance from the centre. To find this function we notice that a density $f(\zeta)$ on the spherical surface corresponds to a density $(2/\zeta)f(\zeta)$ on the disc, where $\rho = a(1-\zeta^2)^{\frac{1}{2}}$.

If, in particular, we write $f(\zeta) = P_{2m}(\zeta)$, we have

$$U = 2Q_{2m}(Z) \text{ when } X = Y = 0.$$

Hence the potential

$$V = 2\pi a \frac{1.3\ldots(2m-1)}{2.4\ldots 2m} P_{2m}(\mu)\, q_{2m}(\nu)$$

is the potential of a disc of radius a whose surface density is

$$(1-p^2)^{-\frac{1}{2}} P_{2m}(\sqrt{1-p^2}),$$

ap being the distance from the centre. On the disc itself we have $\theta = 0$, and since

$$q_{2m}(0) = \tfrac{1}{2}\pi . \frac{1.3 \ldots (2m-1)}{2.4 \ldots 2m},$$

the value of V is V_0, where

$$V_0 = \pi^2 a \frac{1^2 . 3^2 \ldots (2m-1)^2}{2^2 . 4^2 \ldots (2m)^2} P_{2m}(\mu).$$

This method can be used to find the potential of a disc whose density is $(2/\zeta) f(\zeta)$, where $f(\zeta)$ is an even function which can be expanded in a series of Legendre functions in the interval $-1 \leqslant \zeta \leqslant 1$. Sufficient conditions for the uniform convergence of the expansion in the whole of this interval have been obtained in an elementary way by M. H. Stone, *Annals of Math.* vol. XXVII, p. 315 (1926). The requirements are that

$$\int_{-1}^{1} [f''(x)]^2 \, dx$$

should exist and that the equation

$$f(x) = f(-1) + \int_{-1}^{x} du \int_{-1}^{u} f''(v) \, dv$$

should be valid. The coefficients in the expansion are supposed to be derived from $f(x)$ by the usual rule.

For further applications of spheroidal co-ordinates reference may be made to the books of C. Neumann*, E. Mathieu†, M. Brillouin‡ and A. B. Basset§, and to papers by F. Ehrenhaft||, K. F. Herzfeld¶, J. W. Nicholson**, R. Jones†† and K. Sezawa‡‡.

* *Theorie der Elektricitäts- und Wärme-Vertheilung in einem Ringe*, Halle (1864).

† *Cours de Physique Mathématique*, Gauthier-Villars, Paris (1873).

‡ *Propagation de l'Électricité*, Hermann, Paris (1904), Ch. VI.

§ *Hydrodynamics*, vol. II, Cambridge (1888).

|| *Wiener Berichte*, p. 273 (1904).

¶ *Ibid.* p. 1587 (1911).

** *Phil. Mag.* vol. XI, p. 703 (1906); *Phil. Trans.* A, vol. CCXXIV, pp. 49, 303 (1924).

†† *Ibid.* vol. CCXXVI, p. 231 (1926).

‡‡ *Bull. Earthquake Research Inst.* vol. II, p. 29 (1927).

CHAPTER IX

PARABOLOIDAL CO-ORDINATES

§ **9·11**. *Transformation of the wave-equation.* If we write

$$z + i\rho = (\alpha_0 + i\beta_0)^2, \quad \alpha_0^2 = -\alpha, \quad \beta_0^2 = \beta,$$

so that the transformation is

$$z = -\alpha - \beta, \quad \rho = 2(-\alpha\beta)^{\frac{1}{2}},$$

the differential equation

$$\frac{\partial^2 W}{\partial \rho^2} + \frac{2m+1}{\rho}\frac{\partial W}{\partial \rho} + \frac{\partial^2 W}{\partial z^2} - \frac{1}{c^2}\frac{\partial^2 W}{\partial t^2} = 0$$

becomes

$$\alpha\frac{\partial^2 W}{\partial \alpha^2} + (m+1)\frac{\partial W}{\partial \alpha} - \beta\frac{\partial^2 W}{\partial \beta^2} - (m+1)\frac{\partial W}{\partial \beta} - (\alpha - \beta)\frac{1}{c^2}\frac{\partial^2 W}{\partial t^2} = 0,$$

and is satisfied by*

$$W = A(\alpha)\, B(\beta)\, e^{\pm ikct},$$

if

$$\left.\begin{aligned} \alpha\frac{d^2 A}{d\alpha^2} + (m+1)\frac{dA}{d\alpha} - (h - k^2\alpha) A = 0 \\ \beta\frac{d^2 B}{d\beta^2} + (m+1)\frac{dB}{d\beta} - (h - k^2\beta) B = 0 \end{aligned}\right\}, \qquad \text{......(I)}$$

where h is arbitrary.

When $k = 0$ it is simpler to use the variables α_0 and β_0 as independent variables, the differential equations for A and B are then

$$\frac{d^2 A}{d\alpha_0^2} + \frac{2m+1}{\alpha_0}\frac{dA}{d\alpha_0} + 4hA = 0,$$

$$\frac{d^2 B}{d\beta_0^2} + \frac{2m+1}{\beta_0}\frac{dB}{d\beta_0} - 4hB = 0.$$

The inference is that there are simple solutions of Laplace's equation of the types

$$I_m(\lambda\alpha_0)\, J_m(\lambda\beta_0) \cos m(\phi - \phi_0),$$
$$K_m(\lambda\alpha_0)\, J_m(\lambda\beta_0) \cos m(\phi - \phi_0),$$
$$I_m(\lambda\alpha_0)\, Y_m(\lambda\beta_0) \cos m(\phi - \phi_0),$$
$$K_m(\lambda\alpha_0)\, Y_m(\lambda\beta_0) \cos m(\phi - \phi_0),$$

where λ and m are arbitrary constants. The expressions for ρ and z in terms of α_0 and β_0 are

$$z = \alpha_0^2 - \beta_0^2, \quad \rho = 2\alpha_0\beta_0,$$
$$r = \alpha_0^2 + \beta_0^2.$$

* H. J. Sharpe, *Quarterly Journal*, vol. XV, p. 1 (1878); *Proc. Camb. Phil. Soc.* vol. X, p. 101 (1899); vol. XIII, p. 133 (1905); vol. XV, p. 190 (1909); H. Lamb, *Proc. London Math. Soc.* (2), vol. IV, p. 190 (1907).

Many well-known potentials may be expressed in terms of these simple solutions in an interesting way. We shall give here an expression for the inverse distance

$$R^{-1} = [(z+\gamma_0^2)^2 + \rho^2]^{-\frac{1}{2}}$$
$$= [\alpha_0^4 + \beta_0^4 + \gamma_0^4 - 2\beta_0^2\gamma_0^2 + 2\gamma_0^2\alpha_0^2 + 2\alpha_0^2\beta_0^2]^{-\frac{1}{2}}.$$

Let us assume that

$$R^{-1} = \int_0^\infty K_0(\lambda\alpha_0)\, J_0(\lambda\beta_0)\, J_0(\lambda\gamma_0)\, f(\lambda)\, d\lambda,$$

then when $\beta_0 = 0$ we should have

$$(\alpha_0^2+\gamma_0^2)^{-1} = \int_0^\infty K_0(\lambda\alpha_0)\, J_0(\lambda\gamma_0)\, f(\lambda)\, d\lambda.$$

This indicates that perhaps

$$r^{-1} = (\alpha_0^2+\beta_0^2)^{-1} = \int_0^\infty K_0(\lambda\alpha_0)\, J_0(\lambda\beta_0)\, f(\lambda)\, d\lambda,$$

and again, when $\beta_0 = 0$, we should have

$$\alpha_0^{-2} = \int_0^\infty K_0(\lambda\alpha_0)\, f(\lambda)\, d\lambda.$$

This equation is satisfied by $f(\lambda) = C\lambda$, where C is a constant such that

$$C\int_0^\infty K_0(\tau)\,\tau\, d\tau = 1.$$

Now $$K_0(\tau) = \int_0^\infty e^{-\tau\cosh u}\, du,$$

and so $$\int_0^\infty K_0(\tau)\,\tau\, d\tau = \int_0^\infty \tau\, d\tau \int_0^\infty e^{-\tau\cosh u}\, du$$
$$= \int_0^\infty \operatorname{sech}^2 u\,.\,du = 1.$$

Hence $C = 1$, and so the analysis suggests the equation

$$R^{-1} = \int_0^\infty K_0(\lambda\alpha_0)\, J_0(\lambda\beta_0)\, J_0(\lambda\gamma_0)\,\lambda\, d\lambda.$$

This equation may be checked in many ways. In particular, if we make use of the equation

$$\pi J_0(\lambda\beta_0)\, J_0(\lambda\gamma_0) = \int_0^\pi J_0(\beta_0^2+\gamma_0^2 - 2\beta_0\gamma_0\cos\omega)\, d\omega$$

the relation may be deduced from the simpler relation

$$(\alpha_0^2+\delta_0^2)^{-1} = \int_0^\infty K_0(\lambda\alpha_0)\, J_0(\lambda\delta_0)\,\lambda\, d\lambda,$$

which in turn may be checked by substituting the foregoing expression for $K_0(\tau)$. The equation for R^{-1} is a particular case of one given by H. M. Macdonald*. A proof of the formula is given in Watson's *Bessel*

* *Proc. London Math. Soc.* (2), vol. VII, p. 142 (1909).

Functions, p. 412. Some analogous integrals have been evaluated by Watson* at Whittaker's suggestion.

The corresponding expression for R_1^{-1}, where

$$R_1^2 = (z - \gamma_0^2) + \rho^2,$$

is

$$R_1^{-1} = \int_0^\infty J_0(\lambda\alpha_0)\, K_0(\lambda\beta_0)\, J_0(\lambda\gamma_0)\, \lambda\, d\lambda.$$

§ **9·21**. *Sonine's polynomials*. Putting $2ikn = -ik(m+1) - h$ in the equations (I) we find that the differential equations are satisfied by

$$A = e^{-ik\alpha} F_m{}^n(2ik\alpha), \quad B = e^{-ik\beta} F_m{}^n(2ik\beta),$$

where $F_m{}^n(s)$ is a solution of the differential equation

$$s\frac{d^2F}{ds^2} + (m + 1 - s)\frac{dF}{ds} + nF = 0.$$

This is a slight modification of Weiler's canonical form† for an equation of Laplace's type. It is satisfied by a confluent hypergeometric series of type

$$1 - \frac{n}{1\,(m+1)}s + \frac{n\,(n-1)}{1.2\,(m+1)\,(m+2)}s^2 - \dots,$$

which is usually denoted by the symbol $F(-n;\, m+1;\, s)$, but in the British Association report for 1926 the symbol $F(\alpha; \gamma; s)$ is replaced by $M(\alpha; \gamma; s)$.

When n is a positive integer a solution of the equation can be expressed in terms of Sonine's polynomial‡, $T_m{}^n(s)$, which may be defined by means of the expansion

$$(1+t)^{-m-1} e^{\frac{st}{1+t}} = \sum_{n=0}^{\infty} \Gamma(m+n+1)\, t^n\, T_m{}^n(s), \qquad \text{......(A)}$$

where $|t| < 1$. Calculating the coefficients in the expansion of the function on the left-hand side of the equation we find that

$$T_m{}^n(s) = \frac{s^n}{\Gamma(m+n+1)\, n!} - \frac{s^{n-1}}{\Gamma(m+n)\,(n-1)!} + \frac{s^{n-2}}{\Gamma(m+n-1)\, \overline{n-2}!\, 2!} - \dots \qquad \text{......(B)}$$

$$= (-)^n \frac{1}{\Gamma(m+1)\,\Gamma(n+1)} F(-n;\, m+1;\, +s) \qquad \text{......(C)}$$

$$= (-)^n \frac{e^s s^{-m}}{\Gamma(m+n+1)\,\Gamma(n+1)} \frac{d^n}{ds^n}[e^{-s} s^{m+n}] \qquad \text{......(D)}$$

$$= (-)^n \frac{1}{\Gamma(m+n+1)} L_n^{(m)}(s),$$

if we adopt the modern notation for the generalised Laguerre polynomial.

* *Journ. London Math. Soc.* vol. III, p. 22 (1928).

† *Crelle's Journal*, vol. LI, p. 105 (1856).

‡ *Math. Ann.* vol. XVI, p. 1 (1880). The T notation was probably adopted in honour of Tchebycheff who considered particular cases of the polynomial in 1859; *Oeuvres*, t. I, pp. 500–508 (1899).

When $m+n$ is zero or a positive integer there is another formula*

$$T_m{}^n(s) = \frac{(-)^{m+n}e^s}{\Gamma(m+n+1)\,\Gamma(n+1)} \frac{d^{m+n}}{ds^{m+n}}[e^{-s}s^n], \qquad \text{......(E)}$$

which indicates that in this case the polynomial may also be defined by means of the expansion

$$\frac{(1-\lambda)^n}{n!} e^{\lambda s} = \sum_{m=-n}^{\infty} s^m \lambda^{m+n} T_m{}^n(s). \qquad \text{......(F)}$$

A comparison of the formulae (D) and (E) indicates also that in this case

$$s^m T_m{}^n(s) = T_{-m}{}^{m+n}(s). \qquad \text{......(G)}$$

With the aid of the last relation many formulae may be duplicated. Sonine's polynomial satisfies a number of difference equations, most of which are given in the memoir of Gegenbauer†.

$$T_{m-1}{}^n(s) = (s+m)\,T_m{}^n(s) - (m+n+1)\,s\,T_{m+1}{}^n(s),$$

$$n(m+n)\,T_m{}^n(s) - \{s-(m+2n-1)\}\,T_m{}^{n-1}(s) + T_m{}^{n-2}(s) = 0,$$

$$T_m{}^{n-1}(s) = (m+n)\,T_{m+1}{}^{n-1}(s) + T_{m+1}{}^{n-2}(s),$$

$$(n-1)\,T_m{}^{n-1}(s) = \{s-(m+n-1)\}\,T_{m+1}{}^{n-2}(s) - T_{m+1}{}^{n-3}(s),$$

$$(n-1)\,T_m{}^{n-1}(s) = \{s-(m+1)\}\,T_{m+1}{}^{n-2}(s) - s\,T_{m+2}{}^{n-3}(s),$$

$$s\frac{d}{ds}[T_m{}^n(s)] = T_m{}^{n-1}(s) + nT_m{}^n(s),$$

$$\frac{d^p}{ds^p}[T_m{}^n(s)] = T_{m+p}{}^{n-p}(s),$$

$$\frac{d^p}{ds^p}[s^m T_m{}^n(s)] = s^{m-p}T_{m-p}{}^n(s),$$

$$(m+n+1)\frac{d}{ds}[T_m{}^{n+1}(s)] = T_m{}^n(s) - \frac{d}{ds}[T_m{}^n(s)].$$

From these equations we may deduce that

$$\frac{d}{ds}[s^{-m}e^s T_m{}^{n+1}(s)/T_m{}^n(s)]$$

$$= e^s s^{-m-1}[T_m{}^n(s)]^{-2}[m(n+1)\{T_m{}^{(n+1)}(s)\}^2 + \{T_m{}^n(s) + (n+1)T_m{}^{n+1}(s)\}^2].$$

When $m(n+1)$ is positive or zero this equation shows that

$$s^{-m}e^s\,T_m{}^{n+1}(s)/T_m{}^n(s)$$

increases with s except possibly at a place where both $T_m{}^n(s)$ and $T_m{}^{n+1}(s)$ are zero. The roots and poles of this function consequently occur alternately as s increases from $-\infty$ to ∞, and so the roots of the equation $T_m{}^n(s)=0$ separate those of the equation $T_m{}^{n+1}(s)=0$.

Gegenbauer showed that if $m>-1$ the roots of the equation $T_m{}^n(s)=0$, considered as an equation of the nth degree in s, are all real, positive and

* Deruyts, *Liège mémoires* (2), t. XIV, p. 9 (1888).

† *Wiener Berichte*, Bd. XCV, S. 274 (1887).

unequal. This is a generalisation of a result obtained by Laguerre* for the case $m = 0$. It is an immediate consequence of an orthogonal relation which will be obtained presently. A geometrical proof has been given by Bôcher† for the case $m = \frac{1}{2}$. The distribution of the roots is of some physical interest because Lagrange‡ showed that the equation $T_0^n(s) = 0$ gives the possible periods of oscillation of a compound pendulum consisting of equal weights equally spaced on a light string. The transition from the compound pendulum to a continuous heavy flexible chain has been discussed by Suzuki§. Properties of the roots are given by E. R. Neumann‖.

§ **9·22**. *Orthogonal properties of the polynomials.* Let us consider the integral

$$I_{n,\nu} = \int_0^\infty e^{-xs} s^m T_m{}^n(as)\, T_m{}^\nu(bs)\, ds, \quad a > 0, b > 0.$$

If $|t| < 1$ and $|\tau| < 1$ we may write

$$\sum_{n=0}^\infty \sum_{\nu=0}^\infty \Gamma(m+n+1)\,\Gamma(m+\nu+1)\, t^n \tau^\nu I_{n,\nu}$$

$$= \int_0^\infty e^{-xs} s^m ds . [(1+t)(1+\tau)]^{-m-1} \exp\left[\frac{ast}{1+t} + \frac{bs\tau}{1+\tau}\right]$$

$$= \Gamma(m+1)\,[x + (x-a)t + (x-b)\tau + (x-a-b)t\tau]^{-m-1}$$

$$= \sum_{\nu=0}^\infty (-)^\nu \frac{\Gamma(m+\nu+1)}{\Gamma(\nu+1)} \tau^\nu [x - b + (x-a-b)t]^\nu$$

$$\times [x + (x-a)t]^{-m-\nu-1}.$$

The coefficient of $t^n\tau^\nu$ in this expansion is easily obtained, and we find that¶

$$I_{n,\nu} = \frac{(-)^{n+\nu}\,\Gamma(m+n+\nu+1)}{\Gamma(n+1)\,\Gamma(\nu+1)\,\Gamma(m+n+1)\,\Gamma(m+\nu+1)} \frac{(x-a)^n (x-b)^\nu}{x^{m+n+\nu+1}}$$

$$\times F\left(-n, -\nu; -m-n-\nu; \frac{x(x-a-b)}{(x-a)(x-b)}\right),$$

with the usual notation for the hypergeometric function.

When $x = a = b = 1$ the double series has the sum

$$\Gamma(m+1)(1-t\tau)^{-m-1} \quad (m > -1)$$

and so we find that in this case

$$I_{n,\nu} = 0 \quad (n \neq \nu)$$

$$= \frac{1}{\Gamma(n+1)\,\Gamma(n+m+1)} \quad (n = \nu).$$

* *Bull. Soc. Math. de France*, t. VII, p. 72 (1879); *Oeuvres de Laguerre*, t. I, p. 428.

† *Proc. Amer. Acad. of Arts and Sciences*, vol. XL, p. 469 (1904).

‡ *Miscellanea Taurinensia*, t. III (1762–1765); *Oeuvres*, t. I, p. 534.

§ *Proc. Phys. Math. Soc. Japan*, vol. II, p. 185 (1920).

‖ *Jahresbericht deutsch. Math. Verein.* Bd. XXX, S. 15 (1921). See also a paper by A. Milne, *Proc. Edin. Math. Soc.* vol. XXXIII, p. 48 (1915).

¶ This is a simple generalisation of a formula given by P. S. Epstein, *Proc. Nat. Acad. of Sci.* vol. XII, p. 629 (1926).

This orthogonal property of the polynomials was discovered by Abel and Murphy for the case $m = 0$, and by Sonine for the general case. In the cases $m = \pm \frac{1}{2}$, Sonine's polynomial can be expressed in terms of Hermite's polynomial $U_n(x)$, which may be defined by means of the equations

$$e^{-2xt-t^2} = \sum_{n=0}^{\infty} (-)^n \frac{t^n}{n!} U_n(x),$$

$$U_n(x) = (-)^n e^{x^2} \frac{d^n}{dx^n} (e^{-x^2}).$$

We have in fact

$$T_{-\frac{1}{2}}^n(x^2).\sqrt{\pi}.(2n)! = U_{2n}(x),$$

$$xT_{\frac{1}{2}}^n(x^2).\sqrt{\pi}.(2n+1)! = U_{2n+1}(x).$$

This polynomial possesses the orthogonal property

$$\int_{-\infty}^{\infty} e^{-x^2} U_m(x) U_n(x) dx = 0 \quad (m \neq n)$$

$$= 2^n (n!) \quad (m = n),$$

so that the functions $U_n^*(x)$, defined by the equation

$$U_n^*(x) = (2^n n!)^{-\frac{1}{2}} e^{-(x^2/2)} U_n(x),$$

form a normalised orthogonal set for the interval $(-\infty, \infty)$. It seems that these functions first occurred in Laplace's *Théorie Analytique des Probabilités*.

In papers on the new mechanics Hermite's polynomial is usually denoted by $H_n(x)$ instead of $U_n(x)$. A useful bibliography on Hermitian polynomials and Hermitian series is given in a valuable paper by E. Hille, *Annals of Math.* vol. XXVII, p. 427 (1926).

EXAMPLES

1. Prove that $$n! \int_0^{\infty} e^{-xs} s^m T_m^n(s)\, ds = (1-x)^n x^{-m-n-1}.$$ [N. Sonine.]

2. Prove that
$$\Gamma(m+\tfrac{1}{2}).T_m^n(s) = \int_0^{\pi} T_{-\frac{1}{2}}^n(s \cos^2\psi).\sin^{2m}\psi.d\psi \quad (m > -\tfrac{1}{2}).$$

3. Obtain also the more general formula
$$\Gamma(\rho)\, T_{\nu+\rho}^n(x) = \int_0^1 T_\nu^n(xy)\, y^\nu (1-y)^{\rho-1}\, dy$$
$$(\rho > 0, \quad \nu > -1).$$ [N. S. Koshliakov*.]

4. Prove that
$$\Gamma(n+\nu+1)\, s^{m+\mu+1}\, T_{m+\mu+1}^{n+\nu}(s) = \Gamma(n+1)\,\Gamma(\nu+1) \int_0^s (s-t)^m t^\mu\, T_m^n(s-t)\, T_\mu^\nu(t)\, dt.$$

5. Prove that $$T_m^n(x) = \sum_{k=0}^{n} \frac{2^{n-k}}{k!} T_m^{n-k}\left(\frac{x}{2}\right).$$ [B. M. Wilson†.]

6. Prove that $$e^{-x}(i\sqrt{xz})^{-m} J_m(2i\sqrt{xz}) = \sum_{n=0}^{\infty} x^n T_m^n(z).$$ [N. Sonine.]

* *Mess. of Math.* vol. LV, p. 152 (1926).
† *Ibid.*, vol. LIII, p. 159 (1924).

7. Prove that, if m and $n+1$ are positive integers and $s > 0$,

$$| T_m^n (s) | < n!\, m!\, e^{\frac{1}{2}s}.$$ [G. Szegö.]

8. Prove that the equations

$$\frac{d^2 x_n}{dt^2} = n (x_{n-1} - x_n) - (n+1)(x_n - x_{n+1})$$

have a particular set of solutions of type

$$x_n = L_n (u) \cos (t\sqrt{u}),$$

where $$L_n (u) = 1 - \binom{n}{1} \frac{u}{1!} + \binom{n}{2} \frac{u^2}{2!} - \ldots + (-)^n \frac{u^n}{n!}.$$ [J. L. Lagrange.]

9. Prove that $$e^{-\frac{us}{1-s}} = (1-s) \sum_{n=0}^{\infty} s^n L_n (u),$$

and that consequently $L_n(u)$ is the so-called polynomial of Laguerre which possesses the property

$$\int_0^\infty e^{-u} L_m (u) L_n (u)\, du$$
$$= 0 \quad m \neq n$$
$$= 1 \quad m = n.$$ [E. T. Whittaker.]

§ **9·31**. *An expression for the product of two Sonine polynomials.* The analysis of § 9·21 indicates the existence of wave-functions of type

$$\Omega = \sum_{m=-\infty}^{\infty} \sum_{n=0}^{\infty} A_m^n e^{ik(z \pm ct) \pm im\phi} \rho^m T_m^n (2ik\alpha)\, T_m^n (2ik\beta), \quad \ldots(A)$$

where the coefficients A_m^n are suitable constants.

The convergence of a series of this type may be partially discussed with the aid of the equation*

$$T_m^n (ix)\, T_m^n (-ix) = \left[\frac{1}{\Gamma(n+1)\,\Gamma(m+1)}\right]^2 \left[1 + \frac{n(m+n+1)\, x^2}{1\,(m+1)^2 (m+2)} + \frac{n(n-1)(m+n+1)(m+n+2)\, x^4}{1.2\,(m+1)^2 (m+2)^2 (m+3)(m+4)} + \ldots\right],$$

in which the coefficients on the right are all positive. This equation, which will be established presently, shows that the modulus of $T_m^n (ix)$ increases with x^2. Hence if the series (A) converges absolutely for any given value of $| \alpha |$, it converges absolutely for all smaller values of $| \alpha |$.

This equation may be established by first expressing a typical term in the expansion for Ω as a definite integral of type

$$\int_0^{2\pi} f [z - ix \cos \omega - iy \sin \omega,\ \ ct - x \sin \omega + y \cos \omega,\ \omega]\, d\omega.$$

Taking

$$f = e^{ik[z - ct - i(x - iy) e^{i\omega}] - im\omega} F [z - ix \cos \omega - iy \sin \omega],$$

* An analogous equation for the confluent hypergeometric function $F(a; \gamma; s)$ was obtained by S. Ramanujan and extended to the generalised hypergeometric function by C. T. Preece, *Proc. London Math. Soc.* (2), vol. XXII, p. 370 (1924). The present equation was given in the author's *Electrical and Optical Wave Motion*, p. 101 (1915).

we may write the integral in the form

$$e^{ik(z-ct)-im\phi}\int_0^{2\pi} e^{k\rho e^{i\gamma}-im\gamma} F\,[z - i\rho\cos\gamma]\,d\gamma$$

by making the substitution $\gamma = \omega - \phi$. Our aim now is to choose the function F, so that

$$(k\rho)^m\, T_m^{\,n}(2ik\alpha)\, T_m^{\,n}(2ik\beta) = \int_0^{2\pi} e^{k\rho e^{i\gamma}-im\gamma} F\,(z - i\rho\cos\gamma)\,d\gamma.$$

We shall verify that a suitable function F is given by

$$F(s) = \frac{(-)^n}{2\pi\,.\,\Gamma(m+n+1)}\, T_0^{\,n}(-2iks). \qquad \text{......(B)}$$

To do this we multiply both sides of our equation by $e^{-k\xi}$, where ξ is an arbitrary positive quantity greater than ρ, and we then integrate between 0 and ∞. Since each side of our equation is a polynomial in k the equation will be verified if it can be shown that the resulting equation is true.

Making use of the formula of § 9·22 the resulting equation is found to be

$$\int_0^{2\pi} e^{-im\gamma}(\eta + \rho e^{-i\gamma})^n (\xi - \rho e^{i\gamma})^{-n-1}\,d\gamma$$

$$= \frac{\Gamma(m+2n+1)\,2\pi}{\Gamma(n+1)\,\Gamma(m+n+1)}\,\frac{\rho^m(\xi\eta+\rho^2)^n}{\xi^{m+2n+1}}\,F\left(-n, -n; -m-2n; \frac{\xi\eta}{\xi\eta+\rho^2}\right),$$

where $\eta = \xi + 2iz$. Now, if ξ and η are sufficiently large we find by expansion in powers of ρ that the definite integral has the value

$$\frac{2\pi\,\Gamma(m+n+1)}{\Gamma(m+1)\,\Gamma(n+1)}\,\frac{\rho^m\eta^n}{\xi^{m+n+1}}\,F(m+n+1, -n; m+1; -\rho^2/\xi\eta).$$

Putting $u = -\rho^2/\xi\eta$, the equation to be established is

$$[\Gamma(m+n+1)]^2 F(m+n+1, -n; m+1; u)$$

$$= \Gamma(m+2n+1)\,\Gamma(m+1)\,(1-u)^n F\,[-n, -n; -m-2n; (1-u)^{-1}],$$

but this is true on account of a well-known property of the hypergeometric series.

Putting $\beta = -\alpha$, $\rho = 2\alpha$, our formula becomes

$$2\pi\,\Gamma(m+n+1)\,(-)^n(2\alpha k)^m\, T_m^{\,n}(2i\alpha k)\, T_m^{\,n}(-2i\alpha k)$$

$$= \int_0^{2\pi} \exp\,[2\alpha k e^{i\gamma} - im\gamma]\,.\,T_0^{\,n}(-4\alpha k\cos\gamma)\,d\gamma,$$

and from this equation the expression (B) is readily derived. If we write

$$e^{i\gamma} = \tau, \quad -2\alpha k = s,$$

the foregoing equation gives the expansion

$$e^{-s\tau} T_0^{\,n}[s(\tau + \tau^{-1})] = \sum_{m=-n}^{\infty} (-)^{m+n}\,\Gamma(m+n+1)\,(\tau s)^m\, T_m^{\,n}(is)\, T_m^{\,n}(-is).$$

This is a particular case of a more general expansion

$$\exp [(\xi\eta)^{\frac{1}{2}} e^{i\omega}]\, T_0^n [\xi + \eta - 2 (\xi\eta)^{\frac{1}{2}} \cos \omega]$$

$$= \sum_{m=-n}^{\infty} (-)^n \Gamma (m + n + 1) (\xi\eta)^{m/2} e^{im\omega} T_m{}^n (\xi) T_m{}^n (\eta).$$

The differential equation (for F) has been studied for general values of m and n by Pochhammer, Jacobstahl, Whittaker, Barnes and many other writers. Writing it in the canonical form

$$x \frac{d^2y}{dx^2} + (\gamma - x) \frac{dy}{dx} - \alpha y = 0, \qquad \text{......(C)}$$

where α and γ are arbitrary constants which may be complex, the complete solution is

$$y = AF (\alpha; \gamma; x) + Bx^{1-\gamma} F (\alpha - \gamma + 1; 2 - \gamma; x),$$

where
$$F (\alpha; \gamma; x) = 1 + \frac{\alpha}{1.\gamma} x + \frac{\alpha (\alpha + 1)}{1.2\gamma (\gamma + 1)} x^2 + \ldots$$

is the confluent hypergeometric series which is so named because we may write*

$$F (\alpha; \gamma; x) = \lim_{\beta \to \infty} F \left(\alpha; \beta; \gamma; \frac{x}{\beta}\right).$$

When γ is a positive integer the coefficient of B is either infinite or identical with the coefficient of A. In this case the complete solution of (C) is†

$$y = [A + C \log x]\, F (\alpha; \gamma; x) + C \sum_{n=0}^{\gamma-2} (-)^{n+\gamma} B (n+\alpha-\gamma+1, \gamma-n-1) \frac{x^{n+1-\gamma}}{n!}$$
$$+ C \left[\frac{\alpha x}{\gamma} \left(\frac{1}{\alpha} - \frac{1}{\gamma} - 1\right) + \frac{\alpha (\alpha + 1)}{\gamma (\gamma + 1)} \frac{x^2}{2!} \left(\frac{1}{\alpha} + \frac{1}{\alpha + 1} - \frac{1}{\gamma} - \frac{1}{\gamma + 1} - 1 - \frac{1}{2}\right)\right.$$
$$+ \frac{\alpha (\alpha + 1) (\alpha + 2)}{\gamma (\gamma + 1) (\gamma + 2)} \frac{x^3}{3!} \left(\frac{1}{\alpha} + \frac{1}{\alpha + 1} + \frac{1}{\alpha + 2} - \frac{1}{\gamma} - \frac{1}{\gamma + 1} - \frac{1}{\gamma + 2} - 1 - \frac{1}{2} - \frac{1}{3}\right)$$
$+ \ldots$ to infinity.

When m and n are positive integers the equation

$$s \frac{d^2F}{ds^2} + (m + 1 - s) \frac{dF}{ds} + nF = 0$$

possesses only one solution $T_m{}^n (s)$ which can be represented by a convergent power series of integral powers of s. A second solution may be derived by a well-known method by writing $F = uT_m{}^n (s)$. The equation for u is then

$$sT \frac{d^2u}{ds^2} + \left[2s \frac{dT}{ds} + (m + 1 - s) T\right] \frac{du}{ds} = 0,$$

and so
$$F = CT_m{}^n (s) + DT_m{}^n (s) \int^s e^\sigma [T_m{}^n (\sigma)]^{-2} \sigma^{-(m+1)} d\sigma.$$

* E. Kummer, *Crelle's Journal*, vol. xv, p. 138 (1836)

† H. A. Webb and J. R. Airey, *Phil. Mag.* xxxvi, p. 129 (1918). W. J. Archibald, *Phil. Mag.* s. 7, vol. 26, pp. 415–419 (1938), (addition of the second term for $\gamma > 1$).

This solution gives a logarithmic term when the integrand is expanded in powers of σ.

When s is imaginary, or a complex quantity, as it is in our case, use may be made of the solution represented by the definite integral

$$U_m^n(s) = \frac{e^s}{\Gamma(n+1)} \int_0^\infty e^{-\sigma} \sigma^n (s-\sigma)^{-m-n-1} d\sigma.$$

A number of analogous definite integrals are given in papers by Epstein* and Whittaker†.

Whittaker reduces the differential equation to a standard form

$$\frac{d^2W}{dz^2} + \left\{-\frac{1}{4} + \frac{k}{z} + \frac{\frac{1}{4} - m^2}{z^2}\right\} W = 0,$$

and introduces as the principal solution the function

$$W_{k,m}(z) = -\frac{1}{2\pi i} \Gamma(k + \tfrac{1}{2} - n) e^{-\frac{1}{2}z} z^k \int_\infty^{(0+)} (-t)^{-k-\frac{1}{2}+m} \left(1 + \frac{t}{z}\right)^{k-\frac{1}{2}+m} e^{-t} dt,$$

where $\arg z$ has its principal value and the contour is so chosen that the point $t = -z$ is outside it. "The integrand is rendered one-valued by taking $|\arg(-t)| \leqslant \pi$, and taking that value of $\arg(1 + t/z)$ which tends to zero as $t \to 0$ by a path lying inside the contour." When

$$R(k - \tfrac{1}{2} - m) \leqslant 0,$$

$$W_{k,m}(z) = \frac{e^{-\frac{1}{2}z} z^k}{\Gamma(\frac{1}{2} - k + m)} \int_0^\infty t^{-k-\frac{1}{2}+m} \left(1 + \frac{t}{z}\right)^{k-\frac{1}{2}+m} e^{-t} dt.$$

"This formula suffices to define $W_{k,m}(z)$ in the critical cases when $m + k - \frac{1}{2}$ is a positive integer, and so $W_{k,m}(z)$ is defined for all values of z except negative real values."

The relation between this function and Sonine's polynomial is indicated by the relation

$$\frac{1}{n!\,\Gamma(m+n+1)} z^{-\frac{1}{2}(m+1)} e^{\frac{1}{2}z} W_{n+\frac{1}{2}m+\frac{1}{2},\,\frac{1}{2}m}(z) = T_m^n(z).$$

For the properties of the function $W_{k,m}(z)$ and its asymptotic expansion reference must be made to *Modern Analysis* and to some later papers by mathematicians of the Edinburgh school‡. The asymptotic expansion of the function $T_m^n(x)$ for large values of n is discussed by J. V. Uspensky§ and used to obtain sufficient conditions for the validity of the expansion of $f(z)$ in a series of these functions when the coefficients are given by means of the orthogonal relation of § 9·22. The summability of the series

* *Diss. Munich* (1914).

† *Bull. Amer. Math. Soc.* vol. x, p. 125 (1904).

‡ D. Gibb, *Proc. Edin. Math. Soc.* vol. xxxiv, p. 93 (1916); N. M'Arthur, *ibid.* vol. xxxviii, p. 27 (1920); G. E. Chappell, *ibid.* vol. xliii, p. 117 (1924).

§ *Annals of Math.* vol. xxviii, p. 593 (1927). See also O. Perron, *Crelle*, Bd. cli, S. 63 (1921); M. H. Stone, *Annals of Math.* vol. xxix, p. 1 (1927).

has been discussed by E. Hille* and G. Szegö†. The Parseval theorem for the series has been investigated by S. Wigert‡ and M. Riesz §.

EXAMPLES

1. Prove that

$$e^{ikz\cos\omega} J_m(k\rho\sin\omega) = \sum_{n=0}^{\infty} A_m{}^n \rho^m e^{ikz} T_m{}^n(2ik\alpha)\, T_m{}^n(2ik\beta),$$

where
$$A_m{}^n = (-)^n \Gamma(n+1)\,\Gamma(m+n+1)\, k^m \sec^2\frac{\omega}{2}\left(\tan\frac{\omega}{2}\right)^{m+2n}$$

2. If $\phi(x)$ is a continuous function for all real values of x and if, when $|x|$ is very large, $\phi(x) = O(e^{-kx^2})$, where k is a positive constant, the equations

$$\int_{-\infty}^{\infty} \phi(x) H_n(x)\, dx = 0$$

$$n = 0, 1, 2, \ldots,$$

in which $H_n(x)$ denotes Hermite's polynomial, imply that $\phi(x) = 0$ for all real values of x. [M. H. Stone.]

3. If $f(x)$ is integrable for all real values of x and such that

$$\int_{-\infty}^{\infty} e^{-x^2} [f(x)]^2\, dx$$

exists, the quantities

$$c_n = \frac{1}{2^n n! \sqrt{\pi}} \int_{-\infty}^{\infty} e^{-x^2} f(x) H_n(x)\, dx$$

are such that the infinite series

$$\sum_{n=0}^{\infty} 2^n n!\, c_n{}^2$$

converges. [M. H. Stone.]

4. If, in addition, the limits $f(x_0 \pm 0)$ exist absolutely and $f(x)$ is absolutely integrable over any finite interval, the series

$$\sum_{n=0}^{\infty} c_n H_n(x_0) \qquad \text{......(D)}$$

converges and its sum is $\frac{1}{2}[f(x_0+0) + f(x_0-0)]$. [J. V. Uspensky.]

5. If $f(x)$ admits the representation

$$f(x) = f(0) + \int_0^x f'(z)\, dz \qquad (-\infty < x < \infty)$$

and
$$\int_{-\infty}^{\infty} e^{-x^2} [2xf(x) - f'(x)]^2\, dx$$

exists, and if, moreover, for large values of $|x|$

$$f(x) = O(e^{kx^2}) \qquad 0 \leqslant k < 1,$$

the series (D) converges uniformly to $f(x)$ in any finite interval. [M. H. Stone.]

* *Proc. Nat. Acad. Sci.* vol. XII, pp. 261, 265, 348 (1926).
† *Math. Zeits.* Bd. XXV, S. 87 (1926).
‡ *Arkiv för Mat., Astron. och Fysik,* Bd. XV (1921).
§ *Szeged Acta,* t. I, p. 209 (1923).

6. Prove that, if $m > -\frac{1}{2}$, we have for large values of n

$$\Pi_n(x) = \omega_n(x)\left[\cos\theta_n + (nx)^{-\frac{1}{2}} u(x)\sin\theta_n + \frac{1}{n}\phi_n(x)\right]x^{-m/2},$$

$$\Pi_n'(x) = -\omega(x)\left[(n/x)^{\frac{1}{2}}\sin\theta_n - \frac{1}{x}v(x)\cos\theta_n + (nx)^{-\frac{1}{2}}\psi_n(x)\right]x^{-m/2},$$

where
$$\Pi_n(x) = (-)^n[\Gamma(m+n+1)\Gamma(n+1)]^{\frac{1}{2}} T_m^n(x),$$

$$\omega_n(x) = e^{x/2}(\pi^2 nx)^{-\frac{1}{4}}, \quad \theta_n = 2\sqrt{nx} - \frac{2m+1}{4}\pi,$$

$$u(x) = \frac{x^2}{12} - \frac{1+m}{2}x + \frac{1}{16} - \frac{m^2}{4},$$

$$v(x) = \frac{x^2}{12} - \frac{mx}{2} + \frac{1}{16} - \frac{(m+1)^2}{4},$$

and where $\phi(x)$ and $\psi(x)$ remain bounded when x varies in the finite interval

$$0 < a \leqslant x \leqslant b.$$

In this form the asymptotic expression is due to Uspensky. An earlier form, given by Fejér, has been elaborated by Perron and Szegö in papers to which we have already referred. The corresponding asymptotic expression for Hermite's polynomial was obtained by Adamoff, *Ann. Inst. Polytechnique de St Pétersbourg*, t. v, p. 127 (1906). The result was extended to complex values of the variable x by G. N. Watson, *Proc. London Math. Soc.* (2), vol. VIII, p. 393 (1910).

7. Prove that $$sT^n_{m+1}(s) = (n+1)T_m^{n+1}(s) + T_m^n(s).$$

8. Prove that, if $a > 0$,

$$F(a; c; z) = \frac{2^a\Gamma(a+1-c)\Gamma(c)}{2\pi\Gamma(a)}\int_{-\infty}^{\infty}(1+is)^{-c}(1-is)^{c-a-1}e^{\frac{2z}{1+is}}ds.$$

9. Prove that $$F(a;\gamma;-h)e^{hx} = \sum_{m=0}^{\infty}\frac{(hx)^m}{m!}F\left(a, -m; \gamma; \frac{1}{x}\right).$$

[P. Humbert, *Journ. de l'École Polytechnique* (2), Cah. 24, p. 59 (1924).]

CHAPTER X

TOROIDAL CO-ORDINATES

§ **10·1**. *Laplace's equation in toroidal co-ordinates.* If we put

$$x = \rho \cos \phi, \quad y = \rho \sin \phi, \quad z + i\rho = a \cot \tfrac{1}{2} (\psi + i\sigma),$$

$$\rho = \frac{a \sinh \sigma}{\cosh \sigma - \cos \psi}, \quad z = \frac{a \sin \psi}{\cosh \sigma - \cos \psi},$$

the angle ψ is the angle subtended at a point $P(x, y, z)$ by the segment AB which is the diameter of the circle $z = 0$, $x^2 + y^2 = a^2$ in the plane through P and the axis of z. The quantity σ is equal to $\log (PB/PA)$. The surfaces $\psi =$ constant are the spherical caps having the above-mentioned circle in common; the surfaces $\sigma =$ constant are anchor rings. In these co-ordinates

$$dx^2 + dy^2 + dz^2 = \frac{a^2}{(s - \tau)^2} [d\sigma^2 + d\psi^2 + \sinh^2 \sigma . d\phi^2],$$

where $s = \cosh \sigma$, $\tau = \cos \psi$. Laplace's equation consequently takes the form*

$$\frac{\partial}{\partial \sigma} \left(\frac{\sinh \sigma}{s - \tau} \frac{\partial u}{\partial \sigma} \right) + \frac{\partial}{\partial \psi} \left(\frac{\sinh \sigma}{s - \tau} \frac{\partial u}{\partial \psi} \right) + \frac{1}{(s - \tau) \sinh \sigma} \frac{\partial^2 u}{\partial \phi^2} = 0. \quad \text{...(A)}$$

Putting $u = v (s - \tau)^{\frac{1}{2}}$ we find that v satisfies the equation

$$\frac{\partial}{\partial s} \left[(s^2 - 1) \frac{\partial v}{\partial s} \right] + \frac{\partial^2 v}{\partial \psi^2} + \frac{v}{4} + \frac{1}{s^2 - 1} \frac{\partial^2 v}{\partial \phi^2} = 0, \quad \text{......(B)}$$

in which the variables are separated. Hence there are simple toroidal potential functions of type

$$u = (s - \tau)^{\frac{1}{2}} \cos n (\psi - \psi_0) \cos m (\phi - \phi_0) [A P^m{}_{n-\frac{1}{2}} (s) + B Q^m{}_{n-\frac{1}{2}} (s)],$$

where A, B, n, m, ϕ_0 and ψ_0 are arbitrary constants. When $n = \frac{1}{2}$ the typical solutions may be combined so as to give a solution of type

$$u = (s - \tau)^{\frac{1}{2}} \cos \tfrac{1}{2} (\psi - \psi_0) [f (\phi + i\chi) + g (\phi - i\chi)],$$

where

$$\chi = \log (\tanh \sigma/2),$$

and f and g are arbitrary functions. This form is indicated by the solution of Laplace's equation

$$f\left(\frac{x + iy}{z + r}\right) = f\left(\tan \frac{\theta}{2} . e^{i\phi}\right)$$

and is also indicated by the fact that

$$\Gamma (m + 1) . P_0{}^{-m} (\cosh \sigma) = \tanh^m (\sigma/2).$$

* B. Riemann, *Partielle Differentialgleichungen*, Hattendorf's ed. (1861); C. Neumann, *Theorie der Elektricität und Wärme in einem Ringe*, Halle (1864); W. M. Hicks, *Phil. Trans.* vol. CLXXI, p. 609 (1881); A. B. Basset, *Amer. Journ. of Math.* vol. XV, p. 287 (1893); *Hydrodynamics*, vol. II; E. Heine, *Anwendungen der Kugelfunktionen*, 2nd ed. pp. 283–301, Berlin (1881).

We shall now obtain some formulae for the Legendre function of the second kind which will be useful in the subsequent work. It should be remarked that the more general equation

$$\frac{\partial^2 u}{\partial \eta^2} + \frac{\alpha}{\eta}\frac{\partial u}{\partial \eta} + \frac{1}{\eta^2}\frac{\partial^2 u}{\partial \phi^2} + \frac{\partial^2 u}{\partial \zeta^2} + \frac{\beta}{\zeta}\frac{\partial u}{\partial \zeta} + \frac{1}{\zeta^2}\frac{\partial^2 u}{\partial \theta^2} = 0 \qquad \text{......(C)}$$

may be treated by a similar substitution, and it is found that, if

$$\eta + i\zeta = ia \cot \tfrac{1}{2}(\psi + i\sigma), \quad u = v\,(\cosh\sigma - \cos\psi)^{\frac{\alpha+\beta}{2}}, \qquad \text{......(D)}$$

then v satisfies the partial differential equation

$$\frac{\partial^2 v}{\partial \sigma^2} + \alpha \coth\sigma \frac{\partial v}{\partial \sigma} + \frac{\partial^2 v}{\partial \psi^2} + \beta\cot\psi\frac{\partial v}{\partial \psi} - \frac{\alpha^2 - \beta^2}{2} v + \operatorname{cosech}^2\sigma \frac{\partial^2 v}{\partial \phi^2} - \operatorname{cosec}^2\psi\frac{\partial^2 v}{\partial \theta^2} = 0.$$

This equation evidently possesses elementary solutions of type

$$v = S(\sigma)\Psi(\psi)\Phi(\phi)\Theta(\theta).$$

In particular, when $\alpha = \beta = 1$ we have solutions of the type

$$u = (s - \tau)\left[AP_n^m(s) + BQ_n^m(s)\right]\left[CP_n^\kappa(\tau) + DQ_n^\kappa(\tau)\right] e^{ik\theta + im\phi},$$

where A, B, C and D are arbitrary constants and $s = \cosh\sigma$, $\tau = \cos\psi$. The wave-equation may be reduced to the form (C) by writing

$$x = \eta\cos\phi, \quad y = \eta\sin\phi, \quad z = \zeta\cos\theta, \quad ict = \zeta\sin\theta.$$

EXAMPLES

1. Prove that, if $n > -1$, $p > -1$, $m > -1$, $p + m > -1$,

$$\int_0^\infty K_p(\lambda\zeta) J_m(\lambda\eta) J_n(\lambda a) \lambda^{p+m-n+1} d\lambda = 2^{p+m-n-1} a^{n-p-m-2} \frac{\Gamma(p+m+1)\Gamma(p+1)\Gamma(m+1)}{\Gamma(n+1)} \times (s-\tau) P^{-p}{}_{p+m-n}(\tau) P^{-m}{}_{p+m-n}(s),$$

where η and ζ are defined by equation (D). When $m = n$ this equation reduces to one given by H. M. Macdonald, *Proc. London Math. Soc.* vol. VII, p. 147 (1909).

2. Prove that

$$(s - \tau)\tanh^m(\tfrac{1}{2}\sigma)\tan^n(\tfrac{1}{2}\psi) = 2a^{m+2}\int_0^\infty K_n(\lambda\zeta) J_m(\lambda\eta) J_{m+n}(\lambda a)\lambda\, d\lambda,$$

and deduce that

$$x^{-n}(1 + x^2)^{-m-1} = \frac{1}{2^m\Gamma(m+1)}\int_0^\infty K_n(\lambda x) J_{n+m}(\lambda)\lambda^{m+1} d\lambda.$$

3. Prove that, if $m \geqslant 0$,

$$(s-\tau)^{-m-1} P_m\left(\frac{s\tau - 1}{s - \tau}\right) = \sum_{n=m}^{\infty} \frac{(n+m)!}{2^m (n-m)!\, m!\, m!}(2n+1) Q_n(s) P_n(\tau).$$

4. Prove that $\qquad P_p^{-p}(\cos\psi) = \dfrac{1}{\Gamma(1+p)}(\tfrac{1}{2}\sin\psi)^p,$

$$P_p^{-p}(\cosh\sigma) = \frac{1}{\Gamma(1+p)}(\tfrac{1}{2}\sinh\sigma)^p.$$

§ **10·2**. *Jacobi's transformation**. If n is a positive integer we find on integrating by parts that

$$\int_{-1}^{1} \phi^{(n)}(z)\,(1-z^2)^{n-\frac{1}{2}}\,dz = (-)^n \int_{-1}^{1} \phi(z)\,\frac{d^n}{dz^n}\left[(1-z^2)^{n-\frac{1}{2}}\right] dz.$$

Now it follows from the expansion of sin nx in powers of cos x that†

$$\frac{d^{n-1}(1-z^2)^{n-\frac{1}{2}}}{dz^{n-1}} = (-)^{n-1}\,\frac{1.3\ldots(2n-1)}{n}\,\sin(n\cos^{-1}z),$$

$$\frac{d^n}{dz^n}(1-z^2)^{n-\frac{1}{2}} = (-)^n\,\frac{1.3\ldots(2n-1)}{\sqrt{1-z^2}}\,\cos(n\cos^{-1}z).$$

Hence if $\phi^{(n)}(z)$ is continuous in the range $-1 \leqslant z \leqslant 1$,

$$\int_{-1}^{1} \phi^{(n)}(z)\,(1-z^2)^{n-\frac{1}{2}}\,dz = 1.3\ldots(2n-1)\int_{-1}^{1}\phi(z)\cos(n\cos^{-1}z)\,\frac{dz}{\sqrt{1-z^2}}.$$

Putting $z = \cos\theta$ the equation takes the form

$$\int_0^{\pi} \phi^{(n)}(\cos\theta)\sin^{2n}\theta\,d\theta = 1.3\ldots(2n-1)\int_0^{\pi}\phi(\cos\theta)\cos n\theta\,d\theta.$$

There are many applications of this theorem. If we start with the formulae

$$P_n^m(z) = \frac{(n+1)(n+2)\ldots(n+m)}{\pi}\int_0^{\pi}\{z+\sqrt{z^2-1}\cos\phi\}^n\cos m\phi\,d\phi,$$

$$J_m(z) = \frac{1}{i^m\pi}\int_0^{\pi} e^{iz\cos\phi}\cos m\phi\,d\phi,$$

we may derive the formulae

$$P_n^m(z) = \frac{(n+m)!}{(n-m)!}\,\frac{(z^2-1)^{m/2}}{1.3\ldots(2m-1)}\,\frac{1}{\pi}\int_0^{\pi}\{z+\sqrt{z^2-1}\cos\phi\}^{n-m}\sin^{2m}\phi\,d\phi,$$

$$J_m(z) = \frac{z^m}{1.3\ldots(2m-1)}\,\frac{1}{\pi}\int_0^{\pi} e^{iz\cos\phi}\sin^{2m}\phi\,d\phi.$$

The last equation is a particular case of the more general equation‡

$$J_\nu(z) = \frac{z^\nu}{2^\nu\,\Gamma(\nu+\frac{1}{2})\,\Gamma(\frac{1}{2})}\int_0^{\pi} e^{iz\cos\phi}\sin^{2\nu}\phi\,d\phi,$$

which holds when ν is not an integer if its real part is greater than $-\frac{1}{2}$. The first equation may be written in the alternative form

$$P_n^{-m}(z) = (-)^m\,\frac{(z^2-1)^{m/2}}{1.3\ldots(2m-1)}\,\frac{1}{\pi}\int_0^{\pi}\{z+\sqrt{z^2-1}\cos\phi\}^{n-m}\sin^{2m}\phi\,d\phi.$$

* *Crelle's Journal*, vol. xv, p. 1; L. Kronecker, *Berlin. Sitzungsberichte*, S. 539 (1884).

† An elementary proof of the first equation is given by W. L. Ferrar, *Proc. London Math. Soc.* (2), vol. xxii (1924); *Records of Proceedings*, Feb. 14.

‡ Whittaker and Watson, *Modern Analysis*, p. 366.

If in Jacobi's formula we take $\phi(x) = x^m$ we obtain

$$\int_0^\pi \cos^m \theta \cos n\theta \, d\theta = \frac{\pi}{2^m} \frac{\Gamma(n+s+2)}{\Gamma(m)\,\Gamma(s+1)} \quad (m = n + 2s)$$

$$= 0 \quad (m = n + 2s + 1) \qquad (s \text{ an integer})$$

$$= 0 \quad (m < n).$$

This formula is related to the general formula of Poisson*

$$\int_0^{\pi/2} \cos^\nu x \cos(\nu + 2m)\, x \, dx = 0,$$

$$\int_0^{\pi/2} \cos^\nu x \cos \nu x \, dx = \frac{\pi}{2} \cdot \frac{1}{2^\nu},$$

where ν is any positive quantity and m any positive integer.

If we next put
$$\phi(\cos\theta) = (1 - 2a\cos\theta + a^2)^{-\frac{1}{2}},$$

$$\phi^{(n)}(\cos\theta) = 1.3\ldots(2n-1)\, a^n (1 - 2a\cos\theta + a^2)^{-n-\frac{1}{2}},$$

Jacobi's formula gives

$$\int_0^\pi \frac{\cos n\theta . d\theta}{(1 - 2a\cos\theta + a^2)^{\frac{1}{2}}} = a^n \int_0^\pi \frac{\sin^{2n}\theta . d\theta}{(1 - 2a\cos\theta + a^2)^{n+\frac{1}{2}}}.$$

Putting $2s = a + a^{-1}$ and replacing $\cos\theta$ in the last integral by t, we have

$$\int_0^\pi (s - \cos\theta)^{-\frac{1}{2}} \cos n\theta . d\theta = 2^{-n} \int_{-1}^1 1 - t^2)^{n-\frac{1}{2}} (s - t)^{-n-\frac{1}{2}} \, dt$$

$$= 2^{\frac{1}{2}} Q_{n-\frac{1}{2}}(s).$$

Another expression for the Legendre function of the second type is obtained by making use of the transformation

$$\cos\phi = R^{-\frac{1}{2}}(\cos\theta - a), \quad \sin\phi = R^{-\frac{1}{2}} \sin\theta,$$

$$R = 1 - 2a\cos\theta + a^2,$$

$$(1 - a^2 \sin^2\phi)^{-\frac{1}{2}}\, d\phi = R^{-\frac{1}{2}}\, d\theta,$$

$$\int_0^\pi (1 - 2a\cos\theta + a^2)^{-n-\frac{1}{2}} \sin^{2n}\theta . d\theta = \int_0^\pi (1 - a^2\sin^2\phi)^{-\frac{1}{2}} \sin^{2n}\phi . d\phi.$$

Consequently we have the equations

$$\int_0^\pi \frac{\cos n\theta . d\theta}{(1 - 2a\cos\theta + a^2)^{\frac{1}{2}}} = a^n \int_0^\pi \frac{\sin^{2n}\phi . d\phi}{(1 - a^2\sin^2\phi)^{\frac{1}{2}}},$$

$$\int_0^\pi \frac{\sin^{2n}\phi . d\phi}{(1 - a^2\sin^2\phi)^{\frac{1}{2}}} = a^{-n-\frac{1}{2}} Q_{n-\frac{1}{2}} \left[\tfrac{1}{2}(a + a^{-1})\right].$$

* S. D. Poisson, *Journ. de l'École Polytechnique*, Cah. 19, p. 490 (1823).

EXAMPLES

1. Prove that $$\pi [2 (s - \tau)]^{-\frac{1}{2}} = \sum_{n=-\infty}^{\infty} e^{in\psi} Q_{n-\frac{1}{2}} (s),$$
and show that a potential which is constant on the ring $s = s_0$ is given by
$$V = [2 (s - \tau)]^{\frac{1}{2}} \sum_{n=-\infty}^{\infty} e^{in\psi} Q_{n-\frac{1}{2}} (s_0) P_{n-\frac{1}{2}} (s)/P_{n-\frac{1}{2}} (s_0).$$ [Heine.]

2. Prove that the potential of a uniformly charged circular ring coinciding with the fundamental circle $z = 0$, $\rho = a$, is proportional to
$$(\cosh \sigma - \cos \psi)^{\frac{1}{2}} \operatorname{sech} (\tfrac{1}{2}\sigma) . K (\tanh \tfrac{1}{2}\sigma),$$
where $K (k)$ is the complete elliptic integral with modulus k.

3. Prove that Laplace's integral for $P_n (\mu)$ can be used to obtain the formula
$$\tfrac{1}{2}\pi . P_n (\cos \theta) = \int_0^\theta \cos \{(n + \tfrac{1}{2}) \phi\} . \{2 (\cos \phi - \cos \theta)\}^{-\frac{1}{2}} d\phi.$$
[Mehler.]

4. Prove that, when n is a positive integer,
$$\tfrac{1}{2}\pi . P_n (\cos \theta) = \int_0^\pi \sin \{(n + \tfrac{1}{2}) \phi\} . \{2 (\cos \theta - \cos \phi)\}^{-\frac{1}{2}} d\phi.$$

The integrals in Examples 3 and 4 are given in Whittaker and Watson's *Modern Analysis*, p. 315, they have been much used by J. W. Nicholson to evaluate series and integrals involving Legendre functions.

5. Prove that if $f (\alpha)$ is a suitable type of arbitrary function, the differential equation
$$\frac{\partial}{\partial s} \left[(1 - s^2) \frac{\partial v}{\partial s} \right] - \frac{v}{4} - \frac{\partial^2 v}{\partial \psi^2}$$
is satisfied by the integral
$$v = \int_{\psi-\beta}^{\psi+\beta} [\cos (\alpha - \psi) - \cos \beta]^{-\frac{1}{2}} f (\alpha) \, d\alpha, \quad (s = \cos \beta).$$

6. Prove that the differential equation
$$\frac{d}{ds} \left[(1 - s^2) \frac{dv}{ds} \right] - \frac{v}{4} = 0$$
possesses the two particular solutions
$$v_1 = 1 + \frac{1^2}{2!} \left(\frac{s}{2}\right)^2 + \frac{1^2 . 5^2}{4!} \left(\frac{s}{2}\right)^4 + \frac{1^2 . 5^2 . 9^2}{6!} \left(\frac{s}{2}\right)^6 + \ldots,$$
$$v_2 = \frac{s}{2} + \frac{3^2}{3!} \left(\frac{s}{2}\right)^3 + \frac{3^2 . 7^2}{5!} \left(\frac{s}{2}\right)^5 + \frac{3^2 . 7^2 . 11^2}{7!} \left(\frac{s}{2}\right)^7 + \ldots,$$
the series being convergent for $| s | < 1$.

The corresponding solutions of Legendre's equation of order n may be written respectively in the forms
$$v_1 = F (- \tfrac{1}{2}n, \tfrac{1}{2}n + \tfrac{1}{2}; \tfrac{1}{2}; s^2), \quad v_2 = F (- \tfrac{1}{2}n + \tfrac{1}{2}, \tfrac{1}{2}n + 1; \tfrac{3}{2}; s^2).$$
[Heine.]

§ **10·3.** *Green's functions for the circular disc and spherical bowl.* If R is the distance between two points with toroidal co-ordinates (σ, ψ, ϕ), $(\sigma_0, \psi_0, \phi_0)$, we have
$$aR^{-1} = (s - \tau)^{\frac{1}{2}} (s_0 - \tau_0)^{\frac{1}{2}} \{2 \cosh \alpha - 2 \cos (\psi - \psi_0)\}^{-\frac{1}{2}}, \quad \ldots\text{(A)}$$
where $$\cosh \alpha = \cosh \sigma \cosh \sigma_0 - \sinh \sigma \sinh \sigma_0 \cos (\phi - \phi_0).$$

The last factor in (A) may be expanded in a cosine series of multiples of $\psi - \psi_0$ and the coefficient of $\cos m(\psi - \psi_0)$ will be

$$\pi^{-1}\sqrt{2}\int_0^\pi \cos m\psi\,(\cosh\alpha - \cos\psi)^{-\frac{1}{2}}\,d\psi = (2/\pi)\,Q_{m-\frac{1}{2}}(\cosh\alpha).$$

Therefore

$$\pi a R^{-1} = (s-\tau)^{\frac{1}{2}}(s_0-\tau_0)^{\frac{1}{2}}[Q_{-\frac{1}{2}}(\cosh\alpha) + 2Q_{\frac{1}{2}}(\cosh\alpha)\cos(\psi-\psi_0) + \dots].$$

This expansion of the inverse distance was given by Heine.

The series may be summed by writing

$$Q_{m-\frac{1}{2}}(\cosh\alpha) = 2^{-\frac{1}{2}}\int_\alpha^\infty (\cosh u - \cosh\alpha)^{-\frac{1}{2}}\,e^{-mu}\,du.$$

This formula is proved in § 10·5. The method of summation, which is taken from a paper by E. W. Hobson*, leads to the formula

$$\pi a R^{-1} = 2^{-\frac{1}{2}}(s-\tau)^{\frac{1}{2}}(s_0-\tau_0)^{\frac{1}{2}}\int_\alpha^\infty \frac{\sinh u\,du}{\cosh u - \cos(\psi-\psi_0)}(\cosh u - \cosh\alpha)^{-\frac{1}{2}}.$$

In order to obtain the Green's function for a circular disc by an extension of the method of images it is convenient to use an idea originated and developed by A. Sommerfeld†, and to consider two superposed spaces of three dimensions related to one another in much the same way as the sheets of a Riemann surface. In the present case the passage from one space to the other is made when a point "passes through the disc." The two spaces may be distinguished from each other by the inequalities

$$-\pi < \psi < \pi \quad \text{in the first space,}$$
$$\pi < \psi < 3\pi \quad \text{in the second space.}$$

A point P which starts from a place in the first space on the positive side of the disc may pass through the disc into the second space, and ψ will increase continuously to a value greater than its original value π when the point is on the disc. In order that this point after the passage through the disc may return to its original position P_0 it will be necessary for it to pass again through the disc and at the second crossing it returns into the first space.

Corresponding to a point (σ, ψ, ϕ) in the first space $(-\pi < \psi < \pi)$ there is an associated point $(\sigma, \psi + 2\pi, \phi)$ in the second space. The point $(\sigma, \psi + 4\pi, \phi)$ is regarded as identical with the original point in the first space.

We now notice that there is an identity

$$\frac{2\sinh u}{\cosh u - \cos(\psi-\psi_0)} = \frac{\sinh\frac{1}{2}u}{\cosh\frac{1}{2}u - \cos\frac{1}{2}(\psi-\psi_0)} + \frac{\sinh\frac{1}{2}u}{\cosh\frac{1}{2}u + \cos\frac{1}{2}(\psi-\psi_0)},$$

* *Camb. Trans.* vol. XVIII, p. 277 (1899).

† *Math. Ann.* Bd. XLVII, S. 317 (1896); *Proc. London Math. Soc.* (1), vol. XXVIII, p. 395 (1897).

which indicates that we may write

$$R^{-1} = W(\sigma_0, \psi_0, \phi_0) + W(\sigma_0, \psi_0 + 2\pi, \phi_0),$$

where

$$2\pi a W(\sigma_0, \psi_0, \phi_0)$$
$$= 2^{-\frac{1}{2}}(s-\tau)^{\frac{1}{2}}(s_0-\tau_0)^{\frac{1}{2}} \int_\alpha^\infty \frac{\sinh \frac{1}{2}u \, . du}{\cosh \frac{1}{2}u - \cos \frac{1}{2}(\psi - \psi_0)} (\cosh u - \cosh \alpha)^{-\frac{1}{2}}.$$

Performing the integration we find that

$$W(\sigma_0, \psi_0, \phi_0) = R^{-1}\left[\frac{1}{2} + \frac{1}{\pi} \sin^{-1}\{\cos \tfrac{1}{2}(\psi - \psi_0) \operatorname{sech}(\tfrac{1}{2}\alpha)\}\right].$$

It may be shown without much difficulty that this function W is a solution of Laplace's equation when considered as a function of either σ, ψ, ϕ or σ_0, ψ_0, ϕ_0; it is, in fact, a symmetrical function of the two sets of co-ordinates. It is, moreover, a uniform function of σ, ψ, ϕ in the double space since it is unaltered in value when ψ is increased by 4π. It is continuous $(D, 1)$ throughout the double space except at the point $(\sigma_0, \psi_0, \phi_0)$ where it becomes infinite like R^{-1}. It is finite at the point $(\sigma_0, \psi_0 + 2\pi, \phi_0)$ because at this point $\sin^{-1}[\{......\}]$ becomes $-\frac{1}{2}\pi$ instead of $\frac{1}{2}\pi$. It is, indeed, the fundamental potential function for the double space.

Let us now consider the function

$$V = W(\sigma_0, \psi_0, \phi_0) - W(\sigma_0, 2\pi - \psi_0, \phi_0),$$

which is a potential for the double space when there is a charge at the point P with co-ordinates $(\sigma_0, \psi_0, \phi_0)$ and an image charge at a point P' $(\sigma_0, 2\pi - \psi_0, \phi_0)$ which is situated in the second space at the optical image of P in the plane of the disc. This function V is infinite in the first space only at P and is, moreover, zero on the disc.

To see this we note that on the disc R is the same for the point P and its image, consequently it is only necessary to show that when $\psi = \pi$

$$\cos \tfrac{1}{2}(\psi - \psi_0) = \cos \tfrac{1}{2}(\psi - 2\pi + \psi_0),$$

and this is evidently true.

The function V possesses all the characteristics of a Green's function and so we may write

$$V = G(\sigma, \psi, \phi; \sigma_0, \psi_0, \phi_0) = G(Q, P),$$

and regard G as the Green's function of the circular disc. It is evidently a symmetrical function of the two sets of co-ordinates (σ, ψ, ϕ), $(\sigma_0, \psi_0, \phi_0)$; that is, of the points Q and P that have these co-ordinates.

To solve the corresponding problem for the spherical bowl it is convenient to regard the surface of the bowl as the place where a passage is made from one space to the other. If the angle of the bowl is β, so that $\psi = \beta$ is the equation of the bowl, we must suppose that in the first space ψ has values from $\beta - 2\pi$ on the negative side of the bowl up to β on the positive side, and that in the second space ψ increases from β up to $\beta + 2\pi$. If the convexity of the bowl is upwards, $\beta < \pi$; if it is downwards, $\beta > \pi$.

We now need the image of P $(\sigma_0, \psi_0, \phi_0)$ in the spherical surface of the bowl and this must be regarded as a point P' $(\sigma_0, 2\beta - \psi_0, \phi_0)$, which is in the second space if $\beta - 2\pi < \psi_0 < \beta$. If $0 < \psi_0 < \beta$, P is above the bowl and P' below the bowl.

The function

$$G(P, Q) = \frac{1}{PQ}\left[\frac{1}{2} + \frac{1}{\pi}\sin^{-1}\{\cot \tfrac{1}{2}(\psi - \psi_0)\operatorname{sech}\tfrac{1}{2}\alpha\}\right]$$
$$- \frac{1}{P'Q}(\cosh\sigma_0 - \cos\psi_0)^{\frac{1}{2}}[\cosh\sigma_0 - \cos(2\beta - \psi_0)]^{\frac{1}{2}}$$
$$\times\left[\frac{1}{2} + \frac{1}{\pi}\sin^{-1}\{\cos\tfrac{1}{2}(\psi + \psi_0 - 2\beta)\operatorname{sech}\tfrac{1}{2}\alpha\}\right]$$

is seen to satisfy the requirements and is, indeed, the Green's function for the spherical bowl.

§ **10·4**. *Relation between toroidal and spheroidal co-ordinates.* There is a simple relation between toroidal co-ordinates and the spheroidal co-ordinates connected with an oblate spheroid. If we write

$$\rho = a\cos\xi\cosh\eta, \quad z = a\sin\xi\sinh\eta,$$

we have
$$\sin^2\xi = \frac{1 - \cos\psi}{\cosh\sigma - \cos\psi}, \quad \sinh^2\eta = \frac{1 + \cos\psi}{\cosh\sigma - \cos\psi},$$
$$\cos^2\xi = \frac{\cosh\sigma - 1}{\cosh\sigma - \cos\psi}, \quad \cosh^2\eta = \frac{\cosh\sigma + 1}{\cosh\sigma - \cos\psi},$$
$$\tan\xi = \pm\sin\tfrac{1}{2}\psi\operatorname{cosech}\tfrac{1}{2}\sigma, \quad \tanh\eta = \pm\cos\tfrac{1}{2}\psi\operatorname{sech}\tfrac{1}{2}\sigma.$$

With the aid of these formulae we may derive the formulae of Lipschitz for the Green's functions for the circular disc, and spherical bowl from the formulae of Hobson, or *vice versa*. It is necessary, of course, to pay careful attention to the determination of signs where ambiguities are produced by the use of the formulae of transformation.

§ **10·5**. *Spherical lens.* The potential of an insulated electrified conducting lens bounded by two intersecting spherical surfaces $\psi = \alpha - \beta$ and $\psi = -\beta$ has been found by H. M. Macdonald*.

If $n\alpha = \pi$, where n is a positive integer, the problem may be solved by the method of electrical images.

We commence by placing a charge E at the centre of the first sphere, its magnitude being chosen so as to produce a constant potential πU on this sphere. We next introduce a succession of image charges $E_1, E_2, \ldots E_m$, chosen so that for E and E_1 the second sphere is an equipotential for which $V = 0$, for E_1 and E_2 the first sphere is an equipotential for which $V = 0$, and so on. The second step is similar to the first except that we begin by placing a charge E' at the centre of the second sphere, its magnitude being

* *Proc. London Math. Soc.* (1), vol. XXVI, p. 156 (1895); vol. XXVIII, p. 214 (1896). References are given in these papers to some earlier work by W. D. Niven.

chosen so as to produce a constant potential πU on this sphere. We then introduce a succession of image charges E_1', E_2', ... such that E' and E_1' give a potential which is zero over the first sphere, E_1' and E_2' a potential which is zero over the second sphere, and so on.

To express the distance of an arbitrary point from one of the charges in toroidal co-ordinates we notice that

$$r^2 = \rho^2 + z^2 = a^2 \frac{s + \tau}{s - \tau}.$$

Hence

$$(s - \tau)(r^2 + a^2) = 2a^2 s, \quad (s - \tau)(r^2 - a^2) = 2a^2 \tau, \quad (s - \tau) z = a \sin \psi,$$

and so

$$2a^2 [s - \cos(\psi + 2\theta)] = 2(s - \tau)[(z \sin \theta + a \cos \theta)^2 + \rho^2 \sin^2 \theta]. \quad \text{......(A')}$$

Now the z co-ordinates of the different charges are

E	$a \cot(\alpha - \beta)$,	E'	$-a \cot \beta$,
E_1	$-a \cot \alpha$,	E_1'	$a \cot \alpha$,
E_2	$a \cot(2\alpha - \beta)$,	E_2'	$-a \cot(\alpha + \beta)$,
E_3	$-a \cot 2\alpha$,	E_3'	$a \cot 2\alpha$,
....		....	

and in each case the co-ordinate is expressed in the form $z = -a \cot \theta$. Also the equation (A') shows that the function

$$(s - \tau)^{\frac{1}{2}} [s - \cos(\psi + 2\theta)]^{-\frac{1}{2}} = F(\theta), \text{ say,}$$

is a potential which is proportional to the inverse distance from the point $z = -a \cot \theta$. Hence

$F(\beta - \alpha) - F(\alpha)$ is a potential which is zero on the sphere $\psi = -\beta$,

$F(\beta - 2\alpha) - F(\alpha)$ is a potential which is zero on the sphere $\psi = \alpha - \beta$,

and so on. Now $F(\beta - \alpha)$ is a potential which is equal to unity on the sphere $\psi = \alpha - \beta$, and $F(\beta)$ is a potential which is equal to unity on the sphere $\psi = -\beta$. Hence the potential

$$V = U\pi [F(\beta - \alpha) - F(\alpha) + F(\beta - 2\alpha) - F(2\alpha) + \ldots$$
$$+ F(\beta) - F(-\alpha) + F(\beta + \alpha) - F(-2\alpha) + \ldots]$$

is exactly one which is obtained by the method of images. When n is a positive integer we have for $m + k = n$

$$F(\beta - m\alpha) = F(\beta + k\alpha), \quad F(-m\alpha) = F(k\alpha),$$

consequently the charges will repeat themselves unless we take care to stop each series at the proper charge.

The difficulty of knowing when to stop may be avoided by considering the potential

$$V^* = U\pi \sum_1^n [F(\beta + m\alpha) - F(m\alpha)],$$

which is zero over each sphere. When we put $\psi = \alpha - \beta$ the term $F(\beta + m\alpha)$ is cancelled by the term $-F[(k+1)\alpha]$, and when $\psi = -\beta$ the term $F(\beta + m\alpha)$ is cancelled by $-F(k\alpha)$, and $F(\beta + n\alpha)$ by $-F(n\alpha)$.

All the terms in the series except $F(n\alpha)$ are zero at infinity while $F(n\alpha) \to 1$. We therefore write

$$V = V^* + U\pi,$$

and this formula will give us a potential which is zero at infinity and constant over the lens. It should be observed that all the charges $E, E_1, E_2, \ldots E', E_1', E_2', \ldots$ lie within the lens and so our potential is of a form suitable for the representation of the potential of the lens.

Let us now introduce the notation

$$f(\omega, \chi) = \frac{\sinh \omega}{\cosh \omega - \cos \chi}.$$

Since

$$\int_\sigma^\infty (\cosh \omega - \cosh \sigma)^{-\frac{1}{2}} f(\omega, \chi)\, d\omega = \pi (\cosh \sigma - \cos \chi)^{-\frac{1}{2}},$$

we may write

$$F(\theta) = \frac{1}{\pi} \int_\sigma^\infty \left(\frac{\cosh \sigma - \cos \psi}{\cosh \omega - \cosh \sigma}\right)^{\frac{1}{2}} f(\omega, \psi + 2\theta)\, d\omega.$$

Now $$\sum_{s=1}^{n} f(\omega, \psi + 2s\alpha) = nf(n\omega, n\psi),$$

hence we may write

$$V = U\left[\pi - n\int_\sigma^\infty d\omega \left(\frac{\cosh \sigma - \cos \psi}{\cosh \omega - \cosh \sigma}\right)^{\frac{1}{2}} \{f(n\omega, n\psi) - f(n\omega, n\psi + 2n\beta)\}\right].$$

This formula may now be extended to the case in which n is not a positive integer. We must first show that V is a solution of Laplace's equation and to do this we must show that if

$$g(\omega, \psi) \equiv f(n\omega, n\psi) - f(n\omega, n\psi + 2n\beta),$$

the integral

$$v = \int_\sigma^\infty d\omega (\cosh \omega - \cosh \sigma)^{-\frac{1}{2}} g(\omega, \psi)$$

is a solution of the differential equation

$$Dv \equiv \frac{\partial^2 v}{\partial \sigma^2} + \coth \sigma \frac{\partial v}{\partial \sigma} + \frac{1}{4} v + \frac{\partial^2 v}{\partial \psi^2} = 0.$$

To perform the necessary differentiations we first integrate by parts, this gives the equation

$$v = -2\int_\sigma^\infty d\omega (\cosh \omega - \cosh \sigma)^{\frac{1}{2}} \frac{\partial}{\partial \omega} (g \operatorname{cosech} \omega).$$

We may now differentiate with respect to σ and we find that

$$\frac{\partial v}{\partial \sigma} = \int_\sigma^\infty \sinh \sigma (\cosh \omega - \cosh \sigma)^{\frac{1}{2}} \frac{\partial}{\partial \omega} (g \operatorname{cosech} \omega)\, d\omega.$$

Repeating the process and making use of the fact that g is a solution of the equation

$$\frac{\partial^2 g}{\partial \omega^2} + \frac{\partial^2 g}{\partial \psi^2} = 0, \qquad \text{......(B)}$$

we find eventually that*

$$Dv = \int_\sigma^\infty d\omega \frac{d}{d\omega}\left[\frac{\partial g}{\partial \omega} \operatorname{cosech}^2 \omega\,(\sinh^2 \sigma - \sinh^2 \omega)\,(\cosh \omega - \cosh \sigma)^{-\frac{1}{2}}\right.$$

$$\left. + g \operatorname{cosech}^3 \omega\,(\cosh^2 \omega - \tfrac{3}{2} \sinh^2 \omega)\,(\cosh \omega \cosh \sigma)\,(\cosh \omega - \cosh \sigma)^{\frac{1}{2}}\right]$$

$$= 0.$$

A similar result may be obtained with any function g which satisfies the equation (B) and behaves in a suitable manner as $\omega \to \infty$. In particular, if

$$g(\omega, \psi) = e^{-m\omega} \cos m\psi,$$

we have

$$v = 2^{\frac{1}{2}} Q_{m-\frac{1}{2}}(\cosh \sigma) \cos m\psi.$$

The stream-function corresponding to V has been found by Greenhill†. With the notation

$$h(\omega, \chi) = \frac{\sin \chi}{\cosh \omega - \cos \chi},$$

it may be written in the form

$$S = naU \int_\sigma^\infty \left(\frac{\cosh \omega - \cosh \sigma}{\cosh \sigma - \cos \psi}\right)^{\frac{1}{2}} h(\omega, \psi)\,[f(n\omega, n\psi) - f(n\omega, n\psi + 2n\beta)]\,d\omega$$

$$- naU \int_\sigma^\infty \left(\frac{\cosh \sigma - \cos \psi}{\cosh \omega - \cosh \sigma}\right)^{\frac{1}{2}} f(\omega, \psi)\,[h(n\omega, n\psi) - h(n\omega, n\psi + 2n\beta)]\,d\omega.$$

If

$$S = (\cosh \sigma - \cos \psi)^{-\frac{1}{2}} R,$$

the differential equation for R is

$$\frac{\partial^2 R}{\partial \sigma^2} - \coth \sigma \frac{\partial R}{\partial \sigma} + \frac{\partial^2 R}{\partial \psi^2} + \frac{1}{4} R = 0,$$

while the relations between V and S are

$$af(\sigma, \psi)\frac{\partial V}{\partial \psi} = \frac{\partial S}{\partial \sigma}, \quad af(\sigma, \psi)\frac{\partial V}{\partial \sigma} = -\frac{\partial S}{\partial \psi}.$$

When n is a positive integer the expression for S may be verified by noticing that if $\chi = \psi + 2\theta$,

$$\int_\sigma^\infty d\omega \left(\frac{\cosh \omega - \cosh \sigma}{\cosh \sigma - \cos \psi}\right)^{\frac{1}{2}} h(\omega, \psi)\, f(\omega, \chi)$$

$$- \int_\sigma^\infty d\omega \left(\frac{\cosh \sigma - \cos \psi}{\cosh \omega - \cosh \sigma}\right)^{\frac{1}{2}} f(\omega, \psi)\, h(\omega, \chi)$$

$$= \pi \operatorname{cosec} \theta\,[\{\cos \theta \cosh \sigma - \cos(\psi + \theta)\}$$

$$\times (\cosh \sigma - \cos \psi)^{-\frac{1}{2}} (\cosh \sigma - \cos \chi)^{-\frac{1}{2}} - \cos \theta],$$

* The verification is performed in a slightly different manner by A. G. Greenhill, *Proc. Roy. Soc.* A, vol. XCVIII, p. 345 (1921); *Amer. Journ. Math.* vol. XXXIX, p. 335 (1917). The gravitational attraction of a solid homogeneous spherical lens had been worked out previously by G. W. Hill, *ibid.* vol. XXIX, p. 345 (1907) and A. G. Greenhill, vol. XXXIII, p. 373 (1911).

† *Loc. cit.*

while on the other hand

$$z + a \cot \theta = a \operatorname{cosec} \theta . \frac{\cos \theta \cosh \sigma - \cos (\psi + \theta)}{\cosh \sigma - \cos \psi},$$

$$[(z + a \cot \theta)^2 + \rho^2]^{\frac{1}{2}} = a \operatorname{cosec} \theta . \left(\frac{\cosh \sigma - \cos \chi}{\cosh \sigma - \cos \psi}\right)^{\frac{1}{2}}.$$

Hence the foregoing expression is simply proportional to the stream-function for a single point charge and so Greenhill's expression may be derived from the two series of electrical images.

An expression for the capacity of the lens is given by

$$C = \frac{S_1 - S_2}{2\pi U},$$

where S_1 and S_2 are the values of S at the vertices of the lens. At the vertex for which $\sigma = 0$, $\psi = -\beta$, $\psi + 2\beta = \beta$, we have

$$S_1 = 2anU . \sin n\beta . (1 - \cos \beta)^{\frac{1}{2}} \int_0^\infty \frac{(\cosh \omega + 1)^{\frac{1}{2}} \, d\omega}{(\cosh \omega - \cos \beta)(\cosh n\omega - \cos n\beta)},$$

while at the vertex for which $\sigma = 0$, $\psi = \alpha - \beta$, $\psi + 2\beta = \alpha + \beta$, we have

$$S_2 = 2anU . \sin n\beta . [1 - \cos (\alpha - \beta)]^{\frac{1}{2}} \times \int_0^\infty \frac{(\cosh \omega + 1)^{\frac{1}{2}} \, d\omega}{[\cosh \omega - \cos (\alpha - \beta)][\cosh n\omega + \cos n\beta]}.$$

The spherical bowl may be regarded as a particular case of the lens in which $n = \frac{1}{2}$, $\alpha = 2\pi$. The expression obtained for the potential V at a point P is

$$V = U (\pi - \gamma + b\gamma'/r_1),$$

where b is the radius of the sphere $\psi = -\beta$, and r_1 is the distance of P from its centre; furthermore

$$\sin \gamma = 2a/(R_1 + R_2), \quad \cos \gamma' = \operatorname{sech} \tfrac{1}{2}\sigma . \cos (\tfrac{1}{2}\psi + \beta),$$

where R_1 and R_2 are the greatest and least distances of the point P from the rim of the bowl.

The corresponding stream-function is

$$S = - aU (1 - \cos \psi)^{\frac{1}{2}} (\cosh \sigma - \cos \psi)^{-\frac{1}{2}} - b (\omega' \cos \theta_1 - \omega),$$

where θ_1 is the polar angle of the point P when the pole is the centre of the sphere $\psi = -\beta$, and the polar axis the line joining the centres of the two spheres. This result is due to J. R. Wilton* and A. G. Greenhill†.

§ **10·6**. *The Green's function for a wedge.* If we invert the lens from an arbitrary point we may obtain the Green's function for the inverse lens. If, however, the point is on the rim of the lens the surfaces of the lens will invert into planes and we shall obtain the Green's function for a wedge formed from two semi-infinite conducting planes which intersect at an

* *Mess. of Math.* (1914).

† *Loc. cit. ante*, p. 471.

angle π/n. This problem was solved by a direct method of A. Sommerfeld* and H. M. Macdonald†.

If (ρ, z, ϕ) are cylindrical co-ordinates, $\phi = 0$, $\phi = \pi/n$, the equations of the conducting planes, and if an electric charge is placed at the point (ρ', z', ϕ'), the potential V is given by the formula

$$\pi V = n\epsilon\,(2\rho\rho')^{-\frac{1}{2}} \int_\eta^\infty d\zeta\,(\cosh\zeta - \cosh\eta)^{-\frac{1}{2}}\,[f(n\zeta, n\phi - n\phi') - f(n\zeta, n\phi + n\phi')]$$

where

$$2\rho\rho' \cosh\eta = \rho^2 + \rho'^2 + (z - z')^2.$$

The potential can also be expanded in the form of a Fourier series

$$V = 4n\epsilon \sum_{m=1}^{\infty} F_{mn}(\eta) \sin(mn\phi) \sin(mn\phi'),$$

where

$$\pi . F_{mn}(\eta) = (2\rho\rho')^{-\frac{1}{2}} \int_\eta^\infty d\zeta\,(\cosh\zeta - \cosh\eta)^{-\frac{1}{2}}\, e^{-mn\zeta}$$

$$= (\rho\rho')^{-\frac{1}{2}}\, Q_{mn-\frac{1}{2}}(\cosh\eta).$$

An alternative expression for F_{mn} is

$$F_{mn}(\eta) = \int_0^\infty e^{-\kappa|z-z'|} J_{mn}(\kappa\rho)\, J_{mn}(\kappa\rho')\, d\kappa.$$

This may be deduced from a well-known expression for the Q-function‡ or may be obtained directly.

When $n = m + \frac{1}{2}$, where m is a positive integer, the potential can be expressed as a finite sum, for then, if $(2m + 1)\,\alpha = 2\pi$,

$$2\pi V = \epsilon\,(2\rho\rho')^{-\frac{1}{2}} \sum_{s=0}^{2m} \int_\eta^\infty d\zeta\,(\cosh\zeta - \cosh\eta)^{-\frac{1}{2}}\,[f(\tfrac{1}{2}\zeta, \tfrac{1}{2}\phi - \tfrac{1}{2}\phi' + s\alpha)$$
$$- f(\tfrac{1}{2}\zeta, \tfrac{1}{2}\phi + \tfrac{1}{2}\phi' + s\alpha)]$$
$$= 4\epsilon\,(2\rho\rho')^{-\frac{1}{2}} \sum_{s=0}^{2m} \left\{(\eta, \phi - \phi' + 2s\alpha) \tan^{-1} \frac{(\frac{1}{2}\eta, \frac{1}{2}\phi - \frac{1}{2}\phi' + s\alpha)}{[\frac{1}{2}\eta, \frac{1}{2}\phi - \frac{1}{2}\phi' + s\alpha]}\right.$$
$$\left. - (\eta, \phi + \phi' + 2s\alpha) \tan^{-1} \frac{(\frac{1}{2}\eta, \frac{1}{2}\phi + \frac{1}{2}\phi' + s\alpha)}{[\frac{1}{2}\eta, \frac{1}{2}\phi + \frac{1}{2}\phi' + s\alpha]}\right\},$$

where for brevity we have used the notation

$$(a, b) = (\cosh a - \cos b)^{-\frac{1}{2}},$$
$$[a, b] = [\cosh a + \cos b]^{-\frac{1}{2}}.$$

§ **10·7**. *The Green's function for a semi-infinite plane.* When $m = 0$ this expression gives the potential when a point charge ϵ is in the presence of

* *Proc. London Math. Soc.* (1), vol. XXVIII, p. 395 (1897).

† *Ibid.* vol. XXVI, p. 156 (1895).

‡ H. M. Macdonald, *Proc. London Math. Soc.* (2), vol. VII, p. 142 (1909).

a conductor in the form of an infinite half plane. Since $\alpha = 2\pi$ the potential is simply

$$V = \frac{2\epsilon}{\pi}(2\rho\rho')^{-\frac{1}{2}}\left\{(\eta, \phi - \phi')\tan^{-1}\frac{(\frac{1}{2}\eta, \frac{1}{2}\phi - \frac{1}{2}\phi')}{[\frac{1}{2}\eta, \frac{1}{2}\phi - \frac{1}{2}\phi']} - (\eta, \phi + \phi')\tan^{-1}\frac{(\frac{1}{2}\eta, \frac{1}{2}\phi + \frac{1}{2}\phi')}{[\frac{1}{2}\eta, \frac{1}{2}\phi + \frac{1}{2}\phi']}\right\}.$$

§ **10·8**. *Circular disc in any field of force.* H. M. Macdonald* has considered the case in which the potential of the inducing system can be expressed in the form

$$\Sigma\,\Sigma\, A_{n\nu} J_n(\nu\rho)\cos(n\phi + \alpha_\nu),$$

where $A_{n\nu}$, n, ν and α_ν are constants. The solution of the simplified problem in which the series reduces to a single term was given by Gallop for the case $n = 0$, and by Basset for the case $n = 1$. We shall commence the discussion of the general case by considering the formulae

$$(s - \tau)^{\frac{1}{2}}\cos\tfrac{1}{2}\psi\,.\tanh^m(\sigma/2) = a^{\frac{3}{2}}\sqrt{\pi}\,.\int_0^\infty e^{-\kappa z} J_m(\kappa\rho) J_{m-\frac{1}{2}}(\kappa a)\,\kappa^{\frac{1}{2}}\,d\kappa$$

$$m > -\tfrac{1}{2}, \quad z > 0,$$

$$(s - \tau)^{\frac{1}{2}}\sin\tfrac{1}{2}\psi\,.\tanh^m(\sigma/2) = a^{\frac{3}{2}}\sqrt{\pi}\,.\int_0^\infty e^{-\kappa z} J_m(\kappa\rho) J_{m+\frac{1}{2}}(\kappa a)\,\kappa^{\frac{1}{2}}\,d\kappa$$

$$m > -1, \quad z > 0.$$

Each expression multiplied by $\cos m\phi$ represents a solution of Laplace's equation and so each expression divided by ρ^m is a solution of the equation

$$\frac{\partial^2 u}{\partial\rho^2} + \frac{2m+1}{\rho}\frac{\partial u}{\partial\rho} + \frac{\partial^2 u}{\partial z^2} = 0.$$

Since the solution of this equation is determined uniquely by its value on the axis of z it is sufficient to verify the truth of the formulae by making $\rho \to 0$ after dividing by ρ^m. The integrals are uniformly convergent in the neighbourhood of $\rho = 0$ after this operation and so we have merely to verify the equations

$$\Gamma(m+1)\,.\cos\tfrac{1}{2}\psi\,.(1 - \cos\psi)^{m+\frac{1}{2}} = a^{\frac{3}{2}+m}\sqrt{\pi}\int_0^\infty e^{-\kappa a\cot\frac{1}{2}\psi} J_{m-\frac{1}{2}}(\kappa a)\,\kappa^{m+\frac{1}{2}}\,.d\kappa,$$

$$\Gamma(m+1)\,.\sin\tfrac{1}{2}\psi\,.(1 - \cos\psi)^{m+\frac{1}{2}} = a^{\frac{3}{2}+m}\sqrt{\pi}\int_0^\infty e^{-\kappa a\cot\frac{1}{2}\psi} J_{m+\frac{1}{2}}(\kappa a)\,\kappa^{m+\frac{1}{2}}\,.d\kappa.$$

These are easily seen to be true on account of Sonine's formulae

$$\sqrt{\pi}\int_0^\infty e^{-xt} J_{m-\frac{1}{2}}(t)\,t^{m+\frac{1}{2}}\,dt = \Gamma(m+1)\,2^{m+\frac{1}{2}}\,x\,(1 + x^2)^{-m-1} \quad (m > -\tfrac{1}{2}),$$

$$\sqrt{\pi}\int_0^\infty e^{-xt} J_{m+\frac{1}{2}}(t)\,t^{m+\frac{1}{2}}\,dt = \Gamma(m+1)\,2^{m+\frac{1}{2}}\,(1 + x^2)^{-m-1} \quad (m > -1).$$

* *Proc. London Math. Soc.* (1), vol. XXVI, p. 257 (1895).

It may be observed also that the value of the second integral may be deduced from that of the first with the aid of the substitution

$$(\cosh\sigma - \cos\psi)(\cosh\sigma - \cos\chi) = \sinh^2\sigma,$$

which gives
$$a = \frac{\rho\sinh\sigma}{\cosh\sigma - \cos\chi}, \quad z = \frac{\rho\sin\chi}{\cosh\sigma - \cos\chi},$$

$$\cos\tfrac{1}{2}\psi\,(\cosh\sigma - \cos\psi)^{\frac{1}{2}} = \sin\tfrac{1}{2}\chi\,.\sinh\sigma\,(\cosh\sigma + 1)^{\frac{1}{2}}(\cosh\sigma - \cos\chi)^{-1},$$
$$\sin\tfrac{1}{2}\psi\,(\cosh\sigma - \cos\psi)^{\frac{1}{2}} = \cos\tfrac{1}{2}\chi\,.\sinh\sigma\,(\cosh\sigma - 1)^{\frac{1}{2}}(\cosh\sigma - \cos\chi)^{-1}.$$

The value of the first integral for $\rho > a$ can now be deduced from the value of the second integral for $\rho < a$ by interchanging ρ and a and writing $m - \frac{1}{2}$ instead of m.

Let us now consider the potential

$$V = (s - \tau)^{\frac{1}{2}}\cos\tfrac{1}{2}\psi\,.\tanh^m\tfrac{1}{2}\sigma\,.\cos m\phi.$$

When $z = 0$ and $\rho^2 < a^2$ we have $\psi = \pi$, consequently

$$V = 0, \quad a\frac{\partial V}{\partial z} = 2^{\frac{3}{2}}\cosh^3(\tfrac{1}{2}\sigma)\,.\tanh^m(\tfrac{1}{2}\sigma)\,.\cos m\phi.$$

When $z = 0$ and $\rho^2 > a^2$ we have $\psi = 0$, consequently

$$V = 2^{\frac{1}{2}}\cosh\tfrac{1}{2}\sigma\,.\tanh^m(\tfrac{1}{2}\sigma)\,.\cos m\phi, \quad \frac{\partial V}{\partial z} = 0.$$

Next, consider the potential $V = W_m\cos m\phi$, where

$$W_m = \int_0^\infty e^{-\kappa z}\,d\kappa\int_0^c J_{m-\frac{1}{2}}(\nu a)\,J_{m-\frac{1}{2}}(\kappa a)\,J_m(\kappa\rho)\,(\kappa\nu)^{\frac{1}{2}}\,a\,da.$$

Integrating under the integral sign the first formula tells us that when $z = 0$ and $\rho < c$,

$$\begin{aligned} W_m &= \sqrt{(2\nu/\pi)}\int_0^\rho J_{m-\frac{1}{2}}(\nu a)\,\rho^{-m}(\rho^2 - a^2)^{-\frac{1}{2}}\,a^{m+\frac{1}{2}}\,da \\ &= \sqrt{(2\nu\rho/\pi)}\int_0^{\frac{1}{2}\pi} J_{m-\frac{1}{2}}(\nu\rho\sin\theta)\,.\sin^{m+\frac{1}{2}}\theta\,.d\theta \\ &= J_m(\nu\rho). \end{aligned}$$

Again, when $z = 0$ and $\rho < c$,

$$\begin{aligned} \frac{\partial W_m}{\partial z} &= -\int_0^\infty d\kappa\,\sqrt{(\kappa\nu)}\int_0^c J_{m-\frac{1}{2}}(\nu a)\,J_{m-\frac{1}{2}}(\kappa a)\,J_m(\kappa\rho)\,\kappa a\,da \\ &= -\nu J_m(\rho\nu) + \int_0^\infty d\kappa\int_c^\infty \sqrt{(\kappa\nu)}\,J_{m-\frac{1}{2}}(\nu a)\,J_{m-\frac{1}{2}}(\kappa a)\,J_m(\kappa\rho)\,\kappa a\,da \\ &= -\nu J_m(\rho\nu) + \sqrt{(2\nu/\pi)}\int_c^\infty \rho^m(a^2 - \rho^2)^{-\frac{3}{2}}\,J_{m-\frac{1}{2}}(\nu a)\,a^{\frac{3}{2}-m}\,da. \end{aligned}$$

When $z = 0$ and $\rho > c$, we have on the other hand

$$W_m = \sqrt{(2\nu\rho/\pi)}\int_0^{\sin^{-1}(c/\rho)} J_{m-\frac{1}{2}}(\nu\rho\sin\theta)\,.\sin^{m+\frac{1}{2}}\theta\,.d\theta,$$

$$\frac{\partial W_m}{\partial z} = 0.$$

CHAPTER XI

DIFFRACTION PROBLEMS

§ **11·1**. *Diffraction by a half plane.* The problem of diffraction of plane waves by a straight edge parallel to the wave fronts or surfaces of constant phase was shown by Sommerfeld* to be one which could be treated successfully by the methods of exact analysis. The analysis has subsequently been expressed in different forms, and various attempts have been made to make the derivation of the final formulae seem natural and straightforward. It must be confessed, however, that the discovery of any of these methods requires remarkable insight and a very thorough knowledge of the different types of solutions of the wave-equation, and the solution of the diffraction problem must be regarded as a triumph of mathematical ingenuity and experiment.

The method which will be followed here is one which was devised when the solution was known; it has the advantage that it presents the solution in a particularly simple form.

If $F(x, y, z, t)$ is a solution of the wave-equation $\Box^2 F = 0$ it may be easily verified that the definite integral

$$V = \int_0^{\pi/2} F\left[x\tan^2\alpha,\ y\tan^2\alpha,\ z\tan^2\alpha,\ t - \frac{r}{c}\sec^2\alpha\right]\tan\alpha\, d\alpha$$

is a solution of the wave-equation provided that

$$\frac{\partial}{\partial t}F\left[x\tan^2\alpha,\ y\tan^2\alpha,\ z\tan^2\alpha,\ t - \frac{r}{c}\sec^2\alpha\right]\tan^2\alpha$$

has the same value when $\alpha = \frac{\pi}{2}$ as it has when $\alpha = 0$, and that the function F behaves in a suitable manner for values of its arguments which are either zero or infinite.

Again, it is easily verified that the function

$$F(x, y, z, t) = (x + iy)^{-\frac{1}{2}}\, G(\rho, z, t)$$

is a solution of the wave-equation if $G(x, y, t)$ is a solution of the two-dimensional wave-equation

$$\frac{\partial^2 G}{\partial x^2} + \frac{\partial^2 G}{\partial y^2} = \frac{1}{c^2}\frac{\partial^2 G}{\partial t^2}. \qquad \text{......(A)}$$

* *Math. Ann.* Bd. XLVII, S. 317 (1896); *Zeits. für Math. und Phys.* Bd. XLVI, S. 11 (1901). See also H. S. Carslaw, *Proc. London Math. Soc.* (1), vol. XXX, p. 121 (1899); *Proc. Edinburgh Math. Soc.* vol. XIX, p. 71 (1901). An approximate solution was given by H. Poincaré, *Acta Mathematica*, vol. XVI, p. 297 (1892); vol. XX, p. 313 (1896).

When the first theorem can be applied to this function, it tells us that the expression

$$(x + iy)^{-\frac{1}{2}} \int_0^{\pi/2} G\left[\rho \tan^2 \alpha, z \tan^2 \alpha, t - \frac{r}{c} \sec^2 \alpha\right] d\alpha$$

represents a solution of the wave-equation, and since it is of the form †

$$(x + iy)^{-\frac{1}{2}} H(\rho, z, t)$$

the natural inference is that the integral

$$\int_0^{\pi/2} G\left[x \tan^2 \alpha, y \tan^2 \alpha, t - \frac{\rho}{c} \sec^2 \alpha\right] d\alpha$$

must be a solution of (A). The foregoing method of deriving one wave-function from another seems to be applicable, then, to wave-functions in a space of any number of dimensions.

Let us now consider a diffraction problem in which the primary waves are specified in some way by means of the wave-function

$$\phi = \phi_0 = F(ct + y \sin \theta_0 + x \cos \theta_0),$$

where ϕ may be the velocity potential of waves of sound or one of the electromagnetic vectors if we are dealing with waves of light. If these waves are reflected completely at the plane $y = 0$ there is a reflected wave specified by the wave-function

$$\phi = \phi_1 = F(ct + x \cos \theta_0 - y \sin \theta_0),$$

and the complete wave-function is $\phi = \phi_0 + \phi_1$ when the boundary condition is $\frac{\partial \phi}{\partial y} = 0$ for $y = 0$, but is $\phi = \phi_0 - \phi_1$ when the boundary condition is $\phi = 0$ for $y = 0$.

When the primary waves are diffracted by a screen, $y = 0$, $x > 0$, which occupies only half of the plane $y = 0$, it is necessary to divide the xy-plane up into three regions S_1, S_2, S_3, whose boundaries are as follows:

$$S_1, \quad y = 0, \quad x \sin \theta_0 + y \cos \theta_0 = 0;$$
$$S_2, \quad x \sin \theta_0 \pm y \cos \theta_0 = 0;$$
$$S_3, \quad y = 0, \quad x \sin \theta_0 - y \cos \theta_0 = 0.$$

The limits of S_1 are thus the screen and the geometrical limit of the reflected wave, the limits of S_2 are the geometrical limits of the reflected wave and the shadow, the limits of S_3 are the screen and the geometrical boundary of the shadow. In each case the geometrical limit is obtained by the methods of geometrical optics.

To take into consideration the phenomena of diffraction we associate with ϕ_0 a wave-function V_0 defined by the equation

$$V_0 = \frac{1}{\pi} \int_0^{\pi/2} F\left[ct + x \cos \theta_0 + y \sin \theta_0 - (\rho + x \cos \theta_0 + y \sin \theta_0) \sec^2 \alpha\right] d\alpha,$$

† The method is an extension of one used by F. J. W. Whipple, *Phil. Mag.*, (6), vol. 36, p. 420 (1918) and by W. G. Bickley, *ibid.*, (6) vol. 39, p. 668 (1920).

and we associate with ϕ_1 a wave-function V_1 defined by the equation

$$V_1 = \frac{1}{\pi}\int_0^{\pi/2} F\,[ct + x\cos\theta_0 - y\sin\theta_0 - (\rho + x\cos\theta_0 - y\sin\theta_0)\sec^2\alpha]\,d\alpha.$$

We shall try to prove that the conditions imposed by the two boundary problems may be satisfied by using a complete wave-function ϕ which is defined in the different regions as follows:

$$\begin{aligned} \phi &= \phi_0 + \phi_1 - V_0 - V_1 &&\text{in } S_1,\\ \phi &= \phi_0 - V_0 + V_1 &&\text{in } S_2,\\ \phi &= V_0 + V_1 &&\text{in } S_3. \end{aligned}$$

We have to show that this function ϕ and its first derivatives are continuous as a point passes from S_1 to S_2, and as a point passes from S_2 to S_3. Now, when $x = -\rho\cos\theta_0$, $y = -\rho\sin\theta_0$, we have

$$V_0 = \tfrac{1}{2}\phi_0, \quad \frac{\partial V_0}{\partial x} = \frac{1}{2}\frac{\partial\phi_0}{\partial x}, \quad \frac{\partial V_0}{\partial y} = \frac{1}{2}\frac{\partial\phi_0}{\partial y},$$

and when $x = -\rho\cos\theta_0$, $y = \rho\sin\theta_0$, we have

$$V_1 = \tfrac{1}{2}\phi_1, \quad \frac{\partial V_1}{\partial x} = \frac{1}{2}\frac{\partial\phi_1}{\partial x}, \quad \frac{\partial V_1}{\partial y} = \frac{1}{2}\frac{\partial\phi_1}{\partial y};$$

hence the requirement of continuity is seen to be satisfied. The boundary condition is also satisfied on both faces of the screen and ϕ is continuous (D, 1) over the whole plane, when the screen is regarded as a cut, hence it will give the solution of the boundary problem if it has the correct form at infinity.

This is certainly the case if $F(s)$ is zero when $|s|$ is greater than some fixed number, that is, if the incident wave is limited at the front and rear, for then the form of ϕ at infinity is precisely that given by the methods of geometrical optics, V_0 and V_1 being zero.

The interesting case of periodic waves may be discussed by writing $F(s) = e^{iks}$. The expressions for ϕ may then be reduced to those of Sommerfeld by writing

$$\frac{1}{\pi}\int_0^{\pi/2} e^{-iz^2\sec^2\alpha}\,d\alpha = \left(\frac{i}{\pi}\right)^{\frac{1}{2}}\int_{-\infty}^{-|z|} e^{-i\tau^2}\,d\tau.$$

This identity may be derived by writing the right-hand side in the form

$$\frac{2i}{\pi}\int_{-\infty}^{-|z|} d\tau\int_0^{\infty} e^{-i(\sigma^2+\tau^2)}\,d\sigma,$$

a transformation which is possible on account of the well-known equation

$$\int_0^{\infty} e^{-i\sigma^2}\,d\sigma = \frac{1}{2}\left(\frac{\pi}{i}\right)^{\frac{1}{2}}.$$

The repeated integral is now replaced by a double integral and transformed by means of the substitution $\tau = z\cos\alpha$, $\sigma = z\sin\alpha$. Integrating with respect to r we obtain

$$\frac{1}{\pi}\int_0^{\pi/2} e^{-iz^2\sec^2\alpha}\,d\alpha.$$

This supplies the outlines of a proof.

It may be remarked that in the present case it cannot be proved by direct differentiation that the integrals occurring in the solution are wave-functions. A transformation is therefore advantageous.

§ 11·2. The various steps sketched above may be justified without much difficulty, but it is more interesting to examine the steps by which Sommerfeld was led to his famous solution of the problem*. Let

$$f_n(z) = \log\,[e^{iz/n} - e^{i\theta_0/n}],$$

where n is any positive integer, then, if C is a small circle enclosing the point $z = \theta_0$ in the z-plane and no other singularity of the integrand, the contour integral

$$u_1(\rho, \theta, \theta_0) = \frac{1}{2\pi i}\int e^{ik\rho\cos(z-\theta)} f_1'(z)\,dz$$

represents the function

$$e^{ik\rho\cos(\theta-\theta_0)},$$

which is known to be a solution of the equation $\nabla^2 u + k^2 u = 0$. Our representation of this solution still holds when the circle C is replaced by a path $C(\theta)$ consisting of two branches which go to infinity within the shaded strips of breadth π in which $e^{ik\rho\cos(\theta-\theta_0)}$ has a negative real part. These two branches may, in fact, be joined by dotted lines, as indicated in the figure, so as to form a closed contour which can be deformed into the circle C without passing over any singularity of the integrand. The integrals along these dotted lines, moreover, cancel on account of the periodicity of the integrand and so the integral round C is equal to the integral round the path $C(\theta)$ consisting of the two branches.

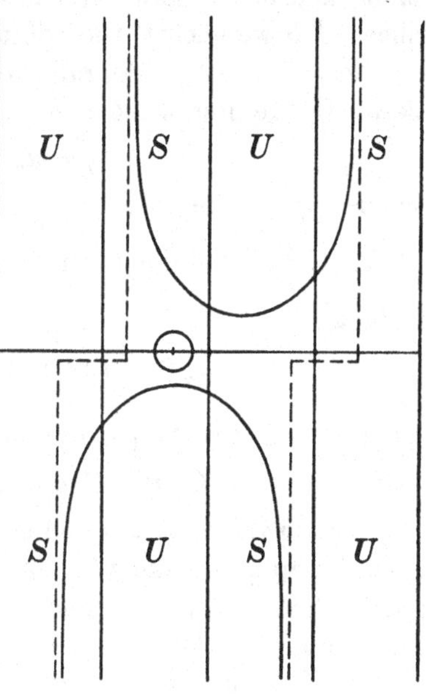

Fig. 29.

S, shaded region; *U*, unshaded region.

The function

$$u_n(\rho, \theta, \theta_0) = \frac{1}{2\pi i}\int_{C(\theta)} e^{ik\rho\cos(z-\theta)} f_n'(z)\,dz$$

is also a solution of $\nabla^2 u + k^2 u = 0$, since the necessary differentiations can

* Our presentation in §§ 11·2–11·4 follows closely that given by Wolfsohn in *Handbuch der Physik*, Bd. xx, pp. 266–277.

be performed under the integral sign. Its period, however, is $2n\pi$, and so to make it uniform we must consider a Riemann surface R with n sheets. When θ increases by $2n\pi$ the path $C(\theta)$ is displaced a distance $2n\pi$ to the right but, since the integrand has the period $2n\pi$, it runs over the same set of values as before.

This function u_n has the following properties:

If $|\theta - \theta_0| < \alpha < \pi$, $u_n \to u_1$ as $\rho \to \infty$.

If $\pi < \beta < |\theta - \theta_0|$, $u_n \to 0$ as $\rho \to \infty$.

Indeed, if $|\theta - \theta_0| < \alpha < \pi$, we may deform the path of integration so that each of the new branches runs in shaded regions only. In the first sheet of R we enclose the point θ_0 by the circle C as before.

As $\rho \to \infty$ the integral over the path in the shaded region tends to zero. In the first sheet of R the integral round C gives u_1, therefore

$$u_1 = u_n^{(1)} + u_n^{(2)} + \dots u_n^{(n)},$$

where $u_n^{(s)}$ is the value of u_n at the point lying over ρ, θ in the sth sheet of R. If $\int_{s+1}$ denotes the integral over the contour $C(\theta + 2s\pi)$, we have in fact

$$u_n^{(1)} + u_n^{(2)} + \dots u_n^{(n)} = \int_1 + \int_2 + \dots \int_n = \int_\Gamma,$$

where Γ denotes the path made up of the branches

$$C(\theta), \quad C(\theta + 2\pi), \dots C[\theta + 2(n-1)\pi].$$

Converting this into a closed contour by broken lines as before and noting that the integrals along the broken lines cancel, we have a contour which can be deformed into C, and so the result follows. The proof for the other case follows similar lines except that now the contour may be reduced to a point by a suitable deformation.

We now put $n = 2$ and write

$$\frac{\partial}{\partial \rho}\left\{\frac{u_2^{(1)} - u_2^{(2)}}{u_1}\right\} = -\frac{k}{2\pi} e^{-2ik\rho\cos^2\frac{\theta-\theta_0}{2}} \int_{C(\theta)} \sin\frac{z + \theta_0 - 2\theta}{2} . e^{2ik\rho\cos^2\frac{z-\theta}{2}} dz.$$

The path of integration can be deformed into the lines

$$z = \theta - \pi + i\tau \text{ and } z = \theta + \pi + i\tau \qquad -\infty < \tau < \infty,$$

after which the integrations can be performed, giving

$$\frac{\partial}{\partial \rho}\{\ \} = \sqrt{(2ik/\pi\rho)} . \cos\frac{\theta - \theta_0}{2} e^{-2ik\rho\cos^2\frac{\theta-\theta_0}{2}} = \sigma\frac{\partial}{\partial \rho} G(T),$$

$$\{\ \} \equiv \frac{u_2^{(1)} - u_2^{(2)}}{u_1} = \sigma G(\tau),$$

where $\sigma = \dfrac{2}{\sqrt{-i\pi}}$, $G(\tau) = \displaystyle\int_0^T e^{-i\tau^2} d\tau$, $T = \sqrt{2k\rho}\cos\dfrac{\theta - \theta_0}{2}$.

Since $u_2^{(1)} + u_2^{(2)} = u_1$, we may write

$$u_2^{(1)} = \tfrac{1}{2}u_1 \{1 + \sigma G(\tau)\} \qquad 0 < \theta < 2\pi,$$
$$u_2^{(2)} = \tfrac{1}{2}u_1 \{1 + \sigma G(-\tau)\} \quad -2\pi < \theta < 0.$$

Also, since $\sigma G(-\infty) = -1$, it is easily seen that

$$u_2^{(1)} = u_2, \quad 0 < \theta < 2\pi; \qquad u_2^{(2)} = u_2, \quad -2\pi < \theta < 0,$$

where
$$u_2 = u_1 . e^{i\pi/4} . \pi^{-\frac{1}{2}} \int_{-\infty}^{T} e^{-i\tau^2} \, d\tau.$$

This is Sommerfeld's expression. When this function u_2 is multiplied by e^{ikct} a wave-function $v = u_2 (\rho, \theta, \theta_0) \, e^{ikct}$ is obtained. If

$$W[\rho, \theta, t, \theta_0, \tau] = k[c(t - \tau) + \rho \cos(\theta - \theta_0)], \quad V = e^{ikc\tau} W^{-\frac{1}{2}},$$

this wave-function can be written in the form

$$v = -\tfrac{1}{2}ck (i/\pi)^{\frac{1}{2}} \int_{-\infty}^{\tau_1} V d\tau, \quad \cos \tfrac{1}{2} (\theta - \theta_0) < 0$$

$$= -\tfrac{1}{2}ck (i/\pi)^{\frac{1}{2}} \left\{ \int_{-\infty}^{\tau_1} V d\tau + 2 \int_{\tau_1}^{\tau_2} V d\tau \right\}, \quad \cos \tfrac{1}{2} (\theta - \theta_0) > 0,$$

$$\tau_1 = t - \rho/c, \quad \tau_2 = t + \frac{\rho}{c} \cos(\theta - \theta_0),$$

where the integrand is in each case a wave-function for all values of the parameter. It should be noticed that τ_2 is a value of τ for which

$$W[\rho, \theta, t, \theta_0, \tau] = 0.$$

§ **11·3**. The chief advantage of the last expression for v is that it suggests the form of the solution for the case in which the waves originate from a line source (ρ_0, θ_0) which may be either at rest or in motion.

Let us take as the potential of our line source the function

$$\phi_0 = \int_{-\infty}^{\tau'} \frac{e^{ick\tau} \, d\tau}{[c^2 (t - \tau)^2 - \rho^2 - \rho_0^2 + 2\rho\rho_0 \cos(\theta - \theta_0)]^{\frac{1}{2}}} = \int_{-\infty}^{\tau'} F(\tau) \, d\tau, \text{ say,}$$

where τ' is a value of τ less than t, for which the denominator of the integrand is zero. When the line source is in motion ρ_0 and θ_0 are functions of τ, we suppose them to be functions such that the velocity of the line source is always less than c. For the reflected wave we write

$$\phi_1 = \int_{-\infty}^{\tau''} \frac{e^{ikc\tau} \, d\tau}{[c^2 (t - \tau)^2 - \rho^2 - \rho_0^2 + 2\rho\rho_0 \cos(\theta + \theta_0)]^{\frac{1}{2}}} = \int_{-\infty}^{\tau''} G(\tau) \, d\tau, \text{ say,}$$

where τ'' is a value of τ less than t, for which the denominator of the integrand vanishes. These integrals are generally wave-functions.

Now let τ_1 be a value of τ defined by the equation*

$$\tau_1 = t - \frac{1}{c} [\rho + \rho_0 (\tau_1)],$$

* Since $\rho^2 + \rho_0^2 - 2\rho\rho_0 \cos(\theta \pm \theta_0) < (\rho + \rho_0)^2$, it is evident that $(t - \tau_1)^2$ is greater than either $(t - \tau')^2$ or $(t - \tau'')^2$, consequently $\tau_1 < \tau'$ and $\tau_1 < \tau''$.

then, if the boundary condition at the surface of the screen is $\frac{\partial \phi}{\partial y} = 0$, the solution of the diffraction problem may be expressed in the form

$$\phi = \tfrac{1}{2}\int_{-\infty}^{\tau_1} F(\tau)\,d\tau + \int_{\tau_1}^{\tau'} F(\tau)\,d\tau + \tfrac{1}{2}\int_{-\infty}^{\tau_1} G(\tau)\,d\tau + \int_{\tau_1}^{\tau''} G(\tau)\,d\tau \text{ in } S_1,$$

$$\phi = \tfrac{1}{2}\int_{-\infty}^{\tau_1} F(\tau)\,d\tau + \int_{\tau_1}^{\tau'} F(\tau)\,d\tau + \tfrac{1}{2}\int_{-\infty}^{\tau_1} G(\tau)\,d\tau \text{ in } S_2,$$

$$\phi = \tfrac{1}{2}\int_{-\infty}^{\tau_1} F(\tau)\,d\tau + \tfrac{1}{2}\int_{-\infty}^{\tau_1} G(\tau)\,d\tau \text{ in } S_3.$$

The integrals with the limit τ_1 being also wave-functions, as may be verified by differentiation.

The space S_1 is bounded by the screen and the limiting surface of the geometrical shadow for the image of the source, the space S_2 is bounded by the limiting surfaces of the geometrical shadows for the source and its image, while S_3 is the space bounded by the screen and the limiting surface of the geometrical shadow for the source.

The boundary of the geometrical shadow for the source is defined by the equation $\tau_1 = \tau'$, while the boundary of the geometrical shadow for the image of the source is defined by the equation $\tau_1 = \tau''$. It is evident, then, that ϕ is continuous at the boundaries of the shadows. That the first derivatives of ϕ are also continuous may be seen by transforming the integrals, in which $F(\tau)$ is the integrand, by the substitution

$$c^2 (t - \tau)^2 = [\rho^2 + \rho_0{}^2 - 2\rho\rho_0 \cos(\theta - \theta_0)] \cosh^2 u,$$

and the integrals in which $G(\tau)$ is the integrand, by the substitution

$$c^2 (t - \tau)^2 = [\rho^2 + \rho_0{}^2 - 2\rho\rho_0 \cos(\theta + \theta_0)] \cosh^2 v.$$

In the case when ρ_0 and θ_0 are constants and the line source is consequently at rest, these substitutions transform the expression for ϕ to Macdonald's form*

$$\phi = \tfrac{1}{2} e^{ikct} \left[\int_{-\infty}^{u_0} e^{-ikR\cosh u}\,du + \int_{-\infty}^{v_0} e^{-ikR'\cosh v}\,dv \right],$$

where

$$R \sinh u_0 = 2(\rho\rho_0)^{\frac{1}{2}} \cos \tfrac{1}{2}(\theta - \theta_0), \quad R' \sinh v_0 = 2(\rho\rho_0)^{\frac{1}{2}} \cos \tfrac{1}{2}(\theta + \theta_0),$$

$$R^2 = \rho^2 + \rho_0{}^2 - 2\rho\rho_0 \cos(\theta - \theta_0), \quad R'^2 = \rho^2 + \rho_0{}^2 - 2\rho\rho_0 \cos(\theta + \theta_0).$$

It should be noticed that when we transform from u to τ in the first integral the path of integration runs from $\tau = -\infty$ to $\tau = \tau_1$ if u_0 is negative, but when u_0 is positive the path runs from $-\infty$ up to a maximum value $\tau = \tau'$ and then back to τ_1. A similar phenomenon occurs in the transformation of the second integral. This accounts for the existence of three different expressions for ϕ when τ is taken as the variable of integration.

* *Proc. London Math. Soc.* (2), vol. XIV, p. 410 (1915).

Let us next consider the case in which the rectilinear source moves with constant velocity along a path which does not intersect the screen and is such that the co-ordinates of a point at time τ are

$$x_0(\tau) = \xi + a\tau, \quad y_0(\tau) = \eta + b\tau,$$

where ξ and η are constants and $a^2 + b^2 < c^2$. In this case we may write

$$(c^2 - a^2 - b^2)\,\tau = c^2 t - a(x - \xi) - b(y - \eta) - S\cosh u,$$

$$(c^2 - a^2 - b^2)\,\tau = c^2 t - a(x - \xi) + b(y + \eta) - S'\cosh v,$$

where

$$S = [\{c^2 t - a(x - \xi) - b(y - \eta)\}^2 - \{c^2 - a^2 - b^2\} \times \{c^2 t^2 - (x - \xi)^2 - (y - \eta)^2\}]^{\frac{1}{2}},$$

$$S' = [\{c^2 t - a(x - \xi) + b(y + \eta)\}^2 - \{c^2 - a^2 - b^2\} \times \{c^2 t^2 - (x - \xi)^2 - (y + \eta)^2\}]^{\frac{1}{2}};$$

then

$$\phi = \tfrac{1}{2} e^{\frac{ikc^3 t}{c^2 - a^2 - b^2}} \left[\int_{-\infty}^{u_0} e^{-ikc\frac{a(x-\xi) + b(y-\eta) + S\cosh u}{c^2 - a^2 - b^2}}\, du + \int_{-\infty}^{v_0} e^{-ikc\frac{a(x-\xi) - b(y+\eta) + S'\cosh v}{c^2 - a^2 - b^2}}\, dv\right],$$

where

$$S\cosh u_0 = \rho c - ax - by + [\{c(ct - \rho) + a\xi + b\eta\}^2 - \{(ct - \rho)^2 - \xi^2 - \eta^2\}\{c^2 - a^2 - b^2\}]^{\frac{1}{2}},$$

$$S'\cosh v_0 = \rho c - ax + by + [\{c(ct - \rho) + a\xi + b\eta\}^2 - \{(ct - \rho)^2 - \xi^2 - \eta^2\}\{c^2 - a^2 - b^2\}]^{\frac{1}{2}}.$$

§ **11·4**. *Discussion of Sommerfeld's solution.* For $T < 0$ we obtain by partial integration

$$\int_{-\infty}^{T} e^{-i\tau^2}\, d\tau = -\frac{1}{2iT} e^{-iT^2} - \int_{-\infty}^{T} \frac{d\tau}{2i\tau^2} e^{-i\tau^2},$$

$$\int_{-\infty}^{T} \frac{d\tau}{2i\tau^2} e^{-i\tau^2} = \frac{1}{4T^3} e^{-iT^2} + \frac{3}{4}\int_{-\infty}^{T} \frac{d\tau}{\tau^4} e^{-i\tau^2}$$

Therefore $$\left|\int_{-\infty}^{T} \frac{d\tau}{2i\tau^2} e^{-i\tau^2}\right| \leqslant \frac{1}{4|T|^3} + \frac{1}{4|T|^3} = \frac{1}{2|T|^3}.$$

Introducing a quantity ϵ to indicate the precision with which measurements may be made, we have for points outside the parabola $\frac{1}{2|\tau|^3} < \epsilon$, the approximation formula

$$\int_{-\infty}^{T} e^{-i\tau^2}\, d\tau = -\frac{1}{2iT} e^{-iT^2} \qquad T < 0.$$

Similarly, when $T > 0$,

$$\int_{-\infty}^{T} e^{-i\tau^2}\, d\tau = \int_{-\infty}^{\infty} - \int_{T}^{\infty} = \sqrt{\pi}\,.\,e^{-i\pi/4} - \frac{1}{2iT} e^{-iT^2}.$$

These results may be used to obtain approximate representations of the light vector when plane waves of light are diffracted at a straight edge. If

$$v_1 = e^{ik\rho \cos(\theta - \theta_0) + i(\pi/4 + kct)} \pi^{-\frac{1}{2}} \int_{-\infty}^{T_1} e^{-i\tau^2} d\tau,$$

where
$$T_1 = \sqrt{2k\rho} \cos \frac{\theta - \theta_0}{2},$$

and
$$v_2 = e^{ik\rho \cos(\theta + \theta_0) + i(\pi/4 + kct)} \pi^{-\frac{1}{2}} \int_{-\infty}^{T_2} e^{-i\tau^2} d\tau,$$

where
$$T_2 = \sqrt{2k\rho} \cos \frac{\theta + \theta_0}{2},$$

the solution of a number of diffraction problems may be expressed in terms of v_1 and v_2. In particular, if plane waves of light are incident upon a screen ($y = 0, x > 0$) which is a perfect conductor, and the waves are polarized so that the electric vector is parallel to the edge of the screen, the electric vector E_z in the total electromagnetic field is given by the expression

$$E_z = R\{v_1 - v_2\}, \qquad \text{......(B)}$$

where the symbol R is used to denote the real part of the expression which follows it. This expression evidently satisfies the boundary condition $E_z = 0$ when $y = 0, x > 0$, i.e. when $\theta = 0$ and $\theta = 2\pi$. On the other hand, if the magnetic vector is parallel to the edge of the screen, the magnetic vector in the total field is given by the equation

$$H_z = R\{v_1 + v_2\}. \qquad \text{......(C)}$$

The boundary condition which is satisfied in this case is $\frac{\partial E_x}{\partial t} = 0$, which, on account of the field equation $\frac{\partial H_z}{\partial y} - \frac{\partial H_y}{\partial z} = \frac{1}{c}\frac{\partial E_x}{\partial t}$, is equivalent to $\frac{\partial H_z}{\partial y} = 0$ or $\frac{\partial H_z}{\partial \theta} = 0$, when $\theta = 0$ and $\theta = 2\pi$.

Let
$$S = \frac{1}{4\pi} \sqrt{(\lambda/\rho)} \cos\left(k\rho - kct + \frac{\pi}{4}\right),$$

then in the region of the geometrical shadow ($T_1 < 0$, $T_2 < 0$) we have respectively in the two cases just considered

$$E_z = S\left(\sec\frac{\theta + \theta_0}{2} - \sec\frac{\theta - \theta_0}{2}\right), \quad H_z = -S\left(\sec\frac{\theta + \theta_0}{2} + \sec\frac{\theta - \theta_0}{2}\right).$$

In the region $T_1 > 0$, $T_2 < 0$ which contains the incident but not the reflected wave, we have the approximate expressions

$$E_z = \cos\{k\rho\cos(\theta - \theta_0) + kct\} + S\left(\sec\frac{\theta + \theta_0}{2} - \sec\frac{\theta - \theta_0}{2}\right),$$

$$H_z = \cos\{k\rho\cos(\theta - \theta_0) + kct\} - S\left(\sec\frac{\theta + \theta_0}{2} + \sec\frac{\theta - \theta_0}{2}\right),$$

and in the region $T_1 > 0$, $T_2 > 0$, which contains the reflected wave, we have approximately

$$E_z = \cos\{k\rho\cos(\theta-\theta_0)+kct\} - \cos\{k\rho\cos(\theta+\theta_0)+kct\} + S\left\{\sec\frac{\theta+\theta_0}{2} - \sec\frac{\theta-\theta_0}{2}\right\},$$

$$H_z = \cos\{k\rho\cos(\theta-\theta_0)+kct\} + \cos\{k\rho\cos(\theta+\theta_0)+kct\} - S\left\{\sec\frac{\theta+\theta_0}{2} + \sec\frac{\theta-\theta_0}{2}\right\}.$$

The disturbance diverging from the edge of the screen like a cylindrical wave whose intensity falls off like $\rho^{-\frac{1}{2}}$ is called the diffraction wave. The phenomena of diffraction may be regarded as the result of an interference between this wave and the incident and reflected waves.

There is no true source of light at the edge of the screen, yet the eye, when it accommodates itself so as to view the edge of the screen, receives the impression that the edge is luminous. Gouy and Wien first observed this phenomenon in the region of the geometrical shadow where the phenomenon is not masked by the incident light.

For the amplitude of the electric vector in the diffracted wave we have in the two cases

$$A_1 = \frac{1}{4\pi}\sqrt{(\lambda/\rho)}\left(\sec\frac{\theta+\theta_0}{2} - \sec\frac{\theta-\theta_0}{2}\right),$$

$$A_2 = \frac{1}{4\pi}\sqrt{(\lambda/\rho)}\left(\sec\frac{\theta+\theta_0}{2} + \sec\frac{\theta-\theta_0}{2}\right),$$

respectively, where in the second case A_2 is the amplitude of the electric vector perpendicular to the edge of the screen, and where in the first case A_1 is the amplitude of the electric vector parallel to the edge of the screen.

If the incident waves are linearly polarized so that they can be resolved into a wave of amplitude a_1 with electric vector parallel to the screen, and a wave of amplitude a_2 with electric vector perpendicular to the screen, the amplitudes of the corresponding components of the diffracted wave are respectively a_1A_1 and a_2A_2. Since these are no longer in the ratio $a_1 : a_2$ there is a rotation of the plane of polarization.

When the screen is illuminated with natural light in which waves with all phases and directions of polarization occur with equal frequency we have $a_1 = a_2$, but since $a_1A_1 \neq a_2A_2$ the diffraction wave is polarized.

It should be noticed that

$$\frac{A_2}{A_1} = \cot\frac{\theta}{2}\cot\frac{\theta_0}{2}.$$

For the case of perpendicular incidence $\theta_0 = \frac{\pi}{2}$, we have

$$\frac{A_2}{A_1} = \cot\frac{\theta}{2} = \tan\left(\frac{\pi}{4}+\frac{1}{2}\delta\right),$$

where δ is the angle of diffraction.

The measurements of W. Wien* with very fine sharp steel blades show a somewhat stronger increase of this ratio than that indicated by this formula. Epstein† attributes this to the finite thickness of the blade.

Raman and Krishnan‡ have recently invented a new method of discussing the influence of the material of the screen in which the solutions taking the place of (B) and (C) are respectively

$$E_z = R\{v_1 - \beta v_2\},$$
$$H_z = R\{v_1 + \gamma v_2\},$$

where β and γ are complex constants depending on the nature of the material and the angle of incidence $\psi = \frac{\pi}{2} - \theta_0$. This amounts to a change in the boundary condition, the solutions adopted being still wave-functions.

If ω denotes the material constant $n(1 - i\kappa)$, where n is the refractive index and κ the index of Absorption, the values adopted for β and γ are respectively

$$\beta = \frac{\alpha - \cos\psi}{\alpha + \cos\psi}, \quad \gamma = \frac{\omega^2\cos\psi - \alpha}{\omega^2\cos\psi + \alpha},$$

where $\alpha^2 = \omega^2 - \sin^2\psi$.

In this way an explanation is obtained of the results of Gouy relating to the effect of the material on the colour and polarization of the diffracted light.

§ **11·5**. *Use of parabolic co-ordinates.* It was shown by Lamb§ and later by Epstein|| and Crudeli¶ that the problems of diffraction can be treated successfully with the aid of the parabolic co-ordinates ξ, η defined by the equation

$$x + iy = (\xi + i\eta)^2, \quad \eta \geqslant 0.$$

In terms of these variables we have

$$\frac{\partial\phi}{\partial x} = \frac{1}{2r}\left(\xi\frac{\partial\phi}{\partial\xi} - \eta\frac{\partial\phi}{\partial\eta}\right), \qquad \xi = r^{\frac{1}{2}}\cos\tfrac{1}{2}\theta,$$
$$\frac{\partial\phi}{\partial y} = \frac{1}{2r}\left(\eta\frac{\partial\phi}{\partial\xi} + \xi\frac{\partial\phi}{\partial\eta}\right), \qquad \eta = r^{\frac{1}{2}}\sin\tfrac{1}{2}\theta.$$

A particular solution of the wave-equation

$$\frac{\partial^2\phi}{\partial t^2} = c^2\left(\frac{\partial^2\phi}{\partial x^2} + \frac{\partial^2\phi}{\partial y^2}\right)$$

is

$$r^{-\frac{1}{2}}\cos\tfrac{1}{2}\theta \, . \, f(ct - r),$$

where $f(ct - r)$ is an arbitrary function with continuous second derivative.

* *Wied. Ann.* Bd. XXVIII, S. 117 (1886).

† *Diss. Munich* (1914); *Encyklopädie der Math. Wiss.* Bd. V. 3, Heft 3 (1915).

‡ C. V. Raman and K. S. Krishnan, *Proc. Roy. Soc. London*, A, vol. CXVI, p. 254 (1927).

§ H. Lamb, *Proc. London Math. Soc.* (2), vol. VIII, p. 422 (1910).

|| P. S. Epstein, *Diss. Munich* (1914).

¶ U. Crudeli, *Il Nuovo Cimento* (6), vol. XI, p. 277 (1916).

Let us denote this function by $\frac{\partial \phi}{\partial x}$, then the equation for ϕ is

$$\xi \frac{\partial \phi}{\partial \xi} - \eta \frac{\partial \phi}{\partial \eta} = 2\xi f\,(ct - \xi^2 - \eta^2),$$

and a solution which satisfies the condition $\frac{\partial \phi}{\partial y} = 0$, or $\frac{\partial \phi}{\partial \eta} = 0$ at the screen $\eta = 0$, is

$$\phi = \int_0^{\xi+\eta} f\,(ct + y - \sigma^2)\,d\sigma + \int_0^{\xi-\eta} f\,(ct - y - \sigma^2)\,d\sigma$$
$$+ \tfrac{1}{2}F\,(ct + y) + \tfrac{1}{2}F\,(ct - y),$$

where F is another arbitrary function. Now when θ is nearly equal to π and r is very great, the upper limits of the integrals become ∞ and $-\infty$, respectively, and the asymptotic form of ϕ is

$$\phi = \int_0^{\infty} f\,(ct + y \quad \sigma^2)\,d\sigma - \int_0^{\infty} f\,(ct - y - \sigma^2)\,d\sigma$$
$$+ \tfrac{1}{2}F\,(ct + y) + \tfrac{1}{2}F\,(ct - y);$$

this is identical with $F\,(ct + y)$ if

$$\int_0^{\infty} f\,(y - \sigma^2)\,d\sigma = \tfrac{1}{2}F\,(y), \quad -\infty < y < \infty.$$

This is an integral equation for the determination of the function f when F is given. Writing $\sigma = (y - v)^{\frac{1}{2}}$ the equation takes the form

$$F\,(y) = \int_{-\infty}^{y} (y - v)^{-\frac{1}{2}} f\,(v)\,dv, \quad -\infty < y < \infty,$$

and the solution given by Abel's inversion formula is

$$f\,(v) = \frac{1}{\pi}\frac{d}{dv}\int_{-\infty}^{v} (v - u)^{-\frac{1}{2}} F\,(u)\,du.$$

If $\lim_{u \to -\infty} F\,(u) = 0$ and $\lim_{u \to -\infty} [(v - u)^{\frac{1}{2}} F\,(u)]$ exists,

$$f\,(v) = \frac{1}{\pi}\frac{d}{dv} \lim_{u \to -\infty} [(v - u)^{\frac{1}{2}} F\,(u)] + \frac{1}{\pi}\int_{-\infty}^{v} (v - u)^{-\frac{1}{2}} F'\,(u)\,du.$$

In particular, if $F\,(u) = (-u)^{\frac{1}{2}} G\,(u)$, where $G\,(-\infty)$ has a finite value different from zero,

$$\lim_{u \to -\infty} [(v - u)^{\frac{1}{2}} F\,(u)] = G\,(-\infty),$$

and we have simply

$$f\,(v) = \frac{1}{\pi}\int_{-\infty}^{v} (v - u)^{-\frac{1}{2}} F'\,(u)\,du. \qquad \text{......(D)}$$

The foregoing conditions imposed on $F\,(u)$ are not necessary for the existence of a solution given by the formula (D), for if $F\,(u) = \cos ku$, we have $F'\,(u) = -k \sin ku$,

$$-\frac{k}{\pi}\int_{-\infty}^{v} (v - u)^{-\frac{1}{2}} \sin ku\,du = -\frac{2k}{\pi}\int_0^{\infty} \sin\{k\,(v - s^2)\}\,ds$$
$$= (k/\pi)^{\frac{1}{2}} \cos\,(kv + \pi/4),$$

and it may be verified directly that $f(v) = (k/\pi)^{\frac{1}{2}} \cos(kv + \pi/4)$ is a solution of the integral equation

$$\cos ky = \int_{-\infty}^{y} (y - v)^{-\frac{1}{2}} f(v)\, dv.$$

§ **11·6.** In the work of Epstein an endeavour is made to allow for the influence of the material of which the screen is made, and so the surface of the screen is taken to be a parabolic cylinder and the material is supposed to have a finite conductivity. The electromagnetic field vectors E and H are regarded as the real parts of the expressions eT, hT, respectively, where T denotes the exponential factor $\exp(int)$. In a material medium with constants κ, θ, μ and σ the equations satisfied by the vectors e and h are

$$\operatorname{curl} h = \alpha e, \qquad \operatorname{curl} e = -\beta h,$$
$$\operatorname{div} e = 0, \qquad \operatorname{div} h = 0,$$

where

$$\alpha = \frac{i\kappa n}{c} + \frac{\sigma}{c}, \quad \beta = \frac{i\mu n}{c}.$$

These equations are transformed by the substitution

$$x = \tfrac{1}{2}(\xi^2 - \eta^2), \quad y = \xi\eta, \quad z = z, \quad s^2 = \xi^2 + \eta^2,$$

into equations connecting the new components e_ξ, e_η, e_z, h_ξ, h_η, h_z,

$$\frac{\partial}{\partial \xi}(sh_\eta) - \frac{\partial}{\partial \eta}(sh_\xi) = \alpha s^2 e_z, \qquad \frac{\partial h_z}{\partial \eta} = \alpha s e_\xi, \qquad \frac{\partial h_z}{\partial \xi} = -\alpha s e_\eta,$$

$$\frac{\partial}{\partial \xi}(se_\eta) - \frac{\partial}{\partial \eta}(se_\xi) = -\beta s^2 h_z, \qquad \frac{\partial e_z}{\partial \eta} = -\beta s h_\xi, \qquad \frac{\partial e_z}{\partial \xi} = \beta s h_\eta.$$

These equations imply that e_z and h_z are solutions of the partial differential equation

$$\frac{\partial^2 V}{\partial \xi^2} + \frac{\partial^2 V}{\partial \eta^2} + k^2(\xi^2 + \eta^2) V = 0,$$

where

$$k^2 = -\alpha\beta = \frac{\kappa\mu n^2 - i\mu\sigma n}{c^2}.$$

The boundary conditions at the surface of the screen are, when $\mu = 1$,

$$V_0 = V_i, \quad (\gamma \partial V/\partial \eta)_0 = (\gamma \partial V/\partial \eta)_i,$$

where $V = e_z$ or h_z, according as the incident wave is polarized perpendicular to the edge of the screen or parallel to the focal axis of the cylinder, and where γ is unity in the first case and k^{-2} in the second.

The differential equation for V may be satisfied by writing

$$V = X(\xi)\, Y(\eta),$$

where X and Y satisfy the differential equations of Weber*

$$\frac{d^2 X}{d\xi^2} + (k^2\xi^2 + A) X = 0, \qquad \frac{d^2 Y}{d\eta^2} + (k^2\eta^2 - A) Y = 0,$$

where A is an arbitrary constant. Writing $A = 2ik(n + \frac{1}{2})$,

$$X = \exp(-ik\xi^2/2)\, U_n[\xi\sqrt{(ik)}], \quad Y = \exp(ik\eta^2/2)\, U_n[\eta\sqrt{(-ik)}],$$

* H. Weber, *Math. Ann.* Bd. I, S. 1 (1869).

we have particular solutions expressed in terms of the polynomials of Hermite

$$n!\,X_n(\xi) = (2ik)^{-n/2}\,e^{ik\xi^2/2}\frac{d^n}{d\xi^n}(e^{-ik\xi^2}),$$

$$n!\,Y_n(\eta) = (2ik)^{-n/2}\,e^{-ik\eta^2/2}\frac{d^n}{d\eta^n}(e^{+ik\eta^2}).$$

These are suitable for the representation of an electromagnetic disturbance in the interior of the parabolic cylinder. Distinguishing the functions associated with the value of k for the interior of the cylinder by a star, we have for the interior of the cylinder

$$V^* = \sum_{n=0}^{\infty} b_n X_n^*(\xi)\,Y_n^*(\eta).$$

To represent the disturbance outside the cylinder it is necessary to use the second solution of the differential equation, and this may be chosen so that for $\eta \gg n$ there is an asymptotic representation

$$Z_n(\eta) \sim i\,(-\,2ik\eta)^{-(n+1)}\,e^{-\frac{1}{2}ik\eta^2},$$

where $Z_n(\eta)$ denotes this second solution. We may now assume for the space outside the cylinder

$$V = e^{-ik(x\cos\theta_0 + y\sin\theta_0)} + \sum_{n=0}^{\infty} a_n X_n(\xi)\,Z_n(\eta),$$

$$e^{-ik(x\cos\theta_0 + y\sin\theta_0)} = \sec(\tfrac{1}{2}\theta_0)\sum_{n=0}^{\infty} n!\,\tan^n(\tfrac{1}{2}\theta_0)\,X_n(\xi)\,Y_n(\eta),$$

where the first term represents the incident wave and the coefficients a_n, like the coefficients b_n, are to be determined by means of the boundary conditions. The analysis has been formally completed by Epstein and some rough calculations made.

In the case of the infinitely thin parabolic cylinder (or half plane) which is a perfect conductor an agreement is found with the formula of Sommerfeld. When the thickness is not quite negligible and the conduction is still perfect, it is concluded that the term representing the correction will increase the amplitude in the case $||$ and decrease it in the case $\perp$, so that the ratio A_2/A_1 is greater than the value $\tan\left(\frac{1}{2}\delta + \frac{\pi}{4}\right)$ found in § 11·4. This is in qualitative agreement with observation. With finite conductivity the parabolic screen gives a selective effect favouring wavelengths which are most strongly reflected by a plane mirror of the same material. This selective effect is extremely weak in the case $\perp$ and quite appreciable in the case $||$. These predictions of theory have been confirmed by recent experiments*.

* F. Jentsch, *Ann. d. Phys.* Bd. LXXXIV, S. 292 (1927–28).

EXAMPLES

1. Prove that if
$$f(x) = \int_x^\infty e^{x-\tau} \frac{d\tau}{\tau},$$
the definite integral
$$V = e^{ct-z} \int_0^{\pi/2} e^{-(r+z)\tan^2 a} \tan a\, da$$
represents the wave-function
$$V = \tfrac{1}{2} e^{ct-r} f(r+z).$$

2. Prove that if $n > -1$,
$$\int_0^{\pi/2} e^{-u \tan^2 a} \tan^n a\, da = \tfrac{1}{2}\Gamma\left(\frac{n+1}{2}\right) \int_u^\infty e^{u-v} v^{-n+1/2}\, dv$$
and write down an expression for a solution of
$$\frac{\partial^2 V}{\partial x_1^2} + \dots \frac{\partial^2 V}{\partial x_n^2} + \frac{\partial^2 V}{\partial z^2} = \frac{1}{c^2}\frac{\partial^2 V}{\partial t^2},$$
which is of the form
$$V = e^{ct-r} f(r+z),$$
where
$$r^2 = x_1^2 + \dots x_n^2 + z^2.$$

3. Prove that if
$$ct + y = b \tan a, \quad ct - y = b \tan \beta, \quad ct - r = b \tan \omega, \quad -\frac{\pi}{2} \leqslant (a, \beta, \omega) \leqslant \frac{\pi}{2}$$
the wave-potential
$$\phi = \frac{1}{2b}\{\cos^2 a + \cos^2 \beta + (\xi + \eta)\, b^{-\frac{1}{2}} \cos^{\frac{1}{2}} \omega \cos a \cos(\tfrac{1}{2}\omega + a + \tfrac{1}{4}\pi)$$
$$+ (\xi - \eta)\, b^{-\frac{1}{2}} \cos^{\frac{1}{2}} \omega \cos \beta \cos(\tfrac{1}{2}\omega + \beta + \tfrac{1}{4}\pi)\}$$
corresponds to a primary wave of type
$$\phi = \frac{b}{b^2 + (ct+y)^2} = \frac{\cos^2 a}{b}.$$
[H. Lamb.]

4. Verify that with the function ϕ in Example 3
$$\int_{-\infty}^{\infty} \phi\, dt = \pi/c,$$
at all points in the field.

[H. Lamb.]

§ 11·7. For a discussion of other diffraction problems reference may be made to G. Wolfsohn's article, "Strenge Theorie der Interferenz und Beugung," *Handbuch der Physik*, Bd. xx, T. 7. In this article accounts are given of the work of Schaefer and others on the diffraction of undamped electric waves by a dielectric circular cylinder and of Schwarzschild's treatment of diffraction by a slit [*Math. Ann.* Bd. LV, S. 177 (1902)]. A study of diffraction by an elliptic cylinder has been commenced by Sieger*. For this study and for an analogous study of diffraction by a hyperbolic cylinder the substitution $x + iy = h \cosh(\xi + i\eta)$ may be used with advantage and then for the representation of divergent waves it is necessary to find solutions of Mathieu's equation which vanish for large values of ξ. Such solutions have been studied by Sieger, Dougall† and Dhar‡.

* B. Sieger, *Ann. d. Phys.* Bd. XXVII, S. 626 (1908).

† J. Dougall, *Proc. Edin. Math. Soc.* vol. XXXIV, p. 191 (1916).

‡ S. Dhar, *Journ. Indian Math. Soc.* vol. XVI, p. 227 (1926); *Amer. Journ. Math.* vol. XLV, p. 208 (1923).

CHAPTER XII

NON-LINEAR EQUATIONS

§ 12·1. *Riccati's equation.* The differential equation

$$\dot{v} = av^2 + bv + c,$$

in which a, b and c are functions of t, is generally known as Riccati's equation; it may be replaced by a linear differential equation of the second order

$$\ddot{s} + p\dot{s} + qs = 0,$$

in which $$s = e^{\int av\,dt}, \quad p = -b - \dot{a}/a, \quad q = c.$$

A very simple Riccatian equation occurs in the theory of motion in a resisting medium when a falling body encounters a resistance proportional to the square of the velocity (the law of Newton). If m is the mass of the body, v the velocity, Kv^2 the drag and g the acceleration of gravity, the equation of motion is

$$m\dot{v} = mg - Kv^2 = mv\,(dv/dx),$$

when the motion is downwards, and

$$m\dot{v} = mg + Kv^2 = mv\,(dv/dx),$$

when the motion is upwards, v being in each case positive when directed downwards.

For downward motion, if $v = 0$ when $t = 0$, we have

$$vV = V^2 \tanh gt,$$

where $$V^2 = mg/K = W/K.$$

Also $$x = \int_0^t v\,dt = \frac{V^2}{2g} \log\left(\frac{V^2}{V^2 - v^2}\right).$$

As the velocity increases from v_0 to v the distance travelled is

$$x - x_0 = \frac{V^2}{2g} \log \frac{V^2 - v_0^2}{V^2 - v^2},$$

and the time taken is given by the equation

$$e^{\frac{2g}{V}(t-t_0)} = \frac{(V+v)\,(V-v_0)}{(V-v)\,(V+v_0)}.$$

For upward motion, if $v = -V\tan\theta$ when $t = 0$, we have

$$v = V \tan\left(\frac{gt}{V} - \theta\right),$$

$$2g\,|\,x\,| = V^2 \log\left[\left(1 + \frac{v^2}{V^2}\right)\cos^2\theta\right]$$

The velocity vanishes when $gt = V\alpha$ and the body is then at a height h above its initial position, h being given by the equation

$$gh = V^2 \log(\sec\theta).$$

In particular, if $v = -V$, we have

$$gh = \cdot 347 V^2.$$

The foregoing analysis has been applied to the case of an airplane in a rectilinear glide on the assumption that the thrust of the propeller may be neglected and that the relative decrease in the airplane drag due to the absence of the slip stream from the propeller can also be neglected*, as one effect will partly compensate the other. The airplane is supposed, moreover, to fly at an angle of attack close to the angle of no lift, this angle being adjusted in flight so as to keep the path rectilinear. The density of the air is supposed to be constant and Kv^2 is regarded as a good approximation to the total drag of the airplane because the drag coefficient is nearly constant at low angles of attack. If α is the angle of descent, the quantity g in the foregoing equations should be replaced by $g \sin\alpha$, and W by $W \sin\alpha$.

Hunsaker considers the case of an airplane which has a normal maximum speed in horizontal flight of 110 ft./sec. and for which

$$W = 2000 \text{ lb.}, \quad K = 0\cdot 031.$$

With $\alpha = 45°$ the limiting speed is given by

$$V^2 = \frac{2000}{(0\cdot 031)(1\cdot 414)},$$

or

$$V = 214 \text{ ft./sec.}$$

With these data it appears that if the airplane starts with a velocity of 110 feet a second, a vertical drop of about 1000 feet is sufficient to give it a velocity of 194 feet a second. This result is obtained by putting $V = 24$, $v_0 = 110$, $(x - x_0)\sin\alpha = 1000$ in the equation

$$2g(x - x_0)\sin\alpha = V^2 \log\frac{V^2 - v_0^2}{V^2 - v^2}.$$

The thrust of the propeller has been taken into consideration by W. S. Diehl†, who assumes that the thrust T is a linear function of v, i.e.

$$T = T_0 - a(v - v_0),$$

where T_0 is the value of the thrust when $v = v_0$. The value of this coefficient is derived from the plausible assumption that T is zero when $v = 2v_0$. This gives $T_0 = av_0$, and

$$T = 2T_0 - T_0 v/v_0.$$

* J. C. Hunsaker, *Aviation and Aeronautical Engineering*, vol. VII, p. 539 (1920).
† *N.A.C.A. Report* 238, Washington (1926).

The equation of motion for a rectilinear path is now

$$m\dot{v} = mg \sin \alpha + 2T_0 - T_0 v/v_0 - Kv^2,$$

the total drag being represented by an expression of type Kv^2. If

$$v + T_0/2Kv_0 = u,$$

this equation takes the form

$$m\dot{u} = mg \sin \alpha + 2T_0 + T_0^2/4Kv_0^2 - Ku^2,$$

and is of the type already considered.

In steady horizontal flight at velocity v_0 we have

$$T_0 = D_0 v_0^2, \quad mg = W = L_0 v_0^2,$$

where L_0 and D_0 are the total lift and drag coefficients for a certain low angle of incidence which gives maximum speed. With a fair approximation we may write $D_0 = K$, and the equation for u becomes

$$m\dot{u} = K (L_0 v_0^2 \sin \alpha/D_0 + qv_0^2/4 - u^2).$$

The limiting velocity V is thus given by the equation

$$(V + \tfrac{1}{2}v_0)^2 = U^2 = qv_0^2/4 + L_0 v_0^2 \sin^2 \alpha/D_0.$$

If $L_0/D_0 = 4{\cdot}5$, $\sin \alpha = \frac{1}{2}$, this equation gives

$$(V + \tfrac{1}{2}v_0)^2 = qv_0^2/2,$$

$$V = 1{\cdot}6v_0.$$

In Diehl's paper curves are given showing the variation of V and its horizontal component as α and L_0/D_0 are changed.

There are many other cases in which the variable motion of a system is governed by an equation of type

$$\dot{v} = A + Bv + Cv^2. \qquad \text{......(A)}$$

For instance, the equation of an unopposed bimolecular reaction of the simplest type is

$$\dot{x} = k (a - x) (b - x),$$

where b may or may not be equal to a. The equation (A) also represents the equation of motion of a motor-coach train when the power has been shut off.

When $A = 0$ the equation can be solved even when B and C are functions of t. The equation is then a particular case of the equation of Bernoulli

$$\dot{x} + Px = Qx^n,$$

in which P and Q are functions of t. An equation of this type with different variables, was obtained by Harcourt and Esson in their work on chemical dynamics*.

The differential equations for the reactions $X \to Y$, $Y + Z \to W$ are

$$\dot{x} = -k_1 x, \quad \dot{y} = k_1 x - k_2 yz, \quad \dot{z} = -k_2 yz, \quad \dot{w} = k_2 yz,$$

where x, y, z, w denote the amounts of the substances X, Y, Z, W present at time t.

* *Phil. Trans.* vol. CLVI, p. 193 (1866).

If initially, $x = z = a$, we have

$$w = a - z = a - x - y.$$

Therefore $y = z - x$, and so

$$\frac{dz}{dx} + Kz(z - x) = 0,$$

where $K = k_2/k_1$. This equation may be solved by dividing by z^2.

An interesting Riccatian equation occurs in the study of the lines of electric force of a moving electric pole. If ξ, η, ζ are the co-ordinates of the pole at time τ, and if

$$x - \xi = lr, \quad y - \eta = mr, \quad z - \zeta = nr, \quad c(t - \tau) = r > 0,$$

$$s = c - l\xi' - m\eta' - n\zeta', \; p = l\xi'' + m\eta'' + n\zeta'', \; q = c^2 - \xi'^2 - \eta'^2 - \zeta'^2,$$

the components of the electric field vector may be expressed in the form

$$E_x = \frac{k}{s^2r^2}[r\{s\xi'' + p(\xi' - cl)\} + q(\xi' - cl)], \text{ etc.},$$

where k is a constant depending on the charge associated with the pole.

If, on the other hand, a particle is emitted from the pole at time τ in the direction with direction cosines (l, m, n) and if this particle travels continually in this direction with speed c, its co-ordinates at time t will be x, y, z. The co-ordinates of a second particle emitted at time $\tau + d\tau$ from the position $\xi + \xi'd\tau$, $\eta + \eta'd\tau$, $\zeta + \zeta'd\tau$ in a direction with direction cosines $l + l'd\tau$, $m + m'd\tau$, $n + n'd\tau$ will be $x + dx$, $y + dy$, $z + dz$, if

$$dx = d\tau[\xi' - cl + rl'], \quad dy = d\tau[\eta' - cm + rm'], \quad dz = d\tau[\zeta' - cn + rn'].$$

It is seen that dx, dy, dz are proportional to E_x, E_y, E_z, respectively, if

$$\left.\begin{aligned} ql' &= s\xi'' + p(\xi' - cl) \\ qm' &= s\eta'' + p(\eta' - cm) \\ qn' &= s\zeta'' + p(\zeta' - cn) \end{aligned}\right\}. \qquad \text{......(B)}$$

These equations indicate the way in which the direction of projection must vary in order that the emitted particles may at each instant t lie on a line of electric force. The equations may be replaced by two Riccatian equations by writing

$$l = \frac{\phi + \psi}{1 + \phi\psi}, \quad m = i\frac{\psi - \phi}{1 + \phi\psi}, \quad n = \frac{1 + \phi\psi}{1 - \phi\psi}. \qquad \text{......(C)}$$

The resulting equation for ϕ is then

$$\begin{aligned} 2(c^2 - \xi'^2 - \eta'^2 - \zeta'^2)\phi' &= [\zeta''(\xi' - i\eta') - (\zeta' + c)(\xi'' - i\eta'')]\phi^2 \\ &+ 2[i(\xi''\eta' - \xi'\eta'') - c\zeta'']\phi + \zeta''(\xi'' + i\eta'') - (\zeta' - c)(\xi'' + i\eta''), \end{aligned} \qquad \text{......(D)}$$

and the equation for ψ may be obtained by changing the sign of i in the last equation.

The general integral of equation (D) involves a single arbitrary constant. By giving different complex values to this constant the different lines of force are obtained.

An important property of the equations (B) is that if we have one set of solutions l, m, n satisfying the relation $l^2+m^2+n^2=1$ which is compatible with the equations, a second set of solutions L, M, N is obtained by writing*

$$(q-2sc)(cL-\xi')=q(cl-\xi'),\quad (q-2sc)(cM-\eta')=q(cm-\eta'),$$
$$(q-2sc)(cN-\zeta')=q(cn-\zeta').$$

The verification is easy because

$$cL'-\xi''=\frac{q}{q-2sc}(cl'-\xi'')+\frac{2c(s'q-sq')}{(q-2sc)^2}(cl-\xi'),$$
$$s'q-sq'=s[\xi'\xi''+\eta'\eta''+\zeta'\zeta''-pc],$$
$$(q-2sc)[cP-\xi'\xi''-\eta'\eta''-\zeta'\zeta'']=q(cp-\xi'\xi''-\eta'\eta''-\zeta'\zeta''),$$
$$2s(\xi'\xi''+\eta'\eta''+\zeta'\zeta''-cp)=c(q-2s)(p-P).$$

Therefore $$cL'-\xi''=\frac{q(cl'-\xi'')+c(p-P)(cl-\xi')}{q-2sc},$$

or $$cL'=-\frac{cs\xi''}{q-2sc}-\frac{cP}{q}(cL-\xi').$$

Now $$S[q-2sc]=-qsc.$$

Therefore $$qL'=\xi''S+P(\xi'-cL).$$

It should be noticed that the relations between the two directions may be written in the form

$$s(cL-\xi')+S(cl-\xi')=0;\quad s(cM-\eta')+S(cm-\eta')=0,$$
$$s(cN-\zeta')+S(cn-\zeta')=0,$$
$$2sSc=q(S+s).$$

This result indicates that there are conjugate directions giving rise to conjugate lines of force.

There is evidently a similar result in a space of n dimensions when the n direction cosines $l_1, l_2, \ldots l_n$ are connected by the relations

$$ql_1'=s\xi_1''+p(\xi_1'-cl_1),$$
$$\cdots\cdots\cdots\cdots\cdots\cdots$$
$$q=c^2-\xi_1^2-\ldots\xi_n^2,\quad s=c-l_1\xi_1'-\ldots l_n\xi_n',\quad p=l_1\xi_1''+\ldots l_n\xi_n''.$$

Kourensky† has noticed that when the factor 2 is dropped from the left-hand side of the Riccatian equation there is a simple general solution

$$\phi=\frac{C(c-\zeta')+(\xi'+i\eta')}{C(\xi'-i\eta')+(c+\zeta')}.$$

He therefore discusses the general problem of the determination of solution of the equation

$$N\Phi'=P\Phi^2+Q\Phi+R,$$

when a solution of the equation

$$\phi'=P\phi^2+Q\phi+R$$

is known.

* F. D. Murnaghan, *Amer. Journ. of Math.* vol. XXXIX, p. 147 (1917).

† *Proc. London Math. Soc.* (2), vol. XXIV, p. 202 (1926).

Another interesting problem is to find the most general set of relations for the rate of variation of direction cosines which are compatible with the relation $l^2 + m^2 + n^2 = 1$ and are reducible to Riccatian equations by means of the substitution (D). If A, B, C, P, Q, R are arbitrary functions of τ, the equations

$$l' = l\,(Al + Bm + Cn) + qn - rm - A,$$
$$m' = m\,(Al + Bm + Cn) + rl - pn - B,$$
$$n' = n\,(Al + Bm + Cn) + pm - ql - C,$$

can be shown to fulfil the conditions.

EXAMPLES

1. Let a be the velocity of sound at the point (x, y, z) of a moving medium, the velocity being measured relative to the medium and the co-ordinates x, y, z being measured relative to some standard set of axes not moving with the medium. Let u, v, w be the component velocities of the medium at (x, y, z) relative to the standard axes. Assuming that u, v, w, a are independent of the time and that the equations of a sound ray are

$$\frac{dx}{dt} = u + la, \quad \frac{dy}{dt} = v + ma, \quad \frac{dz}{dt} = w + na,$$

where l, m, n are the direction cosines of the wave-normal, prove that l, m, n vary along the ray in accordance with the equations

$$\frac{dl}{dt} + l\frac{\partial u}{\partial x} + m\frac{\partial v}{\partial x} + n\frac{\partial w}{\partial x} + \frac{\partial a}{\partial x} = lR,$$

$$\frac{dm}{dt} + l\frac{\partial u}{\partial y} + m\frac{\partial v}{\partial y} + n\frac{\partial w}{\partial y} + \frac{\partial a}{\partial y} = mR,$$

$$\frac{dn}{dt} + l\frac{\partial u}{\partial z} + m\frac{\partial v}{\partial z} + n\frac{\partial w}{\partial z} + \frac{\partial a}{\partial z} = nR,$$

where

$$R = l\left(l\frac{\partial u}{\partial x} + m\frac{\partial v}{\partial x} + n\frac{\partial w}{\partial x} + \frac{\partial a}{\partial x}\right) + m\left(l\frac{\partial u}{\partial y} + m\frac{\partial v}{\partial y} + n\frac{\partial w}{\partial y} + \frac{\partial a}{\partial y}\right) + n\left(l\frac{\partial u}{\partial z} + m\frac{\partial v}{\partial z} + n\frac{\partial w}{\partial z} + \frac{\partial a}{\partial z}\right).$$

[E. A. Milne, *Phil. Mag.* (6), vol. XLII, p. 96 (1921).]

2. Prove that the equations giving the variation of l, m, n are of the type considered at the end of § 12·1 when

$$\frac{\partial u}{\partial x} = \frac{\partial v}{\partial y} = \frac{\partial w}{\partial z}, \quad \frac{\partial w}{\partial y} + \frac{\partial v}{\partial z} = 0, \quad \frac{\partial u}{\partial z} + \frac{\partial w}{\partial x} = 0, \quad \frac{\partial v}{\partial x} + \frac{\partial u}{\partial y} = 0,$$

and the derivatives of u, v, w, a are regarded as known functions of t.

3. Prove that the equations of Ex. 1 are the Eulerian equations for the variation problem $\delta I = 0$, where

$$I = \int \frac{l\,dx + m\,dy + n\,dz + b\,ds\,(1 - l^2 - m^2 - n^2)}{lu + mv + nw + a},$$

and where x, y, z, l, m, n, b are regarded as unknown functions of a variable parameter s. The time t is defined so that when $\delta I = 0$ the integrand is dt.

4. Use the result of Ex. 1 to obtain the results of Ex. 2, p. 337.

5. Obtain the form of Doppler's principle for a steadily moving atmosphere.

§ **12·2**. *The treatment of non-linear equations by a method of successive approximations.* This method has been applied with some success to an equation of type

$$\ddot{x} + k\dot{x} + hx + ax^2 + 2bx\dot{x} + c\dot{x}^2 = f(t), \qquad \text{......(A)}$$

in which a, b, c, h, k are functions of t.

An equation of type

$$\ddot{x} + n^2 x + hx^2 = a \cos pt + b \cos qt,$$

in which n, h, a, b, p, q are constants, was used by Helmholtz to illustrate his theory of combination tones in music, and was solved by a method of successive approximations. The analysis was criticised by some writers partly on account of the fact that no proof was given of the convergence of the series obtained by the method of successive approximations, but this objection is no longer applicable because some general existence theorems in the theory of differential equations* establish the convergence of the series under fairly wide conditions. The analysis, as presented by Schaefer†, is as follows:

Let us seek a solution which satisfies the initial conditions

$$x = c, \quad \dot{x} = 0,$$

when $t = 0$.

We write

$$a = ca_1, \quad b = cb_1,$$

$$x = \Sigma\, c^n \phi_n(t).$$

Substituting in the differential equation and equating coefficients of the different powers of c on the two sides of the equation, we obtain the system of differential equations

$$\ddot{\phi}_1 + n^2\phi_1 = a_1 \cos pt + b_1 \cos qt,$$

$$\ddot{\phi}_2 + n^2\phi_2 + h\phi_1^2 = 0,$$

$$\ddot{\phi}_3 + n^2\phi_3 + 2h\phi_1\phi_2 = 0,$$

$$\cdots\cdots\cdots\cdots\cdots\cdots\cdots\cdots$$

and the supplementary conditions

$$\phi_1(0) = 1, \quad \phi_n(0) = 0, \quad n > 0, \quad \phi_n'(0) = 0.$$

The solution of the first equation may be written in the form

$$\phi_1 = A \cos nt + \alpha \cos pt + \beta \cos qt,$$

where $\quad \alpha(n^2 - p^2) = a_1, \quad \beta(n^2 - q^2) = b_1, \quad A + \alpha + \beta = 1.$

Using this value of ϕ_1 the differential equation for ϕ_2 becomes

$$\ddot{\phi}_2 + n^2\phi_2 = A_1 + B_1 \cos 2nt + C_1 \cos(n+p)t + D_1 \cos(n-p)t$$
$$+ E_1 \cos(n+q)t + F_1 \cos(n-q)t + G_1 \cos(p+q)t + H_1 \cos(p-q)t,$$

* See, for instance, H. Poincaré, *Mécanique céleste*, t. I, p. 58; J. Horn, *Zeits. für Math. u. Phys.* Bd. XLVII, S. 400 (1902); G. A. Bliss, *Amer. Math. Soc. Colloquium Lectures*, vol. XVII, Princeton.

† C. Schaefer, *Ann. d. Phys.* Bd. XXXIII, S. 1216 (1910).

where the coefficients $A_1, \dots H_1$ have values which are easily obtained but need not be written down. The general solution of this equation is of the form

$$\phi_2 = A_2 + B_2 \cos 2nt + C_2 \cos (n+p) t + D_2 \cos (n-p) t + E_2 \cos (n+q) t$$
$$+ F_2 \cos (n-q) t + G_2 \cos (p+q) t + H_2 \cos (p-q) t + \alpha_1 \cos nt + \beta_1 \sin nt$$

and, if we are content with this degree of approximation, the expression for x is

$$x = c\phi_1 + c^2\phi_2,$$

and is seen to contain terms $G_2 \cos (p+q) t + H_2 \cos (p-q) t$, which may be responsible for combination tones with frequencies equal to the sum and differences of the frequencies in the exciting force*.

An equation of type (A) has been obtained and discussed by Galitzin in a study of the free motion of a horizontal pendulum carrying a pin which inscribes a record on smoked paper. His equation is

$$\ddot{x} + 2k\dot{x} + n^2 (x + \rho) + \xi (\dot{x} + \nu x)^2 = 0,$$

where ξ and ρ are two constants depending upon the elements of friction and ν is a constant depending on the rate at which the recording drum turns with the smoked paper attached to it. The quantity ξ is small and so a method of successive approximations may be used in which x is expanded in ascending powers of ξ.

When there is a resistance proportional to the square of the velocity in addition to one proportional to the velocity, the differential equation of motion is

$$\ddot{x} + 2k\dot{x} + n^2x + \mu\dot{x} \mid \dot{x} \mid = f(t), \qquad \text{......(B)}$$

where μ is a positive constant. We write $\dot{x} \mid \dot{x} \mid$ instead of $\dot{x}^2$ because a resistance is always opposite in direction to the velocity. It is tacitly assumed here that the two resistances are of different origin. If we have initially $\dot{x} = c$ when $c > 0$, the equation

$$\ddot{x} + 2k\dot{x} + n^2x + \mu\dot{x}^2 = f(t) \qquad \text{......(C)}$$

will hold for subsequent times up to the instant when $\dot{x}$ changes sign, and then the sign of μ in (C) must be changed until $\dot{x}$ again vanishes and changes sign. If, however, the equation be written in the form

$$\ddot{x} + n^2x + R = f(t),$$

where R is a resistance which can be regarded as of one type physically, the equation (C) may be expected to hold until $2k + \mu\dot{x}$ vanishes.

* Some interesting remarks on this subject will be found in Lamb's *Dynamical Theory of Sound*, p. 294. When $a = b = 0$ a first integral of the equation may be obtained by multiplying by $\dot{x}$ and integrating. The integration may then be completed with the aid of elliptic functions. The exact solution has been discussed recently by H. Nagaoka in a paper on asymmetric vibrations, *Proc. Imperial Acad. Tokyo*, vol. III, p. 23 (1927). When h is small it may be supposed to give a measure of the imperfection of the elasticity of the vibrating system. The effect of h is to increase the period of the vibration by an amount proportional to ah^2x^{-3}, where a is the initial energy.

When $f(t) = 0$ the equation (B) is of a type discussed by Milne*. He has established some general theorems relating to the existence of different types of solution and has constructed a table with the aid of which the equation may be solved numerically. Schaefer has considered the case in which

$$f(t) = a \cos pt + b \cos qt,$$

and points out that the equation will also give combination tones, but the theory is not as simple as in the case of Helmholtz's equation because the coefficients of $\cos(p+q)t$ and $\cos(p-q)t$ change sign when $\dot{x}$ changes sign. He remarks, however, that each of the coefficients may be expressible as a Fourier series of sines and cosines of st, where s is a suitable constant, and that if in this case the constant term is not zero the terms of type $\cos(p+q)t$ and $\cos(p-q)t$ will indeed be present and will give combination tones.

When solid friction which is constant in magnitude but opposite in direction to the velocity of a body modifies the motion of a body acted upon by forces which would ordinarily produce simple damped oscillations, the equation of motion is of type†

$$\ddot{x} + 2k\dot{x} + n^2 x \pm F = 0,$$

where k, n and F are constants and the sign of the frictional term F is the same as that of $\dot{x}$. It is understood here that this equation is valid during motion, that is, when $\dot{x}$ is different from zero. When $\dot{x}$ vanishes it may happen that the friction is sufficient to prevent further motion. This point will be brought out clearly in the analysis.

We commence by considering the solution

$$n^2 x = (-)^s F + A_s e^{-kt} \sin m(t+\tau), \qquad \text{......(D)}$$

$$s\pi < mt < (s+1)\pi,$$

$$m^2 = n^2 - k^2, \quad m \cos m\tau = k \sin m\tau, \quad 0 < m\tau < \pi.$$

The time t is supposed to be measured from an instant at which $\dot{x} = 0$. If $x = a$ when $t = 0$, we have

$$n^2 a = F + A_0 \sin m\tau,$$

and for small positive values of t

$$n^2 \dot{x} = A_0 e^{-kt} [m \cos m(t+\tau) - k \sin m(t+\tau)].$$

The expression chosen for our solution requires that $\dot{x}$ and $(-)^s F$ should have the same sign for small values of t, consequently A_s must be positive and we must have

$$n^2 a > F.$$

The value of A_s changes whenever $\dot{x}$ changes sign. In order that x

* W. E. Milne, *University of Oregon Publications*, vol. II (1923); *ibid. Mathematics Series*, vol. I (1929).

† This equation is discussed by H. S. Rowell, *Phil. Mag.* (6), vol. XLIV, pp. 284, 951 (1922).

may be continuous as t passes through the value given by $mt = s\pi$ we must have, with $k\pi = m\alpha$,

$$(-)^{s-1} F + (-)^s e^{-s\alpha} \sin m\tau\, A_{s-1} = (-)^s F + (-)^s e^{-s\alpha} \sin m\tau\, A_s,$$

so long as the value of A_s given by this equation has the same sign as A_{s-1}. If it had the opposite sign the value of $\dot{x}$ derived from (D) would not change sign as mt passed through the value $s\pi$ and our supposition regarding the sign and magnitude of the frictional force would not be valid. Let us see if this can actually happen.

The difference equation for A_s is

$$A_s = A_{s-1} - 2Fe^{s\alpha} \operatorname{cosec} m\tau,$$

and so the derived value of A_s is

$$A_s = \operatorname{cosec} m\tau \,[n^2 a - F(1 + 2e^{\alpha} + 2e^{2\alpha} + \dots 2e^{s\alpha})]. \quad \text{......(E)}$$

There is clearly some value of s for which A_{s-1} and A_s have opposite signs. To find what happens when $mt = s\pi$ and s has this critical value we examine the value of x. This is given by the equation

$$n^2 x = (-)^s [F + A_s e^{-s\alpha} \sin m\tau].$$

In order that the motion may continue for $mt > s\pi$ the force acting on the body must be sufficient to overcome the solid friction; in other words, $(-)^s n^2 x$ must be greater than F, but this condition is not satisfied when A_s is negative. Hence the motion continues up to a time t given by $mt = s\pi$, where s is the first positive integer for which the value of A_s given by (E) is negative.

Writing $F = n^2 a e^{-\beta}$ the successive swings are of lengths

$$a(1 - e^{\alpha-\beta}), \quad a(1 - e^{2\alpha-\beta}), \quad a(1 - e^{3\alpha-\beta}),$$

respectively; the body comes to rest within the so-called dead region defined by the inequalities

$$-ae^{-\beta} < x < ae^{-\beta}.$$

It should be noticed that the damping of the swings is more rapid than when solid friction does not act, because if $p < s$,

$$\frac{1 - e^{(p+1)\alpha-\beta}}{1 - e^{p\alpha-\beta}} < e^{-\alpha}.$$

This inequality is an immediate consequence of the inequality

$$(1 - e^{-\alpha})(1 - e^{(p+1)\alpha-\beta} - e^{p\alpha-\beta}) < 0.$$

EXAMPLES

1. Prove that if

$$2T = a_1 \dot{\phi}_1^2 + a_2 \dot{\phi}_2^2,$$
$$2V = c_1 \phi_1^2 + c_2 \phi_2^2 + 2[\gamma_1 \phi_1^3 + \gamma_2 \phi_1^2 \phi_2 + \gamma_3 \phi_1 \phi_2^2 + \gamma_4 \phi_2^3],$$

Lagrange's equations give

$$a_1 \ddot{\phi}_1 + c_1 \phi_1 + 3\gamma_1 \phi_1^2 + 2\gamma_2 \phi_1 \phi_2 + \gamma_3 \phi_2^2 = 0,$$
$$a_2 \ddot{\phi}_2 + c_2 \phi_2 + 3\gamma_4 \phi_2^2 + 2\gamma_3 \phi_1 \phi_2 + \gamma_2 \phi_1^2 = 0.$$

Obtain an approximate solution by assuming

$$\phi_1 = H_0 + H_1 \cos nt + H_2 \cos 2nt + H_3 \cos 3nt + \ldots,$$
$$\phi_2 = K_0 + K_1 \cos nt + K_2 \cos 2nt + K_3 \cos 3nt + \ldots.$$

[Rayleigh*.]

2. Prove that when $\pi^2 a \leq 8\omega^2$ the method of successive approximations can be applied to Duffing's equation

$$\frac{d^2x}{dt^2} + ax - \gamma x^3 = k \sin \omega t,$$

in which a, γ, k and ω are real constants. The process leads in fact to a convergent series.

[N. Bogoliouboff, *Travaux de l'Inst. de la mécan. techn. Kiev*, t. II, p. 367.

G. Duffing, *Erzwungene Schwingungen bei veränderlicher Eigenfrequenz und ihre technische Bedeutung* (1918).]

§ **12·3**. *The equation for a minimal surface.* We have already seen that the partial differential equation which characterises a minimal surface is

$$\frac{\partial}{\partial x}(p/H) + \frac{\partial}{\partial y}(q/H) = 0,$$

where
$$p = \frac{\partial z}{\partial x}, \quad q = \frac{\partial z}{\partial y}, \quad H = (1 + p^2 + q^2)^{\frac{1}{2}}.$$

Writing $r = \dfrac{\partial^2 z}{\partial x^2}$, $s = \dfrac{\partial^2 z}{\partial x \partial y}$, $t = \dfrac{\partial^2 z}{\partial y^2}$ and performing the differentiations the equation takes the form

$$(1 + q^2)\, r - 2pqs + (1 + p^2)\, t = 0,$$

and it may be noticed that the coefficients of r, s and t satisfy the inequality

$$(1 + q^2)(1 + p^2) - p^2 q^2 = 1 + p^2 + q^2 > 0,$$

so that the equation is analogous to a linear partial differential equation of elliptic type and does indeed possess some properties in common with such an equation. The equation may be actually replaced by a linear partial equation of elliptic type by using Legendre's transformation

$$x = \frac{\partial \chi}{\partial p}, \quad y = \frac{\partial \chi}{\partial q}, \quad \chi = px + qy - z.$$

The resulting equation is

$$(1 + p^2)\frac{\partial^2 \chi}{\partial p^2} + 2pq\frac{\partial^2 \chi}{\partial p \partial q} + (1 + q^2)\frac{\partial^2 \chi}{\partial q^2} = 0,$$

and the equation satisfied by x, y and z is

$$(1 + p^2)\frac{\partial^2 z}{\partial p^2} + 2pq\frac{\partial^2 z}{\partial p \partial q} + (1 + q^2)\frac{\partial^2 z}{\partial q^2} + 2p\frac{\partial z}{\partial p} + 2q\frac{\partial z}{\partial q} = 0.$$

* *Phil. Mag.* (6), vol. XX, p. 450 (1910); *Papers*, vol. V, p. 611. The theory is developed further by M. Born and E. Brody, *Zeits. f. Phys.* Bd. VI, S. 140 (1921); Bd. VIII, S. 205 (1922). J. Horn, *Crelle*, Bd. CXXVI, S. 194 (1903).

Now this equation may be reduced to the form

$$\frac{\partial^2 z}{\partial\xi\,\partial\eta} = 0,$$

for, if we write $p = w\cos\alpha$, $q = w\sin\alpha$, it becomes

$$(1 + w^2)\frac{\partial^2 z}{\partial w^2} + \left(2w + \frac{1}{w}\right)\frac{\partial z}{\partial w} + \frac{1}{w^2}\frac{\partial^2 z}{\partial\alpha^2} = 0,$$

or

$$w(1 + w^2)^{\frac{1}{2}}\frac{\partial}{\partial w}\left[w(1 + w^2)^{\frac{1}{2}}\frac{\partial z}{\partial w}\right] + \frac{\partial^2 z}{\partial\alpha^2} = 0.$$

The differential equation $d\xi d\eta = 0$ for the characteristics is

$$(1 + q^2)\,dp^2 - 2pq\,dp\,dq + (1 + p^2)\,dq^2 = 0.$$

Let us write $R = \dfrac{\partial^2\chi}{\partial p^2}$, $S = \dfrac{\partial^2\chi}{\partial p\partial q}$, $T = \dfrac{\partial^2\chi}{\partial q^2}$ and multiply the last equation by $RT - S^2$. Since $(1 + p^2)R + 2pqS + (1 + q^2)T = 0$, the equation may be written in the form

$$(Rdp + Sdq)^2 + (Sdp + Tdq)^2 + [(pR + qS)\,dp + (pS + qT)\,dq]^2 = 0,$$

or $dx^2 + dy^2 + dz^2 = 0$. This equation implies that the curves $\xi =$ constant, $\eta =$ constant are minimal curves. Hence the solution of the partial differential equation may be expressed in the form

$$x = X(\xi) + U(\eta), \quad y = Y(\xi) + V(\eta), \quad z = Z(\xi) + W(\eta),$$

where

$$\left.\begin{aligned}[X'(\xi)]^2 + [Y'(\xi)]^2 + [Z'(\xi)]^2 = 0,\\ [U'(\eta)]^2 + [V'(\eta)]^2 + [W'(\eta)]^2 = 0.\end{aligned}\right\} \qquad \text{......(A)}$$

These are the equations of Monge, the method of derivation being that of Legendre.

The equations (A) may be satisfied by writing

$$x = \int(1 - \xi^2)\,F(\xi)\,d\xi + \int(1 - \eta^2)\,G(\eta)\,d\eta,$$

$$y = i\int(1 + \xi^2)\,F(\xi)\,d\xi - i\int(1 + \eta^2)\,G(\eta)\,d\eta,$$

$$z = \int 2\xi\,F(\xi)\,d\xi + \int 2\eta\,G(\eta)\,d\eta,$$

where $F(\xi)$ and $G(\eta)$ are arbitrary integrable functions. These equations give

$$dx^2 + dy^2 + dz^2 = 4(1 + \xi\eta)^2\,F(\xi)\,G(\eta)\,d\xi d\eta.$$

If l, m, n are the direction cosines of the normal at a point (x, y, z) of the surface, l, m, n may be regarded as the co-ordinates of a point on the unit sphere in the spherical representation. Now

$$l = \frac{\xi + \eta}{1 + \xi\eta}, \quad m = \frac{i(\eta - \xi)}{1 + \xi\eta}, \quad n = \frac{\xi\eta - 1}{1 + \xi\eta},$$

and so

$$dl^2 + dm^2 + dn^2 = \frac{4d\xi d\eta}{(1 + \xi\eta)^2} = \frac{1}{F(\xi)\,G(\eta)\,(1 + \xi\eta)^4}\,[dx^2 + dy^2 + dz^2]$$

The surface is thus mapped conformally on the sphere and the magnification factor

$$R = [F(\xi)\, G(\eta)]^{\frac{1}{2}} (1 + \xi\eta)^2$$

is independent of the axes of co-ordinates, being equal, in fact, to the principal radius of curvature at a point on the surface. On the other hand, if we put

$$x_1 + iy_1 = 2 \int f(\xi)\, d\xi, \quad x_1 - iy_1 = 2 \int g(\eta)\, d\eta,$$

we have

$$dx_1^2 + dy_1^2 = 4f(\xi)\, g(\eta)\, d\xi\, d\eta = \frac{f(\xi)\, g(\eta)}{(1 + \xi\eta)^2 F(\xi)\, G(\eta)} (dx^2 + dy^2 + dz^2),$$

and so the surface is mapped conformally on the plane in which x_1 and y_1 are rectangular co-ordinates. The magnification factor is now

$$(1 + \xi\eta)\, [F(\xi)\, G(\eta)]^{\frac{1}{2}}\, [f(\xi)\, g(\eta)]^{-\frac{1}{2}},$$

and is independent of the axes of co-ordinates if

$$f(\xi) = [F(\xi)]^{\frac{1}{2}}, \quad g(\eta) = [G(\eta)]^{\frac{1}{2}},$$

being in this case $R^{\frac{1}{2}}$.

The $x_1 y_1$-plane is now mapped conformally on the unit sphere, and this fact is of great importance for the solution of boundary problems relating to minimal surfaces.

From physical considerations the important problem is to determine a continuous minimal surface which passes through a given curve which either closes or extends to infinity*. When the curve is made up of straight lines the corresponding curve in the spherical representation is made up of arcs of great circles and the curve in the xy-plane is composed of straight lines. A straight line on the minimal surface is, in fact, an asymptotic line and such a line bisects the angle between the lines of curvature at any point. In the mapping of the surface on the $x_1 y_1$-plane, however, the lines of curvature map into lines parallel to the axes of co-ordinates and so the asymptotic lines map into straight lines making equal angles with the axes of co-ordinates.

To prove that the lines of curvature map into the lines $x_1 =$ constant, $y_1 =$ constant, we have to show that along a curve for which $dx_1 = c$ (or $dy_1 = 0$), the normals at (x, y, z) and $(x + dx, y + dy, z + dz)$ intersect. We must therefore show that when

$$[F(\xi)]^{\frac{1}{2}}\, d\xi \pm [G(\eta)]^{\frac{1}{2}}\, d\eta = 0,$$

we have

$$dx \pm R dl = 0, \quad dy \pm R dm = 0, \quad dz \pm R dn = 0.$$

This is easily seen to be true because, for example,

$$dx \pm R dl = (1 - \xi^2)\, F(\xi)\, d\xi + (1 - \eta^2)\, F(\eta)\, d\eta \\ \pm [F(\xi)\, G(\eta)]^{\frac{1}{2}} [(1 - \eta^2)\, d\xi + (1 - \xi^2)\, d\eta].$$

* This is called the problem of Plateau.

Since straight lines on the minimal surface correspond to straight lines in the $x_1 y_1$-plane and to arcs of great circles on the unit sphere, the problem of determining a continuous minimal surface through a contour composed of straight lines reduces to a problem of conformal representation which was discussed by Schwarz.

We shall first discuss the conformal representation on a half-plane of a region bounded by arcs of circles because the rectilinear polygon in the $x_1 y_1$-plane can be mapped on a half w-plane by the method of Schwarz and Christoffel explained in § 4·62, and when the unit sphere is mapped on a plane by stereographic projection, a region bounded by great circles not passing through the point of projection, maps into a region bounded by circular arcs in the z-plane.

Since any two consecutive arcs belonging to the boundary can be transformed into segments of straight lines by a transformation of type

$$z = \frac{a\zeta + b}{c\zeta + d},$$

composed of an inversion with respect to a circle whose centre is at an angular point where the two lines intersect and a reflection in a line, the behaviour of z, considered as a function of w, may be inferred from the discussion given for the case of a polygon with straight sides.

Consequently, at any point of the boundary which is not a vertex the mapping function has the form

$$z = \frac{a\,(w - w_0)\,p\,(w - w_0) + b}{c\,(w - w_0)\,p\,(w - w_0) + d},$$

where $p\,(w - w_0)$ is a power series with real coefficients.

At a vertex where two circles cut at an interior angle $\alpha\pi$

$$z = \frac{a\,(w - w_0)^{\alpha}\,p\,(w - w_0) + b}{c\,(w - w_0)^{\alpha}\,p\,(w - w_0) + d}.$$

At a point of the boundary which corresponds to an infinite value of w

$$z = \frac{\dfrac{a}{w}\,p\left(\dfrac{1}{w}\right) + b}{\dfrac{c}{w}\,p\left(\dfrac{1}{w}\right) + d},$$

if the point is not a vertex, while if it is a vertex of interior angle $\alpha\pi$

$$z = \frac{\dfrac{a}{w^{\alpha}}\,p\left(\dfrac{1}{w}\right) + b}{\dfrac{c}{w^{\alpha}}\,p\left(\dfrac{1}{w}\right) + d}.$$

At an interior point of the region

$$z - z_0 = (w - w_0)\,P\,(w - w_0).$$

Writing

$$\{z, w\} = \frac{d^2}{dw^2}\left(\log \frac{dz}{dw}\right) - \frac{1}{2}\left(\frac{d}{dw}\log \frac{dz}{dw}\right)^2,$$

with Cayley's notation for the Schwarzian derivative, the name usually used for the expression on the right, we have

$$\{z, w\} = \{\zeta, w\},$$

and so the constants a, b, c, d do not enter into the expression for $\{z, w\}$.

At any point of the boundary which is not a vertex, it is found that

$$\{z, w\} = h\,(w - w_0) + k\,(w - w_0)^2 + \ldots.$$

At a vertex of interior angle $\alpha\pi$

$$\{z, w\} = \frac{1}{2}\frac{1 - \alpha^2}{(w - w_0)^2} + \frac{h}{w - w_0} + k + l\,(w - w_0).$$

For a point on the boundary corresponding to an infinite value of w

$$\{z, w\} = hw^{-4} + kw^{-5} + \ldots$$

provided the point is not a vertex; if it is a vertex,

$$\{z, w\} = \frac{1}{2}\frac{1 - \alpha^2}{w^2} + \frac{h}{w^3} + \frac{k}{w^4}.$$

The coefficients in all these expressions are real and so $\{z, w\}$ is real for all real values of w; it can therefore be continued by Schwarz's method and defined analytically in the whole of the w-plane.

Finally, at any point within the region bounded by the circular arcs dz/dw is never zero and $\{z, w\}$ can be expressed as a power series.

Since $\{z, w\}$ has only a finite number of poles and is infinitely small for large values of w, it must be a rational function. Let $a_1, a_2, \ldots a_n$ be the values of w corresponding to the vertices of the region, $\alpha_1\pi, \alpha_2\pi, \ldots \alpha_n\pi$ the associated interior angles, then

$$\{z, w\} - \sum_{s=1}^{n} \frac{1}{2}\frac{1 - \alpha_s^2}{(w - a_s)^2} - \sum_{s=1}^{n} \frac{h_s}{w - a_s}$$

is finite for all finite values of w and becomes infinitely small when $|\,w\,|$ is infinitely great, consequently it must be zero, and so we have the differential equation

$$\{z, w\} = \sum_{s=1}^{n} \frac{1}{2}\frac{1 - \alpha_s^2}{(w - a_s)^2} - \sum_{s=1}^{n} \frac{h_s}{w - a_s}.$$

The constants α_s, h_s are not all independent because the expression of the right-hand side in ascending powers of $1/z$ should start with $1/z^4$. Consequently, we have the relations

$$\sum_{s=1}^{n} h_s = 0,$$

$$\sum_{s=1}^{n}\left(a_s h_s + \frac{1 - \alpha_s^2}{2}\right) = 0,$$

$$\sum_{s=1}^{n}\left[a_s^2 h_s + a_s\,(1 - \alpha_s^2)\right] = 0.$$

If, on the other hand, the point corresponding to $w = \infty$ is a vertex of interior angle $\beta\pi$, the expansion should start with the term $(1 - \beta^2)/2w^2$ and we have only two relations

$$\sum_{s=1}^{n} h_s = 0,$$

$$\sum_{s=1}^{n} \left(a_s h_s + \frac{1 - \alpha_s^2}{2}\right) = \tfrac{1}{2}(1 - \beta^2).$$

The solution of the differential equation

$$\{z, w\} = F(w)$$

may be made to depend on the fact that if W_1 and W_2 are two independent solutions of a linear differential equation

$$\frac{d^2W}{dw^2} + p(w)\frac{dW}{dw} + q(w)W = 0,$$

the function $$z = W_2/W_1$$
satisfies the differential equation

$$\{z, w\} = 2q - \tfrac{1}{2}p^2 - \frac{dp}{dw}.$$

Taking p arbitrarily and choosing q so that

$$2q - \tfrac{1}{2}p^2 - \frac{dp}{dw} = F(w),$$

the solution of $\{z, w\} = F(w)$ is made to depend on that of a linear differential equation of the second order. This equation is of a special type, for all its coefficients and all its singular points are real; moreover, it turns out that all its integrals are regular in the sense in which the word is used in the theory of linear differential equations.

Example $(n = 2)$.

$$\{z, w\} = \frac{1}{2}\left[\frac{1 - \alpha^2}{(w - a)^2} + \frac{1 - \beta^2}{(w - b)^2}\right] + \frac{h}{w - a} + \frac{k}{w - b},$$

$$h + k = 0,$$

$$ah + bk + \frac{1 - \alpha^2}{2} + \frac{1 - \beta^2}{2} = 0,$$

$$a^2h + b^2k + a(1 - \alpha^2) + b(1 - \beta^2) = 0.$$

Therefore $$(a - b)h + 1 - \tfrac{1}{2}(\alpha^2 + \beta^2) = 0,$$

$$(a^2 - b^2)h + a + b - a\alpha^2 - b\beta^2 = 0.$$

Therefore $$2(a\alpha^2 + b\beta^2) = (a + b)(\alpha^2 + \beta^2),$$

$$\alpha^2(a - b) = \beta^2(a - b).$$

If $a \neq b$ we have $$\alpha^2 = \beta^2, \quad h = \frac{\alpha^2 - 1}{a - b},$$

$$\{z, w\} = \tfrac{1}{2}(1 - \alpha^2)\left[\frac{1}{(w - a)^2} + \frac{1}{(w - b)^2} - \frac{2}{(w - a)(w - b)}\right]$$

$$= \tfrac{1}{2}(1 - \alpha^2)\frac{(a - b)^2}{(w - a)^2(w - b)^2}.$$

Hence z is the ratio of two solutions of the differential equation

$$\frac{d^2W}{dw^2} + \frac{W}{4}(1 - \alpha^2)\frac{(a-b)^2}{(w-a)^2(w-b)^2} = 0.$$

Now this equation is known to possess a solution of the form

$$W = A(w-a)^m (w-b)^n + B(w-a)^n (w-b)^m,$$

where m and n are the roots of the equation

$$\rho^2 - \rho + \tfrac{1}{4}(1 - \alpha^2) = 0.$$

Therefore $$m = \tfrac{1}{2}(1+\alpha), \quad n = \tfrac{1}{2}(1-\alpha).$$

A particular value of z is

$$z = (w-a)^{m-n}(w-b)^{n-m} = [(w-a)/(w-b)]^{\alpha}.$$

There is thus a notable reduction in this simple case.

The minimal surface which corresponds to this case is the helicoid. With a special choice of the axis of z the equation is

$$z = c\tan^{-1}(y/x) + b \qquad \text{......(B)}$$

and it is clear that the surface is a ruled surface generated by lines perpendicular to the axis of z.

To determine a minimal surface which passes through two non-intersecting straight lines we take the line of shortest distance between the two lines as axis of z. The equation is then of type (B). If h is the distance between the lines and θ a value of the angle between the lines, the constant c is given by the equation $h = c\theta$.

If the angle between the lines is taken to be $\theta + 2k\pi$ instead of θ the constant c is given by the equation $h = c(\theta + 2k\pi)$.

It thus appears that in this case there is more than one minimal surface through the given contour and the area of the surface is infinite.

Another case of considerable interest is that in which the minimal surface is a surface of revolution and is bounded by two circles having the same axis. Assuming that $z = F(\rho)$, where $\rho^2 = x^2 + y^2$, we obtain the differential equation

$$\rho F''(\rho) + F'(\rho) + [F'(\rho)]^3 = 0.$$

Therefore $$z - b = c\cosh^{-1}(\rho/c).$$

The meridian curve is thus a catenary. It is shown in books on the Calculus of Variations* that it is not always possible to draw a catenary which will pass through two given points and have a given directrix. Under certain conditions, however, two catenaries can be drawn, and in a particular case one only. The conclusion is that it is not always possible to find a continuous one-sheeted minimal surface which is terminated by the two circles. There is, however, a degenerate minimal surface consisting of planes through the two circles and a thin tube joining two points of

* See, for instance, Todhunter, *Researches in the Calculus of Variations*; G. A. Bliss, *Calculus of Variations*, p. 92 (1925).

these planes. In this case the tube is not strictly a minimal surface though its contribution to the total area is negligible.

An interesting inequality for the area of a minimal surface has been obtained recently by Carleman*. He shows that if L is the length of the perimeter of the closed curve limiting the surface, the area A cannot be greater than $L^2/4\pi$. The value $L^2/4\pi$ is attained only when the boundary is a circle.

The method of § 2·33 has been extended by Douglas† so that it gives approximate solutions of Plateau's problem. The partial derivatives in the differential equation are replaced by their finite difference approximations. An important memoir on Plateau's problem has been published recently by Haar‡ who uses a direct method of the Calculus of Variations and establishes the existence of a solution of Plateau's problem for a type of boundary curve C considered previously by Bernstein§ and Lebesgue‖. The curve C is supposed in fact to have a convex projection Γ on the xy-plane and to be such that none of its osculating planes are perpendicular to this plane. Haar also obtains a proof of Radó's theorem¶ that any extremal function which is continuous $(D, 1)$ is analytic. Radó has established moreover the analytical character of any minimal surface represented by a function $z = z(x, y)$ which is continuous $(D, 2)$.

The existence theorem has been simplified by Radó** with the aid of a three-point condition of the following type. Let θ be the acute angle made with the xy-plane by any plane P which meets the boundary curve C in at least three points. If there is a finite positive number Δ which exceeds every value of $\tan\theta$, the curve C is said to satisfy a three-point condition with constant Δ.

Radó has extended the existence theorem so that it is applicable to a variation problem involving an integral

$$I = \iint F(p, q)\, dx\, dy,$$

$$p = \frac{\partial z}{\partial x}, \quad q = \frac{\partial z}{\partial y},$$

where the function $F(p, q)$ is analytic in the arguments p, q and satisfies the inequalities

$$F_{pp} F_{qq} - F_{pq}^2 > 0, \quad F_{pp} > 0,$$

which make the variation problem regular.

* T. Carleman, *Math. Zeits.* Bd. IX, S. 154 (1921).

† J. Douglas, *Trans. Amer. Math. Soc.* vol. XXXIII, p. 263 (1931).

‡ A. Haar, *Math. Ann.* Bd. XCVII, S. 124 (1926).

§ S. Bernstein, *Ann. de l'école normale* (3), t. XXVII, p. 233 (1910), t. XXIX, p. 431 (1912) *Math. Ann.* Bd. XCV, S. 585 (1926).

‖ H. Lebesgue, *Annali di Matematica* (3), t. VII, p. 231 (1902).

¶ T. Radó, *Math. Zeits.* Bd. XXIV, S. 321 (1925).

** T. Radó, *Szeged Acta*, t. II, p. 228 (1926); *Math. Ann.* Bd. CI, S. 620 (1929).

It is shown that if L is any number not less than Δ there is a function $z_L(x, y)$ which satisfies a Lipschitz condition with constant L

$$|z_L(x_2, y_2) - z_L(x_1, y_1)| \leqslant L[(x_2 - x_1)^2 + (y_2 - y_1)^2]^{\frac{1}{2}}$$

for any pair of points (x_1, y_1), (x_2, y_2) in the realm R bounded by the convex projection of C and assumes on this curve a succession of values equal to the z-co-ordinates of points on C. This function z_L satisfies, moreover, a Lipschitz condition with constant Δ.

§ 12·4. *The steady two-dimensional motion of a compressible fluid.* Let ρ be the density, p the pressure and u, v the component velocities at a point (x, y), then the equation of continuity is

$$\frac{\partial u}{\partial x} + \frac{\partial v}{\partial y} + \frac{1}{\rho}\left(u\frac{\partial \rho}{\partial x} + v\frac{\partial \rho}{\partial y}\right) = 0,$$

and if the motion is irrotational with velocity potential ϕ, we have

$$u = \frac{\partial \phi}{\partial x}, \quad v = \frac{\partial \phi}{\partial y}, \quad \frac{\partial v}{\partial x} = \frac{\partial u}{\partial y}.$$

When the pressure and density are connected by the adiabatic law $p = k\rho^\gamma$, Bernoulli's integral takes the form

$$\frac{\gamma k}{\gamma - 1}\rho^{\gamma-1} + \tfrac{1}{2}(u^2 + v^2) = \frac{\gamma k}{\gamma - 1}\rho_0^{\gamma-1} + \tfrac{1}{2}V^2,$$

where V is the velocity at a place where the density is ρ_0. If a is the velocity of sound at this place we have also

$$a^2 = \left(\frac{dp}{d\rho}\right)_0 = k\gamma\rho_0^{\gamma-1},$$

and so $$\left(\frac{\rho}{\rho_0}\right)^{\gamma-1} = 1 - \frac{\gamma - 1}{2a^2}(u^2 + v^2 - V^2) = c^2/a^2, \text{ say.}$$

This equation gives

$$\frac{c^2}{\rho}\frac{\partial \rho}{\partial x} = -\left(u\frac{\partial u}{\partial x} + v\frac{\partial v}{\partial x}\right), \quad \frac{c^2}{\rho}\frac{\partial \rho}{\partial y} = -\left(u\frac{\partial u}{\partial y} + v\frac{\partial v}{\partial y}\right),$$

and so the equation of continuity may be written in the form

$$c^2\left(\frac{\partial u}{\partial x} + \frac{\partial v}{\partial y}\right) = u\left(u\frac{\partial u}{\partial x} + v\frac{\partial v}{\partial x}\right) + v\left(u\frac{\partial u}{\partial y} + v\frac{\partial v}{\partial y}\right),$$

or $$(c^2 - u^2)\frac{\partial^2 \phi}{\partial x^2} - 2uv\frac{\partial^2 \phi}{\partial x \partial y} + (c^2 - v^2)\frac{\partial^2 \phi}{\partial y^2} = 0. \quad \text{......(A)}$$

This equation may be transformed into a linear equation by the method of Legendre, in which u and v are taken as new independent variables and the new dependent variable is a quantity χ defined by the equations

$$\chi = ux + vy - \phi, \quad x = \frac{\partial \chi}{\partial u}, \quad y = \frac{\partial \chi}{\partial v}.$$

The transformed equation is*

$$(c^2 - u^2)\frac{\partial^2\chi}{\partial v^2} + 2uv\frac{\partial^2\chi}{\partial u \partial v} + (c^2 - v^2)\frac{\partial^2\chi}{\partial u^2} = 0, \qquad \text{......(B)}$$

and this is of elliptic type if

$$(c^2 - u^2)(c^2 - v^2) > u^2v^2, \text{ i.e. if } c^2 > u^2 + v^2.$$

A change from the elliptic to the hyperbolic type occurs when

$$q^2 = u^2 + v^2 = c^2, \quad u^2 + v^2 = a^2 - \tfrac{1}{2}(\gamma - 1)(u^2 + v^2 - V^2),$$

that is, when the velocity q is equal to the local velocity of sound c. The critical velocity q_c is given by the equation

$$q_c^2 = \frac{2a^2 + (\gamma - 1)V^2}{\gamma + 1},$$

and may be less than a. For example, if $V = \cdot 6a$ and $\gamma = 1 \cdot 4$, we have

$$q_c^2 = \frac{2 \cdot 144}{2 \cdot 4} a^2.$$

Simple solutions of the differential equation (B) may be obtained by writing $u = q\cos\theta$, $v = q\sin\theta$, $\chi = Q(q)\cos(m\theta + \epsilon)$. The equation for Q is then

$$[a^2 + \tfrac{1}{2}(\gamma - 1)(V^2 - q^2)]\left[\frac{d^2Q}{dq^2} + \frac{1}{q}\frac{dQ}{dq} - \frac{m^2}{q^2}Q\right] + m^2Q = q\frac{dQ}{dq}.$$

Assuming as a trial solution

$$Q = \sum_r^\infty a_n q^n,$$

we find that

$$[a^2 + \tfrac{1}{2}(\gamma - 1)V^2][(n+2)^2 - m^2]a_{n+2} = [\tfrac{1}{2}(\gamma - 1)n^2 + n - \tfrac{1}{2}(\gamma + 1)m^2]a_n,$$

and it is seen that r can be either m or $-m$. Since

$$\operatorname*{Lt}_{n\to\infty} q^2\frac{a_{n+2}}{a_n} = \frac{(\gamma - 1)q^2}{2a^2 + (\gamma - 1)V^2},$$

it appears that the series converge if

$$(\gamma - 1)q^2 < 2a^2 + (\gamma - 1)V^2,$$

but this condition is always satisfied since $c^2 > 0$. The second critical velocity† (for which $c^2 = 0$) is given by the equation

$$(\gamma - 1)q^2 = 2a^2 + (\gamma - 1)V^2,$$

and is evidently greater than the first.

* Some interesting examples of the use of this equation have been given by A. Steichen, *Diss. Göttingen* (1909). A good account of this work is given by J. Ackeret in his article on the dynamics of gases in Bd. VII of the *Handbuch der Physik*.

† The existence of critical velocities was noted by Barré de Saint-Venant and G. Wantzel, *Journ. de l'École Polytechnique*, cah. 16, p. 85 (1839). The fact that changes of pressure cannot be propagated upstream when the velocity of flow exceeds the velocity of sound has been used to account for the existence of the first critical velocity. See especially, O. Reynolds, *Phil. Mag.* vol. XXI, p. 185 (1886); A. Hugoniot, *Ann. de Chim.* t. IX, p. 383 (1886).

EXAMPLES

1. Prove that equation (A) is satisfied by

$$\phi = Cr \sin (\nu\theta), \quad x = r \cos \theta, \quad y = r \sin \theta,$$

where C is an arbitrary constant and ν is a constant connected with γ by the equation

$$\nu^2 (\gamma + 1) = \gamma - 1.$$

The component velocities derived from this velocity potential are useful for the discussion of flow round a corner with straight sides. When the angle of the field of flow is greater than two right angles there is an angular region bounded by lines through the corner in which there is a transitional flow of the present type passing continuously over on each side into a uniform rectilinear flow.

[See L. Prandtl, *Phys. Zeits.* Bd. VIII, S. 23 (1907); Th. Meyer, *Diss. Göttingen* (1908).]

2. Prove that equation (A) possesses a solution of type

$$\phi = A\theta + Bf(r),$$

where A and B are constants if $f(r)$ satisfies the differential equation

$$\nu [\gamma + 1 - \nu (\gamma - 1) - \eta (\gamma + 1)] \frac{d\eta}{d\nu} = \eta [(\gamma + 1) - \nu (\gamma - 3) - \eta (\gamma - 1)],$$

where $$\nu a_0^2 r^2 = A^2, \qquad \eta a_0^2 = B^2 [f'(r)]^2,$$

and a_0 is the velocity of the stream where it attains that of sound.

[G. I. Taylor, *Journ. London Math. Soc.* vol. v, p. 224 (1930).]

3. Prove that the differential equation for η is satisfied by quantities η and ν which are expressed in terms of a parameter, ρ, by the equations

$$\eta + \nu = \frac{\gamma + 1}{\gamma - 1} - \frac{2}{\gamma - 1} \left(\frac{\rho}{\rho_0}\right)^{\gamma - 1},$$

$$\rho^2 \eta = C\nu,$$

where ρ_0 and C are arbitrary constants.

4. Show that when an attempt is made to solve a boundary problem for equation (A) by expanding ϕ in powers of $1/\gamma$, the pressure in the steady stream far from all obstacles being assumed to be constant and independent of γ, the method may fail when, in the corresponding problem for an incompressible fluid, there is some place where the pressure is zero.

APPENDIX

NOTE I. The generality of this result has been questioned by O. Perron, *Gött. Nachr.* (1930), who gives an example in which the theorem fails. Restrictions on $A(t)$ which will make the result valid have been proposed by M. Fukuhara and M. Nagumo, *Proc. Imp. Acad. Tokyo*, vol. VI, p. 131 (1930). The simplest restrictions are $A'(t) > 0$, $A(\infty) = C \neq \infty$.

NOTE II. For the resisted vibrations of a prismatic bar Suyehiro* has derived the equation

$$E\kappa^2 \frac{\partial^4 \eta}{\partial x^4} + \rho \frac{\partial^2 \eta}{\partial t^2} = \rho\kappa^2 \frac{\partial^4 \eta}{\partial x^2 \partial t^2} - \xi\kappa^2 \frac{\partial^5 \eta}{\partial x^4 \partial t},$$

which is a simple extension of Sezawa's equation†

$$E\kappa^2 \frac{\partial^4 \eta}{\partial x^4} + \rho \frac{\partial^2 \eta}{\partial t^2} = -\xi\kappa^2 \frac{\partial^5 \eta}{\partial x^4 \partial t},$$

ξ being a positive constant for the material. In his discussion of this equation Suyehiro shows that it indicates the existence of an upper limit $E/\pi\xi$ to the frequency of the transverse vibrations but finds values of this upper limit which seem too low, thus throwing doubt on the correctness of estimates that have been made of the magnitude of the quantity ξ, which gives a measure of the solid viscosity.

NOTE III. When the functions $\Theta(\mu)$ are properly normalised they form a set of orthogonal functions for the interval $(-1, 1)$. It is customary now to define the Jacobi polynomial $\phi_n(\mu)$ so that

$$\int_{-1}^{1} (1+\mu)^m (1-\mu)^k \phi_n(\mu)\, \phi_{n'}(\mu)\, d\mu = \begin{cases} 0, & n' \neq n \\ 1, & n' = n \end{cases}$$

$$m > -1, \quad k > -1.$$

The convergence of the expansion

$$f(\mu) \sim \sum_{n=0}^{\infty} c_n \phi_n(\mu), \qquad \text{......(A)}$$

where $$c_n = \int_{-1}^{1} (1+\mu)^m (1-\mu)^k \phi_n(\mu) f(\mu)\, d\mu,$$

depends essentially upon the behaviour of $\phi_n(\mu)$ when n is very large. An appropriate asymptotic expression was obtained by Darboux and has been further studied by Stekloff, Ford and others. A simplified derivation has been given recently by Shohat‡ who gives references to the literature.

* K. Suyehiro, *Proc. Imp. Acad. Tokyo*, vol. IV, p. 263 (1928).

† K. Sezawa, *Bull. Earthquake Research Inst.* vol. III, p. 50 (1927).

‡ J. A. Shohat, *American Mathematical Monthly*, vol. XXXIII, p. 354 (1926).

If $\mu = \cos\theta$ and $-1+\epsilon \leqslant \mu \leqslant 1-\epsilon$ the asymptotic expressions are of type

$$\phi_n(\mu) = B_1 \cos n\theta + B_2 \sin n\theta + O(1/n),$$

$$\phi_n(1) = \frac{2^{-\frac{m+k}{2}} n^{k+\frac{1}{2}}}{\Gamma(1+k)} [1 + o(1)],$$

$$\phi_n(-1) = (-)^n \frac{2^{-\frac{m+k}{2}} n^{m+\frac{1}{2}}}{\Gamma(1+m)} [1 + o(1)],$$

where $o(1)$ denotes in each case some quantity which tends to zero as $n \to \infty$. For the Legendre polynomials, which correspond to the case $m = k = 0$, there is the theorem of Stieltjes* that if $0 < \theta < \pi$ the product $(n \sin\theta)^{\frac{1}{2}} P_n(\cos\theta)$ is less, in absolute value, than some fixed number, independent of n and θ. Gronwall† expressed this result in a concise form

$$(2\pi n \sin\theta)^{\frac{1}{2}} |P_n(\cos\theta)| < 4$$

and Fejér‡, by more elementary reasoning, has obtained a similar inequality with the number 8 on the right-hand side in place of 4.

The result has been further extended by Hobson§ who obtains the inequality

$$|P_n^{\pm m}(\cos\theta)| \sin^m\theta < \frac{\Gamma(n \pm m + 1)}{\Gamma(n+1)} \left(\frac{8}{n\pi \sin\theta}\right)^{\frac{1}{2}},$$

or

$$\frac{\Gamma(n \pm m + 1)}{\Gamma(n + \frac{3}{2})} \left(\frac{4\pi}{\sin\theta}\right)^{\frac{1}{2}},$$

the first or second form being used according as m is not restricted to be integral, or is so restricted. The number n is not restricted to be integral but is restricted by the inequalities

$$n \geqslant 1, \quad n - m + 1 > 0, \quad m \geqslant 0.$$

The convergence of the expansion (A) has been much discussed in recent years with the aid of the idea of summability developed by Cesàro. By an extension of the method of § 1·16 the Cesàro sum (C, r) of order r for a series

$$a_0 + a_1 + a_2 + \ldots$$

is said to exist when the Cesàro means $S_n^{(r)}$ of order r tend to a definite finite limit $S^{(r)}$ as $n \to \infty$.

If $A_n^{(r)}$ is the coefficient of z^n in the expansion of $(1-z)^{-r-1}$ in ascending powers of z, the means are defined by the equation

$$A_n^{(r)} S_n^{(r)} = A_n^{(r)} a_0 + A_{n-1}^{(r)} a_1 + \ldots A_0^{(r)} a_n.$$

* T. J. Stieltjes, *Annales de Toulouse*, t. IV, G. 6 (1890).
† T. H. Gronwall, *Math. Annalen*, Bd. LXXIV, S. 221 (1913).
‡ L. Fejér, *Math. Zeitschrift*, Bd. XXIV, S. 290 (1925).
§ E. W. Hobson, *Proc. London Math. Soc.* (2), vol. XXX, p. 239 (1929).

If μ is a point of continuity of a function $f(\mu)$, the associated expansion (A) converges to $f(\mu)$ in the manner (C, r) with $r > 0$ if $-1 < \mu < 1$. When $\mu = \pm 1$ and $m = k > -\frac{1}{2}$ the expansion converges (C, r) to $f(\mu)$ when $r > m + \frac{1}{2}$. This general result, obtained by Kogbetliantz*, includes the case of the Legendre polynomials ($m = k = 0$) studied by A. Haar† and W. H. Young‡ for $-1 < \mu < 1$ and by T. H. Gronwall§ for $\mu = \pm 1$. Further results relating to the summability of series of Legendre functions have been obtained by Chapman ||, who makes use of the idea of uniform summability. The conditions for the convergence of the series have been discussed very thoroughly by Hobson¶.

Series of functions of type $P_n^m(\mu)$ occur most naturally in the double series of functions of Laplace

$$f(\theta, \phi) \sim \Sigma\, c_{m,n} P_n^m(\cos\theta)\, e^{im\phi}$$

mentioned in §§ 6·34, 6·35.

When the function $f(\theta, \phi)$ is continuous at a point (θ, ϕ) of the unit sphere, Fejér** found that the associated series of Laplace is summable $(C, 2)$ with the sum $f(\theta, \phi)$. This theorem was regarded as the analogue of the theorem of § 1·17 for the Fourier series of a function. More general results have been obtained by Gronwall and Fejér in the papers cited in the footnotes on p. 513. The series has also been discussed by MacRobert††.

The first of the orthogonal relations of § 6·28 has been known from the time of Laplace and Legendre. The second one was discovered by Heine‡‡; unlike the first it does not seem to be a particular case of Jacobi's orthogonal relation (§ 6·53).

* E. Kogbetliantz, *Journ. de Math.* (9), t. III, p. 107 (1924).

† A. Haar, *Math. Ann.* Bd. LXXVIII, S. 121 (1917).

‡ W. H. Young, *Comptes Rendus*, t. CLXV, p. 696 (1917); *Proc. London Math. Soc.* (2), vol. XVIII, p. 141 (1919–20).

§ T. H. Gronwall, *Math. Ann.* Bd. LXXIV, S. 213 (1913), Bd. LXXV, S. 321 (1914). See also L. Fejér, *Math. Zeits.* Bd. XXIV, S. 267 (1925–6).

|| S. Chapman, *Quart. Journ.* vol. XLIII, p. 1 (1911); *Math. Ann.* Bd. LXXII, S. 211 (1912).

¶ E. W. Hobson, *Proc. London Math. Soc.* (2), vol. VI, p. 349 (1908), vol. VII, p. 24 (1908).

** L. Fejér, *Comptes Rendus*, t. CXLVI, p. 224 (1908); *Math. Ann.* Bd. LXVII, S. 76 (1909); *Rend. Palermo*, t. XXXVIII, p. 79 (1914).

†† T. M. MacRobert, *Proc. Edin. Math. Soc.* vol. XLII, p. 88 (1924).

‡‡ E. Heine, *Handbuch der Kugelfunktionen*, Bd. I, p. 253.

LIST OF AUTHORS CITED

INDEX